The Minerals, Metals & Materials Series

Shadia Ikhmayies · Bowen Li
John S. Carpenter · Jian Li
Jiann-Yang Hwang · Sergio Neves Monteiro
Donato Firrao · Mingming Zhang
Zhiwei Peng · Juan P. Escobedo-Diaz
Chenguang Bai · Yunus Eren Kalay
Ramasis Goswami · Jeongguk Kim
Editors

Characterization of Minerals, Metals, and Materials 2017

TMS

Springer

Editors
Shadia Ikhmayies
Al Isra University
Amman
Jordan

Bowen Li
Michigan Technological University
Houghton, MI
USA

John S. Carpenter
Los Alamos National Laboratory
Los Alamos, NM
USA

Jian Li
CanmetMATERIALS
Hamilton, ON
Canada

Jiann-Yang Hwang
Michigan Technological University
Houghton, MI
USA

Sergio Neves Monteiro
Military Institute of Engineering, IME
Rio de Janeiro
Brazil

Donato Firrao
Politecnico di Torino
Turin
Italy

Mingming Zhang
ArcelorMittal Global R&D
East Chicago, IN
USA

Zhiwei Peng
Central South University
Changsha
China

Juan P. Escobedo-Diaz
University of New South Wales
Canberra
Australia

Chenguang Bai
Chongqing University
Chongqing
China

Yunus Eren Kalay
Middle East Technical University
Ankara
Turkey

Ramasis Goswami
Naval Research Laboratory
Washington, DC
USA

Jeongguk Kim
Korea Railroad Research Institute
Uiwang
South Korea

ISSN 2367-1181 ISSN 2367-1696 (electronic)
The Minerals, Metals & Materials Series
ISBN 978-3-319-51381-2 ISBN 978-3-319-51382-9 (eBook)
DOI 10.1007/978-3-319-51382-9

Library of Congress Control Number: 2016960729

Printed on acid-free paper

This Springer imprint is published by Springer Nature
The registered company is Springer International Publishing AG
The registered company address is: Gewerbestrasse 11, 6330 Cham, Switzerland

Preface

The characterization of materials is an important step to be taken before utilizing the materials for any purpose. It starts from the production of the material, and continues after each processing and/or engineering stage to explore its influence on the structure and properties of the material. Depending on the purpose, one can subject the material to mechanical, thermal, chemical, optical, electrical, and other characterizations to make sure that the material under consideration can function without failure for the life of the final product. Therefore, advances in the materials science are strongly correlated with advances in the characterization technologies. The Characterization of Minerals, Metals, and Materials symposium sponsored by the Materials Characterization Committee of TMS focuses on material characterization from the bulk down to the nano-scale. All characterization techniques and their applications are covered in this symposium. Developments in these techniques and their application in the quantification of the microstructure of materials are essential facets of this symposium. Specific characterization focus areas include catalyst structure, waste and failure characterization, besides structure-property relationships in metals, minerals and materials.

The characterization symposium is a cornerstone symposium in the TMS annual meeting, which attracts materials scientists, metallurgists, mechanical engineers, microscopists, metallographers, from academia and industry from all over the globe. In the TMS 2017 Annual Meeting & Exhibition held in San Diego, California, USA, the characterization symposium received 229 submissions, of which are 137 oral presentations, and 67 will be presented as posters. Of the presented papers, 93 are published in this book after being peer reviewed. The topics of these papers cover a wide range of materials science, metallurgy, physics, chemistry, and engineering of materials. Minerals, ferrous and nonferrous metals, semiconductors, clays, ceramics, alloys, composites, electronic, magnetic, environmental, soft, and advanced materials are widely covered. In addition, research papers about extraction, processing, welding, solidification, corrosion and method development represent a large portion of the presented papers.

This book features original articles and the state-of-the-art reviews on theoretical and practical aspects of the characterization, extraction, processing, structure, and behavior of minerals, metals, and materials. It is a good reference for academic and industry audiences from advanced undergraduates to seasoned professionals who wish to learn about all types of characterization methods in general, and specifically about real-world applications in the minerals, metals and materials. This book will also be relevant for scientists and engineers engaged in research, development, and production. This book will provide the industry audience with up-to-date information on many types of materials and their characterization with an underlying theme of explaining the behavior of materials using novel approaches.

The reader of this book will learn about all types of characterization methods, their development, and their applications. The reader will enjoy the diversity of topics in this book. He/she will find in this book up-to-date information about bulk materials, thin films, joints and interfaces, powders, slags, micro and nanostructures. The beautiful thing is that this book pays attention to the relationship between production, extraction, processing, recycling, and loading of materials and alloys in practical use. The knowledge gained from this book can be used to prompt innovations in characterization methods and techniques, and to produce new materials with the specific desired properties.

The editors of this book express their sincere thanks and appreciations to the TMS for giving this symposium the opportunity to publish a stand-alone volume. The editors also thank the Materials Characterization Committee for sponsoring this symposium. They also thank the publisher, Springer, for the production of this book, and the authors, who are the core of this scientific work. Finally the editors express their appreciation for the past chairs and members of the Materials Characterization Committee, who built this great symposium and who attracted talented and creative people to the committee, and attracted scientists and research groups from around the world to this symposium.

Shadia Ikhmayies
Bowen Li
John S. Carpenter
Jian Li
Jiann-Yang Hwang
Sergio Neves Monteiro
Donato Firrao
Mingming Zhang
Zhiwei Peng
Juan P. Escobedo-Diaz
Chenguang Bai
Yunus Eren Kalay
Ramasis Goswami
Jeongguk Kim

Contents

About the Editors

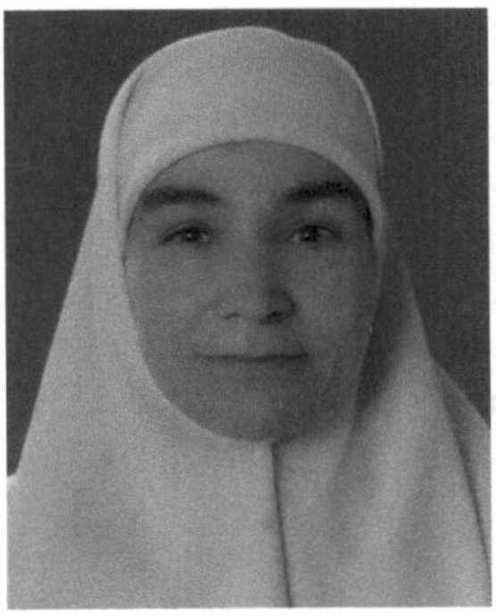

Shadia Ikhmayies had received the B.Sc. degree from the Physics Department in the University of Jordan in 1983, the M.Sc. degree in Molecular Physics from the same university in 1987 and the Ph.D. working on producing and characterizing CdS/CdTe thin film solar cells from the same university in 2002. She worked in the Applied Science University from 2004 to 2009 as Assistant Professor, and now she works in Al Isra University as Associate Professor. Her research is focused on producing and characterizing semiconductor thin films such as SnO_2: F, ZnO, CdS, CdTe, $CuInS_2$, thin film bilayers such as SnO_2:F/CdS:In, and thin film CdS/CdTe solar cells. She also works in characterizing quartz in Jordan for the extraction of silicon for solar cells and characterizing different materials by computation. She published 40 research papers in international scientific journals, three chapters in books, and 62 research papers in conference proceedings. She is the author of two books for Springer —*Silicon for Solar Cell Applications*, and *Performance Optimization of CdS/CdTe Solar Cells* —which are in development.

Shadia is a member of The Minerals, Metals & Materials Society (TMS) and a member of the steering committee of the World Renewable Energy Network (WREN/WREC). She is a member of the international organizing committee and the international scientific committee in the Third and Fourth European Conference on Renewable Energy Systems (ECRES2015 and ECRES2016). She was an associate editor in the journal of *Physics Express* published by Simplex Academic Publishers. She is an associate editor for the journal *Peak Journal of Physical and Environmental Science Research* (*PJPESR*) published by Peak Journals. She is a member of the editorial board of the *International Journal of Materials and Chemistry* for Scientific & Academic Publishing, the editor-in-chief of the book *Advances in II-VI Compounds Suitable for Solar Cell Applications* for the Research Signpost, the editor-in-chief of the book *Advances in Silicon Solar Cells* for Springer, and the eBook series *Material Science: Current and Future Developments* for Bentham Science Publishers, where the last two books are under construction. She was the technical advisor/subject editor for *JOM* as a representative of the Materials Characterization Committee for the year 2014. She is a guest editor for a special section in the *Journal of Electronic Materials*: Third European Conference on Renewable Energy. Shadia is a reviewer in 22 international journals and five international conferences, and she is the 2016–2017 Chair of the TMS Materials Characterization Committee.

Bowen Li is a research professor in the Department of Materials Science and Engineering and Institute of Materials Processing at Michigan Technological University. His research interests include materials characterization, metals extraction, ceramic processing, antimicrobial additives, applied mineralogy, and solid waste reuse. He has more than 100 publications and 14 patents. Bowen Li received a Ph.D. degree in Mineralogy and Petrology from China University of Geosciences Beijing in 1998, and a Ph.D. degree in Materials Science and Engineering from Michigan Technological University in 2008. He has been an active member in TMS, SME, and China Ceramic Society. At TMS, he has served as a member in Materials Characterization Committee, Powder Materials Committee, Biomaterials Committee, EPD Award Committee, and *JOM* Subject Advisor, as well as symposium co-organizer and session chair.

John S. Carpenter is a technical staff member in the Materials Science and Technology Division at the Los Alamos National Laboratory. Dr. Carpenter received his Ph.D. in Materials Science and Engineering from The Ohio State University in 2010 after performing his undergraduate studies at Virginia Tech.

His research interests include the characterization, processing, and mechanical testing of metallic nanocomposites fabricated via severe plastic deformation as well as additive manufacturing. Currently, his work focuses on understanding the relationship between plastic strain, texture, and the mechanical properties of bimetallic nanocomposites fabricated via accumulative roll bonding and joined using friction stir processing. This research involves the use of several characterization techniques including neutron scattering, X-ray synchrotron, PED, TEM, EBSD, and SEM. Mechanical testing for this work includes methods such as micropillar compression, microtension, and nanoindentation. He has more than 45 journal publications, one book chapter, and more than 20 invited technical talks to his credit

With regard to TMS service, Dr. Carpenter currently serves as the past chair for the Materials Characterization Committee, a programming representative for the Extraction and Processing Division (EPD), and the chair for the Advanced Characterization, Testing & Simulation Committee. He is also a participating member of the Mechanical Behavior of Materials, Content Development and Dissemination, and the Nanomechanical Behavior committees. He serves as a Key Reader for *Metallurgical and Materials Transactions A* and has co-edited special sections in *JOM* related to neutron characterization, coherent X-ray diffraction imaging methods, and modeling in additive manufacturing. He is the 2012 recipient of the Young Leaders Professional Development Award for the EPD of TMS. Dr. Carpenter was also awarded an honorable mention for the 2012 Los Alamos National Laboratory Postdoctoral Distinguished Performance Award.

Jian Li is a senior research scientist at CanmetMATERIALS in Natural Resources Canada. He obtained his B.Sc. in Mechanical Engineering from Beijing Polytechnique University; M.Sc. in Metallurgical Engineering from Technical University of Nova Scotia (TUNS) and Ph.D. in Materials and Metallurgical Engineering from Queen's University, Kingston, Ontario. He has broad experience in materials processing and characterization including alloys deformation, recrystallization and micro-texture development. Dr. Li has extensive experience in focused ion beam (FIB) microscope techniques. He is also an expert in various aspects of SEM-EDS and EPMA techniques. Dr. Li holds a patent, authored three book chapters, and published more than 120 papers in scientific journals and conference proceedings.

Jiann-Yang Hwang is a professor in the Department of Materials Science and Engineering at Michigan Technological University. He is also the Chief Energy and Environment Advisor at the Wuhan Iron and Steel Group Company, a Fortune Global 500 company. He has been the Editor-in-Chief of the *Journal of Minerals and Materials Characterization and Engineering* since 2002. Dr. Hwang has founded several enterprises in areas including water desalination and treatment equipment, microwave steel production, chemicals, flyash processing, antimicrobial materials, and plating wastes treatment. Several universities have honored him as a Guest Professor, including the Central South University, University of Science and Technology Beijing, Chongqing University, Kunming University of Science and Technology, Hebei United University, etc.

Dr. Hwang received his B.S. degree from National Cheng Kung University in 1974, M.S. in 1980 and Ph.D. in 1982, both from Purdue University. He joined Michigan Technological University in 1984 and has served as Director of the Institute of Materials Processing from 1992 to 2011 and Chair of the Mining Engineering Department in 1995. He has been a TMS member since 1985. His research interests include the characterization and processing of materials and their applications. He has been actively involved in the areas of separation technologies, pyrometallurgy, microwaves, hydrogen storages, ceramics, recycling, water treatment, environmental protection, biomaterials, and energy and fuels. He has more than 28 patents and has published more than 200 papers. He has chaired the Materials Characterization Committee and the Pyrometallurgy Committee in TMS (The Minerals, Metals & Materials Society) and has organized several symposiums. He is the recipient of TMS Technology Award and the Michigan Tech Bhata Rath Research Award.

Sergio Neves Monteiro graduated as metallurgical engineer (1966) at the Federal University of Rio de Janeiro (UFRJ), and received his M.Sc. (1967) and Ph.D. (1972) from the University of Florida. This was followed by a course (1975) in Energy at the Brazilian War College and postdoctorate (1976) at the University of Stuttgart. He joined (1968) the Metallurgy Department (1977) as full professor of the postgraduation program in engineering (COPPE) of the UFRJ, was elected head of department (1978), coordinator of COPPE (1982) and Under-Rector for Research (1983). Dr. Monteiro was invited as Under-Secretary of Science for the State of Rio de Janeiro (1985) and Under-Secretary of College Education for the Federal Government (1989). He retired in 1993 from the UFRJ and joined the State University of North Rio de Janeiro (UENF), from where he retired in 2012. He is now Professor at the Military Institute of Engineering (IME), Rio de Janeiro. He has published over 1200 articles in journals and conference proceedings and has been honored with several awards including the ASM Fellowship, top researcher (1A) of the Brazilian Council for Scientific and Technological Development (CNPq) and top scientist of State of Rio de Janeiro (FAPERJ). He served as president of the Superior Council of the State of Rio de Janeiro Research Foundation, FAPERJ, (2012) and currently is coordinator of the Engineering Area of this Foundation. Dr. Monteiro is the president of the Brazilian Association for Metallurgy, Materials and Mining—

ABM (2017–2019), a consultant for the main Brazilian R&D agencies and member of the Editorial board of five international journals as well as Associate Editor of the *Journal of Materials Research and Technology*.

Donato Firrao earned his Laurea in chemical engineering at the Politecnico di Torino, Turin, Italy, in 1968 and his M.Sc. in Metallurgical Engineering at The Ohio State University (OSU) on a Fulbright Scholarship in 1970. He began teaching in 1968 as Assistant Professor of ferrous extractive metallurgy and Lecturer of chemistry at the Politecnico di Torino since 1971. In 1983 he became Associate Professor of technology of metallic materials, gaining full professorship in the subject three years later. He also stayed as visiting fellow from 1978 to 1979 at the OSU Materials Science and Engineering Department, where he was named Distinguished Alumnus in 2003. Firrao has authored more than 230 papers, primarily in the fields of physical and mechanical metallurgy and surface heat treatments. He is a member of Associazione Italiana di Metallurgia (AIM), ASTM International, ESIS, TMS, and the Turin Academy of Sciences, Fellow of ASM International (2011) and member of the Failure Analysis Society (FAS). A founding partner of the Italian Group on Fracture (IGF), he was its secretary since its establishment in 1982, and the president between 1988 and 1994. Firrao was co-chair of the ESIS Technical Committee I (Elasto-Plastic Fracture Mechanics) from 1987 to 1996 and was named ESIS Fellow in 2016. He was president of the Federation of European Materials Societies (FEMS) from 2000 to 2001. Since 1993, Firrao has been the president of the board of trustees of the Collegio Universitario di Torino (a private nonprofit university student housing foundation). Firrao served as the Dean of the First College of Engineering at the Politecnico di Torino from 2005 to 2012; he retired in November 2015. He is an expert in failure analysis, and has acted as technical advisor to the judge in national and international trials (such as the Ustica aircraft crash, the Mattei affair, and the Sgrena/Calipari cases).

Mingming Zhang is currently a senior research engineer at ArcelorMittal Global R&D in East Chicago, Indiana. His main responsibilities include raw material characterization and process efficiency improvement in the mineral processing and ironmaking areas. He also leads technical relationships and research consortia with university and independent laboratory members and manages pilot pot-grate sintering test facility at ArcelorMittal Global R&D East Chicago. Dr. Zhang has more than 15 years of research experience in the field of mineral processing, metallurgical and materials engineering. He obtained his Ph.D. degree in Metallurgical Engineering from The University of Alabama and his Master degree in Mineral Processing from General Research Institute for Non-ferrous Metals in China. Prior to joining ArcelorMittal, he worked with Nucor Steel Tuscaloosa, Alabama where he was a metallurgical engineer leading the development of models for simulating slab solidification and secondary cooling process. Dr. Zhang has conducted a number of research projects involving mineral beneficiation, thermodynamics and kinetics of metallurgical reactions, electrochemical processing of light metals, energy efficient and environmental cleaner technologies. He has published more than 30 peer-reviewed research papers and is the recipient of several U.S. patents. Dr. Zhang also serves as editor and key reviewer for a number of prestigious journals including *Metallurgical Transactions A* and *B*, *JOM*, *Journal of Phase Equilibria and Diffusion*, and *Mineral Processing and Extractive Metallurgy Review*. Dr. Zhang has made more than 20 research presentations at national and international conferences including more than 10 keynote presentations. He is the recipient of 2015 TMS Young Leader Professional Development Award. He has been invited by a number of international professional associations to serve as conference organizer and technical committee member. These associations include the Association for Iron & Steel Technology (AISTech) and The Minerals, Metals & Materials Society (TMS).

Zhiwei Peng is Associate Professor in the School of Minerals Processing and Bioengineering at Central South University and Adjunct Assistant Professor in the Department of Materials Science and Engineering at Michigan Technological University. He received his B.E. and M.S. degrees from Central South University in 2005 and 2008, respectively, and his Ph.D. degree in Materials Science and Engineering from Michigan Technological University in 2012. His research interests include heat transfer in microwave heating, dielectric characterization of materials, non-thermal microwave effects, extractive metallurgy, computational electromagnetics, microwave absorbing materials, and biomaterials.

He has published more than 70 papers, including 45 peer-reviewed articles in multiple journals such as *International Materials Reviews*, *Metallurgical and Materials Transactions A*, *JOM*, *Journal of Power Sources*, *Fuel Processing Technology*, *Energy & Fuels*, *IEEE Transactions on Magnetics*, *IEEE Transactions on Instrumentation and Measurement*, *Ceramics International*, and *Annals of Medicine*. He has served as a key reviewer for a number of journals and been on the editorial board of the *Journal of Minerals and Materials Characterization and Engineering* since 2012. He received a TMS Travel Grant Award for the 141st TMS Annual Meeting & Exhibition, the Doctoral Finishing Fellowship and Dean's Award for Outstanding Scholarship of Michigan Technological University in 2012 and the Bhakta Rath Research Award of Michigan Technological University in 2013.

Dr. Peng is an active member of The Minerals, Metals & Materials Society (TMS). He has co-organized five TMS symposia (Characterization of Minerals, Metals and Materials in 2013, 2014, 2015, 2016, and 2017) and co-chaired twelve TMS symposia sessions since 2012. He is a member of the Pyrometallurgy and Materials Characterization committees, the chair of the Continuing Education Sub-Committee of the Materials Characterization Committee, a *JOM* advisor for the Pyrometallurgy Committee, and a winner of the TMS EPD Young Leader Professional Development Award in 2014.

Juan P. Escobedo-Diaz is a senior lecturer in the School of Engineering and Information Technology (SEIT) at the University of New South Wales (UNSW) Canberra. He obtained his doctoral degree in Mechanical Engineering at Washington State University. Prior to taking up this academic appointment he held research positions at the Institute for Shock Physics and Los Alamos National Laboratory.

His main research interests center on the dynamic behavior of materials under extreme conditions, in particular high pressure and high strain rate. His focus has been on investigating the effects of microstructural features on the dynamic fracture behavior of metals and metallic alloys. He has published primarily in the fields of shock physics and materials science.

He has been a member of The Metals, Minerals & Materials Society (TMS) since 2011. During this time he has co-organized more than five symposia at the Annual Meetings including the symposium on Characterization of Minerals, Metals and Materials in 2014. He was awarded a 2014 SMD Young Leader Award.

Chenguang Bai is Professor in the Department of Metallurgical Engineering, School of Materials Science and Engnieering at Chongqing University, China. He received his B.S. in 1982, M.S. in 1987, and Ph.D. in 2003 from Chongqing University. He also continueed his studies in the Department of Metallurgy and Materials, University of Toronto as a visiting scholar between October 1995 to January 1997. Dr. Bai has been actively involved in the teaching and scientific research works in ferrous metallurgy, especially in the field of comprehensive utilization of vanadium-titanium magnetite resources. He has more than 20 patents, published more than 200 research articles, about 60 of which were in the international metallurgical periodicals. He also is Vice Chairman of the Chongqing Society for Metals, and was a member of the Advisory Committee of Experts, Department of Engineering and Materials Science, National Science Foundation of China (NSFC). He was the Vice President from 2009 to 2011, and the Vice Chairman of University Council of Chongqing University from 2011 to 2016.

Yunus Eren Kalay is Associate Professor in the Metallurgical and Materials Engineering Department and assistant to the president at Middle East Technical University (METU) Ankara, Turkey. Dr. Kalay received his Ph.D. degree with Research Excellence award from Iowa State University in 2009. His Ph.D. topic was related to the metallic glass formation in Al-based metallic alloy systems. Following his Ph.D., he pursued postdoctoral research in Ames National Laboratory, where he was given the opportunity to practice "Atom Probe Tomography". In 2011, Dr. Kalay joined the Department of Metallurgical and Materials Engineering (METE) of METU as Assistant Professor and in 2014, he was promoted to Associate Professor. His research interests span microstructural evolution in metallic alloys, rapid solidification of metallic alloys, nanostructured and amorphous alloys, lead-free solders, electronic packaging, and advanced characterization techniques such as scanning and transmission electron microscopy, electron and X-ray spectroscopy, application of synchrotron X-ray scattering in materials research. Dr. Kalay was awarded the METU Prof. Dr. Mustafa Parlar Foundation Research Incentive Award, which is a very prestigious award that recognizes young scientists in Turkey with exceptional achievements and research productivity. He is also an active member of Materials Characterization Committee and Phase Transformations Committee of TMS, and served in organizing committees of three international congresses and one national congress including IMMC, MS&T and TMS. Dr. Kalay has also been involved in many synergistic activities such as being founder editor of Turkey's first undergraduate research journal, *MATTER* (http://matter.mete.metu.edu.tr/), and organizing the Materials Science Camps for K-12 students.

Ramasis Goswami is a scientist with the Multifunctional Materials Branch of the Materials Science and Technology Division at Naval Research Laboratory, Washington, D.C., USA. He obtained his bachelor degree in Metallurgical Engineering from Bengal Engineering College, Shibpur, India. He then earned his Master's and Ph.D. degrees in Materials Engineering from Indian Institute of Science, Bangalore. Dr. Goswami is a recipient of the Alexander von Humboldt fellowship. His current areas of research include the study of dislocation structures ahead of the crack tip, the microstructure and property relationship in metals, alloys and in multilayered thin films, and the study of interfaces and defects in semiconducting thin films. He has published more than 90 peer-reviewed articles in scientific literature.

Jeongguk Kim received his Ph.D. in Materials Science and Engineering at the University of Tennessee, Knoxville, in 2002. The title of his Ph.D. thesis was "Nondestructive Evaluation (NDE) and Mechanical Behavior of Continuous Fiber Reinforced Ceramic Matrix Composites (CFCCs)." Currently, he is a director at the Future Transportation Systems Research Division, Korea Railroad Research Institute (KRRI), Korea. He is also Professor in the Railway Systems Engineering Department, the KRRI campus, at the University of Science and Technology, Korea.

Dr. Kim's research interests include testing and certification of railroad components and systems, failure and safety analyses of railroad materials and systems based on fracture mechanics and several different types of nondestructive evaluation (NDE) techniques including ultrasonic testing, acoustic emission, infrared thermography, magnetic particle testing, etc., and mechanical behavior of advanced railway materials.

His recent research efforts include development of future transportation systems such as rail-canal system based on multi-axle bogies, an innovative train-ferry system, and the smart container lift. He also enlarged his research on the development of effective maintenance technologies for high-speed train systems.

He has been a member of TMS since 1996, and he has been a regular contributor at TMS meetings as an author and session chair at the Characterization of Minerals, Metals and Materials session since 2005.

Part I
Soft Materials

Charpy Toughness Behavior of Fique Fabric Reinforced Polyester Matrix Composites

Artur C. Pereira, Sergio N. Monteiro, Foluke S. Assis and Henry A. Colorado

Abstract The fique fiber is an important natural fiber, originally from Colombia where it is used for sacks and crafts while its plant is used to contain slopes. However, few studies were realized with the fiber obtained from the fique plant leaf, its mechanical properties are superior in many aspects in comparison to some other lignocellulosic fibers. This work investigated the toughness behavior of polyester matrix composites reinforced with up to 30% in volume of a fabric made of fique fiber by means of Charpy impact tests. It was found that the addition of fique fabric results in a marked increase in the absorbed energy by the composites. Macroscopic observation of the post-impact specimens and SEM fracture analysis showed that transversal rupture through the fique fabric interface with the polyester matrix is the main mechanism for the remarkable toughness of these composites.

Keywords Fique fabric · Polyester composites · Charpy impact test · Rupture mechanism

Introduction

The interest of this research is to develop composites with polyester resin matrix reinforced with continuous and aligned fique fibers, for applications in various industries, including construction and automotive industry. Conflicts related to the use of non-renewable forms of energy are increasing the interest to enter the market to replace natural materials; synthetic materials synthetics have a higher power consumption in its manufacture [1–4].

A.C. Pereira · S.N. Monteiro · F.S. Assis (✉)
Military Institute of Engineering, IME, Praça Gen. Tibúrcio, nº80 Urca,
Rio de Janeiro, RJ 22290-270, Brazil
e-mail: foluke.assis@hotmail.com

H.A. Colorado
Universidad de Antioquia, Calle 67 #53 - 108, Medellín, Antioquia, Colombia

S. Ikhmayies et al. (eds.), *Characterization of Minerals, Metals, and Materials 2017*, The Minerals, Metals & Materials Series,
DOI 10.1007/978-3-319-51382-9_1

Therefore, applications of natural lignocellulosic fibers obtained from cellulose-based plants are receiving increased attention as an alternative to replace more environmentally correct non-recyclable materials, energy intensive and glass fiber composites in [5, 6].

The use of composites reinforced with natural fibers is a reflection of the concerns with environmental issues such as pollution caused by waste. That is not biodegradable or cannot be incinerate and climate change due to CO_2 emissions associated with the processes of intensive energy and motivates this work to develop self-sustaining. Since natural fibers generate a source of income, especially in developing countries, where most originate, encouraging the cultivation of non-food agriculture [7–9]. Additionally, it is worth also remembering that these fibers come from renewable sources, in addition to being abundant, inexpensive and have a relevant set of mechanical properties [10].

Then, in order to have a composite rigid enough to compete with conventional products such as sheets of wood, only a limited percentage of fique fabric can be incorporated in the polymeric matrix [11–13]. This means that the final cost of the composite would more depending on its processing and polymer resin used as matrix. Therefore, the aim of this work was to study the mechanical properties of polyester matrix composites reinforced with fique fabric by means of Charpy impact tests.

Experimental Procedure

Composites were prepared with distinct volume fractions, up to 30%, of fique fabric incorporated into a commercial unsaturated polyester resin, already mixed with 0.5 wt% of methyl-ethyl-ketone. The as-received fique fabric were first cleaned in water and then dried at 60 °C for 24 h. Plates of these composites were press molded and allowed to cure at RT for 5 h. Standard notched specimens, 63 × 12.7 × 10 mm for Charpy impact testing according to the ASTM D256 norm, were cut from the plate along the direction of alignment of the fabric. Figure 1 illustrates the schematic specimen with standard dimensions. The notch with 2.54 mm in depth, angle of 45° and a tip curvature radius of 0.25 mm. For each condition, 10 specimens were tested to assure a statistical validation. The specimens were impact tested with an EMIC hammer pendulum. The impact energy was obtained with a 2.7 J hammer for the pure polyester specimens, a 10.8 J hammer for up to 20% fique fabric composites, and a 21.6 J hammer for the 25 and 30% fique fabric composites.

Fig. 1 Standard specimen schematic

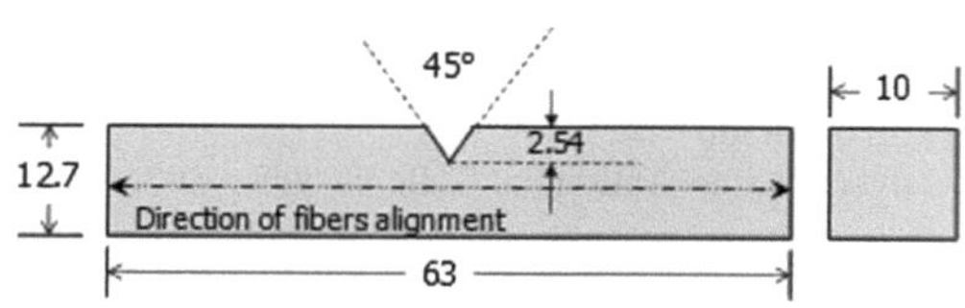

The impact fracture surface of the specimens was analyzed by scanning electron microscopy, SEM, in a model JSM-460 LV Jeol microscope. Gold sputtered SEM samples were observed with secondary electrons imaging at an accelerating voltage of 15 kV.

Result and Discussion

Table 1 presents the results of Charpy impact tests of polyester matrix composites reinforced with different volume fractions of fique fabric.

Based on the results shown in Table 1, the variation of the Charpy impact energy with the amount of fique fabric in the polyester composite is shown in Fig. 2.

In Fig. 2 it should be noticed that the fique fabric incorporation into the polyester matrix significantly improves the impact toughness of the composite. Within the standard deviation, this improvement can be considered as a linear function with respect to the amount of fiber up to 30%. The relatively high dispersion of values, given by the standard deviation associated with the higher fiber percentage points in Fig. 2, is a well known heterogeneous characteristic of the lignocellulosic fibers [5]. The values shown in this figure are consistent with results reported in the literature. The reinforcement of a polymeric matrix with both synthetic [14] and natural [15–17] fibers increases the impact toughness of the composite.

Table 1 Charpy impact energy for polyester composites reinforced with fique fabric

Volume fraction of fique fabric (%)	Charpy impact energy (J/m)
0	23.8 ± 1.2
10	109.44 ± 49.86
20	203.22 ± 43.55
30	293.11 ± 78.14

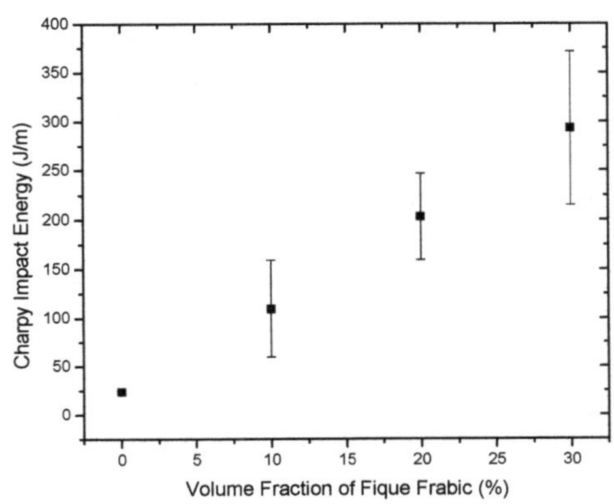

Fig. 2 Charpy impact energy as a function of different volume fractions of fique fabric

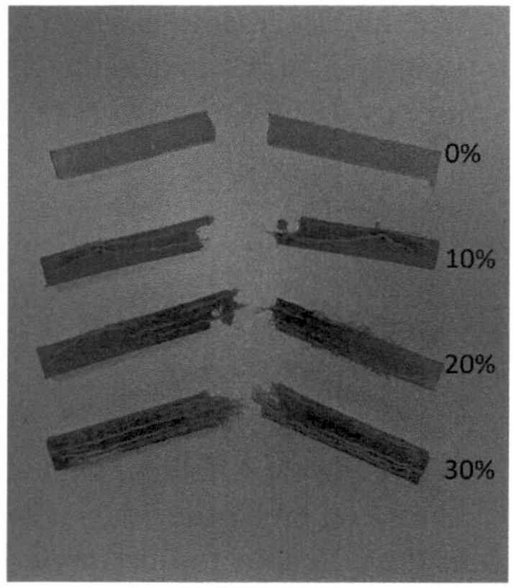

Fig. 3 Typical ruptured specimens by Charpy impact tests

The relatively low interface strength between a hydrophilic natural fiber and a hydrophobic polymeric matrix contributes to an ineffective load transfer from the matrix to a longer fiber. These results in relatively greater fracture surface and higher impact energy needed for the rupture [18]. Another factor is the flexural compliance of a long fiber during the impact test, which will be further discussed. The macroscopic aspects of the typical specimen ruptured by Charpy impact tests are shown in Fig. 3. In this figure it should be noted that the incorporation of aligned fique fabric results in a marked change with respect to pure polyester (0% fiber) in which a totally transversal rupture occurs.

Even with 10% of fabric, the rupture is no longer completely transversal. This indicates that the cracks nucleated at the notch will initially propagate transversally through the polyester matrix, as expected in a monolithic polymer. However, when the crack front reaches a fiber, the rupture will proceed through the interface. As a consequence, after the Charpy hammer hit the specimen, some long fibers will be pulled out from the matrix but, owing to their compliance, will not break but simply bend. In fact, for volume fractions of fiber above 10%, the specimens are not separated at all. For these amounts of fibers, part of the specimen was bent enough to allow the hammer to continue its trajectory carrying away the specimen without breaking it into pieces, which is expected in a Charpy test. The value of the impact toughness in this case cannot be compared with others in which the specimen is totally split apart. Anyway, the fact that a specimen is not completely separated in two parts underestimates the impact toughness. In other words, had all the fibers been broken, the adsorbed impact energy would be even higher.

The SEM analysis of the Charpy impact fracture permitted to have a better comprehension of the mechanism responsible for the higher toughness of polyester composites reinforced with fique fabric. Figure 4 shows the aspect of the fracture surface of a pure polyester (0% fiber) specimen. With lower magnification, the lighter layer in the left side of the fractograph, Fig. 4a, corresponds to the specimen notch, revealing the machining parallel marks. The smoother and gray layer on the right side corresponds to the transversal fracture surface. The fracture in Fig. 4

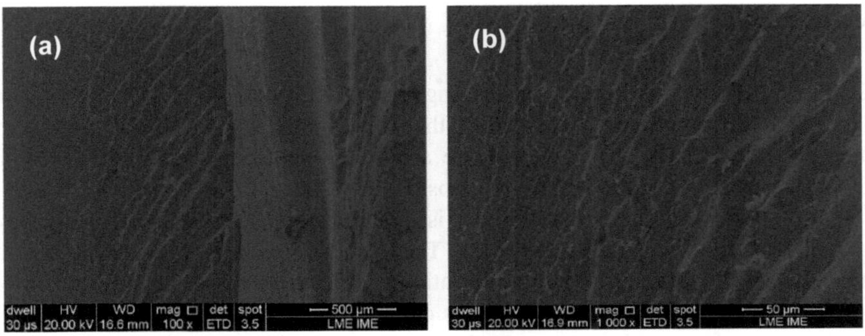

Fig. 4 Charpy impact fracture surface of pure polyester specimen (0% fiber): **a** general view; **b** detail of the polyester transversal fracture

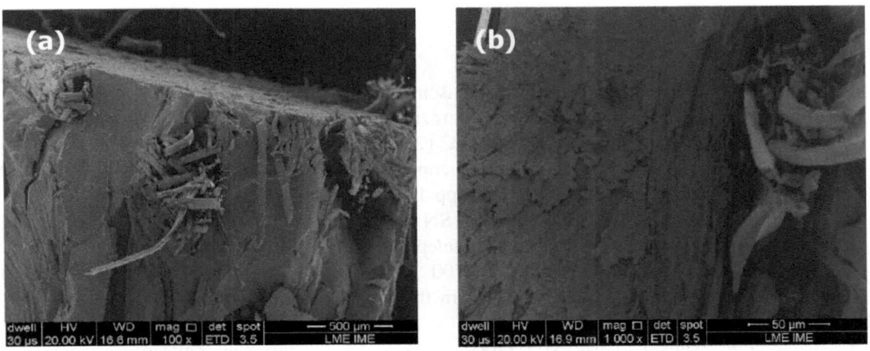

Fig. 5 Impact Charpy fracture surface of a polyester composite reinforced with 30% fique fabric: **a** 30× and **b** 500×

suggests that a single crack was responsible for the rupture with the roughness in Fig. 4b, being associated with voids and imperfections during the processing.

Figure 5 presents details of the impact fracture surface of a polyester composite specimen with 30% of fique fabric. This fractograph shows an effective adhesion between the fibers and the polyester matrix, where cracks preferentially propagate. Some of the fibers were pulled out from the matrix and others were broken during the impact. By contrast, the part of the specimen in which the rupture preferentially occurred longitudinally through the fiber/matrix interface reveal that most of the fracture area is associated with the fiber surface. This behavior corroborates the rupture mechanism of cracks that propagate preferentially in between the fique fabric surface and the polyester matrix due to the low interfacial strength [18]. The greater fracture area, Fig. 5, associated with the aligned fique fabric acting as reinforcement for the composite, justify the higher absorbed impact energy, Fig. 2, with increasing amount of fique fabric.

Conclusion

Composites with fique fabric reinforcing a polyester matrix display a significant increase in the toughness, measured by the Charpy impact test, as a function of the amount of the fiber. The values of the absorbed energy are the highest thus far obtained for lignocellulosic fiber composites.

Most of this increase in toughness is apparently due to the low fique fabric/polyester matrix interfacial shear stress. This results in a higher absorbed energy as a consequence of a longitudinal propagation of the cracks throughout the interface, which generates larger rupture areas, as compared to a transversal fracture.

Amounts of fique fabric above 10% are associated with incomplete rupture of the specimen owing to the bend flexibility, i.e., flexural compliance, of the fique fabric.

References

1. Monteiro SN, Lopes FPD, Ferreira AS, Nascimento DCO (2009) Natural fiber polymer matrix composites: cheaper, tougher and environmentally friendly. JOM 61:17–22
2. Monteiro SN, Nascimento DCO, Motta LC (2008) Effect of jute waste fiber surface treatment on the reinforcement strength in epoxy composites. In: Proceeding of the REWAS 2008 conference, Cancun, Mexico, Oct 2008, pp 1–6
3. Nascimento DCO, Lopes FPD, Monteiro SN (2010) Tensile behavior of lignocellulosic fiber reinforced polymer composites: Part I jute/epoxy. Rev Mater 15(2):199–205
4. Mohanty AK, Misra M, Drzal LT (2002) Sustainable biocomposites from renewable resources: opportunities and challenges in the green material world. J Polym Environ 10: 19–26
5. Wambua P, Ivens I, Verpoest I (2003) Natural fibers: can they replace glass and fibre reinforced plastics? Compos Sci Technol 63:1259–1264
6. Netravali AN, Chabba S (2003) Composites get greener. Mater Today 6:22–29
7. Monteiro SN, Lopes FPD, Ferreira AS, Nascimento DCO (2009) Natural fiber polymer matrix composites: cheaper, tougher and environmentally friendly. JOM 61:17–22
8. Aquino RCMP, D'Almeida JRM, Monteiro SN (2001) Flexural mechanical properties of piassava fibers (Attalea funifera)-resin matrix composites. J Mater Sci Lett 20:1017–1019
9. Kumar AP, Singh RP, Sarwade BD (2005) Degradability of composites, prepared from ethylene-propylene copolymer and jute fiber under accelerated aging and biotic environments. Mat Chem Phys 92:458–469
10. Crocker J (2008) Natural materials innovative natural composites. Mater Technol 2–3: 174–178
11. Monteiro SN, Lopes FPD, Ferreira AS, Nascimento DCO (2009) Natural fiber polymer matrix composites: cheaper, tougher and environmentally friendly. JOM 61(1):17–22
12. Mohanty AK, Khan MA, Hinrichsen G (2000) Influence of chemical surface modification on the properties of biodegradable jute fabrics-polyester amide composites. Compos A 31: 143–150
13. Mohanty S, Verma SK, Nayak SK (2006) Dynamic mechanical and thermal properties of MAPE treated jute/HDPE composites. Compos Sci Technol 66:538–547
14. Fu SY, Lauke B, Mäder E, Hu X, Yue CY (1999) Fracture resistance of short-glass-fiber-reinforced and short-carbon-fiber-reinforced poly-propylene under charpy impact load and dependence on processing. J Mater Process Technol 89(90):501–507

15. Leão AL, Tan IH, Caraschi JC (1998) Curaua fiber—a tropical natural fibre from Amazon—potential and applications in composites. In: Proceedings of the international conference on advanced composites, Hurghada, Egypt, May 1998, pp 557–564
16. Monteiro SN, Aquino RCMP, Lopes FPD, Carvalho EA, d'Almeida JRM (2006) Charpy impact notch toughness of piassava fibers reinforced polyester matrix composites. Rev Mater 11(3):204–210 (in Portuguese)
17. Monteiro SN, Aquino RCMP, Lopes FPD, Carvalho EA, d'Almeida JRM (2006) Charpy impact notch toughness of piassava fibers reinforced polyester matrix composites. Rev Mater 11(3):204–210 (in Portuguese)
18. Yue CY, Looi HC, Quek MY (1995) Assessment of fibre-matrix adhesion and interfacial properties using the pullout test. Int J Adhes Adhes 15:73–80

Comparative Analysis of Curaua Fiber Density Using the Geometric Characterization and Pycnometry Technique

Natália de O.R. Maciel, Carolina G.D. Ribeiro, Jordana Ferreira, Janaina da S. Vieira, Cláudio R. Marciano, Carlos Maurício Vieira, Frederico M. Margem and Sergio N. Monteiro

Abstract One of today's biggest concerns has been environmental issues, which has motivated researches and the development of materials from renewable resources and environmentally friendly. Natural fibers have excelled in replacing synthetic fibers used in composites manufacturing, because natural fibers are biodegradable, abundance in the nature, low cost, low density, high strength, among others. Due to these qualities, some natural fibers are used for many purposes in automobile industry. However, the natural fiber has irregularities and pores in its structure which directly impact the density determination by geometric techniques and, accordingly, the volume of fibers used in composites. Therefore, the main objective is this study is to determine the density of curaua fiber by pycnometry and to compare it with the commonly used geometric technique.

Keywords Curaua fibers · Geometric density · Pycnometry

N. de O.R. Maciel (✉) · C.G.D. Ribeiro · J. Ferreira ·
J. da S. Vieira · C.R. Marciano · C.M. Vieira
Advanced Materials Laboratory, LAMAV, State University of the Northern Rio de Janeiro, UENF, Av. Alberto Lamego, 2000, Campos dos Goytacazes, RJ 28013-602, Brazil
e-mail: nataliaoliveiraroque@gmail.com

F.M. Margem
Redentor, BR 356, 25, Itaperuna, RJ 28300-000, Brazil

S.N. Monteiro
Military Institute of Engineering, IME, Praça General Tibúrcio, 80, Praia Vermelha, RJ 22291-270, Brazil

S. Ikhmayies et al. (eds.), *Characterization of Minerals, Metals, and Materials 2017*, The Minerals, Metals & Materials Series,
DOI 10.1007/978-3-319-51382-9_2

Introduction

Composite materials refer to materials containing strong fibers—continuous or non-continuous—embedded in a weaker material or matrix. The matrix keeps the geometric arrangement of fibers and transmits to these fibers the load acting on the composite component. The composite material shows intermediate mechanical performance, that is, superior to those of the matrix but lower than those of the fibrous reinforcement [1].

The applications to the composite materials include but not limited to electrical-electronics, building and public works, road transports, rail transports and maritime transports, cable transports, air transports, space transports, general engineering sector, sports and leisure, etc. [1].

Synthetic fibers are more common in composite materials but natural fibers (NF composites) offer benefits to the society from different points of view. According Agriculture and Consumer Protection, the benefits are regarded from an Economic, an Ecological and a Technical (E.E.T., see Fig. 1) [2].

In fact, the benefits that motivate the replacement of glass fiber for natural fiber in polymer composites [3], are also technical, economical and societal advantages [4, 5]. There are many reasons that favor the use of natural fibers, mainly those obtained from cellulose-based vegetables, also known as lignocellulosic fibers, such as cotton, flax, sisal, jute, hemp, wood, pineapple and curaua fiber. Actually, it is estimated that more than 500 lignocellulosic fibers are known and have potential to be used in engineering applications. Most of these fibers are native of tropical regions in Africa, South Asia, Central and South Americas. The plant cultivation, extraction and processing of lignocellulosic fibers represent an important source of income for people and countries in these regions.

Curaua fibers were used in this study. This fiber is extracted from the leaves of an Amazonian plant whose family is known as bromeliad and that resembles a pineapple plant. It has specific mechanical properties similar to inorganic fibers,

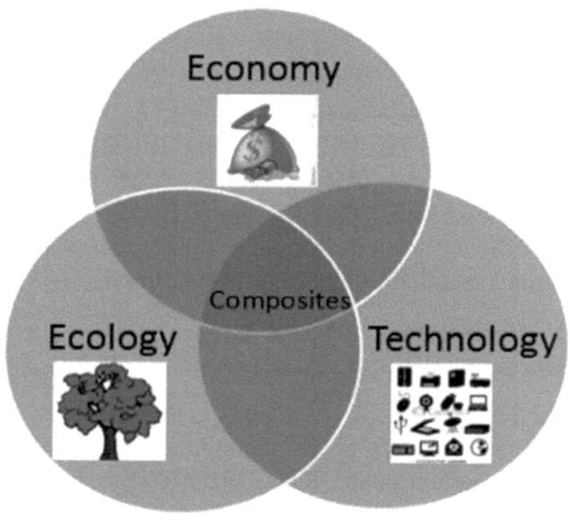

Fig. 1 E.E.T. (Economy, Ecology and Technology) (adapted from Agriculture and Consumer Protection)

low-cost of production, offers relatively a high tensile strength level and is an important renewable raw material [6, 7]. Curaua fiber is composed of lignin (7.5%), glucan (66.4%), xylan (11.6%) and other materials, such as mannan (0.1%), galactan (0.5%) and arabinan (0.5%) [8].

Density is used in many areas of application to designate certain properties of materials or product and it is an important property of the fiber [9]. It is a standard physical term defined as weight (mass) per unit volume. SI units are kg/m^3 but it's common to use g/cm^3 [10]. Normally density of vegetal fibers is known through of the geometric characterization, but these fibers have pores and high density of defects on the surface of the fiber which significantly increase the probability of error in determination the fiber's density. In the other words, it is measured the apparent density of the fiber since apparent density is the mass per unit volume (or the weight per unit volume) of a material, including the voids which are inherent in the material [11, 12].

Density determination by pycnometers a very precise. It uses a working liquid with well-known density as water. It used for determining both the density of liquids and dispersion by simply weighing the defined volume, but especially for determination of the density of powders and granules. Pycnometers can also be used to determine the density of the solid phase in a porous solid [9, 13].

So, thinking about this problem, the aim of this work is compare the analysis of curaua fiber density using the geometric characterization and pycnometry technique.

Experimental Procedure

The curaua was obtained from Amazon Paper. The typical aspect of curaua plantation and a bundle of soft fibers are showed in Fig. 2.

Fig. 2 **a** "White curua" and "purple curaua" plant respectively, **b** bundle curaua fiber [6]

Fiber Surface Analysis

The surface of representative specimen was analyzed by a Scanning Electron Microscope (SEM) SSX-550 Shimadzu.

Geometrical Characterization

The soft fibers were individually separated, for a statistical dimensional analysis. It was separate 100 (hundred) individual fibers and it was weigh each fiber (m_f) in a digital balance with 4 decimal places of the brand SHYMADZU and model AY220.

For determination of fiber diameter (d) was used a profile projector of the brand NIKON with magnification (×50). Each fiber was measured in 10 positions, five in a position and five turning the fiber in 90°.

The length of fiber (L_f) was measured with a metal graduated scale.

After obtaining the three measures described, were used the formulas of the density (1) and volume (2) to each fiber:

$$\rho_f = \frac{m_f}{V_f}\ (\mathrm{kg/m^3}) \tag{1}$$

where:

$$v_f = \frac{\pi d^2}{4} x L_f\ (\mathrm{m^3}) \tag{2}$$

Ultimately, it was calculated average density (3):

$$\bar{\rho} = \frac{1}{n}\sum_{i=1}^{n} \rho_1 \tag{3}$$

Pycnometry Technique

The fibers were dried in kiln at a temperature of 105 °C and weighed (fiber mass - Fm) in a digital balance with 4 (four) decimal places of the brand SHYMADZU and model AY220. Also weighed the empty pycnometer (m_1) and the pycnometer with water (m_2).

In the dry glass pycnometer (50 mL) was inserted the fibers and added distilled water until to reach 90% in volume of the glass pycnometer. The pycnometer with distilled water and fiber was shaken to eliminate the air bubbles. So the pycnometer was put in the desiccator linked with a vacuum pump with 400 mmHg suction applied. After 2 h with suction, it was removed to desiccator and shaken again.

The pycnometer was completed with water and placed the capillary cover. Dried the spare water that leaked through the capillary hole with a filter paper and measured total weight, that is, pycnometer + fiber + water (m_3).

The calculation was made with the density's equation (4):

$$\rho_{fibra} = \frac{m_f}{V_f}\ (\mathrm{kg/m^3}) \tag{4}$$

where, fiber volume (v_f) is equal the displaced water volume (V_d) due the presence of fiber into the fill pycnometer. Temperature depend of distilled water density $\rho_{H_2O}(W_d)$. So, to calculate the fiber volume was used the Eq. (5):

$$v_f = v_d = \frac{[(m_2 - m_1) - (m_3 - m_1 - f_m)]}{w_d} \tag{5}$$

Results and Discussion

Scanning Electron Microscopy—SEM

Figures 3 and 4 show the fiber imperfections. In Fig. 3, there are voids as indicated by the white arrow and in Fig. 4, it shows the irregularities in the all surface of the fiber.

Geometric Characterization

The measured distribution of 100 fibers revealed a dispersion interval in length from 150 to 164 mm, with an average of 157 mm (Fig. 5). Dispersion in diameter from 0.044 to 0.193 mm, with an average of 0.098 mm was also revealed (Fig. 6).

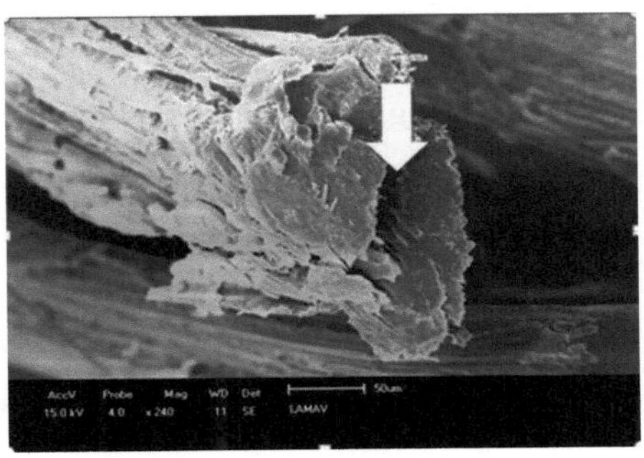

Fig. 3 Curaua fiber surface superior view with magnification (×240)

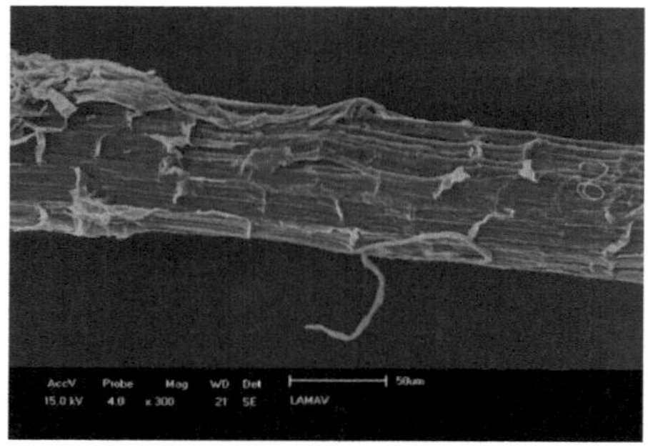

Fig. 4 Curaua fiber lateral view with magnification (×300)

Using Eq. 2 the volume of each fiber was obtained and using the equation, the density of each fiber was obtained. However, the final density was obtained through of Eq. 3:

$$\bar{\rho} = \frac{1}{100} \times 95.77 = 0.9577\ (\text{g/cm}^3) = 957.7\ (\text{kg/m}^3) \tag{6}$$

So, the density obtained was approximately 957.7 kg/m^3.

Pycnometry Technique

It was used 3 samples to do this technique. In Table 1 is showed the fiber mass of each sample submitted to 105 °C to eliminate the humidity, the mass of pycnometer, pycnometer + water and pycnometer + water + fiber respectively.

Using Eq. 5, obtain the displaced volume of fiber to each sample[1]:

$$v_{f1} = \frac{[(80.333 - 28.065) - (80.503 - 28.065 - 0.601)]}{0.9962} = 0.432\ \text{cm}^3$$

$$v_{f2} = \frac{[(86.472 - 37.997) - (86.543 - 37.997 - 0.320)]}{0.9962} = 0.250\ \text{cm}^3$$

$$v_{f3} = \frac{[(90.678 - 37.869) - (90.835 - 37.869 - 0.553)]}{0.9962} = 0.397\ \text{cm}^3$$

So, applying the result in Eq. 4, obtain the density to each sample:

$$\rho_{f1} = \frac{0.601}{0.432} = 1.391\ (\text{g/cm}^3) = 1391\ (\text{kg/m}^3)$$

[1]The water temperature was 28 °C which corresponds to a density 0.9962.

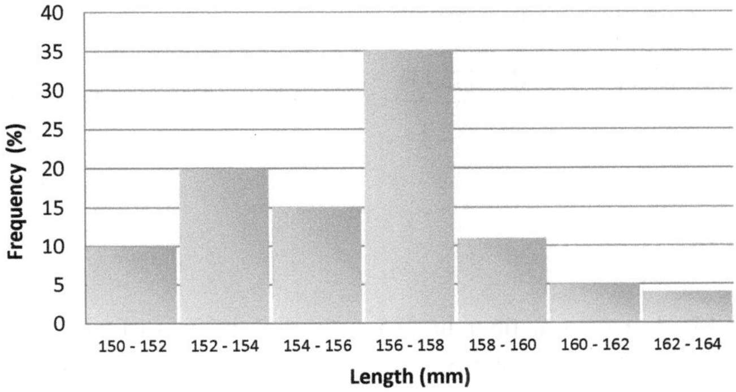

Fig. 5 Fiber length frequency graph

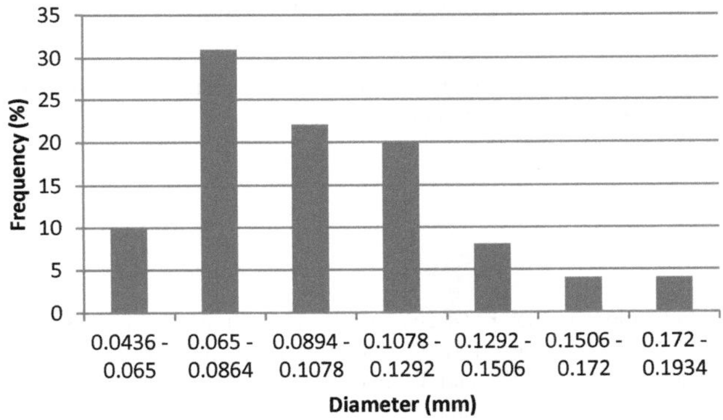

Fig. 6 Fiber diameter frequency graph

Table 1 Mass of the fiber, mass of the pycnometer, mass of the pycnometer with water and mass of the pycnometer with water and fiber

Sample	Dry fiber	Pycnometer (m_1)	Pycnometer + water (m_2)	Pycnometer + water + fiber (m_3)
	g	g	g	g
1	0.601	28.065	80.333	80.503
2	0.320	37.997	86.472	86.543
3	0.553	37.869	90.678	90.835

$$\rho_{f2} = \frac{0.320}{0.250} = 1.280\ (\text{g/cm}) = 1280\ (\text{kg/m}^3)$$

$$\rho_{f3} = \frac{0.553}{0.397} = 1.393\ (\text{g/cm}^3) = 1393\ (\text{kg/m}^3)$$

The average of the density is equal:

$$\bar{\rho} = \frac{(1.391 + 1.280 + 1.393)}{3} = 1.355(\mathrm{g/cm^3}) = 1355(\mathrm{kg/m^3})$$

Conclusions

- From SEM observation, fiber has irregularities in its surface and voids/pores. So, the geometrical technique is not the best method because it does not identify the imperfections of the fiber;
- The fiber has approximately 29% of voids and pores;
- The result of the pycnometer was satisfactory because the density of the fiber is similar to density of the organic matter that ranging from 1300 to 1500 kg/m^3.

Acknowledgements The authors thank the support to this investigation by the Brazilian agencies: CNPq, CAPES, FAPERJ and TECNORTE/FENORTE.

References

1. Gay D (2014) Composite materials: design and applications, 3rd ed. CRC Press, 638 p
2. Wambua P, Ivens I, Verpoest I (2003) Natural fibers: can they replace glass and fibre reinforced plastics? Compos Sci and Technol 63:1259–1264
3. http://www.fao.org/ag/portal/ag-home/en/
4. Crocker J (2008) Natural materials innovative natural composites. Mater Technol 2–3: 174–178
5. Trindade WG, Paiva JMF, Leão AL, Frollini E (2008) Ionized-air-treated curaua fibers as reinforcement for phenolic matrices. Macromol Mater Eng 293(6):521
6. Monteiro SN, Lopes FPD, Ferreira AS, Nascimento DCO (2009) Natural fiber polymer matrix composites: cheaper, tougher and environmentally friendly. JOM 61(1):17–22
7. Spinacé MAS, Lambert CS, Fermoselli KKG, De Paoli MA (2009) Characterization of lignocellulosic Curaua fibres. Carbohydr Polym 77(1):47–53
8. Gomes A, Goda K, Ohgi J (2004) Effects of alkali treatment to reinforcement on tensile properties of Curaua fiber green composites. JSME Int J 47:541–6
9. Kelley SS, Rowell RM, Davis M, Jurich CK, Ibach R (2004) Rapid analysis of the chemical composition of agricultural fibers using near infrared spectroscopy and pyrolysis molecular beam mass spectrometry. Biomass Bioenergy 27:77–88
10. Kelley SS, Rowell RM, Davis M, Jurich CK, Ibach R (1999) Manual of weighing applications: density. Marketing, weighing technology, Dublin City University. http://www.dcu.ie/sites/default/files/mechanical_engineering/pdfs/manuals/DensityDeterminationManual.pdf
11. Lal R (2006) Encyclopedia of soil science. CRC Press, Boca Raton, 1052 p
12. Lal R (2016) Density and porosity Micromeritics Pharmaceutical Services [Internet]. [citado 13 de agosto de 2016]. Available at: http://micrx.com/Analytical-Services/Density-and-Porosity.aspx

13. Lal R (2003) Apparent density. McGraw-Hill dictionary of scientific and technical terms, 6th ed. The McGraw-Hill Companies, Inc., New York, 13 Aug 2016. http://encyclopedia2.thefreedictionary.com/apparent+densit
14. Oremusová J, Vojteková M (1999) Density determination of liquids and solids. Manual for Laboratory Practice, UK, p 121

Izod Impact Test in Polyester Matrix Composites Reinforced with Blanket of the Malva and Jute Fibers

Ygor Macabu de Moraes, Carolina Gomes Dias Ribeiro, Frederico Muylaert Margem, Sergio Neves Monteiro, Jean Igor Margem and João Batista Vasconcelos

Abstract A natural fiber presents interfacial characteristics with polymeric matrices that favor a high impact energy absorption by the composite structure. The objective of this work was then to assess the Izod impact resistance of polymeric composites reinforced with one or two layers of batt malva and jute fibers. The results showed a remarkable increase in the notch toughness with increasing layers of mauve blankets and jute. This can be attributed to a preferential debonding of the fiber/matrix interface, which contributes to an elevated absorbed energy.

Keywords Blanket of the malva and jute · Composite · Polyester matrix · Izod testing · Notch toughness

Introduction

Natural fibers with high cellulose content, known as lignocellulosic fibers, become firmly established as a potential replacement for the search field of synthetic fibers, particularly glass fiber [1].

The use of natural fibers to replace the existing, present especially in aircraft and cars, is motivated by several advantages such as good toughness and less abrasion equipment used in processing composite [2–6].

Y.M. de Moraes · C.G.D. Ribeiro · F.M. Margem
LAMAV, State University of the Northern Rio de Janeiro, UENF,
Av. Alberto Lamego, 2000, Campos dos Goytacazes 28013-602, Brazil

S.N. Monteiro (✉)
Instituto Militar de Engenharia, IME, Praça Gen. Tibúrcio, n°80 Urca,
Rio de Janeiro, RJ 22290-270, Brazil
e-mail: snmonteitro@gmail.com

J.I. Margem · JoãoB. Vasconcelos
Instituto de Ensino Superiores do Censa, ISECENSA, Rua Salvador Correa,
139, Campos dos Goytacazes, Rio de Janeiro 28035-310, Brazil

S. Ikhmayies et al. (eds.), *Characterization of Minerals, Metals, and Materials 2017*, The Minerals, Metals & Materials Series,
DOI 10.1007/978-3-319-51382-9_3

Among these features is the low cost and light weight. In addition, unlike the glass fibers [7] lignocellulosic fibers are relatively flat and the processing procedures produce less wear on equipment. The environmental issue is another point in favor of natural fibers, which are renewable, recyclable, biodegradable and neutral with regard to CO_2 emissions [8, 9].

The incorporation of banana fibers in polymeric matrices of composite was investigated [10–12] and found to have significant properties. These properties are directly related to the microstructure of the fiber as well as the physical and chemical characteristics present in any lignocellulosic fiber [13–17]. The objective of this study was a preliminary assessment through different measures Izod impact energy, together with the micro structural characteristics associated with the fracture of polyester matrix composites with blanket of the malva and jute fibers.

Experimental Procedure

The materials used in this work were blanket of the malva and jute fibers which was Acquired by a producer, Pematec Triangel, from the Southeast region of Brazil (Fig. 1).

The blanket comprises 40% jute fiber and 60% of Malva fiber. The fiber of jute (*Corchorus capsularis*) is a vegetable textile fibers, this woody herb reaches 3–4 m in height and its stalk is approximately 20 mm. It is used mainly in sacks industry, due to resistance and strength of its fiber is also used in the furniture industry. Already Malva fiber belongs to a botanical genus of several species of herbaceous Malvaceae family, is distributed geographically by tropical, subtropical and temperate regions, the leaves are alternate, lobed and usually slaps and measure half to 5 cm. Its raw material is mainly used in the wireless industry and natural fibers packaging.

The polyester resin still liquid, together with 0.5% catalyst based on methyl ethyl ketone, was poured into the one and two layers of the blanket inside the mold. The composite thus formed was allowed to cure for 24 h at room temperature. The plates of each composite were then cut according to the direction of fiber alignment

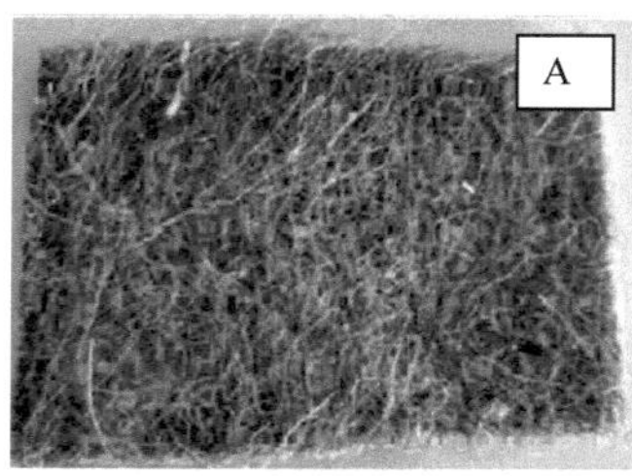

Fig. 1 Blanket of the malva and jute fibers

in bars measuring 10 × 125 × 12.7 mm were used as basis for preparation of test samples Izod impact test according to ASTM D256.

The samples were assayed in a pendulum of the Izod brand PANTEC in configuration belonging to the LAMAV UENF. The impact energy was obtained in power hammer with 11 J for composites. For each condition, relative to a certain fraction of fibers, 10 specimens were used and the results were statistically interpreted.

Results and Discussion

Table 1 shows the results of the values of Izod impact energy with their respective standard deviations for pure polyester and composites with up to two batt layers.

Table 1 Energy impact Izod for polyester matrix reinforced with blanket of the malva and jute fibers

Layers of the blanket	Energy (J/m)
0	104.18 ± 6.19
1	35.06 ± 2.96
2	54.04 ± 2.88

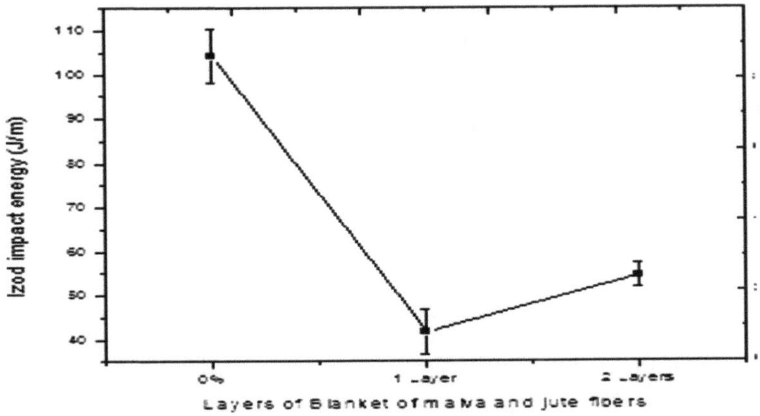

Fig. 2 Izod impact energy as a function of the amount of blanket of the malva and jute fibers

Based on the results of Table 1, the change in Izod impact energy with layers of the blanket this shown in Fig. 2 (Fig. 3).

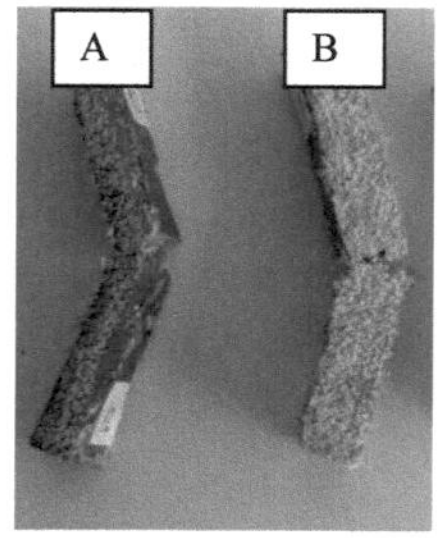

Fig. 3 Macrostructural aspects of Izod impact rupture polyester matrix composites with one layers (**a**) and two layers (**b**) of the blanket

Conclusions

The reduction of the mechanical properties of the hybrid composite with two layers of blanket with a higher content of malva and jute fibers can be attributed to low adherence between fiber/matrix and existing porus in the production of the blanket, added to the lack of chemical compatibility between the fibers naturals and the polyester resin.

- Specimens with two layers of blanket incorporated, however, did not undergo complete rupture. This leads to the decrease in toughness. If all fibers were broken, then the energy absorbed would have been even greater.
- The reason for having a crack nucleated at the notch, changing its trajectory to reach the blanket of malva and jute, and going to propagate through the interface with the matrix is due to the low interfacial resistance.
- As consequence of the incompatibility caused by the fact that lignocellulosic fibers are hydrophilic while the polymer matrix is hydrophobic.

Acknowledgements The authors thank the support to this investigation by the Brazilian agencies: CNPQ, CAPES, FAPERJ and TECNORTE/FERNORTE.

References

1. Bledzki AK, Gassan J (1999) Composites reinforced with cellulose-based fibers. Prog Polym Sci 24:221–274
2. Nabi Sahed D, Jog JP (1999) Natural fiber polymer composites: a review. Adv Polym Technol 18:221–274
3. Mohanty AK, Misra M, Hinrichsen G (2000) Biofibres, biodegradable polymers and biocomposites: an overview. Macromolecular Mater Eng 276:1–24
4. Crocker J (2008) Natural materials innovative natural composites. Mater Technol 2–3(3): 174–178
5. Monteiro SN, Lopes FPD, Ferreira AS, Nascimento DCO (2009) Natural fiber polymer matrix composites: cheaper, tougher and environmentally friendly. JOM 61(1):17–22
6. Gore A (2006) An inconvenient truth. The planetary emergency of global warming and what we can do about it. Rodale Press, Emmaus

7. Peijs T (2000) Natural fibers based composites. Mater Technol 15:281–285
8. Wambua P, Ivens I, Verpoest I (2003) Natural fibers: can they replace glass and fiber reinforced plastic? Compos Sci Technol 63:1259–1264
9. Mohanty AK, Misra M, Drzal LT (2002) Sustainable biocomposites from renewable resources: opportunities and challenges in the green materials world. J Polym Environ 10: 19–26
10. Satyanarayana KG, Sukumaran K, Kulkarni AG, Pillai SGK, Rohatgi PK (1999) Fabrication and properties of natural fibre-reinforced polyester composites. Composites 17:329–333
11. Bledzki AK, Gassan J (1999) Composites reinforced with cellulose-based fibers. Prog Polym Sci 24:221–274
12. Peijs T (2000) Natural fibers based composites. Mater Technol 15:281–285
13. Callister Jr WD (2000) Materials science and engineering—an introduction. Wiley, 5ª Edição, Nova York
14. Fu SY, Lauke B, Mäder E, Hu X, Yue CY (1999) Fracture resistance of short-glass-fiber-reinforced and short-carbon-fiber-reinforced poly-propylene under charpy impact load and dependence on processing. J Mater Process Technol 89–90:501–507
15. Leão AL, Tan IH, Caraschi JC (1998) Curaua fiber—a tropical natural fiber from Amazon—potential and applications in composites. In: International conference on advanced composites, Hurghada, Egito, Maio, pp 557–564
16. Monteiro SN, Costa LL, Lopes FPD, Terrones LAH (2008) Characterization of the impact resistance of coir fiber reinforced polyester composites. In: Mineral, metals and materials characterization symposium—TMS conference, New Orleans, LA, USA, Março, 2008, pp 1–6
17. Yue CY, Looi HC, Quek MY (1995) Assessment of fibre-matrix adhesion and interfacial properties using the pullout test. Int J Adhes Adhes 15:73–80

Tensile Behavior of Epoxy Matrix Composites Reinforced with Eucalyptus Fibers

Caroline G. de Oliveira, Anna C.C. Neves, Gilson V. Fernandes, Marcos V.F. Fonseca, Frederico M. Margem and Sergio N. Monteiro

Abstract Natural fibers represent an economic and environmental motivation for replacing the synthetic fibers, mainly as composite reinforcement material, which is one of their major applications. Some advantages of the natural fibers are the biodegradability, low cost and renewability. Among the natural fibers, the lignocellulosic ones are highlighted for their high resistance. One of the most cultivated lignocellulosic fibers in Brazil is the Eucalyptus fiber, extracted from the bark of *Eucalyptus citriodora* plant. In this work, it was investigated the tensile behavior of the epoxy matrix composites reinforced with different volume fractions of Eucalyptus fibers. The specimens were made by pouring the still liquid resin into the mold and laying the fibers onto the resin. The results show a decrease in the tensile resistance with the increase of volume fraction. It is due to the low adhesion between the fibers and the matrix.

Keywords Eucalyptus fiber · Eucalyptus composites · Natural fibers

Introduction

Economic growth and technological development make it necessary to search for new materials that meet their technical function and, at the same time, preserve the environment. Thus, the search for high-performance and environmentally friendly

C.G. de Oliveira (✉) · A.C.C. Neves · G.V. Fernandes · M.V.F. Fonseca
State University of Northern Rio de Janeiro, Campos dos Goytacazes, RJ, Brazil
e-mail: caroline.oliveyra@gmail.com

F.M. Margem
Faculdade Redentor, Itaperuna, RJ, Brazil

S.N. Monteiro
Military Institute of Engineering, Rio de Janeiro, RJ, Brazil

S. Ikhmayies et al. (eds.), *Characterization of Minerals, Metals, and Materials 2017*, The Minerals, Metals & Materials Series,
DOI 10.1007/978-3-319-51382-9_4

materials has been growing in recent decades. One of the aspects of this research is the study of natural fibers as reinforcement material, due to their low cost, low density (allowing the production of materials with low weight and good mechanical properties), renewability and biodegradability [1–3].

Polymeric matrix composites are one of the most important materials nowadays, with applications ranging from sports to automobile and aeronautic industries. Frequently these composites are reinforced with synthetic fibers, such as glass and aramid ones. However, these materials are obtained from non-renewable resources, including all their inherent environmental disadvantages. Thus, the natural ligno-cellulosic fibers have been studied as a reinforcement alternative, due to their lower density, lower cost, renewability, biodegradability and relevant mechanical properties [4–7].

The use of natural fibers as reinforcement materials to polymeric matrix is of interest of various industries. They are used in automobile parts building, jackets, coatings and furniture, for example [8]. In order to find new possibilities to this research field, the Eucalyptus fiber have been recently studied as an alternative of reinforcement material. This fiber is extracted from the bark of *Eucalyptus citriodora* plant, which is extensively cultivated in Brazil.

The objective of this work is to investigate the tensile behavior of epoxy matrix composites reinforced with continuous and aligned Eucalyptus fibers.

Experimental Procedure

The material used in this work is untreated Eucalyptus fibers. Composites with different volume fractions (30, 40 and 50%) were confectioned by pouring some resin into the mold and then placing the fibers and pouring more resin onto the fibers. The specimens were cured for 24 h at room temperature and posteriorly tested in an universal Instron machine, model 5582, at 25 °C. In Fig. 1 it can be seen the Eucalyptus bark and the fibers.

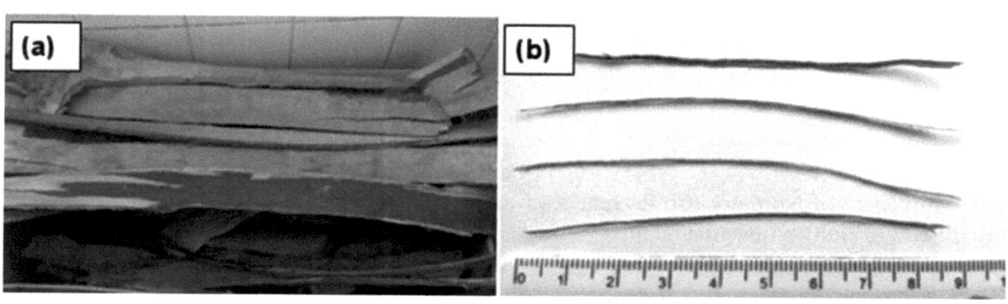

Fig. 1 **a** Eucalyptus plant bark and **b** Eucalyptus fibers

Results and Discussion

In Fig. 2 it can be observed some examples of the load vs. elongation curves obtained from the Instron machine software. There is a typical elastic line followed by a sudden fracture for all the compositions, which evidences the brittle behavior of the matrix as well as of the composites.

From the curves shown in Fig. 2 it was possible to obtain the tensile strength and the elastic modulus for the composites with different volume fractions of fibers. The data is presented in the Table 1.

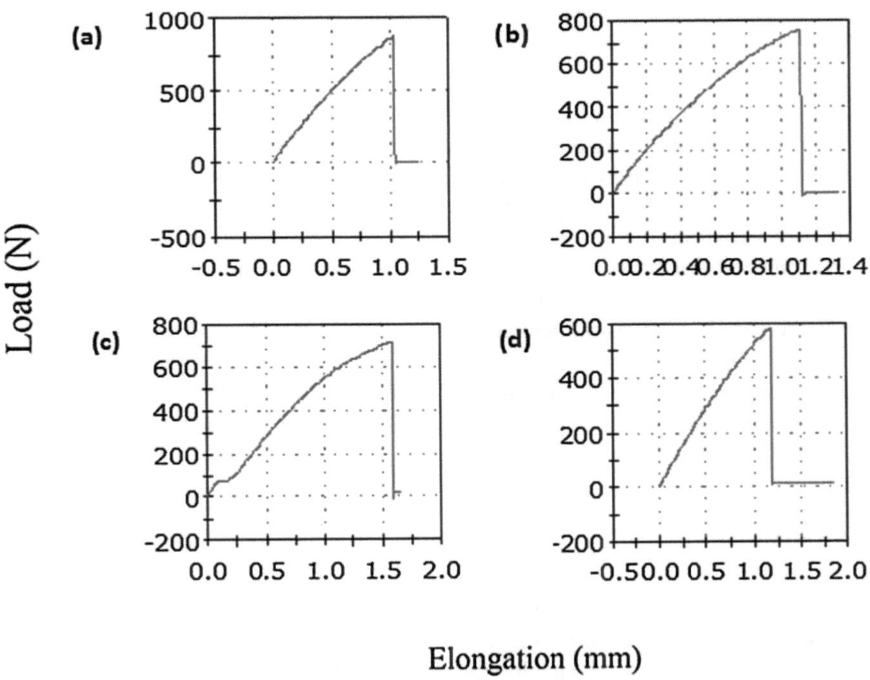

Fig. 2 Load versus elongation curves: **a** pure epoxy, **b** 30% of fiber, **c** 40% of fiber and **d** 50% of fiber

Table 1 Tensile properties of epoxy composites reinforced with different volume fraction of Eucalyptus fibers

Volume fraction of fibers	Tensile strength (MPa)	Elastic modulus (GPa)
0	27.72 ± 7.22	0.31 ± 0.05
30%	24.67 ± 5.97	0.37 ± 0.06
40%	23.72 ± 3.88	0.39 ± 0.06
50%	14.67 ± 6.23	0.46 ± 0.03

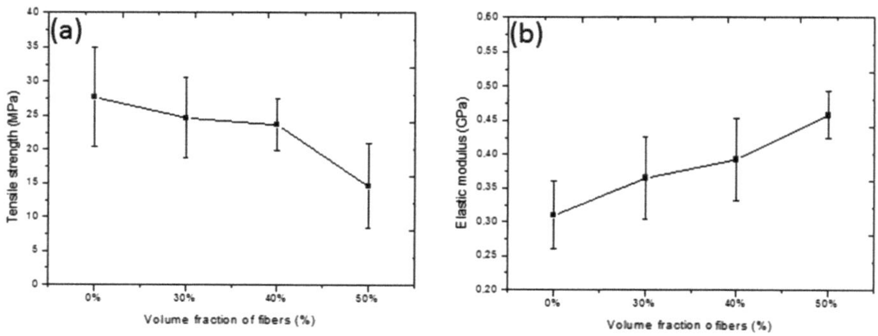

Fig. 3 Graphical results of **a** tensile strength, **b** elastic modulus for composites with different Eucalyptus fiber volume fraction

It is possible to note that the presence of Eucalyptus fibers decreased the tensile strength and increased the elastic modulus of the composite. In fact, the value of tensile strength decreased approximately 45% and the elastic modulus increased approximately 50%. In the Fig. 3 it can be seen the graphical results of the tensile properties variation.

Within the error bars, the variation of both properties are almost linear, which characterizes a material behavior. The worsening observed in the tensile strength is due to the very low adhesion between the matrix and the fibers. It worth noting that the natural fibers, and Eucalyptus particularly, have high quantity of water on their surfaces, aggravating this effect. An alternative to try to solve this issue is to apply a pre-treatment to the fibers, which is out of the objective of this work.

Conclusions

- The introduction of Eucalyptus fibers decreases the tensile strength in approximately 45%, not working as a reinforcement material.
- A pre-treatment can be an alternative to make Eucalyptus fibers better reinforcement material. Although, it implies in higher costs and chemical waste disposal in the nature.
- The introduction of Eucalyptus fibers increases the elastic modulus in approximately 50%.

Acknowledgements The authors thank the support to this investigation by the Brazilian agencies: CNPq, CAPES, FAPERJ.

References

1. Wambua P, Ivens I, Verpoest I (2003) Natural fibers: can they replace glass and fibre reinforced plastics? Compos Sci Technol 63:1259–1264
2. Netravali AN, Chabba S (2003) Composites get greener. Mater Today 6:22–29
3. Crocker J (2008) Natural materials innovative natural composites. Mater Technol 2–3:174–178
4. Chawla KK (1993) Composite materials. Springer, New York
5. Wambua P, Ivens I, Verpoest I (2003) Natural fibers: can they replace glass and fibre reinforced plastics? Compos Sci Technol 63:1259–1264
6. Bledzki AK, Gassan J (1999) Composites reinforced with cellulose-based fibers. Prog Polym Sci 4:201–274
7. Sahed DN, Jog JP (1999) Natural fiber polymer composites: a review. Adv Polym Technol 18:221–274
8. Satyanarayana KG, Guimarães JL, Wypych F (2007) Studies on lignocellulosic fibers of Brazil. Part I: source, production, morphology, properties and applications. Compos A 38:1694–1709

The Dimensional Characterization of Jute Fabric Strips for Reinforcement in Composite Polymeric

Sergio N. Monteiro, Frederico M. Margem, Glenio F. Daniel, Vinícius O. Barbosa, André R. Gomes and Victor B. de Souza

Abstract Nowadays notorious technological growth is encouraging research natural materials in substitution to more usual synthetic ones. One of those initiatives is the vegetable fibers that have been studied for the replacement of glass and carbon fibers. This research addresses broadly the study of the natural fibers mechanics properties, but none study have been made with natural fabrics, such as the pure jute fabric used in Brazil to make ropes and carpets. This work aim to evaluate the mechanical behavior of jute fabric strips with different dimensions to be used as reinforcement into polymeric composites Lots of those fabric strips were subjected to tensile tests in order to obtain the tensile resistance for each strip size. The results obtained during the tests were satisfactory and indicated that 160 × 160 × 1 mm correspond to the best result and the perfect fabric strip for the composite reinforcement.

Keywords Jute fabric · Dimensional characterization · Tensile test · Best strip indication · Green material

S.N. Monteiro · F.M. Margem (✉) · G.F. Daniel
V. O. Barbosa· A.R. Gomes · V.B. de Souza
State University of the Northern Rio de Janeiro, UENF, Av. Alberto Lamego, 2000, Campos dos Goytacazes 28013-602, Brazil
e-mail: fmargem@gmail.com

S.N. Monteiro · F.M. Margem · G.F. Daniel
V.O. Barbosa · A.R. Gomes · V.B. de Souza
Military Institute of Engineer IME/RJ, Praça General Tibúrcio, 80 - Praia Vermelha, Rio de Janeiro 22291-270, Brazil

S.N. Monteiro · F.M. Margem · G.F. Daniel
V.O. Barbosa · A.R. Gomes · V.B. de Souza
Redentor Faculty, Rodovia BR-356, 25, Cidade Nova, Itaperuna, RJ 28300-000, Brazil

S. Ikhmayies et al. (eds.), *Characterization of Minerals, Metals, and Materials 2017*, The Minerals, Metals & Materials Series,
DOI 10.1007/978-3-319-51382-9_5

Introduction

Environmental issues are nowadays assuming a paramount importance in our society owing to the consequences of climate changes and generalized pollution. In fact, any increasing amount of fluid and solid wastes is becoming a worldwide problem with direct connection to global warming [1–3]. Not only gas emission from burning fossil fuels but also oxidation and decomposition of organic matter contribute to increase the temperature of our planet [3]. The simple disposal or land filling of carbon molecule containing products such as food, wood, or other natural fibers will be eventually transformed into CO_2 or CH_4 that are strong greenhouse activation gases. The case of natural fibers deserves special attention. The use of fibers obtained from domestic animals and cultivated plants in handmade or industrialized items such as textile, carpets and roofing has been growing from centuries, following the increase in world population. Both, during fabrication and as end-of-life product, an ever expanding amount of natural fiber wastes is being disposed or burnt, which is threatening the environment.

At this moment that natural fiber wastes are of environmental concern, those lignocellulosic extracted from plants are gaining attention as composite reinforcement [4]. The reasons for this are related to low cost and to the fact that lignocellulosic fibers are renewable and biodegradable. Moreover, the production system of composites reinforced with these lignocellulosic fibers, in comparison with similar composites reinforced with glass fibers, causes low equipment wear, as well as a relative saving in energy. In addition, it has been shown [5–7] that the incorporation of lignocellulosic fibers may significantly improve some mechanical properties of polymeric composites.

Jute is an abundant fiber extracted from leaves of a cultivated bush-like plant (*Chorchorus capsularis*) illustrated in Fig. 1.

Fig. 1 The plant and fibers of jute

In many developing regions, jute fibers are applied in the form of twisted strands and weaved fabrics [8, 9]. In particular, jute fabrics are extensively used around the world to make sackcloth for storing and transport cotton, potatoes, coal, wool, etc. The jute fabric used as sackcloth is underused in its properties and applications; the high capacity of holding the products indicates a good resistance and probably a good reinforcement material for polymer composite.

Composites made from the jute fiber sackcloth or fabric have been investigated for their properties [10, 11] and as possible substitute for conventional building materials [12]. One important mechanical property for these composites subject are the tensile solicitations, but before using the fabric as a composite it is necessary to understand the strip dimension with generates the better composite result. This property has, so far, not been assessed from environmentally correct jute handmade fabric. Therefore, the objective of this work was to investigate the tensile of different dimension jute fabric strips and indicate the better fit to composite fabrication.

Experimental Procedure

Materials

The 3 kg amount of handmade untreated and raw jute fabric was provided by a local producer. Specimens were then fabricated by cutting the fabric in the dimensions 40 × 160 × 1 mm, 80 × 160 × 1 mm, 120 × 160 × 1 mm, 160 × 160 × 1 mm, for each dimensions 15 strips specimens were cut. The jute fabric strips as shown in Fig. 2 were tensile tested in a Kratos machine model Ke Series 500 kgf. The fabric consists of open interlace spaces with more than 5 mm of distance between the twisted strands that were integer in the fabric.

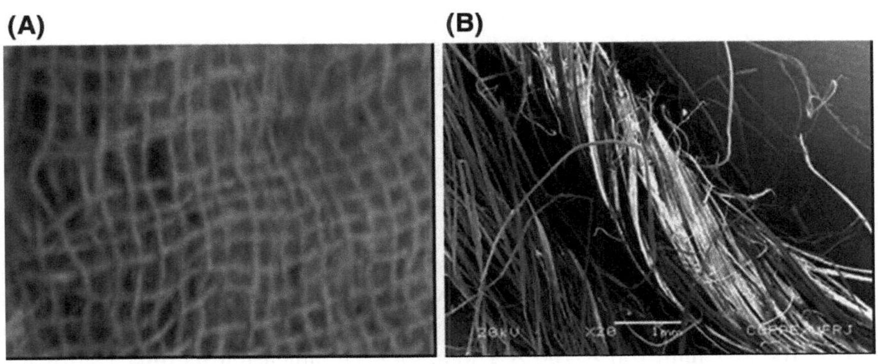

Fig. 2 Jute fabric strips (**a**) and (**b**) twisted strand and interlace space

Fig. 3 Jute fabric strips tested in a Ke series 500 kgf Kratos tensile machine

The tensile machine used to obtain the results for the fabric strips testes is illustrated in Fig. 3.

It should be mentioned that the as-received fabric were relatively uniform, Fig. 2b, and apparently free of contamination from dirt or other imperfections due to the selection carried out by the manufacturing firm.

Methods

Rectangular pieces of jute fabric taken from the lot, Fig. 2, were cleaned and dried in open air. Four sets of jute fabric strips were then fabricated by razor blade cutting process, the used dimensions were 40 × 160 × 1 mm, 80 × 160 × 1 mm, 120 × 160 × 1 mm, 160 × 160 × 1 mm, for each dimension set fifteen strips was cut down. Each specimen was then tensile tested in the Kratos machine and the result for the tensile strength and maximum load plotted for each choose dimension.

Results and Discussion

The variation of the tensile strength and maximum load with the fabric strip dimension is shown in Fig. 4.

In this figure it should be noticed that the jute fabric strips with bigger length obtain a significantly improvement in the maximum load but the tensile strength did not change at all. Within the standard deviation, the improvement in the maximum load can be considered as a linear function with respect to the length of fabric strip up to 160 mm. The relatively high dispersion of values, given by the standard deviation associated with the higher length in Fig. 4, is a well known heterogeneous characteristic of the lignocellulosic fibers [5]. The values shown in this figure are consistent with results reported in the literature.

In this work, using different dimensions of jute fabrics strips, the maximum load and tensile resistance were tested to obtain the better fabric dimension to polymeric composites using this kind of fabric and resins such as polyester and epoxy. The obtain results for maximum load are significantly higher for higher length fabric. So far in the literature was not possible to find other papers that tested the dimensions on natural fiber fabrics to determine the best size to produce specimens. The greater maximum load of the bigger fabric strips in comparison with the smaller ones could be explained by the better woven in the larger fabric strip. However, there are other important factors related to the maximum load of the different dimensions jute fabric strips.

The relatively low fiber density into the woven fabric and the natural fiber characteristic common in the environmentally friendly material contributes to an ineffective load transfer along the fabric. This result indicates that the use of these strips fabrics sizes as composites reinforcement into polymeric resins will probably generate greater fracture surface and higher impact energy needed for the rupture

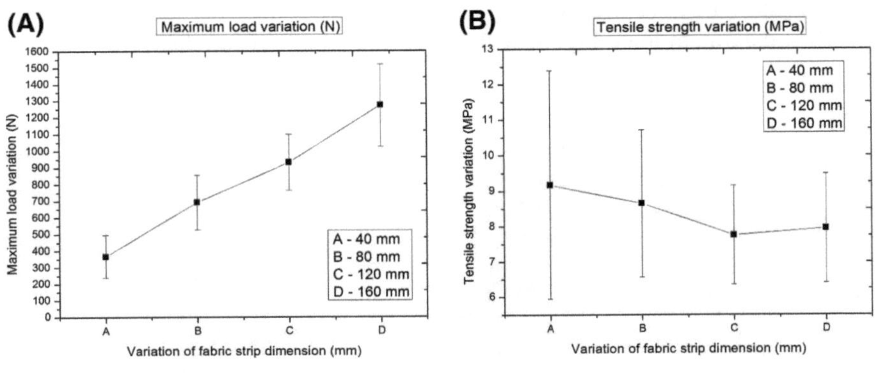

Fig. 4 **a** Maximum load for each jute strip fabric dimension; and **b** tensile strength for each jute strip fabric dimension

[13–16]. Another factor is the flexural compliance of a natural long fiber fabric during the mechanical tests, which will be further discussed.

The higher length (160 mm) of aligned jute fabrics strips results in a marked change with respect to the shorter fabric (40 mm) in which an almost totally transversal rupture occurs. Even with 80 and 120 mm, the rupture is no longer completely transversal. This indicates that the extra length will act as a reinforcement in the material and amplify the path of the rupture, generating a higher difficulty in the transversal propagation on the fabric rupture, as expected in the 40 mm fabric strip. However, when the rupture front reaches a perpendicular fiber, the rupture will proceed through the interface. As a consequence some long fibers will be pulled out from the fabric but, owing to their compliance, will not break but simply stretches. In fact, for fabric strips above 80 mm, the specimens are not separated at all. For these amounts of long jute fabrics, part of the specimen was bent enough to allow the material to be loosed from the tensile machine. The value of the tensile energy in this case cannot be compared with others in which the specimen is totally split apart. Anyway, the fact that a specimen is not completely separated in two parts underestimates the tensile energy. In other words, had all the fibers been broken, the adsorbed tensile energy would be even higher.

The SEM analysis of the fabric woven surface after the tensile tests permitted to have a better comprehension of the mechanism responsible for the higher maximum load of higher length strips of long post-used jute fabric. Figure 5 presents details of the tensile test fracture surface in 160 × 160 × 1 mm of jute fabrics. This fractograph shows an effective adhesion between the fibers in the fabric structure, it is possible to check the perpendicular fiber changing the energy direction during the tensile tests, and bending the material before the rupture. Some of the fibers were pulled out from the fabric mesh and others were broken during the tensile stress.

This behavior corroborates the rupture mechanism of cracks that propagate preferentially in between the woven for the jute fabric due to the low interfacial strength [16, 17]. The greater fracture area, Fig. 5, associate with the aligned and meshed jute fabrics acting as reinforcement for the composite, justify the higher absorbed maximum load, Fig. 4, with increasing length of jute fabrics.

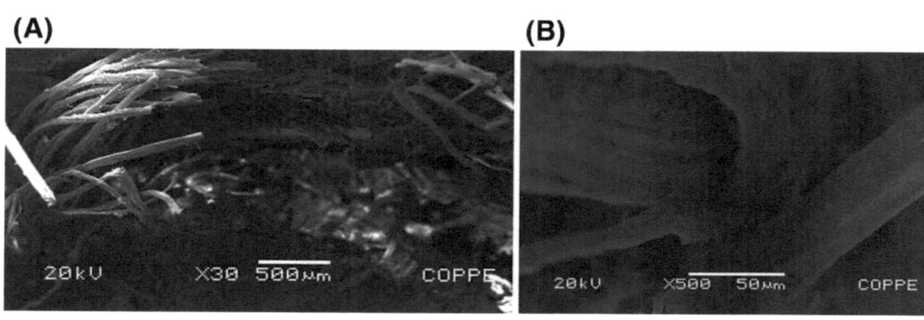

Fig. 5 Fracture surface in a 160 × 160 × 1 mm specimen of jute fabrics: **a** 30× and **b** 500×

Conclusions

- Natural Jute fibers fabric strips with higher length display a significant increase in the maximum load, measures by the tensile strength test, as a function of the amount of the fiber into the specimens.
- Most of this increase in maximum Load is apparently due to the low jute fibers interface interaction between the fibers on the fabric. This results indicates that this fabric strips can create a polymeric composite with a higher absorbed energy as a consequence of a longitudinal propagation of the cracks throughout the interface, which generates larger rupture areas, as compared to a transversal fracture.
- Amounts of jute fabrics strips above 120 mm are associated with incomplete rupture of the specimen owing to the bend flexibility, i.e., flexural compliance, of the jute fabrics.

Acknowledgements The authors thank the support to this investigation by the Brazilian agencies: CNPq, CAPES, FAPERJ and TECNORTE/FENORTE. It is also acknowledged the Redentor Faculty Research Committee for the aid in scientific projects and the UFRJ for the permission to the use of the impact equipment of IQ/UFRJ and the SEM microscope by the PEMM from COPPE/UFRJ.

References

1. Weart SR (2003) The discovery of global warming. Harvard University Press, Boston
2. Houghton JT, (2004) Global warming—the complete briefing 3rd edn. Cambridge University Press, Cambridge
3. Gore, A (2006) An inconvenient truth. In: The planetary emergency of global warming and what we can do about it. Rodale Press, Emmaus
4. Satyanarayana KG, Guimarães JL, Wypych F (2007) Studies on lignocellulosic fibers of Brazil. Part I: source, production, morphology, properties and applications. Compos A 38:1694–1709
5. Bledzki AK, Gassan J (1999) Composites reinforced with cellulose-based fibers. Prog Polym Sci 4:221–274
6. Nabi Sahed D, Jog JP (1999) Natural fiber polymer composites: a review. Adv Polym Technol 18:221–274
7. Mohanty AK, Misra M, Hinrichsen G (2000) Biofibers, biodegradable polymers and biocomposites: an overview. Macormol Mat Eng 276(277):1–24
8. Kumar AP, Singh RP, Sarwade BD (2005) Degradability of composites prepared from ethylene-propylene copolymer and jute fiber under accelerated aging and biotic environments. Mat Chem Phy 92:458–469
9. Doan T-T-L, Gao S-L, Mäder E (2006) Jute/polypropylene composites I. Effect of matrix modification. Compos Sci Technol 66:952–963
10. Monteiro SN, Terrones LAH, Lima AC, Petrucci LJT, d'Almeida JRM (2006) Properties of recycled polyethylene composites reinforced with discarded jute sackcloth (in Portuguese). Rev Mater 11(4):403–411
11. Monteiro SN, Lima AC, Terrones LAH, d'Almeida JRM (2007) Fabric rupture mechanism of jute sackcloth used as reinforcement in polyethylene composites. In: Proceedings of

characterization of minerals, metals and materials—TMS Conference, Orlando, FL, USA, February, 2007, p 1
12. Monteiro SN, Lima AC, Terrones LAH, d'Almeida JRM (2008) Recycled polyethylene composites reinforced with jute fabric from post–used sackcloth: an environmentally correct building material. In: Proceedings of the REWAS 2008, Cancun, Mexico, October, 2008, pp 1–6
13. Fu SY, Lauke B, Mäder E, Hu X, Yue CY (1999) Fracture resistance of short-glass-fiber-reinforced and short-carbon-fiber-reinforced poly-propylene under charpy impact load and dependence on processing. J Mater Process Technol 89(90):501–507
14. Leão AL, Tan IH, Caraschi JC (1998) Curaua fiber—a tropical natural fibr from Amazon—potential and applications in composites, In: Proceedings of the international conference on advanced composites, Hurghada, Egypt, May, 1998, pp 557–564
15. Monteiro SN, Aquino RCMP, Lopes FPD, Carvalho EA, d'Almeida JRM (2006) Charpy impact notch toughness of piassava fibers reinforced polyester matrix composites (in Portuguese). Rev Mater 11(3):204–210
16. Yue CY, Looi HC, Quek MY (1995) Assessment of fibre-matrix adhesion and interfacial properties using the pullout test. Int J Adhes Adhes 15:73–80
17. Monteiro SN, Costa LL, Lopes FDP, Terrones LAH (2008) Characterization of the impact resistance of coir fiber reinforced polyester composites. In: Proceedings of the TMS conference, New Orleans, LA, USA, March, 2008 pp 1–6

Izod Toughness Behavior of Continuous Palf Fibers Reinforced Polyester Matrix Composites

Gabriel O. Glória, Giulio R. Altoé, Maycon A. Gomes, Carlos Maurício F. Vieira, Maria Carolina A. Teles, Frederico M. Margem and Sergio N. Monteiro

Abstract Due to growing concern about the ambient impacts, the society is demanding for materials that are environmentally friendly. The lignocellulosic fibers, like PALF fibers, appears like an option to replace the synthetic ones in composites with polimerics matrices. Besides the environmental advantages, the lignocellulosic fibers have economical and some properties advantages like the low density in comparison with the synthetic fibers. Nowadays the lignocellulosic are used in industries, like the automobile and aerospace. This objective of this study was to analyze the absorbed impact energy of the composites reinforced with PALF fibers. The fibers were incorporated into the polyester matrix with volume fraction from 0 to 30%. The results showed increase in the absorbed impact energy with the increase of the percentage of fiber incorporated. The fracture surface of the specimens were analyzed by SEM (scanning eléctron microscope) and indicates that the PALF fibers act as reinforcement for the polyester matrix.

Keywords PALF fiber · Polyester composite · Izod test

Introduction

Considering the increase of the concern about the ambient impacts, the society is looking for new materials to replace some materials that are harmful. The natural fibers that contains lignin, are known lignocellulosic fibers, can be viable. The application of the lignocellulosic fibers is motivated by others advantages like good

G.O. Glória (✉) · G.R. Altoé · M.A. Gomes
C.M.F. Vieira · M.C.A. Teles · F.M. Margem
State University of the Northern Rio de Janeiro, UENF, LAMAV,
Av. Alberto Lamego, 2000, 28013-602 Campos dos Goytacazes, Brazil
e-mail: gabrieloliveiragloria@hotmail.com

S.N. Monteiro
Instituto Militar de Engenharia, IME, Praça Gen. Tibúrcio,
n°80 Urca, Rio de Janeiro, RJ 22290-270, Brazil

S. Ikhmayies et al. (eds.), *Characterization of Minerals, Metals, and Materials 2017*, The Minerals, Metals & Materials Series,
DOI 10.1007/978-3-319-51382-9_6

toughness and less wear of equipment used in the processing of composites. Nowadays, these fibers are being considered as a substitute for synthetic ones, used by the industry on large scale [1–5]. Besides the advantages of the lignocellulosic fibers are renewable, biodegradable and neutral with respect to CO_2, one of the most responsible for the global warming and climate changes [5, 6].

Despite the advantages the lignocellulosic fibers shows some weaknesses like the dimensional heterogeneity and the difficulty of coupling with polymer matrices, a hydrophobic material contrasting with the hydrophilic character of fiber. These disadvantages are worrisome reasons for the occurrence of reduction on the strengthening capacity, as indicated by the rule of mixtures for composites, a weighted mean used to predict various properties of a composite material [6–8].

According to Wambua et al. [9] the most important problem for the composite strength is the adhesion between the fiber and polymeric matrix. Since the matrix has to transfer an applied load to the reinforcing fibers through interfacial shear stress, a good bond between the matrix and the fiber is required.

Previous works demonstrate that the incorporation of lignocellulosic fibers in polymeric matrix gives rise to composites with mechanical resistance directly proportional of the fiber content [10–12], where these fibers act as reinforcement for matrix due to their high mechanical properties. Therefore, the objective of this study was to evaluate the mechanical properties of polyester matrix composites reinforced with continuous and aligned PALF fibers in the izod impact test.

Experimental Procedure

100 PALF fibers are randomly selected from the bundle in Fig. 1. The equivalent diameter corresponding to the average between the larger and smaller (90° rotation) cross section dimensions at five locations for each fiber was measured in a profile projector Nikon 6C.

The histogram corresponding to the distribution of diameter of the as-received PALF fibers are shown in Fig. 2. The equivalent diameter of each fiber was actually the average value obtained by in a profile projector at five distinct locations (two with 90° rotation at each location).

The histogram in Fig. 2 reveals a big dispersion in the diameter, which is a consequence of the non-uniform physical characteristics of a lignocellulosic fiber, common in natural fibers [1, 2, 7–9]. It should be noticed that the diameter range was 0.10–0.28 mm with an average of 0.20 mm.

Composites with up to 30% in volume of PALF fibers were fabricated by placing the fibers longitudinally aligned inside a steel mold and then pouring the still fluid orthophthalic polyester resin mixed with a methyl-ethyl-ketone hardener into the mold. Standard specimens for Izod impact test, with 62.5 × 12.7 × 10 mm, were prepared according to the ASTM D256 [10, 12] with aligned Fibers along its length. Figure 3 illustrates the Izod impact pendulum and the schematic specimen

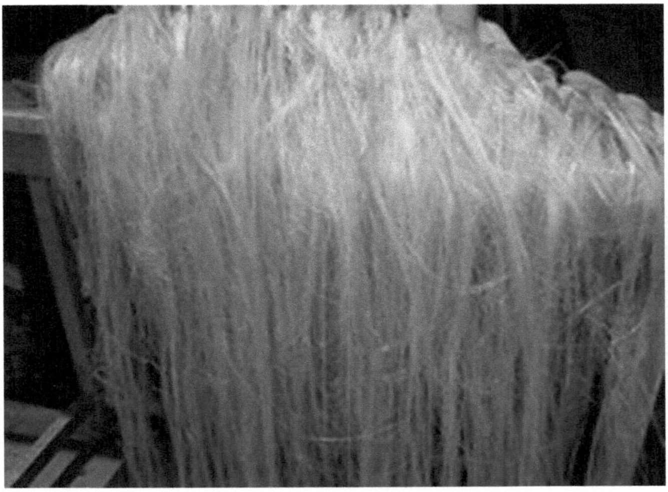

Fig. 1 A small bundle of PALF fibers

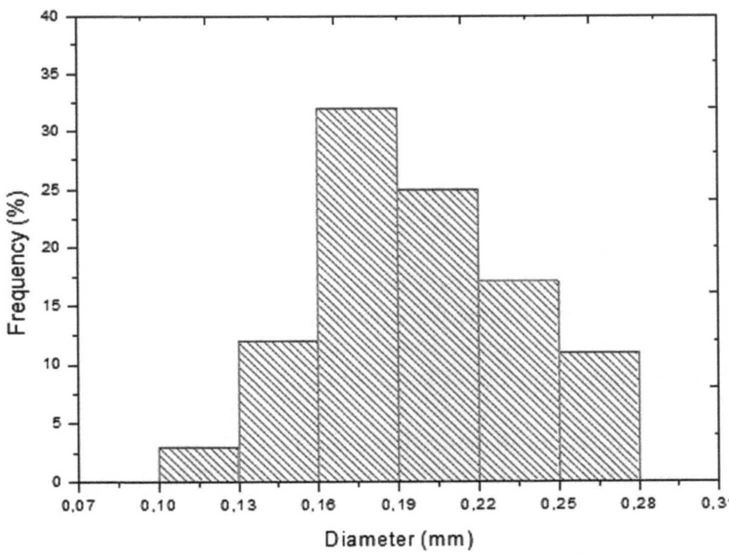

Fig. 2 Histogram for the distribution of diameter of the as-received PALF fibers

with standard dimensions. The notch with 2.54 mm in depth, angle of 45° and a tip curvature radius of 0.25 mm was machined with a DIN 847 milling tool. For each volume fraction of fiber more than 10 specimens were impact tested in a PANTEC hammer pendulum to ensure a statistical validation.

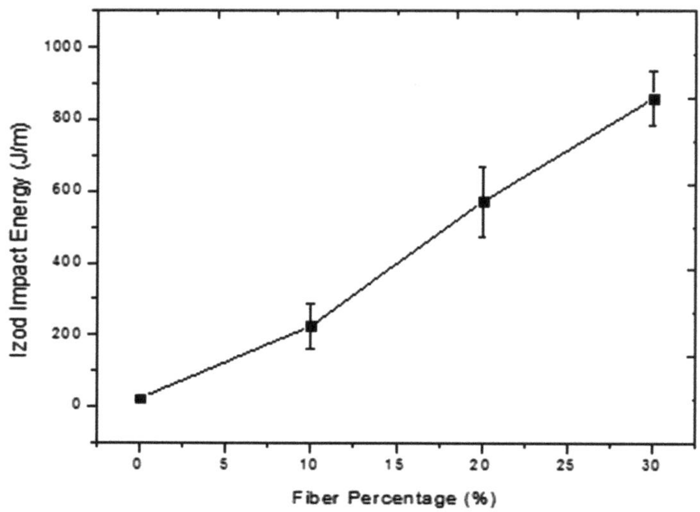

Fig. 3 Izod impact energy as a function of amount PALF fiber

The impact fracture surface of the specimens was gold sputtered and analyzed by scanning electron microscopy, SEM, in a model SSX-500 Shimadzu microscope with secondary electrons imaging at an accelerating voltage of 15 kV.

Results and Discussion

Table 1 presents the results of Izod Impact Tests of polyester matrix composites reinforced with different volume fractions of PALF fibers.

Based on the results shown in Table 1, the variation of the Izod Impact Energy with the corresponding amount of PALF fibers in the polyester matrix is shown in Fig. 3.

In this figure it should be noticed that the PALF fibers incorporation into the polyester matrix significantly improves the impact toughness of the composite. Within the errors bars, this increase can be considered as a linear function with

Table 1 Izod impact energy for polyester composites reinforced with PALF fibers

Amount of PALF fiber (wt%)	Energy (J/m)
0	21.82 ± 2.52
10	224.58 ± 63.76
20	571.33 ± 96.94
30	861.07 ± 76.03

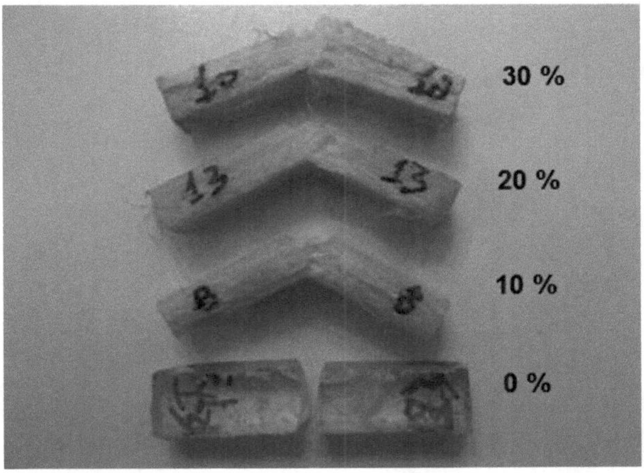

Fig. 4 Specimens for each volume after the test

respect to the amount fiber until 30%. The values shown in this figure are consistent with results of the literature. The reinforcement of a polymeric matrix with lignocellulosic fibers increases the impact toughness of the composite [10, 11] (Fig. 4).

The relatively low interface strength fiber/matrix because of the hydrophilic characteristic of the natural fiber and the hydrophobic characteristic polymeric matrix gives a weak load transfer from the matrix to a longer fiber resulting in a relatively greater fracture surface and higher impact energy is spent for the rupture [13].

The specimens with PALF fibers was not completely broken upon impact. This is due to the relatively high tensile strength characteristic of the PALF fiber, which prevents a total fracture. For these larger fiber fractions, the impacted specimen is bent enough to allow the hammer to continue its trajectory, while the two parts of the folded specimen remained attached. The non-occurrence of rupture upon impact indicates a high toughness of the composite. The incomplete rupture underestimate the results; if it had occurred, the absorbed energy would be even higher.

The SEM analysis (Fig. 5) of the microstructure of the fractured region resulting from Izod impact support a better understanding of the mechanisms responsible for the toughness of PALF fiber reinforced composites as compared to the pure resin. The SEM image shows an effective low adhesion between the fibers and the polyester matrix, where cracks preferentially propagate. This behavior results in a greater fracture area. Some fibers were broken and others were pulled out from the matrix. This greater fracture area, increases energy dissipation area, with the PALF fibers acting as reinforcement for the composite, correlate to the higher absorbed izod impact energy with increasing amount of PALF fiber.

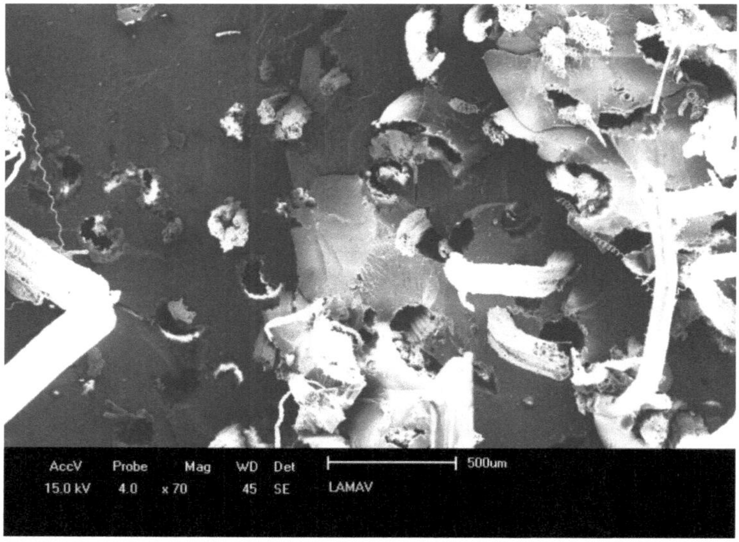

Fig. 5 Impact fractured surface of the composite with 30% in volume of PALF fibers

Conclusions

Polyester matrix composites reinforced with continuous and aligned PALF fibers showed a linear increase within the error bars in notch toughness, measured by the izod impact tests. The incorporation of 30% of PALF fiber in the polyester matrix significantly improves the toughness of the composite, 861.07 J/m, as compared to pure polyester resin, with 21.82 J/m. As a consequence, longitudinal propagation through the weak fiber/matrix interface results in an efficient contribution to the relatively stronger PALF fiber to elevate the izod impact energy and creating difficulty to the final composite rupture.

Acknowledgements The authors thank the support to this investigation by the Brazilian agencies: CNPq, CAPES, FAPERJ and TECNORTE/FENORTE.

References

1. Bledzki AK, Gassan J (1999) Composites reinforced with cellulose-based fibres. Prog Polym Sci 24:221–274
2. Mohanty AK, Misra M, Hinrichsen G (2000) Biofibres, biodegradable polymers and biocomposites: An overview. Macromol Mater Eng 276:1–24
3. Nabi Sahed D, Jog JP (1999) Natural fiber polymer composites: a review. Adv Polym Technol 18:221–274
4. Crocker J (2008) Natural materials innovative natural composites. Mater Technol 2–3: 174–178

5. Monteiro SN, Lopes FPD, Ferreira AS, Nascimento DCO (2009) Natural fiber polymer matrix composites: cheaper, tougher and environmentally friendly. JOM 61:17–22
6. Callister WD Jr (2000) Materials science and engineering—an introduction, 5th edn. Wiley, New York, NY
7. Yue CY, Looi HC, Quek MY (1995) Assessment of fibre-matrix adhesion and interfacial properties using the pullout test. Int J Adhes Adhes 15:73–80
8. Fu SY, Lauke B, Mäder E, Hu X, Yue CY (1999) Fracture resistance of short-glass-fiber-reinforced and short-carbon-fiber-reinforced poly-propylene under charpy impact load and dependence on processing. J Mater Process Technol 89–90:501–507
9. Wambua P, Ivens I, Verpoest I (2003) Natural fibers: can they replace glass and fibre reinforced plastics? Compos Sci Technol 63:1259–1264
10. Monteiro SN, Margem FM, Santos LFL Jr (2008) Izod impact energy in polymeric composites reinforced with ramie fiber. Paper presented at 18th Brazilian congress on materials science and engineering—CBECIMAT, Porto de Galinhas, PE, Brazil, Nov 2008, pp 1–12
11. Monteiro SN, Ferreira AS, Lopes FPD (2009) Izod impact energy of polyester matrix composites reinforced with aligned curaua fibers. Paper presented at characterization of minerals, metals and materials characterization symposium—TMS conference, San Francisco, EUA, Mar 2009, pp 1–8
12. Monteiro SN, Costa LL, Lopes FPD Terrones LAH (2008) Characterization of the impact resistance of coir fiber reinforced polyester composites. Paper presented at characterization of minerals, metals and materials characterization symposium—TMS conference, New Orleans, LA, USA, Mar 2008, pp 1–6
13. Yue CY, Looi HC, Quek MY (1995) Assessment of fiber-matrix adhesion and interfacial properties using the pullout test. Int J Adhes Adhes 15:73–80

Mechanical, Thermal, Morphology and Barrier Properties of Flexible Film Based on Polyethylene-Ethylene Vinyl Alcohol Blend Reinforced with Graphene Oxide

Julyana G. Santana, Angel Ortiz, Rene R. Oliveira, Vijay K. Rangari, Olgun Güven and Esperidiana A.B. Moura

Abstract Ethylene-vinyl alcohol (EVOH) copolymers are widely used in the food packaging industry as gas barrier properties to oxygen, organic solvents, and food aromas. EVOH is very sensitive to moisture and its gas barrier ability deteriorates in high relative humidity conditions. This work aims to prepare flexible films based on melt-blending high density polyethylene (HDPE) and ethylene-vinyl alcohol (HDPE/EVOH/EVA blend) reinforced with graphene oxide (GO). The HDPE/EVOH/EVA/GO flexible films were prepared by twin-screw extrusion and blown film extrusion processing. The flexible films samples were characterized by tensile tests, TG, DSC and FE-SEM analysis and the correlation between properties was discussed. In addition, the oxygen permeability tests were performed at 23 °C, 0 and 90% relative humidity using an OX-TRAN (MOCON Inc.).

Keywords HDPE/EVOH blend · DSC · TG · Graphene oxide · Tensile tests · DSC

J.G. Santana (✉) · A. Ortiz · R.R. Oliveira · E.A.B. Moura
Instituto de Pesquisas Energéticas E Nucleares—IPEN-CNEN/SP,
Av. Prof. L. Prestes, 2242, São Paulo, SP 05508-000, Brazil
e-mail: jugalvaosantana@gmail.com

V.K. Rangari
Department of Materials Science and Engineering,
Tuskegee University, Tuskegee, AL, USA

O. Güven
Department of Chemistry Polymer Chemistry Division,
Hacettepe University, Beytepe, Ankara, Turkey

S. Ikhmayies et al. (eds.), *Characterization of Minerals, Metals, and Materials 2017*, The Minerals, Metals & Materials Series,
DOI 10.1007/978-3-319-51382-9_7

Introduction

Ethylene vinyl alcohol (EVOH) copolymers has been widely applied in food packing, formed by random sequences of ethylene and vinyl alcohol [1]. It is one of the lowest oxygen permeability reported among polymers, commonly used in packaging, due to their excellent barrier properties to oxygen and organic compounds [2, 3]. EVOH copolymers with 32 mol% ethylene is considered to have the superior gas barrier properties when compared to most of the polymeric materials, it is outstanding barrier to gas, odor, and flavor, EVOH presents good chemical resistance, high transparency, low thermal stability and harmlessness toward health [4, 5]. However, EVOH copolymers are severe hygroscopicity, and consequently your applications seriously limited. Such properties are caused by strong hydrogen bond interactions, both inter and intra-molecular, which reduce the free volume of the polymer chains [6–8]. High density polyethylene (HDPE) that is a thermoplastic resulting from the polymerization of ethylene, and is widely used in various industries due to its mechanical properties, chemical resistance, water impermeability, ease of processing and low cost. HDPE is an inexpensive commodity polymer, but its use in some engineering applications may depend on improving its properties by crosslinking [9].

Blending polymer can be an effective alternative for create new materials with enhanced physical properties and polymer diversification. The small quantity of a barrier material into a low-cost matrix material can lead to a low-cost product with a good improvement in barrier properties, where are prominent in the packaging industry. Environmental concerns have thoughts with EVOH application as a thermoplastic, combining high oil resistance with excellent transparency, superior gas barrier properties and easy processability [10]. Its hydrogen-bonded structure, though beneficial to low permeability, causes moisture absorption, which results in the deterioration of the same property. Thus blending with a hydrophobic polymer, like HDPE, may offset this drawback.

These requirements are met through use of multilayer structures consisting of different polymers, each contributing certain specific functions. Because of the stringent requirement of withstanding the rigors of high temperatures and pressures, relatively few polymers are suitable for retort applications. The most common retortable plastic containers are thermoformed from co-extruded multilayer structures consisting of polyolefins (e.g., polypropylene (PP) and polyethylene (PE)), high density (HD) at the surfaces, with an internal oxygen barrier layer. Relevant work reported by Kamal and coworkers on the PP/EVOH blend compatibilized with PP modified with maleic anhydride (PP-g-MA). The influence of various processing parameters on morphology, permeability and impact strength were examined for extruded films. In another work, Prasad and Jackson reported on the mechanical properties and morphology of melt-blended EVOH in an HDPE matrix using PE-g-MA as a compatibilized [11–13]. Nowadays the high-energy radiation is a powerful technique for the modification of polymers, some authors indicates that in certain blends systems the recombination reaction of different macromolecule free radicals induced by irradiation may occur in the interface area, so complex grafting or cross-linking structures may be formed. New structures can improve interphase bonding, and lead to a significant change in mechanical and thermal properties. According with Goulas et al. the effect of radiation on the properties of multilayer films containing EVOH/HDPE blends can be improved by ionizing radiation [14, 15].

Packaging scientists are working on developing solutions like to the moisture sensitivity problem. The incorporation of desiccants in the layers surrounding EVOH to reduce the amount of moisture reaching the EVOH layer, blending of EVOH with other materials to improve properties, use of nanocomposites, etc. [16]. The structural uniqueness of graphite can greatly improve the thermal, mechanical, and gas barrier properties of a polymer if the graphite is well dispersed in the polymer matrix. Graphene oxide is a monolayer of carbon atoms, which may be obtained from exfoliation of graphite and is considered an ultrathin, perfect two dimensional (2D) barrier against gas diffusion [17–20]. Recently, graphene oxide (GO), which have been prepared by chemically oxidizing graphite to graphite oxide with strong oxidants and ultrasonic cleavage for graphene oxide nanosheets, has gained significant attention for incorporation in different polymers for improve the gas barrier and mechanical properties. Because GO has a large number of polar groups such as hydroxyl, ether, and carboxylate groups, intercalation of water-soluble polymers in GO is possible. EVOH/GO nanocomposites can present enhanced molecular ordering and the long tortuous pathway for water and gas molecules may explain the good water/oxygen barrier properties in flexible films [21].

This study aims to evaluate morphology, thermal, mechanical and barrier properties of HDPE/EVOH/EVA/GO flexible films compatibilizer by ionizing radiation.

Experimental

Material

- Graphene oxide
- (GO) nanosheets prepared from conventional flake graphite [15, 16];
- Ethylene vinyl alcohol copolymer (EVOH) with 32% mol/ethylene (EVAL™ manufactured by Kuraray Co. Ltd.);
- High density polyethylene (HDPE);
- Ethylene Vinyl Acetate (EVA).

Preparation of HDPE/EVOH/EVA/GO Flexible Films

The HDPE/EVOH/EVA/GO films were processed by twin-screw extrusion and extrusion blown film processing. The HDPE/EVOH/EVA/GO flexible films with addition of 0.1 wt% of GO nanosheets were prepared by melting extrusion process, using a twin-screw extruder Haake Rheomex P332 with 16 mm and L/D = 25 rate from Thermo Scientific. The temperature profile was of 182/192/197/197/205/205 °C and a screw speed of 30 rpm. The extrudates materials were cooled down for a better dimensional stability, pelletized by a pelletizer, dried again and fed into extrusion blown film, single screw machine (Carnevalli) with 25 mm diameter and flexible film test samples were obtained. The temperature profile used in the blow extrusion process of HDPE/EVOH/EVA and HDPE/EVOH/EVA/GO flexible films were 190/195/210/215/215/220 °C and screw speed was 30 rpm (Table 1).

Electron-Beam Irradiation

Part of HDPE/EVOH/EVA and HDPE/EVOH/EVA/GO flexible films were submitted to electron-beam irradiation at 150 kGy, using a 1.5 meV electron beam accelerator, at room temperature in presence of air.

Characterization of GO Nanosheets and HDPE/EVOH/EVA/GO Flexible Films

Mechanical tests: Tensile tests were determined using an INSTRON Testing Machine model 5564, according to ASTM D882-91 in order to evaluate the mechanical behavior of the materials studied. Each value obtained represented the average of five samples.

Differential scanning calorimetry (DSC): analyses were carried out using a Mettler Toledo DSC 822e from 25 to 250 °C at a heating rate of 10 °C/min under

Table 1 DSC analysis results of EVA/EVOH/HDPE/GO Nanocomposite Flexible Films

Composites	Melting temperature (T_{m1}, °C)	Melting temperature (T_{m2}, °C)	Melting enthalpy (ΔH_{m1}, Jg^{-1})	Melting enthalpy (ΔH_{m2}, Jg^{-1})
HDPE/EVOH/EVA[a]	132.9	178.1	128.0	14.4
HDPE/EVOH/EVA/GO[b]	131.4	178.3	110.7	12.3
HDPE/EVOH/EVA/GO*[c]	131.4	175.1	97.3	8.0

[a]HDPE/EVOH/EVA (70/20/10 wt%); [b]HDPE/EVOH/EVA/GO (70/20/9.99/0.1 wt%); [c]HDPE/EVOH/EVA/GO (70/20/9.99/0.1 wt%)*150 kGy

nitrogen atmosphere (50 ml/min). DSC analyses of the materials were performed on four samples of the materials. DSC were carried out to obtain melt temperature (T_m) and melting enthalpy (ΔH_m).

Thermogravimetric analysis (TG): In this study the TG analyses were made in a Mettler Toledo TGA module "TGA/SDTA851e" from 25 to 500 °C at a heating rate of 10 °C/min under nitrogen atmosphere (50 ml/min).

Field emission scanning electron microscopy (FE-SEM): FE-SEM of irradiated and non-irradiated HDPE/EVOH/EVA and HDPE/EVOH/EVA/GO samples cryofractured under liquid nitrogen were carried out using a JEOL-JSM-6701F, microscope with an accelerating voltage of 1–30 kV, using EDS Thermo-Scientific mod. Noran System Six software, in carbon sputtered samples.

Results and Discussion

DSC analysis results of HDPE/EVOH and HDPE/EVOH/GO flexible films: Fig. 1 shows the DSC analysis results for HDPE/EVOH/EVA and HDPE/EVOH/EVA/GO flexible films.

Thermogravimetric analysis (TG) results: Fig. 2 shows the TG thermograms of HDPE/EVOH/EVA and HDPE/EVOH/EVA/GO flexible films. As can be seen in Fig. 2, the HDPE/EVOH/EVA/GO showed a great reduction in the onset temperature when compared with HDPE/EVOH/EVA without GO addition.

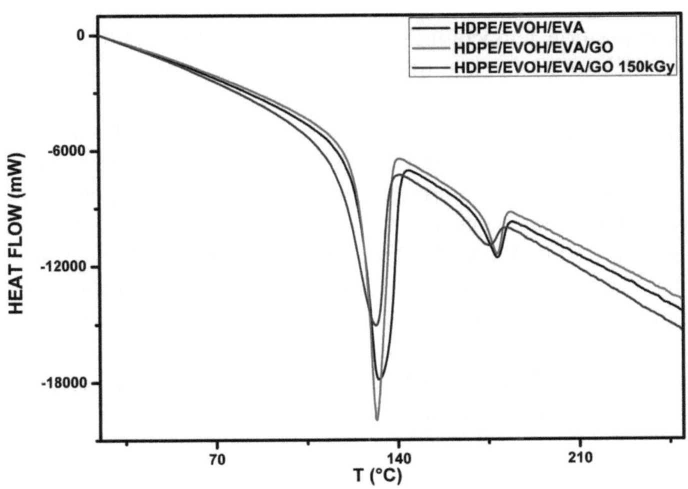

Fig. 1 DSC analysis results for HDPE/EVOH/EVA and HDPE/EVOH/EVA/GO flexible films

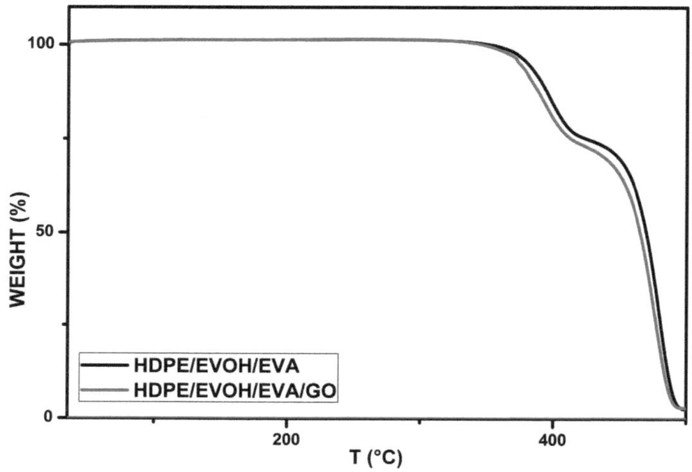

Fig. 2 TG analysis results for HDPE/EVOH and HDPE/EVOH/GO flexible films

Table 2 Decomposition temperature and weight loss of HDPE/EVOH/EVA and HDPE/EVOH/EVA/GO flexible films

Flexible films	T_{onset} (°C)	T_{max} (°C)	Weight loss (%)
HDPE/EVOH/EVA	448.56	457.28	97.41
HDPE/EVOH/EVA/GO	363.86	454.97	96.88

The decomposition temperatures and weight loss for HDPE/EVOH/EVA and HDPE/EVOH/EVA/GO flexible films are presents in Table 2.

Field Emission Scanning Electron Microscopy (FE-SEM) analysis results: FE-SEM micrographs of cryofractured surfaces of irradiated and non-irradiated HDPE/EVOH/EVA blend and HDPE/EVOH/EVA/GO nanocomposite were studied to understand the failure mechanisms and also study possible interaction between different components after electron-beam irradiation. FE-SEM micrographs of HDPE/EVOH/EVA blend and its composite are showed in Fig. 3. The micrographs of irradiated Blend and nanocomposite showed better compatibilization between the different components when compared with non-irradiated samples.

Tensile tests results: Table 3 presents the results of tensile tests of the HDPE/EVOH/EVA and HDPE/EVOH/EVA/GO flexible films. The results presented shows the average values calculated from the data obtained in tests for five test specimens.

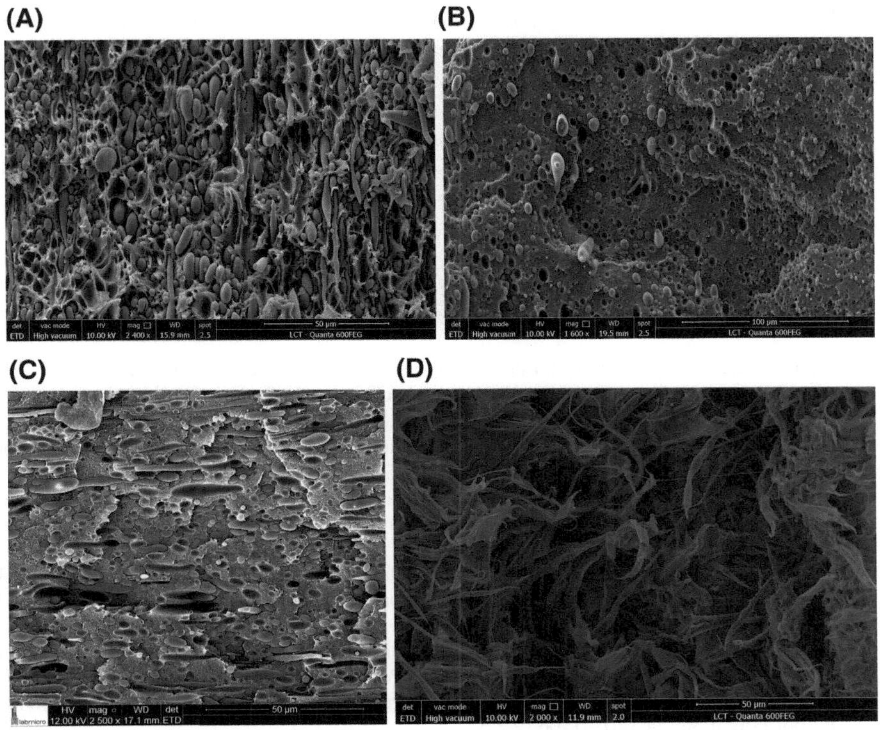

Fig. 3 **a** HDPE/EVOH/EVA blend (2400×), **b** HDPE/EVOH/EVA blend at 150 kGy (1600×), **c** HDPE/EVOH/EVA/GO nanocomposite (2500×), **d** HDPE/EVOH/EVA/GO nanocomposite at 150 kGy (2000×)

Table 3 Tensile test results of the HDPE/EVOH/EVA and HDPE/EVOH/EVA/GO flexible films

Flexible films	Tensile strength at break (MPa)	Elongation at break (%)	Young's modulus (GPa)
HDPE/EVOH/EVA[a]	9.2 ± 0.2	120 ± 10.6	0.54 ± 0.04
HDPE/EVOH/EVA*[b]	10.8 ± 0.5	127 ± 9.3	0.65 ± 0.03
HDPE/EVOH/EVA/GO[c]	13.6 ± 0.2	121 ± 10.4	0.76 ± 0.04
HDPE/EVOH/EVA/GO*[d]	14.3 ± 0.7	122 ± 9.8	0.81 ± 0.06

[a]HDPE/EVOH/EVA; [b]HDPE/EVOH/EVA* irradiated at 150 kGy; [c]HDPE/EVOH/EVA/GO; [d]HDPE/EVOH/EVA/GO* irradiated at 150 kGy

Conclusions

The results showed that the incorporation of GO nanosheets led to an important increase in Tensile strength at break and Young's modulus without an important variation in Elongation at break properties. The FE-SEM micrographs of irradiated

HDPE/EVOH/EVA and HDPE/EVOH/EVA/GO nanocomposite showed better compatibilization between the different components when compared with non-irradiated samples.

Acknowledgements The authors acknowledge Intermarketing Brasil and Kuraray Group to provide for raw materials, IPEN/CNEN-SP, CAPES and IAEA for financial support through IAEA-CRP 17760 RO.

References

1. Franco-Urquiza E et al (2010) Influence of processing on the ethylene-vinyl alcohol (EVOH) properties: application of the successive self-nucleation and annealing (SSA) technique. Express polym Lett 4(3):153–160
2. Mokwena KK, Tang J (2012) Ethylene vinyl alcohol: a review of barrier properties for packaging shelf stable foods. Food Sci Nutr 52:640–650
3. Ramakrishnan S (1991) Well-defined ethylene-vinyl alcohol copolymers via hydroboration: control of composition and distribution of the hydroxyl groups on the polymer backbone. J Macromol 24:3753–3759
4. Lagaron JM, Giménez E, Saura JJ (2001) Degradation of high barrier ethylene–vinyl alcohol copolymer under mild thermal-oxidative conditions studied by thermal analysis and infrared spectroscopy. J Polym Int 50:635–642
5. Lima JA, Felisberti MI (2008) Poly(ethylene-co-vinyl alcohol) and poly(methyl methacrylate) blends: phase behavior and morphology. J Eur Polym 44:1140–1148
6. Aucejo S, Marco C, Gavara R (1999) Water effect on the morphology of EVOH copolymers. J Appl Polym Sci 74:1201–1206
7. Cabedo L et al (2006) The effect of ethylene content on the interaction between ethylene-vinyl alcohol copolymers and water—II: Influence of water sorption on the mechanical properties of EVOH copolymers. Polym Test 25:860–867
8. Kucukpinar E, Doruker P (2004) Effect of absorbed water on oxygen transport in EVOH matrices. A molecular dynamics study. Polymer 45:3555–3564
9. Canevarolo Jr SV (2006) Ciência dos Polímeros. Artliber Ed Ltda 2
10. Samios CK, Kalfoglou NK (1998) Compatibilization of poly(ethylene-co-vinyl alcohol) (EVOH) and EVOH/HDPE blends with ionomers. Structure and properties. Elsevier Science Ltd. Polymer 39(16):3863–3870
11. Kamal MR, Garmabi H, Hozhabr S, Arghyris L (1995) The development of laminar morphology during extrusion of polymer blends. Polym Eng Sci 35(1):41–51
12. Walling N, Kamal MR (1995) Adv Polym Technol 5:269
13. Prasad A, Jackson P (1996) Polym Mater Sci Eng 75:281
14. Goulas AE, Riganakos KA, Kontominas MG (2003) Effect of ionizing radiation on physicochemical and mechanical properties of commercial multilayer coextruded flexible plastics packaging materials. Radiat Phys Chem 68:865–867
15. Elmaghor F, Zhang LY, Li HQ (2002) China Synth Rubber Ind 25:175
16. Mokwena KK, Tang J (2012) Ethylene vinyl alcohol: a review of barrier properties for packaging shelf stable foods. Crit Rev Food Sci Nutr 52:640–650
17. Yang J et al (2013) Thermal reduced graphene based poly(ethylene vinyl alcohol) nanocomposites: enhanced mechanical properties, gas barrier, water resistance, and thermal stability. Ind Eng Chem Res 52:16745–16754
18. Hummers, JWS (1957) Preparation of graphitic acid. US Patent 2798878
19. Geim AK, Novoselov KS (2007) The rise of graphite. Nat Mater 6:183–191

20. Du J, Cheng H-M (2012) The fabrication, properties, and uses of graphene/polymer composites. Macromol Chem Phys 213(10–11):1060–1077
21. Kim D, Kwon H, Seo J (2014) EVOH nanocomposite films with enhanced barrier properties under high humidity conditions. Polym Compos 35:644–654

Radiation Effects on Crosslinking of Butyl Rubber Compounds

Sandra R. Scagliusi, Elizabeth C.L. Cardoso and Ademar B. Lugão

Abstract When butyl rubbers are subjected to high energy radiation, they form easy free radicals that initiate various chemical reactions. These reactions alter the molecular distribution of irradiated rubbers by crosslinking or scission affect their physical and mechanical properties. This work aims to the analysis of effect induced by γ-exposure on the crosslinking density in butyl rubbers by swelling measurements accomplished before and after irradiation at 25, 50, 100 and 200 kGy, with further evaluation of crosslinking density accomplished by Flory-Rehner equation; this is a proper procedure for the qualification of radiation resistance. It can be noticed that changes in material structure was due to build-up of new three-dimensional network in studied rubbers. Changes in crosslinking density of butyl rubber compounds emphasize that degradation mechanism is strongly influenced by gamma-radiation doses higher than 50 kGy, since chain scission process predominates over crosslinking reaction.

Keywords Butyl rubbers · Crosslinking · Degradation · Chain scission · Gamma rays

Introduction

Butyl rubber is a synthetic rubber, a copolymer of isobutylene with isoprene. Butyl rubber is produced by polymerization of about 98% of isobutylene with about 2% of isoprene. Structurally, polyisobutylene resembles polypropylene, having two methyl groups substituted on every other carbon atom [1]. In its hydro-carbonated chain (Fig. 1), unsaturation level is very low and this imparts an excellent resistance to ageing, low permeability to gases, good thermal stability, high resistance to

S.R. Scagliusi (✉) · E.C.L. Cardoso · A.B. Lugão
Nuclear and Energy Research Institute—IPEN/CNEN-SP, Cidade Universitária, Av. Lineu Prestes 2242, São Paulo, SP CEP05508-900, Brazil
e-mail: saresca@ig.com.br

S. Ikhmayies et al. (eds.), *Characterization of Minerals, Metals, and Materials 2017*, The Minerals, Metals & Materials Series,
DOI 10.1007/978-3-319-51382-9_8

oxygen, ozone and solar radiation action and excellent resistance to humidity and to chemical substances attack [2, 3].

Butyl rubber compound acquires mechanical properties due to build-up of crosslinks among polymer macromolecules. These bonds aim to unite elastomer macromolecules, hindering that they flow on each other. This molecular arrangement grants to the elastomer the capacity to support stresses associated to elastic deformations imparting to the material the capacity to recover its original form [4].

The formed intermolecular elastically effective crosslinks are usually classified as carbon-carbon (C–C) crosslinks and sulfidic crosslinks. The C–C crosslinks are created directly between the polymer chains. In contrast, the sulfidic crosslinks are made of a variable number of sulfur atoms. Depending on this number, sulfidic crosslinks are further subdivided into monosulfidic (C–S–C); disulfidic (C–S2–C); and polysulfidic (C–Sx–C, x ≥ 3) crosslinks. All of these crosslinks show varying structure and length, resulting in various characteristics. As a result, they provide different properties to the material. Therefore, it is generally assumed that after the crosslink density, the crosslink structure is the second most important parameter influencing the elastomer properties [5, 6].

Due to vulcanization rubber acquires mechanical resistance contributing for a raise in elasticity modulus and mechanical properties [7]. Elastomers crosslink density determines comprehensively elastomers mechanical properties. In order to improve the performance of these systems it is fundamental to understand relationship among vulcanization process, resulting network structure and their physical properties. Structural characterization of elastomeric networks is generally performed by mechanical essays [5].

Characterization of the network introduced into elastomers by vulcanization has been a fascinating problem since a quite long past. Flory (1943) has developed an equation connecting the equilibrium swelling volume with the degree of cross-links [8].

When polymer is subjected to gamma radiations, many chemicals reactions may occur affecting physical and mechanical properties of material [9–11]. The primary event occurs when a molecule interacts with ionizing radiation, involving the injection of an electron to form a radical, which could lead to macromolecular chain scissions, crosslinking, changes in stereochemistry or formation of grafts through complex chemical reaction process [12].

The major effect of ionizing radiation in butyl rubber is chain scission with a significant reduction of molar mass [13]. Transference of radiation energy to butyl

Fig. 1 Butyl rubber structure illustration

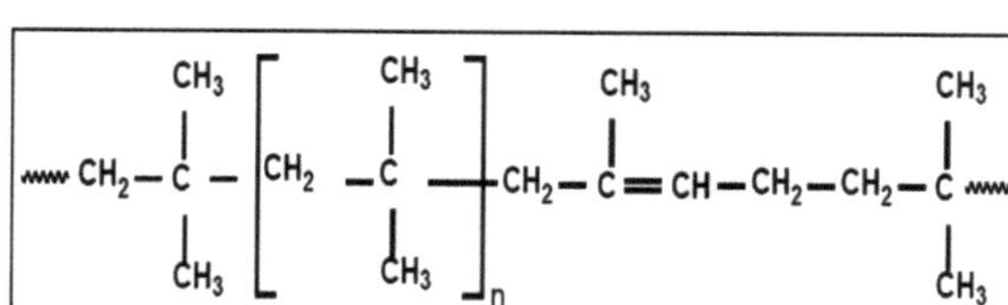

rubber does not occur selectively but the low unsaturation degree of these rubbers favors a quicker scission [14].

The chain scission process usually leads to a reduction of mechanical properties whereas the crosslinking process results in an increase in mechanical properties. Moreover, the materials properties as mechanical for example are a complex function of crosslink chemical density and crosslink nature, both chemical and physical interactions involved [15].

The aim of this paper is the analysis of effect induced by gamma-exposure on the crosslinking density in butyl rubbers by swelling measurements accomplished before and after irradiation.

Experimental

Materials

Elastomeric compounds were prepared using Butyl grade 268, from Exxon Mobil Chemical. Samples were prepared with formulation based on standards commonly used in tires according to automotive industry (Table 1). Parts per hundred rubber (phr) is a measurement unit used in formulation of rubber compounds and refers to the amount of a particular compound in relation to the total amount of rubber used per 100 parts of rubber.

Preparation of Mixtures

Admixtures were manufacturer of the equipment, open roll-mill (Copê), 40 kg capacity according to ASTM D-3182-9 [16]. Samples were cured in an electrically heated HIDRAUL-MAQ at 5 MPa pressure and 180 °C temperature to their optimum cure times (determined from a rheometer Monsanto R-100).

Table 1 Formulation of butyl rubber

Ingredients	Amounts (phr)
Butyl rubber	100
Zinc oxide	5
Stearic acid	1
Paraffinic oil	25
Carbon black	70
Sulfur	2
MBTS	0.5
TMTD	1

Irradiation

Cure sheets in 11.5 × 11.5 × 0.1 cm dimension, 250 g total weight, were irradiated in Embrarad/CBE, gamma rays Cobalt 60 (^{60}Co) in air, at 5 kGy/h rate, within a, 25, 50, 100 and 200 kGy doses range.

Characterization

Characterizations were accomplished before and after irradiation according to following:

- Determination of the swelling rubber

The specimens wirh approximate dimensions 1.5 × 1.5 cm were exposed at room temperature (25 °C), previously weighed and immersed in toluene up to weight stabilization (about five days). Each value was obtained from five samples average, at various time periods. The specimens were removed, gently wiped and paper dried to remove solvent excess in sample surface and reweighted after 30 s. Upon completion of test samples were weighed and then dried at room temperature for 24 h. These analyses were performed in accordance with ASTM D 3616-9 [17].

Degree of swelling was calculated according to Eq. 1.

$$Q = \frac{M - Mo}{Mo} \times 100 \tag{1}$$

where: Mo is the initial weight of sample (g) and
M is the final weight of sample (g)

- Crosslink density

Crosslink density was calculated by equation developed by Flory and Rehner (Eq. 2), based on the equilibrium swelling in organic solvents [18]. Crosslink density [υ] was calculated from the equilibrium swelling.

$$v = \frac{1}{V}\left[\frac{\ln(1 - VR) + VR + \mu VR^2}{VR^{1/3} - \frac{VR}{2}}\right] \tag{2}$$

where:

- Vr is the volume fraction of swollen rubber;
- μ is the parameter Rubber—solvent interaction (IIR toluene—0.5) [18];
- V is the molar ratio of the solvent volume in $cm^3 . mol^{-1}$ (toluene = 105.91 cm^3 mol^{-1}) [18].

Samples were weighed daily at that extent no more weight change was verified within 24 h. Then, samples were dried in a vacuum oven at 60 °C for 4 h. Specimens dimensions used were 2.0 × 2.0 × 0.2 cm and the representative one was the result arithmetic mean of five determinations.

Results and Discussions

Applications of radiation processing for the manufacture of new products or for polymeric recycling are based on the instant bonds breakage.

The penetration of solvent molecules depends on many factors, such as free space between molecules, size of molecules and crosslinking degree, all of them of relevant significance. The exposure to γ-radiation causes a reduction in molecular characteristics: molecular weight, size and distribution; the nature of bonds existing in irradiated materials defines the amplitude of modifications.

In Fig. 2 swelling degrees corresponding to longer immersion times (72, 96 and 120 h) for irradiated and non-irradiated butyl rubber. At 200 kGy evaluation was prejudiced because samples were completely dissolved.

The radiation stability of butyl rubber allows the easy diffusion of foreign molecules when irradiation dose becomes considerable. At very low dose, 25 kGy, butyl rubber does not change dimensional features according to negligible increase of swelling degree. The presence of double bond in the backbone of butyl rubber determines similar values of swelling degree for low irradiation doses For exposures exceeding 50 kGy it is noticed an advanced level of degradation.

In Fig. 3 the values of free volumes describe similar tendencies of degradation, according to swelling determinations. The decrease of free volumes from

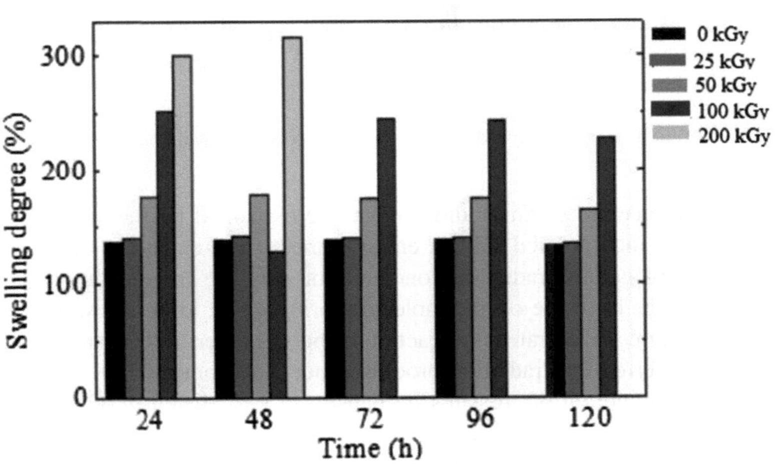

Fig. 2 Swelling degree of butyl rubber after and before gamma radiation

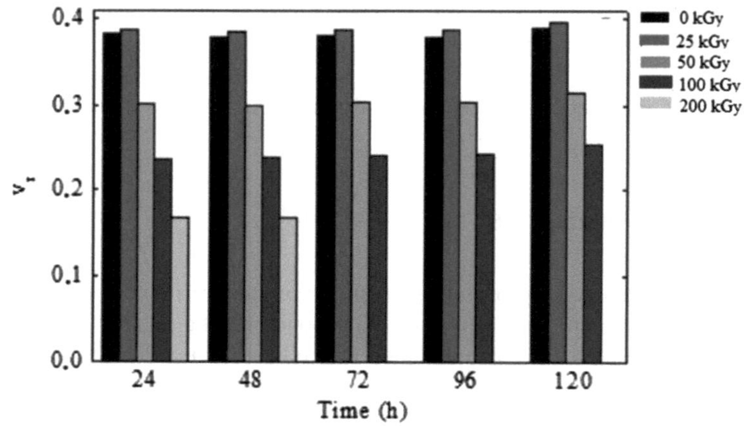

Fig. 3 Modification in the free volumes of butyl rubber irradiated and non-irradiated

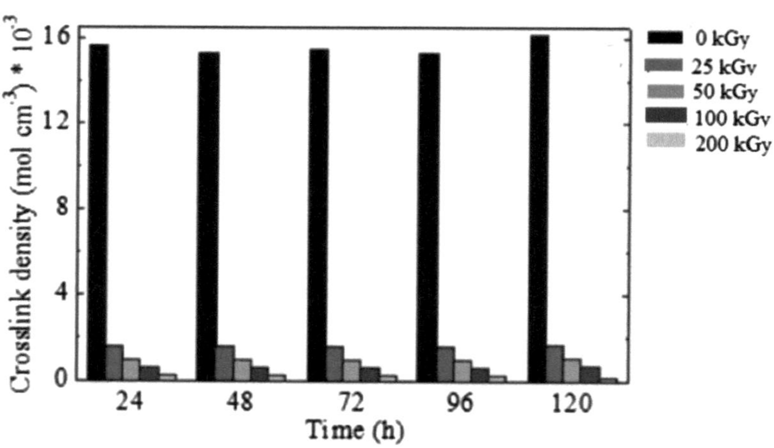

Fig. 4 Modification in crosslinking density of butyl rubbers irradiated and non-irradiated

fragmentation occurred during radiolysis can be explained by the advanced swelling when more incorporated solvent enlarges the size of entangled zones.

The action of gamma radiation concerns not only the fragmentation of molecules, but also the cleavage of intermolecular bridges that joint macromolecules in gel islands. Figure 4 illustrates the reaction of butyl rubber to the damaging action of energy transferred into radiation processed butyl elastomers.

The strike diminution of crosslinking density in γ-irradiated butyl rubber is the effect of the presence of high insoluble fraction in raw butyl rubber. Butyl rubber exhibits a sharp dropping in crosslinking density because the scission of C-C, where involved energy is lower than in C-H bonds on forming intermolecular network.

Conclusions

Swelling of butyl rubber showed a comprehensive predominance of chain-scission with molar mass reduction. Another relevant phenomenon is a higher fragility of butyl rubber to ionizing radiation with a complete solubilisation after 48 h for irradiated rubber at 200 kGy.

The availability of butyl rubbers to the diffusion of foreign entities, either molecules or free radicals is demonstrated by the adequate swelling properties, which can be considered as the basic information for their radiation processing for other applications including the reclaiming polymer wastes.

Acknowledgements The authors would like to thank CBE, Flexlab, Basile Química, Capes and IPEN, for all their support for this work.

References

1. Fusco JV, Hous P (1990) Butyl and halobutyl rubbers. In: The Vanderbilt rubber handbook, 13th edn. R. T. Vanderbilt Company, Inc., USA, p 2
2. Dubey V, Pandey SK, Rao NBSN (1995) Research trends: the degradation of butyl rubber. J Anal Appl Pyrol 34:111–125
3. Teinov AV, Zavyalov NV, Khokhlov YA, Sitnikov NP, Smetanin ML, Tarantasov VP, Shadrin DN, Shorikov IV, Liakumovich AL, Miryasova FK (2002) Radiation degradation of spent butyl rubbers. Radiat Phys Chem 63:245–248
4. Coran AY, Eirich FR (1978) Science and tecnology rubber. Academic Press, New York, p 282
5. Hertz DL (1984) Theory & practise of vulcanization. In: Seals Eastern Inc., Communication Chanel Inc., Red Bank 07701, New Jersey
6. Dogadkin BA (1976) Chemistry of elastomers, 1st edn; WNT, Warsaw, Poland, p 201
7. Oliveira MG, Soares BG (2002) Influencia do Sistema de Vulcanização nas Propriedades da Mistura de NBR/EPDM. UFRJ: Instituto de Macromoléculas Professora Eloisa Mano, 12 (1):11–19
8. Flory P, Rehner J (1943) Statistical mechanics of cross-linked polymer networks II. Swelling. J Chem Phys 11:521–526
9. Bhagawan SS, Kuriakose B, De SK (1987) Mechanical properties and failure surfaces of gamma-ray irradiated systems based on thermoplastic 1, 2 polybutadiene. Int J Radiat Appl Instrum Part C Radiat Phy Chem 30:95–104
10. Zaharescu T, Cazac C, Jipa S, Setnescu R (2001) Assessment on radiochemical recycling of butyl rubber. Nucl Instrum Methods Phys Res: Sect B, 185: 360–364
11. Hill DJT, O'Donnell JH, Perera MCS, Pomery PJ (1992) Determination of scission and crosslinking in gamma irradiated butyl rubber. Int J Radiat Appl Instrum Part C Radiat Phy Chem 40:127–138
12. Makuuchi K, Cheng S (2012) Radiation processing of polymer materials and its industrial applications. Wiley, Nova York, NY, p 174
13. Chandra R, Subhas HV, Verm AK (1982) Changes in physical properties and molecular structure of butyl rubber during radiation. Polymer 23:1457–1460
14. Scagliusi SR, Cardoso ECL, Parra DF, Lima LSCE, Lugão AB (2013) Evaluation of "Payne Effect" in radiation-induced modification of chlorobutyl rubber. Radiat Phys Chem 84:42–46

15. Chandra R, Subhas HV, Verm AK (1982) Changes in physical properties and molecular structure of butyl rubber during radiation. Polymer 23:1457–1460
16. Annual Book of ASTM Standards (2009) Standard practice for rubber—materials, equipment, and procedures for mixing standard compounds and preparing standard vulcanized sheets. vol. 09.01, (ASTM D-3182)
17. Annual Book of ASTM Standards (2009) Standard test method for rubber, raw—determination of gel, swelling index, and dilute solution viscosity, vol. 09.01, (ASTM D-3616)
18. Flory F (1953) Principles of polymer chemistry, Ed. Cornell University Press, London
19. Brandrup J, Grulke HE (1999) Polymer handbook. Wiley-Interscience Publication, New York, pp 621–730

Part II
Clays and Ceramics

Effect of Skin-Core Hierarchical Structure on Dielectric Constant of Injection Molded and Cast Film Extruded Liquid Crystalline Polymer

Mark H. Shooter, Michael A. Zimmerman and Anil Saigal

Abstract Liquid Crystalline Polymers (LCPs) provide favorable dielectric properties for extremely high frequency (30–100 GHz) radio frequency device packaging and printed circuit board substrates. The material studied in this work is a proprietary, thermotropic LCP composite developed and manufactured by iQLP (Woburn, MA, USA). Dielectric constants measured in the flow direction of the post-processed materials differ between the two processes, 3.51 in injection molded plaques, and 4.23 in cast films. This investigation sought to determine the cause of process-induced variation in the dielectric constant between injection molded and cast film extruded LCP. The complex thermomechanical environment present during processing of LCPs creates a through-thickness hierarchical structure, simplified in this study as a skin and core layer. The volume fraction of skin layer of the post-processed LCP was determined to be the differentiating factor, and responsible for the difference in dielectric constant between the injection molding and cast film extrusion processes. The resulting dielectric constant is a function of this volume fraction with a near linear relationship.

Introduction

Liquid crystalline polymers (LCPs) form a unique class of performance materials with specialized properties that make them ideal candidates for various engineering applications. Their long-range molecular order offers high mechanical strength at high temperatures and high chemical resistance, as well as flame retardancy and good weatherability [1]. The influential molecule in LCPs is

M.H. Shooter · M.A. Zimmerman · A. Saigal (✉)
Department of Mechanical Engineering, Tufts University, Medford, MA 02155, USA
e-mail: Anil.Saigal@Tufts.edu

M.H. Shooter
e-mail: Mark.Shooter@Tufts.edu

M.A. Zimmerman
e-mail: Michael.Zimmerman@Tufts.edu

S. Ikhmayies et al. (eds.), *Characterization of Minerals, Metals, and Materials 2017*, The Minerals, Metals & Materials Series,
DOI 10.1007/978-3-319-51382-9_9

p-hydroxybenzoic acid (HBA), a bulky aromatic molecule located in the main chain. HBA provides main chain rigidity and reduces segmental motion, allowing high degrees of crystallinity and the unique properties seen in LCP in comparison to amorphous polymers [2]. In addition, LCPs exhibit good electrical behavior and possess low and stable dielectric constant throughout the extremely high frequency range (EHF) (30–100 GHz) [3]. Researchers have taken advantage of these favorable properties to improve the performance of commercial polymers by adding LCPs as reinforcements [4]. The directional molecular order which drives the favorable material properties leads to inherent anisotropic behavior in the bulk material. The material studied in this work is a proprietary, thermotropic LCP composite developed by iQLP (Woburn, MA, USA).

LCPs favorable mechanical, chemical, and electrical properties have driven research in many niche electronic industry applications, including, circuit substrates, flexible RF electronics and antennas, high power and high frequency air cavities, surface mount packages for Ka-band, and hermetic packaging [2, 3, 5–7]. Current packaging technology for EHF range includes polytetrafluoroethylene (PTFE) composites, multilayer ceramic, metal, and glass materials that provide adequate hermeticity and low loss [8].

LCPs are investigated in high frequency and other electronic applications in part due to their low relative permittivity. Permittivity is related to electric susceptibility, a measurement of a material's capability to polarize in an externally applied electric field. Relative permittivity is comprised of a complex expression when used in an alternating electric field, and determines the material's capability to store and dissipate energy in the presence of an applied external electric field compared to the capability of a vacuum. The real part of permittivity, dielectric constant (D_k), is the materials ability to store energy in an applied electric field. It is driven by the materials electric susceptibility, and is often a non-linear relationship with frequency. The imaginary part measures the materials dissipation factor in the applied field, and is often called the loss tangent. Complex permittivity is shown below in Eq. 1.

$$\varepsilon = \varepsilon_r' + i\varepsilon_r'' \tag{1}$$

where ε is complex permittivity, ε_r' is the real part called dielectric constant, and ε_r'' is the imaginary part called loss factor. Complex permittivity can be illustrated as vectors 90° out of phase (Fig. 1—Complex permittivity expressed in vector form) [9].

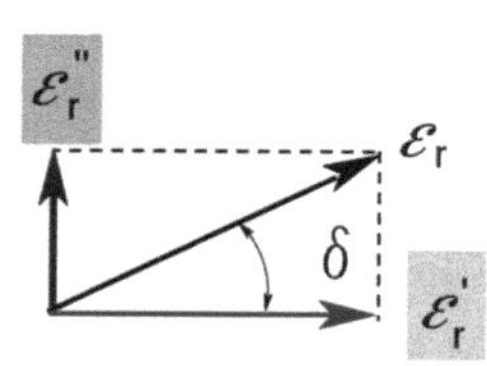

Fig. 1 Complex permittivity expressed in vector form

Materials that demonstrate low D_k are important electronic insulators used to insulate signal carrying devices, and reduce crosstalk and power dissipation [10]. In the 30–100 GHz range, polarization is dominated by dipolar orientation. As frequency increases, the field shifts orientation at a rate at which dipoles cannot reorient simultaneous. This weakens the polarization of the dipoles, storing less energy in the dielectric material [11].

Evolution of crystal domain anisotropy in the LCP during manufacturing can adversely affect the macroscopic properties. Therefore, it is critical to understanding of which processing parameters control the macroscopic crystal orientations to enable the production of more isotropic LCPs. Unlike amorphous polymers which become isotropic in the molten state, LCPs retain crystallinity and orientation at high temperatures. The interaction and orientation of the crystal domains play an important role in the material properties [12]. During manufacturing processes, LCP crystals align parallel to the direction of shear and maintain their orientation even after processing. The initial location of crystallization creates a region called the skin layer, which possesses a different mesoscopic structure than the bulk polymer, and is found to be highly oriented. As polymer melt continues to flow past this newly crystallized skin layer, the interaction causes shear stresses, creating a new intermediate region with its own morphology. In the center of the processed polymer resides a core region that is less affected by the shear stresses, and therefore a less oriented layer (Fig. 2—Hierarchical structure of post-processed) [13, 14].

During injection molding of LCP, a skin layer is formed as the polymer melt freezes upon initial contact with the mold cavity walls [15–17]. As the melt advances into the mold, a parabolic or fountain flow front is formed. At the foremost boundary of the flow front, the crystal domains experience elongational stresses. The crystal domain orientation developed from the elongational stress is immediately frozen against the cavity walls as more polymer melt enters the mold. Melt entering the mold then comes into contact with the frozen skin layer, which causes shearing of the melt also orienting the crystal domains (Fig. 3—Fountain flow front, elongation stresses, and skin layer formation during injection molding) [13, 18].

The resulting morphology from injection molded LCP has been well studied and a through-thickness hierarchical structure has been discovered. The thickness of the regimes within the hierarchy are effected by processing conditions and part thickness. As injection speeds increase, mold cavity temperature decreases, and part thicknesses increases, not only are the layer regimes thicker, four regimes have been identified [19–23]. Crystal domain orientation has been found to be highest in the

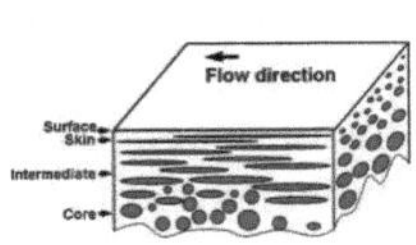

Fig. 2 Hierarchical structure of post-processed

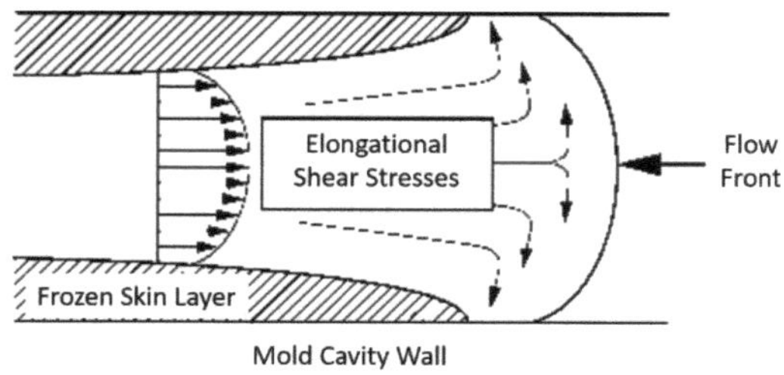

Fig. 3 Fountain flow front, elongation stresses, and skin layer formation during injection molding

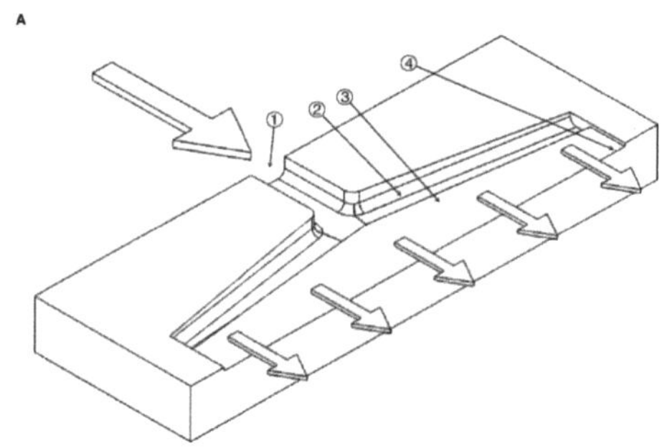

Fig. 4 Coat hanger die design

skin layer, lower in the transitional layers, and slightly transverse to the flow direction in the core [19].

In cast film extrusion, the polymer melt is sheared and thinned in the z-direction in a coat hanger design die (Fig. 4—Coat hanger die design) [24]. Described as the coat hanger-type die concept (A): (1) central inlet port; (2) manifold (distributes melt); (3) island (along with manifold, provides uniform pressure drop from inlet to die lip; (4) die lip (die exit forms a wide slit). The LCP crystal domains are sheared as they are forced through the die, and that shear induces a crystal orientation which is immediately frozen as the polymer melt exits the die and contacts chill rolls [24, 25]. iQLP is a principal manufacturer of cast film LCP, but there is limited available literature on resulting morphology or dielectric constant. This investigation is also a principal study of LCP cast film morphology and hierarchical structure. Throughout this study an assumption was made that the hierarchical structure from both processes was a two-layer system, consisting of a skin and core layer.

Experimental Methods

Several experimental methods were utilized in determination of the dielectric constant of LCPs in relation to skin layer thickness. First, a method of dielectric measurement via resonant cavity perturbation will be discussed, followed by an explanation of the processing parameters studied in a design of experiments using the injection molding machine, and finally a description of the process used to remove the skin layer from both injection molded plaques and cast films.

Resonant Cavity Perturbation Method Dielectric Measurement

The polymer dielectric constant, D_k, was measured using an Agilent Technologies (Boston, MA, USA) Vector Network Analyzer (VNA) and a non-contact dielectric measuring technique known as the Resonant Cavity Perturbation method with the in-waveguide measurement (Fig. 5—Resonant cavity, coaxial cables, and Agilent 8722ES Vector Network Analyzer). Resonant cavity perturbation is the preferred method due to its ability to test small samples in a designed frequency range. Each cavity will have a set of modes in certain frequency ranges based on the geometry of the cavity. iQLP's in house testing method and cavity geometry allows for testing between 7–9 GHz. This system was designed by Chao et al. in partnership with Agilent Technologies according to ASTM D2520, and was repeated, validated, and expanded on up to 100 GHz by DuPont (Wilmington, DE, USA) [9, 26–28]. Electromagnetic waves enter the cavity though the waveguide, and a resonant frequency is measured by the VNA. This resonant frequency is perturbed by the insertion of the sample, and as a result, the resonant peak shifts as energy is stored in the dielectric sample (Fig. 6—Inserting an LCP film sample into the resonant cavity and waveguides). This shift in resonant peak is used to calculate the dielectric constant of the sample, shown in Eq. 2

$$\varepsilon' = \left(1 - \frac{f_s}{f_0}\right)\frac{t_c}{t_s} + 1 \tag{2}$$

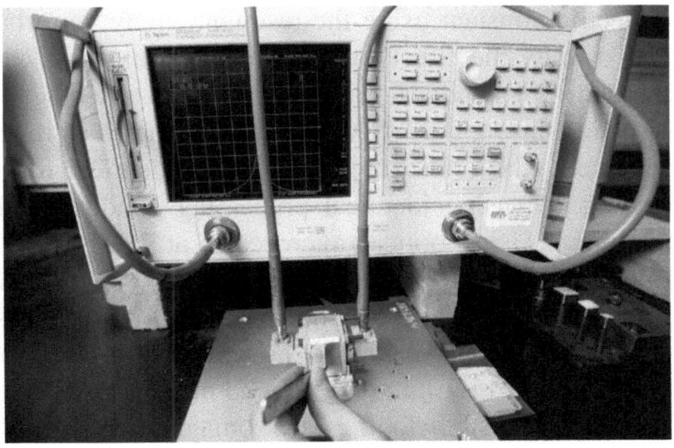

Fig. 5 Resonant cavity, coaxial cables, and agilent 8722ES vector network analyzer

Fig. 6 Inserting an LCP film sample into the resonant cavity and waveguides

where ε' is the dielectric constant, f_0 is the empty resonant frequency, f_s is the resonant with sample, t_c is the thickness of the cavity, and t_s is the thickness of the sample.

Hierarchical Structure Examination

To study the relationship between the hierarchical structure and D_k anisotropy, the highly oriented skin layer was removed from injection molded and cast film samples by an end mill and sanding, respectively, to varying depths and tested for dielectric constant using the resonant cavity perturbation method and VNA.

To study the effects of the skin layer on the dielectric constant of the injection molded LCP plaques, ten samples were tested for D_k in the flow and transverse directions using the VNA. The same samples were retested after the skin layer was removed. These ten samples were of identical formulation, and produced in the same batch using identical processing conditions. The samples were 1.6 mm in thickness. Using optical microscopy at 2.5X the skin layer thickness was determined to be 0.10 mm (100 μm) thick (Fig. 7—LCP injection molded plaque sample cross-section polished and skin layer measured under a microscope at 2.5X). The skin layer was removed using a Bridgeport end mill. Two orthogonal passes with a 12.7 mm (0.5 in.) end mill were needed to remove the skin layer from the surface visible to the electromagnetic waves during dielectric testing. Two depth-of-cut investigation were done to compare the dielectric constant of the skin

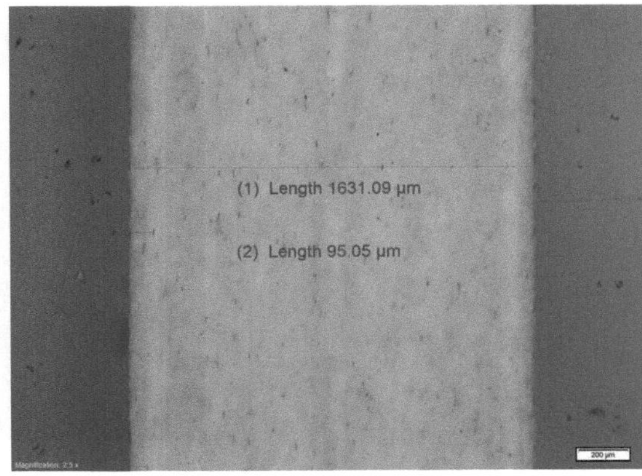

Fig. 7 LCP injection molded plaque sample cross-section polished and skin layer measured under a microscope at 2.5X

Table 1 Skin layer removal experimental procedures for cast films

Removal test	Description	Avg. amount removed per side from each sample (μm)
Bulk	Skin layer removed from both sides—target 15 μm	13
One side	Skin layer removed from one side—target 15 μm	15
Incremental removal	Skin layer removed in 6 total increments, 3 per side—target 5 μm per increment	18

and core layers. The average depth of cut of the skin layer and core examinations were 0.072 and 0.27 mm, respectively.

To study the effects of the skin layer on the dielectric constant of LCP cast film, ten samples were tested for D_k in the flow and transverse directions using the VNA. The samples were then retested with the skin layer removed. Three removal experiments were conducted; bulk skin layer removal, half skin layer removal, and a 5 μm incremental skin layer removal (Table 1).

These ten samples were of identical formulation, and produced in the same batch using identical processing parameters. The samples were 0.115 mm in thickness. Using optical microscopy at 10X and color adjustment to emphasize the morphology, the skin layer thickness was determined to be 15 μm (Fig. 8—Cast film LCP cross section at 10X with hierarchical structure thicknesses labeled).

The hierarchical structure is less pronounced in the extruded film. This follows previous research by investigators which finds the boundary layers become more pronounced as thickness increases [19, 22, 23]. However, a simple skin-core morphology is clearly seen in these extruded film samples. Shown in Fig. 8, the seemingly hollow region on each surface of the sample is representative of the skin

Fig. 8 Cast film LCP cross section at 10X with hierarchical structure thicknesses labeled

layer. A level block and 180 grit sandpaper using a Buehler Ecomet IV polisher (Lake Bluff, IL, USA) were used to remove material.

Results and Discussion

The results from the skin layer removal experiments are shown and discussed in this section. The dielectric constant values were measured using the vector network analyzer, and analysis of variance was calculated using a statistical software package Minitab (State College, PA, USA).

Injection Molded Plaque Skin Layer Removal and Dielectric Constant

The thicker injection molded plaques were dielectric properties showed no effect to both the 0.07 and 0.27 mm removal. Microscopy cross sectional images at the intersection of the depth of cut and intact skin layer at 2.5X magnification are shown (Fig. 9—Cross section of an LCP plaque at the intersection of the removal cut of 0.07 mm and intact skin layer at 2.5X and Fig. 10—Cross section of an LCP plaque at the intersection of the removal cut of 0.27 mm and intact skin layer at 2.5X).

As it can be seen in Fig. 9, it is unclear if the skin layer was completely removed, or if the remaining lighter colored boundary sections are an optical property of the sample under light. However, in Fig. 10, it is clearly visible that the entire skin layer has been removed, and the core is exposed.

The dielectric results of the ten control and experimental samples from the skin layer removal test are presented (Table 2).

Fig. 9 Cross section of an LCP plaque at the intersection of the removal cut of 0.07 mm and intact skin layer at 2.5X

Fig. 10 Cross section of an LCP plaque at the intersection of the removal cut of 0.27 mm and intact skin layer at 2.5X

Table 2 LCP injection molded plaque dielectric constant 0.07 mm skin layer removal results

Measurement	Control avg. of 10 samples	Experimental avg. of 10 samples
D_k in MD	3.51	3.50
D_k in TD	3.38	3.37
D_k MD/TD Ratio	1.04	1.04

As shown on the averages to the right of each table, D_k showed little effect to the 0.07 mm removal in both the flow (MD) and transverse (TD) directions. A similar result is presented in (Table 3) after the 0.27 mm removal seeking to expose the core layer.

The larger depth of cut changed the average D_k and MD/TD ratio in this sample of ten by 0.01. This is not a large enough variation to overcome experimental error, and is certainly not industrially significant. No meaningful variation in dielectric constant was observed in the removal of the skin layer in injection molded LCP plaques of 1.60 mm thickness.

Cast Film Skin Layer Removal and Dielectric Constant

In contrast to the injection molded plaques, the thinner film extruded samples demonstrated a much larger variation in dielectric constant when the skin layer was removed. These results were originally observed during the bulk removal test, which removed on average 13 microns of skin layer on both sides. The tabulated results for the bulk skin layer removal of a sample size of ten is presented (Table 4).

With an average removal of 13 μm on each side, the dielectric constant demonstrated a sizeable reduction, but only a modest reduction in MD/TD anisotropic ratio. ANOVA was used in the statistical software Minitab to determine the significance of the variance. A p-value of 0.000 was calculated, which is less than then the alpha used $\alpha = 0.05$ deeming the variation between the two groups to be statistically significant [29]. Figures 11, 12 and 13 are the individual value plots of MD, TD, and MD/TD ratio, respectively, of the control and experimental samples.

Table 3 LCP injection molded plaque dielectric constant 0.27 mm skin layer removal results

Measurement	Control avg. of 10 samples	Experimental avg. of 10 samples
D_k in MD	3.49	3.48
D_k in TD	3.37	3.40
D_k MD/TD Ratio	1.03	1.02

Table 4 LCP cast film dielectric constant bulk skin layer removal results

Measurement	Control avg. of 10 samples	Experimental avg. of 10 samples
D_k in MD	4.23	3.61
D_k in TD	3.81	3.32
D_k MD/TD Ratio	1.11	1.09

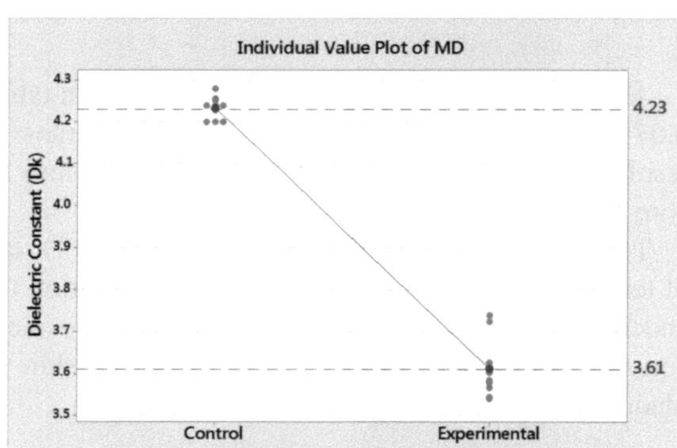

Fig. 11 Individual value plot of the cast film bulk removal dielectric constant in MD of the control and experimental sample groups

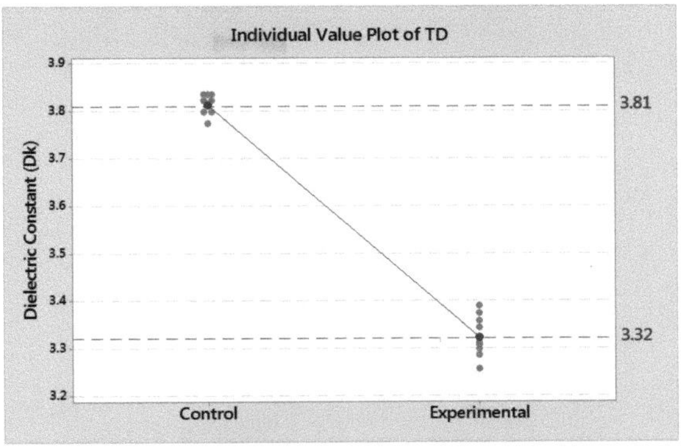

Fig. 12 Individual value plot of the cast film bulk removal dielectric constant in TD of the control and experimental sample groups

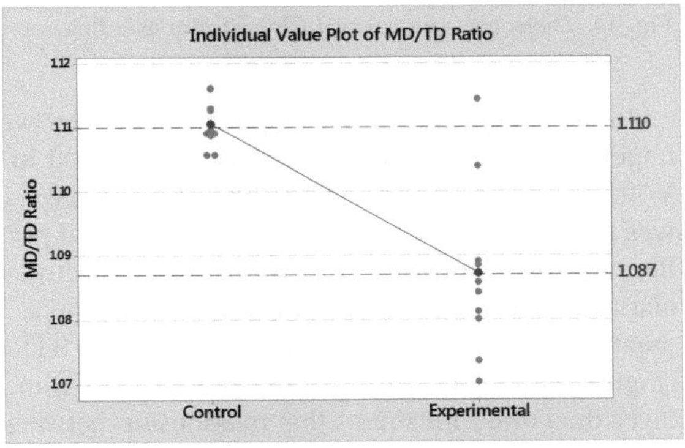

Fig. 13 Individual value plot of the cast film bulk removal dielectric constant of MD TD ratio of the control and experimental sample groups

Table 5 LCP cast film dielectric constant half skin layer removal results

Measurement	Control avg. of 10 samples	Experimental avg. of 10 samples
D_k in MD	4.23	3.84
D_k in TD	3.79	3.49
D_k MD/TD Ratio	1.12	1.10

To investigate observation further, two additional skin layer experiments were conducted; a half skin layer removal, and a 5 μm incremental removal. The skin layer was completely removed (15 μm) from one side of ten cast film samples. Table 5 presents the dielectric constant values between the control and experimental groups.

This experiment yielded results with an average reduction in D_k in MD and TD of 0.039 and 0.030, respectively, demonstrating that removing half of the skin layer resulted in near half the seen reduction in D_k when compared to removing the entire skin layer. These experimental results indicate a near linear relationship between D_k and skin layer thickness in cast films.

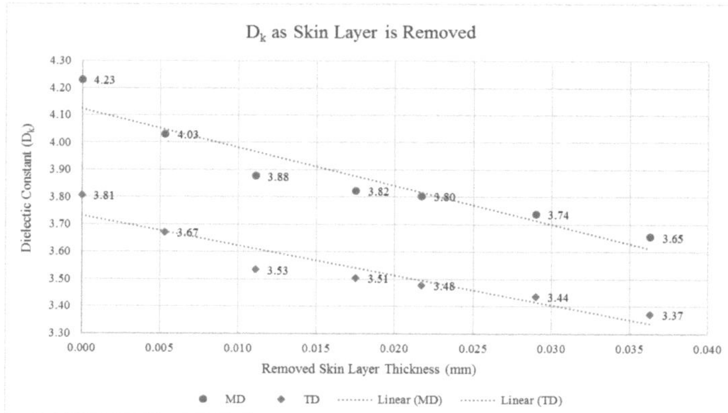

Fig. 14 Dielectric constant of LCP cast film as a function of removed skin layer thickness

In the incremental step experiment, skin layer was removed in increments with a target of 5 μm per side and dielectrically tested in the VNA after each increment. With the control data included, this provided seven data points. An average of 6 μm was removed per step, totaling 18 μm removed per side after. The results validated the trend as skin layer is removed, the dielectric constant is reduced, and added clarity to the relationship with more data points. An average D_k for each incremental step was calculated in both MD and TD, and fit with a linear equation. (Figure 14—Dielectric constant of LCP cast film as a function of removed skin layer thickness) illustrates this relationship between D_k and skin layer thickness.

Linear equations were used to fit trend lines to the reduction in D_k as skin layer was removed in LCP film samples in MD and TD. In MD, D_k fit the equation

$$y = -14.135x + 4.1238$$

and correlated to the data with an R^2 value of 0.887. In TD, D_k fit the equation

$$y = -10.866x + 3.731$$

and correlated to the data with an R^2 value of 0.8898.

Volume Fraction

A possible explanation for the reduction in dielectric constant only demonstrated in cast films is volume fraction of skin layer present in the sample. The volume fraction was determined by dividing the volume of the skin layer by the volume of the sample. Injection molded plaques had a thicker skin layer, but a smaller volume

Table 6 Volume fraction of skin layer and average change in D_k after skin layer removal in injection molded plaques and films

Sample	Volume fraction of skin layer	$\bar{\Delta} D_k$
Plaque	0.0625	0.01
Film	0.2610	0.55

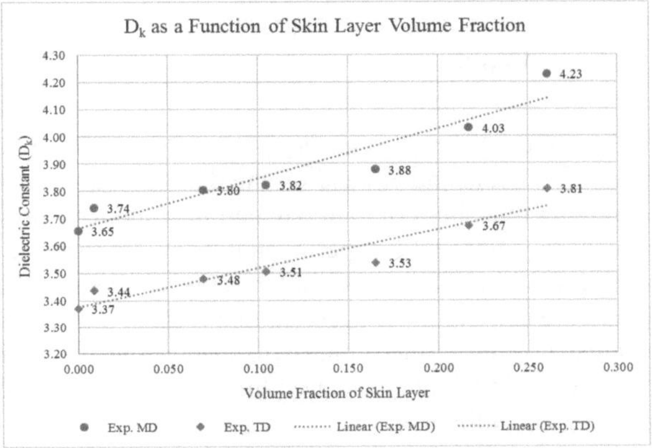

Fig. 15 Dielectric constant of LCP cast film as a function of skin layer volume fraction

fraction of skin layer of 0.0625, in comparison to that of the cast films which had a skin layer volume fraction of 0.261 (Table 6).

With a volume fraction of skin layer about four times smaller in plaques, it is conceivable that the effect of removing the skin layer is negligible on the overall dielectric constant. Conversely, because the volume fraction of skin layer is four times greater in cast film, removing it caused a larger effect on the dielectric constant. The results from the incremental removal of skin layer of the cast film is also displayed in respect to volume fraction of skin layer. A near linear relationship is also demonstrated (Fig. 15—Dielectric constant of LCP cast film as a function of skin layer volume fraction).

The volume fraction of the skin layer formed between the two processing methods is presented here to be the determining factor responsible for the variation between the resulting dielectric constants.

Conclusions

The investigation presented here studied the effect of skin layer on the dielectric constant of post-processed LCP from injection molding and cast film extrusion. Included in this effort was an examination and removal of the skin layer of both injection molded plaque and cast film samples. It was discovered that the hierarchical structure and volume fraction of skin layer are responsible for the difference in dielectric constant of injection molding (3.51) and cast film extrusion (4.23). Cast films contain a larger volume fraction of skin layer (0.261) than injection molded plaques (0.0625), and therefore have a significant effect when removed from the sample. The dielectric constant of cast films was reduced from 4.32 to 3.65 in the flow direction after the skin layer was removed. Dielectric constant and volume fraction of skin layer were demonstrated to follow a near-linear relationship.

Future Work

To continue this research, more accurate skin layer removal techniques should be used to validate and expand on the data presented here. Microtomes are capable of removing thin slices of material, and may induce less mechanical surface conditions which may introduce experimental error. Fixturing the thin LCP plaque and film samples for surface removal presents challenges. This would be a worthwhile pursuit to achieve more accurate removals of skin layer, resulting in a more accurate results. Use of a microtome will also allow for an investigation deeper into the thickness of LCP plaques and films with reduced experimental error due to uneven sanding or milling.

Extruded films consistently result in higher D_k as well as higher anisotropic ratio when compared to injection molded plaques. Developing a plaque with a similar D_k as the film would prove processing control over dielectric constant, and may aid in relating crystal domain orientation to D_k. An injection molding design of experiments attempting to replicate the dielectric constant of film should use a lower temperature chilled mold cavity and faster fill times to induce more elongation and shear stresses and faster freezing of the material. This could result in more oriented and thicker skin and subskin layers. The thicker skin and subskin layer may increase the skin volume fraction enough to effect D_k. Thinner plaques may also increase the skin volume fraction, and could show similar reduction properties when the skin layer is removed. Creating an injection molded plaque with the same properties of the cast film will provide insight to what parameter most significantly affect dielectric constant, and will more clearly illuminate the relationship between skin layer and dielectric constant.

References

1. Cox M (1987) The application of liquid crystal polymer properties. Mol Cryst Liq Cryst 153(1):415–422
2. Zimmerman M (2004) Low cost air cavity liquid crystal polymer packaging. Quantum Leap Packaging
3. Thompson DC et al (2004) Characterization of liquid crystal polymer (LCP) material and transmission lines on LCP substrates from 30 to 110 GHz. IEEE Trans Microw Theory Tech 52(4):1343–1352

4. Chinsirikul W, Hsu T, Harrison I (1996) Liquid crystalline polymer (LCP) reinforced polyethylene blend blown film: effects of counter-rotating die on fiber orientation and film properties. Polym Eng Sci 36(22):2708–2717
5. Dougherty D et al (2009) Multi-lead organic air-cavity package for high power high frequency RFICs. In: 2009 IEEE MTT-S International microwave symposium digest, MTT'09. IEEE
6. Aihara K (2008) Development of liquid crystalline polymer (LCP) surface mount packages for Ka-band applications. ProQuest
7. Zimmerman M et al (2014) New liquid crystal polymer substrate for high frequency applications. In: International symposium on microelectronics. International Microelectronics Assembly and Packaging Society
8. Salmela O, Ikalainen P (1997) Ceramic packaging technologies for microwave applications. In: Proceedings of the wireless communications conference 1997. IEEE
9. Technologies A (2006)
10. Tummala R, Rymaszewski EJ, Klopfenstein AG (2012) Microelectronics packaging handbook: technology drivers. Springer Science & Business Media
11. Ahmad Z (2012) Polymeric dielectric materials. Dielectric material, pp 3–26
12. Donald AM, Windle AH, Hanna S (2006) Liquid crystalline polymers. Cambridge University Press
13. Zhong GJ, Li ZM (2005) Injection molding-induced morphology of thermoplastic polymer blends. Polym Eng Sci 45(12):1655–1665
14. Schrauwen B et al (2004) Structure, deformation, and failure of flow-oriented semicrystalline polymers. Macromolecules 37(23):8618–8633
15. Todd RH, Allen DK, Alting L (1994) Manufacturing processes reference guide. Industrial Press Inc
16. Malloy RA (1994) Plastic part design for injection molding. Hanser Publishers, New York
17. Rockey B (2009) Injection Molding, U.o.A.I.D. Created by Brendan Rockey, for Injection Molding Wikipedia article, Editor. Injection Molding Machine how to work injection moulding
18. Partlow W (2015) 543 Injection molding. In: Civil engineering handbook
19. Hsiung C, Cakmak M (1991) Characterization of structural gradients in an injection molded thermotropic liquid crystalline polymer. In: ANTEC
20. Ide Y, Ophir Z (1983) Orientation development in thermotropic liquid crystal polymers. Polym Eng Sci 23(5):261–265
21. Fritch L (1979) Injection-molding abs-mold in quality with the machine. Plast Eng 35(5): 68–72
22. Hedmark PG et al (1988) Microstructure in injection molded samples of liquid crystalline poly (P-hydroxy-benzoic acid-Co-ethylene terephthalate). Polym Eng Sci 28(19):1248–1259
23. Cakmak M, Cronin S (2000) The effect of composition and processing conditions on the structure development in injection molded dynamically vulcanized PP/EPDM blends. Rubber Chem Technol 73(4):753–778
24. Fundamentals of Cast Film Extrusion Technology. 2015, Macro Engineering and Technology Inc
25. Kostic M, Reifschneider L (2006) Design of extrusion dies. Encyclopedia of Chemical Processing DOI, 10
26. Chao L et al (2013) Techniques to identify the dielectric properties and morphology of LCP Composites in different directions
27. Chao L et al (2012) Permittivity and permeability measurement of spin-spray deposited Ni-Zn-ferrite thin film sample. IEEE Trans Magn 48(11):4085–4088
28. Chao L et al (2013) Non-contact dielectric characterization of lithium ionic solid electrolyte polymer. In: 2013 IEEE International instrumentation and measurement technology conference (I2MTC). IEEE
29. Montgomery DC (2013) Introduction to statistical quality control, 7th edn. Wiley

Aging Behaviour in $Ni_{0.5}Co_xMn_{2.5-x}O_4$ (x = 0.5, 0.8 and 1.1) Thermistors

Gökhan Hardal and Berat Yüksel Price

Abstract Negative temperature coefficient (NTC) thermistors are required to have good electrical stability for many domestic and industrial applications. The aging phenomenon of NTC thermistors can be described as the change in electrical resistance due to thermal stress with time. In this study, $Ni_{0.5}Co_xMn_{2.5-x}O_4$ (x = 0.5, 0.8 and 1.1) ceramics were fabricated by the conventional solid-state reaction method. The powder mixtures of manganese, cobalt and nickel oxides were ball-milled for 5 h. The powders were calcinated at 900 °C for 2 h. The samples were sintered at 1100 °C for 5 h. To investigate the aging behaviour, the samples were held at 150 °C for 400 h. Aging in the samples were calculated by the following equation; $\Delta R/R_0 = (R - R_0)/R_0$ in which R_0 is the resistance at 25 °C before the aging test, and R is the resistance at 25 °C after the aging test.

Keywords Aging · Electrical properties · NTC thermistors

Introduction

Nickel manganite based materials are used as negative temperature coefficient (NTC) thermistors. They are widely used in air conditioners, refrigerators, medical equipment etc. as temperature sensors due to their interesting electrical properties [1]. Transport phenomena in these materials are frequently explained by phonon-assisted jump of carriers among localized states, this is known as hopping conductivity. Their electrical resistivity (ρ) varies exponentially with temperature (T) by the well-known Arrhenius equation $\rho = \rho_o \exp(B/T)$, where ρ_o is the resistivity of the material at infinite temperature and B is the material constant

G. Hardal (✉) · B.Y. Price
Metallurgical and Materials Engineering Department,
Istanbul University, Istanbul, Turkey
e-mail: gokhan.hardal@ogr.iu.edu.tr

B.Y. Price
e-mail: berat@istanbul.edu.tr

S. Ikhmayies et al. (eds.), *Characterization of Minerals, Metals, and Materials 2017*, The Minerals, Metals & Materials Series,
DOI 10.1007/978-3-319-51382-9_10

which is a measure of the sensitivity of the material over a given temperature [2]. The material constant (B) can be calculated by Eq. (1).

$$B_{T1/T2} = \frac{ln\,\rho_1 - ln\,\rho_2}{\frac{1}{T_1} - \frac{1}{T_2}} \tag{1}$$

ρ_1 and ρ_2 being the electrical resistivity at temperature T_1 and T_2, respectively. Nickel manganite based thermistors exhibit the spinel-type crystal structure with the general formula AB_2O_4. In the spinel structure, there are two sites available for the cations; (1) the tetrahedral site is known as A-site and (2) the octahedral site is known as B-site. The distribution of the ions is as follows in nickel manganite: Mn^{3+} will predominantly occupy the B-site, while Mn^{2+} can be found on the A-site and the majority Ni^{2+} can be found on the B-site [3].

The electrical stability of NTC thermistors is very important. However, the drift in resistance ($\Delta R/R$) of a thermistor occurs due to thermal stress with time. The long-term stability of the electrical properties depends on many factors such as the chemical composition, crystal structure (cubic or tetragonal) and the heat treatments applied to NTC thermistors. The chemical composition in thermistor alters its microstructural properties such as crystal structure, density, grain size etc. The atomic diffusion in the spinel lattice can be altered by various heat treatments, the intergranular defects could act as barriers against ion mobility thus explaining the better thermal stability of the thermistors [1].

It is generally considered that the aging is connected with ion exchange between tetrahedral and octahedral sites. The ionic diffusion on octahedral sites may give rise to ionic migration between octahedral and tetrahedral sites. The redistribution of cations occurs owing to cation vacancies and a cationic vacancy migration take places from grain boundary to grains during aging [4].

In this study, we aim to investigate the aging behaviour of $Ni_{0.5}Co_xMn_{2.5-x}O_4$ ($x = 0.5$, 0.8 and 1.1) thermistors.

Experimental Procedures

NiO (99% purity, Alfa Aesar), Co_3O_4 (99.5% purity, Sigma-Aldrich) and Mn_2O_3 (99% purity, Sigma-Aldrich) powders were weighed according to the compositions of $Ni_{0.5}Co_xMn_{2.5-x}O_4$ (where $x = 0.5$, 0.8 and 1.1). The raw powder mixture was ball-milled using ZrO_2 balls as a grinding media with ethyl alcohol in a jar for 5 h. The obtained slurries were dried and powders were calcinated at 900 °C for 2 h. The powders were pressed to form disc shaped specimens and then sintered at 1100 °C for 5 h in air employing a 360 °C/h heating rate then cooled naturally in the furnace.

The phases in the sintered samples were determined by X-ray diffraction (XRD, Rigaku D/Max-2200/PC) analysis using CuK_α radiation at 60 kV/2 kW. The microstructure of samples was observed using a scanning electron microscope (SEM, JEOL, JSM 5600) on fracture surfaces. The sintered samples were coated with silver paste to form electrodes. The electrical resistance was measured in a temperature programmable furnace between 25 and 85 °C in steps of 0.1 °C. The material constant (B, K), the activation energy (E_a, eV), and the sensitivity coefficient (α, %/K) values were calculated for the NTC thermistors.

The samples were held at 150 °C for 400 h in order to age the samples. The drift in resistance was calculated by equation;

$$\% \Delta R = \frac{R - R_0}{R_0} \times 100 \qquad (2)$$

in which "R" is the resistance at 25 °C after aging for 400 h, R_0 is the resistance at 25 °C before aging.

Results

Figure 1a shows the variation of electrical resistance with Co content at 25 °C before and after the aging process. The resistance of samples decreased with the increasing Co content. For the $Ni_{0.5}Co_{0.5}Mn_2O_4$ (A05) sample, the resistance was found as 298 Ω, it decreased to 136 Ω for the $Ni_{0.5}Co_{0.8}Mn_{1.7}O_4$ (A08) sample. A further decrease in resistance to 88 Ω was observed for the $Ni_{0.5}Co_{0.8}Mn_{1.7}O_4$ (A11) sample. In our previous work, the resistivity, $B_{25/85}$ constant and activation energy of samples decreased when Co content increased from 0.5 to 1.1 [5]. This can be explained by the increase in Co^{2+} and Co^{3+} ions, which are responsible for the hopping mechanism on octahedral sites, due to increased Co content.

Muralidharan et al. [6] reported that the resistivity, B constant, the activation energy and temperature coefficient of resistance decreased with the increasing Co content for $Ni_{0.7}Mn_{2.3-x}Co_xO_4$ ($0 \le x \le 0.7$) NTC thermistors. This observation was explained by the Co^{2+} and Co^{3+} ions also occupying the octahedral sites and contribute to the electrical conductivity along with Mn^{3+}/Mn^{4+} ion pairs in the octahedral sites.

Park et al. [7] reported that the resistivity of samples increased with the addition of Cr_2O_3 in $Mn_{1.1}Ni_{1.4}Co_{0.5-x}Cr_xO_4$ ($0 \le x \le 0.35$) NTC thermistors. They reported the resistivity of the thermistors increased as the Cr content increased. There are two possible reasons for the increase in the resistivity with increasing Cr content. (1) Both the grain size and density of the as-sintered samples decreased with an increase in Cr content, decreasing the time between electron scattering events of charge carriers and thus increasing the resistivity. (2) The amount of Co_3O_4 for the thermistors decreases with increasing Cr content, decreasing Co^{2+}/Co^{3+} ions on octahedral sites. As a result, the number of Mn^{3+}/Mn^{4+} ions on octahedral sites decreases to preserve the overall electrical neutrality of the system.

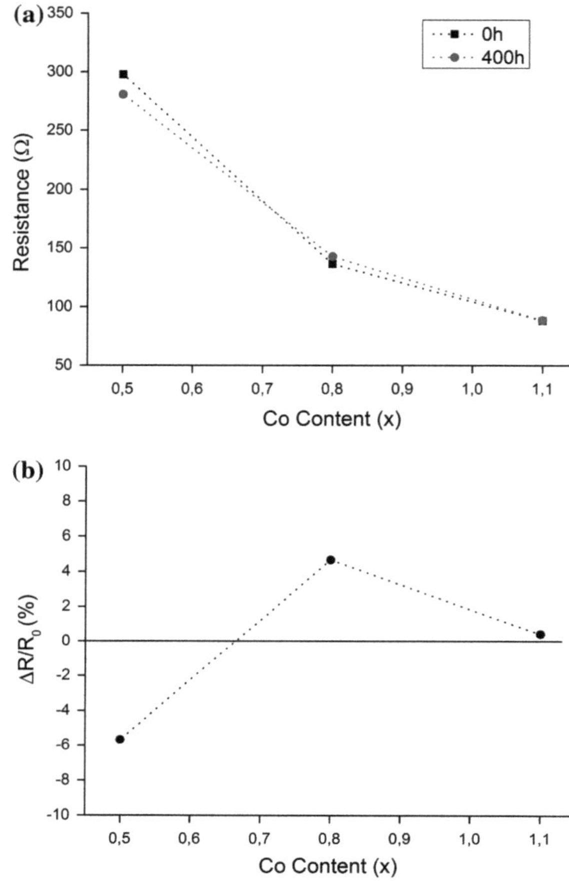

Fig. 1 **a** Variation of electrical resistance at 25 °C **b** Variation of $\Delta R/R_0$ as a function of Co content (x)

Figure 1b shows the variation of $\Delta R/R_0$ as a function of Co content for $Ni_{0.5}Co_xMn_{2.5-x}O_4$ (where x = 0.5, 0.8 and 1.1) ceramic system. The A05 sample aged negatively, but A08 and A11 samples behaved unlike A05 sample. The drift in resistance of A05 sample was calculated as $\approx -6\%$. For A08 sample, the aging value was a positive change in resistance of $\approx 5\%$. Then, it decreased sharply to +0.4% with increasing Co content from 0.8 to 1.1.

Conclusion

Aging phenomenon of nickel manganite based NTC thermistors thermally stressed at 150 °C for 400 h was investigated. The drift in resistance ($\Delta R/R$) of $Ni_{0.5}Co_xMn_{2.5-x}O_4$ (where x = 0.5, 0.8 and 1.1) thermistors varied in the range from −5.7% to +0.4%. The drift in resistance decreased with increasing Co content from 0.8 to 1.1. In this study, the best electrical stability was observed for the $Ni_{0.5}Co_{1.1}Mn_{1.4}O_4$ thermistor with a positive change in resistance of 0.4%.

Acknowledgements This study is supported by TÜBİTAK (The Scientific and Technical Research Council of Turkey), Project number 3001-114M860. We would like to thank TÜBİTAK for its financial support.

References

1. Battault T, Legros R, Brieu M, Couderc J-J, Bernard L, Rousset A (1997) Correlation between microstructure and ageing of iron manganite thermistors. J de Physique 3(7):979–992
2. Wang J, Zhang J (2012) Effects of Mg substitution on microstructure and electrical properties of $NiMn_{2-x}Mg_xO_4$ NTC ceramics. J Mater Res 27:928–931
3. Park K (2005) Fabrication and electrical properties of Mn–Ni–Co–Cu–Si oxides negative temperature coefficient thermistors. J Am Ceram Soc 88:862–866
4. Li DF, Zhao SX, Xiong K, Bao HQ, Nan CW (2014) Aging improvement in Cu-containing NTC ceramics prepared by co-precipitation method. J Alloy Compd 582:283–288
5. Hardal G, Price BY Influence of nano-sized cobalt oxide addition on the structural and electrical properties of nickel manganite based NTC thermistors. In: Paper presented at the 23rd international conference on materials and technology, Portorož, Slovenia, 28–30 Sept 2015, p 256
6. Muralidharan MN, Rohini PR, Sunny EK, Dayas KR, Seema A (2012) Effect of Cu and Fe addition on electrical properties of Ni–Mn–Co–O NTC thermistor compositions. Ceram Int 38:6481–6486
7. Park K, Han IH (2006) Effect of Cr_2O_3 addition on the microstructure and electrical properties of Mn-Ni-Co oxides NTC thermistors. J Electroceram 17:1069–1073

Adsorption of Lead from Aqueous Solutions to Bentonite and Composite

Shujing Zhu and Ying Qin

Abstract The characterization of adsorption of natural and composite bentonite in lead aqueous solution has been studied. The results showed that adsorption of both natural and composite bentonites strongly depended on the pH values of the solution. At lower pH values, the mechanisms that govern the adsorption of bentonites were dissolution of crystal structure and competition between the metal ions and the H^+. To composite bentonite, the lead ion removal capacity was close to 250 mg/g. At higher pH values, formation of lead hydroxyl species might lead to either participation to the adsorption or precipitation onto the bentonites. Hence, a rapid increase in the equilibrium removal capacity of lead was obtained when pH > 7.0. An increase in the initial lead ion concentration led to the increase in the equilibrium adsorption capacity increased to a certain degree; and then, a plateau was obtained at higher concentrations. As a result, the adsorption performance of composite bentonite was better than the natural bentonite in both physical and chemical changes. The data fitted well with both Langmuir and Freundlich isotherms.

Keywords Amorphous hydrous oxides (AHO) · Bentonite · Ion exchange · Lead ion · Adsorption

S. Zhu (✉)
Department of Resource and Environment Science, R&D Center of WISCO,
Wuhan 430079, People's Republic of China
e-mail: shujingz@mtu.edu

Y. Qin
School of Science, Wuhan University of Technology,
Wuhan 430070, People's Republic of China

S. Ikhmayies et al. (eds.), *Characterization of Minerals, Metals, and Materials 2017*, The Minerals, Metals & Materials Series,
DOI 10.1007/978-3-319-51382-9_11

Introduction

Releasing industrial wastewaters and flooding of ore mines into the environment results in contamination of soils and groundwater. The remediation technologies used for clean up processes of polluted areas were time consuming and expensive. Despite their high cost, clay-based barriers were built up at the waste disposal areas to protect the migration of hazardous contaminants through the environment. The natural materials widely used in liner application were clays and bentonites mixtures. The utilization of clays as a liner material has been applied for the last few decades [1]. Recent studies showed that bentonite/zeolite mixtures could be used in a barrier system as well [2]. On account of higher surface area and cation exchange capacity, bentonite (mainly montmorillonite) was the most preferable clay mineral for barrier applications. Most of these studies considered the adsorption characteristics of bentonite for some toxic elements. It is well documented that bentonite was an efficient adsorbent for some heavy metals, especially for lead, copper, cadmium and zinc [3–5].

Details of adsorption studies for bentonite were given in the literature for some heavy metals, i.e. lead, copper and nickel [4, 5]. However, studies of its behavior in the presence of lead ions under different pH, heavy metal initial concentrations and composite bentonite pretreated by fly ash were very scarce. The main purpose of this study was to fill the gap for lead uptake for bentonite and composite bentonite under different physical and chemical conditions. For this purpose, a detailed investigation was performed to determine the effect of pH and initial adsorbate concentration on adsorption characterization of natural and composite bentonites. Langmuir and Freundlich models were used to describe the adsorption isotherm.

Materials and Methods

Materials

The natural bentonite was obtained from Huagu Co., Hubei, P.R.China. Bentonite was in the clod sized forms when first received. Later, the fly ash and natural bentonite were mixed together to get the composite bentonite. The ratio of fly ash and nature bentonite in the mixture was 80:20%. The components mixed together and was ground by a vertical planetary lab ball mill (QM-1SP2, Nanjing University Chemical Company Ltd. in Nanjing, China) for four hours to achieve fine powders (75 μm). Powdered bentonites were dried for 1 week in an oven at 60 °C before the experiments. The mineralogical and chemical compositions of the materials were determined by X-ray diffractometer (Siemens D5000, Siemens Company Ltd., Germany). The cation exchange capacity was obtained by the Na-method [6]. The results of the mineralogical and chemical compositions of the bentonite are presented in Tables 1 and 2, respectively.

Table 1 Propertity of nature bentonite

Samples	Gelatine (mL/15 g)	Dilatability (mL/2 g)	AMBA (g/100 g)	CEC (mmoL/100 g)
Natural bentonite	26	8.9	30.47	80.5

Note AMBA and CEC represent Amount of methylene blue absorption and Cation-Exchange Capacity respectively

Table 2 Chemical compositions of nature bentonite and fly ash (m/m%)

Materials	SiO_2	Al_2O_3	Fe_2O_3	MgO	CaO	Na_2O	K_2O	LOI
Natural bentonite	47.28	10.97	1.28	6.81	7.90	2.81	0.22	–
Fly ash	36.50	18.58	6.90	5.88	14.02	3.32	5.35	10.81

Methods

For determination of adsorption capacity of lead ion, 1.0 g samples were firstly dissolved in concentrated nitric acid solution at 24.8 °C. When most of the samples were added into the reactor (Micromeritics Model ASAP-2000, Nanjing University Chemical Company Ltd. in Nanjing, China) to fix the dimension with 100 mL. The concentration of lead ion in the filtrate was analyzed with atomic absorption spectrometry (TAS-990, Shanghai Chemical Company Ltd. in Shanghai, China), and thus the contents of lead ion removal was calculated. The samples were respectively denoted as NBPb and CBPb.

Puls and Bohn [7] stated that lead adsorption onto the calcium saturated bentonite was higher for Cl^- solution than that of ClO_4^- and SO_4^{2-} because of the formation of limited lead–nitrate complexes resulting from the incompatibility of the harder Pb^{2+} with the soft NO^{3-}. Therefore, $Pb(NO_3)_2$ solution was used in the batch experiments. At the beginning of the experiment, a volume of 100 mL of lead solutions with various amounts of bentonites was placed in an Erlenmeyer flask. For both adsorbents, the initial metal ion concentration used in the tests ranged from 50 to 1000 mg/L. Since adsorbent dosage was required for 10.0 g/L, adsorbents were weighed for 1 g. The batch adsorption experiments were also performed at different pH levels (i.e. pH 3–11) and solution pHs were adjusted by adding diluted NaOH and HNO_3 solutions. NaOH and HNO_3 were prepared in 0.01 moL/L stock solutions and diluted with de-ionized water. Note that the pHs of the solutions were adjusted before adding the bentonites. This was because bentonite and its composite had a tendency to increase the solution pH after adding it into the solution. The shaking speed was constant and it kept 200 rpm during the test. Contact time was 24 h. Most of the lead ions were adsorbed by the bentonites in the first 1 h. After centrifugation, the supernatant liquid was used in atomic adsorption spectrophotometry for determination of the equilibrium lead concentration after adsorbed.

Results and Discussion

Determining the Amount of Adsorbed Lead Ions, Qe

The interaction between the lead ion and stuffing can be given by the equilibrium:

$$\mathrm{A(liquid)} + \mathrm{B(solid)} \rightarrow c\mathrm{C(liquid)} + d\mathrm{D(solid)} \quad (1)$$

where A is the aqueous solutions of lead ion and B is the materials, in which the amorphous hydrous oxides (AHO) composition contains exchangeable cation valence Na, K, Ca, Mg etc. The amount of adsorbed lead ions by any material can be given by:

$$Q_e = (C_i - C_e)/S \quad (2)$$

$$R_e = (C_i - C_e)/C_i \quad (3)$$

where Qe, C_i, C_e, S and R_e are the adsorbed metal ions onto the bentonites (mg/g), initial metal ion concentration (mg/L), equilibrium metal ion concentration (mg/L), slurry concentration (g/L) and lead removal ratio (%), respectively. The change in adsorption characteristics of bentonites against lead ion during the testing period can be seen in Fig. 1. And here C_i = 200 mg/L.

Influence of Reacting Time

Figure 1 Adsorption characteristics of materials during 1 day.

Figure 1 showed that the adsorption of the lead ions onto the natural bentonite and composite bentonite reached equilibrium approximately at the end of 4 h. For natural bentonite, it may due to the limited replacement of alkaline and alkaline earth metals with lead ions. The gradual increase in the lead uptake by composite bentonite pointed out the continuity of the hydraulic reaction of AHO in the

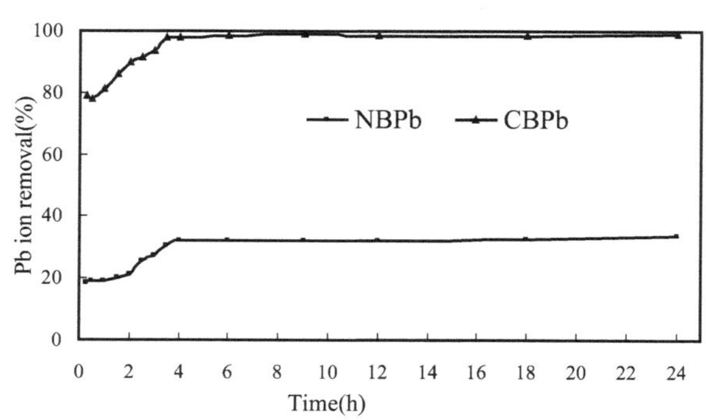

Fig. 1 Adsorption characteristics of materials during 1 day

aqueous solution during the whole adsorption; then, it became constant at the end of another 20 h. At the first 4 h, the curve of composite bentonite followed a lateral path within a few minutes just as that of natural bentonite. Then, a gradual increase was observed from 1 to 4 h that may because the AHO hydraulic products calcium silicate hydrates(C-S-H) has large number of micropores and very little crystal particles, which leads to high specific surface area (50 m^2/g), and so the lead ions can be adsorbed on the crystal particles interlayer. Therefore, the initial stage of the lead uptake curve of composite bentonite was similar to the natural bentonite. However, adsorption ability for composite bentonite was three times higher than that of natural bentonite at the end of the experiments. Triantafyllou et al. [8] stated that the time required to reach steady state of adsorption differed with the sample characteristics, metal ion and pH. Thus, we chose following factors as the key factors that would affect the adsorption characteristics of natural and composite bentonites.

Influence of Solution pH

To determine the pH effect on adsorption capacity of materials, solutions were prepared at different pH levels ranging from 3 to 11 and the initial concentration was 200 mg/L before adding bentonites. Figure 2 showed the change in lead uptake by both bentonites at different initial pH levels. As can be seen from it, adsorption capacity of composite bentonite increased when the initial pH of the solution increased from 3 to 7. And the increase of adsorption capacity in this region was remarkable for the composite bentonite, while such change was not marked for natural bentonite. Then, Pb^{2+} removal by both bentonites suddenly increased beyond 90% when the initial pH of the solution was higher than 8. These results are in agreement with other studies performed with montmorillonite [9, 10]. The mechanisms that influence the adsorption characteristics of bentonites can be given by dissolution, ion exchange/adsorption, and precipitation [10]. From Fig. 2, the lowest Pb^{2+} sorption rates were obtained at pH 3. This may due to the increasing competing for adsorption sites by H^+ [9, 10]. Altin et al. [10] revealed that the

Fig. 2 Changes in the equilibrium removal of Pb^{2+} between pH 3 and 11

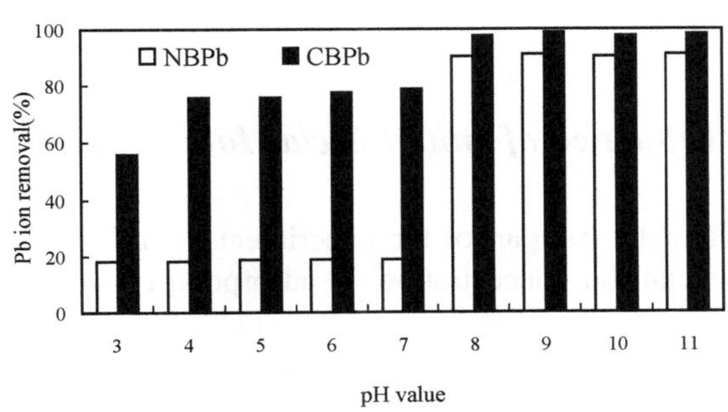

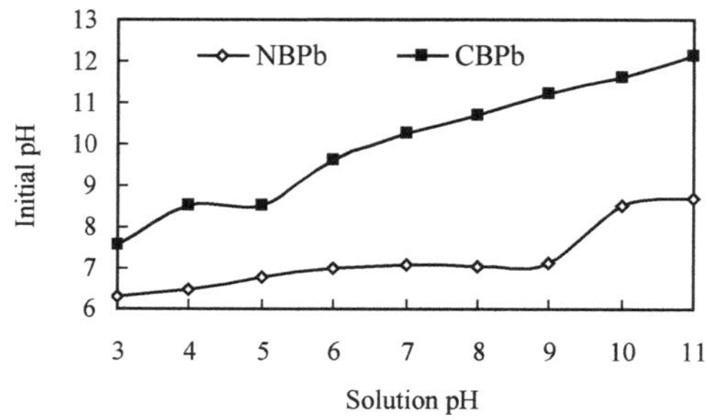

Fig. 3 Change in initial solution pH after the experiments

removal efficiency of Pb^{2+} by montmorillonite decreased at low pH values (pH 2–4). The similar results were obtained by other researchers for some heavy metals such as Zn, Cd, Cu, Ni and Cr when pH = 3 [9]. With the increase of pH value from 4 to 7, the basic mechanism that governs the adsorption characteristics of bentonites is adsorption and ion exchange. Sharp increase was observed for both bentonites at higher pH level (i.e. pH 8 $\sim$ 11) and then a decrease of lead ion removal, which was the result of the formation of lead hydroxyl species. With different pH values and metal concentration, lead may form complexes with OH^-, for example, $Pb(OH)_2$, $Pb(OH)_3^-$ and $Pb(OH)_4^{2-}$ at higher pHs and as a result, lead hydroxyl species may participate in the adsorption and precipitate onto the bentonite structure [3]. Previous studies had showed a similar performance for other heavy metals. For example, Barbier et al. (2001) reported that lead sorption suddenly increased between pH 5.5 and 6.5 for montmorillonite and pH 6 and 7 for commercial bentonite; whereas cadmium sorption increased abruptly between pH 7 and 8 for both clays.

Measuring the final pHs of the suspension after a contact time of 50 h could explain such experiment phenomena. Figure 3 shows the final pH values at the end of the experiments. Treating the natural bentonite at different initial pHs place the final pH value around 7 until pH 11; then, a sudden increase noted at pH 9. Unlike nature bentonite, initial pH of the composite bentonite did not change significantly when the initial pH of the solution was 5, however the final pH value raised with the initial one. As seen from this figure, increase of the final suspension pH might be attributed to the hydrolic products of AHO in the composite bentonite with lead ions existed in the solutions.

Influence of Initial Metal Ion Concentration

For the rest part of the experiments, pH 4 was set to evaluate the effect of initial metal ion concentration on adsorption characteristics of bentonites. Figure 4 represents the change in adsorption behavior of both materials under different metal ion concentrations. As seen from Fig. 4, increasing the metal ion concentration leaded

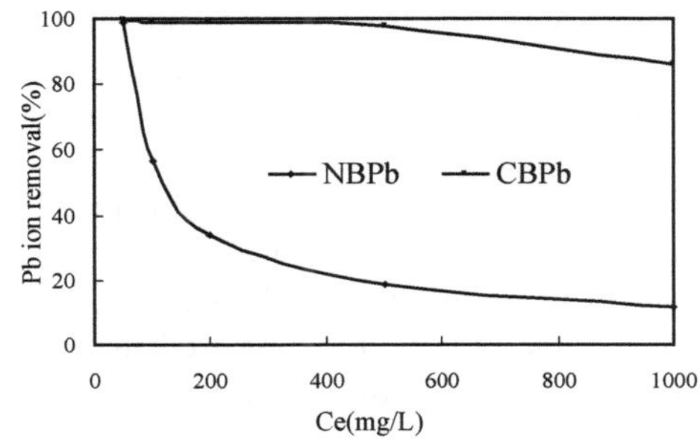

Fig. 4 Relationship between initial metal ion concentration and Pb^{2+} adsorption

to increase the Pb^{2+} uptake onto the materials to a certain point; then, a plateau occurs for both clays that indicate unavailability of adsorption sites on the bentonites for adsorption of lead. The maximum adsorption capacity of Composite bentonite and natural bentonite was obtained 250 and 60 mg/g, respectively. The difference in the lead adsorption onto the bentonites might be due to the difference in the mineralogical compositions and associated cations in the exchangeable sites. Sheta et al. [11] investigated the sorption characteristics of natural zeolite and bentonite via zinc and iron and concluded that heavy metal uptake strongly depended on the mineralogical compositions of materials and kinds of heavy metal used in the tests. They found that one of two clinoptilolites had lower sorption ability for Pb^{2+} in comparison with the bentonite and other had higher ability than those of bentonite.

Comparison of Adsorption Isotherms

Adsorption isotherms of bentonites for lead ion were expressed mathematically in terms of the Langmuir and Freundlich models. The obtained experimental data fitted the Langmuir (Eq. 4) and Freundlich (Eq. 5) models well:

$$Q_e = K_L C_e/(1 + aC_e) \tag{4}$$

$$Q_e = K_F \cdot C_e^n \tag{5}$$

Following equations were liner express based on above equations:

$$C_e/Q_e = 1/K_L + aC_e/K_L \tag{6}$$

$$\text{Lg}\, Q_e = \text{Lg}\, K_F + n\text{Lg}\, C_e \tag{7}$$

where K_L, a and K_F, n are the constants for Langmuir and Freundlich models, respectively. In addition to the experimental data, the linearized form of Langmuir and Freundlich isotherms using Eqs. (6) and (7) for lead ion removal by bentonites can be seen in Figs. 5 and 6, respectively.

As it was seen from Figs. 5 and 6, the experimental data for both clays fitted the liner Langmuir isotherm well. Except for composite bentonite, liner Freundlich isotherm corresponded with the experimental data as well. The values of K_L, a, K_F and n obtained from both models were presented in Table 3.

The comparison of the experimental values with the values of Qe obtained by both models was shown in Fig. 7. As it is seen from figure, Langmuir isotherms usually fitted better with the experimental data rather than Freundlich isotherms. Also, some other studies showed that Langmuir and Freundlich isotherms corresponded well with the experimental results of some heavy metals [11].

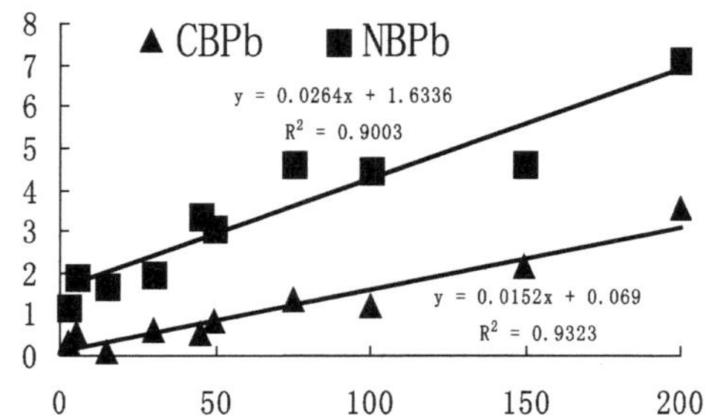

Fig. 5 Linearized Langmuir isotherms for Pb^{2+}removal by bentonites

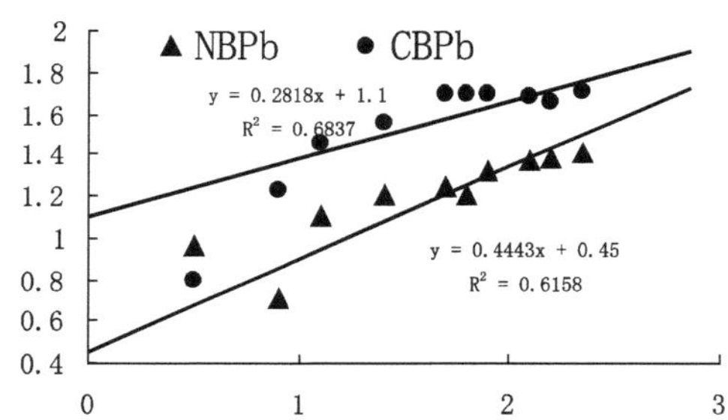

Fig. 6 Linearized Freundlich isotherms for Pb^{2+} removal by bentonites

Table 3 The parameters for the isotherms

Isotherm parameters	Composite bentonite	Natural bentonite
Langemuir		
$1/K_L$	0.069	1.6336
a/K_L	0.0152	0.0264
Freundlich		
Log K_F	1.1	0.45
n	0.282	0.444

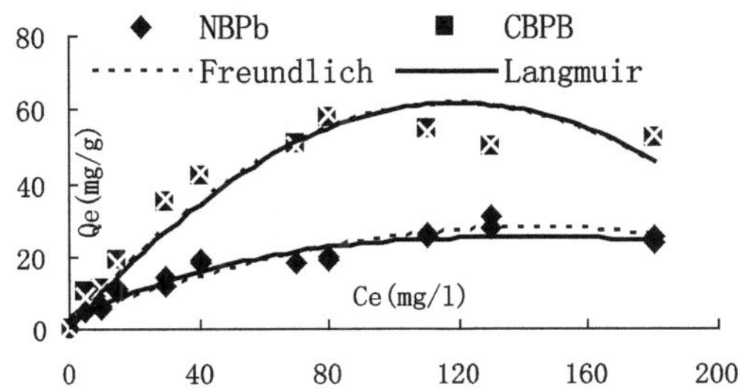

Fig. 7 Comparison of the experimental results with the Q_e values obtained by Freundlich isotherms

Conclusions

The obtained results can be summarized:

1. Results obtained from the effect of pH on the adsorption capacity of the bentonites indicated that the major factor which affected this parameter were competition between the H^+ ions and Pb^{2+} ions (under pH 4), ion exchange (pH 4–7), and participation of the heavy metal species to the adsorption and sedimentation of some onto the bentonites (pH 8–11).
2. Increase in the initial lead ion concentration leads to increase of the adsorption capacity of the bentonites.
3. The experimental results fitted well with the Langmuir and Freundlich isotherms. Compared to Freundlich isotherms, the Langmuir isotherms corresponded better with the experimental findings.

References

1. Fenanadez F, Quigley RM (1985) Hydraulic conductivity of natural clays permeated with simple liquid hydrocarbons. Can Geotech J 22:205–214
2. Kaya A, Durukan S (2004) Utilization of bentonite-embedded zeolite as clay liner. Appl Clay Sci 25:83–91
3. Mellah A, Chegrouche S (1997) The removal of zinc from aqueous solutions by natural bentonite. Water Res 31(3):621–629
4. Gonzales-pradas E, Villafranca-Sanchez M, Canton-Cruz F, Socias-Viciana M, Fernandez-Perez M (1994) Adsorption of cadmium and zinc from aqueous solution on natural and activated bentonite. J Chem Tech Biotechnol 59:289–295
5. Naseem R, Tahir SS (2001) Removal of Pb(II) from aqueous/acidic solutions by using bentonite as an adsorbent. Water Res 35(16):3982–3986
6. Chapman HD (1965) Cation exchange capacity. In: Black JA (ed) Methods of soil analysis. No. 9, in the Series, Agronomy, American Institute of Agronomy, Madison, WI, pp 891–901
7. Puls RW, Bohn HL (1998) Sorption of cadmium, nickel, and zinc by kaolinite and montmorillonite suspensions. Soil Sci Soc Am J 52:1289–1292
8. Triantafyllou S, Christodolou E, Neou-Syngouna P (1999) Removal of nickel and cobalt from aqueous solutions by Na-activated bentonite. Clays Clay Miner 47(5):567–572

9. Alvarez-Ayuso E, Garcia-Sanchez A (2003) Removal of heavy metals from waste waters by natural and Na-exchanged bentonites. Clays Clay Miner 51(5):475–480
10. Altin O, Ozbelge OH, Dogu T (1999) Effect of pH, flow rate and concentration on the sorption of Pb and Cd on montmorillonite. I. Experimental. J Chem Tech Biotechnol 74:1131–1138
11. Sheta AS, Falatah AM, Al-Swailem MS, Khaled EM, Salam ASH (2003) Sorption characteristics of zinc and iron by natural zeolite and bentonite. Micropor Mesopor Mater 61:127–136

Part III
Electronic, Magnetic, Environmental, and Advanced Materials

Characterization of Low-Zinc Electric Arc Furnace Dust

Zhiwei Peng, Xiaolong Lin, Jiaxing Yan, Jiann-Yang Hwang, Yuanbo Zhang, Guanghui Li and Tao Jiang

Abstract Electric arc furnace (EAF) dust is an important secondary resource that should be recycled to enhance its considerable economic value and potential environmental benefit. In this study, a low-zinc EAF dust was characterized by various techniques, including chemical titration, X-ray diffraction, granulometric analysis, scanning electron microscopy and thermogravimetry. It is shown that the dust contains 2.08 wt% Zn, 23.16 wt% Fe and 19.84 wt% Ca, accompanying small amounts of Cr, Pb, etc. Magnetite, calcium ferrite and zinc ferrite are the main phase constituents. The majority (90%) of particles have size less than 137.862 μm. According to these characteristics, it is expected that the use of microwave energy for intensification of the reduction of EAF dust in the presence of biochar will succeed in the dust recycling by promoting the processing efficiency with elimination of secondary hazardous pollutants.

Keywords EAF dust · Characterization · Composition · Microstructure · Thermal stability · Recycling

Introduction

The rapid growth of iron and steel industry has led to huge increases in the use of energy and other resources. It also caused severe environmental degradation because of considerable generation of dust, waste water and exhaust gas which must be treated appropriately. Electric arc furnace (EAF) dust is an important by-product of the steelmaking industry with output accounting for approximately

Z. Peng (✉) · X. Lin · J. Yan · Y. Zhang · G. Li · T. Jiang
School of Minerals Processing and Bioengineering, Central South University, Changsha 410083, Hunan, China
e-mail: zwpeng@csu.edu.cn

J.-Y. Hwang
Department of Materials Science and Engineering, Michigan Technological University, Houghton, MI 49931, USA

S. Ikhmayies et al. (eds.), *Characterization of Minerals, Metals, and Materials 2017*, The Minerals, Metals & Materials Series,
DOI 10.1007/978-3-319-51382-9_12

1–2% of the charge in a usual EAF operation [1, 2]. In 2014, its output reached 8.764 million tons. Because of its high contents of zinc and iron (up to 40 wt% Zn [3] and 50 wt% Fe [4]), EAF dust is often deemed to be an industrial waste having high recycling value. However, this dust also contains minor amounts of harmful heavy metals, e.g., Pb and Cr [5–8]. Hence, it is commonly categorized as a hazardous waste [5]. For disposal of the dust, it is necessary to gain a comprehensive understanding of the characteristics of EAF dust.

This paper offers a systemic study on the characteristics of low-zinc EAF dust by examining its chemical and phase compositions, size distribution, particle morphology and thermal property by chemical titration, X-ray diffraction, granulometric analysis, scanning electron microscopy and thermogravimetry, respectively. Based on the characteristics, possible approaches for treatment of this dust are discussed and a promising method involving use of microwave energy for processing of the dust is proposed.

Experimental

Material

The EAF dust, collected from Baosteel Group Corporation, Shanghai, China, was used for the characterization in this study.

Experimental Methods

The chemical composition of the EAF dust was determined by X-ray fluorescence spectrometry (XRF-Axios^{mAX}) with Rh$K\alpha$ radiation. This technique can measure the content of the elements between O^8 and U^{92} with good reproducibility, fast test speed and high sensitivity. The phase composition of EAF dust was acquired by a D/Max-2500 diffractometer at 2° min^{-1} from 0 to 90° with Cu$K\alpha$ radiation. The morphology of the sample was examined using an environmental scanning electron microscope (Quanta-200) and the particle size distribution of the dust was determined using a laser particle size analyzer (LPSA-Mastersize 2000) with the granularity range from 20 nm to 2 mm. The thermal property of the dust was characterized using a thermal gravimetric analyzer (TGA851) at a ramp rate of 15 °C min^{-1} under nitrogen atmosphere.

Results and Discussion

Chemical Composition

Table 1 shows the chemical composition of the EAF dust. Iron and calcium are the main elements of EAF dust. The contents of Fe and Ca reach 23.16 wt% and 19.84 wt%, respectively. Because the content of zinc is only 2.08%, it is categorized as a typical low-zinc dust (<4 wt%) [9, 10]. The dust also contains hazardous species, including Cr, Pb and halogens, constituting threat to environment though their contents are less than 1 wt%. Particularly, the content of Mn reaches 3.03% in the EAF dust, representing a potential manganese resource.

Phase Composition

The phase composition of the dust is shown in Fig. 1. The main phases are Fe_3O_4, $CaFe_2O_4$ and $Zn_{0.35}Fe_{2.65}O_4$. Fe is primarily enriched in the dust in the form of Fe_3O_4 and $CaFe_2O_4$ with relatively strong magnetism. This finding indicates the possibility of recovering Fe from the dust by magnetic separation. $Zn_{0.35}Fe_{2.65}O_4$ may decompose into ZnO and Fe_xO_y under certain metallurgical conditions, causing separation of iron and zinc. The other less important phases include Mn_3O_4, $CaAl_8Fe_4O_{19}$, $Mg_{0.64}Fe_{2.36}O_4$, $MgAl(SiO_4)_3$ and SiO_2.

Particle Size Distribution

The particle size distribution of EAF dust is displayed in Fig. 2. It is observed that the dust has two main size fractions, namely a very fine-grained portion (0.1–1 μm) and a coarser portion (1–138 μm). 50% of the dust particles are smaller than 19.661 μm and 90% of the particle sizes are less than 137.862 μm.

Table 1 Chemical composition of EAF dust

Element	wt%	Element	wt%	Element	wt%	Element	wt%
Zn	2.08	Cl	0.69	F	0.98	Nb	0.01
Fe	23.16	K	0.35	Ni	0.01	Mo	0.11
Ca	19.84	Mg	2.83	Cu	0.02	Ba	0.27
Al	6.21	Ti	0.15	O	30.53	Pb	0.05
Si	3.58	V	0.02	Br	0.003	loss	4.44
P	0.05	Cr	0.26	Sr	0.02		
S	1.30	Mn	3.03	Zr	0.01		

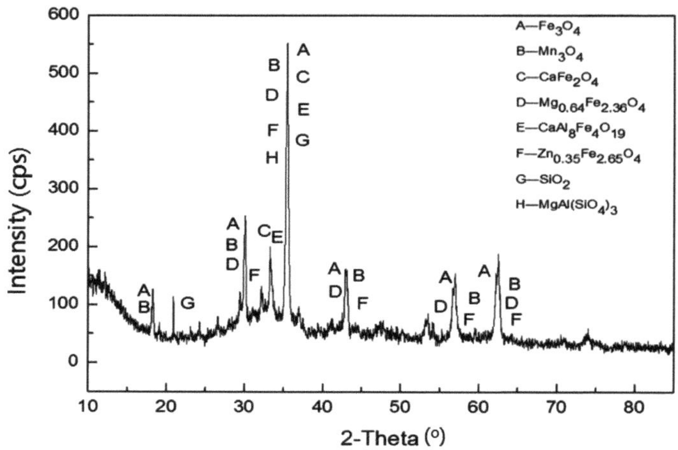

Fig. 1 XRD pattern of EAF dust

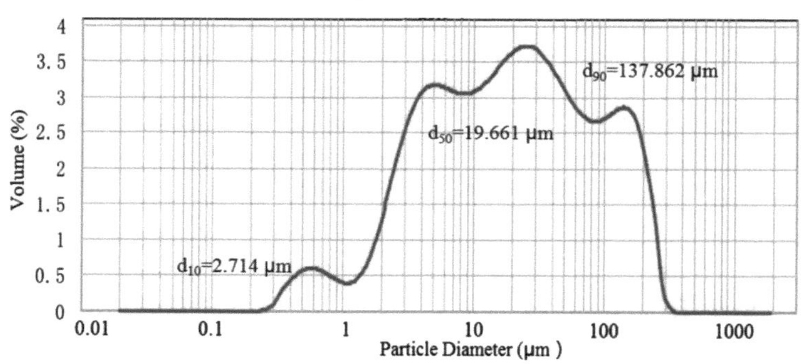

Fig. 2 Particle size distribution of EAF dust

Microstructure

Figure 3 shows scanning electron microscope images of the microstructure of EAF dust at different magnifications. It is observed that most particles appear in the sphere-like shape. The small particles accumulate together or adhere to the surface of large particles, forming large agglomerates.

Thermal Property

The thermal stability of EAF dust was revealed by the thermogravimetric analysis. Figure 4 shows the thermogravimetric profile of the EAF dust at a ramp rate of

Fig. 3 Scanning electron microscope images of EAF dust

Fig. 4 Thermogravimetric profile of EAF dust

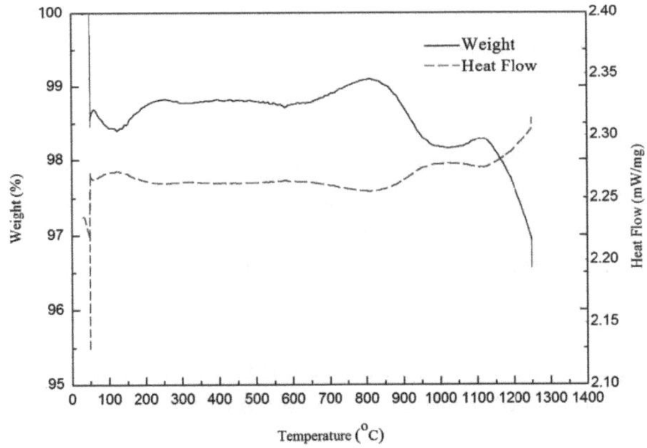

15 °C min^{-1} under N_2 atmosphere. It is seen that there exist three steps for the mass loss, consistent with the findings in literature [9]. The first mass loss event starts at 50 °C with the maximum mass loss rate occurring at 130 °C, corresponding to evaporation of entrained free water and chemically adsorbed water (e.g., water bound to the metal chlorides). The second mass loss has an onset temperature of 800 °C with the mass loss occurring in the temperature range up to 1000 °C, possibly attributed to the vaporization of halides. The final mass loss event initiates at a temperature of 1100 °C, and the loss may be assigned to the vaporization of the elements with low boiling points in the dust. It is clear that additional

characterizations of the dust are required to identify the specific reactions involved in the process.

The above results show that the dust is characterized by low content of zinc in the form of zinc ferrite, and high content of iron and calcium in the form of magnetite and ferrites. The main hazardous species include Pb, Cr and Cl. It also has small particle size and relatively low thermal stability which has been verified by the three main steps of mass loss in the temperature range up to 1250 °C. These features are in favor of use of pyrometallurgical approaches (e.g., carbothermic reduction) in separating these elements. The zinc and iron species are capable of being separated and recovered from the dust by controlling the processing temperature and atmosphere properly. When the zinc constituents are volatilized, the Fe-bearing residue can be used as a material for ferrous metallurgy after further purification by removing the major impurities, especially Ca. As the dust has high content of ferrites which have strong microwave absorption, it is anticipated that the use of microwave energy (featured by volumetric and selective heating [10, 11]) and biomass-derived char (biochar, as reducing agent and microwave absorber [12, 13]) for treatment of this dust will be highly promising from the perspectives of energy conservation and environmental protection.

Conclusions

EAF dust is an important secondary resource enriching a variety of metallic elements. In this study, a low-zinc EAF dust was characterized by a variety of techniques, including chemical and phase compositions, particle morphology, size distribution and thermal property. The dust contains 2.08 wt% Zn and 23.16 wt% Fe, accompanying a number of harmful heavy metals, e.g., Pb and Cr. Magnetite, calcium ferrite and zinc ferrite are the main phase constituents. Further, approximately 90% of particles have size less than 137.862 μm. According to these characteristics, it is anticipated that the use of microwave energy for intensifying the reduction of EAF dusts in the presence of the reducing agents originated from sustainable biomass resources, such as biochar, may succeed in the dust recycling by elevating the processing efficiency with elimination of secondary hazardous pollutants.

Acknowledgements This work was supported by the Innovation-Driven Program of Central South University under Grant 2016CXS021, the Shenghua Lieying Program of Central South University under Grant 502035001, and the Fundamental Research Funds for the Central Universities of Central South University under Grant 2016zzts464.

References

1. Dutra AJB, Paiva PRP, Tavares LM (2006) Alkaline leaching of zinc from electric arc furnace steel dust. Miner Eng 19:478–485
2. Suetens T, Klaasen B, Van Acker K, Blanpain B (2014) Comparision of electric arc furnace dust treatment technologies using energy efficiency. J Clean Prod 65:152–167
3. Nolasc-Sobrinho PJ, Espinosa DCR, Tenorio JAS (2003) Characterization of dusts and sludges generated during stainless steel production in Brazilian industries. Ironmaking Steelmaking 30:11–17
4. Orhan G (2005) Leaching and cementation of heavy metals from electric arc furnace dust in alkaline medium. Hydrometallurgy 78:236–245
5. Pickles CA (2008) Thermodynamic analysis of the selective carbothermic reduction of electric arc furnace dust. J Hazard Mater 150:265–278
6. Lis T, Nowacki K, Żelichowska M, Kania H (2015) Innovation in metallurgical waste management. Metalurgija 54:283–285
7. Sofilić T, Rastovčan-Mioč A, Cerjan-Stefanović S, Novosel-Radovi V, Jenko M (2004) Characterization of steel mill electric-arc furnace dust. J Hazard Mater 109:59–70
8. Suetens T, Guo M, Van Acker K, Blanpain B (2015) Formation of the $ZnFe_2O_4$ phase in an electric arc furnace off-gas treatment system. J Hazard Mater 287:180–187
9. Al-Harahsheh M, Kingman S, Al-Makhadmah L, Hamilton IE (2014) Microwave treatment of electric arc furnace dust with pvc: dielectric characterization and pyrolysis-leaching. J Hazard Mater 274:87–97
10. Peng Z, Hwang JY, Mouris J, Hutcheon R, Huang X (2010) Microwave penetration depth in materials with non-zero magnetic susceptibility. ISIJ Int 50:1590–1596
11. Peng Z, Hwang JY (2015) Microwave-assisted metallurgy. Int Mater Rev 60:30–63
12. Tang H, Qi T, Qin Y (2015) Production of low-phosphorus molten iron from high-phosphorus oolitic hematite using biomass char. JOM 67:1956–1965
13. Motasemi F, Afzal MT (2013) A review on the microwave-assisted pyrolysis technique. Renew Sustain Energy Rev 28:317–330

Gamma-Radiation Effect on Biodegradability of Synthetic PLA Structural Foams PP/HMSPP Based

Elizabeth Carvalho L. Cardoso, Sandra R. Scagliusi and Ademar B. Lugão

Abstract This research investigated how gamma-radiation affected PP/HMSPP structural foams filled with biodegradable synthetic polyester PLA in terms of thermal properties, biodegradability and infrared spectrum. Polymers are used in various applications providing enormous quantities of wastes in environment and contributing with 20–30% of total volume of solid residues. Besides, shortage of plastics resins obtained from fossil base is addressing research and development toward alternative materials; environmental concerning in litter reduction is being directed to renewable polymers for manufacturing of polymeric foams. Biodegradable polymers, a new generation of polymers produced from various natural resources, environmentally safe and friendly, can contribute for pollution reduction. PLA (poly(lactic) acid)) was chosen to be blended with PP/HMSPP. Blends thereof were subjected to gamma radiation at 50, 100, 150, 200 and 500 kGy doses. Soil burial test pointed toward a partial biodegradation of irradiated PLA-PP/HMSPP foams; other characterizations involved TG/DTG technique and FTIR assessments.

Keywords PLA · Gamma-radiation · Biodegradability · PP/HMSPP

Introduction

Structural foams have a comprehensive application field; they are especially used to improve structures appearance or to reduce costs of involved material, besides their applications in construction market. Most of thermoplastics can be extruded in order to obtain structural foams; commercial activities are concentrated in thermoplastic of lower cost, as Polypropylene (PP) [1]. Polypropylene suffers simultaneously chain-scission and crosslinking, when subjected to ionizing radiation;

E.C.L. Cardoso (✉) · S.R. Scagliusi · A.B. Lugão
Instituto de Pesquisas Energéticas e Nucleares-IPEN/CNEN-SP, Av. LineuPrestes 2242, Cidade Universitária, São Paulo, SP CEP05508-900, Brazil
e-mail: eclcardo@ipen.br

S. Ikhmayies et al. (eds.), *Characterization of Minerals, Metals, and Materials 2017*, The Minerals, Metals & Materials Series,
DOI 10.1007/978-3-319-51382-9_13

both chemical phenomena are able to induce changes in polymers properties capable to enhance their use in various applications and in different processes. Polypropylene is a linear hydrocarbon polymer worldwide used; its versatility is linked to its origin from a cheap petrochemical material and its easy processing for obtaining various products. Unfortunately, the enormous production of fossil based polymers leads to an accumulation in environment, because their difficulty to be degraded by microorganisms, presenting a serious pollution risk to flora and fauna. These polymers are very bio-resistant, once they have just carbon atoms in backbone chain, with non-hydrolysable functional group. Non-degradable plastics accumulate in environment, at a rate of 25 million per year; they can remain in the ocean for hundreds of years in their original form and even longer in small particles, which means that the amount of plastic in the ocean cumulates over time. The best research currently available estimates that there are over 150 million tonnes of plastic waste in the ocean today. A few possibilities have being considered in order to minimize environmental impact caused by the use of conventional polymers [2–4].

Some reasons are responsible for continuous growth of foams, especially related to excellent resistance/weight relation, that endows to foamed product a 30% reduction in comparison with non-foamed one. In our study, it was developed structural foams (320–800 kg m^{-3} density range), from a PP/HMSPP 50% structural foams, considering the poor viscosity of linear PP homopolymer [5]. HMSPP is obtained from PP modification imparted by gamma-radiation, acetylene environment, in order to increase PP melt strength [6–8]. Nevertheless, branched polymers show a low productivity during the processing, with a lower concentration of cells when compared to linear polymers. In order to solve this processing problem, a 50% PP/HMSPP mixture is used as structural foam based.

Polymers from renewable resources have attracted an increasing amount of attention over the last two decades, predominantly due to two major reasons: firstly environmental concerns, and secondly the depletion of our petroleum resources. There are many sources of biodegradable plastics, from synthetic to natural polymers [9]. Over recent years biodegradable products are attracting interest in function of the expansion of sustainable development politics, with a decrease in fossil combustible supply and the raising concern about environment. Biodegradable polymers impart a significant contribution toward sustainable development due to a more comprehensive range of disposal options with a lesser environmental impact.

Poly(lactic acid) (PLA) is an aliphatic polyester that can be derived from 100% renewable resources [10]. It is a synthetic polymer based on lactic acid and produced from the fermentation of agricultural resources, such as corn. PLA polymer is compostable, as it easily degrades by simple hydrolysis under the appropriate conditions [11, 12]. The PLA polymer demonstrates a clear decrease in molecular weight on heat treatment above 190 °C, at which the thermal degradation of PLA starts to take place [13]. The production of PLA blends with other polymers aims to improving their physical properties, processing and reducing costs. Among these blends some of them are biodegradable and other ones partially biodegradable depending on the polymer combined with PLA. When polymers are selected for production of a blend, two factors should be considered: miscibility and compatibility in order to enhance the development and application of new polymeric materials [14, 15].

Factors responsible for accelerating polymers degradation can be resumed in: a more hydrophilic backbone chain, more hydrophilic groups in backbone chain extremities, more reactive groups in backbone chain, lower crystallinity and higher porosity. Due to polymers water insolubility and molecular size, the microorganisms are not capable to a direct attack in cells to actuate as carbon and energy source. All degradable polymers share surface degradation property, considered as conclusive parameters for biodegradable materials. At initial degradation stage, biodegradable sample is penetrated by water that breaks off chemical bonds and convert polymers backbone chain into lower chains; this process is called hydrolysis. Gamma radiation [16–18] will be imparted to PLA/PP foams, within 50–500 kGy in order to accelerate biodegradability.

Imparted gamma-radiation modifications will be pointed out by TG/DTG, soil burial test and FTIR assessments.

Experimental

Materials

PLA, *Ingeo Biopolymer 3251 D*, from Natureworks.
Polypropylene (PP): H-503 from Braskem (or H-550H from Quattor), 3.5 g (10 min)$^{-1}$ melt index.
High Melt Strength Polypropylene (HMSPP): PP subjected to gamma-radiation, 12.5 kGy dose, acetylene environment.
PP/HMSPP blends, 50% weight.
CO_2 (carbon dioxide): physical blowing agent.

Blends Preparation

Blends were prepared according to Table 1; they were first homogenized by melting extrusion process, using a co-rotating twin-screw extruder (HAAKE Rheomex 332p, 3.1 L/D, 19/33 compression ratio), by using a 175–200 °C temperature profile and 50 rpm:

Table 1 Blends prepared in extruder

	PP/HMSPP, 50%	PLA
	90	10
%, wt/wt	85	15
	70	30
	50	50

Foaming

Blends prepared according to Table 1 were extruded foams within a 175–200 °C temperature profile, by using a specific mono screw for foaming, under CO_2 physical blowing agent, 30 bar pressure. Samples were collected directly from die for further investigations.

Gamma-radiation

Foams prepared were gamma radiated at 50, 100, 150, 200 and 500 kGy doses, ^{60}Co source, in air, at a 5 kGy/h rate.

Characterization

Thermogravimetry (TG)

It was used Thermogravimetric Analyzer TGA/SDTA 851 Mettler-Toledo, with thermo-balance, for evaluating sample mass variation in function of temperature, according to ASTM E1641 [22].

Essays were programmed according to: nitrogen environment, 50 mL min^{-1} flow rate, within 25–600 °C, 10 °C min^{-1} heating rate. Samples were inserted in aluminum pan, 40 μL capacity, nitrogen used with 99.999% purity degree and oxygen level lower than 1 ppm (White Martins). Thermogravimetric analyzer was used aiming to obtaining following parameters:

- T_{onset} = thermal degradation initial temperature, in each stage;
- T_{endset} = thermal degradation final temperature, in each stage.

Soil Burial Test

Soil burial test methods in plastics were established and standardized for assessing plastics resistance to microorganisms, in function of mass change [23–25]. The material is buried under controlled conditions of temperature (24 °C) and relative humidity (80%) and mass loss is determined after soil burying. Foamed samples aliquots were buried in 1000 ml capacity beckers containing gardening specific inoculum with water content approximately 20%, w/w. Recipients were kept in specific conditions of temperature and humidity (24 °C ± 1 °C/80% relative humidity). Evaluations were fortnightly accomplished up to one year of soil burying, with samples carefully cleaned with brush and towel, before weighing in analytical digital balance BP210D Sartorius AG, RFA model. Mass variation rate was assessed in function of time, according to Eq. (1):

$$T(\%) = \frac{m_o - m_t}{m_o} \times 100 \quad (1)$$

where m_o is sample initial mass, at t_o time, and m_t is sample mass at t time, after soil burying.

Infrared Region Absorption Spectroscopy with Fourier Transformed (FTIR)

Infrared Spectroscopy technique is adequate for qualitative analysis of polymers and their blends; it is reliable, quick and low cost [26]. Infrared spectrum derives from light absorption that excites molecular vibrations. Sites in absorption bands in spectrum provide information on presence or absence of specific functional groups in a molecule; as a whole, the spectrum acts as a fingerprint used for determining sample identity. A difference between two samples indicates that they are composed by different constituents. The major concern in bands investigation to detect biodegradability evidences is tied up to two regions: 3100–3700 cm^{-1} corresponding to OH (hydroxyl) and 1600–1800 cm^{-1}, corresponding to C=O (carbonyl). Absorption spectra in infrared region shown by analysed materials were registered within 400–4000 cm^{-1} spectral range, in Nexus 6700, Thermo Nicolet MID-FTIR spectrophotometer, using Attenuated Total Reflectance (ATR). Before each analysis, baseline is automatically established, with a blank run, in air, 32 scanning. Samples are analyzed in direct contact with diamond tip and spectra are obtained after 64 scanning.

Results and Discussion

Thermogravimetry (TG/DTG)

Thermal decomposition profiles for non-irradiated blends of foamed samples are shown in Fig. 1:

Initial and final degradation temperature are practically coincident: t_{onset} = 225 °C and t_{endset} = 400 °C.

10, 15, 30 and 50% of PLA (mix PLA) in PP/HMSPP were blended, foamed and further gamma-radiated at 50, 100, 150, 200 and 500 kGy and presented TG behavior similar to shown in Fig. 1, proving that gamma-radiation did not affect thermal samples resistance.

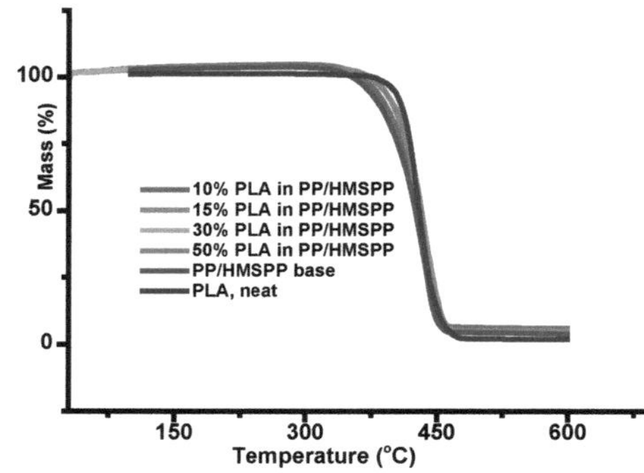

Fig. 1 TG/DTG curves for non-irradiated foamed blends: 10, 15, 30 and 50% of PLA in PP/HMSPP; PP/HMSPP; PLA, neat

Soil Burial Test

Foamed samples of PP/HMSPP, 10, 15, 30 and 50% of PPLA in PP/HMSPP were previously gamma-radiated at zero, 50, 100, 150 and 200 kGy and further soil buried for 12 months. After this period, mass loss attained a negligible value around 1% for all buried samples.

PP/HMSPP and PLA in PP/HMSPP foamed samples were gamma-radiated at 500 kGy and after 12 month-period soil buried presented behavior shown in Table 2. Even subjected to 500 kGy, PP/HMSPP presented a low and negligible mass loss after 12 month-period soil buried: 1.9%. 500 kGy gamma-radiated PLA in PP foamed samples showed low values for mass loss variation: within 3.7–10.7 in function of a higher PLA content.

Infrared Region Absorption Spectroscopy with Fourier Transformed (FTIR)

Prior to soil burial treatment PLA in PP/HMSPP foamed samples were subjected to gamma radiation at zero, 50, 100, 150 and 200 kGy. In Fig. 2 it is shown behavior

Table 2 Mass loss variation (%) of 500 kGy gamma-radiated samples of PP/HMSPP and 10, 15, 30 and 50% PLA in PP/HMSPP foamed samples after a 12 month-period soil buried

500 kGy	Mass loss (%)
PP/HMSPP	1.9
PLA in PP/HMSPP 10%	3.7
PLA in PP/HMSPP 15%	4.5
PLA in PP/HMSPP 30%	6.3
PLA in PP/HMSPP 50%	10.7

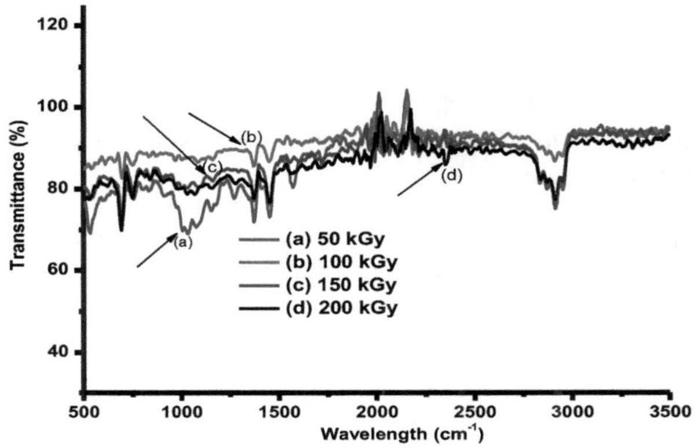

Fig. 2 Infrared-spectroscopy for PLA foamed samples in function of gamma-radiation doses: 50, 100, 150 and 200 kGy, prior to soil buried

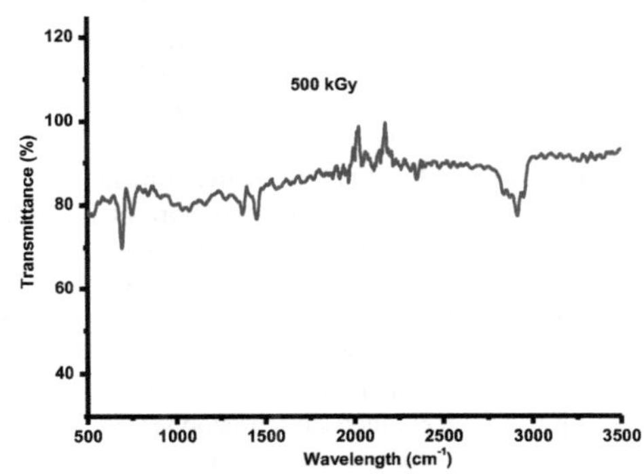

Fig. 3 Infrared-spectroscopy for MIX PLA in PP/HMSPP foamed samples subjected to 500 kGy and soil buried for 12 months

presented by gamma-radiated samples, prepared each one from a mixture among 10, 15, 30 and 50% of PLA in PP/HMSPP:

A major concern in peaks investigation, in favor of biodegradability is associated to:

3100–3700 cm^{-1}: OH (hydroxyl) and
1600–1800 cm^{-1}: C=O (carbonyl).

In Fig. 2 it was not detected none of these peaks.

Mix PLA (mixture among 10, 15, 30 and 50% of PLA in PP/HMSPP) in PP/HMSPP foamed sample was subjected to 500 kGy and further soil buried for a 12 month-period; behavior is shown in Fig. 3.

Even after a 12 month-period of soil burying mix PLA in PP/HMSPP subjected to 500 kGy did not present a visible carbonyl band.

Conclusions

PLA did not present changes when subjected to gamma-radiation doses, within 50–500 kGy, according to TG and FTIR tests. Low results presented by soil burial test confirmed this behavior.

Acknowledgements The authors wish to thank Braskem and Nature Works for raw materials, CBE-Embrarad for irradiation process, IPEN/CNEN-SP and CAPES for their financial support.

References

1. Longo C, Savaris M, Zani M, Brandalise RM, Gris AMC (2011) Degradation study of polypropylene(PP) and bi-oriented polypropylene(BOPP) in the environment. Mater Res 14 (4):442–448
2. Arutchelvi J, Sudhakar M, Arkatkar A, Doble M, Bhaduriandp S, Uppara V (2008) Biodegradation of polyethylene and polypropylene. Indian J Biotechnol 7(1):9–22
3. Chandra R, Rustgi R (1997) Biodegradation of maleated linear low-density polyethylene and starch blends. Polym Degrad Stab 56(2):185–202
4. Tokiwa Y, Calabia BP, Ugwu CU, Aiba S (2009) Biodegradability of plastics. Int J Mol Sci 10(9):3722–3742
5. Cardoso ECL (2009) Desenvolvimento de espumas a partir de misturas poliméricas de polipropileno linear (PP) e polipropileno de alta resistência do fundido (HMSPP), máster thesis, IPEN, Instituto de Pesquisas Energéticas e Nucleares, São Paulo
6. Lugão AB et al (2004) Process for preparing high melt strength polypropylene and crosslinked prepared therewith. U.S. Patent 20040171712
7. Lugão AB et al (2004) Production of high melt strength polypropylene by gamma irradiation. Radiat Phys Chem 76(11–12,13):464–529 issues
8. Spitael Pieter, Macosko Christopher W (2004) Strainhardening in polypropylenes and its role in extrusion foaming. Polym Eng Sci 44(11):2090–2100
9. Ghanbarzadeh B, Almasi H (2013) Biodegradable polymers.In Dr. Chamy R (ed) Biodegradation—Life of Science, InTech, doi:10.5772/56230. Available from: http://www.intechopen.com/books/biodegradation-life-of-science/biodegradable-polymers
10. Drumright RE, Gruber PR, Henton DE (2000) Polylactic acid technology. Adv Mater 12 (23):1841–1846
11. Tsuji H, Ikada Y (1998) Blends of aliphatic polyesters. II. Hydrolysis of solution-cast blends from poly (L-lactide) and poly (E-caprolactone) in phosphate-buffered solution. J Appl Polym Sci 67(3):405–415
12. Palade LI, Lehermeier HJ, Dorgan JR (2001) Melt rheology of high L-content poly (lactic acid). Macromolecules 34(5):1384–1390
13. Jamshidi K, Hyon SH, Ikada Y (1998) Thermal characterization of polylactides. Polymer 29 (12):2229–2234
14. http://www.ecycle.com.br/component/content/article/37/738-pla-o-plastico-compostavel.html . Acessado em 07/06/2012
15. Cabedo L, Plackett D, Gimenez E, Andlagaron JM (2009) Studying the degradation of polyhydroxybutyrate-co-valerate during processing with clay-based nanofillers. J Appl Polym Sci 112:3669–3676
16. Calvo WAP et al (2012) Electron beam accelerators—trends in radiation processing technology for industrial and environmental applications in Latin America and Caribbean. Radiat Phys Chem 81:1276–1281

17. Güven O (2005) New opportunities for the utilization of electron accelerators in polymer processing industries. IAEA-CN-115-68, pp 67–68
18. Bhattacharya A (2000) Radiation and industrial polymers. Prog Polym Sci 25:371–401
19. ASTM E 1641–07–Standard test method for decomposition kinetics by thermogravimetry
20. Naime N (2010) Embalagens ativas de fonte renovável. Dissertação, IPEN – Instituto de Pesquisas Energéticas e Nucleares, São Paulo
21. Lardjane N et al Soil burial degradation of new bio-based additives, Part I rigid poly (vinylchloride) films
22. Vogelsager N Jr et al (2004) Acompanhamento da Biodegradação em Solo por Propriedades Térmicas e Morfológicas. Revista Matéria 9(4):370–377
23. Yang Charles (2011) Chemical analysis of polymer materials using infrared spectroscopy. University of Zagreb, Croatia
24. Sherman Hsu C-P Infrared spectroscopy. Chapter15, Separation sciences research and product development, Mallinckrodt, Inc, UK
25. Bucci DZ, Tavares LBB, Sell I (2005) PHB packaging for the storage of food products. Polym Testing 24(5):564–571
26. Wu CS (2006) Assessing biodegradability and mechanical, thermal and morphological properties of an acrylic acid modified poly(3-hydroxybutyric acid) wood flours biocomposite. J Appl Polym Sci 102(4):3565–3574

Study of Flexible Films Prepared from PLA/PBAT Blend and PLA *E*-Beam Irradiated as Compatibilizing Agent

Elizabeth Carvalho L. Cardoso, René R. Oliveira, Glauson Aparecido F. Machado and Esperidiana A.B. Moura

Abstract Plastics global annual production exceeds 250 million tons and more than 40% of the total plastic production is used as packaging materials. In recent years polymeric materials derived from biomass have received great attention due to scarce petroleum resources and environmental concerns. PLA is one of the most extensively studied bio-based and bio-compostable aliphatic polyesters; nevertheless, high brittleness and low toughness limits its application. Blending PLA with PBAT (Poly(butylene adipate-*co*-terephthalate)), an aliphatic-aromatic copolyester, was the solution found due to its high toughness and biodegradability. Differences between PLA and PBAT solubility parameter values lead to the formation of immiscible blends as well as reduction in their mechanical performance besides a poor morphology; so, 5% of pre-irradiated PLA and a previous compatibilized PLA/PBAT (Ecovio^{TM}) were added to PLA/PBAT blends in order to improve miscibility phases. Investigations were accomplished on phase morphology (SEM), mechanical properties, thermal behavior and X-ray diffractions (XRD).

Keywords PLA/PBAT · Electron-beam radiation · PLA/PBAT (Ecovio^{TM})

Introduction

Owing to the increasing of waste, oil price and general environmental awareness in our society, the industries are leaded to pack their products by a more eco-friendly and sustainable way [1]. In the near future, renewable resources from vegetable, animal elements or microbial production could replace conventional fossil resources in the production of plastics [2]. Both governments and industries are awareness of the need to develop the use of sustainable resources. Actually, resources and

E.C.L. Cardoso (✉) · R.R. Oliveira · G.A.F. Machado · E.A.B. Moura
Instituto de Pesquisas Energéticas e Nucleares-IPEN/CNEN-SP, Av. LineuPrestes 2242, Cidade Universitária, São Paulo, SP CEP05508-900, Brazil
e-mail: eclcardo@ipen.br

S. Ikhmayies et al. (eds.), *Characterization of Minerals, Metals, and Materials 2017*, The Minerals, Metals & Materials Series,
DOI 10.1007/978-3-319-51382-9_14

environment should be preserved for future generations. Waste products are a recurring problem: certain wastes can no longer be deposited in landfills and landfill fees are on the rise. Very high proportion of flexible packaging is sourced from oil-based derivatives [3, 4]. Flexible packaging is defined as a package or container made of flexible materials that can be readily changed in shape. The construction may be of paper, plastic film, foil or any combination of these. Similarly to other areas, plastics industry started looking for alternative sources of raw materials in the last few decades and considerable interest is shown in natural, renewable solutions. So, much endeavor has been committed to developing biodegradable and renewable resource–derived polymers; as the world moves toward the production of sustainable environmentally friendly plastics (bio-plastics), the production of biodegradable polymers has become more and more an area of relevant scientific interest [5].

Poly(lactic acid) (PLA), a linear aliphatic polymer, is a biodegradable thermoplastic polymer with widely potential applications, due to a number of interesting properties including biodegradability, biocompatibility, high strength and high modulus. For these reasons, PLA is a strong candidate for producing packaging materials; however, its high brittleness and low toughness limit its application. PLA can be manufactured from biomass materials, such as cornstarch or beet sugar; nevertheless, its elongation ability, impact strength, shock absorbance, toughness, physical aging and heat deflection properties are unsatisfactory for practical applications [6–9].

Blending PLA with other biodegradable polymers presents a more practical and economic measure to obtain a final product with properties not currently attainable. PLA/Poly(butylene-adipate-co-terephthalate) (PBAT) blends have been extensively investigated by many researchers [10, 11] especially due to PBAT flexibility. In addition, PBAT is considered a good candidate for improving PLA toughness, besides being fully biodegradable and a flexible aliphatic-aromatic copolyester designed for film extrusion and extrusion coating. Its most attractive feature is its high elongation at break [12].

Inherent immiscibility of PLA/PBAT neat blends led to the addition of compatibilizer agents. A minor addition of other biopolymers (around 5 weight—%) enhances the compatibility of PLA and PBAT [13]. In this study it was used 5% (w/w) of Ecovio^{TM}. Another compatibilizer used was PLA e-beam pre-irradiated at 150 kGy, 5% (w/w) [14, 15].

Characterizations accomplished in neat PLA/PBAT blends and composites with Ecovio^{TM} and PLA e-beam irradiated comprised: SEM, mechanical properties, DSC, TG and XRD.

Experimental

Materials

PLA: GS PLA AD 92 W, from Mitsubishi, dried at 60 °C, for 12 h, before processing.

Table 1 Designation and composition of blends and composites

Materials	PLA (wt%)	PBAT (wt%)	Ecovio^{TM} (phr)	PLA eb (phr)
PLA/PBAT	50.0	50.0	–	–
PLA/PBAT/Ecovio	50.0	50.0	5.0	–
PLA/PBAT/Ecovio/PLA eb	50.0	50.0	5.0	5.0
PLA/PBAT	30.0	70.0	–	–
PLA/PBAT/Ecovio	30.0	70.0	5.0	–
PLA/PBAT/Ecovio/PLA eb	30.0	70.0	5.0	5.0

phr = per hundred resin

PBAT: Ecoflex—FS, Basf, dried at 60 °C, for 12 h, before processing.

Ecovio FS Paper C1500, Basf, dried at 60 °C, for 12 h, before processing.

PLA e-beam irradiated at 150 kGy, in air, at room temperature, in electrons accelerator model Dynamitron II, Dynamics Inc.

Preparation of Blends and Composites

Composite materials were prepared according to Table 1. They were first processed by melting extrusion process, using a twin-screw extruder HaakeRheomex L/D-25, temperature profile from feed zone to die 120–145 °C, 50 rpm screw speed and extrudates air-cooled before pelletizing.

Films Preparation

Films were produced via single screw extruder Haake Rheomex, 48 mm annular die, 140 °C barrel temperature, BUR 2.0: 1, 14 rpm, films average thickness = 167 microns.

Characterization

Mechanical tests: Samples were analyzed in an Instron Equipment model 5567, 1 kN load cell, 500 mm min^{-1} cross-head speed, room temperature, specimen dimension 130 × 25 mm^2 (ASTM D882-02) [16].

Differential Scanning Calorimetric Analysis: Thermal behavior was examined in a DSC Mettler Toledo apparatus, ASTM D3418-08 [17]. Samples were heated from 30 to 300 °C, kept at 300 °C for 10 min and after cooled down to 30 °C,

10 °C min^{-1} and further heated back to 30 °C at 10 °C min^{-1}. The percent crystallinity was calculated upon the second heating by using Eq. 1 based on Hermans and Weidinger equation [18]:

$$X_c(\%) = 100 * \frac{\Delta H_m - \Delta H_{cc}}{f^* \Delta H_{m\infty}} \quad (1)$$

where:

ΔH_m is the measured heat of fusion, ΔH_{cc} the cold crystallization enthalpy (j g^{-1}),"*f*" is the weight fraction of PLA or PBAT in the blend and "$\Delta H_{m\infty}$ "is the enthalpy of fusion for a crystal having infinite crystal thickness (93 j g^{-1} for PLA and 114 j g^{-1} for PBAT).

Thermogravimetric analyses (TG): TG were performed using a DSC Mettler Toledo apparatus, according to ASTM—E 1641-07 [19], by using 5–9 mg of sample, within a 25–600 °C program, at a 10 °C min^{-1}, in a nitrogen flow of 50 ml min^{-1}.

XRD (X-RAY Diffraction Analysis): It was used a X-ray diffractometer, RigakuMultiflex, 40 kV, 20 mA, scanning 5–60°, 2θ, step 0.02°, 2θ, 1 °C min^{-1}, continuous analysis. It provides, among other, information on samples crystallinity, via diffractograms.

Scanning Electron Microscopy: Investigations were accomplished in a Hitachi-table top TM 300 microscope, for an investigation of morphology, in various magnification micrographs. Acceleration voltage of 10 kV was used to collected SEM images of sample. The samples were freeze-fractured in liquid nitrogen and coated with gold before analysis.

Results and Discussion

Mechanical tests: Mechanical properties of PLA/PBAT blends and composites are summarized in Table 2.

From Table 2, when comparing PLA/PBAT neat blends, higher results for Tensile Strength and Young Modulus were presented by PLA/PBAT (50.0/50.0); a higher PBAT concentration in PLA/PBAT (30.0/70.0) just increased slightly elongation at break, as expected.

Addition of 5% of Ecovio and 5% of PLA eb in both PLA/PBAT neat blends reduced tensile strength and elongation at break. Low values obtained for Young Modulus in all PLA/PBAT with a higher percentage of PBAT just proved lower mechanical resistance imparted to PBAT.

Differential Scanning Calorimetric Analyses (DSC): DSC is an useful technique to reveal the variation in the thermal flow and to obtain the thermal critical points like melting point, enthalpy and specific heat. DSC analyses results and the degree of crystallinity reported as percent crystallinity for PLA/PBAT neat blends and its composites are listed in Table 3. As can be seen in Table 3, a higher

Table 2 Tensile properties of PLA/PBAT blends and its composites

Blends and composites	Tensile strength (MPa)	Elongation at break (%)	Young's modulus (MPa)
PLA/PBAT (50.0/50.0)	18.0 ± 0.7	880.0 ± 26.1	68.6 ± 1.5
PLA/PBAT/Ecovio (50.0/50.0/5.0)	16.9 ± 1.1	792.1 ± 32.1	73.4 ± 0.9
PLA/PBAT/Ecovio/PLA eb (50.0/50.0/5.0/5.0)	13.0 ± 1.4	748.7 ± 27.8	72.0 ± 1.1
PLA/PBAT (30.0/70.0)	16.0 ± 0.3	890.0 ± 54.7	42.5 ± 0.8
PLA/PBAT/Ecovio (30.0/70.0/5.0)	11.0 ± 1.8	694.3 ± 34.7	15.5 ± 0.6
PLA/PBAT/Ecovio/PLA eb (30.0/70.0/5.0/5.0)	12.7 ± 2.5	795.1 ± 46.0	23.1 ± 0.9

Table 3 DSC analysis results and percent crystallinity of PLA/PBAT blends and composites

Blends and composites	$T_c^{(a)}$ (°C)	$\Delta H_{cc}^{(b)}$ (J g^{-1})	$T_m^{(c)}$ (°C)	$\Delta H_m^{(d)}$ (J g^{-1})	$X_C^{(e)}$ (%)
PLA/PBAT (50.0/50.0)	71.7	7.2	86.9	21.1	29.9
PLA/PBAT/Ecovio (50.0/50.0/5.0)	67.7	8.8	90.2	17.7	<1
PLA/PBAT/Ecovio/PLA eb (50.0/50.0/5.0/5.0)	68.2	5.7	91.3	18.1	<1
PLA/PBAT (30.0/70.0)	62.1	9.8	88.3	10.7	3.4
PLA/PBAT/Ecovio (30.0/70.0/5.0)	71.2	11.5	96.5	11.6	<1
PLA/PBAT/Ecovio/PLA eb (30.0/70.0/5.0/5.0)	92.1	16.0	134.2	12.0	<1

(a) cold crystallization temperature; (b) crystallization enthalpy; (c) melting temperature; (d) melting enthalpy; (e) crystallinity

addition of PBAT in PLA/PBAT blend contributed for a strong reduction in crystallinity. Addition of Ecovio^{TM} and PLA eb imparted strong reduction in crystallinity in both PLA/PBAT neat blends.

DSC thermograms for blends and composites were typical of partially crystalline materials, showing an endothermic peak, related to the melting (T_m) of the crystals, according it is shown in Fig. 1.

An addition of 20% of PBAT in PLA/PBAT (50.0/50.0) blend reduced drastically enthalpy of fusion and contributed for a very low crystallinity; the same behavior was shown after addition of Ecovio^{TM} and PLA *eb*.

Thermogravimetric analyses (TG): As can be seen in Fig. 2, PLA/PBAT neat blends and their composites presented the same initial degradation temperature (t_{onset} = 325 °C) and the same final degradation temperature (t_{endset} = 380 °C). None mass loss was registered in function of temperature increase.

XRD (X-RAY Diffraction Analysis): In Fig. 3, The large peak shown within 2θ equal 17° and to 25° is typical of PLA/PBAT blends, with a higher intensity at 2θ equal 22°. Several small diffraction peaks appeared in all samples and might be associated with PLA micro-crystallinity. Ecovio^{TM} and PLA *eb* added in neat PLA/PBAT blends showed peaks of higher intensity.

Fig. 1 DSC thermograms of blends and composites

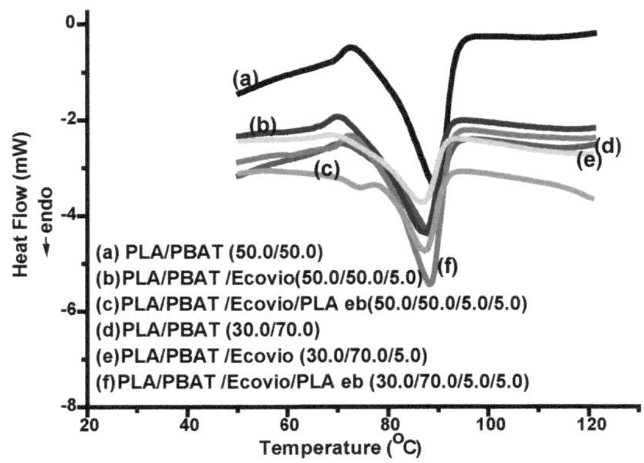

Fig. 2 TG thermograms for PLA/PBAT neat blends and their composites

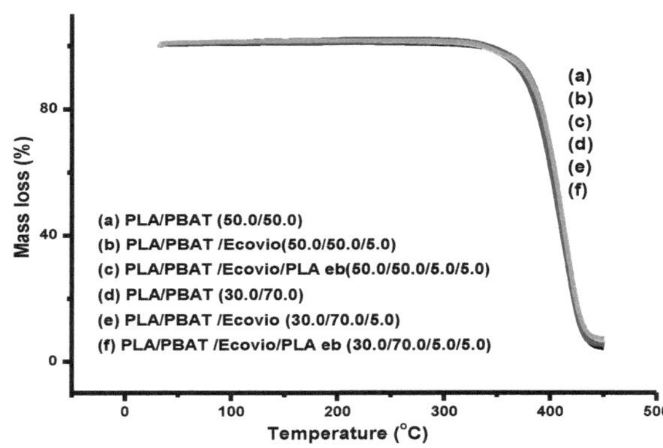

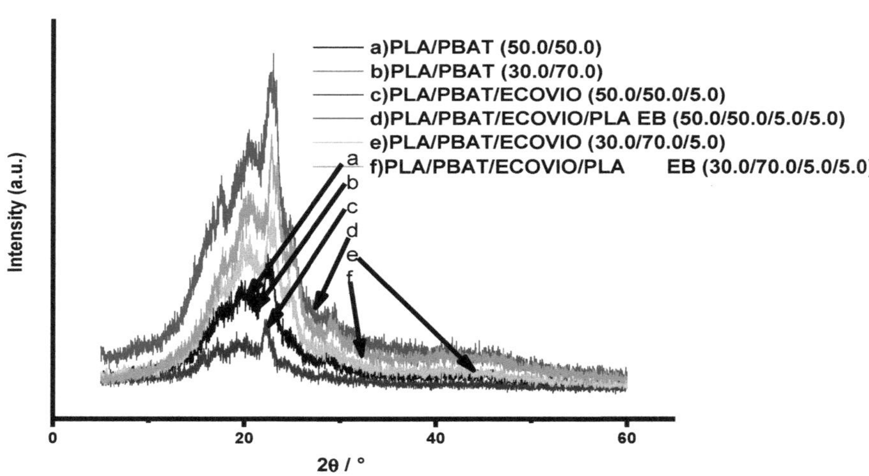

Fig. 3 XRD patterns pf PLA/PBAT blends and composites

Fig. 4 SEM micrographs of **a** PLA/PBAT (50.0/50.0) and **b** (30.0/70.0)

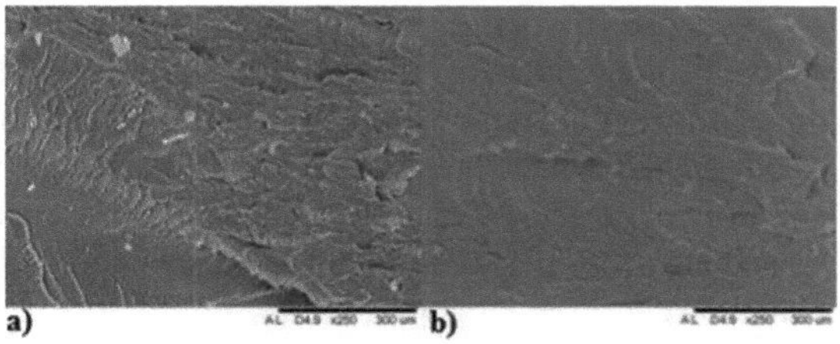

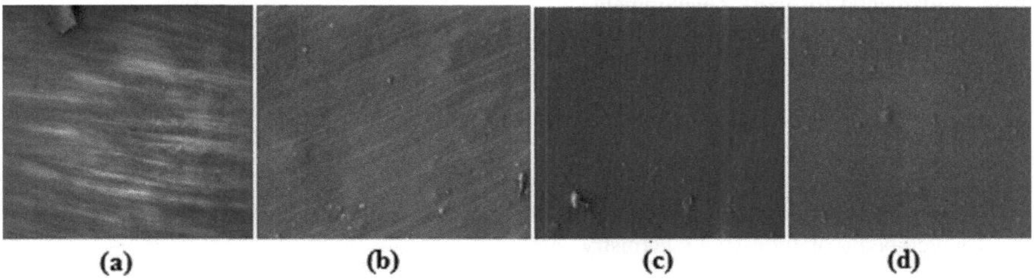

Fig. 5 SEM micrographs of: **a** PLA/PBAT/Ecovio^{TM} (50.0/50.0/5.0); **b** PLA/PBAT/Ecovio^{TM}/PLA *eb* (50.0/50.0/5.0/5.0); **c** PLA/PBAT/Ecovio^{TM} (30.0/70.0/5.0); **d** PLA/PBAT/Ecovio^{TM}/PLA *eb* (30.0/70.0/5.0/5.0)

Scanning Electron Microscopy (SEM): Morphologies of all PLA/PBAT neat blends and their composites were examined by a scanning electron microscope Table-top Hitachi, acceleration voltage of 10 kV, used to collected SEM images of sample. The samples were freeze-fractured in liquid nitrogen and coated with gold before analysis. In Fig. 4 are shown micrographs related to neat PLA/PBAT blends.

PLA shows an irregular dispersion and PBAT a continuous phase in blends; that is, PLA has a typical and irregular morphology *island-phase* type and the PBAT a *sea-phase* type morphology [20], as can be observed in Fig. 3a and b.

After addition of Ecovio^{TM} and PLA eb 150 kGy, morphologies presented homogeneous and smooth surface as can be seen in Fig. 5.

A higher amount of PBAT in PLA/PBAT blends contributed for a better dispersion, improved after addition of Ecovio^{TM}; it should be emphasized that further addition of PLA *eb* enhanced this good performance. All micrographs were enlarged about 200 times.

Conclusions

A higher amount of PBAT in PLA/PBAT neat blends and composites was responsible for reducing crystallinity, tensile and elongation at break. Insertion of 5.0% of Ecovio^{TM} and PLA *eb* did not significantly affected mechanicals tests. All tested samples presented the same behavior in thermogravimetric analyses and consequently the same initial and final degradation temperature. DRX analyses showed higher intensity peaks after addition of PLA *eb* confirming surface and miscibility optimization from SEM analyses. Future works will comprise the study of PLA/PBAT neat blends varying PBAT level within a 65–75% range as well addition of Ecovio^{TM} and PLA *eb,* in a percentage higher than 5%. This will contribute for improving phase dispersion without prejudicing mechanicals behavior.

Acknowledgements The authors acknowledge CBE, Capes, Ipen, Mitsubishi and Basf, all of them fundamental for work development.

References

1. State of order of polymers, chapter 1 (1990) Plastics, Properties and Testing (Ullmann's Encyclopedia of Industrial Chemistry, A21)
2. Mishra N et al (2014) Route to generate fuel oil via catalytic pyrolysis of waste polypropylene bags: towards waste management of >20 μm plastic bags. J Fuels
3. Endres H-J, Siebert-Raths A (2009) Technische Biopolymere – Rahmen bedingungen, Markt situation, Herstellung, Aufbau und Eigenschaften. Hanser, München
4. Fakirov S, Bhatacharyya D (2007) Handbook of engineering biopolymers. Homopolymers, blends and composites. Hanser, München
5. Zweifel H, Maier RD, Schiller M (2009) Plastics additives handbook. Hanser, München
6. Jiang L, Wolcott MP, Zhang J (2006) Study of biodegradable polylactide/ poly(butylene adipate co-terephthalate) blends. Biomacromolecules 7(1):199–207
7. Nofar M et al (2015) Interfacial and rheological properties of PLA/PBAT and PLA/PBSA blends and their morphological stability under shear flow. J Rheol 59:317–333
8. Quero E et al (2012) Isothermal cold-crystallization of PLA/PBAT blends with and without the addition of acetyl tributyl citrate. Macromol Chem Phys 213:36–48
9. Al-Itry R, Lamnawar K, Maazouz A (2014) Reactive extrusion of PLA, PBAT with a multifunctional epoxide: physico-chemical and rheological properties. Eur Polym J 58:90–102
10. Quero E et al (2012) Isothermal cold-crystallization of PLA/PBAT blends with and without the addition of acetyl tributyl citrate. Macromol Chem Phys 213:36–48
11. Signori F, Coltelli MB, Bronco S (2009) Thermal degradation of poly(lactic acid) (PLA) and poly(butylene adipate-co-terephthalate) (PBAT) and their blends upon melt processing. Polym Degrad Stab 94:74–82
12. Jiang L, Wolcott MP, Zhang J (2006) Study of biodegradable polylactide/poly(butylene adipate-co-terephthalate) blends. Biomacromolecules 7(1):199–207
13. Pivsa-Art W et al (2013) Preparation of polymer blends between poly(lactic acid) and poly (butylene adipate-co-terephthalate) and Biodegradable polymers as compatibilizers. Energy Procedia 34:549–554
14. Goulas E, Riganakos KA, Badeka A, Kontominas MG (2002) Effect of ionizing radiation on the physicochemical and mechanical properties of commercial monolayer flexible plastics packaging materials. Food Addit Contam 19(12):1190–1199
15. Oliveira VM, Nogueira BR, Ortiz AV, Moura EAB (2009) Mechanical and thermal properties of commercial multilayer flexible plastics packaging materials irradiated with electron beam. Nukleonika (D-ruk) 53:S141–S144
16. ASTM D-882-02, Standard test method for tensile properties of thin plastic sheeting
17. ASTM D-3418-08, Standard test method for transition temperatures and enthalpies of fusion and crystallization of polymers by differential scanning calorimetry

18. Hermans PH, Weidinger (1949) Estimation of crystallinity of some polymers from X-ray intensity measurements. J Polym Sci 4:709–723
19. ASTM E-1641-07, Standard test method for decomposition kinetics by thermogravimetry
20. Chiu HT et al (2013) Heat treatment effects on the mechanical properties and morphologies of poly(lactic acid)/poly(butylene adipate-co-terephthalate) blends. Int J Polym Sci 11 pages (Hindawi Publishing Corporation)

Synthesis of ZnO Micro Prisms on Glass Substrates by the Spray Pyrolysis Method

Shadia Ikhmayies

Abstract ZnO micro prisms in thin film form were produced by the low cost spray pyrolysis technique at a substrate temperature of 350 ± 5 °C on glass substrates. Scanning electron microscopy (SEM) and energy dispersion X-ray spectroscopy (EDS) were used to explore morphology and composition of the films respectively. The films were found to be oxygen rich and contain chlorine. ImageJ software was used to analyze the size of the prisms, where the diameter and length of the micro prisms were estimated.

Keywords II-VI compound semiconductors · Transparent conducting oxide · ZnO micro prisms · Solar cells · Spray pyrolysis

Introduction

Wurtzite ZnO with space group P63mc or C6v4 is an important II–VI compound semiconductor material with wide direct bandgap of 3.37 eV at room temperature, and large exciton binding energy of 60 meV at room temperature [1–4]. ZnO is a transparent conductive oxide (TCO) that has promising advantages in optoelectronic devices, solar cells and gas sensors. Nowadays, ZnO is one of the most used TCO materials for thin film silicon solar cell development, and it is of potential use in dye sensitized solid-state solar cells, which are currently the most stable and efficient excitonic solar cells for large-scale solar energy conversion [5–7]. In these cells, 1D ZnO nanorods with high carrier mobility can serve as direct conduction pathways for the excitons [8]. In addition, it is the compound of choice to achieve low-cost, reproducible electrode layers on large-scale production lines. Various physical and chemical techniques have been employed to synthesize ZnO nano/micro structures, such as hydrothermal method [3], chemical vapor deposition

S. Ikhmayies (✉)
Faculty of Science, Department of Physics,
Al Isra University, Amman 11622, Jordan
e-mail: shadia_ikhmayies@yahoo.com

S. Ikhmayies et al. (eds.), *Characterization of Minerals, Metals, and Materials 2017*, The Minerals, Metals & Materials Series,
DOI 10.1007/978-3-319-51382-9_15

(CVD) [4], metalorganic chemical vapor deposition (MOCVD) [9–11], sputtering [12], molecular beam epitaxy [13], sol–gel [14], and spray pyrolysis (SP) [15] method. However, for low cost, large-scale production, and intentional doping purposes, the SP method is preferred.

A lot of experimental work was conducted to prepare and characterize ZnO prisms. For example, Ekthammathat et al. [3] produced ZnO prisms on Zn foils by a simple hydrothermal method. Kao et al. [16] deposited ZnO hexagonal prisms on indium titanium oxide-coated glass substrates by using the solution phase deposition method. Hamada et al. [17] deposited these prisms using the electrochemical deposition method on a graphite substrate, and Liu et al. [18] grown them from polyvinylpyrrolidoned-assisted electrochemical assembly onto *p*-type Si (111) substrate. In this work ZnO hexagonal micro prisms were produced on glass substrates using the spray pyrolysis method. These micro prisms are of important use in solar cells, gas sensors, and optoelectronics industry.

Experimental Procedure

ZnO thin films were deposited on glass substrates of dimensions of $60 \times 26 \times 1$ mm^3 by the spray pyrolysis technique at a substrate temperature of 350 ± 5 °C. The starting material was zinc chloride ($ZnCl_2$) of purity 99%. The precursor solution was prepared by dissolving 2.5 g of $ZnCl_2$ in 60 ml of distilled water to get a 0.03 M solution. This solution was sprayed vertically and intermittently on the hot glass substrates, and oxygen was used as the carrier gas. After finishing the deposition process, the heater was turned off, and the samples were left on the heater to cool naturally to room temperature. The morphology of the films was examined by a FEI scanning electron microscope (SEM) (Inspect F 50) operating at 5.0 kV. The composition of the samples was analyzed by the same SEM system, which is supported by energy dispersion X-ray spectroscopy (EDS). The thickness of the films was measured using SEM microscope too. ImageJ software was used to analyze the size of the micro prisms, where both the diameter and length were estimated.

Results and Discussion

Figure 1 displays the SEM images of two ZnO thin films of hexagonal prism micro structure. The film in (a) has a thickness of 1.622 μm, while that in (b) has a thickness of 1.277 μm. The films show high density of ZnO micro prisms where the density in Fig. 1a appears larger than that in Fig. 1b. The difference in density may be due to the larger thickness of the film in Fig. 1a. In this film, three or four positions appear with collections of high density of micro prisms of approximate sizes. In both films, ZnO prisms have hexagonal cross-section, and they appear approximately perpendicular to the surface of the substrate.

Fig. 1 SEM micrographs of two ZnO films of thickness t = 1.622 μm (**a**), and t = 1.277 μm (**b**)

Figure 2 displays the EDS spectrum of the film in Fig. 1b, in addition to an image showing the point at which measurement was performed. From the EDS report it is found that the films contain Zn, O, and Cl, where the presence of chlorine is due to the use of $ZnCl_2$ as a starting material. The ratio of Zn: O in the film is 37.93: 59.58, or 0.637, while the stoichiometric ratio is 1. This means that the films are oxygen rich. This may be due to the use of oxygen as the carrier gas besides the long period (more than one month) of exposure to air. The EDS report also showed that the ratio of Cl: O in the film is 2.49: 59.58 or 4.18%. When Cl is incorporated in the ZnO crystal lattice, it replaces oxygen, and hence it increases the

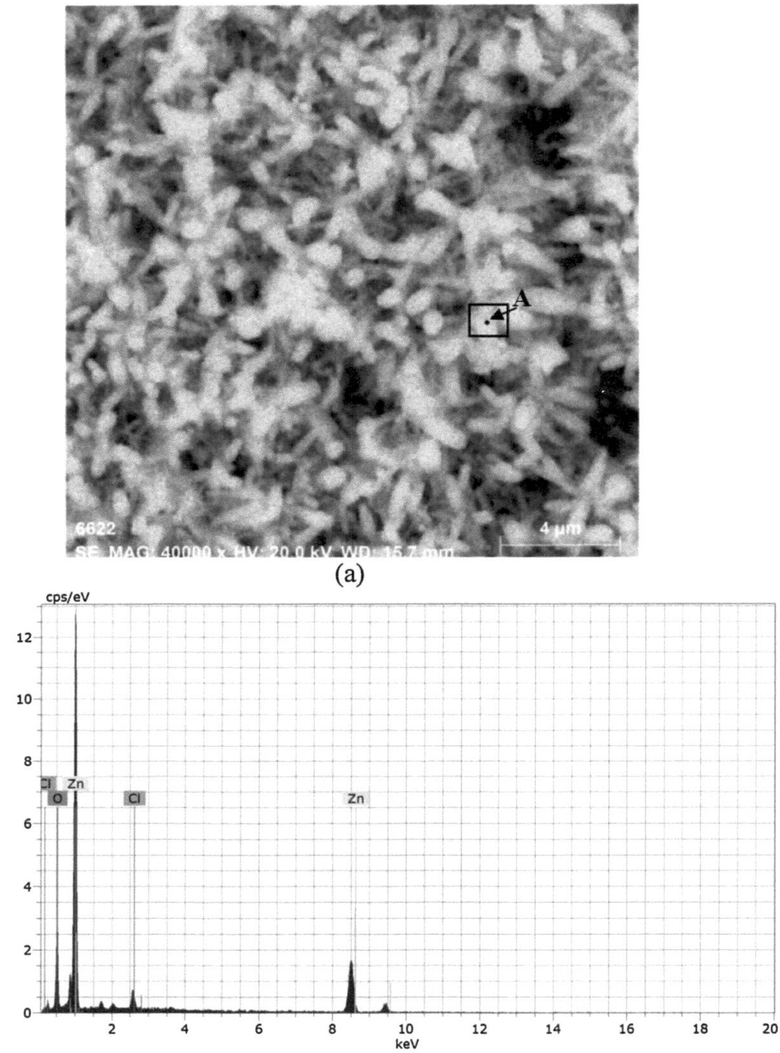

Fig. 2 SEM image shows the point A at which measurement was performed (**a**), and EDS spectrum for the film of SEM shown in Fig. 1b (**b**)

n-type conductivity of the films, which is preferred in the case of the use of ZnO as a fore contact or a window layer in thin film solar cells.

Figure 3 shows the histograms of the diameter of ZnO micro prisms appearing in the SEM micrographs in Fig. 1a and b, and Fig. 4 shows the histograms of the length distribution for the same micrographs in Fig. 1 too. For the film in Fig. 1a,

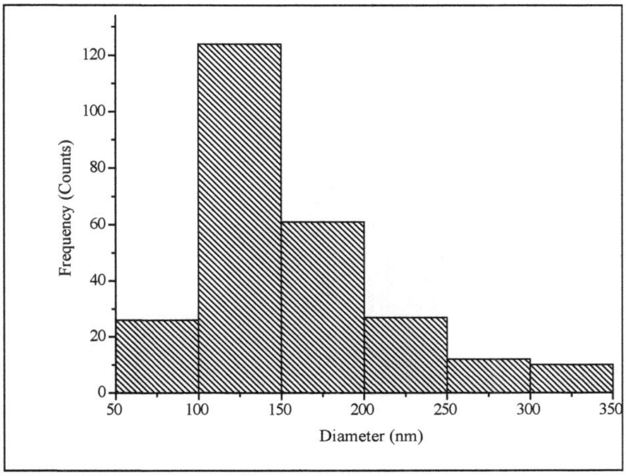

(a)

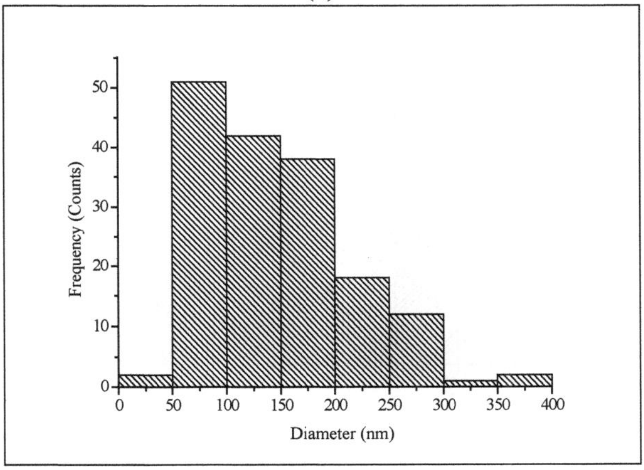

(b)

Fig. 3 The histograms of the diameter distributions of ZnO micro prisms in SEM micrographs in: **a** Fig. 1a, and **b** Fig. 1b

the average diameter estimated by imageJ software is $D \pm \Delta D = 156.2 \pm 57.7$ nm, and the average length of the prisms is $L \pm \Delta L = 433.0 \pm 134.6$ nm. Hence the aspect ratio of the prisms in this film is about 2.77. For the film in Fig. 1b, the average diameter is $D \pm \Delta D = 145.3 \pm 68.8$ nm, and the average length is $L \pm \Delta L = 222.1 \pm 74.1$ nm. In this film the ZnO prisms have an aspect ratio of about 1.53. So the prisms in (a) have larger diameter, larger length, and larger aspect ratio. These differences are most probably related to the difference in film thickness in the two cases. The large values of the uncertainties are expected in the case of

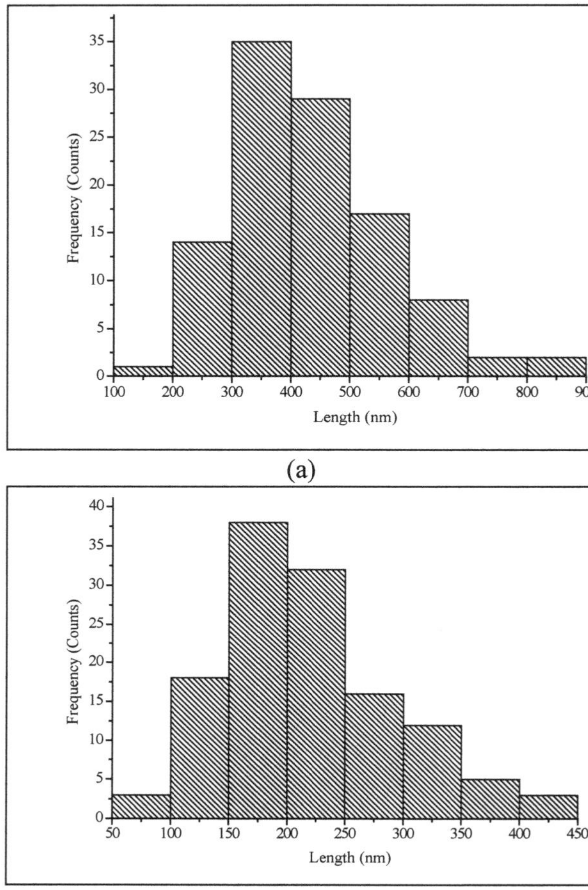

(b)

Fig. 4 The histograms of the length distributions of ZnO micro prisms in SEM micrographs in: **a** Fig. 1a, and **b** Fig. 1b

using the SP method to produce micro and nano structures. There are several factors that determine the size and shape of the produced structures, and to control all of them is a challenge.

Such micro prisms provide larger surface area, which is necessary for several applications such as the dye sensitized solar cells (DSSC). To make a rough estimation of the area of the whole array of micro prisms in the film of Fig. 1a, two assumptions were made. First; no overlap of the micro prisms, and second; the length of the micro prism equals the thickness of the film, where both assumptions are not found in reality. The second assumption may be accepted if there are several layers of the micro prisms are stacked over each other, where the thickness of the film d = NL, and N is the number of layers, while L is the average length of the

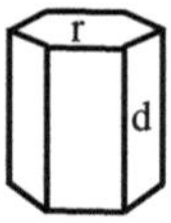

Fig. 5 Prism of hexagonal cross section, radius r and length d = NL

prisms calculated before. Dividing d by L to get N ≈ 4 for the film of Fig. 1a. Hence each four ZnO micro prisms each of length L and diameter D stacked vertically on each will be replaced by one prism of the same diameter and length d as shown in Fig. 5. The area of the portion of such ZnO prism where the dye in DSSC will be adsorbed is $A_{Dye-prism} = \frac{3\sqrt{3}}{2}r^2 + 6rd$, where r is the radius of the hexahedron, and d the thickness of the film in Fig. 1a, which is 1.622 μm. The radius $r = \frac{D}{2} = 78.1$ nm. So $A_{Dye-prism} = 7.6 \times 10^5$ nm^2. If the density of prisms per nm^2 in Fig. 1a is approximated by 10 prisms/nm^2, and the area of the substrate = 60 × 26 mm^2 = 1.56 × 10^{15} nm^2, then the total area of the array ≈ 1.19 × 10^{16} nm^2, which is more than 7 times the area of the substrate.

Conclusions

ZnO Micro prisms were produced as thin films by the low cost spray pyrolysis method on glass substrates. The SEM images showed that the prisms have hexagonal cross sections and they are approximately perpendicular to the substrates. The average diameter, and average length of the micro prisms, besides their distributions were calculated using imageJ software. The EDS reports revealed the composition of the films, and they are found to be oxygen rich and contain chlorine. The estimated surface area of prism arrays on one of the films was found to be seven times that of the substrate. These results are important for the use of such ZnO micro prisms in solar cells and optoelectronic devices.

References

1. Yan C, Xue D (2008) Solution growth of nano- to microscopic ZnO on Zn. J Cryst Growth 310:1836–1840
2. Yang JH, Zheng JH, Zhai HJ, Yang LL, Lang JH, Gao M (2009) Growth mechanism and optical properties of ZnO nanosheets by the hydrothermal method on Si substrates. J Alloy Compd 481:628–631
3. Ekthammathat N, Phuruangrat A, Thongtem S, Thongtem T (2015) Hydrothermal-assisted synthesis and photoluminescence of ZnO microrods. Digest J Nanomaterials Biostructures 10 (1):149–153
4. Kang DS, Seok KH, Sang MY, Kim JG, Hwang WJ, Hong SK (2008) Growth and characterization of zinc oxide nanostructures on (111) silicon substrates with aluminum compound layer. J Korean Phys Soc 53(1): 292–298

5. Galoppini E, Rochford J, Chen H, Saraf G, Lu Y, Hagfeldt A, Boschloo G (2006) Fast electron transport in metal organic vapor deposition grown dye-sensitized ZnO nanorod solar cells. J Phys Chem B 110(33):16159–16161
6. Law M, Greene LE, Johnson JC, Saykally R, Yang P (2005) Nanowire dye-sensitized solar cells. Nat Mater 4:455–459
7. Gregg BA (2003) Excitonic solar cells. J Phys Chem B 107:4688–4698
8. Lupan Oleg, Chow Lee, Chai Guangyu, Roldan Beatriz, Naitabdi Ahmed, Schulte Alfons, Heinrich Helge (2007) Nanofabrication and characterization of ZnO nanorod arrays and branched microrods by aqueous solution route and rapid thermal processing. Mater Sci Eng B 145:57–66
9. Nicolay S, Benkhaira M, Ding L, Escarre J, Bugnon G, Meillaud F, Ballif C (2012) Control of CVD-deposited ZnO films properties through water/DEZ ratio: decoupling of electrode morphology and electrical characteristics. Sol Energy Mater Sol Cells 105:46–52
10. Faÿ S, Steinhauser J, Nicolay S, Ballif C (2009) Polycrystalline ZnO: B grown by LPCVD as TCO for thin film silicon solar cells. Thin Solid Films 518:2961–2966
11. Hongsingthong A, Yunaz IA, Miyajima S, Konagai M (2010) ZnO films prepared by two-step MOCVD process for use as front TCO in silicon-based thin film solar cells. In Proceedings of the 35th IEEE photovoltaic specialists conference, pp 1508–1511
12. Kluth O, Schöpe G, Hüpkes J, Agashe C, Müller J, Rech B (2003) Modified Thornton model for magnetron sputtered zinc oxide: film structure and etching behavior. Thin Solid Films 442:80–85
13. Chen Y, Ko H-J, Hong S-K, Segawa Y, Yao T (2002) Morphology evolution of ZnO (0001) surface during plasma-assisted molecular-beam epitaxy. Appl Phys Lett 80:1358–1360
14. Ohyama M, Kozuka H, Yoko T (1998) Sol-gel preparation of transparent and conductive aluminum-doped zinc oxide films with highly preferential crystal orientation. J Am Ceram Soc 81:1622–1632
15. Ikhmayies Shadia (2016) Synthesis of ZnO microrods by the spray pyrolysis technique. J Electron Mater 45(8):3964–3969
16. Kao Ming-Cheng, Chen Hone-Zern, Young San-Lin, Lin Chen-Cheng, Kung Chung-Yuan (2012) Structure and photovoltaic properties of ZnO nanowire for dye-sensitized solar cells. Nanoscale Res Lett 7(1):260
17. Hamada Takahiro, Fujii Eiji, Chu Dewei, Kato Kazumi, Masuda Yoshitake (2011) Aqueous synthesis of single-crystalline ZnO prisms on graphite substrates. J Cryst Growth 314: 180–184
18. Liu YL, Liu YC, Feng W, Zhang JY, Lu YM, Shen DZ, Fan XW, Wang DJ, Zhao QD (2005) The optical properties of ZnO hexagonal prisms grown from polyvinylpyrrolidone-assisted electrochemical assembly onto Si (111) substrate. J Chem Phys 122:174703

Electrical and Microstructural Investigation of $Ni_{0.5}Co_{0.5}Cu_{0.3}Zn_{0.3}Mn_{1.4}O_4$ Temperature Sensors

Gökhan Hardal and Berat Yüksel Price

Abstract Nickel manganite based negative temperature coefficient (NTC) thermistors have good temperature sensor characteristics and electrical stability for temperature measurement applications. In this study, the electrical and microstructural properties of $Ni_{0.5}Co_{0.5}Cu_{0.3}Zn_{0.3}Mn_{1.4}O_4$ NTC thermistor were investigated. The samples were fabricated by the conventional solid-state reaction method. The powders were calcinated at 900 °C for 2 h and the samples were sintered at 1200 and 1300 °C for 5 h in air. The bulk density of samples was 5.07 and 4.96 g/cm^3, respectively. The microstructure of samples was observed using a scanning electron microscope. The electrical resistance was measured in a temperature programmable furnace between 25 and 85 °C. The electrical resistivity of the $Ni_{0.5}Co_{0.5}Cu_{0.3}Zn_{0.3}Mn_{1.4}O_4$ sample increased from 52 Ω cm to 95 Ω cm when the sintering temperature increased from 1200 to 1300 °C. The material constant "B", sensitivity coefficient "α" and activation energy "E_a" values were also calculated for the NTC thermistors.

Keywords Copper oxide · Microstructure · Electrical properties · NTC thermistors

Introduction

Negative temperature coefficient (NTC) thermistors are widely used for temperature sensing. Nickel manganite ($NiMn_2O_4$) is the most investigated NTC material due to its low cost and ease of manufacturing [1]. Most NTC materials are based on Mn_3O_4 ceramics, where Mn^{3+} is partially replaced by other transition-metals such as Ni, Co, Fe, and Cu in order to induce semiconductivity in this rather resistive

G. Hardal (✉) · B. Yüksel Price
Metallurgical and Materials Engineering Department, Istanbul University, Istanbul, Turkey
e-mail: gokhan.hardal@ogr.iu.edu.tr

B. Yüksel Price
e-mail: berat@istanbul.edu.tr

S. Ikhmayies et al. (eds.), *Characterization of Minerals, Metals, and Materials 2017*, The Minerals, Metals & Materials Series,
DOI 10.1007/978-3-319-51382-9_16

material. These substitutions lead to the formation of Mn^{3+}/Mn^{4+} pairs on the octahedral sites, which are essential for electrical conductivity [2]. These materials crystallize in the spinel structure, that is a cubic close packing of oxygen anions in which the cations are located on the tetrahedral sites (A) and the octahedral sites (B) [3]. The semiconducting properties of nickel manganite based NTC thermistors can be explained by a phonon-assisted jump of carriers between Mn^{3+} and Mn^{4+} ions present in octahedral B-sites. The conductivity reaches maximum when the number of Mn^{3+} ions equals that of Mn^{4+} ions [4].

The aim of this work is to understand the effect of sintering temperature on the microstructural and electrical properties of the $Ni_{0.5}Co_{0.5}Cu_{0.3}Zn_{0.3}Mn_{1.4}O_4$ sample.

Experimental Procedure

NiO (99% purity, Alfa Aesar), Co_3O_4 (99.5% purity, Sigma-Aldrich), CuO (≥99% purity, Sigma-Aldrich), ZnO (≥99% purity, Merck) and Mn_2O_3 (99% purity, Sigma-Aldrich) powders were weighed according to the compositions of $Ni_{0.5}Co_{0.5}Cu_{0.3}Zn_{0.3}Mn_{1.4}O_4$. The raw powder mixture was ball-milled using ZrO_2 balls as a grinding media with ethyl alcohol in a jar for 6 h. The obtained slurries were dried and powders were calcinated at 900 °C for 2 h. The powders were pressed to form disc shaped specimens and then sintered at 1200 and 1300 °C for 5 h in air employing a 360 °C/h heating rate then cooled naturally in the furnace.

The bulk density (d, g cm^{-3}) of sintered samples was calculated using the Archimedes principle. The phases in the sintered samples were determined by X-ray diffraction (XRD, Rigaku D/Max-2200/PC) analysis using CuK_α radiation at 60 kV/2 kW. The microstructure of samples was observed using a scanning electron microscopy (SEM, JEOL, JSM 5600) on fracture surfaces. The sintered samples were coated with silver paste to form electrodes. The electrical resistance was measured in a temperature programmable furnace between 25 and 85 °C in steps of 0.1 °C.

The material constant (B, K) can be calculated by Eq. 1.

$$B_{25/85} = \frac{\ln \rho_{25} - \ln \rho_{85}}{\frac{1}{T_{25}} - \frac{1}{T_{85}}} \tag{1}$$

ρ_1 and ρ_2 being the resistivity at temperature T_1 and T_2, respectively. The temperature coefficient of resistance or sensitivity coefficient (α, %/K) can be obtained by using Eq. 2.

$$\alpha = \frac{1}{R}\frac{dR}{dT} = -\frac{B}{T^2} \tag{2}$$

$$E_a = Bk_b \tag{3}$$

The activation energy (E_a, eV) can also found by the Eq. 3, where k_B is the Boltzmann constant [1].

Results

The XRD pattern of the $Ni_{0.5}Co_{0.5}Cu_{0.3}Zn_{0.3}Mn_{1.4}O_4$ sample sintered at 1300 °C for 5 h is shown in Fig. 1. The cubic spinel phase was observed as a main phase in the sintered sample. In addition, the NiO-rich and tetragonal spinel phase was also found in this sample. The decomposition of spinel $NiMn_2O_4$ phase is responsible for the presence of NiO-rich and tetragonal spinel phase [5–8].

The bulk density of the $Ni_{0.5}Co_{0.5}Cu_{0.3}Zn_{0.3}Mn_{1.4}O_4$ samples sintered at 1200 and 1300 °C was 5.07 and 4.96 g cm^{3-}, respectively. In our previous work, the bulk density of $Ni_{0.5}Co_{0.5}Cu_{0.3}Mn_{1.7}O_4$ sintered at 1100 °C was found as 5.1 g cm^{3-}. When comparing these two results, it was seen that the bulk density was quite similar, irrespective of ZnO addition. A similar observation was reported by Park et al. [9].

SEM micrographs of the $Ni_{0.5}Co_{0.5}Cu_{0.3}Zn_{0.3}Mn_{1.4}O_4$ sample sintered at 1200 °C was shown in Fig. 2. The NiO rich phases and the "floret" structure which are surrounded by NiO rich phases were observed. The surrounding matrix was a NiO rich rock-salt structured material and the florets were manganese rich [10]. These results are in good agreement with the results of Györgyfalva and Reaney [10].

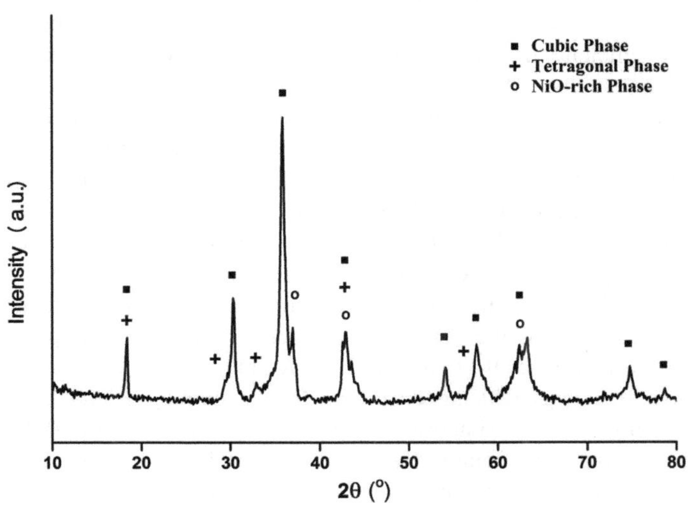

Fig. 1 XRD pattern of the $Ni_{0.5}Co_{0.5}Cu_{0.3}Zn_{0.3}Mn_{1.4}O_4$ sample sintered at 1300 °C

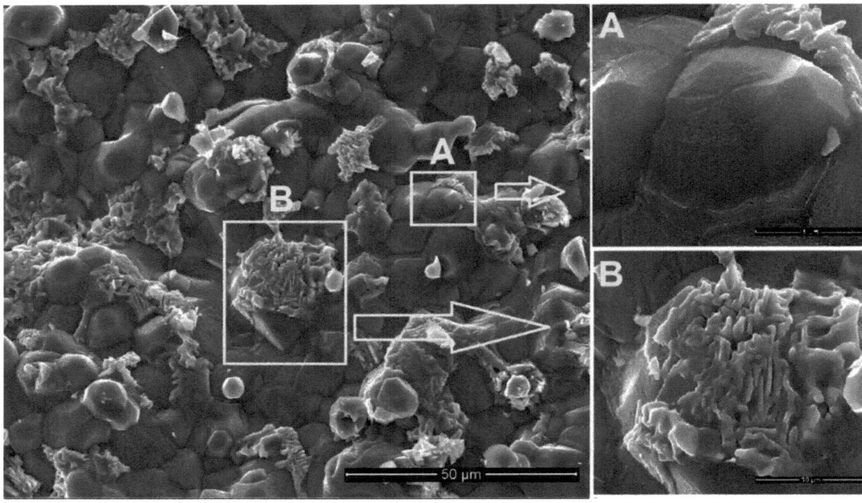

Fig. 2 SEM micrographs of the sample sintered at 1200 °C

The SEM micrograph and EDS analysis of the sample sintered at 1300 °C is shown in Fig. 3. The grain size of the sample was found ≈20 μm. EDS analysis results from four different points of the sample show solid solutions of the constituent oxides. The NiO-rich phases and Mn-rich phases were also observed in the structure. The analysis of EDS results was confirmed in comparison with XRD result of same sample.

The decomposition of the $Ni_{0.5}Co_{0.5}Cu_{0.3}Zn_{0.3}Mn_{1.4}O_4$ sample sintered at 1300 °C was more obvious than the sample sintered at 1200 °C.

The electrical resistivity at room temperature (ρ_{25}), material constant ($B_{25/85}$), sensitivity coefficient (α_{25}) and activation energy (E_a) values were tabulated in Table 1. The value of electrical resistivity was 52 Ω cm for the $Ni_{0.5}Co_{0.5}Cu_{0.3}Zn_{0.3}Mn_{1.4}O_4$ sample sintered at 1200 °C. It increased to 95 Ω cm when the sintering temperature increased to 1300 °C. Similarly, the increase was observed between the $B_{25/85}$ values of samples sintered at 1200 and 1300 °C, and was calculated as 2825 and 3226 K, respectively.

Park and Lee found that the electrical resistivity and material constant of $Ni_{0.45}Co_{0.15}Cu_{0.45}Zn_{0.3}Mn_{1.65}O_4$ sample sintered at 1075 °C for 3 h were 20.2 Ω cm and 2523 K [5]. In another work, they reported that the electrical resistivity and material constant of the $Ni_{0.45}Co_{0.15}Cu_{0.45}Zn_{0.3}Mn_{1.65}O_4$ sample sintered at 1100 °C for 24 h were 22.9 Ω cm and 2791 K [11].

The activation energy which is the energy for the hopping mechanism was found as 0.243 eV in the sample sintered at 1200 °C. It was increased to 0.278 eV with increasing sintering temperature. The sensitivity coefficient value of 3.181 -%/K was calculated for the sample sintered at 1200 °C. This value increased to 3.632 -%/K when sintering temperature increased to 1300 °C.

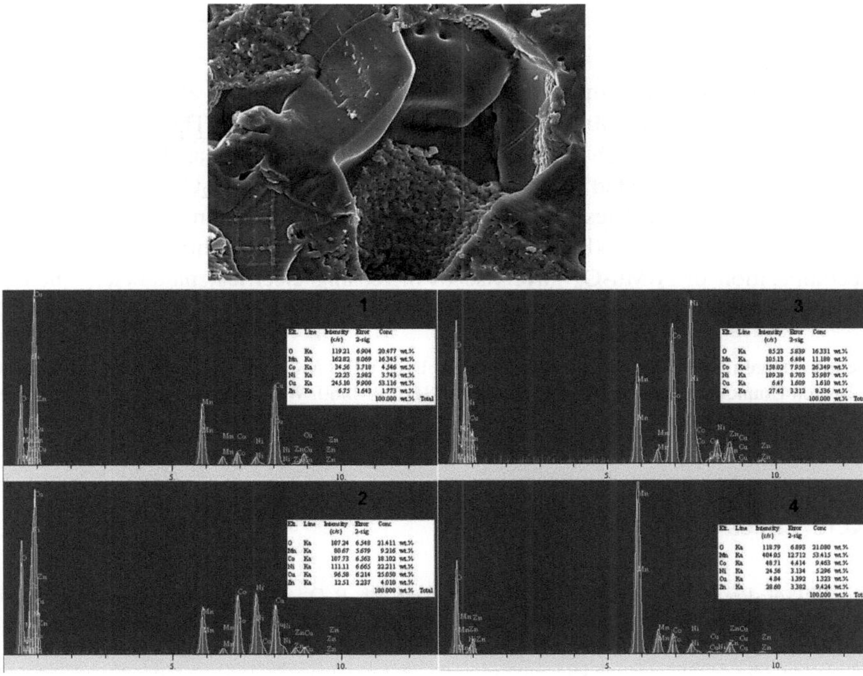

Fig. 3 The SEM micrograph and EDS analysis of sample sintered at 1300 °C

Table 1 The electrical parameters of samples sintered at different sintering temperatures

Sintering temperature (°C)	ρ_{25} (Ω cm)	$B_{25/85}$ (K)	E_a (eV)	α_{25} (-%/K)
1200	52	2825	0.243	3.181
1300	95	3226	0.278	3.632

As can be seen in all these results, the electrical properties of NTC thermistors altered due to the difference in sintering temperature. It is well-known that the ratio of the Mn^{3+}/Mn^{4+} ions in the octahedral sites influences the electrical resistivity of samples. The increase of electrical resistivity, material constant, activation energy and sensitivity coefficient with increasing sintering temperature is related to the decomposition of $NiMn_2O_4$. Gao et al. [4] reported that the number of Mn^{3+}/Mn^{4+} couples in the octahedral sites altered resulting in variations in the resistivity. Additionally, some Mn^{4+} ions reduced to Mn^{3+} due to the occurrence of decomposition.

Conclusion

The effect of sintering temperature on the microstructural and electrical properties of $Ni_{0.5}Co_{0.5}Cu_{0.3}Zn_{0.3}Mn_{1.4}O_4$ sample was investigated. The XRD analysis demonstrated the NiO-rich and tetragonal spinel phases as well as the main phase of cubic spinel. It indicated that the decomposition of $NiMn_2O_4$ occurred in all samples. The electrical resistivity, material constant, activation energy and sensitivity coefficient increased with increasing sintering temperature. The increasing decomposition of $NiMn_2O_4$ gave rise to increasing electrical parameters such as electrical resistivity, material constant etc.

Acknowledgements This study is supported by TÜBİTAK (The Scientific and Technical Research Council of Turkey), Project number 3001-114M860. We would like to thank TÜBİTAK for its financial support.

References

1. Yüksel Price B, Hardal G (2016) Influence of B_2O_3 addition on the electrical and microstructure properties of $Ni_{0.5}Co_{0.5}Cu_xMn_{2-x}O_4$ ($0 \leq x \leq 0.3$) NTC thermistors without calcination. J Mater Sci Mater Electron 27:9226–9232
2. Feteira A (2009) Negative temperature coefficient resistance (NTCR) ceramic thermistors: an industrial perspective. J Am Ceram Soc 92:967–983
3. Fritsch S, Sarrias J, Brieu M, Couderc JJ, Baudour JL, Snoeck E, Rousset A (1998) Correlation between the structure, the microstructure and the electrical properties of nickel manganite negative temperature coefficient (NTC) thermistors. Solid State Ionics 109:229–237
4. Gao H, Ma C, Sun B (2014) Preparation and characterization of $NiMn_2O_4$ negative temperature coefficient ceramics by solid-state coordination reaction. J Mater Sci Mater Electron 25:3990–3995
5. Park K, Lee JK (2009) The effect of ZnO content and sintering temperature on the electrical properties of Cu-containing $Mn_{1.95-x}Ni_{0.45}Co_{0.15}Cu_{0.45}Zn_xO_4$ ($0 \leq x \leq 0.3$) NTC thermistors. J Alloy Compd 475:513–517
6. Feltz A, Pölzl W (2000) Spinel forming ceramics of the system $Fe_xNi_yMn_{3-x-y}O_4$ for high temperature NTC thermistor applications. J Eur Ceram Soc 20:2353–2366
7. de Györgyfalva GDCC, Nolte AN, Reaney IM (1999) Correlation between microstructure and conductance in NTC thermistors produced from oxide powders. J Eur Ceram Soc 19:857–860
8. Macklen ED (1986) Electrical conductivity and cation distribution in nickel manganite. J Phys Chem Solids 47:1073–1079
9. Park K, Lee JK, Kim S-J, Seo W-S, Cho W-S, Lee C-W, Nahm S (2009) The effect of Zn on the microstructure and electrical properties of $Mn_{1.17-x}Ni_{0.93}Co_{0.9}Zn_xO_4$ ($0 \leq x \leq 0.075$) NTC thermistors. J Alloy Compd 467:310–316
10. de Györgyfalva GDCC, Reaney IM (2001) Decomposition of $NiMn_2O_4$ spinel: an NTC thermistor material. J Eur Ceram Soc 21:2145–2148
11. Park K, Lee JK (2007) Mn–Ni–Co–Cu–Zn–O NTC thermistors with high thermal stability for low resistance applications. Scripta Mater 57:329–332

Part IV
Nano Materials

Enhanced Physical Properties of Thin Film Nanocomposites

T. Thuy Minh Nguyen, Sathish K. Lageshetty and Paul Bernazzani

Abstract The addition of nanoparticles to epoxy resins often results in changes in the physical properties of the material particularly when dealing with thin coatings. By characterizing the mechanism of interactions between nanoparticles and polymer chains the material could be tailored for specific applications or properties such as enhanced biodegradability. This study compared the characteristics of common DGEBA resin coated on a silicon substrate and embedded with nanoparticles of TiO_2 and Fe_2O_3. Specifically, we investigated the rate of curing, and the thermal stability of these nanocomposites. Results show changes in the stability for compounds containing TiO_2, while Fe_2O_3 significantly enhanced the curing rates of the nanocomposites while maintaining similar physical properties.

Keywords Epoxy · Nanocomposites · Thermal analysis · Reaction rates

Introduction

The development of improved material, i.e. stronger, lighter more heat and impact resistant, has been a general pursuit of mankind. Recently, the development of nanocomposites and their possible applications [1] has generated an enormous amount of interest in both academia and industry. In fact, a "Google Scholar" search for the term "nanocomposite" over the last ten years, resulted in about 260,000 hits [accessed 09/01/16]. While the dominating driving force of these studies is the development of materials with improved properties, the process by which nanocomponents modify the physical and thermal properties of the polymer matrix is still far from understood. The present study attempts to shed some light on the mechanism of interaction between nanocomponents and matrix.

T.T.M. Nguyen (✉) · S.K. Lageshetty · P. Bernazzani
Department of Chemistry and Biochemistry, Lamar University,
77710 Beaumont, TX, USA
e-mail: ttnguyen15@lamar.edu

S. Ikhmayies et al. (eds.), *Characterization of Minerals, Metals, and Materials 2017*, The Minerals, Metals & Materials Series,
DOI 10.1007/978-3-319-51382-9_17

In recent years, a large variety of materials have been investigated as nanocomposites. The nanoparticles themselves may be inorganic or organic compounds. Common examples of inorganic nanoparticles are, ZnSe, AgCl, Iron oxides and Silicates such as clay which are usually aluminosilicates [2] to name some of the most commonly used ones. TiO_2, is one of the most commonly studied nanoparticles. In its natural state, titanium dioxide is available in three major crystalline forms: rutile, anatase and brookite. Because rutile is the most stable form, anatase and brookite can be converted into it upon heating. Nanocomposites containing titanium dioxide nanoparticles have been found to display unusual characteristics mainly been centered on surface optical such as an increase in the refractive index, mechanical and thermal properties [3–6].

A large variety of polymer matrices have been reported [7]. These include vinyl polymers such as polymethyl methacrylate [8, 9], polyacrylamide [10], polytetrafluoro ethylene [11], and polystyrene [12, 13]; polyolefins such as polypropylene [14, 15] and polyethylene [16, 17]; several others such as polyamides [18], polyethylene oxide [19, 20], and epoxidized natural rubber [21, 22].

Epoxy resins have also been investigated as polymer matrix for nanocomposites. Epoxy resins are polymers, which contain more than one unreacted oxirane groups. They are classified into two main types: glycidal and non-glycidal epoxies. Diglycidal ether of bisphenol-A is the most commonly available commercial product of the glycidal epoxies. It is usually synthesized by the reaction between bisphenol-A and epichlorohydrin in the presence of catalyst or accelerators. The final properties of the cured resin depend on the chemical structure of the starting compounds. Hence, characterization of the epoxy resin is critically important to the preparation of a cured epoxy with tailored properties [23]. The use of epoxy resins as nanocomposite matrices is because of the fact that epoxy resins are known for their heat resistance and excellent adhesion properties for which they are widely used in various applications that include in aerospace industry, coatings and adhesives. However, their applications are limited because of their brittle nature and poor thermal conductivity. A material's response to failure is described as its toughness. Two important approaches to increase the toughness of epoxy resins are the modification of the epoxy backbone, and incorporation of inorganic fillers or more recently nanoparticles.

The most often encountered epoxy resin is composed of diglycidal ether of bisphenol A, commonly known as DGEBA epoxies. This is a matrix of choice and is the one used in our investigation because of its existing high thermal stability and physical properties and large chemical group diversity. The preparation of this matrix involves several other compounds used as plasticizer (for increase in malleability), hardener (for crosslinking purposes), and accelerator to increase the rate and efficiency of crosslinking. Figure 1 presents the chemical structures of these compounds used in this study. While several compounds are involved in the preparation of an epoxy resin, the hardener is the most important. Curing is the process in which the three-dimensional crosslinked network is formed from a linear polymer when reacted with a hardener. During the curing process the reaction proceeds by the opening of epoxide group, which leads to the formation of

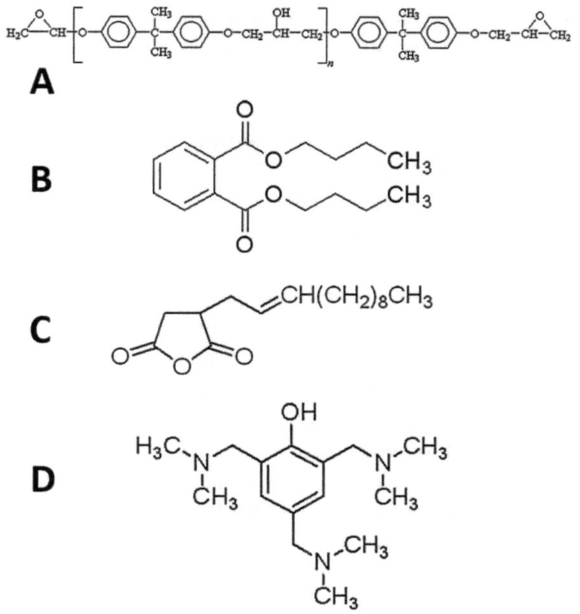

Fig. 1 Chemical structures of the compounds involved in the preparation of the epoxy resins matrix. **a** Diglycidal ether of bisphenol A (DGEBA); **b** the plasticizer, Dibutyl phthalate (DBP); **c** the hardener, dodecenyl succinic anhydride (DDSA); **d** the accelerator, 2,4,6-tris (dimethylaminomethyl) phenol (DMP-30)

hydroxyl groups. The chemical reaction, which takes place during curing, determines the properties of the cured network [24]. The main function of the hardening agent is to transform the epoxy resin into a rigid matrix hence the final properties, such as the glass transition temperature and the kinetics of the curing process depend on the structure of the hardening agent. The most commonly used hardening agents for epoxies include: phenolic resins, amines, polyamides, isocyanates and anhydrides. Dodecenyl succinic anhydride (DDSA) is a very common liquid anhydride-curing agent for epoxy resins.

Studies on epoxy resins showed that temperature, type of epoxy, and the epoxy/hardener ratio affected the cure kinetics [25–27], while the addition of nanoparticles had an effect on the fatigue behavior of composites [28]. Attempts have been made to understand the role of dipole interactions on tensile properties, glass transition temperature, fracture toughness and impact strength of epoxy resins [29]. These studies have led to our hypothesis that the cause of any change in behavior with the addition of nanoparticles is either the existence of chemical interactions between the particles and the polymer chains or physical interactions i.e. the particles act defects at the molecular level.

The cause of the variety of the changes in properties of a material with the addition of the nanoparticles has been the object of some investigations. While the variations in modulus have been attributed to good dispersion and higher interfacial adhesion between nanoparticles and the polymer matrix [30], the contribution of the shape of the nanoparticles in the mechanical and the processing behavior of polymer nanocomposites was also shown to be significant [31]. The type and amount of interactions between nanoparticles and polymer chain has been the subject of several studies. Mohamed et al. [32] suggested that a decrease in crosslinking density of the nanocomposites could be attributed to physical interactions between the polymer matrix and nanocomposites. An increase in the glass transition temperature of AuPS 481 and poly (methyl methacrylate)/C60 nanocomposite systems was attributed to interactions between the polymer matrix and nanoparticles system [33] without explicit details but implying chemical interactions. However, Sanchez and others suggested that an observed increase in the glass transition temperature of polystyrene nanocomposites was caused by the chemical interactions between TiO_2 nanoparticles and polymer chains [34]. Studies on the increase in stress of polymer chains in nanocomposites suggested that chemical interactions were present between the polymer chains and the nanoparticles [35]. The addition of TiO_2 nanoparticles was reported to shown a significant effect on the curing process of the vinyl ester resin, which associated to both physical and chemical (physicochemical) interactions between the polymer matrix and the nanoparticles [36].

Other studies have shown that the main factors that affect the properties of nanocomposites include morphology, surface treatments of nanoparticles and type of polymer matrix [37]. The high surface to volume ratio exhibited by the nanoparticles in the nanocomposites differentiates it with the conventional composite material. Hence, increases in the thermal stability and mechanical properties were attributed to the uniform particle dispersion and strong chemical bonds between nanoparticles and polymer-matrix [38], while the mechanical properties of the nanocomposites mainly depend on the proper linkage or bonding between the particles and the polymer matrix [38, 39]. Hence the type of interactions either physical or chemical or both that is responsible for the enhanced properties of nanocomposites, is under debate.

Experimental

Materials

The epoxy system used in this study is an Araldite 502 (a Diglycidal ether of bisphenol A), Poly Bed 812 cured with DDSA and catalyzed by a phenol,2,4,6-tris (dimethyl amino) methyl (DMP-30). The above kit was purchased from Polysciences, Inc. Titanium dioxide anatase particles were obtained from Aldrich.

Sample Preparation

In a beaker first weighed 5 g of Poly/Bed 812 and 3 g Araldite, stirred them with little heat up to 10 min under a fume hood until a homogeneous mixture is obtained. To this mixture 0.9 g of Dodecenyl succinic anhydride (curing agent) was added maintaining the same conditions, stirred for 35 min. Then after 45 min. of total mixing 0.26 g catalyst 2, 4, 6-tris (dimethylaminomethyl) phenol (DMP-30) was added to the above mixture and continued stirring for more 15 min. Finally, after obtaining a homogeneous solution it is transferred into a Petri dish which is wrapped with aluminum foil and cured for 24 h at 60 °C. The structures of the products used are reported.

The different concentrations of nanoparticles ranging from 0.01 to 0.03% w/w were dispersed in the above solution, after 10 min. of addition of catalyst and stirred for five more minutes.

Curing Procedure

TiO_2 nanoparticles in the anatase crystalline form were dispersed in amounts ranging from 0.01 to 0.03%, in the liquid resin mixture and were kept in an isothermal oven for about 24 h at 60 °C. The kinetics of the curing was determined by FTIR Spectroscopy, employing a Nicolet Nexus 470. The sample at different time intervals was placed on the ZnSe crystal of ATR and absorbance was recorded. Following the absorbance of epoxy region at 915 cm^{-1} monitored the conversion of the epoxy group.

UV-Visible Spectroscopy

A Cary 50-Bio UV-Visible spectrophotometer was used to determine the rate and extant of curing process of the epoxy system and epoxy system with nanoparticles. A scan rate of 2400 nm/min was used and the scan range was 200–800 nm. Absorbance was observed at 425 nm. Scans were recorded every 20 min.

FTIR Spectroscopy

A Nicolet Nexus 470 FTIR with attenuated total reflectance (ATR) sampling was used to determine the curing process and the interactions of nano particles with the epoxy system. Spectra were recorded using 32 scans at a resolution of 8 cm^{-1}. The final format used was absorbance and the correction was ATR.

Differential Scanning Calorimetry

TA Instrument Q20 was used to determine the changes in the enthalpy and glass transition temperature associated due to the presence of interactions with the TiO_2

nanoparticles. The sample time was 0.10 s/point and the equilibrium temperature was −40 °C. The heating ramp temperature was −40–300 °C, followed by an isothermal step maintained for 5 min. and a cooling ramp to −40 °C and again followed isothermal for 5 min. and a heating ramp again to 300 °C under nitrogen flux. Indium was used as the calibration standard.

Results and Discussion

To investigate the nature of the interactions of nanoparticles with polymer chains, three aspects of the properties of nanocomposites were evaluated. The first of these investigated aspects concerns the cure kinetics of the epoxy resin matrix and any changes in rate of curing caused by the addition of nanoparticles. The second aspect involves the thermodynamic characteristics of the nanocomposite including variations in glass transition temperatures and thermal stability. The third aspect pertains to the physical properties of the material namely the creep and stress relaxation behavior dependence of nanocomposites as a function of nanoparticle content. These aspects will be discussed in the following sections.

Kinetics of Epoxy Resins

The kinetics of epoxy resins were followed using two spectroscopic techniques: Fourier Transform Infrared (FTIR) and UV-Visible spectroscopy. Both of these methodologies have been used in the past to follow the completion of the cure reaction in crosslinked systems. The curing reaction consists in the opening of the epoxide ring, which results in the formation of hydroxyl groups.

The infrared spectrum of the reaction mixture contains several peaks of interest: a broad peak formation around 2900 cm^{-1} is attributed to C–H stretching under aromatic and non-aromatic conditions and therefore includes CH_2 symmetric and asymmetric stretching vibrations. A strong absorption peak at 1500 cm^{-1} is associated with C=C aromatic ring stretching and represents the p-phenylene structure. A strong absorption peaks at 1250 and 1038 cm^{-1} represent the ether groups on the DGEBA molecule i.e. various stretching of C–O and C–O–C groups with a peak at 1250 cm^{-1} being associated with the aromatic carbon-oxygen stretching. These bands are not expected to be significantly affected by the curing reaction since it does not involve these chemical groups. The curing process consists in the opening of the epoxide ring followed by the formation of hydroxyl groups. A band around 915 cm^{-1}, assigned to the epoxide ring group, is therefore expected to decrease in proportion to the degree of curing while another band corresponding to the formation of hydroxyl groups should be observed around 3400 cm^{-1} attributed to the O–H group stretching.

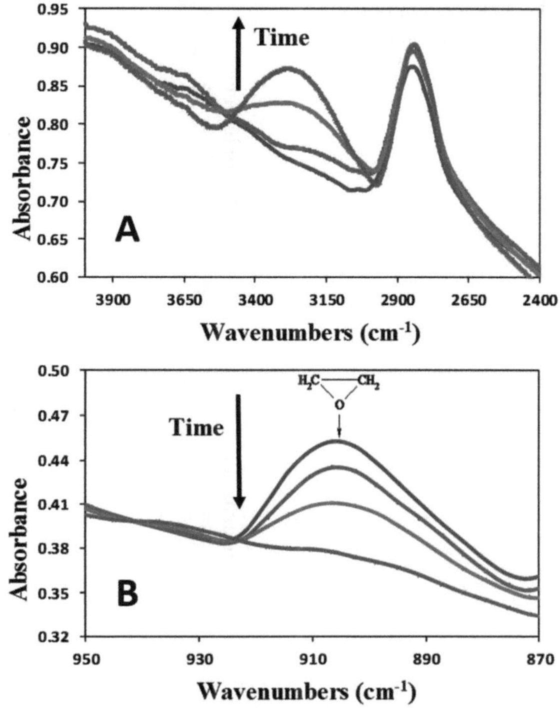

Fig. 2 FTIR spectra of epoxy resins at different times. Panel A shows the spectra in the hydroxyl region while panel B shows the spectra in the epoxide region

For a proper investigation of the kinetics of the curing process, FTIR spectra were recorded as a function of time. Figure 2 shows the spectra of epoxy resins taken at different times during the curing process at 60 °C in the regions of interest namely the hydroxyl region (panel A) and the epoxide band region (panel B). Since the amount of reagent (the epoxide ring) and product (the crosslinked epoxy containing a hydroxyl group) can be followed over the time of the reaction, the order of the reaction and the rate constants associated with it can be determined by plotting the absorbance of the bands in Fig. 2 as a function of time.

Figure 3 presents kinetic plots of the evolution of the curing reaction at 60 °C by showing the evolution of the corrected absorbance of bands at position 3400 cm^{-1}, (panel A) and 915 cm^{-1} (panel B) with time. The correction involved subtraction of a linear baseline and its offset. Assuming first order kinetics and starting from the disappearance of the reagents these plots were fitted to the exponential Eq. 1.

$$A = A_0 e^{-kt} \tag{1}$$

In this equation A represents the corrected absorbance of the reagent at time t, A0, the absorbance at the start of the reaction and k is first order rate constant. At such low temperature, the rate constants are very small and found to be

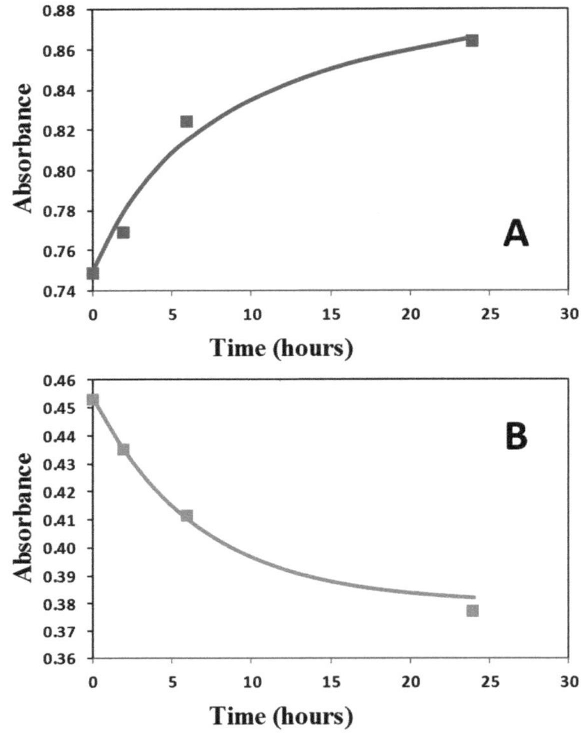

Fig. 3 Evolution of the absorbance at 3400 cm^{-1} (panel A) and at 915 cm^{-1} (panel B)

12.6×10^{-4} min^{-1} and 5.6×10^{-4} min^{-1} for the epoxide and hydroxyl bands respectively. The apparent discrepancy between the two values is caused by the differences in absorptivity coefficient of the respective chemical groups. Figure 3 shows the fitted curves as solid lines. The reasonably good convergence of the fits within the error of the data points indicates that the curing reaction proceeds through first order kinetics as expected. However, one drawback of this technique is the transparency of the epoxy system and the inaccurate measurements that will result when nanoparticles are added to the system. Hence the investigation of the cure kinetics of the material in the presence of nanoparticles was investigated using another spectroscopic method, UV-Visible spectroscopy.

To investigate such effect TiO_2 and Fe_2O_3 nanoparticles of similar dimensions were incorporated into the Araldite 812 epoxy resin mixture and the rate of the curing process was evaluated by following the increase in the absorbance at 425 nm. Figure 4 presents the evolution of the absorbance at this wavelength of the epoxy resin and of epoxy resins containing dispersions of nanoparticles during a curing procedure maintained at 150 °C. Since this temperature is significantly higher than that used previously for IR study, higher rates are expected. The data points were fitted to Eq. 1 by subtracting the recorded absorbance value from that of the completely cure epoxy system containing no nanoparticles. The fits obtained

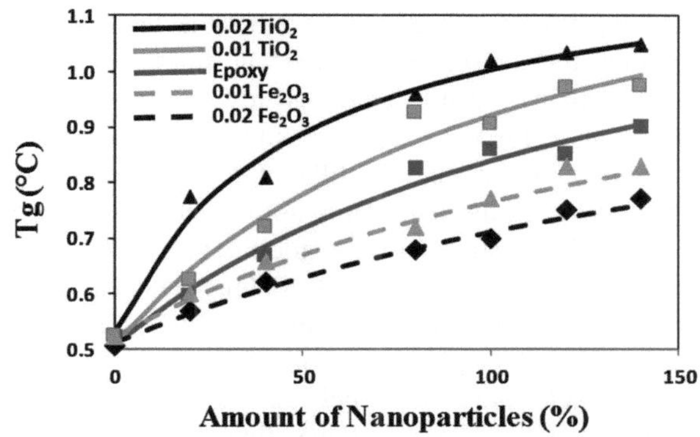

Fig. 4 Evolution of the absorbance at 425 nm of epoxy and epoxy/nanoparticles composite during the curing at 150 °C

Table 1 Kinetic rate constants for the curing process using FTIR and UV-visible spectroscopy for different material

FTIR epoxy resin	Rate constant (10^{-4} min^{-1})	Standard error ($\pm 10^{-4}$ min^{-1})	Goodness of fit R^2
Hydroxyl group	5.6	3.0	0.88
Epoxide group	12.6	4.4	0.83
UV-visible			
Epoxy Resin	190.0	0.6	0.99
Resin with 0.01% TiO_2	134.9	0.4	0.96
Resin with 0.02% TiO_2	94.4	1.6	0.97
Resin with 0.01% Fe_2O_2	210.2	0.5	0.97
Resin with 0.02% Fe_2O_2	236.1	1.1	0.95

were significantly better than those from the FTIR data probably because of the smaller signal to noise ratio in FTIR, as well as the strong dependence of the absorbance on the frequency, which is a factor of the geometry of the thin epoxy films.

Table 1 summarizes the values of the rate constants for the different experiments. These values are averages of several experiments. The error column represents calculated standard deviations. The table presents the values of k for both the FTIR and the UV-Visible methods. However, since these methods have not been normalized using relative concentrations, the values of k from different methods and even different bands cannot be compared. However, one should note that the calculated error is significantly higher for the FTIR methodology. The table also shows that upon the addition of a small amount, 0.01% (W/W), of TiO_2 nanoparticles, the rate of the reaction decreases by about 30%. Doubling this amount, 0.02% (W/W), resulted in a further decrease of the rate, which then becomes 50% its original rate. On the other hand, the addition of iron oxide nanoparticles resulted in an observable increase in the rate of the reaction, suggesting the presence of chemical interactions between the reactants and the nanoparticles. Using amounts lower than 0.01% of nanoparticles had little observable effect. A caution about this statement must be mentioned as such a small amount of compound is difficult to measure. Also worthy of notice is that the mechanism of the reaction is not affected by the presence or absence of the particles, as demonstrated by the good values of R^2, the goodness of the fit, using first order kinetic equation. Figure 4 along with the values presented in Table 1 clearly show that incorporation of nanoparticles leads to a decrease in curing rate.

Results of the investigation of the cure kinetics show the strong impact the nanoparticles have on the behavior of the epoxide resin films as well as the different behavior of the material depending on the nature of the particles. However, the mechanism of this change in behavior is unclear. Since the order of the reaction was left unchanged, the nanoparticles may not have a direct chemical interaction with the polymer chains. In fact they may simply act as obstacles for the diffusion or reptation of the molecules or macromolecules involved in the reaction. If the latter is the case, we expect a decrease in the thermal characteristics of the nanocomposite as well as in the physical properties of the material. To investigate this further, the thermal analysis of the system was undertaken.

Thermal Analysis

Thermal analysis and more specifically the determination of the glass transition temperature, can help shed light on the issue concerning the presence or absence of interactions between polymer chains and nanoparticles. The purpose of investigating the thermal properties of nanocomposites resides in the fact that the behavior of polymer chains under the influence of matter acting as defects is well modeled. For example, the change in the glass transition temperature (Tg) of a material in the presence of an impurity with a different Tg is given by the Fox equation;

$$\frac{1}{T_g} = \frac{x}{T_g^1} + \frac{(1-x)}{T_g^2} \tag{2}$$

In this equation, T_g^1 and T_g^2 represent the glass transition temperatures of the pure compound acting as a defect and the pure polymer respectively, and x represents the fraction of the first compound. This equation shows that the addition of a compound that does not have any interactions either physical or chemical with the polymer chains will result in a depression of the Tg of the material relative to the Tg of the pure polymer. Following Tg of the epoxy resins as a function of the amount of nanoparticle will indicate if the Fox equation as followed.

Figure 5 shows the typical differential scanning calorimetry (DSC) traces of four epoxy resins containing different amounts of TiO_2. The thermal stability of the

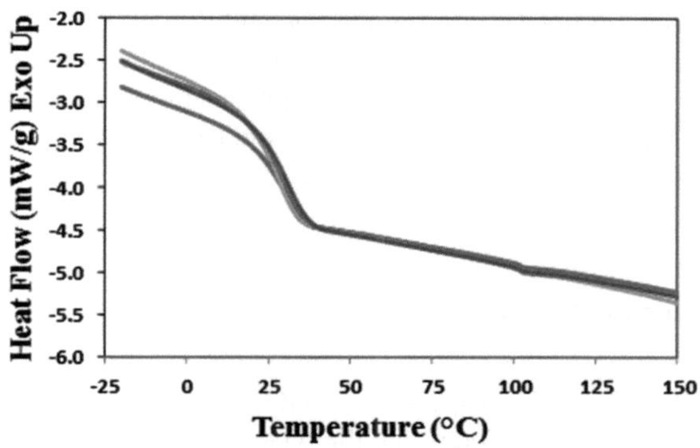

Fig. 5 DSC traces for epoxy resin nanocomposites with various concentrations of TiO_2 nanoparticles

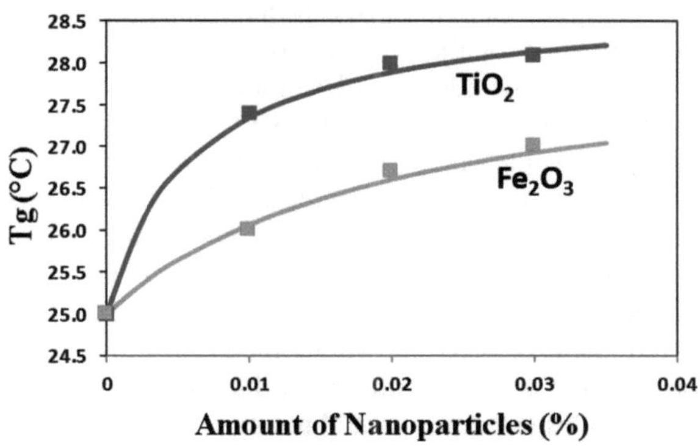

Fig. 6 Variation of the glass transition temperature with amount of nanoparticles

sample is invariable and the epoxy resin is stable until 300 °C. The clear endothermic drop in heat flow around 25 °C is typical of a glass transition. However, the non-zero value of the slope of the glassy lines (the baseline at temperatures above Tg) indicates the presence of a more or less extensive semi-rigid phase in the sample, which is usually found in crosslinked polymers. The figure also shows that the Tg increases for some samples. This could be attributed to variations in degree of curing as the more crosslinks occur in a sample the more the chains are confined leading to an increase in Tg. Since the crosslinking reaction is strongly exothermic, heating the sample at higher temperatures would result in the reaction of any remaining epoxide groups. However, the DSC traces' lack of an exothermic peak at 150 °C indicates that all samples are fully cured. The observed variations in Tg can than only be attributed to the effect of the presence of nanoparticles.

Figure 6 shows the variation of the glass transition temperature as a function of the amount of nanoparticle dispersed in the polymer. The data represent the values (with a standard deviation of about 0.3 °C), while the solid line is represents a logarithmic trendline of the data with a satisfactory goodness of fit R2 of 0.94. Contrary to the prediction that non-interacting particles would automatically decrease the Tg of a sample, the addition of TiO_2 to epoxy resins increases the Tg. This result indicates that the TiO_2 nanoparticles have an interaction. At this time, it is not clear if this interaction is physical or chemical however, the fact that both the FTIR and the thermal stability indicate no clear changes, suggests that physical interactions are present.

Conclusions

The effect of a dispersion of TiO_2 nanoparticles on the behavior epoxy resins was investigated to determine the type, if any, of interactions between the nanoparticles and polymers. The goal was to establish if chemical or physical interactions existed. The study focused on three aspects of the properties of nanocomposites. Firstly, the determination of the cure kinetics and especially the rate of the curing reaction showed that the addition of nanoparticles to the epoxy resin decreased the rate of curing. This suggests that the TiO_2 particles may act as obstacles to the proper reptation of the chains and are not involved in either physical or chemical interactions with the macromolecules. This was supported by the absence of variations observed on the FTIR spectra comparing nanocomposite and epoxy resin. Secondly, the thermodynamic characteristics of the nanocomposite including variations in glass transition temperatures (Tg) were determined. The Tg of fully cured material was shown to increase with the amount of nanoparticles. This observation was in conflict with the previous idea that TiO_2 nanoparticles only act as obstacles since this should result in a decrease of Tg. The thermal characteristics of the addition of TiO_2 therefore suggest that interactions between nanoparticles and polymer chains exist. Thirdly the physical properties of the material were followed, namely changes in storage and loss shear modulus as well as the creep and stress relaxation behavior dependence of nanocomposites as a function of nanoparticle content. Frequency sweeps at constant stress were performed at different temperatures. Results indicate that G' and especially G'' show a larger stability of the nanocomposite compared to the epoxy resin. Creep and relaxation measurements showed that elastic fluctuations were almost eliminated and recovery to stress was much faster in nanocomposite materials. These observations were attributed to the hypothesis that nanoparticles migrate into preferred locations driven by reduced chemical potential and strongly influence the behavior of the material. The mechanism of this influence may be through the chemical binding of TiO_2 with some part of the polymer chains. The added interactions may help store the energy necessary for recovery. The release of this extra stored energy may enable the relaxation of the material in a more rapid time scale. Hence some chemical interactions may be present which are only observable under conditions of high temperature.

Acknowledgements The partial support of the Welch Foundation (grant V-0004) is gratefully acknowledged.

References

1. Lebrum L, Bruzaud S, Grohens Y, Langevin D (2006) Elaboration and characterisation of PDMS-HTiNbO$_5$ nanocomposite membranes. Eur Polymer J 42:1975–1985
2. Kotsilkova (2006) Rheological approach to prognostic design of polymer nanocomposites. In: Proceedings of the 11th international conference, Sofia, Bulgaria, October 2–4, 2006
3. Sangermano M, Malucelli G, Amerio E, Bongiovanni R, Priola A, Gianni AD, Voit B, Rizza G (2006) Preparation and characterization of nanostructured TiO$_2$/epoxy polymeric films. Macromol Mater Eng 291:517–523
4. Lin J, Siddique JA, Ottenbrite RM (2001) Surface modification of inorganic oxide particles with silane coupling agent and organic dyes. Polym Adv Technol 12:285–292
5. Cossagneau T, Fendler JH, Johnson S, Mallouk TE (2000) Self-Assembled diode junctions prepared from a ruthenium tris(bipyridyl) polymer, n-Type TiO$_2$ nanoparticles, and graphite oxide sheets. Adv Mater 12:1363–1366
6. Chang CC, Chen WC (2001) High-refractive-index thin films prepared from aminoalkoxysilane-capped pyromellitic dianhydride–titania hybrid materials. J Polym Sci Part A Polym Chem 39:3419–3427
7. Ray SS, Okamoto M (2003) Polymer/layered silicate nanocomposites: a review from preparation to processing. Prog Polym Sci 28:1539–1641
8. Zeng C, Lee LJ (2001) Poly(methyl methacrylate) and polystyrene/clay nanocomposites prepared by in-situ polymerization. Macromolecules 34: 4098–4103
9. Su S, Wilkie CAJ (2003) Exfoliated poly(methyl methacrylate) and polystyrene nanocomposites occur when the clay cation contains a vinyl monomer. Polym Sci Part B Polym Phys 41:1124–1135
10. Gao D, Heimann RB, Williams MC, Wardhaugh LT, Muhammad M (1999) Rheological properties of poly(acrylamide)-bentonite composite hydrogels. J Mater Sci 34:1543–1552
11. Wheeler A (1958) Poly(vinyl alcohol)/clay nanocomposites. U.S. Patent 2, 847, 391
12. Zax DB, Yang DK, Santos RA, Hegmann H, Giannelis EP, Manias E (2000) Dynamical heterogeneity in nanoconfined poly(styrene) chains. J Chem Phys 112:2945–2951
13. Gelfer MY, Hyun HS, Liu L, Hsiao BS, Rafailovich M, Si M, Zaitsev VJ (2003) Effects of organoclays on morphology and thermal and rheological properties of polystyrene and poly (methyl methacrylate) blends. Polym Sci Part B Polym Phys 41:44–54
14. Somwangthanoroj A, Lee EC, Solomon MJ (2003) Early stage quiescent and flow-induced crystallization of intercalated polypropylene nanocomposites by time-resolved light scattering. Macromolecules 36:2333–2342
15. Lele A, Mackley M, Galgali G, Ramesh C (2002) In situ rheo-x-ray investigation of flow-induced orientation in layered silicate—syndiotactic polypropylene nanocomposite melt. J Rheol 46:1091–1110
16. Rong J, Jing J, Li H, Sheng M (2001) A polyethylene nanocomposite prepared via in-situ polymerization. Macromol Rapid Commun 22:329–334
17. Alexandre M, Dubois P, Sun T, Graces JM, Jerome R (2002) Polyethylene-layered silicate nanocomposites prepared by the polymerization-filling technique: synthesis and mechanical properties. Polym J 43:2123–2132
18. Dabrowski F, Bourbigot S, Delbel R, Bras ML (2000) Kinetic modelling of the thermal degradation of polyamide-6 nanocomposite. Eur Polym J 36:273–284
19. Hyun YH, Lim ST, Choi HJ, Jhon MS (2001) Rheology of poly(ethylene oxide)/organoclay nanocomposites. Macromolecules 34:8084–8093
20. Chen HW, Chang FC (2001) The novel polymer electrolyte nanocomposite composed of poly (ethylene oxide), lithium triflate and mineral clay. Polym J 42:9763–9769
21. Manna AK, Tripathy DK, De PP, De SK, Chatterjee MK, Pfeiffer DG (1999) Bonding between epoxidized natural rubber and clay in presence of silane coupling agent. J Appl Polym Sci 72:1895–1903

22. Vu YT, Mark JE, Pham LH (2001) Clay nanolayer reinforcement of cis-1,4-polyisoprene and epoxidized natural rubber. J Appl Polym Sci 82:1391–1403
23. Xu K, Chen M, Zhang K, Hu J (2004) Synthesis and characterization of novel epoxy resin bearing naphthyl and limonene moieties, and its cured polymer. Polymer 45:1133–1140
24. Lin B, Gelves GA, Haber JA, Sundararaj U (2007) Electrical, rheological, and mechanical properties of polystyrene/copper nanowire. Ind Eng Chem Res 46:2481–2487
25. Karayannidou EG, Achilias DS, Sideridou ID (2006) Cure kinetics of epoxy—amine resins used in the restoration of works of art from glass or ceramic. Eur Polym J 42:3311–3323
26. Wu P, Siesler HW (2003) Water diffusion into epoxy resin: a 2D correlation ATR-FTIR investigation. Chem Phys Lett 374:74–78
27. Guo Q (1995) Effect of curing agent on the phase behaviour of epoxy resin/phenoxy blends. Polymer 36(25):4753–4760
28. Akinyede O, Mohan R, Kelkar A, Sankar J (2009) Static and fatigue behavior of epoxy/fiberglass composites hybridized with alumina nanoparticles. J Compos Mater 43 (7):769–781
29. Zhang W, Lu W, Wang S, Zhou H (2003) Influence of dipole interactions on mechanical behavior of modified epoxy resins. Polym J 35:470–475
30. Yasmin A, Luo JJ, Abot JL, Daniel IM (2006) Mechanical and thermal behavior of clay/epoxy nanocomposites. Compos Sci Technol 66(14):2415–2422
31. Knauert ST, Douglas JF, Starr FW (2007) The effect of nanoparticle shape on polymer-nanocomposite rheology and tensile strength. J Polym Sci Part B Polym Phys 45:1882–1897
32. Mohomed K, Gerasimov TG, Abourahma H, Zaworotko MJ, Harmon JP (2005) Nanostructure matrix interactions in methacrylate composites. Mater Sci Eng A 409:227–233
33. Oh H, Green PF (2009) Polymer chain dynamics and glass transition in athermal polymer/nanoparticle mixtures. Nat Mater 8(2):139–143
34. Bernazzani P, Sanchez RF (2008) Effect of substrate interactions on the melting behavior of thin polyethylene films. Eur Phys J Soft Matter 26:427–434
35. Bernazzani P, Pandi HKR, Peyyavula VK (2008) Structural changes associated with interactions between starch and particles of TiO_2 or ZnSe. Sci J Chem Biochem Mol Biol 2:1–13
36. Guo Z, Ng HW, Yee GL, Hahn HT (2009) Differential scanning calorimetry investigation on vinyl ester resin curing process for polymer nanocomposite fabrication. J Nanosci Nanotechnol 9:3278–3285
37. Koo JH (2006) Polymer nanocomposites: processing, characterization, and applications. McGraw-Hill, New York
38. Guo Z, Kim TY, Lei K, Pereira T, Sugar JG, Hahn HT (2008) Strengthening and thermal stabilization of polyurethane nanocomposites with silicon carbide nanoparticles by a surface-initiated-polymerization approach. Compos Sci Technol 68:164–170
39. Zhang X, Simon LC (2005) In situ polymerization of hybrid polyethylene-alumina nanocomposites. Macromol Mater Eng 290:573–583

A Study on the Size and Type of Inclusions in Si–Mn Combined Deoxidated Low Carbon Steel Strip

Ting Wang, Wenqiang Bao, Shaobo Zheng, Qijie Zhai and Huigai Li

Abstract The type of inclusions with different sizes in as-cast low carbon steel strip which is deoxidized by Si–Mn has been characterized by scanning electron microscope (SEM) and transmission electron microscopy (TEM). SEM observation and analysis suggested that the type of inclusions larger than 1 μm is manganese silicate. TEM analysis was carried out to investigate the composition and structure of nano-inclusions. Inclusions with diameter about 100 nm are manganese silicate, MnS–Si(Mn)O_x complex particles and isolated cubic Al_2O_3. Furthermore, HREM observations have revealed that there were two kinds of ultrafine particles in size range of 5–10 nm. One is spherical monoclinic Cu_2S; the other is silicon-iron oxide particles with morphology of long strip. In addition, some thermodynamic calculations were performed to explain the corresponding relation between the type of inclusions and their size.

Keywords As-cast steel strip · Si–Mn combined deoxidation · Nano-inclusions · TEM · HREM

Introduction

Low carbon steel has relatively low strength and good plasticity. The relatively low strength of low carbon steel limits its wide application. Thin-strip casting is a technique in which strip (2–3 mm thickness) is cast directly from liquid steel to a final production of strip steels with very fast cooling rates (10^2–10^3 K/s) [1, 2]. So far, the technology has already used for production of low carbon steel [3]. The strength-elongation balance of as-cast low-carbon steel strip can be comparable to that of strip by the conventional process with rolling and heat treatment, which can

T. Wang · W. Bao · S. Zheng · Q. Zhai · H. Li (✉)
State Key Laboratory of Advanced Special Steel, Shanghai Key Laboratory of Advanced Ferrometallurgy, School of Materials Science and Engineering, Shanghai University, Shanghai 200072, China
e-mail: lihuigai@shu.edu.cn

S. Ikhmayies et al. (eds.), *Characterization of Minerals, Metals, and Materials 2017*, The Minerals, Metals & Materials Series,
DOI 10.1007/978-3-319-51382-9_18

refine the microstructure. Inclusion is the important factor that affect the properties and surface quality of steel strip. Inclusions in the steel strips are distributed in a diameter of less than 5 μm due to the fast cooling rate [4]. Finely dispersed inclusions are beneficial in mechanical, chemical and thermal properties of steel strip. Liu et al. [5] concluded that a large number of nano-scale oxide and sulfide located at the grain boundaries can play an important role in the grain refinement of low carbon steel. Intragranular acicular ferrite which nucleated at inclusions with featured sizes and types is known to provide a good combination of high strength and toughness [6, 7]. For these purposes, the controls of particle size and type are of crucial importance to obtain the strip with excellent property.

Inclusions with different sizes are corresponding to different types in the same steel. In this paper, the intrinsic relation between factors on steel strip by Si–Mn combined deoxidation was studied with SEM and TEM. Some thermodynamic calculations were performed to explain the corresponding relationship between the types of inclusions and their size.

Experiment Produce

Industrial pure iron was melted in induction melting furnace. 80% ferromanganese and 75% ferrosilicon are sequentially added to the melt for deoxidization and composition adjustment. A steel strip with 2.5 mm thickness were prepared by copper mold suction casting method. The major components such as Si, Mn and Ti were analyzed by an optical emission spectrometry (PMI-MASTER). In addition, the levels of total oxygen were analyzed via inert gas fusion-infrared absorptiometry (TC-436, LECO). The chemical composition of the as-cast steel is shown in Table 1.

Two plates with size of 10 × 5 × 2.5 mm^3 were cut from the as-cast strip and polished with typical metallographic polishing routine, and analyzed with a SEM (JSM-6700F) to identify the type of micro-inclusions. The other plate was prepared with carbon replica method for crystal-structure observation of inclusions. Nickel net was used to scoop the carbon film containing inclusions. Finally, the inclusions on carbon replica were characterized with JEM-2010F TEM equipped with an energy-dispersive X-ray spectrometer (EDS). The composition of inclusions was analyzed by EDS and the crystal-structures were determined by selected area electron diffraction patterns (SADPs) and the FFT (Fast Fourier Transform) patterns of HREM lattice images.

Table 1 The chemical composition of the steel strip (wt%)

Element	$[O]_{Tot}$	$[O]_{Sol}$	C	Si	Mn	Al	Cu	P	S	Fe
Content	0.044	0.007	0.037	0.078	0.28	0.0036	0.0034	0.016	0.02	Balance

Results

Inclusions Larger Than 1 μm

Inclusions with diameter larger than 1 μm were observed by SEM. The morphologies, size and composition of typical particles are shown in Fig. 1. Four inclusions with size of 5, 3, 1.2, and 1 μm are found to be manganese silicate (Mn–Si–O) by the analysis of EDS spectra as shown in Fig. 1c. The atomic ratio between Mn and Si is about 1:16 without absorption correction. Based on the SiO_2–MnO binary phase diagram, the oxide contained tridymite and a small quantity of manganese silicate.

Inclusions with Diameter About Tens to Hundreds Nanometers

Some inclusions with size range of tens to hundreds nanometers were apparent at low magnification in many regions as shown in Fig. 2a. It can be found that they

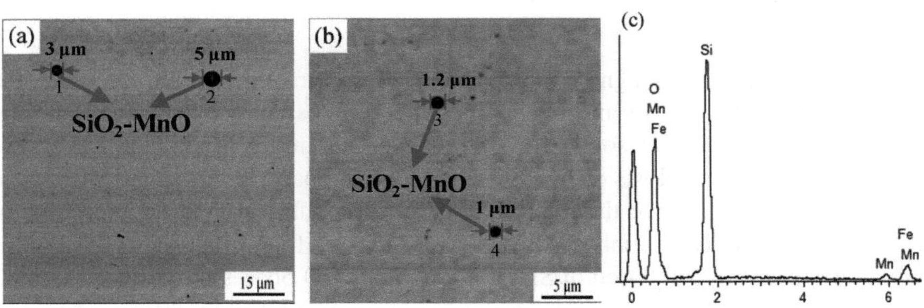

Fig. 1 SEM analysis for four inclusions with a size more than 1 μm: **a**, **b** the morphology of inclusions; **c** the composition of inclusions

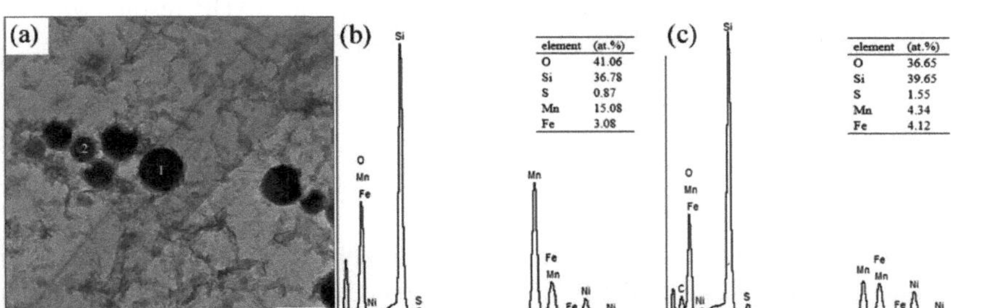

Fig. 2 **a** TEM photograph of inclusions with size range of hundreds nanometer, **b**, **c** EDS spectra showing the distribution of elements of inclusions 1 and 2, respectively

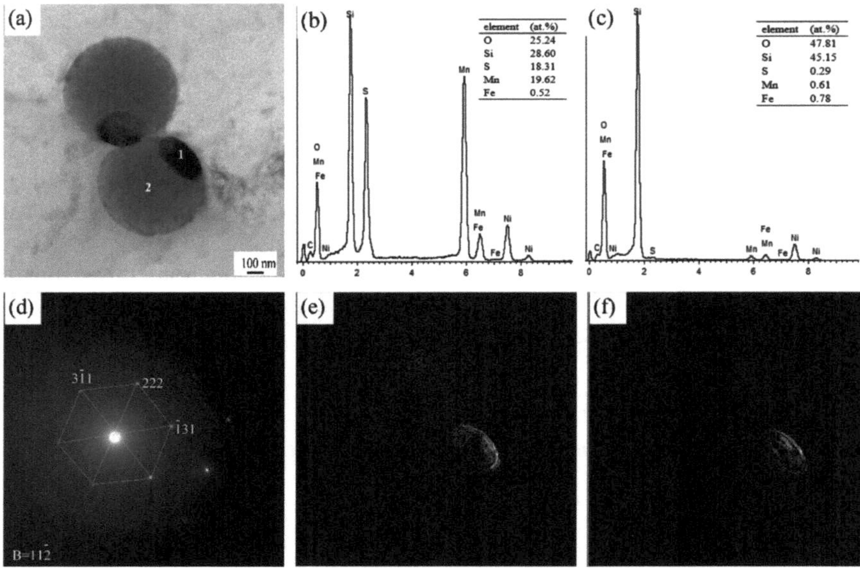

Fig. 3 Compound inclusion analysis by TEM: **a** BF images of inclusions; **b** EDS spectra of 1; **c** EDS spectra of 2; **d** SADP of the White block area of (**a**); **e**, **f** DF images

are spherical at such magnification. The EDS spectra in Fig. 2b, showing the presence of Si, O, S, Mn and Fe, implies the composition of inclusions may be silicon-manganese salt or solid SiO_2 and a small amount of MnS. Figure 3 shows the results of TEM analysis of two typical compound particles. A spherical particle could be divided into two different regions by their different contrast presented in Fig. 3a. From the EDS analysis, Fig. 3b, c, it can be found that the typical compound particles are composed of Si–O–S–Mn and Si–O phases for particles 1 and 2 (marked in Fig. 3a), respectively. The SADP of the Si–O–S–Mn phase is shown in Fig. 3d, indicating that it is cubic MnS. Figure 3e and f are the dark field images when trapping spots of (d), confirming the phase is cubic MnS. Figure 4 represents a typical particle in size of 50 nm. Comparing EDS and SADP results with the crystallographic data, it could be concluded that the Al oxide is cubic Al_2O_3.

Ultrafine Particles

The composition and crystal structure of particles with a diameter less than 10 nm was difficult to study due to the limit of test methods and the few fraction in the matrix. In this study, carbon extraction replica method was used to prepare the TEM samples by extracting inclusions on the carbon film.

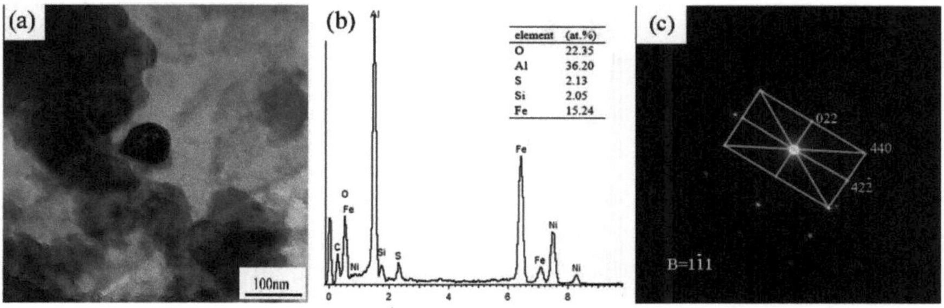

Fig. 4 Al_2O_3 in low carbon steel strip: **a** BF images of inclusions; **b** EDS spectra; **c** SADP of the particle

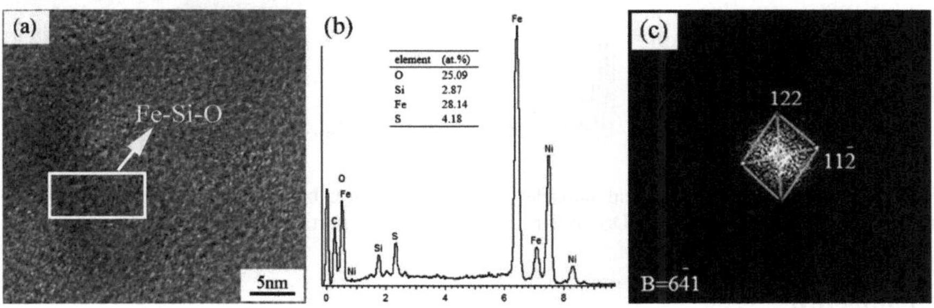

Fig. 5 Analysis of an anisometric ultra-fine particle: **a** HREM image; **b** EDS spectra; **c** FFT image of particle

The HREM image of an isometric ultra-fine particle with a length of 10 nm is presented in Fig. 5a. From EDS spectra, Fig. 5b, it is found that the particle is mainly composed of Fe–O–S, and considerably little content of Si is detected. This result is in good agreement with the previous report [4]. The FFT analysis in Fig. 5c reveals that it have an orthogonal structure Fe_2O_3, observed along the [6$\bar{4}$1] axis.

The characterization of other two particles with a diameter around 5 nm is presented in Fig. 6. It is evident that the two particles marked in this figure were of spherical, which is significantly different from the particle appeared in Fig. 5. The EDS spectra show that both of two particles are composed of Cu–S–Fe–Si–O. The content of S is decreased with the decrease of Cu. The FFT analysis of the two particles in Fig. 6c and f reveals that this kind of spherical particles should be Cu_2S which have a monocline structure, observed along the [721] axis.

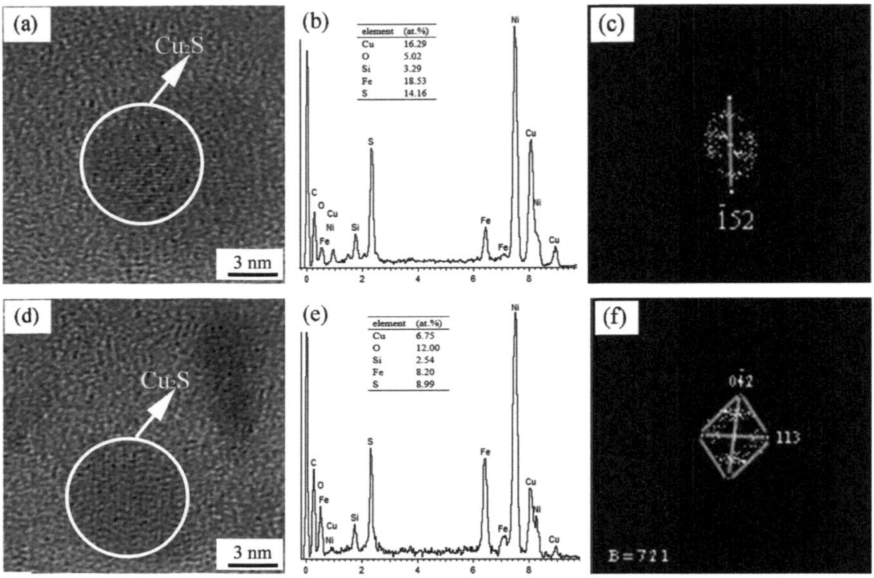

Fig. 6 Analysis of two spherical particles: **a** HREM image; **b** EDS spectra; **c** FFT image of particle 1; **d** HREM image; **e** EDS spectra; **f** FFT image of particle 2

Discussion

The types of inclusions with different size are different. According to the early reports [8, 9], inclusions larger than 1 μm formed during refining in the ladle before solidification. In the case of Si–Mn combined deoxidation, ferromanganese and ferrosilicon were sequentially added to the melt. According to their oxide-forming capacity, oxides originating during deoxidation represented by the following reactions:

$$[\mathrm{Mn}] + [\mathrm{O}] = \mathrm{MnO} \tag{1}$$

$$\mathrm{Si} + 2\mathrm{MnO} = 2\mathrm{Mn} + \mathrm{SiO_2} \tag{2}$$

$$K_{\mathrm{Mn,Si}} = \frac{[\%\mathrm{Mn}]^2}{[\%\mathrm{Si}]} * \frac{a_{\mathrm{SiO_2}}}{(a_{\mathrm{MnO}})^2} \tag{3}$$

$$\log K_{\mathrm{Mn,Si}} = \frac{1510}{T} + 1.27 \tag{4}$$

where a is activity, it depends on the interaction coefficient between the elements and the solute content. $a_{\mathrm{SiO_2}}$ and a_{MnO} are the activity of pure solid oxide, and equal to 1. For a given temperature, the deoxidation product is solid $\mathrm{SiO_2}$ when

$[\%Mn]^2/[\%Si]$ exceeds the critical value. These inclusions formed in the melt have enough time for growth, hence, they performed larger size more than 1 μm.

The precipitation temperature of various inclusions is important for judging precipitation sequence and their growth time. A series of thermodynamic calculations can provide a great reference. The liquidus temperature (T_L) and the solidus temperature (T_S) can be calculated by Eqs. (5) and (6), respectively.

$$\begin{aligned} T_L = 1811 - \{&55[C] + 80[C][C] + 13[Si] + 4.8[Mn] \\ &+ 1.5[Cr] + 4.3[Ni] + 30[P] + 30[S]\} \end{aligned} \tag{5}$$

$$\begin{aligned} T_S = 1495 - \{&20.5[Si] + 6.5[Mn] + 2.0[Cr] + 11.5[Ni] \\ &+ 500[P] + 700[S] + 5.5[Al]\} \end{aligned} \tag{6}$$

The Gibbs free energy of the reaction (Eq. 7) in the liquid steel can be calculated with Eqs. (8) and (9). Considering the segregation of solute during solidification, the concentration of element is a function of initial concentration (C_0), solidification fraction (f_s) and equilibrium distribution coefficient (k). The standard Gibbs free energy for the reactions in the liquid steel and solubility product of sulfide are listed in Table 2.

$$\mathrm{x}[M] + \mathrm{y}[N] = (M_xN_y) \tag{7}$$

$$\Delta G = \Delta G^0 + RT\ln K \tag{8}$$

$$K = \frac{a_{M_xN_y}}{a^x_{[M]}a^y_{[N]}} = \frac{a_{M_xN_y}}{(f_M \cdot [\%M])^x \cdot (f_N \cdot [\%N])^y} \tag{9}$$

$$C = C_0 \cdot (1 - f_s)^{(k-1)} \tag{10}$$

Figure 7a shows the Gibbs free energy-temperature curves of SiO_2, MnO, Al_2O_3 and MnS. It can be found that precipitation temperature of Al_2O_3 is above 2000 K, which is really higher than the liquidus temperature (T_L = 1805 K). Above the T_L, the Gibbs free energy of other three compounds are still greater than 0, so the SiO_2, MnO and MnS cannot precipitate before solidification. Figure 7b shows the Gibbs free energy values the solid fraction during solidification. It can be known that the

Table 2 The Gibbs free energy (ΔG^0) of reactions and solubility product of inclusions [10, 11]

Equation for the reaction	ΔG^0 (J/mol)	Solubility product
[Si] + 2[O] = SiO_2	−591990 + 229.4 T	–
[Mn] + [O] = MnO	−244316 + 106.84 T	–
2[Al] + 3[O] = Al_2O_3	−1225000 + 393.8 T	–
[Mn] + [S] = MnS	−167000 + 88.68 T	lg([Mn] [S]) = 2.929 − 9200/T
2[Cu] + [S] = Cu_2S	–	lg($[Cu]^2$ [S]) = 6.082 − 8966/T

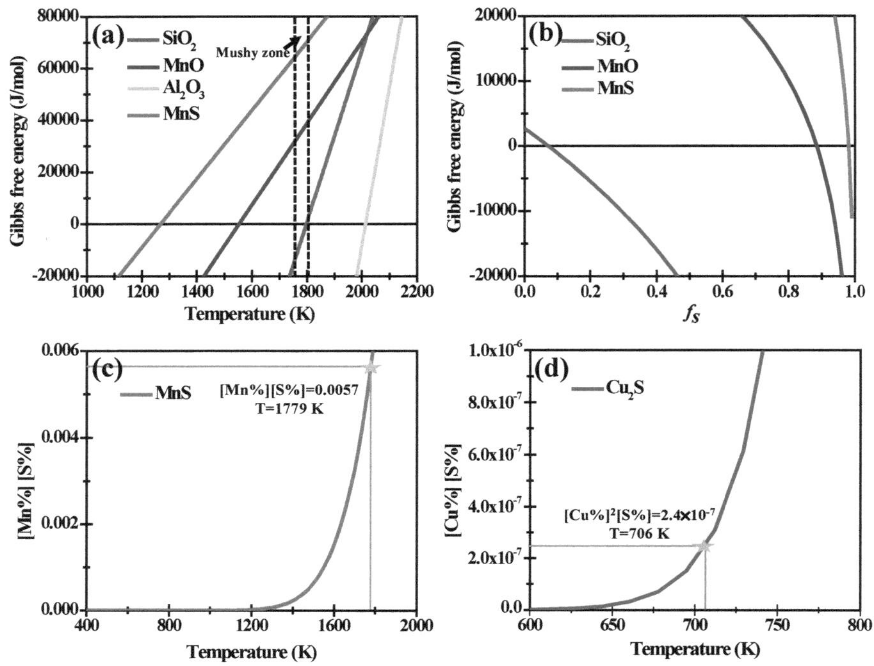

Fig. 7 The results of thermodynamic calculations: **a** the Gibbs free energy-temperature curves before solidification; **b** the Gibbs free energy versus the solid fraction during solidification; **c**, **d** the solubility product of MnS, Cu_2S-temperature curves, respectively

precipitation order is from SiO_2, MnO to MnS during solidification. The precipitation of MnS starts from the end of solidification, and the precipitation temperature is 1779 K (approach Ts = 1767 K) as shown in Fig. 7c. MnS can be precipitated on oxide in the end of solidification to form complex inclusion with larger size. From Fig. 7d, the precipitation temperature of Cu_2S is 706 K. Obviously, there is no enough time for growth of Cu_2S.

The formation of ultrafine particles containing Si, Fe and O, can be attributed to the processing technology. The high total oxygen and the very high cooling rate may result in the precipitation of such particles.

Conclusion

(1) Inclusions larger than 1 μm are manganese silicate, they are the deoxidation product formed in the melt.

(2) Inclusions with tens to hundreds nanometers are oxide (Al_2O_3, SiO_2, MnO) and complex inclusion (MnS distributed on part of the surface of oxide). Oxides have higher precipitation temperature than sulfide.
(3) Ultrafine particles (few nanometers) are Cu_2S and silicon-iron oxide. The precipitation temperature of Cu_2S is 706 K.

Acknowledgements This work was supported by 973 Project (No. 2011CB012902) and National Natural Science Foundation of China (No. U1460103). We express our grateful thanks to the Instrumental Analysis & Research Center of Shanghai University for the instrument supports.

References

1. Xiong ZP, Kostryzhev AG, Stanford NE (2016) Effect of deformation on microstructure and mechanical properties of dual phase steel produced via strip casting simulation. Mater Sci Eng, A 651:291–305
2. Ge S, Chang S, Wang T, Calzado LE, Isac M, Kozinski J, Guthrie RI (2016) Mathematical modeling and microstructure analysis of low carbon steel strips produced by horizontal single belt casting (HSBC). Metall Mater Trans B 47:1893–1904
3. Edelman DG, Campbell PC, Killmore CR, Carprnter KR, Kau HR, Williams JG, Blejde WN (2009) Recent developments with ultra-thin cast strip products produced by the CASTRIP® process. Iron Steel Technol 6:47–58
4. Killmore CR, Phillips A, Kaul HR, Campbell P, Schueren M, Blejde W (2008) Development of ultra-thin cast strip products by the CASTRIP process. Mater Forum 32
5. Liu DL, Wang YL, Huo XD, Chen NJ, Shao WR, Fu J, Kang YL (2002) Grain refinement and strengthening of low carbon steel by CSP technology. Acta Metall Sin 38:647–651
6. Gregg JM, Bhadeshia H (1997) Solid-state nucleation of acicular ferrite on minerals added to molten steel. Acta Mater 45:739–748
7. Byun JS, Shim JH, Cho YW, Lee DN (2003) Non-metallic inclusion and intragranular nucleation of ferrite in Ti-killed C-Mn steel. Acta Mater 51:1593–1606
8. Yu D, Barbaro FJ, Chandra T, Dunne DP (1996) Effect of large particles and fine precipitates on recrystallization and transformation behaviour of Ti treated low carbon TiO steel. ISIJ Int 36:1055–1062
9. Ohta H, Suito H (2006) Effects of dissolved oxygen and size distribution on particle coarsening of deoxidation product. ISIJ Int 46:42–49
10. Huang XG (2008) Iron and steel metallurgy. Metallurgical Industry Press, Beijing
11. Yong QL (2006) The second phase in iron and steel materials. Special Steel Branch of the Chinese Society of Metals, Beijing

Effect of Argon Gas Purging of Spark Plasma Sintered ZrB_2+SiC Nano-Powder Composites

Naidu Seetala, Owen Reedy, Lawrence Matson, HeeDong Lee and Thomas Key

Abstract Spark Plasma Sintering (SPS) consolidated ZrB_2+SiC composites using nano-powders (around 40 nm) showed smaller grains compared to those using micron size powders and segregation of SiC into islands is minimal but with higher oxidation of ZrB_2 to form ZrO_2 in nano-composites. Argon-gas purging prior to SPS consolidation at around 2000 °C and 40 MPa of ZrB_2+20vol.%SiC nano-powders was used to minimize the oxidation and obtain fine granules with high densification. The densification of the Argon-gas purged nano-composites is higher compared to those consolidated without Argon gas purging. The EDX analysis showed a strong reduction in the oxygen peak for the Argon gas purged composites. The XRD spectra also support this observation with less ZrO_2 phase composition in Argon gas purged composites. The Vickers micro-hardness showed slightly lower values for Argon gas purged composites though they have higher densification.

Keywords Ultra-high-temperature ceramics · Spark plasma sintering · Nano-composites · Micro-hardness · Grain growth · Oxidation

N. Seetala (✉) · O. Reedy
Department of Mathematics and Physics,
Grambling State University, Grambling, LA 71245, USA
e-mail: naidusv@gram.edu

L. Matson
Air Force Research Laboratory, Wright-Patterson Air Force Base,
Dayton, OH 45433, USA

H. Lee · T. Key
UES, Inc., Dayton, OH 45432, USA

S. Ikhmayies et al. (eds.), *Characterization of Minerals, Metals, and Materials 2017*, The Minerals, Metals & Materials Series,
DOI 10.1007/978-3-319-51382-9_19

Introduction

Ultra-High-Temperature Ceramics (UHTCs), such as ZrB_2 and HfB_2 with incorporation of SiC nano-filler, are useful as structural materials for applications in propulsion and thermal protection systems (such as turbine-engine hot section components) and the leading edge of hypersonic vehicles where extremely high heat fluxes generate very high temperatures and steep temperature gradients [1–3]. Several synthesis and sintering routes were explored to obtain dense ZrB_2+SiC, HfB_2+SiC, and HfB_2+HfC+SiC UHTCs starting from composite powders [4–6]. We made a comparative study of spark plasma sintered composites ZrB_2-SiC and HfB_2-SiC made using coarse-powders (0.7–1.5 μm) and ~40 nm nano-powders at high temperatures and pressures [7]. Large segregates of SiC were observed in coarse-powder consolidated composites (coarse-composites) which were almost nil in the nano-powder consolidated composites (nano-composites). Nano-composites showed remarkable ZrB_2 oxidation during Spark Plasma Sintering (SPS), which is minimal for the coarse-composites. Here, we tried Argon-gas purging prior to SPS consolidation at around 2000 °C and 32–40 MPa of ZrB_2-SiC nano-powders to minimize the oxidation and obtain fine granular and high dense nano-composites.

Experimental

The coarse (0.7–1.5 μm) precursor powders (99.9% pure) of ZrB_2 and SiC were obtained from Electronic Space Products International (ESPI), Ashland, OR, and the nano-powders of ZrB_2 (43 nm hexagonal 99% pure) and SiC (40 nm cubic 99% pure) from US Research Nanomaterials, Inc., Houston, TX. For the coarse-powders and the first set of nano-powders we mixed ZrB_2 with 20 vol.%SiC using ultra-sonication and rotary evaporation, followed by spark plasma sintering to high temperatures (1800 or 2100 °C) and pressures (32 MPa) at Wright-Patterson Air Force Base, Dayton, OH. For the second set of nano-powders, the spark plasma sintering was performed at SPS NanoCeramics, LLC, Morton Grove, IL using Argon-gas purging prior to SPS sintering to high temperatures (1850 or 1950 °C) and pressures (35 or 40 MPa). We used SEM/EDXS for granular and elemental analysis, Archimedes method for density and open porosity measurements, and a micro-hardness tester for Vickers hardness. An FEI Quantum-2 SEM was used to analyze the particle sizes and structures of the precursor powders, and an EDX system was used to perform elemental analyses of the different types of granules found on spark plasma heat treated sample surfaces. Bruker D2 X-ray Diffractometer (XRD) was used to study phase compositions. Material Analysis Using Diffraction (MAUD) computer code was used to obtain the phase compositions from the XRD spectra. The Vickers Hardness Number (VHN) was measured using a micro-hardness tester with 4.9 N force and Dwell time of 20 s.

Results and Discussion

Some of the coarse-powder composites results have been presented previously [8]. Here, first we compare the coarse-composite results with nano-composites.

The SEM and EDX analysis of the spark plasma heat treated ZrB_2+20vol.%SiC coarse-composites is presented in Figs. 1 and 2. As seen in EDX spectrum Fig. 2b, the dark spots in Fig. 1a are SiC rich, and the remaining white regions (as magnified in Fig. 1b) have less SiC.

a) Overall area 37X mag. (Dark regions SiC rich) *b) White region 1,250X mag. (Av. grain size 9.7 μm)*

Fig. 1 SEM images of SPS sintered ZrB_2+20vol.%SiC coarse-composites

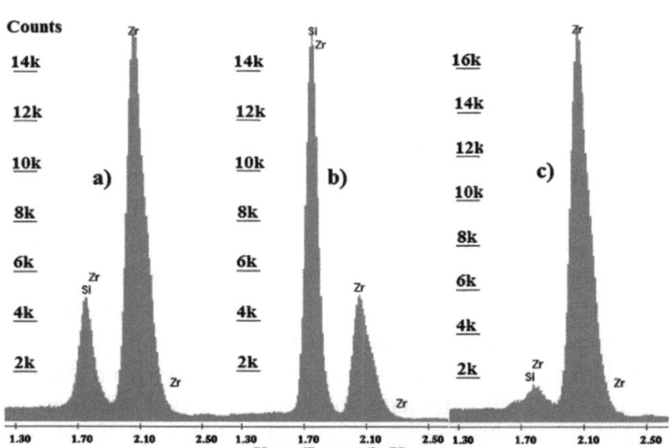

Fig. 2 EDX spectra for ZrB_2+20vol.%SiC coarse-composites shown in Fig. 1, **a** overall area, **b** dark spots, and **c** white area

a) Overall area 500X mag *b) Magnified 5,150X (Av. grain size 1.6 μm)* *c) Magnified 6,042X (Av. grain size 2.1 μm)*

Fig. 3 SEM images of SPS sintered ZrB_2+20vol.%SiC nano-composites: **a** and **b** are without Argon-gas purging, and **c** is with Argon-gas purging prior to SPS consolidation

The observations show that the ZrB_2+SiC coarse-composites have segregation of SiC into a few micron-sized dark spots, and the rest of the area is dominated by ZrB_2 with small amounts of SiC (2–6 wt%, varying from grain to grain). The average grain size is 9.65 μm, which is about 14 times the average particle size for 0.7 μm ZrB_2 precursor particles sintered at 2100 °C (note: the larger precursor particles may consist of few smaller grains).

The segregation of SiC is minimal in nano-composites as seen in Fig. 3 with an average grain size of 1.6 μm. This is about 37 times the grain growth of the 43 nm ZrB_2 particles sintered at 2100 °C. Lowering the sintering temperature to 1800 °C further reduced the grain growth to 575 nm ($\sim$13.4 times the grain growth of the ZrB_2+20vol.%SiC nano-composite), and 492 nm ($\sim$11.4 times the grain growth of the ZrB_2+15vol.%SiC nano-composite). Though the overall grain size is much smaller in nano-composites, the grain growth is not remarkably reduced at the same sintering temperature by moving from coarse- to nano-sized precursors.

The density (and the densification compared to theoretical density) and open porosity measured by the Archimedes method are presented in Table 1 for nano-composites and coarse-composites prepared by SPS sintering at 1800 or 2100 °C and 32 MPa pressure. One set of coarse-powder HfB_2+SiC composites is also included in the table. Since no supplier is available for nano-powders of HfB_2, we are unable to provide nano-powder HfB_2+SiC composite for comparison. The nano-composites show lower open-porosity ($\sim$0.1%) compared to coarse-powder composites ($\sim$9%). The nano-composites sintered at 1800 °C showed poor densification, but strangely have the lowest (0.005%) open-porosity.

Figure 4 shows a comparison between the X-ray Diffraction (XRD) spectra analysis for coarse- and nano-composites of ZrB_2+20vol.%SiC using Material Analysis Using Diffraction (MAUD) program [9]. XRD peak broadening results (table not included) corroborate with the SEM observations that the ZrB_2+SiC nano-composites have smaller grains compared to coarse-composites. The striking difference between the nano- and coarse-composites is the ZrO_2 phase, where nano-composites showed remarkable oxidation of ZrB_2 to form ZrO_2 during SPS

Table 1 Density and open porosity of nano- and coarse-composites. The errors in the measurements are below 5%

Sample	Final temp. (°C)	Final pressure (MPa)	Density (g/cm^3)	Open-porosity (%)
ZrB_2+20vol.%SiC-nano	2100	32	4.81 (87% th.)	0.113
ZrB_2+20vol.%SiC-nano	1800	32	4.18 (76% th.)	0.005
ZrB_2+15vol.%SiC-nano	1800	32	4.89 (75% th.)	0.001
ZrB_2+20vol.%SiC-coarse	2100	32	4.90 (89% th.)	9.73
HfB_2+20vol.%SiC-coarse	2100	32	8.82 (95% th.)	4.36

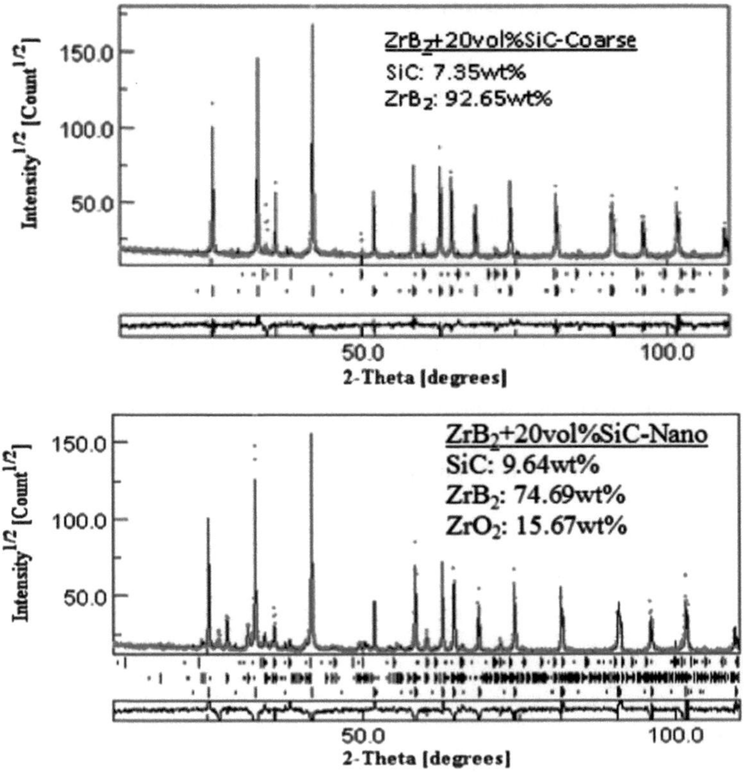

Fig. 4 XRD phase composition comparison between coarse- and nano-composites of ZrB_2+20vol.%SiC

consolidation (15.67 wt% as shown in Fig. 4). A good match of XRD peaks with the ZrO_2 phase is observed easily in nano-composites, but these ZrO_2 phase peaks are almost absent in the coarse-composites. Some additional peaks are observed in coarse-composites which could not be identified with any phase and did not match with ZrO_2 (or HfO_2) peaks.

MAUD analysis has also been performed on the XRD spectrum of ZrB_2 nanoparticles as supplied by the vendor and found no ZrO_2 phase, while a good match was found with 100% ZrB_2 phase. This suggests that the oxidation of ZrB_2 occurs during SPS (a small amount of trapped surface moisture could contribute to oxidation during SPS in the nano-powder case as the surface area is large). The EDX spectra were also compared between: (a) precursor ZrB_2 nanoparticles and (b) SPS consolidated ZrB_2+SiC nano-composite. The SPS consolidated ZrB_2+SiC nano-composite had a large oxygen peak at around 0.5 keV which was almost negligible in the precursor nanoparticle spectrum. This verifies that the oxidation of ZrB_2 occurred during the SPS consolidation. We made attempts to find ways to minimize the availability of oxygen to nano-particles during high temperature SPS consolidation.

In order to minimize the oxidation of ZrB_2 during SPS of nano-powders, a set of samples was SPS consolidated after purging the SPS chamber with Argon gas starting with vacuuming to 10 Pa, purging to 23 kPa with Argon gas, and following with vacuuming to 8 Pa prior to heat treatment.

Figure 3c shows the SEM viewgraph of ZrB_2+20vol.%SiC nano-composite purged with Argon-gas prior to SPS consolidation, which shows higher densification than unpurged nano-composite with average grain size of 2.1 μm compared to 1.6 μm for the unpurged composite. The densification of the nano-composites that were Argon-gas purged prior to SPS consolidation is higher at all SiC percentages compared to those consolidated without Argon gas purging (Table 2).

Comparing the Energy Dispersive X-ray (EDX) spectra for ZrB_2+20vol.%SiC nano-composites with and without Argon gas purging prior to SPS consolidation showed a strong distinction in the oxygen peak that is very much reduced for the sample SPS consolidated after purging with Argon gas (Fig. 5). The minimized

Table 2 Micro-hardness of ZrB_2-SiC nano-composites. The errors in the measurements are about 5%

Sample	Micro-hardness (VHN) (%Densification)	
	Without Ar_2-purging	With Ar_2-purging
ZrB_2+20vol.%SiC-nano SPS @ 1950 °C, 40 MPa		1242 (88.2%)
ZrB_2+20vol.%SiC-nano SPS @ 1850 °C, 35 MPa		1455 (86.2%)
ZrB_2+20vol.%SiC-nano SPS @ 2100 °C, 32 MPa	2304 (87.2%)	
ZrB_2+20vol.%SiC-nano SPS @ 1800 °C, 32 MPa	1815 (75.9%)	
ZrB_2+15vol.%SiC-nano SPS @ 1800 °C, 32 MPa	2094 (75.0%)	
ZrB_2+15vol.%SiC-nano SPS @ 1850 °C, 35 MPa		1819 (89.7%)
ZrB_2+15vol.%SiC-nano SPS @ 1850 °C, 35 MPa		1845 (92.2%)

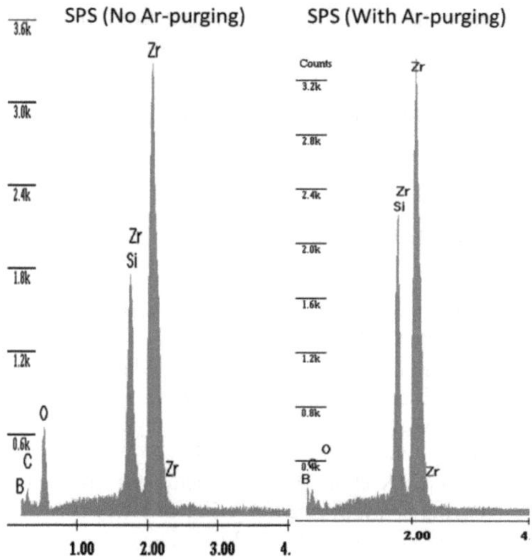

Fig. 5 EDX spectra of ZrB_2+20vol.%SiC nano-composites with and without Argon gas purging prior to SPS consolidation

oxidation of ZrB_2 might lead to the higher densification in Argon-gas purged nano-composites. The Argon-gas purged densification values are comparable with the coarse-composites which showed no oxidation of ZrB_2, and a higher densification compared to nano-composites SPS consolidated without Argon-gas purging.

The micro-hardness results of ZrB_2+SiC nano-composites with and without Argon-gas purging prior to SPS consolidation are presented in Table 2. The results indicate that the micro-hardness is slightly lower for Argon-gas purged nano-composites though they have higher densification compared to unpurged nano-composites. The reason could be the larger grain size ($\sim$2.1 μm) in purged nano-composites compared to the grains ($\sim$1.6 μm) in unpurged nano-composites [10], probably due to the differences in SPS parameters.

Conclusions

A comparative study is made between spark plasma sintered coarse- and nano-composites of ZrB_2+SiC made using ($\sim$1 μm) coarse-powders and ($\sim$40 nm) nano-powders. Large segregates of SiC are observed in coarse-composites those are almost nil in the nano-composites. Nano-composites showed remarkable ZrB_2 oxidation during SPS which is minimal for the coarse-composites. Argon-gas purging was used prior to SPS consolidation of ZrB_2-SiC nano-powders to minimize the oxidation. The EDX and XRD analysis showed that Argon-gas purging prior to SPS consolidation remarkably reduced the oxidation of ZrB_2 and resulted in higher densification of the nano-composite. The micro-hardness is slightly lower for Argon-gas purged nano-composites which may be as a result of larger size grains in purged nano-composites compared to the unpurged nano-composites, probably due to the differences in SPS parameters.

Acknowledgements This work is supported by Air Force Contract FA8650-13-C-5800.

References

1. Fahrenholtz WG et al (2007) Refractory diborides of zirconium and hafnium. J Am Ceram Soc 90:1347–1364
2. Squire TH, Marschall J (2010) Material property requirements for analysis and design of UHTC components in hypersonic applications. J Eur Ceram Soc 30(11):2239–2251
3. Van Wei DM et al (2004) The hypersonic environment: required operation conditions and design challenges. J Mater Sci 39(19):5915–5924
4. Enneti RK et al (2012) Taguchi analysis on the effect of process parameters on densification during spark plasma sintering of HfB_2-20SiC. Int J Refract Metal Hard Matter 31:293–296
5. Wang Z, Wu Z, Shi G (2010) The oxidation behaviors of ZrB_2-SiC-ZrC ceramic. Solid State Sci 13(3):534–538
6. Han J et al (2008) Oxidation-resistant ZrB_2-SiC composites at 2200 °C. Comp Sci Technol 68:799–806
7. Seetala N et al (2016) Spark plasma heat treated coarse- and nano-powder ZrB_2-SiC and HfB_2-SiC composites. TMS annual meeting & exhibition, Nashville, TN, 14–18 Feb 2016
8. Seetala NV, Webb M (2014) Spark plasma heat treated ZrB_2-SiC and HfB_2-SiC composites. Mater Sci Forum 783–786:1176–1782
9. Lutterotti L et al (1999) MAUD: a friendly Java program for material analysis using diffraction. IUCr Newsl CPD 21:14–15
10. Seetala N et al (2015) Densification and microhardness of spark plasma sintered ZrB_2+SiC nano-composites. Microsc Microanal 21(3):1051–1052

Part V
Alloys

Investigating the Anisotropic Behaviour of Lean Duplex Stainless Steel 2101

A.A.H. Ameri, J.P. Escobedo-Diaz, M. Ashraf and Md. Z. Quadir

Abstract The quasi-static mechanical response of hot rolled lean duplex stainless steel 2101 (LDSS 2101) in rolling, transverse and normal directions under compression loading has been investigated for its perceived anisotropic behaviour. Microstructural characterization is performed by optical microscopy and Electron Back Scatter Diffraction (EBSD) to qualitatively define the influence of the texture by using Taylor factor and grain sizes, and phase boundaries by utilizing the modified Hall-Petch formula. It was observed that LDSS2101 does not show anisotropic behaviour in terms of the mechanical strength but it has a clear anisotropy in the lateral strains; the microstructural explanation of this response may result from the phase boundaries strengthening.

Keywords LDSS2101 · Anisotropy · Microstructure · Compression

Introduction

Lean duplex stainless steel (LDSS2101) is a new type of duplex stainless steel (DSS) alloys offering an excellent combination of strength and corrosion resistance at a reduced cost compared with conventional duplex stainless steel DSS2205, for instance the alloy surcharge, which covers the cost of metals in stainless steel alloys, is about 70% higher for DSS2205 compare with LDSS2101 for this year as reported by Outokumpo® [1]. LDSS2101 is designed by partially substituting

A.A.H. Ameri (✉) · J.P. Escobedo-Diaz · M. Ashraf
School of Engineering and Information Technology,
The University of New South Wales, Canberra, ACT 2600, Australia
e-mail: Ali.Ameri@student.adfa.edu.au

Md.Z. Quadir
Faculty of Science and Engineering, Curtin University,
Perth, WA 6102, Australia

S. Ikhmayies et al. (eds.), *Characterization of Minerals, Metals, and Materials 2017*, The Minerals, Metals & Materials Series,
DOI 10.1007/978-3-319-51382-9_20

nickel with nitrogen, and molybdenum with manganese and chromium; this provides a more stable and affordable price tag for LDSS2101 making it more attractive than its other conventional counterparts [2–4]. The attractive mechanical properties of LDSS2101 results from its unique crystallography, which consists of two phases ferrite (α) and austenite (γ), and the proportion of these phases are approximately equal as in DSS2205 [5, 6]. However, at the present time, the usage of LDSS2101 is limited because this material is relatively new and its deformation mechanisms is complex [6, 7].

The manufacturing process (hot and cold rolling) of DSS causes changes in the orientation of both phases, which results an anisotropic behaviour [8]. Many studies [8–12] investigated the anisotropic behaviour of DSS2205 under tension and fatigue loading by testing samples parallel to: Rolling Direction (RD), Transverse Direction (TD) and 45° with respect to the rolling direction; it was reported that DSS has anisotropic mechanical properties in these directions, and TD is the strongest direction in terms of fatigue resistance, yield stress and ultimate tensile strength. Mateo et al. [9] and Hutchinson et al. [10] accounted that to the texture effect. Because the chemical composition of LDSS2101 is different from DSS2205 [13], it is not expected that these materials will have the same response under the same loading conditions.

Investigating the possible anisotropy for hot rolled LDSS2101 is important prior to moving to cold rolling process; to the best of our knowledge, there is no information about the mechanical anisotropy of LDSS2101 under compression. This paper aims to explore the behaviour of LDSS2101 in three directions (RD, TD and ND) under quasi-static condition, and presents microstructural analysis to justify the observed discrepancies in mechanical properties.

Materials and Experimental Methods

A 20 mm thick hot rolled LDSS2101 plate produced by Outokumpo® was used in this study. The plate was hot rolled, heat treated at 1050 °C and then quenched in water.

Samples were machined for compression test with loading axes parallel to three main directions (RD, TD and ND). Three sections of the plate were characterised, as shown in Fig. 1, by using Beraha etchant after preparing samples by following the standard metallographic procedure up to vibratory polishing with 0.02 μm non-crystallised colloidal silica suspension.

Compression samples in three directions had the same dimensions (7.5 mm diameter and 7.5 mm length), and three samples were tested for each direction at a quasi-static strain rate by using a Shimadzu universal testing machine.

EBSD was conducted on the three main planes of the as-received material by using field emission scanning electron microscope (FESEM), Zeiss UltraPlus and Carl Zeiss AURIGA, operated at 20 kV.

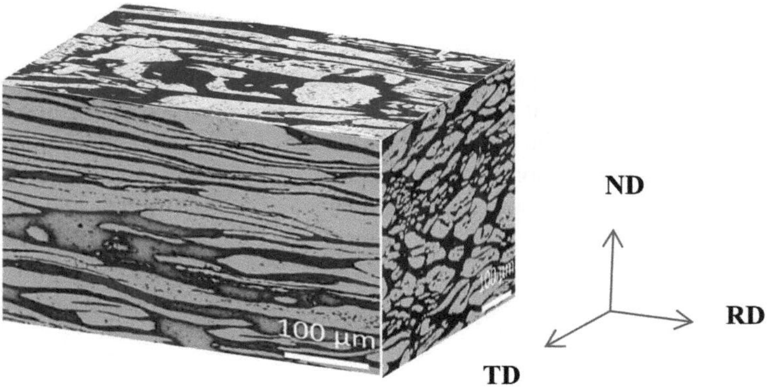

Fig. 1 Three dimensional representation of the microstructure of the studied LDSS2101, the *light colour* refers to austenite phase and the *dark colour* represents ferrite matrix (Color figure online)

Results and Discussion

Figure 2 and Table 1 show the experimental results of quasi-static compression tests. LDSS2101 shows negligible difference in the mechanical properties but RD shows some higher stress values at true strain larger than 20%. These results are consistent for all attempts which make it represents the material response, where all

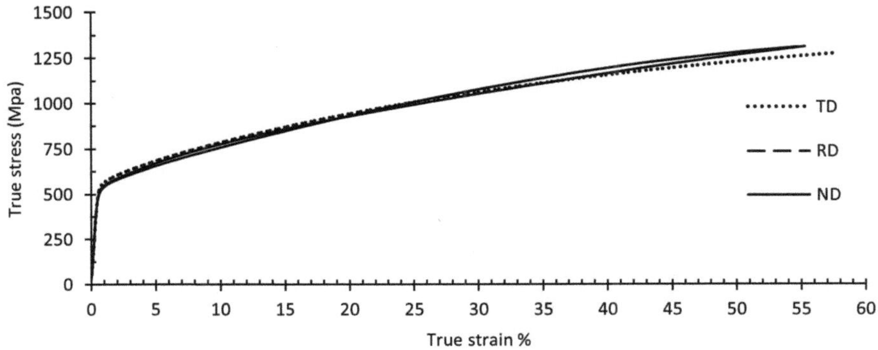

Fig. 2 True stress- true strain diagrams of three directions

Table 1 Compression tests results (average true stress at different true strains) for the studied LDSS2101 at different directions

Loading direction	$\sigma_{0.2}$ (MPa)	σ_4 (MPa)	σ_{25} (MPa)
RD	419 ± 8	640 ± 2	1027 ± 13
TD	398 ± 11	670 ± 1	1005 ± 11
ND	404 ± 12	655 ± 2	996 ± 9

tests were performed at the same conditions using the same lubricant (MOS2) and with the same sample geometry.

However, the cross sectional area of the samples experienced a clear anisotropic response, where the initial circular cross-sectional area was deformed into an elliptical section during the test as shown in Fig. 3. This response varies clearly between the tested directions where samples with loading axis parallel to RD experienced the highest anisotropic effect compared to those tested in TD and ND Fig. 4.

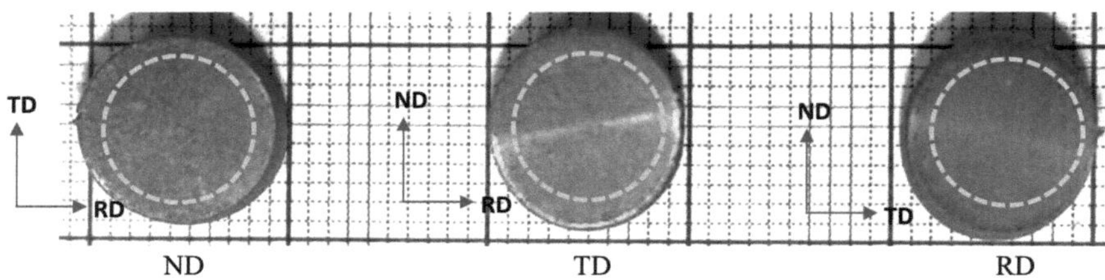

Fig. 3 The deformed samples at the tested directions; *red dotted circle* represents the original cross-sectional area (Color figure online)

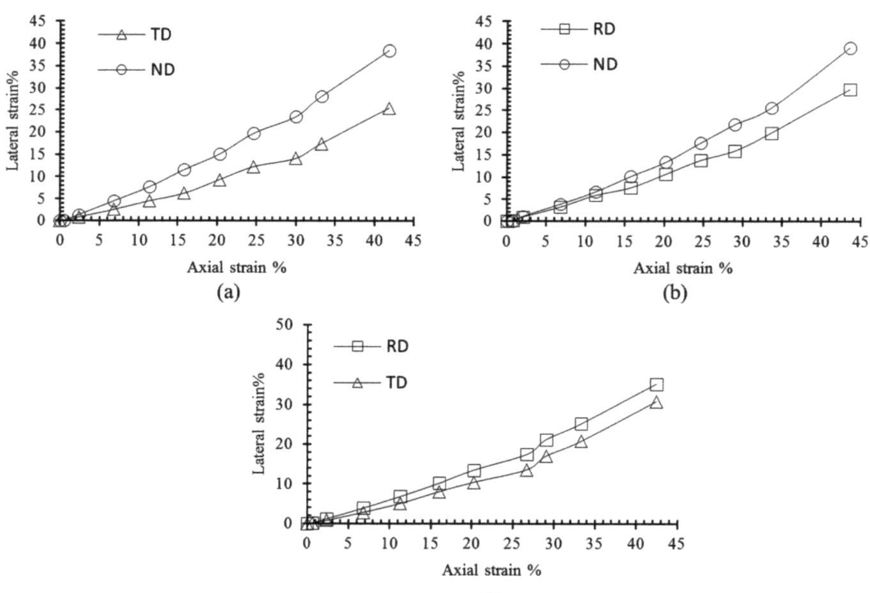

Fig. 4 Lateral strains of the cross sectional areal as a function of the axial strain for LDSS2101 when the load is applied in different directions **a** sample loaded in RD **b** sample loaded in TD **c** sample loaded in ND

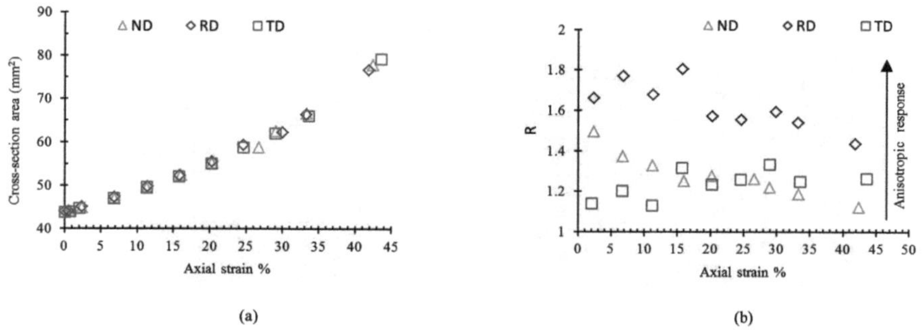

Fig. 5 **a** Cross sectional area of samples tested at different directions as a function of strain **b** R value for samples tested at different directions as a function of strain

To explain the change in lateral strains and the shape of the cross sectional area as a function of strain, the test was stopped at specific intervals to measure the changing dimensions of the cross section. Figure 4 shows the development of lateral strains of samples tested at different directions.

It is important to mention that this anisotropy at the cross section level was not visible in the flow stress (Fig. 2) because the cross sectional areas during the test for the all directions were similar as shown in Fig. 5a. Plastic strain ratio (R) is often used to express the anisotropy observed at the cross-sectional level and can be computed using Eq. (1) [14]:

$$R = \ln\frac{b}{d} \Big/ \ln\frac{a}{d} \quad (1)$$

where a, b are the short and long axes of the elliptical shaped sample and d is the initial diameter of the sample. R value was higher than one for all considered directions (R > 1 means the loaded area deforms anisotropically) but it was almost constant for each direction during the test as shown in Fig. 5b.

In order to understand the reason for the anisotropic response of a material, it is important to know the strengthening components of the material at micro-level; Eq. (2) shows the strengthening components of the flow stress [15]:

$$\sigma_f = M\tau_{CRSS} + \sigma_{ss} + \sigma_{pre} + M\alpha Gb\sqrt{\rho} + \sigma_g \quad (2)$$

where τ_{CRSS} is the critical shear stress, σ_{ss} is the strengthening from solid solution, σ_{prec} is the strengthening from precipitation effect, α is a constant, G is the shear modulus, b is Burger vector, and ρ is dislocation density; all of these quantities are independent for the loading direction of the sample. Whereas, Taylor factor M and grain boundary strengthening σ_g are direction dependent and hence they are the main factors causing the anisotropic behaviour. M and σ_g depend on grain orientation and grain size, respectively.

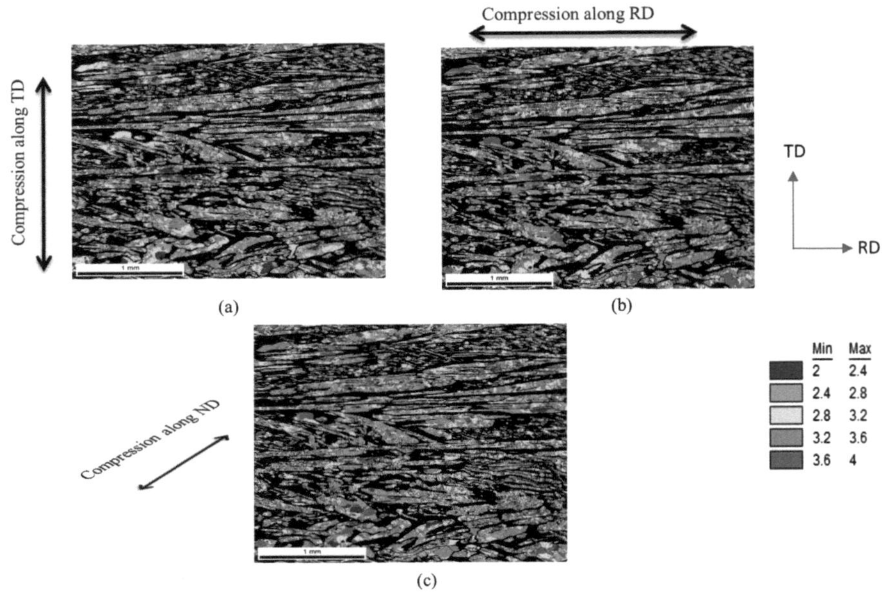

Fig. 6 Taylor maps of austenite phase for ND plane **a** compression in TD M = 3.03 **b** compression in RD M = 3.09 **c** compression in ND M = 3.08

Table 2 M values of the as-received material for both phases at different directions of compression loading

Phase	Loading direction	Plane			
		RD	TD	ND	Average value
Austenite	RD	3.07	3.09	3.09	3.08
	TD	3.11	3.09	3.03	3.08
	ND	3.13	3.04	3.08	3.08
Ferrite	RD	2.64	2.89	2.58	2.70
	TD	2.70	2.60	2.64	2.65
	ND	2.82	2.83	2.76	2.80

EBSD maps were used to define M values for all planes (RD, TD and ND) and for three directions of loading for each plane as shown in Fig. 6. Table 2 summarize the values of M for both phases of LDSS2101 for the main planes under compression loading in main directions. The average values of M for austenite phase are similar in all directions, and hence the texture effect of austenite phase can be neglected. Mateo et al. [9] accounted that to the behaviour of austenite islands during hot rolling, and reported that they only rotate inside ferrite matrix to be parallel to the rolling direction without changing grain orientations. On the other hand the variation of M for ferrite phase is considerable (about 12.00%), and therefore, texture might have an effect on its deformation [9].

Regarding the grain boundary strengthening, the modified Hall-Petch formula for dual phase material can be utilized, the yield stress of the material can be computed by Eq. (3) [16]:

$$\sigma_y^C = \sigma_y^\alpha f_{\alpha c} + \sigma_y^\gamma f_{\gamma c} + \sigma_y^{\alpha\gamma} \mathrm{F_s} \tag{3}$$

the contribution of grain and phase boundaries in the flow stress is represented by Eq. (4):

$$\sigma_\mathrm{g} = f_{\gamma c} \frac{\mathrm{K}_\mathrm{HP}^\gamma}{\sqrt{\mathrm{d}_\gamma}} + f_{\alpha c} \frac{\mathrm{K}_\mathrm{HP}^\alpha}{\sqrt{\mathrm{d}_\alpha}} + \mathrm{Fs} \frac{\mathrm{K}_\mathrm{HP}^{\alpha\gamma}}{\sqrt{\bar{\mathrm{d}}_{\alpha\gamma}}} \tag{4}$$

where $\mathrm{K}_\mathrm{HP}^\gamma$, and $\mathrm{K}_\mathrm{HP}^\alpha$ are Hall-Petch constants for austenite and ferrite (as single phases), $\mathrm{K}_\mathrm{HP}^{\alpha\gamma}$ is Hall-Petch constant for dual phase matrix, d $_\gamma$ and d $_\alpha$ are grain sizes of austenite and ferrite, $\bar{\mathrm{d}}_{\alpha\gamma}$ is a volume fraction-weighted grain size of dual phase matrix and $f_{\gamma c}$,$f_{\alpha c}$, Fs are defined in [16, 17].

Grain sizes for all planes for both phases (Table 3) were calculated from EBSD maps of the three planes. As the shape and orientation of grains are random, as shown in Fig. 7, utilizing intercept length approach gives better representation to the grain size effect than assuming grains are spheres and defining grain size by their diameter.

By comparing grain sizes in Table 3, it is clear that ND plane has the smallest grain size for both phases and the largest grain size belongs to RD plane. By considering the combined effect of texture strengthening and grain strengthening, loading the material in ND direction should have the highest flow stress in comparison to loading it in TD and RD but this is not the case as shown in Fig. 2. Moreover, the reason for the observed anisotropy in the lateral strains cannot be explained by using the results in Table 3. For instance, loading the sample in RD direction produce more lateral strain in ND than TD as shown in Fig. 4, but results in Table 3 show that the lateral strain in ND should be less than TD because grain sizes in ND are smaller. Phase boundary strengthening could be responsible for this observed discrepancy.

Dual phase materials are designed to get material properties better than single phase materials, and these good properties originate from the interaction between

Table 3 Grain size distribution of both phases using intercept length method

Phase	Intercept length of ND plane H: parallel to RD V: parallel to TD (μm)	Intercept length of TD plane H: parallel to RD V: parallel to ND (μm)	Intercept length of RD plane H: parallel to TD V: parallel to ND (μm)
Austenite	8.14(H)-3.40(V)	11.1(H)-6.00(V)	11.3(H)-8.1(V)
Ferrite	7.14(H)-4.62(V)	11.23(H)-6.35(V)	20.4(H)-8.6(V)

Fig. 7 Inverse pole figure of ferrite and austenite phases of ND plane

phases and this happens through phase boundaries [17, 18]. For LDSS2101, Liu et al. [18] explained that the initial plastic deformations take place at the soft phase (ferrite phase), and then transfer to the austenite phase through phase boundaries. Moreover, this inhomogeneous plastic deformation between the phases causes local plastic deformation at phase boundaries which forms Geometrically Necessary Dislocations (GNDs) [17]. Consequently, phase boundary strengthening plays a very important role in the deformation mechanism and the mechanical response of LDSS2101. The phase boundary effect for loading the material in TD direction happens mainly in ND direction because austenite phase aligned in RD as shown in Fig. 1 which makes the movement of dislocations in ferrite phase in ND easier than RD. This result from the fact that phase boundaries is weaker than grain boundaries of ferrite phase and this explains why the sample deforms in ND more than RD (Fig. 4) [17]. For the other two planes (RD and ND) explaining the effect of phase boundaries on the mechanical anisotropy is not easy because there is no uniform distribution for austenite islands inside the ferrite matrix as shown in Fig. 1, and using the modified Hall-Petch formula for dual phase is not suitable because it is not considering the effect of phase boundary orientations [16]. Therefore, more investigation is required to understand the contribution of phase boundaries on the anisotropic response of sample loaded in RD and ND directions.

Conclusions

This paper has discussed the mechanical response of LDSS2101 under compression loading and investigated the effects of microstructural parameters on the observed anisotropic behaviour of LDSS2101. The following conclusions can be drawn in the light of this study:

1. The flow stresses of LDSS2101 in all directions (RD, TD and ND) are almost similar.
2. The lateral strains of the loaded cross sectional areas showed a clear anisotropy where the lateral strain in ND is the highest while it is the lowest in TD. The strain ratio (R) is constant for each of the considered loading directions; the cross sectional areas for all loading directions were similar during the test which made the observed anisotropic behaviour not visible in the flow stresses.
3. The microstructural explanation for the anisotropic response of LDSS2101 is complex but it is envisaged that the phase boundaries could have played the main role in this response.

Acknowledgements The authors would like to thank Dr. Frank Brink and Dr. Hua Chen in the Centre for Advanced Microscopy at the Australian National University in conducting some of EBSD scans.

References

1. Outokumpo (2016) Stainless steel alloy surcharges Europe. [Online]. Available: http://www.outokumpu.com/en/pricing-aaf/stainless-steel-alloy-surcharges/Pages/default.aspx
2. Liljas M, Johansson P, Liu HP, Olsson COA (2008) Development of a lean duplex stainless steel. Steel Res Int 79(6):466–473
3. Strubbia R, Hereñú S, Marinelli MC, Alvarez-Armas I (2012) Short crack nucleation and growth in lean duplex stainless steels fatigued at room temperature. Int J Fatigue 41:90–94
4. Sieurin H, Sandström R, Westin EM (2006) Fracture toughness of the lean duplex stainless steel LDX 2101. Metall Mater Trans A 37(10):2975–2981
5. Bassani P, Breda M (2013) Characterization of a cold-rolled 2101 lean duplex stainless steel. Microsc Microanal 19(4):988–995
6. Strubbia R, Hereñú S, Alvarez-Armas I, Krupp U (2014) Short fatigue cracks nucleation and growth in lean duplex stainless steel LDX 2101. Mater Sci Eng A 615:169–174
7. Huang Y, Young B (2014) Experimental investigation of cold-formed lean duplex stainless steel beam-columns. Thin-Walled Struct 76:105–117
8. Fargas G, Akdut N, Anglada M, Mateo A (2008) Microstructural evolution during industrial rolling of a duplex stainless steel. ISIJ Int 48(11):1596–1602
9. Mateo A, Llanes L, Akdut N, Stolarz J, Anglada M (2003) Anisotropy effects on the fatigue behaviour of rolled duplex stainless steels. Int J Fatigue 25(6):481–488
10. Hutchinson W, Ushioda K, Runnsjö G (1985) Anisotropy of tensile behaviour in a duplex stainless steel sheet. Mater Sci Technol 1(August):728–736
11. Ul-Haq A, Weiland H, Bunge HJ (1994) Textures and microstructures in duplex stainless steel. Mater Sci Technol 10(4):289–298
12. Fargas G, Akdut N, Anglada M, Mateo A (2011) Reduction of anisotropy in cold-rolled duplex stainless steel sheets by using sigma phase transformation. Metall Mater Trans A Phys Metall Mater Sci 42(11):3472–3483
13. ASTM A 240/A 240 M—04a (2004) Standard specification for chromium and chromium-nickel stainless steel plate, sheet, and strip for pressure vessels and for general applications. Standards i
14. Standard test method for plastic strain ratio r for sheet metal (2010) ASTM E517
15. Liang ZY, Wang X, Huang W, Huang MX (2015) Strain rate sensitivity and evolution of dislocations and twins in a twinning-induced plasticity steel. Acta Mater 88:170–179

16. Fan Z, Tsakiropoulos P, Smith PA, Miodownik AP (1993) Extension of the Hall-Petch relation to two-ductile-phase alloys. Philos Mag A 67(2):515–531
17. Chen L, Yuan FP, Jiang P, Xie JJ, Wu XL (2014) Simultaneous improvement of tensile strength and ductility in micro-duplex structure consisting of austenite and ferrite. Mater Sci Eng A 618:563–571
18. Liu Y, Yan H, Wang X, Yan M (2013) Effect of hot deformation mode on the microstructure evolution of lean duplex stainless steel 2101. Mater Sci Eng A 575:41–47

Microstructural Investigation and Impact Testing of Additive Manufactured TI-6AL-4V

D.C. Austin, M.A. Bevan, D. East, A.D. Brown, A.A.H. Ameri, P.J. Hazell, A. Chen, S.L.I. Chan, M.Z. Quadir and J.P. Escobedo

Abstract The microstructure and mechanical response of additive manufactured Ti64 under compression and tension has been investigated. Quasi-static and dynamic experiments were conducted on electron beam powder bed Arcam A1 (EBM) and blown powder laser Optomec LENS MR-7 (LMD) additive manufactured specimens. Hot isostatic pressing (HIP) was conducted on both of the additive manufactured specimens and results were compared with as-received specimens. Digital image correlation (DIC) was utilized to observe the in situ development of deformation fields during tensile loading. Optical microscopy was conducted on pristine and deformed specimens to examine the microstructure of the specimens. HIP treatment reduced the porosity by $\sim$90% with a concomitant increase in hardness by 40%. The results from tensile experiments show that additive manufactured Ti64 display a higher values for the yield stress, hardness, and ultimate tensile strength than conventionally manufactured Ti64. Under quasi-static compression, it was observed that EBM specimens exhibit a higher yield strength and ultimate tensile strength than LMD specimens.

Keywords Titanium · Porosity · Compression · Tension · Additive manufacture

---

D.C. Austin · M.A. Bevan · A.D. Brown · A.A.H. Ameri · P.J. Hazell · J.P. Escobedo (✉)
School of Engineering and Information Technology, UNSW Australia,
Australian Defence Force Academy, Canberra, ACT 2600, Australia
e-mail: j.escobedo-diaz@adfa.edu.au

D. East
Commonwealth Scientific and Industrial Research Organisation,
Clayton, VIC 3168, Australia

A. Chen · S.L.I. Chan
School of Materials Science, UNSW Australia, Sydney, NSW 2052, Australia

M.Z. Quadir
Microscopy and Microanalysis Facility (MMF), John de Laeter Centre (JdLC),
Curtin University, Perth, WA 6102, Australia

S. Ikhmayies et al. (eds.), *Characterization of Minerals, Metals, and Materials 2017*, The Minerals, Metals & Materials Series,
DOI 10.1007/978-3-319-51382-9_21

Introduction

Titanium's high corrosion resistance, environmental resistance and high strength to density ratio have led the non-ferrous material to be highly desired in the automotive, biomedical and aerospace industry. Ti-6Al-4V (or Ti64) is particularly desirable due to its significant strength at low temperatures and good formability and weldability [1–3] Extensive work has been conducted on the microstructure and mechanical behaviour of Ti64 under quasi-static and dynamic compression and tension [1–7]. However very little work has been conducted on the mechanical responses and microstructural investigation on additive manufactured Ti64 [8–11].

Electron beam additive manufacturing is a method for making metal components. Ti64 powder is used to build components by layering the powder into a set formation and welding it together using an electron beam as a heat source. This is referred to as Electron Beam Melting (EBM). EBM works by fully melting the metal powder layers with an electron beam in a high vacuum chamber. This is crucial for Ti64 as it has a high affinity for oxygen. This process produces fully dense metal parts or lattices with characteristics of the metal powder. The EBM operates at temperatures up to 2000 °C which can cause differences in phase formation through solidification which in turn might result in materials with significant amount of porosity.

An alternative method is the blown powder laser, also known as Laser Melting Deposition (LMD), in this case the metal powder is blown coaxially into a melt pool created by the laser beam. The laser beam is able to melt the powder onto the previous layer which is usually around 300 μm. Once cooled, this process forms the metallurgical bond [12]. Both manufacturing methods use the same Ti64 powder and print each specimen individually. LMD is not as accurate as EBM as the layer is prone to inaccuracy during formation.

This paper examines the differences in mechanical response of additive manufactured Ti64 (EBM vs. LMD) to quasi-static and dynamic loading. A further comparison is drawn between as-received and heat treated specimens, via hot isostatic pressing (HIP), to investigate the effect of porosity.

Material Characterisation

For this project, the specimens were manufactured using the Optomec LENS MR7 and Arcam A1 housed at CSIRO Manufacturing-Lab 22. The EBM and LMD specimens are machined after their manufacturing process in order to have an acceptable surface finish for testing.

The porosity in the initial specimens was measured to be 0.26% (EBM) and 0.18% (LMD). These amounts of porosity might cause a low and stochastic mechanical performance of the material, thus, a set of specimens was HIP treated. During HIP treatment specimens are hold at elevated temperatures, generally half of the materials melting temperature, and high isostatic gas pressures in a high pressure compartment [13]. This process reduces the porosity in the microstructure of the material by eliminating internal voids via a combination of creep, plastic deformation and diffusion bonding [14]. Specimens were HIP treated in a vacuum sinter at 1100 °C for 90 min followed by HIP at 100 MPa at 1000 °C for 90 min in an Argon atmosphere. After HIP treatment the porosity reduced to 0.015% (EBM) and 0.016% (LMD), a relative decrease of ~92% for both cases.

Experimental Procedure

Specimens for compression were manufactured to comply with the ASTM E9-09 standard, 7.5 mm for height and diameter for both quasi-static and dynamic tests. The tensile specimens for quasi-static experiments followed the AS 1391 standard, 25 × 6 × 3 mm gauge area. In all cases, the specimens were manufactured so that the layer deposition direction is perpendicular to the load direction.

In preparation for quasi-static tensile experiments, the specimen gauge areas were sprayed with white paint with a black speckle pattern over the top for digital image correlation (DIC) measurements. Hardness testing was conducted using the Vickers Hardness test under a 5 kg load for 10 s. Quasi-static compression and tensile testing was conducted in a Shimadzu 100 kN universal testing machine. Dynamic compression testing was conducted on a Split Hopkinson Pressure Bar (SHPB) at a strain rate of ~1300 s^{-1}.

Results and Discussion

Quasi-static Mechanical Response

The Vickers hardness of the as-received Ti64 specimens was measured as 380HV (EBM) and 385HV (LMD). These values are higher than the reported for conventionally manufactured Ti64 349HV found in ASTM B265 [15]. After HIP treatment, the hardness increased to 530HV (EBM) and 540HV (LMD). This is an increase of ~40% for both additive manufacturing processes.

Results from quasi-static compression, at a strain rate of 10^{-3}/s, are shown in Fig. 1 and the most salient features are reported in Table 1. The reported response of a standard Ti64 specimen ASTM B265 [15] is provided for comparison. The EBM specimens display a higher yield stress and ultimate compressive stresses than the LMD specimens. The modulus of elasticity and strain hardening are very similar between the two manufacturing methods.

It is observed that the yield stress remains similar in HIP treated specimens but there is a significant reduction of ductility for both manufacturing processes. This

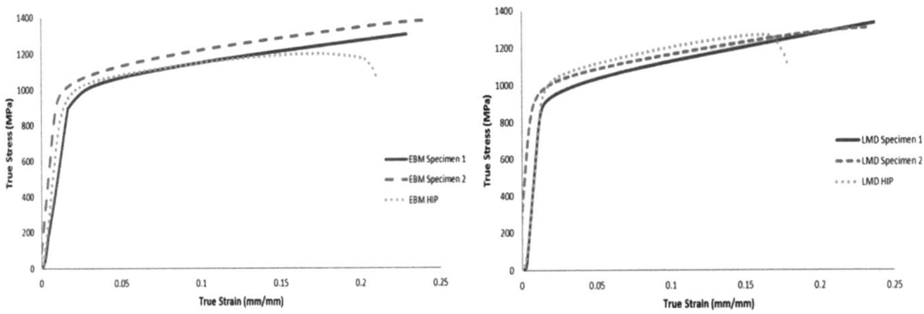

Fig. 1 QS compression results for **a** EBM and **b** LMD specimens

Table 1 QS compression results

Manufacture method	Modulus of elasticity (GPa)	Yield stress (MPa)	Ultimate compressive strength (MPa)	Strain hardening rate (GPa)
ASTM B265	114	970	1860	
EBM	92 ± 2	950 ± 47	1826 ± 73	3.895 ± 0.067
LMD	96 ± 3	803 ± 18	1792 ± 23	4.849 ± 0.138
EBM-HIP	91	968	1508	2.764
LMD-HIP	91	985	1546	3.387

observation correlates with the reduction in porosity and microstructural changes in the materials due to the HIP treatment. It is likely that as the specimens being to plastically deform, shear banding, dislocation slip and multiplication is hindered by pores, which in essence act as barriers for damage development. As these pores are removed via HIP treatments, there are less barriers for damage development which results in faster deformation rates when compared with as-received specimens. Examination of microstructural changes due to HIP treatment, e.g. changes in grain size or texture development, is currently being conducted via EBSD measurements and the results will be reported in due time. These results will provide a full view of how the HIP treatment changes the microstructure which in turn, affects the mechanical response.

When comparing as-received and HIP specimens with the ASTM B265 standard, it is observed that the EBM displays a yield stress and UCS closer to the ASTM B265 standard (~2% difference) whereas the LMD differs 10% with respect to the standard. This difference could be attributed to LMD manufacturing of the specimens as each layer is prone to inaccuracy during formation.

Results from quasi-static tension, at a strain rate of $\sim 10^{-3}$/s, are shown in Fig. 2 and the most salient features are reported in Table 2. The DIC for the QS EBM and LMD tension tests are shown in Fig. 3a and b, respectively. As before, the response of a conventional is provided for comparison. AR EBM specimens exhibit higher yield stress and UTS than conventionally manufactured specimens whereas the

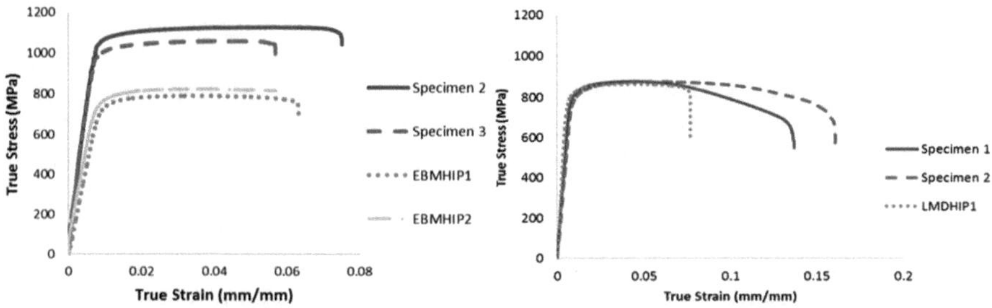

Fig. 2 QS tension results for **a** EBM specimens, **b** LMD specimens

Table 2 QS tension result

Manufacture method	Modulus of elasticity (GPa)	Yield stress (MPa)	Ultimate tensile strength, UTS (MPa)	Elongation (%)
ASTM B265	114	880	950	14
EBM	113.3 ± 1	901.7 ± 146.7	1001.6 ± 159.2	6.5 ± 0.91
LMD	112.4 ± 0.5	842.4 ± 16.4	876.6 ± 9.5	14.8 ± 1.7
EBM HIP	109.4 ± 6.8	784.6 ± 15	796.4 ± 13	5.9 ± 0.42
LMD HIP	114.9	854.1	864.1	7.7

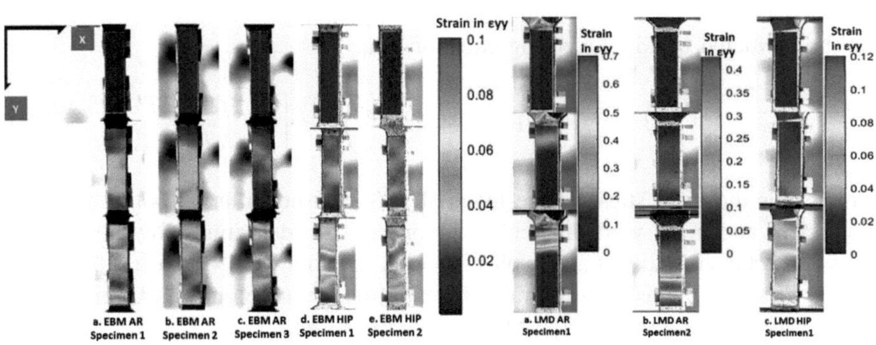

Fig. 3 DIC QS tension **a** EBM specimens, **b** LMD specimens

modulus of elasticity is very similar to the standard. However, the EBM specimens show less ductility, approximately 50% less elongation than the ASTM B265 standard.

Furthermore, the three EBM specimens tested display a scattered response in terms of yield stress, UTS and ductility. This heterogeneous plastic response can be attributed to the porosity in each specimen. The measured porosity for each specimen was 0.562% (specimen 1), 0.182% (specimen 2) and 0.373% (specimen 3). A lower yield stress and UTS correlate with a higher porosity.

The AR LMD specimens exhibit slightly lower yield stress and UTS than conventionally manufactured specimens whereas the modulus of elasticity is very similar to the ASTM B265 standard. However, the AR LMD specimens show greater ductility than EBM specimens and have similar elongation to the ASTM B265 standard. It is observed that the yield stress and ultimate tensile strength in the EBM HIP specimens is reduced in the QS tension results but with similar elongation values compared to AR EBM. However, the HIP specimens are more consistent. LMD HIP results show similar results for yield stress and ultimate tensile strength compared to AR LMD, however the elongation is reduce dramatically by 50%. The QS tension results demonstrate how HIP treatment creates consistency with the mechanical behaviour of the material.

The DIC plots illustrate how the AR specimens break due to porosities in the specimen, shown through the different breaking points on the specimens in the gauge area. After HIP treatment, fracture occurs generally in the middle of the gauge area indicating that the material failed due to reaching its mechanical limit rather than an inconsistency due to proposity.

Dynamic Response: Johnson-Cook Model

The Johnson-Cook (J-C) [16] model is able to predict the plastic behaviour of materials under a variety of strain rates and temperatures. The mechanical response can be predicted by the governing equation (Eq. 1):

$$\sigma_{JC} = \left[A + B \times \varepsilon_{pl}^{n}\right] \times \left[1 + C \times \ln\left(\frac{\dot{\varepsilon}_{pl}}{\varepsilon_0}\right)\right] \times \left[1 - \left(\frac{T - T_r}{T_m - T_r}\right)\right]^m \tag{1}$$

where A is the yield stress of the material at 0.2% true strain; B is the hardening coefficient and n the work hardening exponent; C is determined by the dynamic stress to static stress ratios at 1% plastic strain, σ_f: Johnson Cook Yield Stress; ε_{pl}: Plastic Strain; $\dot{\varepsilon}_{pl}$: Plastic Strain rate; T: Room temperature during experiment; T_r: Reference Temperature and T_m: Melting Temperature.

The first set of brackets in the J-C model describes the stress as a function of strain when the strain rate is equal to 1 s^{-1}. The second set of brackets expresses the strain rate effect at room temperature. The third set of brackets describes the temperature effects during the experiment, however this experiment was run at room temperature therefore the third set of brackets simplifies to 1. For this study quasi-static experimental data was used to determine the constants, A, B and n.

The results from dynamic compression testing conducted on EBM and LMD AR and HIP specimens at room temperature and at a strain rate of 1200 s^{-1} are shown in Figs. 4 and 5, respectively. The J-C fitting obtained from our data as well as the best fitting model found in Shivpuri et al. [17] are also reported. The varying factors between the three different models are the JC parameters listed in Table 3.

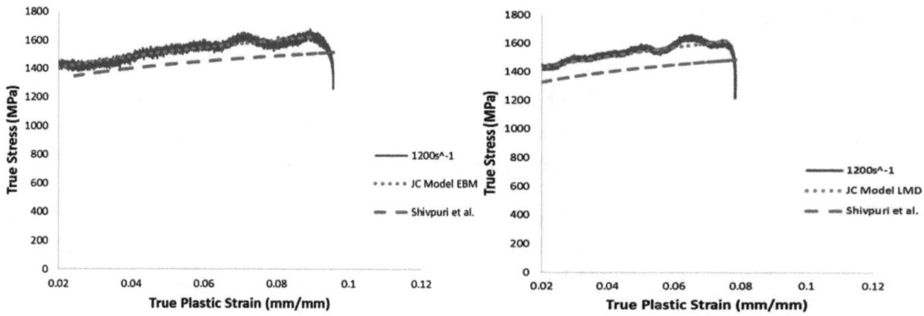

Fig. 4 Experimental data and J-C fittings for **a** EBM, **b** LMD Ti-6Al-4V

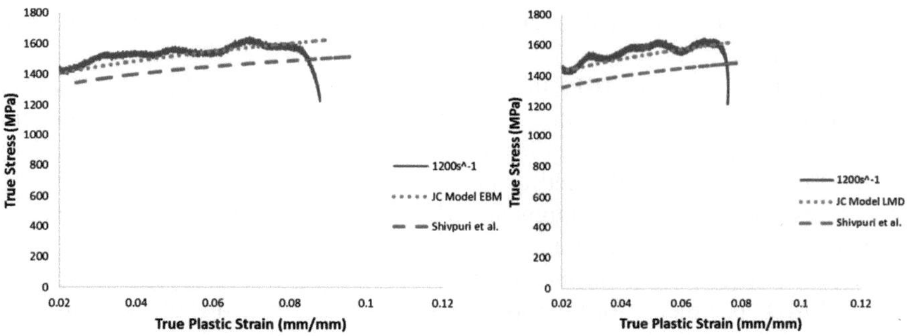

Fig. 5 Experimental data and J-C fittings for **a** EBM HIP, **b** LMD HIP Ti-6Al-4V

Table 3 Johnson Cook parameters

Material coefficient	EBM JC fit	LMD JC fit	Shivpuri et al.
A (MPa)	950	871	870
B (MPa)	1046	1026	990
C	0.068	0.06	0.011
n	0.52	0.54	0.25

As aforementioned, the J-C parameters from [17] of a conventional Ti64 alloy are the closest to fit our data on the EBM and LMD specimens. Main differences are on the C parameter (strain-rate effects) and value and strain hardening coefficient, n. The JC model for EBM and LMD was used against HIP specimens that were tested in the same conditions. These results can be seen in Fig. 5.

Conclusions

The mechanical response of Ti64 specimens produced by two additive manufacturing processes, electron beam melting (EBM) and laser metal deposition (LMD), has been investigated and our main findings are as follows:

- EBM and LMD specimens display a higher hardness than conventional Ti64.
- As received EBM display a scatter tensile response, due to porosity, which in average exhibits higher yield stress and UTS but lower ductility than conventional Ti64 in tension.
- As received LMD display a lower yield stress and UTS but have similar ductility to conventional Ti64 in tension.
- Both EBM and LMD have a lower yield stress and UCS in compression.
- EBM has a higher yield strength and ultimate tensile strength than LMD specimens.
- Hot isostatic pressing (HIP) treatment reduced the porosity by $\sim$90% with a concomitant increase in hardness by 40%. This improved the yield strength and UCS in compression but significantly reduced the ductility. The latter is likely due to failure being enhanced as pores were removed which resulted in fewer barriers for damage development. HIP treatment causes specimens exhibit consistent behaviour therefore making the material more reliable.
- There was a good agreement of Johnson-Cook fittings and the response of EBM and LMD specimens, for both as-received and heat treated (HIP) specimens.

Acknowledgements The authors would like to thank Thomas d'Ews Thomson assisting considerably in the manufacturing of specimens in a timely manner and David Sharp for assisting with sample metallographic preparation. Peter Nicholson is thanked for his assistance in performing the HIP treatment. Pat Noland is also thanked for his assistance in conducting the experiments.

References

1. Ivasishin O, Teliovych R, Ivanchenko V (2008) Processing, microstructure, texture, and tensile properties of the Ti-6Al-4V-1.55B eutectic alloy. Metall Mat Trans 39:402
2. Khan AS, Suh YS, Kazmi R (2004) Quasi-static and dynamic loading responses and constitutive modelling of titanium alloys. Int J Plast 20:2233
3. Lee W, Lin C (1998) High temperature deformation behaviour of Ti-6Al-4V alloy evaluated by high strain-rate compression tests. J Mater Process Technol 57:127
4. Nemat-Nasser S, Guo W, Nesterenko VF, Indrakanti SS, Gu Y (2001) Dynamic response of conventional and hot isostatically pressed Ti-6Al-4V alloys: experiments and modeling. Mech Mater 33:425
5. Millett J, Whiteman G, Bourne N, Gray G (2008) Role of anisotropy in response of Ti-6Al-4V to shock loading. J Appl Phys 104:073531
6. Osovski S, Rittel D, Landau P, Venkert A (2012) Microstructural effects on adiabatic shear band formation. Scripta Mater 66:9

7. Biswas N, Ding JL, Balla VK, Field DP, Bandyopadhyay A (2012) Deformation and fracture behavior of laser processed dense and porous Ti-6Al-4V alloy under static and dynamic loading. Mater Sci Eng A 549:213
8. Alcisto J, Enriquez A, Garcia H, Hinkson S, Steelman T, Silverman E, Valdovino P, Gigerenzer H, Foyos J, Ogren J, Dorey J, Karg K, McDonald T, Es-Said OS (2011) Tensile properties and microstructures of laser-formed Ti-6Al-4V. JMEP 203–212
9. Murr LE, Esquivel EV, Quinones SA, Gaytan SM, Lopez MI, Martinez EY, Medina F, Hernandez DH, Martinez E, Martinez JL, Stafford SW, Brown DK, Hoppe T, Meyers W, Lindhe U, Wicker RB (2009) Microstructures and mechanical properties of electron beam rapid manufactured Ti-6Al-4V biomedical prototypes compared to wrought Ti-6Al-4V. Mater Characteristics 2009:96–105
10. Kobryn PA, Semiatin SL (2011) The laser additive manufacturing of Ti-6Al-4V. JOM 53:40
11. Parthasarathy J, Starly B, Raman S, Christensen A (2010) Mechanical evaluation of porous titanium (Ti-6Al-4V) structures with electron beam melting (EBM). J Mech Behav Biomed Mater 3:249
12. Selcuk C (2011) laser metal deposition for powder metallurgy parts. Powder Metall 2:94
13. Delo D, Piehler H (1999) Early stage consolidation mechanisms during hot isostatic pressing of Ti-6Al-4V powder compacts. Acta Mater 47:2841
14. Atkinson H, Davies S (2000) Fundamental aspects of hot isostatic pressing: an overview. Metall Mater Trans A 31:2981
15. ASTM B265-15 (2013) Standard specification for titanium and titanium alloy strip, sheet and plate
16. Johnson G, Cook W (1983) A constitutive model and data for metals subjected to large strains, high strain rates and high temperatures. In: Proceedings of the 7th international symposium on ballistics, vol 31
17. Shivpuri R, Hua J, Mittal P, Srivastava A, Lahoti G (2002) Microstructure-mechanics interactions in modelling chip segmentation during titanium machining. CIRP Ann Manuf Technol 51(1):71–74

Part VI
Powders and Foams

Synthesis of TiN Nano-Composite Powder by High-Energy Ball Milling of TiH_2 Under Nitrogen Atmosphere

Xiaolong Wu, Xuewei Lv, Xuyang Liu, Chunxin Li and Yu Zhang

Abstract Titanium nitride is a broad-spectrum specific functional material because of its many superior properties. Therefore, it is necessary to investigate synthesis methods of TiN and alloys containing TiN reinforcement. In the current study, with the increase of nitrogen gas pressure and the optimization of routine ball milling parameters, the conventional mechanical ball-milling method can be improved, the nanometers TiH_2-TiN composite powder can be synthesized successfully in a shorter time. Nitrogen gas filled in ball mill jar in which the pressure was 0.6 MPa. TiH_2 powder was milled under nitrogen atmosphere at a spinning rate of 250 rpm. The results show that the stable TiN is formed after grinding for 2.5 h according to the X-ray diffraction pattern. DSC (differential scanning calorimetry) curves also indicate that milled compound powder adsorbs certain nitrogen gas. The gas adsorbed in TiH_2-TiN composite can exist stably below 200 °C.

Keywords TiH_2 · Ball milling · TiH_2-TiN composite powder

Introduction

Melt and metalloid nitrides have considerable interests due to their desirable properties including high hardness, high temperature stability, high corrosion resistance and attractive color [1]. In addition, many ultra-fine particles of excellent properties such as TiC, SiC, TiB, TiN, Al_2O_3 were added into titanium matrix to enhance material properties. The titanium matrix composites were reinforced by adding TiN particulate [2]. This process was generally carried out under hot press sintering furnace or vacuum induction furnace. However, the final product was often inhomogeneous and may had the problem of interfacial reaction.

X. Wu · X. Lv (✉) · X. Liu · C. Li · Y. Zhang
College of Materials Science and Engineering, Chongqing University, Chongqing 400044, China
e-mail: lvxuewei@163.com

S. Ikhmayies et al. (eds.), *Characterization of Minerals, Metals, and Materials 2017*, The Minerals, Metals & Materials Series,
DOI 10.1007/978-3-319-51382-9_22

Mechanical milling method is often used to previous process of synthesis titanium matrix composite. Reinforcement prepared by method of ball milling. And then, processed powder was added into reinforcement synthesis alloys in vacuum furnace by hot forming. Calka et al. [3] made a large number of investigations of metallic powder ball milling in gas atmosphere. Block titanium preparation by synthesis of titanium powder in liquid nitrogen and reactive ball milling of Ti in nitrogen followed by sintering. In follow-up work, ball milling metallic powder was obtained from titanium in NH_3 atmosphere. Results show synthesis of metal nitride become more rapid in NH_3 atmosphere than in nitrogen atmosphere. Reaction with NH_3 in the sealed mill chamber results in a rapid loss in initial pressure, due to the combination of absorption and reaction with NH_3. The methods for the synthesis of bulk titanium matrix material included synthesis of Ti-TiN nanostructure composites powder by ball milling Ti in nitrogen and hot press sintering compacted powder preparation of reinforcement homogeneous material [4]. Because ball milling powder surface free energy is higher than before, making use of nanostructure of milled powder to synthesize bulk material has eliminated successfully wetting problems [1]. The synthesis of TiH_2-TiN composites by TiH_2 powder ball milling in nitrogen gas was presented in the current experiment.

Experimental Procedure

Titanium hydride powder of particle size 44 μm and purity of 99.8% was used. Ball-milling experiments were carried out in planetary ball-mill device of PM100 type made in Germany. Ball mill jar is stainless steel, speed of main-wheel is 100–600 r/min, speed ratio of main-wheel and mill cell is 1:2. The sample of about 10 g was placed in the sealed mill chamber containing a few zirconia balls. And the initial pressure of anhydrous N_2 (99.99%) was kept as 0.6 MPa. Nitrogen gas was introduced into mill cell to get the air out of the cell after placing samples. The optimization rotational speed was 250 r/min. After a series of exploration, the ball and powder weight ratio is 20:1 [5]. In this experiment, maximum time of ball milling is 4 h. When ball-milling was completed, milled powders were taken out from chamber in vacuum glove box to prevent as-obtained powders oxidization. The structure of as-milled samples were monitored by X-ray diffraction (XRD) using Cu Kα radiation. X-ray diffraction patterns were taken in the range of 2θ from 10° to 90°. Differential scanning calorimetry (DSC) and thermal gravity (TG) were preformed to analyze the adsorbed nitrogen of milled powder.

Results and Discussion

The XRD diffraction and microstructure pattern of raw material TiH_2 are shown in Fig. 1. In this experiment, a shearing milling mode was employed [6]. TiN synthetized in mill chamber after the crushing, grinding and rolling of original powders were performed [7]. The ball mill chamber pressure declined continuously with reaction proceeding. It illustrates nitrogen was absorbed by TiH_2 powder and then reacted with Ti powder for 2.5 h to produce TiN.

The XRD pattern of the as-obtained product is shown in Fig. 2. When ball-milling time increases, grain refinement leads to characteristic peaks wider and shorter gradually. The broadening of the characteristic peaks is due to the very fine grain size [8]. It is typical of materials to be milled in this way. The Scherrer formula was used to calculate grain size [9]. The calculated grain size of original materials is 293 nm after deconvolution. Grain size diminished from 166 to 135 nm, when milled time increased from 0.5 to 1.0 h. Furthermore, the diffraction peaks associated with the TiN phase were observable after 2.5 h and grain size diminished to 86 nm. This result indicated TiH_2-TiN composite powder of nanostructure was obtained by ball-milled reaction. In addition, α-Ti did not appear in Fig. 2. This result show that TiH_2 didn't decompose in ball milling process [10]. Furthermore, it can be concluded that H atoms of TiH_2 were substituted by N atoms in synthesis of TiN based on the above results.

Figure 3 shows the microstructure of TiH_2 powder after ball-milling (a-0.5 h, b-1.0 h, c-2.0 h, d-2.5 h). Powder size becomes more and more fine with ball milling time going by. In the beginning of ball milling, grain size decreases remarkably. Prolongation of ball milling time facilitates the formation of nanocomposite. In addition, milled powder gathered together spontaneously, and thus formed aggregation agent because fresh interface has higher surface energy [11].

Fig. 1 Microstructure and XRD pattern of TiH_2 powder

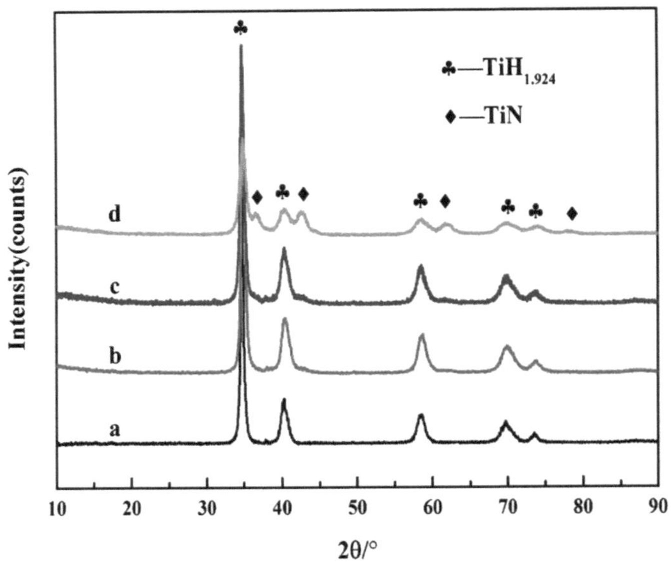

Fig. 2 XRD patterns of samples milled respectively for 0.5, 1.0, 2.0, 2.5 h

Fig. 3 Microstructure of TiH_2 powder after ball-milling. **a** 0.5 h, **b** 1.0 h, **c** 2.0 h, **d** 2.5 h

Figure 4 shows the DSC and TG thermal curves under nitrogen(N_2) atmosphere of milled TiH_2 powder. The DSC and TG curves of 2.0 and 2.5 h milled powder are represented as c and d respectively. The c, d DSC curves of c and d indicate a smooth descending endothermic peak at about 350 and 300 °C, respectively. Both the curves exist relatively obvious endothermic peak and endothermic interval span

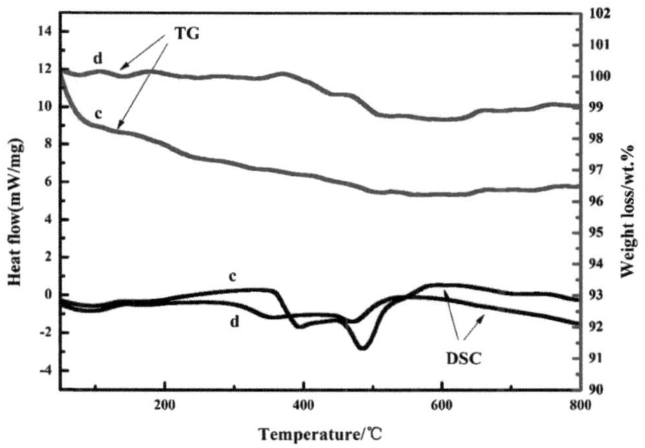

Fig. 4 DSC and TG thermal curves under nitrogen atmosphere (c-2.0 h, d-2.5 h)

is about 240 °C. Endothermal values of the first endothermic peaks of c, d curves were 190.194 and 196.335 J by integral method.

Metallic titanium especially ultra-fine titanium particle is an excellent getter. It has particularly strong ability of gas adsorption. In zhu paper, DSC curves showed that there were gas desorption exothermic peak of the nano-solid titanium at about 340 °C [12]. According to Figs. 2 and 4, first endothermic peaks of c, and d DSC curves show that the increasing of a portion of energy causes variation of moles number and state of adsorption of gas when as-milled powders were heated in argon atmosphere. The changing rate of c DSC curve is larger than d DSC curve at beginning declining stage in first peaks, so as-milled powders adsorption of gas of d DSC curve is more stable than that of c DSC curve. The second peak representing the decomposition of milled TiH_2 powders results from raising temperature.

As shown in Fig. 4, the initial decomposition temperature and finish temperature of milled TiH_2 powders decline with the increasing of ball milling time. Endothermal values of the second endothermic peaks of d, and c DSC curves are 276.258 and 599.244 J, respectively. A portion of TiH_2 translates into TiN after 2.5 h. The synthesis reactions of TiN, in reality, H atoms were substituted by N atoms in synthesis process of TiN. Therefore, endothermal value of second peak of c DSC curve is lager than d.

Conclusions

The synthesis of TiH_2-TiN composites by TiH_2 ball milling in nitrogen gas is experimentally demonstrated, and the conclusions can be obtained as follows:

(1) The nanometers TiH_2-TiN composite powder can be synthesized successfully in a shorter time which is about 2.5 h.
(2) There is no α-Ti in the ball milling process of TiH_2, so the H atoms are substituted by N atoms in synthesis process of TiN.
(3) As-milled powders adsorb a certain amount of N_2, and the desorption happens over about 300 °C.

Acknowledgements This work was supported by the National Natural Science Foundation of China (Grant No.51374262).

References

1. Wexler D, Calka A, Mosbah AY (2000) Ti–TiN hard metals prepared by in situ formation of TiN during reactive ball milling of Ti in ammonia. J Alloy Compd 309(1):201–207
2. Cao J, Li C, Qi J et al (2015) Combustion joining of carbon–carbon composites to TiAl intermetallic using a Ti–Al–C powder composite interlayer. Compos Sci Technol 115:72–79
3. Calka A, Williams J, Millet P (1992) Synthesis of silicon nitride by mechanical alloying. Scripta Metall Et Mater 27(12):1853–1857
4. Chin Z-H, Perng T-P (1997) In situ observation of combustion to form TiN during ball milling Ti in nitrogen. Appl Phys Lett 70(18):2380–2382
5. Yadav TP, Yadav RM, Singh DP (2012) Mechanical milling: a top down approach for the synthesis of nanomaterials and nanocomposites. Nanosci Nanotechnol 2(3):22–48
6. Murty B, Ranganathan S (2013) Novel materials synthesis by mechanical alloying/milling. Int Mater Rev
7. Koch C, Cho Y (1992) Nanocrystals by high energy ball milling. Nanostruct Mater 1(3):207–212
8. Gotor F, Alcala M, Real C et al (2002) Combustion synthesis of TiN induced by high-energy ball milling of Ti under nitrogen atmosphere. J Mater Res 17(07):1655–1663
9. Bolokang A, Phasha M (2010) Formation of titanium nitride produced from monocrystalline titanium powder under nitrogen atmosphere. Int J Refract Metal Hard Mater 28(5):610–615
10. Kennedy A, Lopez V (2003) The decomposition behavior of as-received and oxidized TiH_2 foaming-agent powder. Mater Sci Eng A 357(1):258–263
11. Mcmahon BW, Yu J, Boatz JA et al (2015) Rapid aluminum nanoparticle production by milling in NH3 and CH3NH2 atmospheres: an experimental and theoretical study. ACS Appl Mater Interfaces 7(29):16101–16116
12. Zhu Y(1994) DSC analysis nanostructure of MoN, Mo_2N, Ti, TiN. Chin Sci Bull 39(5):469–470

Application of AFM in Morphology Determination of Powder Material

Jian Wu, Ping Long and Yaochun Yao

Abstract Atomic force microscope (AFM) is widely used for determining of surface performance of materials with smoother surface. In this research, atomic force microscope was used in morphology determination of powder material. Samples were dispersed in aqueous system by ultrasound wave and prepared on mica sheet. Multiple mappings of different powder samples, such as sintered products of ferrous oxalate, ferrous oxalate with lamella structure, lithium carbonate, boron powder, ferric oxide and zinc oxide, were investigated and analyzed in detail. Results showed as follow: ferrous oxalate sintered in different processes and conditions presented different morphology characteristics both in shape and size. Ferrous oxalate obtained from special process presented two-dimension lamella structure and self-assembly phenomenon. The shape and size of lithium carbonate particles were uniform. Defects of mappings on boron powder, ferric oxide and zinc oxide reflected problems of the preparation process before scanning. Determination experiences of powder material in AFM technology could be obtained.

Keywords AFM · Morphology determination · Powder samples

J. Wu · P. Long · Y. Yao (✉)
State Key Laboratory of Complex Nonferrous Metal Resources Clean Utilization/National Engineering Laboratory for Vacuum Metallurgy, Kunming University of Science and Technology, Kunming 650093, Yunnan, China
e-mail: yaochun9796@163.com

J. Wu
e-mail: wujian3756@163.com

P. Long
e-mail: 2463047321@qq.com

J. Wu · P. Long · Y. Yao
Key Laboratory for Nonferrous Vacuum Metallurgy of Yunnan Province, Kunming 650093, Yunnan, China

J. Wu · Y. Yao
Engineering Laboratory for Advanced Batteries and Materials of Yunnan Province, Kunming 650093, Yunnan, China

S. Ikhmayies et al. (eds.), *Characterization of Minerals, Metals, and Materials 2017*, The Minerals, Metals & Materials Series,
DOI 10.1007/978-3-319-51382-9_23

Introduction

In the field of surface science, atomic force microscope (AFM) is an advanced technique and widely used for characterization of surface performance of materials [1]. The macroscopic properties or behavior of the material correlate to the microscopic or even nano-level features in a quantitative way, which could be directly determined by AFM [2]. But AFM mappings with high resolution and quality are usually obtained from parts of materials with smoother surface, such as bulk solid, film and biological sample [3–10]. It is hard to use AFM for characterizing the surface morphology or any other properties of powder samples directly. The reasons were reported as follow [3]: submicron or nano particles are easy to adhere to the tip of probe because of electrostatic attraction and other viscosity force arising from their high surface energy and low mass, which lead to the hardship of scanning and the damage of probe. In addition, particles are likely to be moved by interaction force existed between probe and particles, resulting in the unstability of scanning. Relevant researches reported that the surface morphology of particles could be obtained from AFM, however, most of their preparation methods seemed complicated and unpractical [11–15]. Nowadays, the preparation methods of powder samples are proposed to be dispersed uniformly in liquid and prepared on smooth sheet, such as mica slice, silicon slice, cover glass and so on [9, 16], but these methods were only limited to prepare and determine one or two samples related to the researchers.

In this research, atomic force microscope was used in morphology determination of diverse powder samples encountered in our test work, such as ferrous oxalate, lithium carbonate, barium carbonate, boron powder, ferric oxide, and zinc oxide. Preparation methods for powder samples mentioned above were attempted. Multiple mappings of these powder samples were investigated and parts of their surface performances and special information were analyzed in detail, combining with their preparation process. Lots of useful experiences on preparation and Determination of powder material in AFM technology could be obtained and summarized.

Experimental

Experimental Materials and Reagents

Powder samples involved in this research are explained as follow: ferrous oxalate powder was prepared from ferrous sulfate and oxalic acid by liquid phase precipitation method, then, sintered in two different processes and conditions, the two final sintered products will be determined. Ferrous oxalate powder with two-dimension lamella structure will be determined similarly. Lithium carbonate (purity more than 99.5%) was purified from industrial-grade lithium carbonate by carbonation-decomposition method. Barium carbonate, boron powder, ferric oxide and zinc oxide were provided by clients who need to test their powder samples by AFM. Deionized water was used as dispersion medium for the preparation of powder. Mica sheets were used as matrix for loading powder samples.

Experimental Methods

About 10 mg powder samples were weighed accurately and poured into a 50 ml beaker which was filled with 50 ml deionized water previously, and then, dispersed in aqueous system for 5 min by ultrasound wave. About 20 μl turbid liquid was taken by micro pipette tip and dropped onto mica sheet. Liquid on mica sheet was vaporized naturally at room temperature. The powders' morphology could be observed by AFM when mica surfaces are thoroughly dry.

Experimental Instruments

Atomic force microscope (AFM, Agilent-5500) was purchased from Agilent Technologies in USA. AC mode was used in testing process. Silicon probes (NANOSENSORS-NCL-20) with force constant of 21–98 N/m were used to san. Deionized water was prepared by Exceed-E-UP ultra-pure water system.

Results and Discussion

Sintered Products of Ferrous Oxalate

Ferrous oxalate powder was divided into two parts. One was taken into TG/DTA device and sintered at 800 °C under argon atmosphere and another was sintered in vacuum oven at 900 °C. These two kinds of sintered products were separately prepared on two mica sheets and scanned by AFM at the series range of 10 × 10 μm, and 2 × 2 μm, respectively. The morphology images were showed in Figs. 1 and 2.

According to these images, both of them have the similar morphology and their shapes look like snowflake. But the particles of products sintered in TG/DTA at 800 °C under argon atmosphere are larger and looser than that of sintered in vacuum oven at 900 °C. It is obvious from Fig. 1b at much higher resolution of 2 × 2 μm that particles have net structure with uniform holes, but particles in Fig. 2b have few hole and seem dense. AFM images presented these special differences in morphology characteristics clearly.

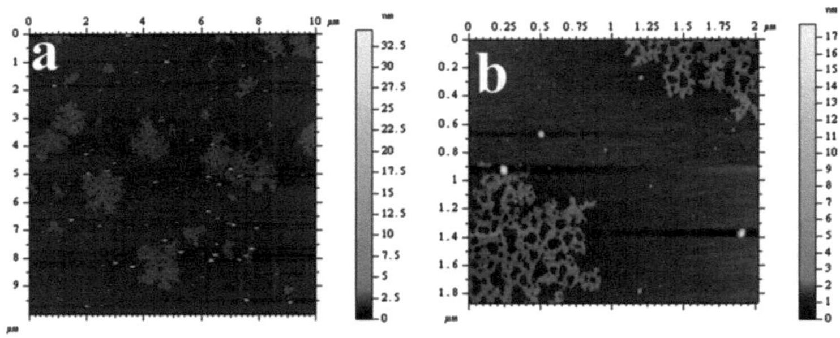

Fig. 1 Sintered products of ferrous oxalate in TG/DTA at 800 °C under argon atmosphere (scanning area: **a** 10 × 10 μm; **b** 2 × 2 μm)

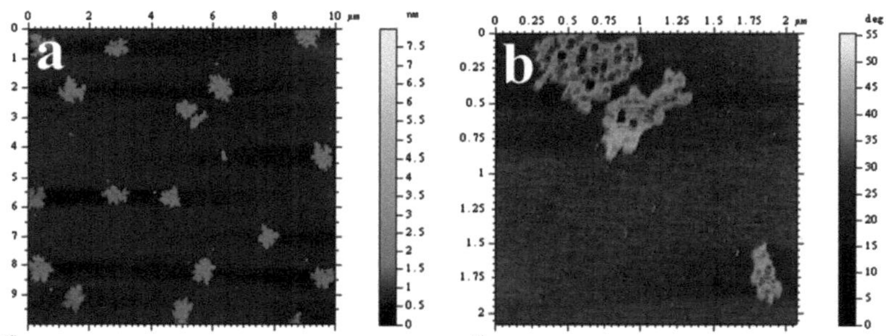

Fig. 2 Sintered products of ferrous oxalate in vacuum oven at 900 °C (scanning area: **a** 10 × 10 μm; **b** 2 × 2 μm)

Ferrous Oxalate with Two-Dimension Lamella Structure

According to the morphology and section profile mappings in Fig. 3, AFM images present Two-dimension lamella structure of ferrous oxalate in morphology, contrast and thickness at higher resolution. The shapes of the lamella crystal units are similar. The size and thickness of them are homogenous and seem about 200 and 2 nm, respectively. Ferrous oxalate is one of the key and common material for cathode materials of lithium-ion battery. Researches showed that the electrochemical performance of lithium iron phosphate could be improved greatly by ferrous oxalate with two-dimension lamella structure.

It is important to note that aggregate structure is observed in the lower left corner of Fig. 3b. The aggregates are presented obviously in the view of Fig. 3c. The profile mappings in Fig. 3b, c show that the aggregate have uniform shape and thickness of 2–4 nm, which might be formed from the 2 nm-thick lamella crystal units. Self-assembly phenomenon of ferrous oxalate lamella crystal is observed by AFM unexpectedly.

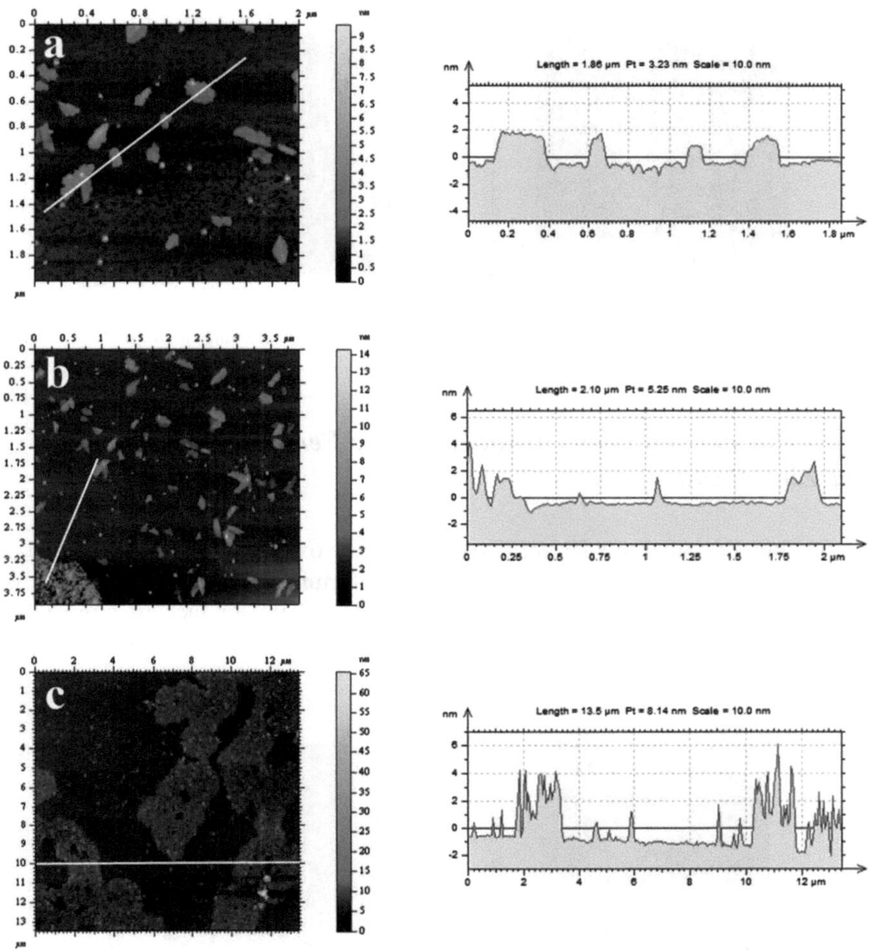

Fig. 3 AFM morphology and profile images of ferrous oxalate with lamella structure (scanning area: **a** 2 × 2 μm; **b** 4 × 4 μm; **c** 13 × 13 μm)

Battery-Grade Lithium Carbonate

Battery-grade lithium carbonate was purified from industrial-grade lithium carbonate by carbonation-decomposition method. AFM images show the morphology of battery-grade lithium carbonate in Fig. 4. In the view of the images, it is not hard to observe that particles have similar spherical shape and uniform size of 1 μm. Improved proposes on purification process could be offered to promote the electrochemical performance of electrode material for lithium-ion battery by referring to AFM morphology images.

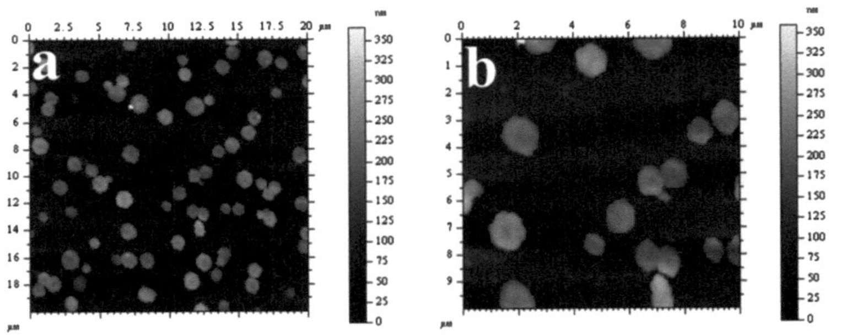

Fig. 4 AFM morphology of battery-grade lithium carbonate (scanning area: **a** 20 × 20 μm; **b** 10 × 10 μm)

Extra Powder Samples (Boron Powder, Ferric Oxide and Zinc Oxide)

Morphology of boron powder, ferric oxide and zinc oxide, which were provided by related researchers, are shown in Fig. 5. Unfortunately, detailed analysis on preparation technology of them might be not done here for lack of more information. However, problems and experiences of the preparation process for AFM

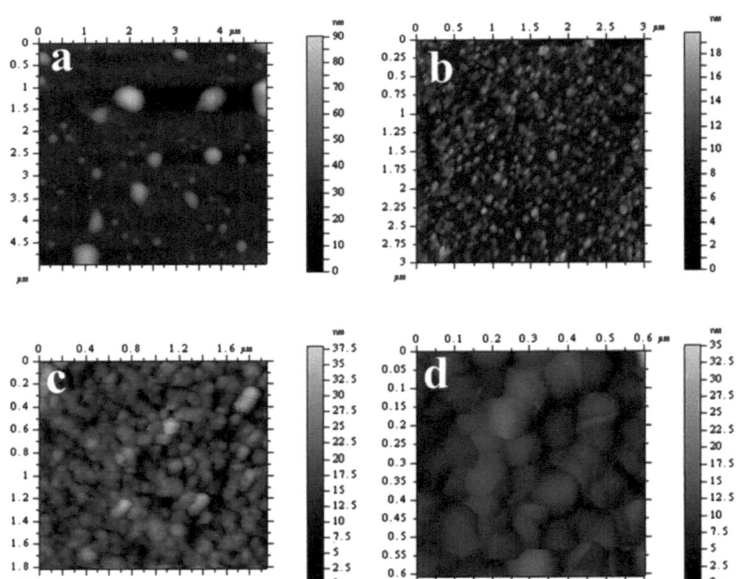

Fig. 5 AFM morphology of extra powder samples (**a** boron powder, scanning area of 5 × 5 μm; **b** ferric oxide, scanning area of 3 × 3 μm; **c** zinc oxide, scanning area of 2 × 2 μm, select section; **d** zinc oxide, scanning area of 0.6 × 0.6 μm)

determination could be reflected from the defects of AFM images. It is easy to see from Fig. 5a that boron particles depart from each other and profiles of them seem clear. On the contrary, particles of ferric oxide and zinc oxide accumulate together, leading to the difficult of observing the single particle morphology and size.

Images results are affected by lots of factors, such as excess powder in aqueous system, insufficient of dispersion time, drying process of turbid liquid on mica sheet, etc. According to different properties of powder, appropriate concentration of turbid liquid should be tested repeatedly and determined finally. Dispersion time should be controlled as long as possible. However, all of the preparing conditions are variable, which need the experiences from operating AFM for a long time.

As shown in Fig. 5c, d, zinc oxide particles have smaller size and special morphology. Their shapes seem as peanuts and have core-shell structure. The significant information on microscopic morphology may be useful for related technology and properties. But unfortunately, the detail information of single particle has not been presented greatly due to accumulation of particles which resulted from the preparation problem of sample before scanning.

Conclusions

In this research, atomic force microscope was applied to determine morphology of powder material. Powder samples, such as sintered products of ferrous oxalate, ferrous oxalate with special structure, lithium carbonate, barium carbonate, boron powder, ferric oxide and zinc oxide, were investigated and analyzed in detail. The conclusions are summarized as follow:

1. Sintered products of ferrous oxalate have snowflake morphology and net structure. The particles of products sintered in TG/DTA at 800 °C under argon atmosphere are larger and looser than that of sintered in vacuum oven at 900 °C.
2. Ferrous oxalate with two-dimension lamella structure was observed from high-resolution AFM technology. Self-assembly phenomenon of ferrous oxalate lamella crystal is also observed in view of AFM mappings unexpectedly.
3. The morphology of battery-grade lithium carbonate is shown by AFM. Particles have similar spherical shape and uniform size of 1 μm. These microscopic properties provide effective information for improving purification process.
4. Defects of mappings on boron powder, ferric oxide and zinc oxide reflect problems of the preparation process before scanning. Preparation and determination experiences of powder material in AFM technology are obtained from determination of multiple powder samples.

Acknowledgements This work was financially supported by the subject of State Key Laboratory of Complex Nonferrous Metal Resources Clean Utilization (CNMRCUTS1505), Chinese National Science Foundation (51364021), and the Natural Science Foundation of Yunnan Province (2014FA025).

References

1. Wu X, Li X, Mansour HM (2010) Surface analytical techniques in solid-state particle characterization for predicting performance in dry powder inhalers. KONA Powder Part J 28:3–19
2. Bonyár A (2016) AFM characterization of the shape of surface structures with localization factor. Micron 87:1–9
3. Luo C, Yang B (2006) Study on the powder imaging method of atomic force microscopy. J Chin Electron Microsc Soc 25:384–389
4. Wang X, Bai C, Zhu C (1996) Study on the DNA imaging method of atomic force microscope. Acta Biophys Sin 12:538–543
5. Li W, Liu S, Lu T, Huang Y (1999) The endothelial cell surface change in different concentrations of homocysteine by atomic force microscope. Acta Biophys Sin 15:214–219
6. Wang X-P, Shi Q-W (2003) Bundle structure formation on polymer film by varying temperature AFM scanning. Acta Phys Sin 52:656–663
7. Zhang J, Wang Y-L, Gu L, Pan J (2003) Atomic force microscopy of actin. Acta Biochem Biophys Sin 35:489–494
8. Guo H-M, Liu H-W, Wang Y-L, Xie H-M, Dai F-L, Gao H-J (2003) Moire fringes of HOPG and mica in scanning probe microscopy. Acta Phys Sin 52:2514–2519
9. Liu L, Yu L, Li X-l, Wang Z-b, Liang Q (2015) Structure and optical properties of Cu-doped SnS thin films prepared by PLD. Chin J Lumin 36:1311–1319
10. Guo L-f, Huang H, Luo C, Yang X-h (2010) Preparation and characterization of superparamagnetic Fe_3O_4 chitosan nanoparticles. J Chongqing Med Univ 35:690–1692
11. Jiang D-s, Yu H-h, Zhou L-d (2004) AFM studies on electrostatically self-assembled films of gold nanoparticles. J Chin Electron Microsc Soc 23:177–182
12. Ramesh S, Cohen Y, Aurbach D, Gedanken A (1998) Atomic force microscopy investigation of the surface topography and adhesion of nickel nanoparticles to submicrospherical silica. Chem Phys Lett 287:461–467
13. Zhao K, Wu X-k, Guo J-y, Sui L, Mei J-p, Ni M-n (2003) Application of atomic force microscope in cluster physics reseach. J Chin Electron Microsc Soc 22:219–222
14. Zhang X-y, Ying W-j, Wang Y, Zhang Y-f (2003) SPM study on FePt magnetic nanoparticle film. J Chin Electron Microsc Soc 22:64–167
15. Zhang J, Xu D, Cao Q, Li J (2012) Study on hydroxyapatite measured by atomic force microscope. Chin Powder Sci and Technol 18:73–75
16. Zhu X-r, Xu Z-p, Zhu Z-g, Chen C (2013) Preparation and characterization of $Ni_{0.5}Zn_{0.5}Fe_20_4$ nanomaterials via sol-gel technique. J Shanghai Second Polytech Univ 30:290–295

Effects of Thermal Processing on Closed-Cell Aluminium Foams

A.D. Brown, W.D. Hutchison, M.A. Islam, M.A. Kader, J.P. Escobedo and P.J. Hazell

Abstract The effects of post-foaming thermal processing on the quasi-static compressive strength of closed-cell aluminium foam with an average relative density of 0.56 g/cc have been investigated. Samples were subject to one of four material conditions: as-received (AR), warm aged (WA), solution treated (ST), and annealed (AT). X-ray diffraction (XRD) was performed to determine a relationship between each material condition and macro-crystallographic texture of the dendritic α-Al phase. It was found that all heat treated samples contained clear deviations from random texture towards the exterior regions of the samples, indicating non-uniform recrystallization kinetics. The ST and WA samples exhibited the highest yield stresses and energy absorption, with the AT samples exhibiting the lowest. Additionally, it was found that the AR and WA conditions lost 12.8 and 15.6% of their total mass from brittle cell wall crumbling, whereas the AT and ST conditions lost 2.5 and 3.0% of their total mass.

Keywords Aluminium foams · Energy absorption · Heat treatments · XRD · Mechanical properties

Introduction

Metallic foams are a highly desirable engineering material in the automotive, aerospace, civil structures, and defense industries due to their ability to absorb large amounts of energy for various loading conditions [1–5]. Metallic foams subjected

A.D. Brown (✉) · M.A. Islam · M.A. Kader · J.P. Escobedo · P.J. Hazell
SEIT, UNSW Canberra at the Australian Defence Force Academy,
Canberra, ACT 2600, Australia
e-mail: A.Brown@adfa.edu.au

W.D. Hutchison
PEMS, UNSW Canberra at the Australian Defence Force Academy,
Canberra, ACT 2600, Australia

S. Ikhmayies et al. (eds.), *Characterization of Minerals, Metals, and Materials 2017*, The Minerals, Metals & Materials Series,
DOI 10.1007/978-3-319-51382-9_24

to an external load results in a mechanical response that can typically be broken down into three distinct regimes: (1) elastic loading, (2) plastic yielding followed by a decrease in load to a plateau stress level that may be held to upwards of 20–40% plastic strain, and (3) densification [1, 2]. The ability of aluminium alloy foams to deform to high plastic strains has been widely studied, however; the effects post foaming thermal processing have on aluminium alloy foam's microstructure [6–9], crystallographic texture, and mechanical properties [3, 6–12] require further investigation to be comprehensively linked.

The elemental composition for some closed-cell aluminium foams are manufactured from an Al–Cu–Si or Al–Mg–Si base alloy containing a high weight percent (6–12%) of silicon [3–5, 9]. Optimization of mechanical properties by solution treatments [13] and ageing treatments [14] are well known for Al–Cu–Si and Al–Mg–Si cast alloys. Unlike solutions treatment processes, warm aging time durations have shown to be independent on the coarseness of the dendritic α-Al phase microstructure [13, 14]. Intermetallic particle precipitation and phase coarseness are heavily dependent on thermal processing [15]. High Si wt% content aluminium alloys have been found to contain platelet-shaped β-Al_xFeSi intermetallic phases. Jeenager et al. [7] reported coarsening of the intermetallic β-phase from over-aging of aluminium foams lead to the formation of micro-cracks leading to brittle fracture.

Although it has been shown that various post-foaming thermal processing can lead to the improvement in the energy absorption of metallic foams, it is important to delve deeper into the microstructural analyses of these processes and is the aim of this study. It may be possible to further enhance the mechanical properties of any given open or closed-cell metallic foam via introduction of crystallographic texture, as metals are anisotropic, and optimization of the intermetallic β-phase(s).

Experimental Procedures

A plate of closed-cell aluminium foam with a density of ~0.56 g/cc (relative density 20.5%) was obtained from CYMAT™. Additional information on the foam composition may be found in [4, 5]. Quasi-static compression samples measuring 25 × 25 × 26 mm (±0.5 mm) were machined from the plate using a low-speed band saw. The plate contained a thin exterior skin (~0.5 mm thick) layer on the top and bottom faces which were left on for testing. Four samples underwent one of four thermal processing schedules shown in Table 1; one characterization sample for XRD and Vickers hardness testing and three samples for mechanical testing. The heat treatment schedule was based on the manufacturer's suggestions.

Characterization samples were sectioned using a diamond wafer blade to study the effects of the heat treatment on phase distributions and crystallographic macro-texture of the primary dendritic α-Al phase. Square sections measuring ~12.5 mm^2 × ~1.5 mm in height were produced from near the outer edges and halfway through the thickness of the samples to fit on a single crystalline silicon

Table 1 Heat treatment schedule for all samples tested

Heat treatment condition	Temperature (°C)	Hold time (h)	Cooling method	Mean Vickers hardness; std. deviation
As-received (AR)	–	–	–	85.0; $\sigma = 12.1$
Warm aging (WA)	155	1.0	Air	104; $\sigma = 12.5$
Annealing (AT) + WA	415	1.0	Furnace	67.0; $\sigma = 9.45$
Solutionising (ST) + WA	525	8.0	Water quench	102.1; $\sigma = 18.0$

holder. A PANalytical Empyrean diffractometer with a copper radiation source was used for XRD analysis. Samples were rotated at one revolution per second with 2θ acquisition every 49.7 s. The α-Al phase intensity peaks were isolated and underwent a LeBail method refinement to obtain (111), (220), and (200) intensity peaks for crystallographic macro-texture analysis.

Similar to the XRD samples, a diamond wafer blade was used to exhume sections from each heat treatment conditions $\sim$25 mm^2 × 5 mm in height for Vickers hardness testing. A minimum of ten Vickers hardness measurements were performed using a diamond indenter at 1 kgf load and a hold time of 10 s; these measurements are shown in Table 1. Indentations were taken along cell wall junctions and thicker cell walls spread out across the surface cross-section.

Quasi-static compression testing (10^{-3} 1/s) was conducted using a Shimadzu universal test machine equipped with a 50 kN load cell and a laser extensometer for additional accuracy of cross-head displacement data. The overall dimensions and mass of each sample were recorded prior to and after testing. The energy absorption per unit volume of each sample up to a strain of ε_i was calculated by taking the integral of the stress-strain data:

$$W = \int_0^{\varepsilon_i} \sigma \mathrm{d}\varepsilon \tag{1}$$

Additionally, a Nikon DSLR camera recorded each experiment at 60 fps to further study pore collapse mechanisms.

Results and Discussions

Macro-texture analysis of the α-Al phase and phase composition of the aluminium SAFs are investigated from XRD. Table 2 shows intensities of the primary planes in which aluminium produces Bragg scattering for the entire heat treatment schedule listed in Table 1. The closer the intensities and plane ratio intensities are to the aluminium powder standard the closer the crystallographic texture is to purely random. The average intensities of five AR samples taken from various locations in

Table 2 Texture analysis for the dendritic α-Al phase from XRD (values marked with * are an average of 4 scans from multiple plate locations; ratios for the AR condition also include a 5th scan of higher resolution/counts)

Heat treatment condition	(111) Intensity	(200) Intensity	(220) Intensity	(111)/(200)	(111)/(220)	(200)/(220)
Al powder	**1000**	**485**	**310**	**2.06**	**3.23**	**1.56**
AR	930*	451*	352*	2.23	2.72	1.27
WA—inner	1240	608	385	2.04	3.22	1.58
WA—outer	1763	585	536	3.01	3.29	1.09
AT—inner	1491	705	467	2.11	3.19	1.51
AT—outer	1379	628	338	2.19	4.08	1.86
ST—inner	1089	568	376	1.92	2.89	1.51
ST—outer	1715	584	424	2.94	4.05	1.38

the bulk plate indicate that the AR material contains a fairly random crystallographic orientation distribution, with a slight excess in out of plane [110]. Interestingly, the interior regions of all heat treated samples maintained a close to random crystallographic texture, with the ST condition containing slight deficiencies in out of plane [111], however; regions near the sample exterior contained elevated deviations from the powder standard. The outer regions of the WA and AT samples contained deficiencies in out of plane [100] and [110], respectively. The ST condition appears to contain excess out of plane [111] and slight deficiency in out of plane [110]. These results indicate non-uniform recrystallization kinetics through the thickness of closed-cell aluminium foams during post-foaming thermal processing and warrant further microstructural investigation.

The raw XRD intensity plot for the AR material condition along with the three primary phase constituents is plotted in Fig. 1a. Accurate refinement analysis requires further information on the atomic positions of the monoclinic β-Al–Fe–Si intermetallic phase. The raw XRD intensity plots for scanned interior and exterior sections for the ST foam condition are shown in Fig. 1b. This clearly illustrates a deviation from a closely random crystallographic texture towards the exterior regions of the sample with approximately double the intensity of (111) Bragg reflection. Further thermal processing is required in order to attempt to put a uniformly global texture into aluminium SAFs. Depending on the crystallographic orientation of the texture with respect to a given loading direction, it may be possible to utilize anisotropy to create a higher strength and/or more ductile mechanical response depending on the desired material application.

The quasi-static compressive response of the closed-cell aluminium foams varied slightly with the post foaming thermal processing conditions. Average yield strengths for the AR, WA, AT, and ST conditions were 5.96, 7.49, 5.53, and 6.45 MPa, respectively. Figure 2a, b show the average stress-strain response of each thermal processing condition. It was observed that the WA process contained the highest hardness and yield strength, though both statistically similar to the ST condition. As expected from similar studies in the literature [10, 12], the AT

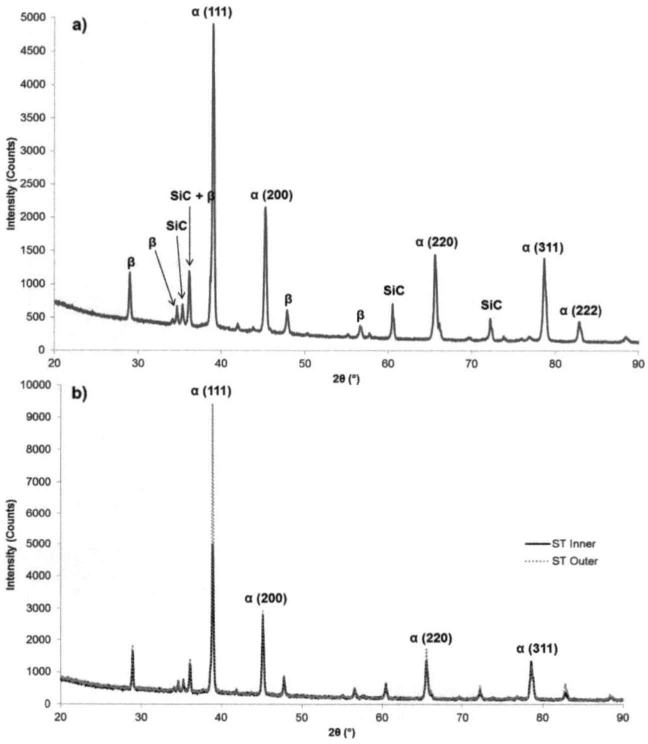

Fig. 1 X-ray diffraction patterns from **a** bulk average as-received aluminium foam and **b** solution treated foam from the interior (*black*) and exterior (*dotted red*) regions of a sample (Color figure online)

condition decreased in overall strength from the AR condition. Figure 2c–e illustrates how the collapse of the SAFs begins at weak, low density regions within the material at the initial stages of deformation. This region was found to be a function of local material strength only; independent of its location with respect to the platens.

Figure 3 shows the average absorbed energy for each material condition and indicates that at about 40% total strain the ST condition begins to absorb more energy than the WA condition. The AT condition results in an overall decrease in absorbed energy compared to the AR material. The final mass of each compacted sample was measured after each test, with any loose material discarded. It was found that the AR and WA material conditions lost 12.8 and 15.6% of their total masses from brittle crumbling of the material during densification. Alternatively, the AT and ST material conditions only lost 2.5 and 3.0% of their mass from crumbling. The AT and the ST processes have stabilized the foam and made the material more ductile than the AR and WA conditions, consistent with reports by Jeenager et al. [7]. Considering all the aforementioned data, it is expected that the ST condition may be most desirable for this material for applications where near

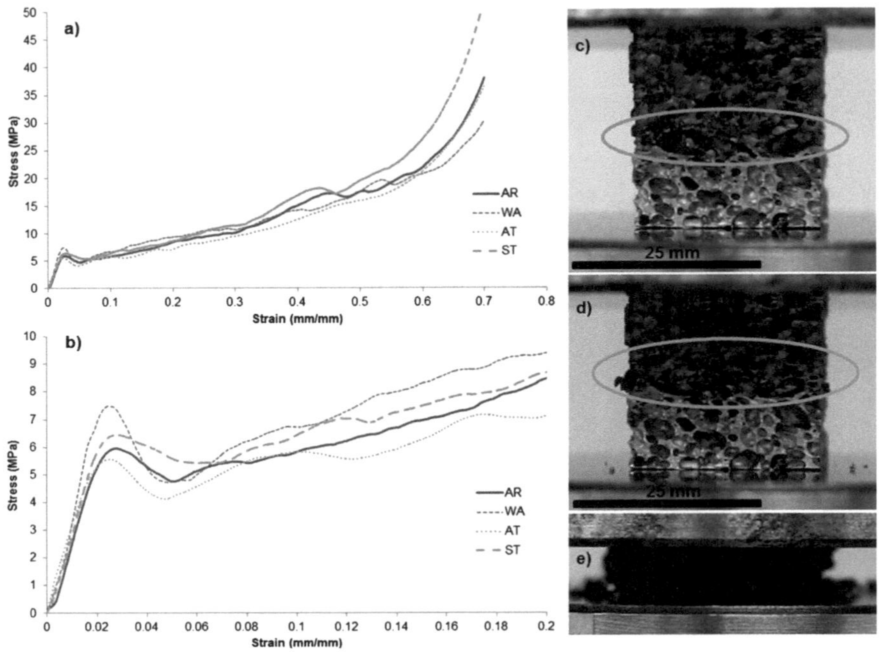

Fig. 2 Average stress-strain response for each thermal process tested for **a** full densification up to 70% strain and **b** enhanced view of the yield, relaxation, and plastic region up to 20% strain. Additionally, the collapse of an as-received sample is shown in **c** elastic regime at $\varepsilon = 1\%$, **d** pore collapse of the lowest density region at $\varepsilon = 10\%$, and **e** full compaction at a $\varepsilon = 73\%$

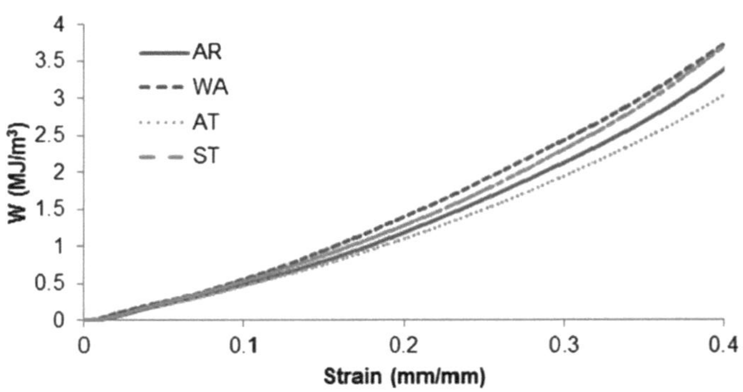

Fig. 3 Average absorbed energy versus strain for each thermal processing condition tested

full compaction is expected. Optical microscopy and further material characterization is required to determine if these heat treatments are stabilizing the dendritic α-Al phase by reducing the thickness of the intermetallic β-Al–Fe–Si phase and/or the structure surrounding the SiC particles.

Conclusions

It has been demonstrated that various thermal processing conditions may change the mechanical properties and total energy absorption of high density closed-cell aluminium foam. X-ray diffraction performed on each material condition suggests deviation from an expected random macro-crystallographic texture as a function of sample through-thickness, indicating non-uniform kinetics of recrystallization. Solution treating and warm aging increased the overall stress response and energy absorption of the aluminium foam, whereas annealing resulted in a decrease of each. Mass loss from material crumbling during the compaction to full densification indicates brittle structural failure of the cell walls for the as-received and warm aged conditions and an increase in ductile failure for the annealed and solution treatment conditions; likely due to coarseness of the intermetallic β-phase. Additional mechanical testing of an expanded heat treatment schedule and more in-depth analysis of the microstructure are required to quantitatively elucidate the roles phase structure and crystallographic texture on the mechanical response of closed-cell aluminium foams.

Acknowledgements The authors gratefully acknowledge UNSW Canberra's Defence Related Research program that partly funded this work. Patrick Nolan of the UNSW Canberra technical support group is thanked for his help on the mechanical testing performed for this work.

References

1. Ashby MF, Medalist RM (1983) The mechanical properties of cellular solids. Metall Trans A 14(9):1755–1769
2. Banhart J (2001) Manufacture, characterisation and application of cellular metals and metal foams. Progress Mater Sci 46(6):559–632
3. Wang Z et al (2009) Effect of heat treatments on the crushing behaviour and energy absorbing performance of aluminium alloy foams. Mater Des 30(4):977–982
4. Islam MA, et al (2016) Investigation of microstructural and mechanical properties of cell walls of closed-cell aluminium alloy foams. Mater Sci Eng A 666:245–256
5. Kader MA et al (2016) Modelling and characterization of cell collapse in aluminium foams during dynamic loading. Int J Impact Eng 96:78–88
6. Jeenager VK, et al (2012) The effect of aging on energy absorption capability of closed cell aluminium foam. Adv Mater Res 585
7. Jeenager VK, Pancholi V (2014) Influence of cell wall microstructure on the energy absorption capability of aluminium foam. Mater Des 56:454–459
8. Kanahashi H et al (2001) Improvement of crashworthiness in ultra lightweight metallic foam by heat-treatment for microstructural modification of base material. Mater Trans 42(10): 2087–2092
9. Lehmhus D, Banhart J (2003) Properties of heat-treated aluminium foams. Mater Sci Eng A 349(1):98–110
10. Campana F, Pilone D (2009) Effect of heat treatments on the mechanical behaviour of aluminium alloy foams. Scripta Mater 60(8):679–682
11. Yamada Y (2000) Effects of heat treatment on compressive properties of AZ91 Mg and SG91A Al foams with open-cell structure. Mater Sci Eng A 280(1):225–228

12. Zhou J, Soboyejo WO (2004) Compression–compression fatigue of open cell aluminium foams: macro-/micro-mechanisms and the effects of heat treatment. Mater Sci Eng A 369(1): 23–35
13. Sjölander Emma, Seifeddine Salem (2010) Optimisation of solution treatment of cast Al–Si–Cu alloys. Mater Des 31:S44–S49
14. Sjölander E, Seifeddine S (2011) Artificial ageing of Al–Si–Cu–Mg casting alloys. Mater Sci Eng A 528(24):7402–7409
15. Liu L et al (2009) Precipitation of β-Al5FeSi phase platelets in Al-Si based casting alloys. Metall Mater Trans A 40(10):2457–2469

Experimental Investigation of Mechanical Behaviour of Closed-Cell Aluminium Foams Under Drop Weight Impact

M.A. Islam, M.A. Kader, A.D. Brown, P.J. Hazell, J.P. Escobedo and M. Saadatfar

Abstract The dynamic pore collapse mechanisms and energy absorption capacity of closed-cell aluminium foams subjected to low velocity impacts have been investigated. Impact experiments were carried out using an instrumented drop-tower with two impactor geometries: a 50 mm square flat steel impactor to investigate pore collapse mechanisms and a 20 mm diameter hemispherical steel impactor to study indentation resistance. X-ray computed tomography was utilized to generate views of the deformation field within the impacted specimen. The results show that the strain rate and incident impact energy influence the energy absorption capacities of closed-cell aluminium foams. Furthermore, damage initiation, propagation and cell collapse mechanisms under drop weight impact have been elucidated.

Keywords Aluminium foams · Drop impact · Compression deformation · Closed-cell aluminium foams · Energy absorption · Cell collapse mechanism · Low velocity impacts

Introduction

The extraordinary energy absorption capability of closed-cell aluminium foam in impact loading is well established [1–5]. The applications of these materials are promising in automotive, aerospace engineering and defence industries [6]. Furthermore, application as a sacrificial cladding is also encouraging as they deform to very high plastic strains [7]. Efficient use of these foams in structural applications

M.A. Islam (✉) · M.A. Kader · A.D. Brown · P.J. Hazell · J.P. Escobedo
SEIT, UNSW Canberra at the Australian Defence Force Academy,
Canberra, ACT 2600, Australia
e-mail: Md.Islam4@student.adfa.edu.au

M. Saadatfar
Department of Applied Mathematics, Australian National University,
Canberra, ACT 0200, Australia

S. Ikhmayies et al. (eds.), *Characterization of Minerals, Metals, and Materials 2017*, The Minerals, Metals & Materials Series,
DOI 10.1007/978-3-319-51382-9_25

demands meticulous characterization of their deformation behaviour and energy absorption capabilities for various loading rates. Collision and localized impact at low velocity are the most common phenomena these materials are likely to experience, hence; it is necessary to fully understand the mechanical response for metallic foams at these loading rates.

One of the most widely used methods to observe low velocity impact behaviour is an instrumented drop weight impact. A number of studies on indentation impact of aluminium foams were presented [8–10] where test results described mechanical behaviour of aluminium foams used in a core sandwich. In those cases, the top plate absorbed most of the energy. Normalized plastic strength variation on strain rate of closed-cell aluminium foam with a volume mass of 0.21 g/cm^3 has been studied by Ramachandra et al. [11]. Castro et al. presented the drop weight test on aluminium syntactic foams (ASF) with a hemispherical impactor of diameter 4.45 mm [12]; whereas, Shen et al. compressed ALPORAS foams at strain rates from 10^{-3} to 2.2×10^2 s^{-1} with a constant cross-head velocity [13, 14]. The result shows that the ALPORAS foams exhibit noticeable strain rate sensitives to the plateau stress as well as the densification strain. For many practical applications the impactor, or projectile, is likely smaller than the metallic foam energy absorber. Thus, a non-constant deformation rate arises as the impactor is slowed by the metallic foam during penetration and may behave differently than when subject to a constant loading rate as in [13, 14]. Furthermore, the cell collapse mechanism during dynamic loading is not clearly established in the open literature.

In this work, we investigated dynamic compression response of CYMAT closed-cell aluminium foams of volume mass 0.51 g/cm^3 with a drop tower using a flat-ended impactor to observe pore collapse mechanisms. Furthermore, indentation experiments were carried out using both square and hemispherical shaped steel impactors. Micro-structural analysis during the impact has been described using computed X-ray tomography. In summary, this study describes failure mechanisms and structural performance of closed-cell aluminium foams subject to low velocity impacts and penetration.

Experimental Procedures

The performance of commercially available stabilized aluminium foam (SAF) of nominal volume mass 0.51 g/cm^3 supplied by CYMAT corp. was used for this work. A 0.5–1 mm thick exterior skin layer was removed from the top and bottom of the as received foam, resulting in a final nominal volume mass of $\sim$0.48 g/cm^3 for all samples tested. Sample dimensions for studying pore collapse measured $50 \times 50 \times 25$ mm, while flat and hemispherical indentation samples measured $90 \times 90 \times 25$ and $40 \times 40 \times 25$ mm, respectively. This foam's elemental composition, microstructural composition, and mechanical properties have been reported in previous work [15, 16].

Drop tower tests were carried out at impact velocities ranging from 2.5 to 10 m/s with an instrumented drop tower (CEAST 9350) equipped with a 90 kN force sensor. The total weight of the free-fall mass was kept within a range of 5.85–12.85 kg to obtain the desired impact energies and velocities. The tests were performed following ASTM D5628 FA standard. Samples were clamped to a flat rigid plate using G-clamps to ensure rigid boundary conditions throughout the duration of the experiments. Additionally, a Phantom v710 high-speed camera and a HIVE plasma lamp were used to capture video of the impacts at a frame rate of $\sim 11 \times 10^3$ fps. The high-speed video and laser diode velocimetry were used to determine the velocity of the impactors just prior to making contact with the samples. Other transient responses of the test such as impact energy, load, and displacement were automatically calculated by the system's software from the force sensor data.

Compaction experiments were conducted using a flat steel impactor with a square cross-section of 50 mm on the samples of the same dimensions. Pore collapse mechanisms were investigated for this configuration. Indentation experiments were conducted using the aforementioned square impactor and a 20 mm diameter hemispherical steel impactor. To ensure bulk material response to the dynamic indentation tests the samples had lateral dimensions almost double that of the impactors. Furthermore, the μ-CT data has been utilized to analyse the deformation. The specimen was used to image using μ-CT before impact loading and later one post-impacted specimen was also imaged again to analysis the pore collapse.

Results and Discussions

The force-time curves for closed-cell aluminium foams with different impact velocities have been shown in Fig. 1. The trend of force-time curves can be generally described as the load reaching a maxima followed by subsequent drop due to the collapse of a band of cells resulting in localised densification followed by an increase in measured force. Thus the area covered by the curve is the impulse given by the

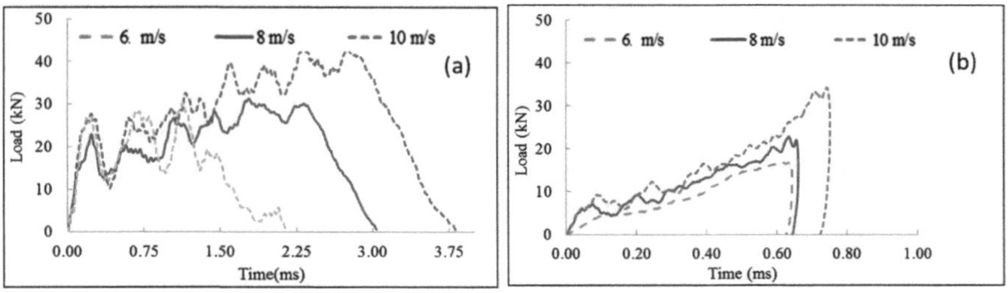

Fig. 1 Load-time curves of the closed-cell aluminium foam of volume mass 0.51 g/cm^3 using: **a** flat and, **b** hemispherical indenter

Table 1 Energy absorption per unit deformed volume (MJ/m^3) by the flat and hemispherical indenter

Indenter	$V_0 = 2.5$ m/s	$V_0 = 3.5$ m/s	$V_0 = 6.0$ m/s	$V_0 = 8.5$ m/s	$V_0 = 9.5$ m/s
Flat	5.896	10.23	14.22	16.88	20.63
Hemispherical	9.85	14.17	17.78	20.22	23.89

drop weight impactor. It is noticed that the load is increased with time linearly up to the initial peak load (elastic limit) which is considered as the resistance of the foam structure to indentation. After the elastic compression, the rest of the plastic-compression curve follows a non-linear path. Similar trends of load versus time curves were found using the hemispherical tup end which covered around 7–9 cells of the foam specimen under the contact surface. Resistance for hemispherical indentation is lower than that to the flat punch indentation. The most remarkable distinction between these two types of test was that the hemispherical indentation did not show a sharp rise of load as the area of the indenter gradually increased as it went through the specimen. However it showed the three stages of deformation similar to the quasi-static compaction because of increase in contact area. It is also noticed that large oscillations have been followed when the tests were carried out using a flat impactor. This was probably due to several crush (shear) bands formations during the compression. Similar oscillations were observed by Deshpande and Fleck [17] from strain rates above 50 s^{-1}. The energy absorption per unit volume deformation for different drop impact velocities is shown in Table 1. It demonstrates that energy absorbed per unit deformed volume (MJ/m^3) increases with the increase of impact velocity. Moreover, the increase of energy absorption with the increase of impact velocity is evidence of strain-rate sensitivity of the closed-cell aluminium foam that has also been claimed by other studies [18, 19]. It was noticed that energy absorption in the hemispherical indenter was higher than that of the flat indenter. The possible reasons of higher energy required in hemispherical indenter are: (i) constantly increasing crushed zone of compressed material underneath the hemispherical tup-end and (ii) extra work required to tear its peripheral cells.

To observe the deformation behaviour and pore collapse, high-speed Phantom camera recorded video has been analysed. Also, correlation between the load-time curves was assessed. The visual deformation of foams by a flat indenter was illustrated with a test at an impact velocity of 6.83 m/s with an impact energy of 291 J. Figure 2a shows the test sample just before the impact (t = 0). At the initial peak load, pressure build up within the cells exceeds the fracture strength of the cell-wall which results in pore collapse. The initial band of cell collapse appeared along the middle dotted line indicated in Fig. 2b. The high-speed video shows that the distortion and twisting went together with the progression of a localized deformation band. Further deformation continued with time due to the subsequent increase of load.

A rapid deformation along the weak band of the specimen was also observed; that is, a series of cell walls simultaneously collapsed during loading. However, in quasi-static loading [15], it was found that deformation process was sequentially

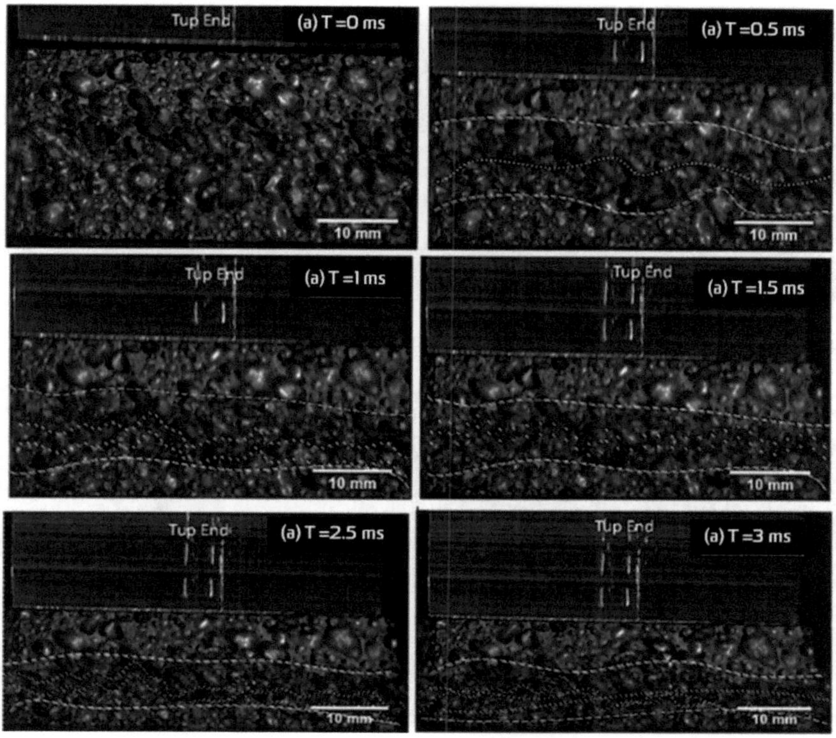

Fig. 2 Load-time curve of the drop tower impact performed at an impact velocity of 6.85 m/s. Stages **a–f** are the moments of impact between the impactor and specimen with an interval of approximately 0.5 ms

followed by consecutive cell collapse. There were similarities between the drop impact deformation and quasistatic deformation in that pore collapse was observed in the area where there were big pores within a cell wall. This region was usually in the middle region of the specimen. Then the X-ray density histogram was constructed from the datasets. Figure 3 shows the density map for 2-D slice obtained from the uncompressed specimen (Fig. 3a) and compressed specimen (Fig. 3b).

It is observed that cell distribution in CYMAT foams was not uniform over the cellular structure along the height. Cell size at the middle of the foam is larger than that of the top and bottom region of the foam. Figure 3b shows in CT image of the deformed specimen which was impacted with energy 321.3 J ($V_0 = 6.85$ m/s). Most of the deformation remains in the middle of the specimen.

The micro-graph of deformed sample tested in quasi-static ($\sim 10^{-3}$ s^{-1}) [15] and drop impact (4.8×10^2 s^{-1}) loading shows that the deformation pattern is distinct from each other (see Fig. 4). In the quasi-static compression, less cell crumbling was observed in contrast to drop tower compaction. This perhaps means that during quasi-static compaction, cell collapse is sequential (bending, buckling and flattening

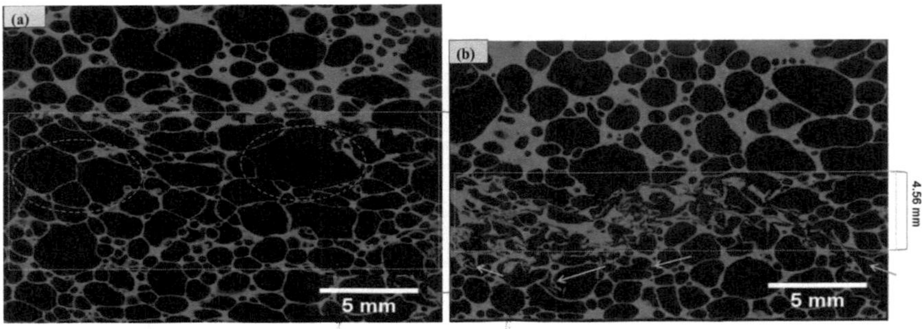

Fig. 3 X-ray density map showing a undeformed specimen and b deformed specimen

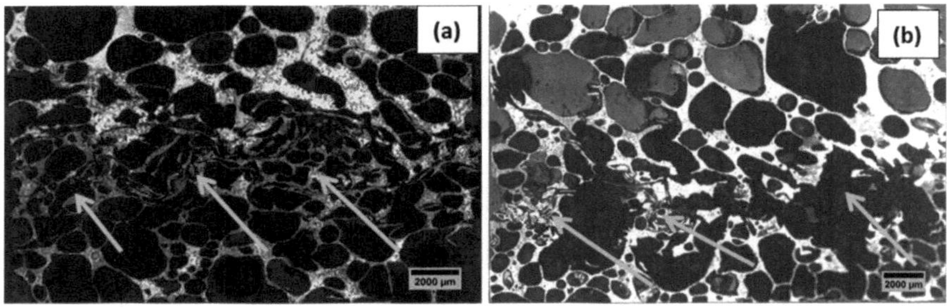

Fig. 4 Micro-graph of a quasi-statically compressed (ε = 0.56) and b drop compacted sample (ε = 0.6)

due to densification); whereas at the higher strain rate compaction the cell wall break in several places simultaneously. The breakage of cells disjointed further with the interaction of wave reflection and frictional interaction by the rigid bottom surface of the specimen.

From the observation of micro-graph and 2D X-ray tomography data, an initial cell-collapse mechanism is proposed both in quasi-static and during drop-weight impact. Let us consider each closed-cell of aluminium foam cell as polyhedral. This consideration of polyhedral shape can be fairly accepted for foams with low density [5]. Half of the unit cell has been taken into consideration for two varying orientations with respect to the loading direction, as shown in the Fig. 5. The collapse mechanism is comparable to the previous investigation by Shen et al. [14]. The model is substantiated by μ-CT tomo image of deformed samples.

In the quasi-static condition, it is shown that cell wall starts to bend from the position **1–2**. Then with an increase in load, the cell wall bent further and became flat. This process of pore collapse in static condition is corroborated by the micro-graph. Cell collapse due to dynamic loading was accomplished in a very short time and multiple fragmentations in the cell wall appear to result in movement

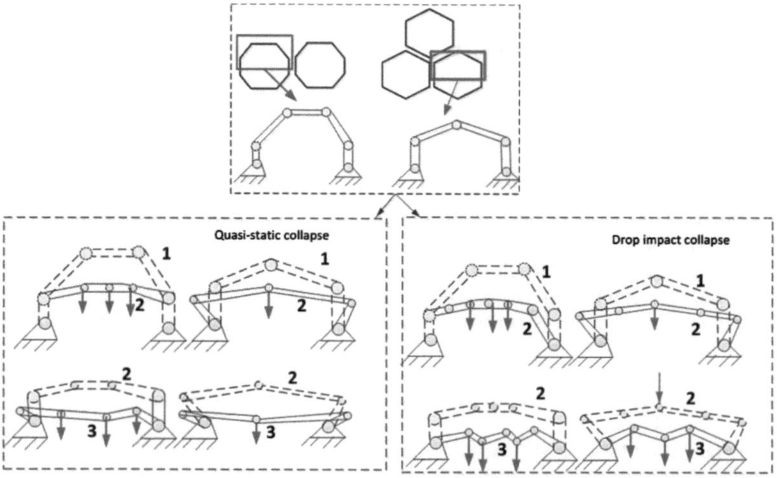

Fig. 5 Pore collapse mechanism of closed-cell aluminium foam under quasi-static and drop-weight impact

from position **2–3**. Furthermore, the cell walls do not have enough time for systematic bending, buckling and flattening.

Conclusions

Dynamic compaction and indentation responses of CYMAT closed-cell aluminium foams have been tested to investigate cell collapse mechanisms and energy absorption capacity. It can be concluded that:

- Energy absorption per unit deformed volume increased with impact velocity. Also it was found that energy absorption during hemispheric indentation was higher than flat punch indentation for the same impact velocity.
- X-ray tomography of deformed and undeformed specimen shows that the cell collapse of CYMAT foams initiates in the weak band of the specimen.
- Finally, a cell collapse mechanism during the quasi-static compression and drop impact has been suggested for these foams. Moreover, it is established that pore collapse mechanism in drop-weight impact loading does not follow the similar failure pattern as in quasi-static compression (bending, buckling and flattening). In fact, the cell-walls fragmented at multiple places simultaneously during impact loading.

Acknowledgements The authors gratefully acknowledge UNSW Canberra's Defence Related Research program that part-funded this work. M.A. Islam also likes to thank Jill Middleton for her technical support in X-ray image analysis in the Department of Applied Mathematics, The Australian National University, Canberra, Australia.

References

1. Banhart J (2001) Manufacture, characterisation and application of cellular metals and metal foams. Prog Mater Sci 46:559–632
2. Islam MA, Escobedo JP, Hazell PJ, Appleby-Thomas GJ, Quadir MZ (2015) Characterisation of closed-cell aluminium foams subjected to compressive loading. In: Proceedings of characterization of minerals, metals and materials, 2015 TMS annual meeting and exhibition, 15–19 Mar 2015, Orlando, FL, USA
3. Kadar MA, Islam MA, Escobedo JP, Hazell PJ, Saadatfar M, Brown AD (2016) Modelling and characterization of cell collapse in aluminium foams during dynamic loading. Int J Impact Eng 96:78–88
4. Saadatfar M, Mukherjee M, Madadi M, Schröder-Turk GE, Garcia-Moreno F, Schaller FM, Hutzler S, Sheppard AP, Banhart J, Ramamurty U (2012) Structure and deformation correlation of closed-cell aluminium foam subject to uniaxial compression. Acta Mater 60:3604–3615
5. Ruan D, Lu G, Chen FL, Siores E (2002) Compressive behaviour of aluminium foams at low and medium strain rates. Compos Struct 57:331–336
6. Hazell PJ (2015) Armour: materials, theory, and design. CRC Press, Boca Raton
7. Hanssen AG, Enstock L, Langseth M (2002) Close-range blast loading of aluminium foam panels. Int J Impact Eng 27:593–618
8. Qin Q, Zhang J, Wang Z, Li H-m, Guo D (2014) Indentation of sandwich beams with metal foam core. Trans Nonferrous Met Soc China 24:2440–2446
9. Shazly M, Bahei-El-Din Y, Salem S (2013) Characterization of sandwich panel for indentation and impact. J Phys (conf ser) 451:012001
10. Yahya MA, Ruan D, Lu G, Dargusch MS (2015) Response of aluminium honeycomb panels subjected to foam projectile impact—an experimental study. Int J Impact Eng 75:100–109
11. Ramachandra S, Sudheer Kumar P, Ramamurty U (2003) Impact energy absorption in an Al foam at low velocities. Scr Mater 49:741–745
12. Castro G, Nutt SR, Wenchen X (2013) Compression and low-velocity impact behaviour of aluminium syntactic foam. Mater Sci Eng A
13. Lu G, Shen J, Hou W, Ruan D, Ong LS (2008) Dynamic indentation and penetration of aluminium foams. Int J Mech Sci 50:932–940
14. Shen J, Lu G, Ruan D (2010) Compressive behaviour of closed-cell aluminium foams at high strain rates. Compos B 41:678–685
15. Islam MA, Hazell PJ, Escobedo JP, Saadatfar M (2014) In-situ quasistatic compression and microstructural characterization of aluminum foams of different cell topology. Int J Chem Nucl Mater Metall Eng 8:1193–1198
16. Islam MA, Kader MA, Hazell PJ, Brown AD, Saadatfar M, Quadir MZ, Escobedo JP (2016) Investigation of microstructural and mechanical properties of cell walls of closed-cell aluminium alloy foams. Mater Sci Eng, A 666:245–256
17. Deshpande VS, Fleck NA (2000) High strain rate compressive behaviour of aluminium alloy foams. Int J Impact Eng 24:277–298
18. Tan PJ, Reid SR, Harringan JJ, Zou Z, Li S (2005) Dynamic deformation of aluminium foams. Part I—Experimental data and observations. J Mech Phys Solids 53:2174–2205
19. Paul A, Ramamurty U (2000) Strain rate sensitivity of a closed-cell aluminium foam. Mater Sci Eng, A 281:1–7

Deformation Mechanisms of Closed Cell-Aluminium Foams During Drop Weight Impact

M.A. Kader, M.A. Islam, A.D. Brown, P.J. Hazell, M. Saadatfar and J.P. Escobedo

Abstract The present study investigates the dynamic deformation mechanisms of closed-cell aluminium foams during low velocity drop weight impact with experimental analyses and finite element (FE) simulation. The evolution of foam collapse was explored with FE simulation using ABAQUS/Explicit. X-ray computed tomography (XRT) based geometry was reconstructed to understand the actual microstructural changes during impact. The experimental stress-strain response was compared with FE simulation with reasonably good agreement observed between FE prediction and experimental data. A vertical cross-sectional XRT slice was analysed at different strains from FE simulation to understand the pore collapse mechanisms.

Keywords Aluminium foam · X-ray computed tomography · Pore collapse · Drop impact

Introduction

Closed-cell aluminium foams offer promising energy absorption capacity for impact loading conditions by deforming to large amounts of plastic strain. Recent studies have found that the complex structures of foam play important role on the energy absorption performance during dynamic deformation [1, 2]. Islam et al. [3] carried out an experimental investigation on dynamic collapse behaviour of aluminium foams and reported that the pore collapse mechanisms at drop impact loading significantly affect the mechanical behaviour of foam. Therefore, investigating the

M.A. Kader · M.A. Islam · A.D. Brown · P.J. Hazell · J.P. Escobedo (✉)
School of Engineering and Information Technology,
The University of New South Wales, Canberra, Australia
e-mail: Diaz@adfa.edu.au

M. Saadatfar
Department of Applied Mathematics, Australian National University (ANU), Canberra, Australia

S. Ikhmayies et al. (eds.), *Characterization of Minerals, Metals, and Materials 2017*, The Minerals, Metals & Materials Series,
DOI 10.1007/978-3-319-51382-9_26

dynamic collapse mechanisms have become an important issue for fully characterizing the mechanical behaviour of metallic foams. A number of previous experimental and numerical studies have been performed to investigate the deformation behaviour of foams at low velocity drop impact [4–6]. However, it is difficult to explore the dynamic deformation and pore collapse mechanisms in-situ and limited to the exterior surfaces of the foam. Finite element simulations are able to further elucidate volumetric collapse of foam structures subject to a wide range of strain rates. Most previous numerical analyses on drop impact were based on continuum geometry, thus making it impossible to explain the collapse process [7–9].

In the present study, we have investigated the deformation behaviour of aluminium foams during drop impact loading. A number of low velocity drop impact experiments and FE simulations with XRT based geometry have been performed to explore the foam collapse mechanism throughout the dynamic deformation process.

Materials

The dynamic behaviour of closed-cell aluminium foams (nominal density of 0.51 g/cc) have been investigated in the present study. The details of morphological and topological features and characterization of this foam has been carried out by XRT which has been provided in our previous work [10, 11]. The supplied foam panel had solid layers on top and bottom surfaces. These solid layers have been removed to obtain structural homogeneity across all samples. Twelve samples of 15 mm × 15 mm × 8 mm dimensions have been prepared for experimental observation. The supplied foam panel and two prepared samples are shown in Fig. 1.

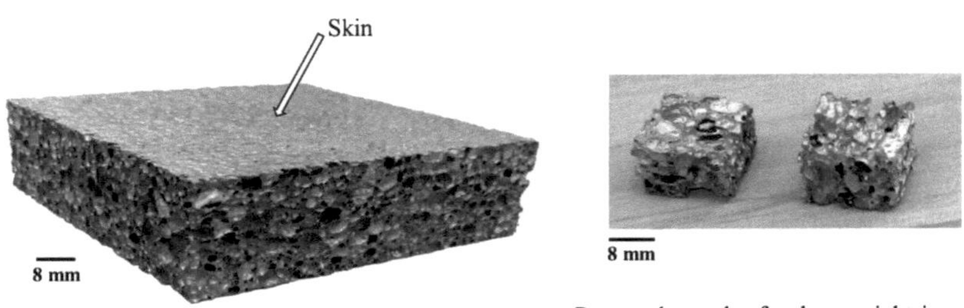

Foam Panel (with skin)

Prepared samples for drop weight impact (without skin)

Fig. 1 Foam panel and prepared samples

Experimental

A CEAST 9350 instrumented drop tower was used for conducting low velocity (1.5–3 m/s) impact experiments. A flat square shaped steel impactor was attached to the end of the free-falling instrumented 90 kN load cell-tup design. The tests were carried out at room temperature, a constant impactor mass (5.639 kg), and with the sample resting on a flat and rigid steel platen inside of the test chamber. A laser-diode velocity sensor was positioned such that the velocity just before impact was recorded.

The foam was imaged with XRT to reconstruct the full 3D structure. A flat panel detector and micro-focus X-ray source of 100 kV were used for imaging. The foam specimen was placed on a motor controlled rotating stage and radio-scope projections were taken after each degree of rotation. The voxel size of the X-ray tomography was 39 μm.

Finite Element Simulation

The foam geometry for FE simulation was reconstructed using XRT data. This process allows for non-destructive investigation of the complex internal foam structure. The grey scale computed tomography was converted to binary images with Medial Axis Geometry Development software (developed at Australian National University). Then the segmented geometry was meshed and imported to FE software ABAQUS/Explicit. The reconstructed geometry and FE assembly are shown in Fig. 2. Tetrahedral elements were used to mesh the real geometry because

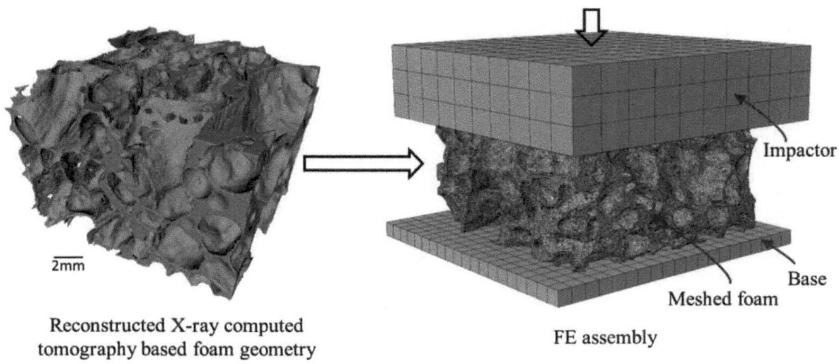

Fig. 2 Foam geometry rendering and FE assembly for drop impact experiment

of its morphological complexity. The top and bottom platen were modelled as rigid plate. The elastic-plastic material model with strain hardening was used for foam material. The ductile damage criteria were applied to the model.

Result and Discussion

Table 1 shows the variation of yield strength of aluminium foams for different strain rates. It shows that the yield strength of the foam increases with increasing the strain rate. Therefore, it clearly reveals that the closed-cell aluminium foam is a strain rate sensitive material.

A numerical simulation of a 1.5 m/s impact velocity has been performed for this study. The comparison of stress-strain response between experiments and simulation is shown in Fig. 3. A reasonably good agreement is found between the

Table 1 Variation of yield strength with strain rate

Type of loading	Impact velocity (m/s)	Strain rate (s^{-1})	Yield strength, σ_y (MPa)
Quasi-static [10]	–	10^{-3}	5.20
Drop impact	1.50	170	7.35
	2.00	230	8.07
	2.50	295	8.20
	3.00	305	8.60

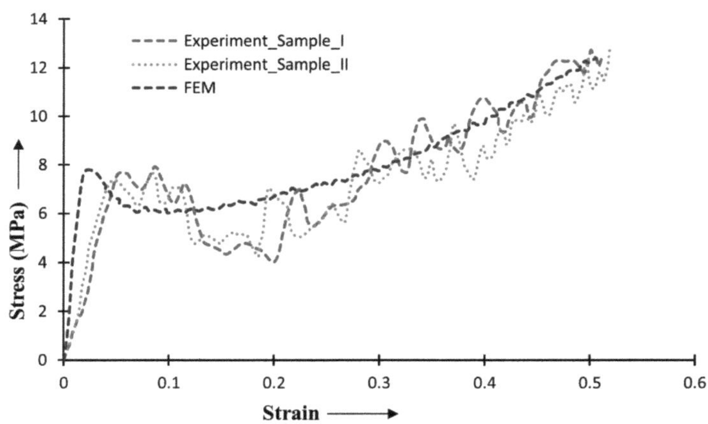

Fig. 3 Comparison of stress-strain response between experiments and FE modelling for an impact velocity 1.5 m/s

$\varepsilon = 0$ $\varepsilon = 0.52$

Fig. 4 Experimental sample **a** before compaction and **b** after compaction at 1.5 m/s

experimental results and FE simulation except for minor discrepancy when capturing the initial elastic loading. The possible underlying reason behind this is the initial deformation of the foam's uneven top surface and possible minor tilt of the experimental samples with respect to the loading direction. It produces the initial plastic deformation and consequently affects the elastic response. Figure 4 shows the sample before and after drop impact compaction. However, it is difficult to understand the intermediate states of collapse through post impact observation of samples. Although it is possible to visualise the deformation of surface collapse with high resolution live camera, but it is impossible to understand the in-situ collapse evolution of the foam's interior [3]. On the other hand, the present FE simulation with XRT-based geometry is able to show the whole volume at any particular instant.

A single interior cross-section has been selected for analysis at varying strains and is presented in Fig. 5. It is shown that deformation initiates at thin cell-walls and the load is subsequently transmitted to surrounding cell-walls and plateau borders (junctions) where the next weakest region will begin to collapse, and so-forth. The deformation is a continuous process of cell-wall rotation, bending, buckling and tearing. This process continues until the structure undergoes full densification. Additionally, the maximum amount of deformation occurs at low solid fraction regions, which are marked with two dotted lines in the XRT segments shown in Fig. 5. The Von Mises stress distributions show that the plateau borders are able to carry more stress whilst undergoing comparatively less deformation than cell-walls. Collapsing cell-walls occupy the interior of neighbouring voids rather than contributing to bulk lateral expansion; resulting in a negligible amount of lateral expansion up to $\sim 30\%$ axial strain. The collapse mechanisms of a selected pore, **A**, are also shown in Fig. 5. The top cell-wall collapses by first rotating at hinges located at regions of high curvature angles, then undergoing buckling at the hinges, followed by bending of the top cell-wall, and lastly by rupture of the hinges.

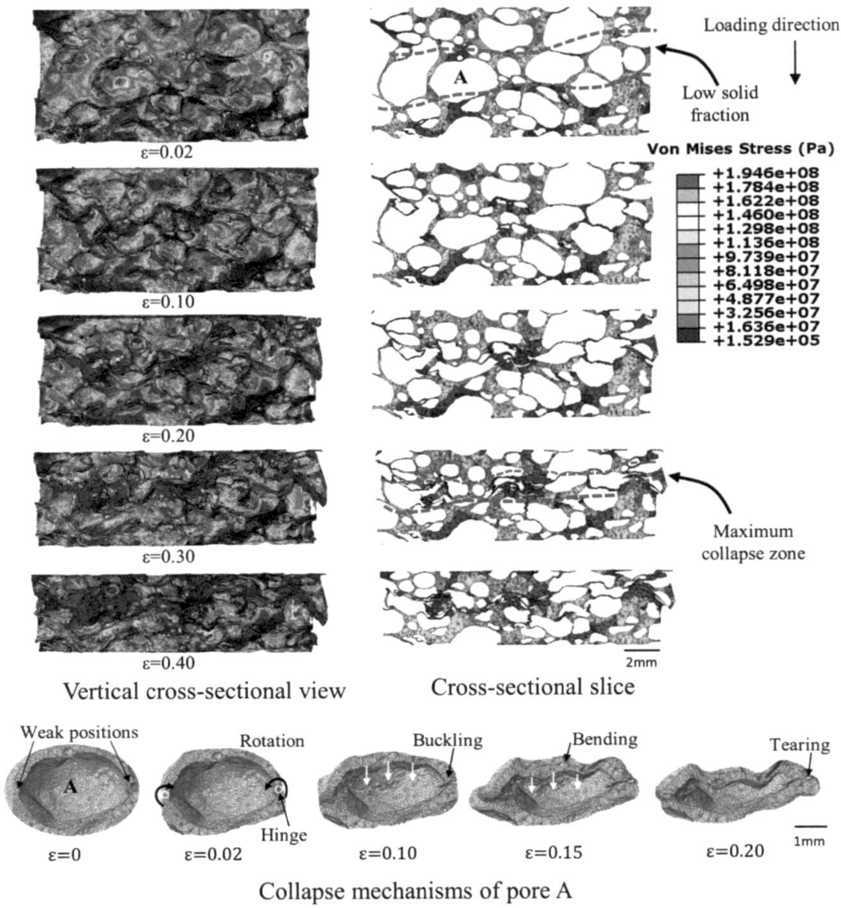

Fig. 5 Deformation mechanisms of closed-cell aluminium foam at drop impact loading from meso-scale modelling at various total strain increments

Conclusions

The present numerical approach is able to replicate the detailed deformation behaviour of closed-cell aluminium foams. The experimental results and FE simulation exhibit similar stress-strain responses. The FE simulation with X-ray computed tomography based reconstructed geometry is able to capture the entire volume at any particular instant and explore the full 3D collapse mechanisms. The thinner cell-walls predominantly deform and transfer the load to surrounding cell-walls and plateau borders. The observed modes of deformation are cell-walls' rotation by forming one or more hinges, bending, buckling and tearing. Moreover, the observed foam shows variation of yield strength at different impact velocities, suggesting that this aluminium foam is a strain rate sensitive material. The observed Poisson's ratio of foam is almost negligible at lower strains (up to ~30% strain).

Acknowledgements The authors gratefully acknowledge the laboratory facility provided by Department of Applied Mathematics, Australian National University. We also gratefully acknowledge the UNSW Canberra Defence Related Research Program that part-funded this work.

References

1. Kader MA, Islam MA, Hazell PJ, Escobedo JP, Saadatfar M, Brown AD, Appleby-Thomas GJ (2016) Modelling and characterization of cell collapse in aluminium foams during dynamic loading. Int J Impact Eng 96:78–88
2. Hazell PJ (2016) Armour: materials, theory, and design, 1st edn. CRC Press, Boca Raton
3. Islam MA, Kader MA, Hazell PJ, Escobedo JP, Brown AD, Saadatfar M (Submitted) Deformation and failure mechanism of closed-cell aluminium foams under drop weight impact. In: Proceeding of characterisation of minerals, metals and materials, 2017 TMS annual meeting & exhibition, 26 Feb–02 Mar, California, USA
4. Shim V, Yap K (1997) Modelling impact deformation of foam-plate sandwich systems. Int J Impact Eng 19:615–636
5. Mills N, Stämpfli R, Marone F, Brühwiler P (2009) Finite element micromechanics model of impact compression of closed-cell polymer foams. Int J Solids Struct 46:677–697
6. Xie B, Tang L, Liu Y, Liu Z, Jiang Z (2014) Research on the energy absorption properties of aluminum foam composite panels with enhanced ribs subjected to uniform distributed loading. J Sandw Struct Mater. doi:10.1177/1099636214555046
7. Xu A, Vodenitcharova T, Kabir K, Flores-Johnson EA, Hoffman M (2014) Finite element analysis of indentation of aluminium foam and sandwich panels with aluminium foam core. Mater Sci Eng A 599:125–133
8. Lu G, Shen J, Hou W, Ruan D, Ong L (2008) Dynamic indentation and penetration of aluminium foams. Int J Mech Sci 50:932–943
9. Hanssen A, Hopperstad O, Langseth M, Ilstad H (2002) Validation of constitutive models applicable to aluminium foams. Int J Mech Sci 44:359–406
10. Islam MA, Hazell PJ, Escobedo JP, Saadatfar M (2014) In-situ quasistatic compression and microstructural characterization of aluminium foams of different cell topology. In: XII international conference on applied mechanics and mechanical engineering (ICAMME 2014), 11–12 Dec 2014, Melbourne, Australia
11. Islam MA, Kader MA, Hazell PJ, Brown AD, Saadatfar M, Quadir MZ, Escobedo JP (2016) Investigation of microstructural and mechanical properties of cell walls of closed-cell aluminium alloy foams. Mater Sci Eng A 666:245–256

Part VII
Minerals

Industrial Use of Brazilian Bentonite Modified by Mild Acid Attack

C.G. Bastos Andrade, D.M. Fermino, M.G. Fernandes and F.R. Valenzuela-Diaz

Abstract Treatments with high concentration solution of inorganic acid at temperatures under boiling point are a usual method for clay minerals structure modification. For many industrial uses bentonites must be cleared of impurities. In the present paper, a sample of Brazilian bentonite from Vitoria da Conquista, Bahia was modified by mild acid attack using hydrochloric acid under moderate conditions (90 °C, reaction times of 1, 6, 12, 18 and 24 h in close reactor, concentration of the aqueous solution of hydrochloric acid 1.5 M, acid solution/clay ratio of 1 g/10 mL). The attacked samples were characterized by CEC, XRF, SEM, EDS and Stereomicroscopy. The purpose of these attacks is to reduce the concentration of impurities with minimal damage in clay minerals structure. The modified bentonite presented good results and tends to be a good economic and environmental alternative to manufacturing products with high added value such as polymer/clay nanocomposites, cosmetics and medicines.

Keywords Bentonite · Clay · Industrial use · Mild acid attack

Introduction

The Brazilian bentonite from Vitoria da Conquista, Bahia´s State, is a smectite clay naturally calcic, containing low iron concentration and presenting a green color [1–3].

The classification of bentonites is determinate according the geologic origin of clay, but if the smectite clays presents the same properties of traditional bentonites and/or is commercialized to the same use, those clays, by a common agreement, can be classified as bentonites, which presents similar properties such as colloidal material, absolving capacity and activation capacity with high grades [2–5].

C.G. Bastos Andrade (✉) · D.M. Fermino · M.G. Fernandes · F.R. Valenzuela-Diaz
Department of Metallurgical and Materials Engineering, EPUSP, Escola Politécnica
Universidade São Paulo, São Paulo, SP 05508-900, Brazil
e-mail: gianesic@usp.br

S. Ikhmayies et al. (eds.), *Characterization of Minerals, Metals, and Materials 2017*, The Minerals, Metals & Materials Series,
DOI 10.1007/978-3-319-51382-9_27

The bentonite products are vast, mostly with medium added value. For example, mineral binders and foundry molds, drilling fluids and adsorbents. In last case is used an activated clay with strong acid. To industrial application, bentonites must be cleaned of mineral impurities. Strong acid attack is a common method that provides a good cleaning of mineral impurities. Other benefit of acid attack is improving of acid sites with more porosity, excellent properties when apply in catalysis [6–11].

Several groups have been studying treatments to modified clays using organic acid with highly concentrations and temperatures, aiming clays bleaching e purifying, for posterior use mostly in food industry to bleaching process. The industrial use for clays is also based on exchangeable cations and clay minerals properties [12–14].

Calcic bentonites are wide used in modification process since, in essence, are montmorillonite clays were the predominate cation is calcium. Otherwise, exist an another type of montmorillonite with a negative response to acid attack, that is the case of sodic bentonite [15–17].

Studies of brazilian bentonite modified by mild acid attack demonstrated that could be a good alternative to material raw with economic advantages over imported clays used in processes to obtain products with high value as cosmetics and nanocomposites.

Materials and Methods

Start Materials

The calcic bentonite, from Vitoria da Conquista, Bahia´s State, Brazil in its natural form, was submitted to mild acid attack using a concentration of the aqueous solution of hydrochloric acid 1.5 M, clay/acid solution ratio of 1 g/10 mL, at 90 °C under bellow boiling temperature and at short times of reaction 1, 6, 12, 18 and 24 h in close reactor.

The attacked clay was washed, by filtration, with distilled water until pH 7–8, and then subjected to drying at 60 °C for 24 h.

After drying, the clay was grounded using a manual mortar and vibratory ball mill until completely pass through #200 mesh sieve.

Materials Characterization

The modified samples were characterized by X-ray fluorescence (XRF) and Energy Dispersive X-ray Detector (EDS) to observe the chemical composition after treatment. Another important clay property was verified by Cation Exchange Capacity

(CEC) which is indicative of the crystalline structure preservation. The images were obtained by Stereomicroscopy and Scanning Electron Microscopy (SEM) to observe impurities reduction, decolorizing, particles size and distribution.

The scanning electron microscopy (SEM) and energy dispersive X-ray detector (EDS) were performed on scanning electron microscopy, Philips - EDAX INSPECT 50 with energy dispersive X-ray detector (EDS).

To observe the clay was used a stereomicroscopy Zeiss, model Stemi 2000C.

CEC was performed using the ammonium acetate method.

The XRF was performed, using TBL Rock standard tablets as a parameter, on X-ray fluorescence spectrometer Panalytical model Axios Advanced with loss on ignition.

Results and Discussion

Figure 1 present the EDS specter of sample treated with water for 24 h at 90 °C. It is possible to observe the presence of Si, O, Al, Fe and Mg. For this test was used a thin layer of gold over the sample.

Figure 2 present the EDS specter of sample submitted to mild acid attack for 24 h at 90 °C. It is possible to observe the peak reduction of metals, remaining a minor peak of Al. This indicates a good result of bentonite purification.

Table 1 presents the XRF results of bentonite submitted to treatment, a quantitative analysis where the values are expressed in percentage of oxides. The presence of aluminum and iron predominates among other elements, and after treatment, presenting a significant reduction. It is possible to observe the oxides reduction due to the increase of the attack time.

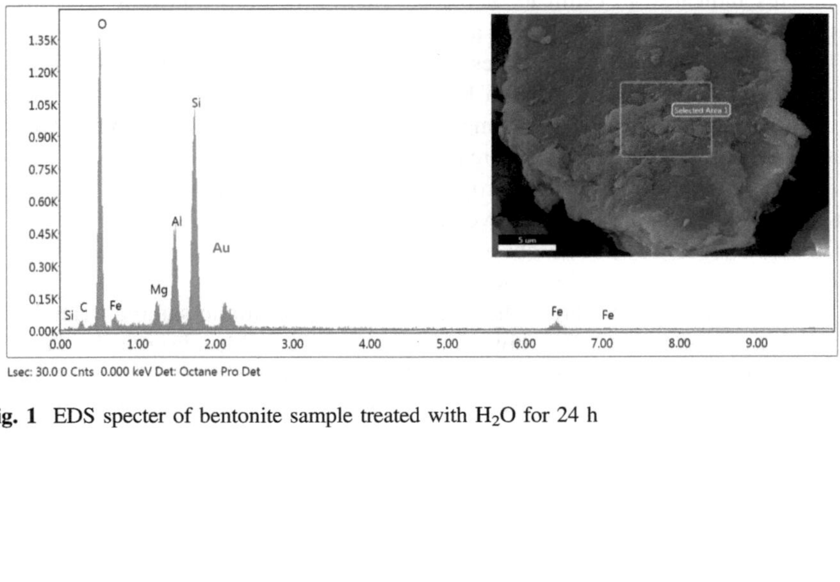

Fig. 1 EDS specter of bentonite sample treated with H_2O for 24 h

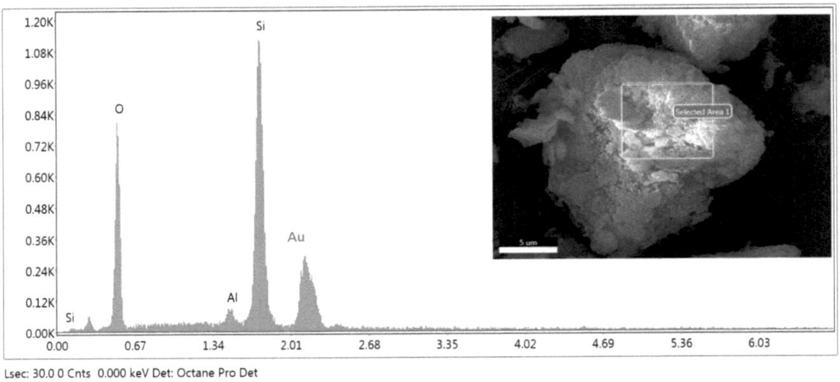

Fig. 2 EDS specter of bentonite sample treated with HCl for 24 h

Table 1 XRF results of bentonite submitted to treatment, quantitative analysis, values in percentage of oxides

Sample	SiO_2	Al_2O_3	Fe_2O_3	MnO	MgO	CaO	Na_2O	K_2O	TiO_2	P_2O_5	Li
H_2O 24 h	54.7	19.8	8.06	<0.10	3.73	0.20	0.10	<0.10	0.36	<0.10	13.5
HCl 1 h	57.7	20.6	7.82	<0.10	3.01	<0.10	<0.10	<0.10	0.39	<0.10	11.0
HCl 6 h	60.1	19.0	6.97	<0.10	2.71	<0.10	<0.10	<0.10	0.39	<0.10	11.2
HCl 12 h	64.1	15.9	5.08	<0.10	2.11	0.18	<0.10	<0.10	0.43	<0.10	13.0
HCl 18 h	65.9	15.0	4.65	<0.10	1.93	0.10	<0.10	<0.10	0.43	<0.10	12.6
HCl 24 h	63.2	16.1	5.31	<0.10	2.22	0.12	<0.10	<0.10	0.42	<0.10	13.3

In Fig. 3, it is possible observe a good reduction in the concentration of iron oxides and aluminum concentration. For aluminum a reduction of 18% and to 34% iron. This indicates an estimate of the destruction of the octahedral layer where aluminum and iron, hypothetically are located.

Table 2 presents the CEC values of bentonite samples submitted to mild acid attack. The results are indicative of low destruction of the crystalline clay mineral structure. A reduction of 40% compared with sample treated only with purified water was observed after 24 h of treatment.

The Fig. 4 shows the stereomicrography of bentonite sample treated with water for 24 h where it is possible observe several impurity particles.

In Fig. 5, it is possible to observe a good bleaching and cleaning of impurity particles at bentonite sample treated with HCl for 24 h.

Figures 6, 7, 8 and 9 shows the SEM images of bentonite sample treated with purified water and with HCl, both for 24 h. It is possible observe the lamellar structure agglomerated in layers, mostly presenting same size and irregular forms. This could indicate a good preservation of smectite crystal structures after acid attack.

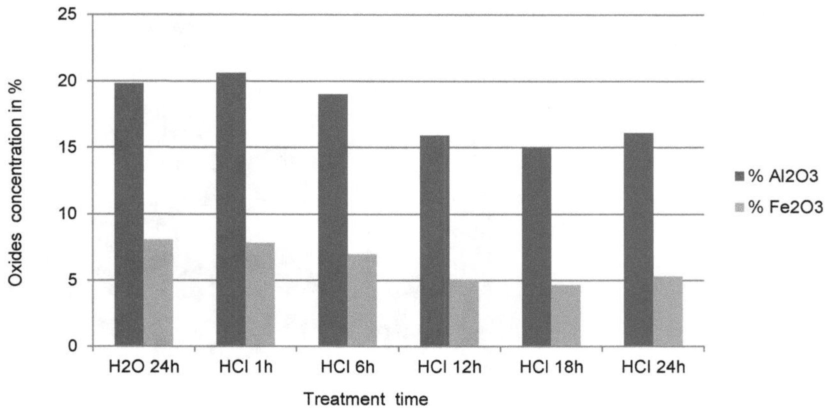

Fig. 3 Percentages of aluminum and iron oxides obtained by X-ray fluorescence of bentonite sample submitted to mild acid attack with treatment time

Table 2 CEC results of bentonite treated

Sample	CEC (meq/100 g)
H_2O 24 h	92
HCl 1 h	79
HCl 6 h	77
HCl 12 h	74
HCl 18 h	72
HCl 24 h	52

Fig. 4 Stereomicrography of sample treated with water for 24 h

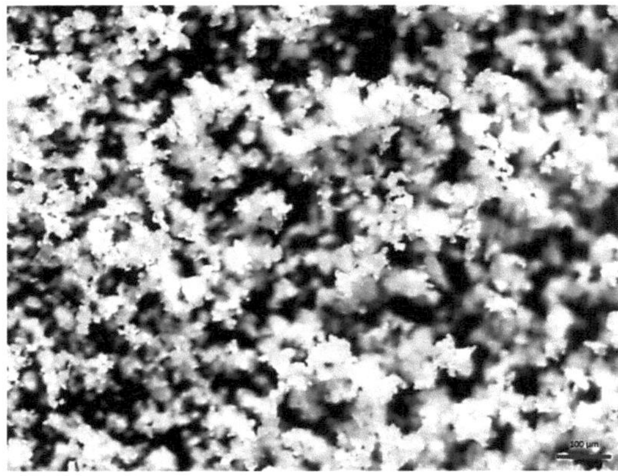

Fig. 5 Stereomicrography of bentonite sample treated with HCl for 24 h

Fig. 6 SEM image of sample treated with water for 24 h (crystals dispersion)

Fig. 7 SEM image of sample treated with HCl for 24 h (crystals dispersion)

Fig. 8 SEM image of sample treated with water for 24 h (crystal detail)

Fig. 9 SEM image of sample treated with HCl for 24 h (crystal detail)

Conclusions

The brazilian bentonite sample from Vitoria da Conquista submitted to mild attack acid under moderated conditions demonstrated a good response to proposed process.

The bleaching and purification were obtained with a good preservation of crystalline structure which was observed in Stereomicroscopy and SEM images.

A good purification of metallic oxides and other elements was observed in XRF.

The most evident modifications in decolorizing and CEC occurred in the first 12 h of acid treatment.

The results acquired presented a good concentration reduction of impurities with no significant change in crystalline clay mineral structure.

According the methodology presented, the proposed process permits new possibilities to industrial use of bentonite, in products of high value such as cosmetics and polymer/clay nanocomposites and could be an efficient alternative to economy and environment.

Acknowledgments São Paulo Research Foundation, FAPESP.

References

1. Valenzuela-Diaz FR (1994) Preparação a nível de laboratório de algumas argilas esmectíticas organofílicas. Tese (Doutorado). Departamento de Engenharia Química da Escola Politécnica da Universidade de São Paulo, SP
2. Souza Santos P (1992) Ciência e Tecnologia de Argilas, vol 1, 2a ed, Edgar Blücher, S. Paulo, SP
3. Grim RE (1968) Clay mineralogy, 2nd edn. McGraw-Hill Book, New York
4. Valenzuela-Díaz RF, Souza Santos P, Souza Santos H (1992) A Importância das argilas industriais brasileiras. Química Industrial 42:33–37
5. Abreu SF (1973) Recursos minerais do Brasil, Ed. Edgard Blucher, 2ª ed, V.1, SãoPaulo, SP
6. Bastos Andrade CG (2016) Bentonitas Brasileiras submetidas à ataques ácidos leves visando uso industrial. Tese (Doutorado), São Paulo, SP
7. GomesCF (1988) Argilas: o que são e para que servem, 1a ed, Fundação Calouste Gulbenkian, Lisboa, Portugal Ed., Fundação Calouste Gulbenkian, Lisboa, Portugal
8. Murray HH (2000) Appl Clay Sci 17:207–221
9. Andrade CGB, Fermino DM, Gonzales Fernandes M, Valenzuela M das G da S, Volzone C, Valenzuela Díaz FR (2015). Utilização de bentonita brasileira modificada na manufatura de produtos com alto valor agregado. In: Paulo AS, ABC, 2015. São Paulo, ABC
10. Amorim LV, Farias KV, Viana JD, Barbosa MIR, Pereira E, França KB, Lira HL, Ferreira HC (2005) Cerâmica 51:128
11. Vaccari A (1999) Appl Clay Sci 14:161–198
12. Souza Santos P (1975) Ciência e Tecnologia de Argilas, Aplicada às Argilas Brasileiras e Aplicações, vols 1–2, Edgar Blücher, São Paulo, SP
13. Valenzuela-Diaz FR, Souza Santos P (2001) Studies on the acid activation of Brazilian smetctitic clays. Química Nova, vol 24(3)
14. Grim RE (1978) Bentonites. Elsevier, Amsterdam
15. Grim RE (1962) Applied clay mineralogy. McGraw-Hill, New York
16. Nutting PG (1943) Adsorbent clays, their distribution, properties, production and uses. U. S. Geological Survey Bulletin 928C, United States Government Printing Office, Washington
17. Coelho ACV, Santos PS, Santos HS (2007) Argilas especiais: argilas quimicamente modificadas – uma revisão. Quim Nova 30(5):1282–1294

Mullitization Characteristics and Sinterability of Kyanite in Ceramic Preparation

Huaguang Wang, Bowen Li, Mingsheng He and Jiann-Yang Hwang

Abstract Kyanite is one of the major high-alumina minerals for producing refractory ceramics. At high temperature, kyanite can be decomposed into mullite and amorphous silica. This change is an irreversible decomposition and provides potentials for fiber reinforcement of ceramic matrix. In this study, kyanite powder was compressed into a cylinder green body of ceramic and heated at 1150–1450 °C. The phase transformation of kyanite and sinterability were investigated by X-ray diffraction and scanning electron microscopy. The results showed that the phase transformation of kyanite started at 1150 °C when rising the temperature. At 1350 °C, kyanite is completely converted into mullite and silica. The compressive strength of the ceramics reached 73.3–134.8 MPa, while the bulk density was 1.95–2.47 g/cm^3. The microstructure indicated the correspondence of the compressive strength to mullitization degree of kyanite.

Keywords Kyanite · Mullitization · Phase transformation · Sinterability

Introduction

Mullite ($3Al_2O_3{\cdot}2SiO_2$) is a refractory mineral with attractive properties such as high refractoriness, low thermal conductivity, low thermal expansion, excellent chemical stability and mechanical properties at high temperatures [1–4]. It has been extensively used for producing mullite ceramics or to improve strength and toughness of other ceramic products [5].

Kyanite ($Al_2O_3{\cdot}SiO_2$) is one of the major raw materials can be used for producing mullite [6]. At high temperature, kyanite is decomposed into mullite and

B. Li (✉) · M. He
Advanced Materials R&D Center of WISCO, Beijing 102211, China
e-mail: boli@mtu.edu

H. Wang · B. Li · J.-Y. Hwang
Department of Materials Science and Engineering, Michigan Technological University, 1400 Townsend Dr, Houghton, MI 49931, USA

S. Ikhmayies et al. (eds.), *Characterization of Minerals, Metals, and Materials 2017*, The Minerals, Metals & Materials Series,
DOI 10.1007/978-3-319-51382-9_28

amorphous silica. This phase transformation is an irreversible decomposition [7]. The main reaction is:

$$3(Al_2O_3 \cdot SiO_2) \rightarrow 3\,Al_2O_3 \cdot 2SiO_2 + SiO_2 \quad (1)$$

When additional alumina is provided, it will be reacted with the amorphous silica to form more mullites:

$$3Al_2O_3 + 2SiO_2 \rightarrow 3Al_2O_3 \cdot 2SiO_2 \quad (2)$$

For this phase transformation, the prior researches mainly focused on the mullitization and sinterability of kyanite [8, 9]. Meanwhile, there was less researches on the structural characteristics during and after the phase transformation. However, the microstructure change during the phase transformation from kyanite into mullite plays a critical role to the properties of mullite ceramics. In this study, kyanite powder was used as the raw material and the research was aimed at the microstructure of kyanite-mullite phase transformation at high temperature.

Materials and Methods

The raw materials used in this study included kyanite and kaolin powders. The kyanite powder (below 325 mesh) was provided by Xinguo Kyanite Ltd., Xintan, China. The chemical compositions of kyanite are (wt%): Al_2O_3 58.55, SiO_2 37.02, Fe_2O_3 1.39, MgO 0.02, CaO 1.78, TiO_2 0.05, Na_2O + K_2O 0.03, and LOI 1.00. The kaolin powder (below 325 mesh) was from Old Hickory Clay Co., Mayfield, KY USA. The chemical compositions of kaolin are (wt%): Al_2O_3 38.6, SiO_2 45.1, Fe_2O_3 0.43, TiO_2 1.49, CaO 0.06, MgO 0.14, Na_2O + K_2O 0.22, and LOI 13.5.

The kyanite ceramic was prepared by mixing kyanite powder with kaolin powder and water in certain ratio in a blender, then compressed the mixture into a cylinder green body (ϕ15 mm, h = 15 mm). There were three copies of green bodies made for each formula of the mixtures. The green ceramic bodies were then transferred in a box furnace, heated to designed temperature, maintained the temperature for 2 h, then naturally cooled down to room temperature for characterization. The specific processing parameters were showed in Table 1.

After heating at the desired temperature, compressive strength of ceramics was measured with a universal tester (Model: Instron 4206). Bulk density wes determined referred to ASTM C373-14a. The test for compressive strength and bulk density were triplicated and the reported data were in average of the testing results. The XRD analysis was carried out with a Scintag XDS2000 Powder Diffractometer (CuK, Step 0.030°, count time 0.900 s, scanning range 5.00–59.99 °, scan rate 2 °/min.). The field emission scanning electron microscope (FE-SEM, Hitachi S-4700) was employed for microstructure analysis.

Table 1 The preparation of mullite ceramics under different conditions

Number	Kyanite, %	Kaolin, %	Water, %	Sintering temperature, °C	Forming pressure, MPa
K1	90	10	5	1150	20
K2	90	10	5	1200	20
K3	90	10	5	1250	20
K4	90	10	5	1300	20
K5	90	10	5	1350	20
K6	90	10	5	1400	20
K7	90	10	5	1450	20

Results and Discussion

Sinterability of Kyanite Ceramic

The kyanite-mullite ceramic, prepared by kyanite associated with kaolin, has a bulk density of 1.95–2.47 g/cm^3 and a compressive strength of 73.3–134.8 MPa. Figure 1 showed the affection of sintering temperature to compressive strength and bulk density of the ceramics. The green ceramic cylinders were sintered at 1150–1450 °C. It can be seen, when elevated the temperature from 1150 to 1200 °C, compressive strength of the ceramic increased from 115.6 to 134.8 MPa. Further rising sintering temperature to 1350 °C, the compressive strength of the ceramic decreased from 134.8 to 49.9 MPa. In the sintering temperature range 1150–1450 °C, the compressive strength of the ceramic increased slightly.

Figure 1 also revealed the relationship between sintering temperature and bulk density in preparation of mullite ceramics. On the whole, the change trend of bulk density was basically corresponded with the change of compressive strength. From 1150 to 1350 °C, it showed that the bulk density decreased obviously as the

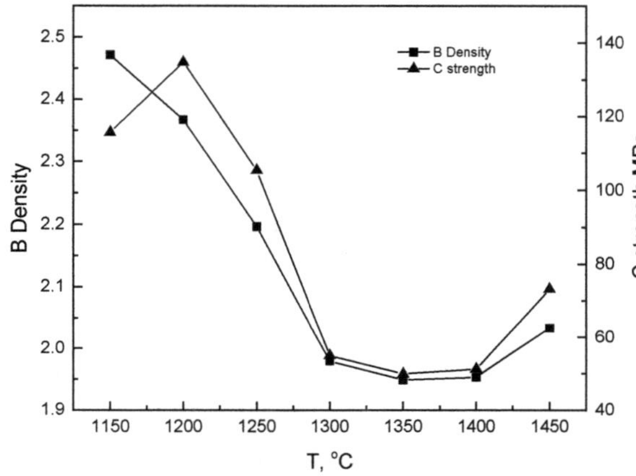

Fig. 1 The affection of sintering temperature to compressive strength and bulk density of ceramics

swelling of the phase transformation from kyanite to mullite. When the sintering temperature was raised to above 1350 °C, the bulk density of the ceramics increased while the compressive strength increased.

Mullitization Characteristics of Kyanite Ceramic

The green ceramic bodies were sintered at 1150, 1200, 1250, 1300, 1350, 1400 and 1450 °C. The XRD patterns of the ceramic sintered at these temperatures were showed in Fig. 2. As it can be seen, for raw materials (Formulated by kyanite and kaolin powders, but without firing), there were only two crystal mineral phases found, kyanite (K) and kaolinite (K1). After sintered at 1150 °C, there were two mineral phases mullite (M) and quartz (S) emerged, while the characteristic peaks of kaolinite disappeared. When sintering temperature was increased to 1300 °C, the intensities of kyanite were decreased constantly, meanwhile, mullite was increased. When the sintering temperature was elevated to above 1350 °C, the major mineral components of ceramic became mullite and quartz and the characteristic peaks of kyanite disappeared. These results indicated that the mullitization of kyanite started

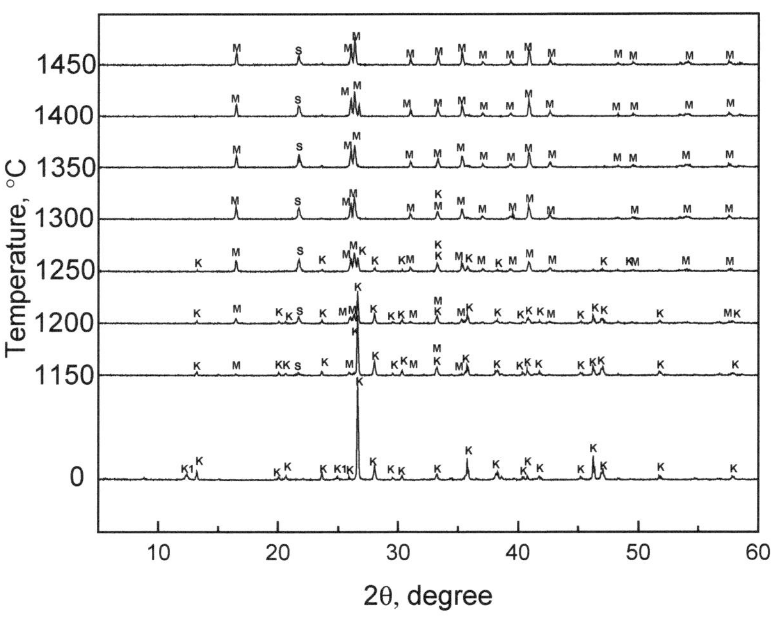

Fig. 2 XRD patterns of ceramic sintered at different temperature

Fig. 3 FE-SEM images of sample sintered at 1150 °C (K1) ►

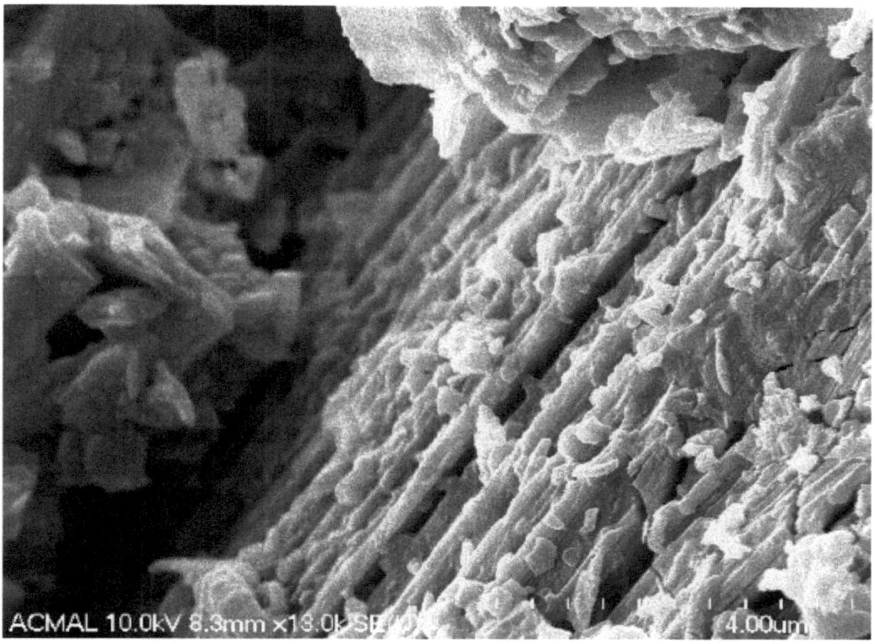

(a)

(b)

(a)

(b)

◀ **Fig. 4** FE-SEM images of sample sintered at 1300 °C (K4)

at 1150 °C, while it completed when the sintering temperature was raised to above 1350 °C, which were corresponding to the changes of compressive strength and bulk density of the ceramics.

The phase transformation of kyanite to mullite and the sinterability were also investigated by field emission scanning electron microscopy. Figures 3, 4, and 5 showed the SEM images of the microstructures in the ceramics.

Figure 3a indicated that fibrous mullite crystals have been formed within the original kyanite grains. The mullite fibers arranged parallelly. This type of crystal arrangement is easily lead to cracks along the crystalline boundaries of single crystals, which will affect the mechanical strength of ceramics. Figure 3b also showed that there was few mullite fibers formed among the grains in the ceramic body, which may be formed from the sintering of kaolinite, and it explained that sample K1 had a high compressive strength.

From 1150 to 1300 °C, the majority of kyanites were transformed into mullite. This was combined with volume expansion (Theoretically, 18% of volume swelling). In this condition, the existent form of mullite phase was fibrous aggregate. Meanwhile, there was no enough bond force among the grains (Fig. 4). It was clear that the structure of sample K4 was easily to be cracked, which also indicated the correspondence of the low compressive strength in Fig. 1.

As Fig. 2 showed, kyanite has been completely converted into mullite and silica at 1350 °C. Figure 5 demonstrated that all the shapes of kyanite grains have been changed. The well-crystallized mullite fibers and rods have been created and grow up locally. The fibrous and rod-shaped mullites were homogeneously distributed and formed with an interlock network instead of parallel mullite crystals. The spaces among the large mullite grains were filled with small mullite grains and amorphous silicate phases. The matric phases have well bonded with the large mullite crystals which made the ceramic body dense. This explained that when the sintering temperature was raised to 1450 °C, the bulk density increased and the compressive strength also increased.

Conclusion

The ceramics made from kyanite powder combined with kaolin was prepared and the phase transformation of kyanite into mullite were investigated when heated the ceramics at 1150–1450 °C. The results showed that the phase transformation of kyanite started at 1150 °C when rising the temperature. At 1350 °C, kyanite is completely converted into mullite and silica. The compressive strength of the ceramics reached 73.3–134.8 MPa, while the bulk density was 1.95–2.47 g/cm^3. The microstructure indicated the correspondence of the compressive strength and bulk density to mullitization degree of kyanite.

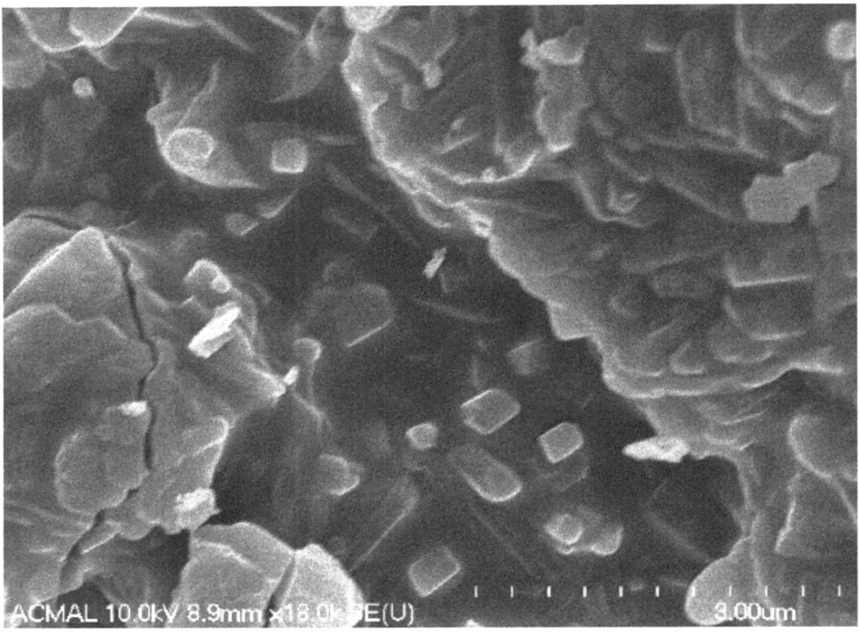
ACMAL 10.0kV 8.9mm x18.0k SE(U)
3.00um

(a)

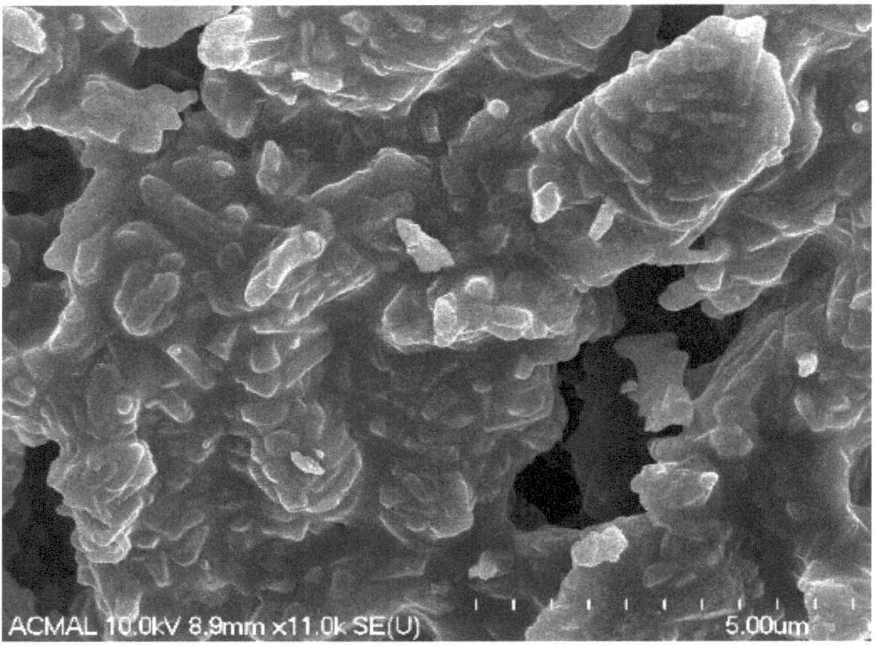
ACMAL 10.0kV 8.9mm x11.0k SE(U)
5.00um

(b)

◀ **Fig. 5** FE-SEM images of sample sintered at 1450 °C (K7)

References

1. Schneider H et al (2008) Structure and properties of mullite—a review. J Eur Ceram Soc 28 (2):329–344
2. Clarke DR (2003) Materials selection guidelines for low thermal conductivity thermal barrier coatings. Surf Coat Technol 163:67–74
3. Barea R et al (2005) Thermal conductivity of highly porous mullite material. Acta Mater 53 (11):3313–3318
4. Dokko PC et al (1977) High-temperature mechanical properties of mullite under compression. J Am Ceram Soc 60(3-4):150–155
5. Chen CY et al (2000) Preparation of mullite by the reaction sintering of kaolinite and alumina. J Eur Ceram Soc 20(14):2519–2525
6. Anggono J (2005) Mullite ceramics: its properties structure and synthesis. Jurnal Teknik Mesin 7(1):1
7. Li BW et al (2000) Aluminosilicate high-performation ceramic prepared from mineral micripowder. Earth Sci Front 7(2):535–539
8. Hülsmans A et al (2000) The transformation of andalusite to mullite and silica: part I. Transformation mechanism in [001] A direction. Am Mineral 85(7–8):980–986
9. Hülsmans A et al (2000) The transformation of andalusite to mullite and silica: part II. Transformation mechanisms in [100] A and [010] A directions. Am Mineral 85(7–8):987–992

Ore Dressing and Technological Characterization of Palygorskite from Piauí/Brazil for Application as Adsorbent of Heavy Metals

Karla M.A. Simões, Bruna L. Novo, Adriana A.S. Felix, Julio C. Afonso, Luiz C. Bertolino and Fernanda A.N.G. Silva

Abstract Palygorskite $(Mg,Al)_5Si_8O_{20}(OH)_2(OH_2)_4.4H_2O$, is a lamellar mineral with a fibrous morphology. The micropores and canals in the structure give a high specific surface and sorption capacity of various kinds of species. Guadalupe's palygorskite (Piauí/Brazil) run of mine (ROM) was prepared (PA) and characterized by XRD and XRF. PA is composed by palygorskite, kaolinite, quartz and goethite and has 15.8, 58.0, 4.2 and 7.6% of SiO_2, Al_2O_3, MgO and Fe_2O_3, respectively. Wet magnetic separation tests were conducted with PA obtaining the magnetic and non-magnetic fractions (PAM and PANM, respectively). PANM was characterized by XRD and XRF and the results of the XRF indicated that the iron content in the PA decreased 8.0% in the non-magnetic product. A new wet sieving was conducted with PANM to obtain a purer fraction (−20 μm) (PANMP) which was characterized by XRD, XRF, CEC, Zeta potential, SEM, TG/DTG. The result showed that this mineral could be used as adsorbent of heavy metals.

K.M.A. Simões · B.L. Novo · J.C. Afonso · F.A.N.G. Silva (✉)
Instituto de Química, Universidade Federal do Rio de Janeiro,
Avenida Athos da Silveira Ramos, 149, Bloco a, sala 630,
Cidade Universitária, Rio de Janeiro, RJ 21941-909, Brazil
e-mail: fnogueira@iq.ufrj.br

K.M.A. Simões
e-mail: karla.carestiato@gmail.com

B.L. Novo
e-mail: brunalemosnovo@gmail.com

J.C. Afonso
e-mail: julio@iq.ufrj.br

K.M.A. Simões · B.L. Novo · L.C. Bertolino
Centro de Tecnologia Mineral/CETEM-MCTI, Avenida Pedro Calmon,
900, Cidade Universitária, Rio de Janeiro, RJ 21941-908, Brazil
e-mail: lcbertolino@cetem.gov.br

A.A.S. Felix
Campus Maracanã, Instituto Federal de Educação Ciência e Tecnologia do Rio de Janeiro,
Rua Senador Furtado, n°121/125, Maracanã, Rio de Janeiro, RJ CEP:20270-021, Brazil
e-mail: adriana.soeiro@ifrj.edu.br

S. Ikhmayies et al. (eds.), *Characterization of Minerals, Metals, and Materials 2017*, The Minerals, Metals & Materials Series,
DOI 10.1007/978-3-319-51382-9_29

Keywords Palygorskite · Ore dressing · Characterization

Introduction

Palygorskite is a lamellar clay mineral with a hydrous fibrous morphology, whose unit cell is $(Mg,Al)_5Si_8O_{20}(OH)_2(OH_2)_4.4H_2O$. Mineralogically, it belongs to the group of phyllosilicates 2:1 with an aluminum octahedral layer, which can undergo isomorphic substitutions Mg^{2+} or Fe^{2+} between two layers of silicon tetrahedrons, exchanged for Al^{3+} or Fe^{3+} [1]. Cations substituents have the substituted lower valence, which leads to a deficiency of positive charge and negative charge excess.

The presence of micropores and canals in the structure and the elongated nature of the particles and their fine particle size, give a high specific surface area at this clay mineral and sorption capacity of different species [2].

Industrialization and population growth are identified as key factors responsible for the increase in pollution by heavy metals. Treatment of effluents containing such polluting species became indispensable, and metal adsorption processes are studied to reduce the environmental impacts. However, traditional adsorbents such as activated carbon and ion exchange resins are considered high cost, which prevents their use in the industrial sector.

The purpose of this work is to characterize and processing of palygorskite from Guadalupe in Piauí, Brazil, for their potential application as heavy metals'adsorber like lead and cadmium in aqueous solutions.

Materials and Methods

Sample Preparation

Palygorskite's sample was collected from Guadalupe deposits in Piauí/Brazil. In Center of Mineral Technology (CETEM) was accomplished the ore dressing. About 7 kg of run of mine's sample (ROM) was crushed on high frequency crusher jaws with, approximately, 0.4 mm of aperture, for the breakdown, and then, the sample was homogenized in a prismatic pile. Sample of the final product of the preparation stages was wet ground, in a stainless steel mill bar, with pulp (1:1, sample:water) and followed was carried out a wet size classification, according to the damp method. A vibratory siever (684.5 rpm), equipped with a sieves with openings of 300–45 μm, where the fraction below 45 μm (PA) was filtered and dried in an oven at, approximately, 50 °C.

Wet magnetic separation tests were performed with the PA sample in BOXMAG RAPID magnetic separator, with high intensity field in the order of 15 kGauss and ferromagnetic matrix. With the non-magnetic fraction obtained (PANM), a new wet size classification was performed with deionized water and sieve with opening 20 μm, the sample below 20 μm was called PANMP.

Characterization

Samples from preparation stages and wet magnetic separation were characterized by X-ray diffraction and X-ray fluorescence (XRF and XRD). PANMP sample was characterized by XRD and XRF techniques, cation exchange capacity (CEC), surface charge measurements (Zeta potential), scanning electron microscopy (SEM) and thermal differential and thermogravimetric analysis (DTG - TG).

X-Ray Diffraction (XRD)

XRD patterns samples were obtained by the powder method in the equipment Bruker D4 Endeavor, Co Kα radiation (35 kV/40 mA), goniometer velocity of 0.02° (2θ) per step with 1 s per step and collected from 5 to 80° (2θ). The qualitative interpretation of the spectra was made by comparison with standards contained in PDF02 database (ICDD 2006) on Bruker AXS software DiffracPlus.

X-Ray Fluorescence (XRF)

Samples were pressed at automatic squeezer VANEOX (20 mm of mould, P 20 ton and t 30 s), using as binder the boric acid (H_3BO_3) 1:0.2 of proportion (0.6 g of boric acid and 3 g of dried sample at 100 °C). The contents were verified and expressed in percentage of their oxide forms, by semi quantitative analysis (Standardless Method mode) in X-ray fluorescence spectrometer (Panalytical WDS), AXIOS MAX model.

Cation Exchange Capacity (CEC)

CEC was determined by adsorption technique of methylene blue with PANMP sample. Then was added, approximately, 2 g of sample in a beaker containing 30 mL of deionized water in order to promote a stirring on magnetic stirrer. pH's solution was adjusted to 3.0 with H_2SO_4 0.1 mol/L and started the titration with methylene blue solution, 0.01 mol/L, for the saturation of the sample to be observed in a qualitative filter paper.

Zeta Potencial

Surface charge (Zeta potential) of PANMP was determined by equipment Zetasizer Nano ZS from Malvern. 0.5 g of sample solutions in 10 ml of KCl 10^{-3} mol/L (electrolyte) were prepared to carry out the measures. pH values were adjusted using HCl or KOH in a range from 1.5 to 11.0 with the aid of a potentiometric titrator coupled to the system.

Scanning Electron Microscopy (SEM)

Analysis in scanning electron microscopy were performed in microscope equipment, HITACHI TM303 Plus. The samples were metalized with gold, using the equipment BAC-TEC-SCD005 Sputter Coater. The modules secondary electrons, backscattered electrons and energy dispersive X-ray spectrometry (EDS) were analyzed for the qualitative determination of chemical elements.

Thermogravimetric Analysis (TG-DTG)

The thermodifferential and thermal gravimetric analysis were performed on the equipment METTLER TOLEDO TGA model/DSC1 STAR^e System. The thermogravimetric analysis curves were obtained in the temperature range of 25–1000 °C, heating rate 10 °C/min, platinum sample holder and nitrogen atmosphere with a flow rate of 50 ml/min.

Results and Discussion

The XRD spectrum shown in Fig. 1 reveals the sample is mainly composed of clay mineral palygorskite, kaolinite, quartz and goethite.

XRD shows the presence of a peak relating to palygorskite diffraction at 2θ = 9.77 and 2θ = 14.33, 31.05, 4.18 to kaolinite, quartz and goethite, respectively, these minerals are consider impurities present in the clay mineral. Furthermore, it was found that there was no change in the crystalline structure of

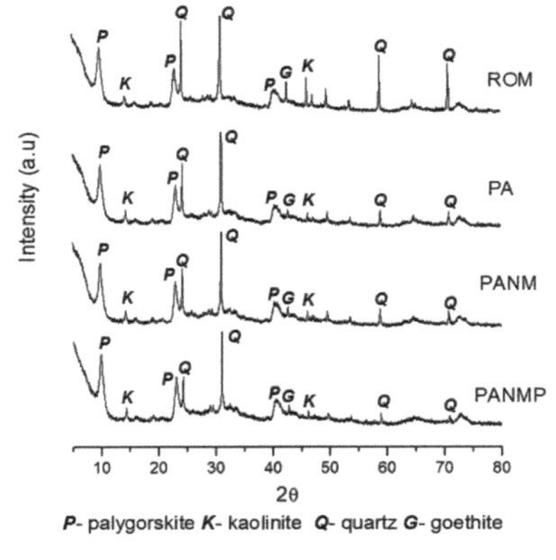

Fig. 1 X-ray diffraction pattern of ROM, PA, PANM and PANMP samples

palygorskite in all samples, indicating that the preparation stages does not provide major changes or alterations in the crystalline structure of phases present in palygorskite ore.

XRF analysis was performed to determine the levels of the main oxides that compound the sample, so it was possible to obtain the chemical composition of samples. Table 1 shows the contents of oxides expressed in wt% of the samples ROM, PA, PANM and PANMP, which were determined by semiquantitative analysis.

Results related to the chemical composition (XRF) corroborate with those obtained by XRD, once Al_2O_3, SiO_2, MgO and Fe_2O_3 to PA and PANMP were 15.80 and 15.50, 58.00 and 50.50, 4.20 and 4.50, 7.60 and 7.10%, respectively. Thus, the results indicate that magnetic separation associated to wet size classification, below 20 μm, promotes an increased degree of palygorskite's purity.

Cation exchange capacity (CEC) of PANMP sample was determined by calculating the methylene blue index, obtaining the result of 35.5 meq/100 g of sample. The higher the value of the cation exchange capacity of the clay, greater could be the number of cations which it can adsorb. The obtained value is within the prescribed by the literature, where the values of CEC for palygorskite is estimated between 20 and 50 meq/100 g of sample [3].

Zeta potencial's result indicated that the surface charges of PANMP sample is negative in throughout the pH range (from 1.5 to 11.0), Fig. 2. According to the results, the PANMP sample could be an excellent cation adsorbent.

Table 1 Chemical composition of ROM, PA, PANM and PANMP samples (wt%)

Samples	SiO_2	Al_2O_3	Fe_2O_3	MgO	K_2O	TiO_2	LOI
	(wt%)						
ROM	56.20	16.40	6.50	4.40	2.10	0.69	12.70
PA	58.00	15.80	7.60	4.20	2.10	0.75	10.40
PANM	54.30	15.40	7.00	4.10	1.90	0.68	15.60
PANMP	50.50	15.50	7.10	4.50	2.20	0.62	18.60

LOI Loss On Ignition

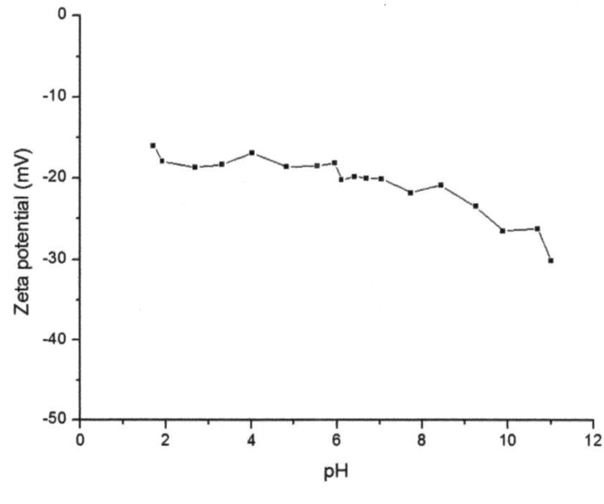

Fig. 2 Zeta potential versus pH of PANMP sample

Figure 3 shows the fibrous structure of palygorskite characterized by SEM. According to the images of PANMP, it could be observed that the stages of crushing and grinding did not affect the tubular fibrous morphology of this material, an important characteristic for its use as adsorbent. It is important to emphasize that the crushing and grinding process could destroy the morphology of some clay minerals, if did not performed in mild conditions.

Associated with SEM, the elemental qualitative composition of PANMP was determined by EDS. It can be observed, Fig. 4, that PANMP is mainly composed by Si, Al, Mg and O. K could be considered as compensation cation of the crystal lattice of palygorskite, and Ti and Fe are impurities, as previously discussed according to results of XRD and XRF.

Figure 5 shows the thermogravimetric and thermo differential (TG/DTG) curves obtained in order to evaluate the thermal behavior of PANMP sample, highlighting three important events of mass loss. The first dehydroxylation is related to loss of physically adsorbed water on the surface of palygorskite and occurs at 74 °C with mass loss of 7.21%. The second one, mass loss of 0.92% was observed at 212 °C

Fig. 3 Images of scanning electron microscopy of PANMP sample

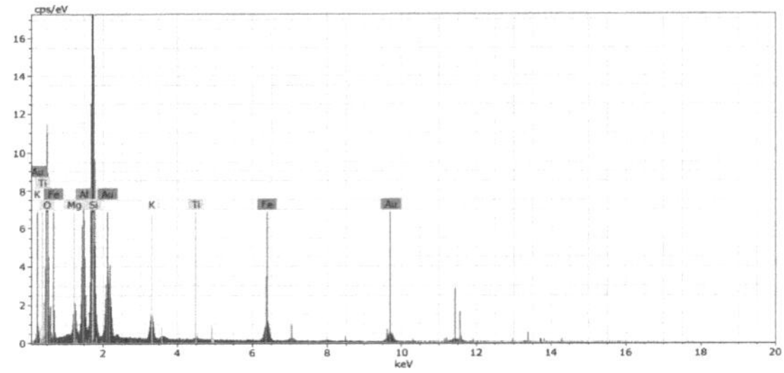

Fig. 4 EDS of PANMP sample

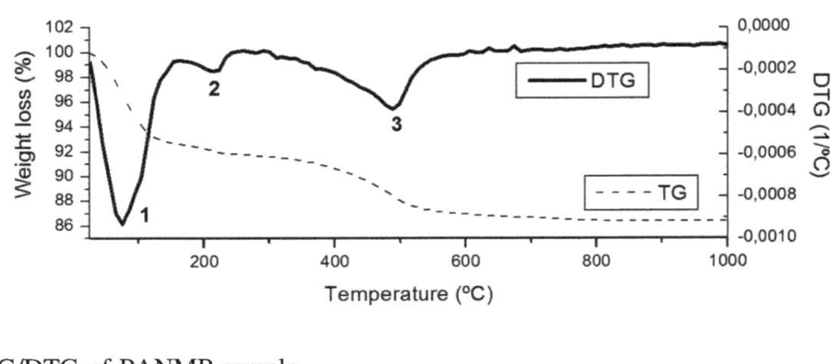

Fig. 5 TG/DTG of PANMP sample

and is attributed to the release of the zeolitic water, present in the channels, of the clay mineral. The third dehydroxylation occurs due to irreversible mass loss of structural water palygorskite, which event occurs at 487 °C, with mass loss of 4.92%.

Conclusions

Based on the observation, Guadalupe's palygorskite is composed by quartz, kaolinite and goethite. The chemical contents (XRF) of SiO_2, Al_2O_3, MgO and Fe_2O_3 in pure sample (PANMP), after the ore dressing, are 50.5, 15.5, 4.5, and 7.1%, respectively.

XRD has shown that the wet magnetic separation followed by wet size classification were efficient on the removal of impurities, since the intensity of the peaks related to quartz and iron decreased while the intensity of the peak related to palygorskite increased, thus raising its purity level. The images obtained by SEM allowed the recognition of the fibrous structure of the clay mineral, which was preserved after the preparation and ore dressing stages.

Cation exchange capacity results with a measure of surface charge gives to PANMP a high adsorption capacity, which associated to the results obtained in the characterization and preliminary tests of lead and cadmium adsorption in synthetic wastewater. Therefore, according to the results, palygorskite can be applied as heavy metals'adsorber.

Acknowledgements We would like to thanks IQ/UFRJ and COAM/CETEM for analysis and lab infrastructure, to CNPq and CAPES their financial support and scholarship.

References

1. Post JL, Crawford S (2007) Varied forms of palygorskite and sepiolite from different geologic systems. Appl Clay Sci 36:232
2. Murray HH (2000) Traditional and new applications for kaolin, smectite, and palygorskite: a general review. Appl Clay Sci 17:207–221
3. Luz, AB, Almeida SLM (2008) Capítulo 10: Argila—Atapulgita e Sepiolita. Rochas e Minerais Industriais, Usos e Especificações, Adão B. Luz e Fernando F. Lins, 2° edição (eds), p 225, Rio de Janeiro

Technological Characterization of Waste from Gold Mining Dam

Vanessa P.R. Silva, Fabiano A.C.M. Passos, Lillian M.B. Domingos, Roberto B. Faria, Zuleica C. Castilhos and Fernanda A.N.G. Silva

Abstract Wastes from gold mining need to be monitored from the beginning of mining activities until to the mine closure. Wastes from tailings dam were characterized by XRD, XRF, elemental analysis of carbon and sulfur and ICP-OES. In flotation waste (FW) were observed quartz, dolomite, pyrite, muscovite and chamosite were observed. Meanwhile, in cyanidation waste (CW), besides those described in FW, minerals such as sphalerite, galena, pyrrhotite, microcline and albite were detected, indicating a higher concentration of sulfide minerals in the cyanidation waste. In relation to C and S content, to FW and CW, they were 2.5 and 0.0% and 0.64 and 15.4%, respectively, confirming the majority of sulfides at CW sample and indicating the absence of the flotation organic reagents at the

V.P.R. Silva · R.B. Faria · F.A.N.G. Silva (✉)
Instituto de Química, Universidade Federal do Rio de Janeiro, Avenida Athos da Silveira Ramos, 149, Bloco a, Cidade Universitária, Rio de Janeiro, RJ 21941-909, Brazil
e-mail: fnogueira@iq.ufrj.br

V.P.R. Silva
e-mail: vanessa.prs89@gmail.com

R.B. Faria
e-mail: faria@iq.ufrj.br

F.A.C.M. Passos · L.M.B. Domingos · Z.C. Castilhos
Centro de Tecnologia Mineral/CETEM-MCTI, Avenida Pedro Calmon Avenue, 900, Cidade Universitária, Rio de Janeiro, RJ 21941-908, Brazil
e-mail: passosfabiano_vl@hotmail.com

L.M.B. Domingos
e-mail: lilliandomingos@uol.com.br

Z.C. Castilhos
e-mail: zcastilhos@cetem.gov.br

F.A.C.M. Passos
Escola de Química, Universidade Federal do Rio de Janeiro, Avenida Horácio Macedo, 2030, Bloco E, Cidade Universitária, Rio de Janeiro, RJ 21941-909, Brazil

Z.C. Castilhos
Instituto de Química, Universidade Federal Fluminense, Campus do Valonguinho, Niterói, RJ 24210-130, Brazil

S. Ikhmayies et al. (eds.), *Characterization of Minerals, Metals, and Materials 2017*, The Minerals, Metals & Materials Series,
DOI 10.1007/978-3-319-51382-9_30

cyanidation waste. The analysis by ICP-OES indicated As, Zn, Pb, Cr and Cd content for FW and CW: 13.30 and 343 mg/kg; 399 and 7,900 mg/kg; 136 and 13,000 mg/kg; 130 and 260 mg/kg; 3.60 and 84.60 mg/kg, respectively. It was also verified significant levels of heavy metals, which indicates the necessity of performing wastes toxicity tests.

Keywords Gold · Flotation · Cyanidation · Waste · Characterization

Introduction

The mining activity has been playing an important role at society through time. Nevertheless, if poorly planned, it can affect the life quality of future generations. The prevention and minimization of the environmental impacts are of great significance in terms of sustainability, and it can be achieved with good planning and environment impact assessment both at the beginning and at the end of the mining exploitation [1].

Gold is present mostly at its elemental state, free or forming alloys not only with silver, copper, bismuth, elements of platinum group, mercury, etc. The second most important occurrence of gold is as substitute on the crystalline reticule of sulfides, mostly in pyrite (FeS_2), arsenopyrite (FeAsS) and chalcopyrite ($CuFeS_2$), and it is known as invisible or occluded gold. Moreover, gold can form series of minerals with sulfur (S), antimony (Sb) and selenium (Se), generally rare and of less economic importance. The mineralogical form of gold is a decisive factor for its processing, and its determination is not simple [2].

The principal wastes generated by the industrial gold mining process come from the flotation and the cyanidation stages. Tailings deposits can be metal sources environmental contamination sources due to the presence of metals, including arsenic, mainly when these materials have high metal levels and sulfide minerals. These lasts can be oxidized when in contact with water and air realising very strong acid solution, which may dissolve these minerals acidifying surrounding water bodies and soil [3].

Gold ore processing has physical and chemical separation stages aiming at the principal mineral species. It usually involves nine steps: crushing, milling, classification, gravity concentration, flotation, regrinding, cyanidation, electrolytic extraction and foundry [4].

Flotation is a solid particle separation method, which involves superficial difference based on polar (hydrophilicity) or nonpolar (hydrophobicity) [4]. The added reagents in this stage increase the hydrophobic characteristics of particles associated with gold, which, in contact with air bubbles, are carried to surface of the solution, creating a spume concentrated in gold particle and gold associated minerals, which, in turn, is removed from the flotation tank. The remaining particles in the water go to the waste dam and are namely as Flotation Waste (FW). The surfactants used in flotation as foaming usually come from alcohols or ethers. One of the most used on the sulfide flotation is the methyl-isobutyl-carbinol [5].

The gold particle concentrate spume is sent to the hydrometallurgical process. In cyanidation, the leaching cycle consists of a pre-aeration tank and Carbon In Leach (CIL) tanks. The first step combines of an oxidizing agents, such as oxygen, with complexing agents, such as salt cyanides, in order to stabilize gold ions in the solution. After the complexation of the contained metals, the adsorption step begins. The gold complex $[Au(CN)_2]^{1-}$ is selectively adsorbed by the activated carbon and then separated from the leached solution. The remaining species in that solution are disposed at the waste dam and named as Cyanidation Waste (CW).

Therefore, this work aims to characterize the flotation and the cyanidation wastes, in order to understand their composition and provide the first steps regarding their potential environmental risks.

Materials and Methods

Sample Preparation

Two samples of wastes, FW and CW were studied. The first was solid, wrapped in plastic bags (60 kg), while the other was in pulp form within barrel (50 L). The second had to be homogenized for 30 min with 10 L of water, from which 6 L of this solution were collected.

The waste samples were disposed in stainless steel trays and dried within kilns at 50 °C. After that, the samples were manually disaggregated with ceramic smooth roll and screen (2 mm), followed by homogenizing at conic and prismatic pile and quartering.

Technological Characterization

Aliquots of FW and CW were characterized using laser granulometry distribution. Other aliquots were ground with ring mill, followed by sieving in screen (104 μm), and the passing fraction was sent to characterization for X-ray diffraction, X-ray fluorescence spectrometry, C and S elemental analysis and optical emission spectrometry with inductively coupled plasma.

Laser Granulometry Distribution

At the laser granulometry distribution, the samples were disaggregated in deionized water and disposed at the equipment Mastersize 2000 (Malvern). This technique scan the volume of the sample and transfer the data to a computer, which collect and organize the data is displayed in a graphic whose profile shows what is the sample main particle size.

X-ray Diffraction (XRD)

XRD spectra were obtained by the powder method in the equipment Bruker-AXS D5005, Co Kα radiation (35 kV/40 mA), goniometer velocity of 0.02° (2θ) per step with 1 s per step and collected from 5 to 80° (2θ). This is a qualitative analysis, that allows to estimate the mineralogical composition, or in other words, the crystalline mineral species contained on the sample.

X-ray Fluorescence Spectrometry (XRF)

Samples were pressed at automatic squeezer VANEOX (20 mm of mould, P 20 ton and t 30 s), using as binder the boric acid (H_3BO_3) 1:0.2 of proportion (0.6 g of boric acid and 3 g of dried sample at 100 °C). The contents were verified and expressed in percentage of their oxide forms, by semi quantitative analysis (Standardless Method mode) in X-ray fluorescence spectrometer (Panalytical WDS), AXIOS MAX model.

Carbon and Sulfur Elemental Analysis

The samples were pressed at automatic squeezer, and then inserted on the Carbon/Sulfur Determining VANEOX (SC632 model), with pure oxygen atmosphere at 1350 °C. The combination between the oven temperature and the oxygen flow induce the sample combustion. All the contained materials at the sample passes by a redox process, which enables the carbon and sulfur liberation. Carbon is oxidized to CO_2 and sulfur to SO_2, which are detected by infrared cells, which measure de CO_2 and SO_2 concentrations. The equipment converts the measured concentrations in percentage values using an equation which counts samples weigh, calibration and standard sample value.

Optical Emission Spectrometry with Inductively Coupled Plasma (ICP-OES)

Samples were inserted on the equipment ULTIMA 2 (Horiba Scientific), after acid digestion process (hydrochloric acid, nitric acid, hydrofluoric acid and perchloric acid) in order to remove the organic compounds, and release all inorganic elements to the solution, followed by the sequential analysis.

Results and Discussion

Laser Granulometry Distribution

According to the granulometry distribution graphic of FW, Fig. 1, 61% of sample has, approximately, 148 μm of particle size, while to CW, Fig. 2, 46.3% has 6 μm, approximately.

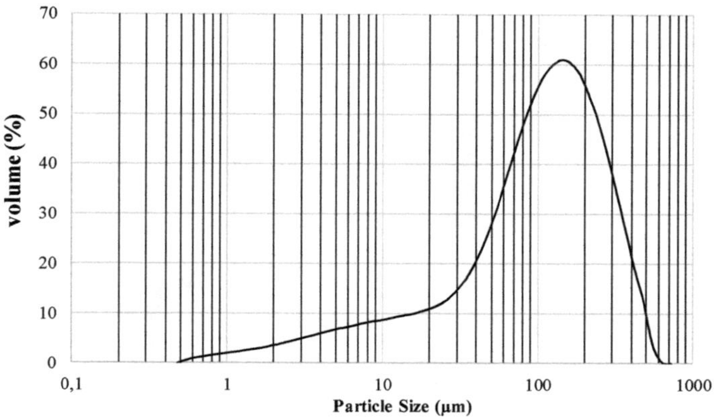

Fig. 1 Particle size distribution graphic of FW sample

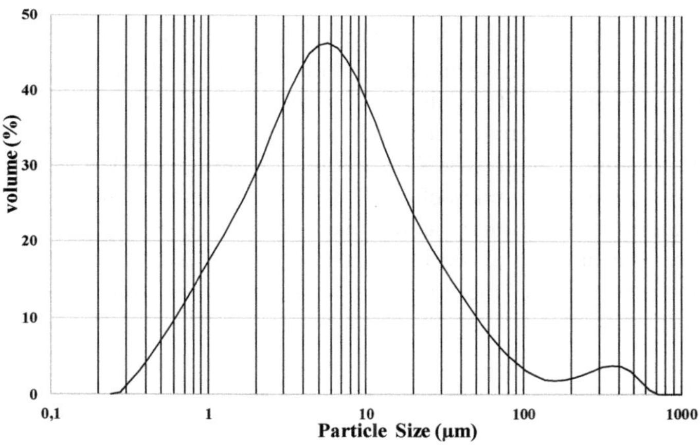

Fig. 2 Particle size distribution graphic of CW sample

Thereby, the particle size distribution of FW is smaller than that of CW. It is justified due to the regrind step between flotation and cyanidation processes. In other words, gold concentrate and the gold associated minerals collected from flotation tank pass by a comminuting step where these materials turn to a smaller granulometry, and then are under cyanidation cycle.

X-ray Diffraction (XRD)

According to FW's XRD graphic, there are characteristic peaks of quartz (SiO_2), pyrite (FeS_2), muscovite [$KAl_2(Si_3Al)O_{10}(OH,F)_2$], dolomite [$CaMg(CO_3)_2$] and

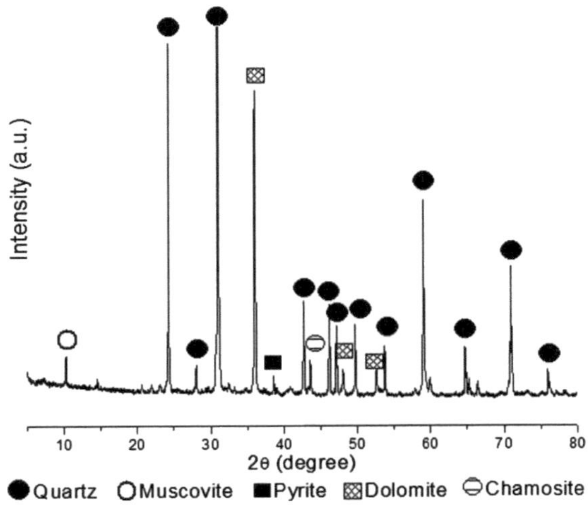

Fig. 3 X-ray diffraction pattern of FW sample, Co Kα Radiation

chamosite [$(Fe^{2+}, Mg, Fe^{3+})_5AI(Si_3Al)O_{10}(OH,O)_8$], with quartz as the main peaks, Fig. 3. The crystallinity associated with the majority of quartz in the sample explains the greater intensity of its peaks with respect to other peaks.

Meanwhile, in CW sample, there is noticed characteristic peaks of quartz, muscovite, dolomite, calcite ($CaCO_3$), pyrite, sphalerite (ZnS), galena (PbS), pyrrhotite (FeS), intermediary microcline ($KAlSi_3O_3$), ordered albite ($NaAlSi_3O_8$) and traces of chamosite, with quartz as the main peaks, Fig. 4.

Therefore, the number of sulfide minerals at CW is higher than that at FW, featuring the association of these minerals with gold concentrate collected from the flotation tank.

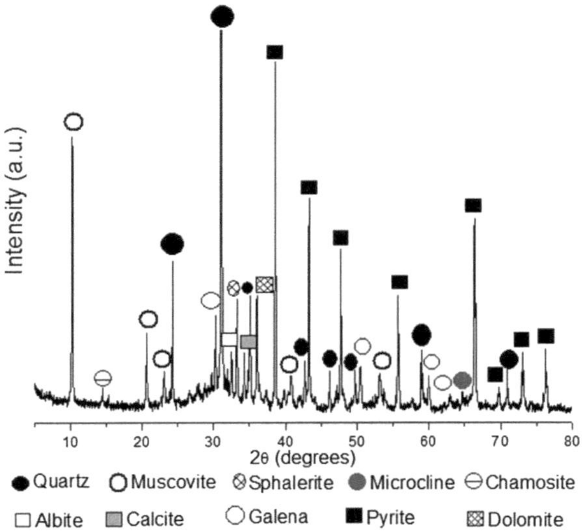

Fig. 4 X-ray diffraction pattern of CW sample, Co Kα Radiation

Nevertheless, peaks associated with arsenopyrite (FeAsS) were absent in XRD graphics, mineral usually associated with gold. This is can be justified by arsenopyrite oxidation during the flotation process, whose agitation induces a higher contact between sulfide minerals, water and oxygen of the air. Since arsenopyrite is a less noble mineral in comparison to pyrite, that mineral possibly was completely oxidized, what means that its crystalline structure has been broken, turning to arsenate, an amorphous species [6]. Thus, this species can not be identified by X-ray diffraction, since this technique only detects crystalline materials.

Carbon and Sulfur Elemental Analysis

Table 1 shows the results for elemental analysis, to carbon and sulfur, expressed in percentage by weigh (%w/w).

In relation to C and S content, to FW and CW, they were 2.5 and 0% and 0.64 and 15.4%, respectively, confirming the majority of sulfides at CW sample and indicating the absence of the flotation organic reagents and the low amounts of carbonate minerals at the cyanidation waste.

According to XRD results, this sulfur content can be associated with the minerals pyrite, galena, sphalerite, pyrrhotite and possibly to arsenopyrite.

X-ray Fluorescence Spectrometry (XRF)

XRF, quite as Carbon/Sulfur Elemental Analysis, allowed a semi quantitative analysis of chemical elements contained at FW and CW samples. Results are expressed, as well as in the previous item, in percentage by weigh (%w/w), but the elements are converted to their respectively oxides. LOI (Loss On Ignition) is the loss of sample weight during the calcination, in other words, the loss of all species which can be burned to the gas phase, Table 2.

According to the results of Table 2, it is visible the silicon majority in both samples (71.3% to FW and 24.7% to CW), confirming the qualitative results shown by XRD. Furthermore, there is a larger amount of metals associated with gold and

Table 1 Percentages (% w/w) of carbon and sulfur for FW and CW samples

Sample	C	S
	%w/w	
FW	2.5	0.64
CW	–	15.4

Table 2 Results of XRF and LOI for FW and CW samples

	Samples		
Elements/Oxides (%w/w)	Species	FW	CW
	MgO	3.7	1.6
	Al_2O_3	5.1	10.7
	SiO_2	71.3	24.7
	SO_3	1.4	–
	CaO	6.1	1.6
	K_2O	0.86	1.9
	TiO_2	0.2	–
	Fe_2O_3	3.0	8.7
	CdO	–	1.8
	PbO	–	1.6
	LOI	7.9	29.4

LOI Loss On Ignition

the sulfide minerals (Fe-pyrite and pyrrhotite; Pb-galena) at CW in comparison to FW sample.

Optical Emission Spectrometry with Inductively Coupled Plasma (ICP-OES)

ICP-OES has been done for the detection of elements, mainly arsenic and gold. In the gold case, it is to verify the efficiency of the gold processing, and, in the arsenic case, it is due to its high toxicity.

Table 3 shows the results of the mainly analyzed elements by ICP-OES for the FW, CW solid samples and CW aqueous sample.

According to the ICP-OES results, the presence of arsenic in FW sample, 13.30 mg/kg, and CW samples, 343 mg/kg and 0.34 mg/L, for solid and aqueous samples respectively, indicates its existence as arsenopyrite before the mining process and the gold processing.

Turning to the gold issue, its presence in the samples (2.70, 7.20 mg/kg and 1.10 mg/L to FW, CW solid and aqueous, respectively) is significant, indicating gold loss by current industrial process. Hence, the efficiency of the processing steps has to be verified to reduce the process loss, due to the waste amounts of 700,000 and 27,000 tons of flotation and cyanidation, respectively. Bases on the relation waste/gold had at directly proportional, the amount of gold within each waste would be, approximately, 1.89 and 0.19 tons, to flotation and cyanidation, respectively. According to quotation for one gramme of gold is sold for, approximately, USD 43 (Dow Jones 2016). Bases on the high amount of heavy metals, besides arsenic, it means that their environmental potential risks should be evaluate as well.

Table 3 Concentration of main elements contained at FW and CW solid samples, in mg/kg, and at CW aqueous sample, in mg/L

Elements	FW (solid)	CW (solid)	CW (aqueous)
	mg/kg		mg/L
Gold	2.70	7.20	1.10
Arsenic	13.30	343	0.34
Zinc	399	7900	0.014
Lead	136	13,000	< 0.020
Chrome	130	260	< 1.0
Copper	126	3100	183
Cadmium	3.60	84.60	< 1.0
Iron	111,000	18,000	2.6

Conclusions

Flotation and cyanidation wastes are composed mainly of silica associated with aluminossilicate crystalline structures, with the presence of metals such as sulfide minerals mostly from cyanidation waste. Cyanidation waste has a smaller granulometry in comparison with flotation waste. Gold was found in the samples, according to ICP-OES, which brings the necessity to calculate the gold metallurgical balance, in terms of verifying the gold loss at the ore processing.

ICP-OES results indicated the presences of high levels of heavy metals which could have potential environmental implications. Thus, toxicity tests, as dangerousness and solubility tests provided by the Brazilian Standard, NBR 10005 and NBR 10006, respectively, and invertebrates ecotoxicity tests, are being performed. In order to better classify the potential risk of these gold mining wastes to ecosystems are being developed as the complementary characterization techniques, such as Zeta potential, thermogravimetry, scan electronic microscopy, and vibrational infrared spectroscopy (IR) are being carried out as well.

Acknowledgements We would like to thanks Odilon da Silva and Gilvan for their help in the pilot plant processes; to IQ/UFRJ and COAM/CETEM for the analysis and lab infrastructure; the gold mining company; for the samples and CNPq for their financial support.

References

1. Muniz, DHF, Oliveira-Filho EC (2006) Metais Pesados Provenientes de Rejeitos de Mineração e seus Efeitos sobre a Saúde e o Meio Ambiente, Universitas: Ciências da Saúde, 4(1/2): 83–100
2. Neumann VH, Rocha DEGA, Moraes AS, Sial AN, Tabboada-Castro MT, Barbosa JA, Fambrini GL, Carvalho RR (2010) Microfácies Carbonáticas e Comportamento Isotópico de C e O nos Calcários Laminados Aptianos Lacustres da Serra Negra. Bacia do Jatobá. Estudos Geológicos 20:89–100
3. Silva SR, Procópio SO, Queiroz TFN, Dias LE (2004) Caracterização de Rejeito de Mineração de Ouro para Avaliação de Solubilização de Metais Pesados e Arsênio e Revegetação Local. Revista Brasileira de Ciência do Solo 28:189–196

4. Peres AEC, Chaves AP, Lins FAF, Torem ML (2002) Beneficiamento de Minérios de Ouro. In: Trindade RBE, Filho OB (eds) Extração de Ouro: Princípios, Tecnologia e Meio Ambiente. Rio de Janeiro. CETEM/CNPq, 2002. 2, pp 15–50
5. Peres AEC (2004) Flotação, Apostila do Curso de Flotação – Pós Graduação UFMG
6. Junior CRFA, Dutra AJB, Monte MBM (2004) Eletroquímica da Pirita e da Arsenopirita na Presença de Amil Xantato de Potássio. Revista Escola de Minas, Ouro Preto 57(4):297–302

Synthesis and Characterization of Sodalite and Cancrinite from Kaolin

Fabiano A.C.M. Passos, Danielle C. Castro, Karoline K. Ferreira, Karla M.A. Simões, Luiz C. Bertolino, Carla N. Barbato, Francisco M.S. Garrido, Adriana A.S. Felix and Fernanda A.N.G. Silva

Abstract Cancrinite and sodalite are characterized by high capacity of adsorption due to its porous structure. The aim of this work was to use kaolin from Borborema-Seridó/Brazil as a source of aluminium and silicon to hydrothermal synthesis of sodalite and cancrinite and to evaluate the best conditions to the

D.C. Castro · K.M.A. Simões · F.M.S. Garrido · F.A.N.G. Silva
Instituto de Química, Universidade Federal do Rio de Janeiro, Avenida Athos da Silveira Ramos, 149, Bloco A, sala 630, Cidade Universitária, Rio de Janeiro, RJ 21941-909, Brazil
e-mail: daniellecostaldecastro@gmail.com

K.M.A. Simões
e-mail: karla.carestiato@gmail.com

F.M.S. Garrido
e-mail: chico@iq.ufrj.br

F.A.N.G. Silva
e-mail: fnogueira@iq.ufrj.br

F.A.C.M. Passos · K.K. Ferreira
Escola de Química, Universidade Federal do Rio de Janeiro, Avenida Horácio Macedo, 2030, Bloco E, Cidade Universitária, Rio de Janeiro, RJ 21941-909, Brazil
e-mail: passosfabiano_vl@hotmail.com

K.K. Ferreira
e-mail: karolinekf@gmail.com

A.A.S. Felix (✉)
Campus Maracanã, Instituto Federal de Educação Ciência e Tecnologia do Rio de Janeiro, Rua Senador Furtado, 121/125, Maracanã, Rio de Janeiro 20270-021, Brazil
e-mail: adriana.soeiro@ifrj.edu.br

F.A.C.M. Passos · K.M.A. Simões · L.C. Bertolino
Centro de Tecnologia Mineral/CETEM-MCTI, Avenida Pedro Calmon, 900, Cidade Universitária, Rio de Janeiro, RJ 21941-908, Brazil
e-mail: lcbertolino@gmail.com

C.N. Barbato
Campus Duque de Caxias, Instituto Federal de Educação Ciência e Tecnologia do Rio de Janeiro, Avenida República do Paraguai, 120, Sarapuí, Duque de Caxias RJ 25050-100, Brazil
e-mail: carla.barbato@ifrj.edu.br

S. Ikhmayies et al. (eds.), *Characterization of Minerals, Metals, and Materials 2017*, The Minerals, Metals & Materials Series,
DOI 10.1007/978-3-319-51382-9_31

formation of these phases. Therefore, kaolin was heated at 700 and 900 °C for 2 h in order to transform kaolinite in metakaolin. Thereafter, experiments with the calcined kaolin's were carried out between 150 and 230 °C for 30–270 min on a stainless steel pressure vessel using NaOH, Na_2CO_3 and H_2O. The samples were characterized by X-ray diffraction, nuclear magnetic resonance, scanning electronic microscopy, thermogravimetry, and Zeta potential. The most pure sodalite phase was obtained at 230 °C, with the kaolin/NaOH ratio of 1:1, low concentration of Na_2CO_3 and smaller time of synthesis (30 min). On the other hand, for the synthesis of cancrinite phase at 230 °C, it was necessary to use kaolin heated at 900 °C, a longer time of synthesis (270 min) and a higher concentration of Na_2CO_3.

Keywords Kaolin · Sodalite · Cancrinite · Hydrothermal synthesis

Introduction

Anthropogenic release of heavy metals in rivers and lakes exceeds the amount that usually enters the aquatic environment by the action of weathering. Even in the oceans is considerable the resulting contribution of human activity. Thus, it is extremely important the treatment of effluents to remove toxic metals [1].

Adsorption has been used as a tertiary treatment of industrial waste, due to its high efficiency, low cost, easy operation and applicability in low concentrations. Clay minerals occur abundantly in nature, and act as natural removers of cationic and anionic pollutants by ion exchange and/or adsorption. Moreover, they have shown to be excellent adsorbent materials because of its high surface area, large exchange capacity and chemical and mechanical stability [1].

Among those clays, we can highlight kaolinite, from group of phyllosilicates, raw material used in this work in order to obtain zeolites sodalite and cancrinite. Both are natural aluminosilicates with same chemical formula ($Na_6[Al_6Si_6O_{24}].2NaX.6H_2O$) but with different three-dimensional structures. Cancrinite is characterized by hexagonal structure, while sodalite is characterized by cubic one [2]. They are made up of tetrahedral sites $[AlO_4]^{1-}$ and $[SiO_4]^{1-}$ alternating and connected by the vertices from the oxygens [3].

The Bayer process is the main industrial method to obtain alumina (Al_2O_3) from bauxite. In this process, bauxite is digested by NaOH at a temperature of, approximately, 170 °C. The reactive silica (kaolinite and/or quartz), mineral gangue, reacts with NaOH and with the ionic species in sodium aluminate solution ($Na_2OAl_2O_3$), forming insoluble desilication product (DSP). Residue from Bayer process, besides containing DSP, also contains iron oxides, titanium oxides, and the zeolitic phases sodalite and cancrinite, is denominated of red mud. Sodalite and cancrinite minerals, due to its porous structure, have adsorption capacity, giving them the possibility of application in adsorption processes [3, 4].

Studies have shown the possibility to synthesize sodalite from Borborema-Seridó's kaolin (PB/RN-Brazil), previously calcined [5, 6]. Within this context, the aim of this research was to assess the reaction conditions required for the synthesis of sodalite phase and its transformation to cancrinite phase, for future application on heavy metals adsorption processes.

Materials and Methods

Sample Preparation

Kaolin ore from Borborema/Seridó region (RN/PB—Brazil) was submitted to wet sieving, using a vibrate sieving (684.5 rpm) equipped with opening screens from 74 to 37 μm. This procedure was developed in order to concentrate kaolinite at finer fractions (37 μm). The fraction below 37 μm was filtered (filter press) and dried in an oven at, approximately, 80 °C.

The dry kaolin was grinded and quartered into aliquots, which were calcined at 700 and 900 °C for 2 h in a muffle furnace. After the calcination, the samples were cooled, by a natural process, and used on the synthesis.

Sodalite and Cancrinite Synthesis

Reagents used in order to react with the calcined kaolin were NaOH, Na_2CO_3 and H_2O, substances present in the red mud (Bayer process's residue). The influence of variables were evaluated to the synthesis of sodalite and cancrinite, thus a set of experiments were performed and the conditions are summarized in Table 1.

For carrying out the synthesis, reagents were weighed at an analytical balance, while the distilled water, 50 mL, was measured in a graduate cylinder. These reagents were placed in Teflon recipient, which, in its turn, were put into in a stainless steel pressure vessels (autogenous pressure). These vessels were heated in

Table 1 Conditions used in the synthesis experiments to available the influence of variables on cancrinite and sodalite synthesis

Experiment	m_{Kaolin} (g)	$m_{Na_2Co_3}$ (g)	m_{NaOH} (g)	V_{H_2O} (ml)	T (°C)	t (min)
A	5.0	5.0	5.0	15.0	230	270
B	0.1	5.0	0.1	5.0	230	30
C	5.0	0.1	5.0	5.0	230	30
D	5.0	5.0	0.1	15.0	150	30
E	5.0	0.1	0.1	5.0	150	270
F	0.1	0.1	0.1	15.0	230	270
G	0.1	5.0	5.0	5.0	150	270
H	0.1	0.1	5.0	15.0	150	30
Central Point	2.5	2.5	2.5	10.0	190	150

m mass; *V* volume; *T* temperature; *t* time

an oven at 10 °C/min of heating rate until the final temperature during the time of the respective experiment (T = 150 or 230 °C, t = 30 or 270 min).

After the heating process, the stainless steel pressure vessels were cooled with water stream until room temperature. The products were vacuum filtered and washed with distilled water in a vacuum system kitassato-Bünchner funnel, for at least 1 h and 30 min, in order to ensure the removal of NaOH excess. The pH of the washing water, from filtration step, was measured until the value of 7.5.

Characterization

The synthetized products were characterized by X-ray diffraction (XRD), nuclear magnetic resonance (NMR), scanning electronic microscopy (SEM), the-mogravimetry analysis (TG/DTG) and Zeta potential.

The samples' XRD patterns were obtained by the powder method in the equipment Bruker D4 Endeavor, Co Kα radiation (35 kV/40 mA), goniometer velocity of 0.02° (2θ) per step with 1 s per step and collected from 5 to 80° (2θ). The qualitative interpretation of the spectra was made by comparison with standards contained in PDF02 database (ICDD 2006) on Bruker AXS software DiffracPlus.

NMR spectrums were obtained into Bruker Avance III Spectrometer, model 400 W B, 9.4 T wide bore, operating at 5 and 10 kHz for ^{29}Si and ^{27}Al NMR spectrums, respectively.

The analyses in scanning electron microscopy were performed in microscope equipment, HITACHI TM303 Plus. The samples were metalized with gold, using the equipment BAC-TEC-SCD005 Sputter Coater. The modules secondary electrons, backscattered electrons.

The thermal gravimetric and gravimetric differential analyzes were performed on the equipment METTLER TOLEDO TGA model/DSC1 STAR^e System. The thermogravimetric analysis curves were obtained in the temperature range of 25–1.000 °C, heating rate 10 °C/min, platinum sample holder and nitrogen atmosphere with a flow rate of 50 ml/min.

The surface charge (Zeta potential) of samples were determined by equipment Zetasizer Nano ZS from Malvern. To carry out the measures were prepared 0.5 g of sample solutions in 10 ml of KCl 10^{-3} mol.L^{-1} (electrolyte). The pH values were adjusted using HCl or KOH in a range from 2.5 to 12.0 with the aid of a potentiometric titrator coupled to the system.

Results and Discussion

The zeolites synthesis and its crystallization, depends on the value of reaction medium's pH, which varies between 8 and 13. For this reason, the alkaline reagents were used. It is important to emphasize that the alkaline medium promotes the dissolution of amorphous aluminosilicates, thereby making them more reactive [6].

NaOH reagent was chosen in order to simulate the Bayer process, once sodalite and cancrinite are synthetized during the bauxite digestion. Meanwhile, the reagent Na_2CO_3 was used as template for the cavities formation contained on the zeolites, besides being a reagent used on the Bayer process [5, 6].

According to literature, the theoretical composition of kaolinite is 46.54 and 39.50% to SiO_2 and Al_2O_3, respectively. In this context, the Kaolin from Borborema-Seridó (PB/RN-Brazil) was chosen due to its chemical composition which is close to theoretical, i.e., 45.9 and 39.2% to SiO_2 and Al_2O_3, respectively [5–7]. This kaolin is composed by kaolinite, muscovite and quartz [8].

After the calcination step, the Al and Si sites of the minerals that compound this kaolin are more available to reaction with the reagents, once the crystalline structure of kaolinite and impurities, muscovite and quartz, were destroyed [6].

X-ray Diffraction (XRD)

According to the diffractograms shown in Fig. 1a, peaks associated with sodalite can be observed in the synthesis C and F (Table 1). In these synthesis were not observed peaks associated with impurity (muscovite), reagent (Na_2CO_3) and mixture of sodalite and cancrinite, contrary to what is shown in diffractograms of experiments A and Central Point (Table 1).

Analyzing the results of XRD for the kaolin calcined at 900 °C, Fig. 1b, is possible to observe that the syntheses C and F (Table 1) produced sodalite. These results are similar to those obtained from kaolin calcined at 700 °C. However, the synthesis A and Central Point produced a mixture of cancrinite and sodalite without impurities and reagents. The peaks of impurity and reagent have been disappeared

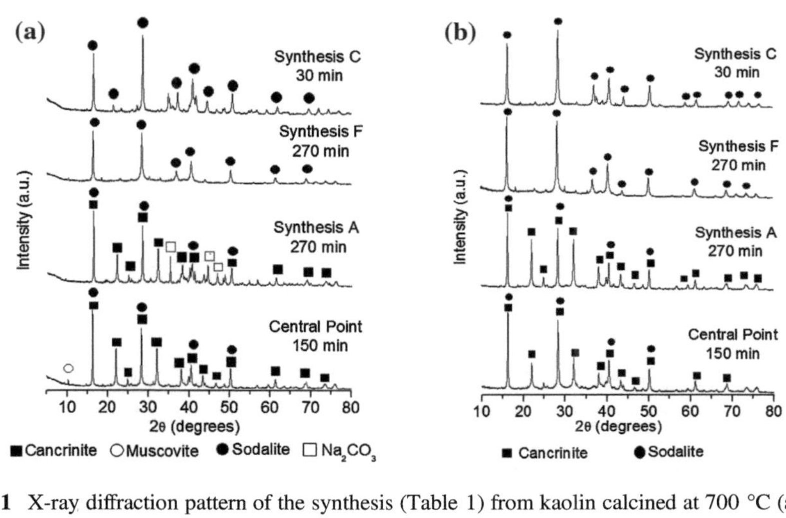

Fig. 1 X-ray diffraction pattern of the synthesis (Table 1) from kaolin calcined at 700 °C (**a**) and 900 °C (**b**)

from synthesis A and Central Point because the higher calcination temperature was able to disrupt the crystalline structure of impurity, as well as allowing more contact between reagents and kaolin active sites.

Due to overlap of the peaks of sodalite and cancrinite, it was not possible to confirm the complete phase transition between from sodalite to cancrinite, therefore, other characterization technics were used.

Nuclear Magnetic Resonance (NMR)

The NMR was used as a complementary technique to XRD in order to identify the difference between synthesis A and Central Point (Table 1) from the kaolin calcined at 900 °C. Figure 2a shows the ^{29}Si NMR spectrum of Central Point synthesis. According to result, product obtained from Central Point synthesis was characterized for two chemical shifts, at −83.61 and −86.77 ppm, assigned to sodalite and cancrinite, respectively, indicating a phase mixture, therefore this synthesis did not satisfy the complete transition from sodalite to cancrinite [2].

On the other hand, the spectrum of ^{29}Si, Fig. 2b, of synthesis A was characterized for one chemical shift, at −87.30 ppm, representative to cancrinite [2]. Therefore, according to NMR spectrums, while the Central Point's product is characterized by a mixture of zeolite phases (sodalite and cancrinite), the product of synthesis A possibly represents the complete transition from sodalite to cancrinite.

Scanning Electron Microscopy (SEM)

The SEM technique was used as visual observation of morphology of the synthesized zeolites. According to Fig. 3a, synthesis C sample, from kaolin calcined at 900 °C, has a rounded format, as "wool skeins", characteristics of polycrystalline

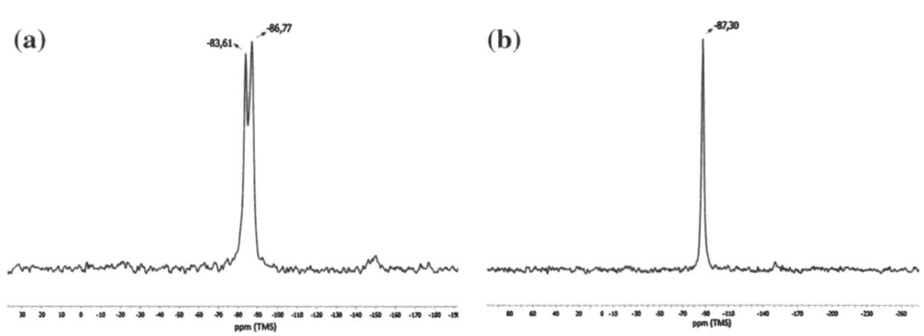

Fig. 2 ^{29}Si NMR spectrum of the Central Point synthesis sample (**a**) and the A synthesis sample (**b**) from kaolin calcined at 900 °C

Fig. 3 Photomicrography by SEM of the synthesis C sample (**a**) and the synthesis A sample (**b**) of kaolin calcined at 900 °C

clusters of sodalite. Figure 3b illustrates the crystalline structure of synthesis A sample, from kaolin calcined at 900 °C, in a stick form, representative of cancrinite hexagonal crystalline structure phase.

Thermogravimetry Analysis (TG/DTG)

Thermogravimetry analysis were used in order to verify the thermal stability of synthesized zeolite phases. According to Barnes et al. [9], the transition from sodalite to cancrinite is a process, which involves four mechanism: sodalite's dissolution; cancrinite's supersaturation resulting from sodalite's dissolution; cancrinite's nucleation; and cancrinite's core growth simultaneously to sodalite dissolution.

The hydrothermal transformation, from metakaolin in sodalite and cancrinite reveals a similar sequence of reactions proposed by Barnes et al. [9], starting with an amorphous material by transformation from kaolin to metakaolin (after calcination), followed to dissolution of the amorphous aluminosilicates, synthesis of LTA zeolite as the first and metastable aluminosilicate phase, sodalite structure, a mixture of sodalite and cancrinite phases and finally the cancrinite (more stable than sodalite) as a final stable synthesis product. Figures 4 and 5 represent the TG/DTG graphics of the synthesis F sample (sodalite) and synthesis A sample (cancrinite), both from kaolin calcined at 900 °C.

The dehydroxylation events, between 50 and 300 °C, for both samples (synthesis F and A), are related to loss of physically adsorbed water on the surface of zeolites and the release of the zeolitic water, present in the channels, of these zeolites.

The dehydroxylation events at 800 and 900 °C, for the synthesis A sample, and at 800 and 950 °C, for the synthesis F sample, are associated with carbonate release from zeolites structure [2]. Therefore, sodalite and cancrinite are stable at a wide range of temperature, since the zeolite phases did not suffer collapse neither disruption.

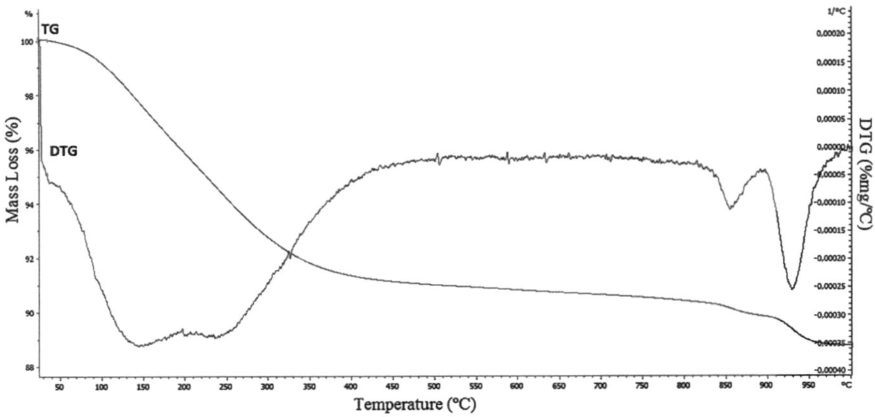

Fig. 4 TG/DTG of the synthesis F sample from kaolin calcined at 900 °C (sodalite)

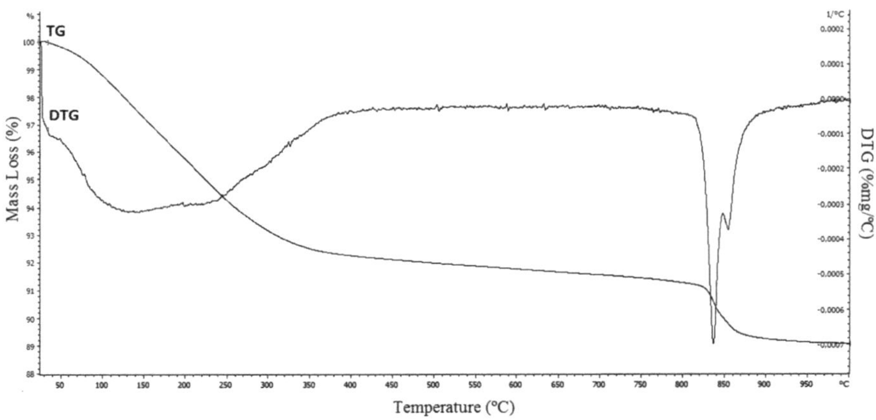

Fig. 5 TG/DTG of the synthesis A sample from kaolin calcined at 900 °C (cancrinite)

Zeta Potential

Zeta potential's results, Fig. 6, to sodalite (synthesis C sample) and cancrinite (synthesis A sample), from kaolin calcined at 900 °C, indicated that the isoelectric point of both samples is, approximately, in pH 6. The surface charges are positives below the isoeletric point and negatives above its. According to the results, these zeolites could be an excellent cation adsorbent (from pH 6 to 12) and anion adsorbent (from pH 3 to 6). It is important to emphasize that these phases are not chemically associated, which favors their separation.

Fig. 6 Zeta potential versus pH of sodalite (syntheis C sample) and cancrinite (synthesis A sample) from kaolin calcined at 900 °C

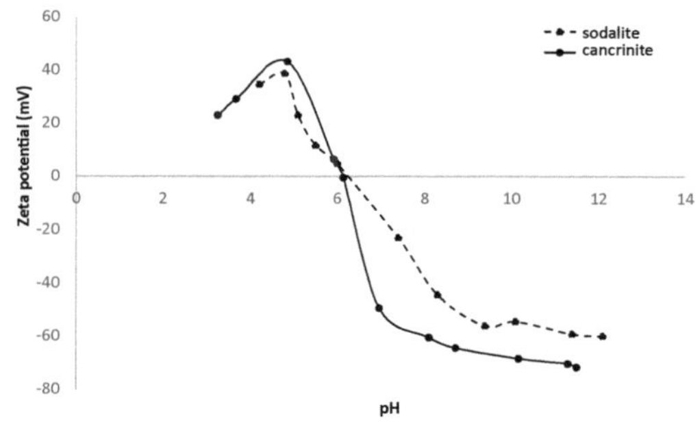

Conclusions

According to results, the study showed that sodalite and cancrinite phases could be synthesized efficiently using calcined kaolin from Borborema-Seridó (PB/RN-Brazil), as well as its crystallization, are dependent on the pH of the reaction medium.

Sodalite was synthesized in accordance with the 1:1 ratio between kaolin and NaOH, at 230 °C, and with an inverse relationship between the time of the synthesis and the concentration of NaOH. Besides that, the formation of cancrinite phase also depends on the ratio 1:1 between kaolin and NaOH, since it is formed after the sodalite phase as well as the major conditions as temperature, conditioning time and concentration of NaOH.

The characterization techniques allowed the verification of mineralogical composition with lower amount of impurities, as well as, to identify the crystalline structure of the zeolites phases. NMR spectrums permitted the possible identification of the complete transition from sodalite to cancrinite phase, in Synthesis A.

The thermogravimetric measurement confirmed the thermal stability of the synthesized samples under a wide range of temperature, since there were no exothermic peaks representing disruption of their crystalline structure. The results of surface charge measures (Zeta potential) indicated that the surface of synthesized phases are negatives for pH higher than 6. Therefore, these particles are not physically associated, facilitating their separation, since the red mud is found in basic medium.

Based on this study, further investigation will be conducted on adsorption of heavy metals using the synthesized zeolite phases.

Acknowledgements We would like to thanks João Alves Sampaio (*in memorian*) for their technical support; to IQ/UFRJ and COAM/CETEM for analysis and lab infrastructure; to CETEM for Fabiano A. C. M Passos's internship, and FAPERJ for their financial support (Process: APQ1 - E-26/110.419/2014).

References

1. Alves JPH, Aragão DM, Arguelho MLPM, Prado CMO (2013) Estudo Comparativo da Adsorção de Pb (II), Cd (II) e Cu (II) em Argila Natural Caulinítica e Contendo Montmorilonita, Orbital: The Electronic. J Chem 5(3):157–163
2. Reys CAR, Williams C, Alarcón OMC (2013) Nucleation and growth process of sodalite and cancrinite from kaolinite-rich clay under low-temperature hydrothermal conditions. Mater Res 16(2):424–438
3. Grader C, Buhl JC (2013) The intermediate phase between sodalite and cancrinite: synthesis of nano-crystals in the presence of Na_2CO_3/TEA and its thermal- and hydrothermal stability. Microporous Mesoporous Mater 171:110–117
4. Maldonado M, Oleksiak MD, Chinta S, Rimer JD (2013) Controlling crystal polymorphism in organic-free synthesis of Na-Zeolites. J Am Chem Soc 135:2641–2652
5. Santos CGM, Lana PMPB, Silva AAS, Barbato CN, Sampaio JA, Garrido FMS, Silva FANG (2012) Sintese da Fase Sodalita com o Uso do Caulim da Região Borborema-Seridó (RN/PB) Ativado Termicamente. HOLOS 28(5):52–59
6. Santos CGM, Passos FACM, Silva AAS, Barbato CN, Sampaio JA, Garrido FMS, Silva FANG (2014) Estudo de Síntese e Transição entre as Fases Zeolíticas Sodalita e Cancrinita. HOLOS 30(3):299–307
7. Silva FANG, Garrido FMS, Medeiros ME, Sampaio JA, Luz AB, Mello LS, Silva FT (2011) Chemical bleaching of the Brazilian kaolin: the effect of the pulp eletrochemical potential and pH adjustment. Quim Nova 34(2):262–267
8. Silva FANG, Luz AB, Sampaio JA, Bertolino LC, Scorzelli RB, Duttine M, Silva FT (2009) Technological characterization of kaolin: study of the case of the Borborema Seridó Region (Brazil). Appl Clay Sci 44:189
9. Barnes MC, Addai-Mensah J, Gerson AR (1999) The mechanism of the sodalite-to-cancrinite phase transformation in synthetic spent bayer liquor. Microporous Mesoporous Mater 31:287–302

Part VIII
Ferrous Metals

Effect of Alumina and Magnesia on Microstructure and Mineralogy of Iron Ore Sinter

Mingming Zhang and Marcelo W. Andrade

Abstract The evolution of microstructure and mineral phases in iron ore sinters have been investigated by scanning electron microscope and QEMSCAN (Quantitative Evaluation of Minerals by Scanning Electron Microscopy). It was found that MgO, basicity and Al_2O_3 content play an important role in the formation of mineral phases and microstructure in sinter. The low Al_2O_3 (<2.0 wt%), medium MgO (2.0–2.5 wt%) and high basicity (>3.0) sinter showed higher low temperature degradation index (LTD) due to the presence of silico ferrites of calcium and aluminium (SFCA), and magnetite phase, and lower pore phase as compared to low basicity iron ore sinter.

Keywords Sinter · Mineralogy · Microstructure · Metallurgical properties

Introduction

Although the annual worldwide production of sinter has been decreasing in the past several decades and alternative ironmaking processes haven been emerging, blast furnace production is still the most economic ironmaking process which utilizes sinter as an important burden materials. In a typical iron ore sintering process, the temperature of a micropelletized sinter raw mix is raised to achieve partial fusion and product some molten material which, on cooling, crystallizes or solidifies into mineral phases that bond the structure together. The properties of the iron ore sinter are controlled by the composition and distribution of these mineral phases [1–3].

Three types of phases, i.e. iron oxides, calcium ferrites and silicates in crystalline and glassy forms, are the major constituent of iron ore sinter. The crystal size and related surface area, the crystal structure (bridged, bonded or loose matrix), and the gangue constituents and their dispersion all have an effect on sintering rate and the

M. Zhang (✉) · M.W. Andrade
ArcelorMittal Global R&D, East Chicago, IN 46312, USA
e-mail: Mingming.Zhang@arcelormittal.com

S. Ikhmayies et al. (eds.), *Characterization of Minerals, Metals, and Materials 2017*, The Minerals, Metals & Materials Series,
DOI 10.1007/978-3-319-51382-9_32

physical-chemical characteristics of the resulting sinter product. Dolomite and other MgO-bearing materials are being increasingly used as a basic flux constituents for production of fluxed sinter. Low-grade iron ores, and recycled materials containing high percentage of Al_2O_3 are also being used more and more to reduce raw material costs of sintering. Addition of such materials influences the resultant sinter mineralogy. It has been known that the sinter mineralogy has a considerable influence on the properties of the sinter. The work pertaining to the mineralogy of MgO and Al_2O_3-bearing sinters is limited [4–7]. In this study a systematic investigation has been made on the influence of MgO and Al_2O_3 addition on sinter mineralogy for sinters in a widely used industrially important basicity range and its possible effect on the sinter properties.

Experimental

Materials

Raw materials used in this study consisted of iron ore blend (concentrates or pellet fines), sinter blend (in-plant recycled materials), fluxes (limestone and dolomite), and solid fuel (flue dust) as well as return fines as shown in Table 1. The MgO and Al_2O_3 contents of the sinter was varied through the addition of fluxes and recycled materials. The sinter basicities (mass percent CaO/mass per cent SiO_2) of the sinters ranged between 2.1 and 3.5.

Methods

The sintering experiments were carried out in a laboratory pot-grate sinter test station. The test station is capable of simulating industrial sintering process with high vacuum (up to 4800 mm W.G., or 47 kPa) and high temperature ignition (up to 1350 °C).

Table 1 Chemical analyses of raw materials used for sintering tests

Raw materials	Total Fe	SiO_2	CaO	Al_2O_3	MgO
Sinter blend	45.6	5.5	12.3	4.5	5.6
Iron ore blend	60.0	4.9	6.0	0.4	1.7
Fluxes	0.1	1.3	39.2	0.4	13.3
Solid fuel	9.9	4.3	2.7	2.7	0.8
Return fines	52.8	4.7	14.0	1.6	3.3

The pot-grate sintering test include proportioning, mixing, loading, ignition, sintering, cooling and conditioning. The raw materials were weighed into the required proportions and mixed in a cement mixer. Then the raw mix was loaded into the sinter pot (400 mm in diameter and 500 mm in height). The finished sinter cake was conditioned and screened to simulate the actual condition in a typical industrial sinter plant. The sinter above ¼ in (6.35 mm) was considered as sinter product and that of minus ¼ in was used as return fines for next test.

Determination of Chemical and Metallurgical Properties

The total oxides of sinter samples were determined by X-ray fluorescence spectrometry (XRF). The FeO content of the sinter samples was determined separately by a wet chemical method (redox titration with potassium dichromate). Metallurgical properties of the sinter samples were determined by ASTM E279-97, and ISO 4696-1 testing methods for Tumble Index (TI), and Low Temperature Degradation (LTD), respectively.

Microstructure and Mineralogical Examination

For each sinter sample, at least five sinter pieces in the granulometric fraction of 10–15 mm were examined. A reflected light metallurgical microscope was used for the microstructure examination. A JEOL-JXA 50 scanning electron microscope (SEM) with energy dispersive spectrometer was used to obtain additional elemental composition information.

Mineralogical examination was conducted using QEMSCAN pseudo-image and semi-quantitative (wt%) normalized analyses with Backscatter electron images (BSE). The mineral compounds were categorized based on their elemental chemistry and the main phases as per the X-ray diffraction analysis. The mineral list is shown in Table 2 which also provides the density applied for the QEMSCAN minerals distributions and elemental deportment data.

X-ray Diffraction (XRD) Analysis

The XRD analyses were conducted on powdered sinter samples. The sinter samples were ground down to 100% passing 100 mesh. The diffraction patterns were obtained employing an X'pert Pro Diffractometer by PANalytical using Ni filtered Cu Kα radiation at 35 kV and 50 mA. The samples were scanned at room

Table 2 Mineral group list (QEMSCAN modal list)

Modal list	Chemical composition	Density(g/cm^3)
Wustite	FeO	5.9
Iron oxides	hematite (Fe_2O_3) and magnetite (Fe_3O_4)	5.0
Iron oxides (Low Ca)	hematite (Fe_2O_3) and magnetite (Fe_3O_4) bearing calcium	5.2
Calcium ferrite/magnetite mix	Dominant Ca-Ferrite, Ca-silicates and Fe-oxides interlayers	5.2
Calcium ferrite	Dominant Fe and Ca	3.5
Silico-ferrite of calcium	Dominant Si and Fe	3.3
Calcium alumino-ferrite	Dominant Ca, Al and Fe	3.6
Calcium silicates	Dominant Ca and Si	2.8
Alteration silicates	Mixed altered silicates (Chlorite, plagioclase, quartz etc.)	2.7
Others	Trace amounts of other minerals	2.8

Table 3 Sinter sample chemistry

Sample ID	RM 3055	RM 3059	RM 3062	RM 3065	RM 3075
SiO_2, wt%	5.1	5.2	5.1	5.0	5.0
Fe_2O_3, wt%	76.9	76.5	77.0	77.9	78.6
MgO, wt%	3.4	3.3	3.3	3.1	2.1
Al_2O_3, wt%	1.6	1.4	1.3	1.1	1.0
CaO, wt%	13.8	14.2	13.8	12.9	13.3
Fe^{2+} as FeO, wt%	8.9	9.4	9.7	9.5	10.3
$(CaO + MgO)/SiO_2$	3.4	3.4	3.4	3.2	3.1
Tumbler Index (wt% + 6.35 mm)	75.7	74.2	72.8	73.5	78.5
Sinter LTD (wt% + 6.35 mm)	84.6	80.1	79.9	75.4	74.0

temperature from 20° to 130° at a step size of 0.10°, with a recording time of 4 s for each step.

Results and Discussion

The detailed chemistry from five different sinter samples are presented in Table 3. The MgO and Al_2O_3 contents of the sinter samples were varied in order to understand their effects on sinter mineralogy and properties.

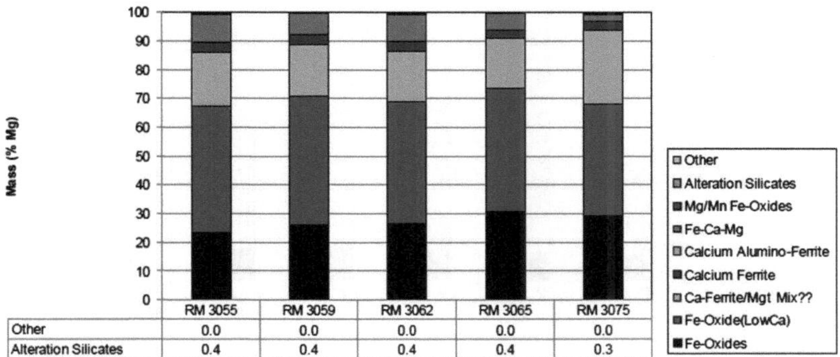

	RM 3055	RM 3059	RM 3062	RM 3065	RM 3075
Other	0.0	0.0	0.0	0.0	0.0
Alteration Silicates	0.4	0.4	0.4	0.4	0.3
Mg/Mn Fe-Oxides	0.5	0.3	0.3	0.1	0.5
Fe-Ca-Mg	9.5	7.0	9.3	5.7	2.3
Calcium Alumino-Ferrite	0.0	0.0	0.0	0.0	0.0
Calcium Ferrite	3.2	3.8	3.5	2.7	2.9
Ca-Ferrite/Mgt Mix??	19.1	17.6	17.8	17.4	26.0
Fe-Oxide(LowCa)	43.9	44.8	42.1	42.8	38.7
Fe-Oxides	23.4	26.1	26.6	30.9	29.3

Fig. 1 Distribution of magnesium in mineral phases in sinter samples

Elemental Distribution of Magnesium

The distribution of magnesium (normalized) in mineral phases is presented in Fig. 1. The graph reveals that magnesium is mainly hosted by the Fe-oxides and Fe-oxides (low Ca) ranging from 67.3 and 73.7%; followed by Ca-ferrite/Mgt mix (17.4 and 26.0%). Minor amounts are carried by calcium ferrite (2.7 and 3.8%) and Fe–Ca–Mg (2.3 and 9.5%), while other magnesium bearing minerals host less than 1%.

Elemental Distribution of Aluminum

The distribution of aluminum (normalized) in mineral phases is presented in Fig. 2. The graph reveals that aluminum is dominantly hosted by the silico-ferrite of calcium between 33.3 and 38.8%; followed by Fe-oxides/Fe-oxides (low Ca) (35.8 and 38.1%), Ca-silicates (14.6 and 22.5%), calcium ferrite (6.9 and 9.6%), while other aluminium bearing minerals host less than 2%.

Influence of Sinter MgO and Al_2O_3 on Low and High Basicity Sinter Microstructure

The change of sinter basicity and MgO/Al_2O_3 can cause major changes in sinter mineralogical phases and its properties. With the increase in basicity the formation

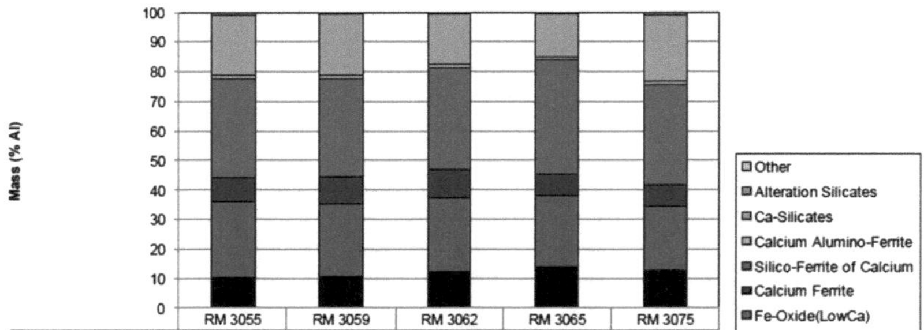

	RM 3055	RM 3059	RM 3062	RM 3065	RM 3075
Other	0.0	0.0	0.0	0.0	0.1
Alteration Silicates	0.8	0.5	0.4	0.4	0.7
Ca-Silicates	20.5	20.5	17.1	14.6	22.5
Calcium Alumino-Ferrite	1.0	1.3	1.4	1.1	1.3
Silico-Ferrite of Calcium	33.6	33.3	34.1	38.8	33.7
Calcium Ferrite	8.3	9.2	9.6	6.9	7.3
Fe-Oxide(LowCa)	25.3	24.2	25.1	24.4	21.7
Fe-Oxides	10.5	11.0	12.4	13.7	12.8

Fig. 2 Distribution of aluminium in mineral phases in sinter samples

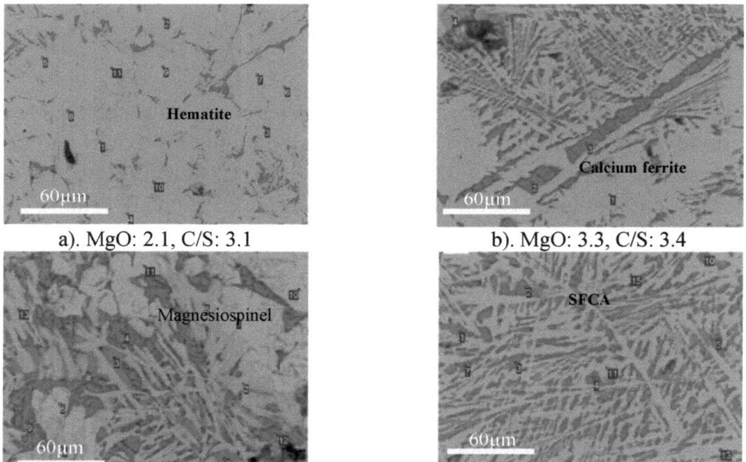

a). MgO: 2.1, C/S: 3.1

b). MgO: 3.3, C/S: 3.4

c). MgO: 3.1, C/S: 3.2

d). MgO: 3.4, C/S: 3.4

Fig. 3 SEM-BSE micrographs of sinter at different MgO levels and basicity

of melt phase increases and it causes significant changes in sinter mineralogy. Slag volume in the mineral phase varies as per change of sinter basicity. The mineralogical correlation with basicity and MgO/Al_2O_3 is more essential to understand the sinter properties.

SEM micrographs of low and high basicity sinter at different MgO levels are shown in Fig. 3a–d. It can be seen that the high level MgO sinter consists of higher SFCA phase in low and high basicity sinter. In all basicity level sinter with the increase in MgO addition the needle like SFCA phase increased. Compared with high basicity sinter, low basicity sinter consists of bigger size pores. At higher MgO level, the low and high basicity sinter consist of higher magnetite (magnesiospinel) phase. In low and high basicity sinter with the increase in magnesiospinel the sinter porosity decreased and slag phase increased. Mineralogical studies on low and high basicity sinter showed that the amount of total hematite phase, silico ferrites of calcium and aluminium, and pore phase increased with the increase in MgO addition. The SFCA phase increased with the increase in MgO addition. Low basicity sinter consists of higher hematite and lower SFCA phases compared to high basicity sinter. The magnesiospinel phase increased with the increase in MgO addition in low and high basicity iron ore sinter. Low basicity sinter consists of higher magnesiospinel and slag phases compared to high basicity sinter. High basicity sinter consists of a large amount of acicular/columnar SFCA phases compared to low basicity sinter at higher MgO level. This may be due to the association of high alumina in high basicity sinter.

Influence of MgO and Al_2O_3 on Sinter Tumble and LTD

Generally low reactivity of alumina and high viscosity of slag containing alumina can lead to a weaker sinter structure with more irregular interconnected pores. During sintering process, alumina also requires large amount of heat to assimilate thus slowing sintering process. Furthermore, un-reacted alumina powders in the sinter may form centers of weakness and result in weaker sinter. Typically alumina exists in the iron ores in the forms of alumino-silicates like kaolinite ($Al_2O_3{\cdot}2SiO_2{\cdot}2H_2O$), or gibbsite ($Al(OH)_3$). Almost all the alumina is segregated in the hematite crystals. During sintering significant diffusion of alumina in hematite crystals was observed. In recycled materials, Aluminum mainly exists in the forms of alumina (Al_2O_3) which comes from refractory linings or bauxites.

FeO content in the sinter decreases as the alumina content increases because of the decrease in magnetite phase as shown in Table 3. With increasing alumina content in sinter, a higher sintering temperature is required due to its refractory nature. At constant sintering temperature and coke breeze rate the FeO content in the sinter decreases as the alumina content increases.

Sinter LTD properties are more important to maintain the blast furnace upper stack permeability to reduce the coke consumption while increasing the productivity of blast furnace. Table 3 shows the influence of MgO addition on LTD of the sinter. The LTD of low and high basicity sinter decreased with the increase in MgO addition. The sinter LTD mainly depends on the microstructural phases present in the sinter. The weakening and degradation of sinter is associated with volume increase owing to phase transformation of hematite to magnetite occurring in the sinter. With the addition of MgO, hematite content decreased and magnetite content increased. Because of less availability of hematite less reduction takes place at lower temperature and less strain is generated in the matrix structure in the upper shaft of the blast furnace, hence LTD decreased. In addition, sinter LTD decreased with the increase in glassy silicate phase. The LTD of high basicity sinter was higher compared to low basicity sinter. This may be due to the presence of higher percentage of hematite and lower percentage of slag/silicate and magnetite phase in the high basicity sinter. The presence of MgO in the sinter matrix may be attributed to the production of slag phase with some precipitation of Mg ions. These Mg ions reduce the cracks during reduction.

The factors affecting sinter LTD have been discussed in a great number of papers [1–3]. The parameters affecting the LTD can be categorized as sintering parameters, properties of raw materials, and the chemical composition of a sintering mix. For example, increasing the MgO content, basicity (CaO/SiO_2), and fuel rate improves the reduction degradation property of the sinter. On the other hand, increasing the content of Al_2O_3 has a negative effect on LTD. Moreover, mineralogy of sinter is a significant factor of reduction strength. One controlling factor of mineralogy is the chemical composition, especially the CaO/SiO_2 ratio and even small changes in the content of minor components such as MgO and Al_2O_3, have a clear effect on the sinter mineralogy. An increase in the alumina content has been shown to cause more calcium ferrites or SFCA-phase (silico-ferrite of calcium and aluminium) in sinter. Similarly, MgO has been shown to stabilize magnetite in sinter.

Conclusions

Low Al_2O_3 (<2.0 wt%), medium MgO (2.0–2.5 wt%) and high basicity (>3.0) sinter showed higher low temperature degradation index (LTD) due to the presence of silico ferrites of calcium and aluminium (SFCA), and magnetite phase, and lower pore phase as compared to low basicity iron ore sinter.

The results of the QEMSCAN analysis yielded a relatively complex mineral assemblage which consist of major amounts of Fe-oxides (including Fe-oxides and Fe-oxides (Low Ca)) varying between 42.2 and 48.8%; followed by Ca-Ferrite/Mgt Mix (22.0–31.2%), and silico ferrite of calcium (13.8–16.5%). Minor amounts of wustite (5.2–7.8%) and calcium ferrite (3.4–4.9%); and trace amounts of other minerals were also detected.

Magnesium is mainly hosted by Fe-oxides ranging from 67.3 to 73.7%; followed by Ca-ferrite/Mgt mix (17.4 and 26.0%). Minor amounts are carried by calcium ferrite (2.7 and 3.8%) and Fe–Ca–Mg (2.3 and 9.5%). Aluminum is dominantly carried by the silico-ferrite of calcium between 33 and 39%; followed by Fe-oxides/Fe-oxides (low Ca) (36 and 38%), Ca-silicates (15 and 22%), calcium ferrite (7 and 10%), while other aluminium bearing minerals host less than 2%.

Acknowledgements The authors would like to thank ArcelorMittal Global R&D management for permission to submit this paper for publication.

References

1. Zhang M, Andrade M (2016) Effect of MgO and basicity on microstructure and metallurgical properties of iron ore sinter. In: Ikhmayies SJ, Li B, Carpenter JS, Hwang J, Monteiro SN, Li J, Firrao D, Zhang M, Peng ZW, Escobedo JP, Bai C (eds) Characterization of minerals, metals, and materials. TMS (The Mineral, Metals and Materials Society), Feb 2016, pp 167–174
2. Zhang M, Coe M, Andrade M (2015) Effect of sinter basicity on sinter productivity and quality with high rate of recycled materials. In: Battle TP, Downey JP, May LD, Davis B, Neelameggham NR, Sanchez-Segado S, Pistorius PC (eds) Drying, roasting, and calcining of minerals. TMS (The Mineral, Metals and Materials Society), Feb 2015, pp 259–267
3. Kalenge MK et al (2010) Investigation into how the magnesia, silica, and alumina contents of iron ore sinter influence its mineralogy and properties. J South Afr Inst Min Metall 110: 447–456
4. Umadevi T et al (2009) Influence of alumina on iron ore sinter properties and productivity in the conventional and selective granulation sintering process. Steel Res Int 80(9):10–15
5. Lu L, Holmers RJ, Manuel JR (2007) Effects of alumina on sintering performance of hematite iron ores. ISIJ Int 47(3):349–358
6. Yamaguchi K et al (1994) Influence of Al_2O_3 on reduction-meltdown behavior of sinter in blast furnace, ISIJ Int 34(12): 964–972
7. Kumar M, Choudhary et al (2008) Effect of variation of alumina on development of phases during iron ore sintering. ISIJ Int 48(12):1804–1806

Isothermal Reduction Kinetics of $CaO{\cdot}2Fe_2O_3$ by Thermogravimetric Analysis

Chengyi Ding, Xuewei Lv, Senwei Xuan, Kai Tang, Yun Chen and Jie Qiu

Abstract Isothermal reduction kinetics of CF_2 by 30% CO and 70% N_2 at 1123, 1173 and 1223 K was investigated by TGA in this study. The reduction rate analysis indicated that reduction of CF_2 comprises three stages during the whole reduction process, Apparent activation energy of samples CF_2 reduction is 34.37 kJ/mol. The ln-ln analysis implied the reduction of CF_2 was expressed initially by a 2D subsequently by 3D shrinking layer reaction.

Keywords CF_2 · Isothermal reduction kinetics · Apparent activation energy · Model function

Introduction

It is necessary to clarify the reducibility in more detail basicity content for the fluxed iron ore sinters which are the significant burden in a blast furnace. As the main bonding phases in fluxed sinters, investigation of reducibility on $CaO–Fe_2O_3–SiO_2–Al_2O_3$ quaternary calcium ferrite (SFCA) was performed by many previous studies. In the 1960s and 1970s [1], $CaO–Fe_2O_3$ system was found as the highlight of bonding phase in sintering process to instead of silicate system. The studies on $CaO–Fe_2O_3$ system has undergone from composition characterization, macroscopic properties to micro-morphology [2].

The reduction of bonding phases CF_2 ($CaO{\cdot}2Fe_2O_3$) in iron ores sinters in a blast furnace is typical gas-solid reaction. Maeda [3] clarified reduction equilibria of ternary calcium ferrite with $CO–CO_2$ gas mixture, reduction sequence and reaction equilibria in different CO/CO_2 ratio was fully investigated. Ganguly [4] examined that CF_2 is reduced in a number of reaction steps, the investigations also indicated the product morphologies and reduction kinetics are dependent on the rates of

C. Ding · X. Lv (✉) · S. Xuan · K. Tang · Y. Chen · J. Qiu
College of Materials Science and Engineering,
Chongqing University, Chongqing 400044, China
e-mail: lvxuewei@163.com

S. Ikhmayies et al. (eds.), *Characterization of Minerals, Metals, and Materials 2017*, The Minerals, Metals & Materials Series,
DOI 10.1007/978-3-319-51382-9_33

chemical reactions, rates of calcium and iron ion diffusion (surface and bulk), oxygen potential of the reducing gas and inherent structural and surface defects of the samples. Du [5] studied reduction kinetics of C_2F, CF, and CF_2 and compared their reduction rates at elevated temperatures. Taguchi [6] discussed the reduction process of C_2F, CF, and CF_2 by CO–CO_2 mixed gas and the accompanied expansion in connection with the behavior during reduction of lime self-fluxing sinter, the study found that the expansion of binary calcium ferrite increased with increasing Fe_2O_3 content in ferrite. The previous work was short of the macroscopic description of reduction mechanism, that is, the model function to reveal the reduction process and its relation to reduction rate were not involved in their research.

Experimental

Materials

The samples were prepared from $CaCO_3$ (≥99.99%) and Fe_2O_3 (α-Fe_2O_3, ≥99.99%) with 1:2 molar ratio. The powdery raw materials were uniformly mixed and then pressed to cylindrical shape samples. The samples were roasted in furnace with heating element of $MoSi_2$, the samples were first heated to 1173 K (900 °C) and kept for 1 h for fully decomposition of $CaCO_3$ to CaO and then increased to 1473 K (1200 °C) for 20 h to allow the complete formation of C_2F, CF and CF_2. The entire process was carried out in the air with 1 atmosphere. The samples were ground into powder (<74 μm) for next investigation.

X-Ray Diffraction (XRD) (Model D/max2500/PC (Cu Kα)) analysis was carried out to ensure the phase composition of samples. Scanning was carried out at an angular range of 10°–90° and scan rate of 4°/min. MDI Jade 6.5 was utilized to analyze the intensity data obtained by XRD measurement. The patterns of samples and standard patterns of CF_2 were shown in Fig. 1. The results indicate that samples mainly comprise CF_2.

TG Analysis

TG measurement of CF_2 reduction was conducted by using a Setaram analyzer (Model Setsys Evo TG-DTA 1750, Setaram Instrumentation) shown in Fig. 2. Samples (20 mg) were heated from room temperature to 1123 K (850 °C), 1173 K (900 °C) and 1223 K (950 °C) in N_2 (≥99.999%) atmosphere at a heating rate of 15 K/min, then switching to gas mixtures of 30% CO (≥99.999%) and 70% N_2 at a flow rate of 20 ml/min for 70 min to enable complete reaction with the samples at

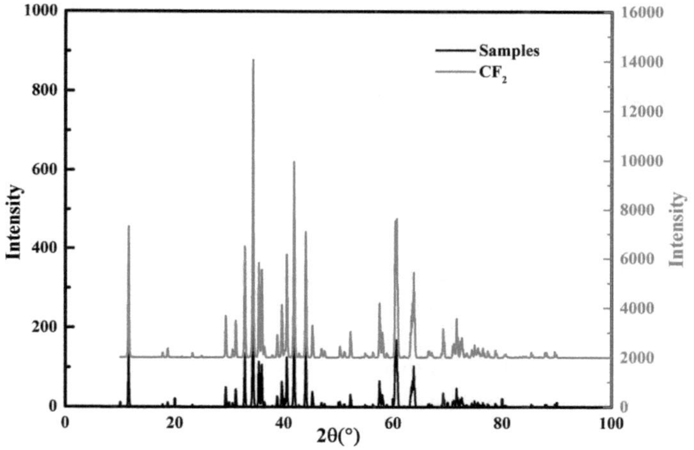

Fig. 1 XRD patterns of sample and standard patterns of CF_2

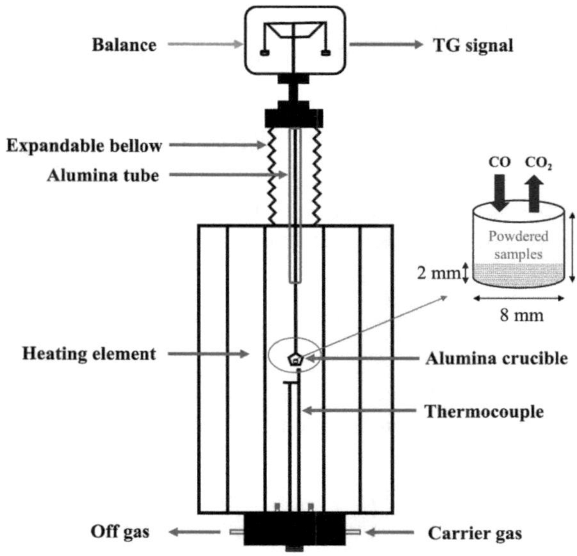

Fig. 2 Schematic of the TG analyzer

the isothermal stage. To exclude the influence of the system error from the thermal analyzer and the buoyance force from the gas mixtures, a blank test was conducted under the same elimination conditions with only empty alumina crucibles. Weight loss were obtained during the isothermal reduction stage, from which the TG data of the blank test were deduced.

Thermal Analysis Kinetics

Reduction degree is defined as the ratio of removed oxygen mass at a fixed time t to the theoretically removed oxygen mass from iron oxide; it can be expressed as

$$\alpha = \frac{\Delta m_t}{\Delta m_0}, \tag{1}$$

where α is the reduction degree and Δm_t and Δm_O refer to the removed oxygen mass at fixed time t and the theoretically removed oxygen mass from iron oxide, respectively.

The basic kinetic equation [7] that describes the relationship between reduction rate and time can be expressed as

$$\frac{d\alpha}{dt} = k(T)f(\alpha), \tag{2}$$

where dα/dt is the reduction rate and $k(T)$ and $f(\alpha)$ are the rate constant and model function of the reduction reaction, respectively. $f(\alpha)$ is influenced by the reaction mechanism. $k(T)$ is determined by the Arrhenius equation as follows:

$$k(T) = A\ \exp\left(\frac{-E}{RT}\right), \tag{3}$$

where A is the pre-exponential factor, E is the apparent activation energy, and R is the gas constant [8.314 J/(mol·K)]. Equation (2) can be further expressed as

$$\frac{d\alpha}{dt} = A\ \exp\left(\frac{-E}{RT}\right)f(\alpha). \tag{4}$$

Given that reduction degree α is a constant, ln $f(\alpha)$ remains unchanged. Therefore, activation energy can be calculated as

$$E = -R\frac{d\left(\ln\frac{d\alpha}{dt}\right)}{d\left(\frac{1}{T}\right)}. \tag{5}$$

The method to calculate activation energy can eliminate the limit from the model function and is called the model-free method [8]. $f(\alpha)$ is usually not easily obtained. $G(\alpha)$, which is the integral function of $f(\alpha)$, can be described as

$$G(\alpha) = \int_0^{\alpha} \frac{d\alpha}{f(\alpha)} = \int_0^{t} A\ \exp\left(\frac{-E}{RT}\right)dt = k(T)t. \tag{6}$$

Results and Discussion

Reduction Degree and Reduction Rate

The reduction degree of samples CF_2 with time during the isothermal reaction stage is shown in Fig. 3, maximum reduction degree of samples and its corresponding time are shown in Table 1. The results indicated that reduction degree is obviously promoted and reduction time decreases as temperature increases.

Reduction rates $d\alpha/dt$ CF_2 at 1123, 1173, and 1223 K are expressed in Fig. 4. Fe_2O_3 is assumed to be the essential reduced composition existing in C_2F, CF, and CF_2, therefore, reduction routes of $CaO–Fe_2O_3$ system can be simply expressed as

$$Fe_2O_3(C_2F, CF, CF_2) \rightarrow Fe_3O_4 \rightarrow FeO \rightarrow Fe. \quad (7)$$

Figure 4 clearly describes that the reduction processes CF_2 was observably "three–steps" reaction. The theoretical reduction degree is 0.11 and 0.33 when Fe_2O_3 fully reduces to Fe_3O_4 and Fe_3O_4 fully reduces to FeO. Figure 5 shows peak fitting based on Gauss rule of the reduction rate for samples C_2F, CF, and CF_2 at 1173 K.

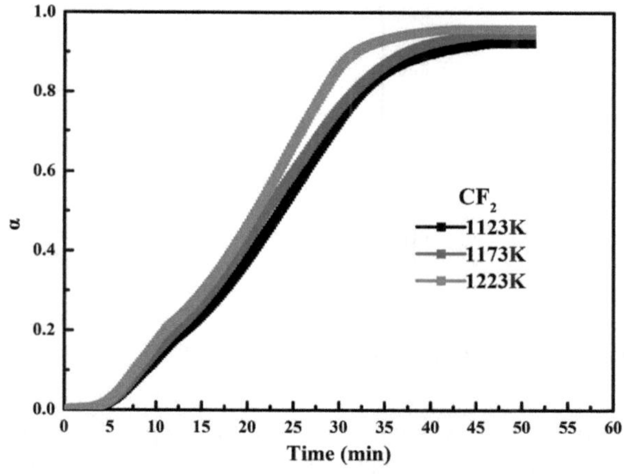

Fig. 3 Reduction degree of sample CF_2 at 1123, 1173, and 1223 K

Table 1 Maximum reduction degree of sample CF_2 at 1123, 1173, and 1223 K and the corresponding time

Sample	α_m			t_m/min		
	1123 K	1173 K	1223 K	1123 K	1173 K	1223 K
CF_2	0.92	0.94	0.96	46.6	46.4	42.5

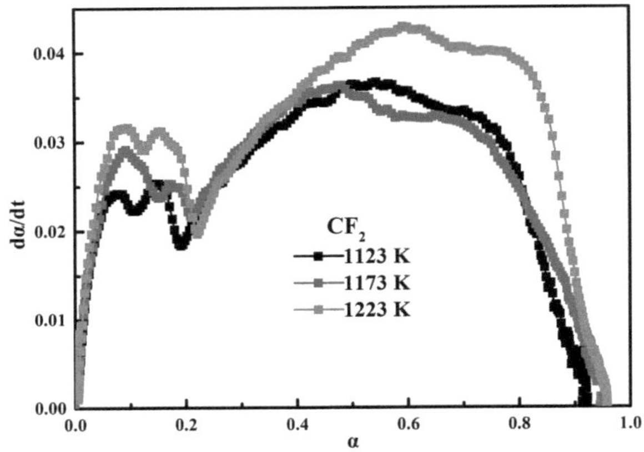

Fig. 4 Reduction rate of sample CF_2 at 1123, 1173, and 1223 K

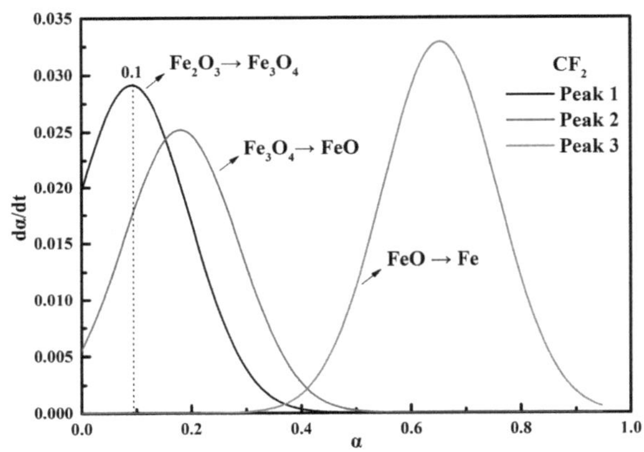

Fig. 5 Peak fitting of the reduction rate for sample CF_2 at 1173 K

Apparent Activation Energy for Reduction

Apparent activation energy was calculated by the slope of the plots of ln(dα/dt) against 1/T shown in Fig. 6. The results shown in Table 2 indicated that apparent activation energy of CF_2 reduction lies higher which as reaction proceeds, and the average value of apparent activation energy is 34 kJ/mol. The average fitting correlation coefficient R^2 is 0.92 for the activation energy values calculated at different reduction degree. The low value of R^2 was mainly led by the multi-steps in the reduction process, that is, several stages such as gas diffusion limited or chemical reaction exist in the reduction process, therefore the activation energy is the apparent value in this study. The liner fitting coefficient could not be higher.

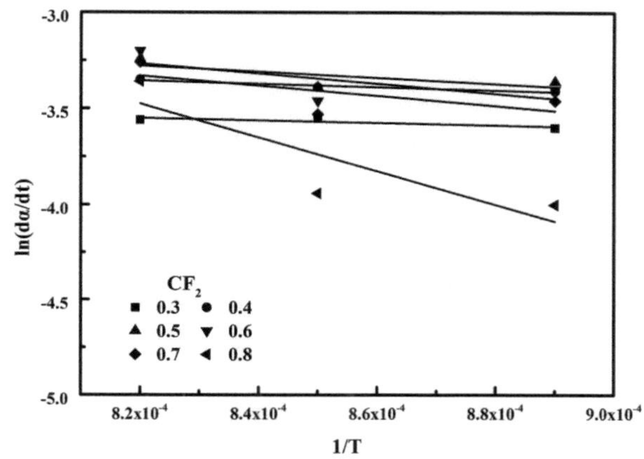

Fig. 6 ln(dα/dt) against 1/T of samples CF_2

Table 2 Apparent activation energy of sample CF_2

$E/(kJ \cdot mol^{-1})$							
α	0.3	0.4	0.5	0.6	0.7	0.8	Avg
CF_2	5.06	6.97	13.14	21.8	72.58	86.68	34.37

Model Function Results

ln-ln analysis [9] is applied to target $G(\alpha)$.

According to the Avrami-Erofeev [10–12] equation, the reduction degree α can be expressed as the relationship of time t and rate constant k:

$$1 - \alpha = \exp^{-kt^n}, \tag{8}$$

where n refers to the Avrami exponent. Avrami analysis, derived from process of copolymer, mineral, or metal crystallization, was also widely used in the description of solid-gas reaction. Obtaining the double natural logarithm on either side of the equal sign of Eq. (8), we can formulate the following equation:

$$\ln[-\ln(1-\alpha)] = n\ \ln t + \ln k. \tag{9}$$

The Avrami exponent n can be obtained by the slope of the fitted line of $\ln[-\ln(1-\alpha)]$ against $\ln t$, i.e.,

$$n = \frac{d\{\ln[-\ln(1-\alpha)]\}}{d\ln t}. \tag{10}$$

The value of n is directly related to $G(\alpha)$, which means that $G(\alpha)$ can be determined by first obtaining the value of n. Table 3 [9] shows the corresponding

Table 3 Relationship between $G(\alpha)$ and n for normal solid reactions and its reaction mechanism

Function	$G(\alpha)$	n	Mechanism
$D_1(\alpha)$	$\alpha^2 = kt$	0.62	One-dimensional diffusion
$D_2(\alpha)$	$(1-\alpha)\ln(1-\alpha)+\alpha = kt$	0.57	Two-dimensional diffusion (bidimensional particle shape)
$D_3(\alpha)$	$\left[1-(1-\alpha)^{1/3}\right]^2 = kt$	0.54	Three-dimensional diffusion (tridimensional particle shape) Jander equation [13]
$D_4(\alpha)$	$(1-2/3\alpha)-(1-\alpha)^{2/3} = kt$	0.57	Three-dimensional diffusion (tridimensional particle shape) Ginstling-Brounshtein equation [14]
$F_1(\alpha)$	$-\ln(1-\alpha) = kt$	1	Bimolecular decay law (instantaneous nucleation and unidimensional growth)
$R_2(\alpha)$	$1-(1-\alpha)^{1/2} = kt$	1.11	Phase boundary controlled reaction (contracting area, i.e., bidimensional shape)
$R_3(\alpha)$	$1-(1-\alpha)^{1/3} = kt$	1.07	Phase boundary controlled reaction (contracting volume, i.e., tridimensional shape)
$A_2(\alpha)$	$[-\ln(1-\alpha)]^{1/2} = kt$	2	Random instant nucleation and two-dimensional growth of nuclei (Avrami-Erofeev equation [9–11])
$A_3(\alpha)$	$[-\ln(1-\alpha)]^{1/3} = kt$	3	Random instant nucleation and three-dimensional growth of nuclei (Avrami-Erofeev equation)

relationship between $G(\alpha)$ and n and its reaction mechanism. The Avrami exponent n is related to the functions corresponding to nine solid-state reactions which can be expressed mechanisms as diffusion controlled, reaction-rate controlled, first order kinetics controlled, or following the equations of Avrami and Erofeev.

Figure 7 shows the relationship between $\ln[-\ln(1-\alpha)]$ and $\ln t$ for samples CF_2, and Avrami exponent n was calculated in Table 4. The results revealed that the n corresponding to samples CF_2 reduction at 1123, 1173 and 1223 K lie at 2.41, 2.52, and 2.53, respectively, indicating reduction of sample CF_2 is described by function A_2 tending to A_3.

Functions A_2 and A_3 were obtained from the nucleation process (crystallization) and can also be applied to the shrinking layer reaction. Functions A_2 can expresses 2D shrinking layer reaction and A_3 can expresses a 3D reaction, moreover, to be better understood, the reaction determined by A_2 appears proceeding in a plane whereas in a cylinder based on A_3.

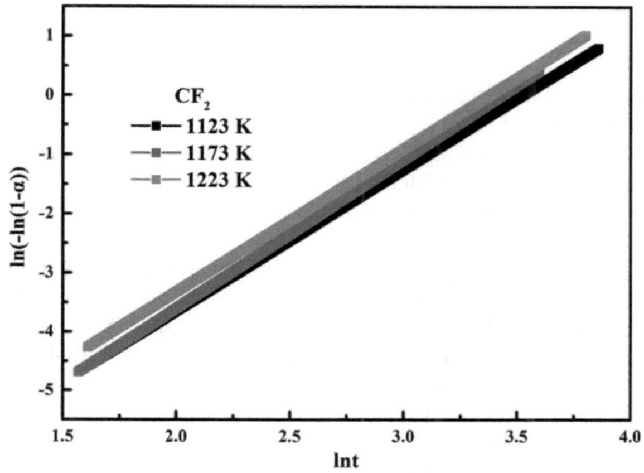

Fig. 7 ln[−ln(1−α)] versus ln$t$ at 1123, 1173 and 1223 K based on ln-ln analysis for sample CF_2

Table 4 Avrami exponent n at 1123, 1173 and 1223 K for sample CF_2

Temperature/K	CF_2
1123	2.41
1173	2.52
1223	2.53
Function	$A_2 \rightarrow A_3$ (more)

Conclusions

The isothermal reduction kinetics of CF_2 was investigated via TG measurement with 30% CO and 70% N_2 gas mixtures. The reducibility CF_2 was implied by revealing the reduction rate and apparent activation energy. The reduction mechanisms were examined through ln-ln and Sharp analysis. The following conclusions were summarized.

- The reduction rate analysis indicated that reduction of samples CF_2 comprises three stages during the whole reduction degrees.
- Apparent activation energy of samples CF_2 reduction process lies 34.37 kJ/mol.
- The ln-ln analysis implied the reduction of CF_2 was expressed initially by a 2D subsequently by 3D shrinking layer reaction.

Acknowledgements The authors are grateful for the financial support provided by the Natural Science Foundation of China (51544203).

References

1. Phillips B, Muan A (2006) Phase Equilibria in the System CaO-Iron Oxide-SiO_2, in Air. J Am Ceram Soc 42(9):413–423
2. Ding C, Lv X, Chen Y et al (2015) Non-isothermal crystallization kinetics for CaO–Fe_2O_3 system. J Therm Anal Calorim 124(1):509–518
3. Maeda T, Ono Y (1994) Reduction equilibria of ternary calcium ferrite with CO–CO_2 Gas Mixture. Tetsu-to-Hagane, vol 80
4. Ganguly S (1991) A study of gaseous reduction of calcium ferrites. Reduction
5. Du S, Sichen N, Staffansson LI (1988) Reduction of calcium ferrites by hydrogen in the temperature interval 1191–1426K. Scand J Metall 17(5):232–238
6. Taguchi N, Otomo T, Tasaka K (1983) Reduction process of CaO–Fe_2O_3 binary calcium ferrite and resultant expansion. Tetsu-to-Hagane 69:1409–1416
7. Mccune RC, Wynblatt P (1983) Calcium segregation to a magnesium oxide (100) surface. J Am Ceram Soc 66(2):111–117
8. Vyazovkin S, Wight CA (1999) Model-free and model-fitting approaches to kinetic analysis of isothermal and non-isothermal data. Thermochim Acta 340:53–68
9. Hancock JD, Sharp JH (1972) Method of comparing solid-state kinetic data and its application to the decomposition of kaolinite, brucite, and $BaCO_3$. J Am Ceram Soc 55(55):74–77
10. Avrami M (1939) Kinetics of phase change. I general theory. J Chem Phys 7(12):1103–1112
11. Avrami M (1940) Kinetics of phase change: II. transformation—time relation for random distribution of nuclei. J Chem Phys 8(2):212–224
12. Avrami M (1941) Granulation, phase change, and microstructure kinetics of phase change. III. J Chem Phys 9(2):177–184
13. Tamhankar SS, Doraiswamy LK (1979) Analysis of solid-solid reactions: a review. AIChE J 25(4):561–582
14. Ginstling AM, Brounshtein BI (1950) On diffusion kinetics in chemical reactions taking place in spherical powder grains. Zhur. Priklad. Khim., vol 23

Phase Transformation of MnO_2 and Fe_2O_3 Briquettes Roasted Under $CO–CO_2$ Atmospheres

Bingbing Liu, Yuanbo Zhang, Zijian Su, Guanghui Li and Tao Jiang

Abstract With the depletion of rich manganese ores, high-Fe content manganese ores are becoming the mainstream resources for manganese extraction. Reduction roasting-magnetic separation process is reported as a relatively effective route to realize the separation of manganese and iron. This study investigated the phase transformation of MnO_2 and Fe_2O_3 briquettes (molar ratio 1:1) roasted under various temperatures and $CO–CO_2$ atmospheres. Results showed that when the CO content was in the range of 0–10 vol%, $Mn_xFe_{3-x}O_4$ ($x = 0-1$) phase was formed and its content increased with the rising of temperature. The formation of ferro-magnetic $Mn_xFe_{3-x}O_4$ with strong magnetism resulted in an inferior separation of manganese and iron by magnetization roasting-magnetic separation process. With the increase of CO content to over 25 vol%, the $Mn_xFe_{3-x}O_4$ content decreased while the content of feebly magnetic substance $(MnO)_y(FeO)_{1-y}$($y = 0-1$) increased. Further increasing the CO content to over 80 vol%, $(MnO)_y(FeO)_{1-y}$ was reduced to MnO and metallic iron.

Keywords MnO_2 · Fe_2O_3 · Reduction roasting · Phase transformation

Introduction

With the depletion of rich manganese ores, high-Fe content manganese ores are becoming the mainstream resources for manganese extraction. Previous investigations indicated that it's very difficult to effectively separate iron oxides from manganese oxides only by using mechanical beneficiation methods, such as gravity, magnetic or flotation processes, due to the closely disseminated relationship between iron oxides and manganese oxides [1–3].

B. Liu · Y. Zhang (✉) · Z. Su · G. Li · T. Jiang
School of Minerals Processing and Bioengineering,
Central South University, Changsha 410083, Hunan, China
e-mail: zybcsu@126.com

S. Ikhmayies et al. (eds.), *Characterization of Minerals, Metals, and Materials 2017*, The Minerals, Metals & Materials Series,
DOI 10.1007/978-3-319-51382-9_34

Pyrometallurgical processes for separation of manganese and iron, such as magnetic roasting-magnetic separation process and direct reduction roasting-magnetic separation process, are reported as effective routes to handle the high-Fe content manganese ores [4–8]. Pyrolusite (MnO_2) and hematite (Fe_2O_3) are the main valuable minerals in the manganese ores. During the reduction roasting process, the dominating reduction reactions include the reduction of MnO_2, Fe_2O_3 and the reciprocal influence between them. During the magnetic roasting process, the Fe_2O_3 is selectively reduced to Fe_3O_4 by CO while the MnO_2 was reduced to nonmagnetic MnO. Then, the iron-rich component was then separated from MnO by magnetic separation [4–6]. Direct reduction roasting process was conducted under strong reduction atmosphere, in which the Fe_2O_3 was converted to metallic iron and MnO_2 was reduced to MnO, and then followed by magnetic separation to realize the separation of manganese and iron [7, 8]. It's well known that the MnO_2 and Fe_2O_3 were reduced step by step to low-valence oxides or metals. There are some mutual influences existing among MnO_2 and Fe_2O_3 as well as the reduced products. Zhang et al. [9] investigated the consolidation behavior of high-Fe content manganese ore fines with natural basicity. It's found that the optimal coke breeze dosage for the sintering was 9.9 wt% and in this case the CO content in the outlet flue gas was in the range of 7.5–8.0 vol%. The main bonding phase of the sintered products was $Mn_xFe_{3-x}O_4$ which was formed from the reaction of MnO_2 and Fe_2O_3 under reduction atmosphere. On the other hand, extensive investigations [10–12] reported that the manganese ferrites ($Mn_xFe_{3-x}O_4$) are one kind of excellent magnetic materials. Therefore, it's essential to study the reactions between MnO_2 and Fe_2O_3 during the reduction process under CO–CO_2 atmosphere and the magnetic characterization of the roasted products, which can be acted as a good understanding of the reduction roasting-magnetic separation process for high-iron content manganese ores.

In the current research, phase transformation of pure MnO_2 and Fe_2O_3 briquettes with molar ratio of 1:1 was conducted under different temperatures and various CO/(CO + CO_2) atmospheres by using XRD, VSM, optical microscopy and SEM-EDS analyses. Eventually, the main possible reactions during the reduction roasting process of MnO_2 and Fe_2O_3 were discussed.

Experimental Procedure

Materials

The tested MnO_2 powder was spectrum pure reagent and Fe_2O_3 powder was chemically pure reagent. Both MnO_2 and Fe_2O_3 were with purity of 99.5 wt%. Particle size distributions of MnO_2 and Fe_2O_3 powders examined by Malvern Laser Particle Size Analyzer (Mastersizer 2000, Britain), and the average grain size of the MnO_2 and Fe_2O_3 powders are 20.1 μm and 69.6 μm, respectively. The purity of gases, including CO, CO_2 and N_2, are more than 99.99 vol%.

Roasting Procedure

Reduction roasting experiment was conducted in a horizontal resistance furnace, and the schematic diagram experimental of roasting system is illustrated in Fig. 1. MnO_2 and Fe_2O_3 powders with molar ratio of 1:1 were evenly mixed and then the mixture was briquetted by a mold into a cylindrical briquette with a diameter of 10 mm and height of 15 mm at a pressure of 10 MP, and then the cylindrical briquettes were dried at 110 °C for 4 h. After that, the dried briquettes were placed on a corundum substrate (size: 80 mm × 10 mm) and loaded into a heat resistant corundum tube. The corundum substrate carrying with the dried briquettes was pushed towards the constant roasting zone located in a central area of an electrically heated horizontal tube furnace. Beforehand, the furnace was heated to a preset temperature. The roasting temperature was measured by a Pt–Rh thermocouple and controlled using a digital temperature controller (accuracy ±1 °C). Pure N_2 gas was introduced into the corundum tube until the temperature reached a constant value. Next, the N_2 was immediately replaced by the mixed CO–CO_2 gas. The CO content in the CO–CO_2 mixed gas was precisely adjusted by the flowmeters. The inlet gas flow was fixed at 4.0 dm^3/min. After roasted at given temperatures for certain time periods, the samples were cooled in N_2 atmosphere for the subsequent characterization tests.

Characterization

XRD patterns of the roasted samples were detected by a diffractometer (RIGAKU D/Max 2500, Japan) under the conditions of radiation: Cu Kα, tube current and voltage: 250 mA, 40 kV, scanning range: 10°–80° (2θ), step size: 0.02° (2θ) and

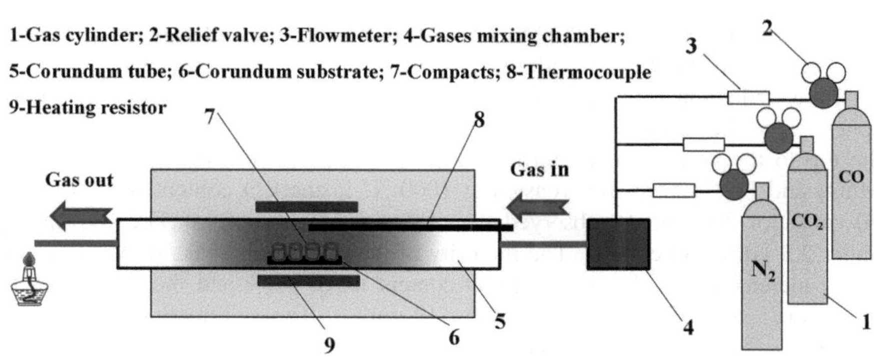

Fig. 1 Schematic diagram of experimental equipment for roasting

scanning speed: 8°/min. Some representative briquettes were heat mounted in resin and polished to mirror finish. Then the polished sections were examined by an optical microscopy (LEICA MDI5000 M, Germany) and scanning electron microscopy (SEM, TESCAN, MIRA3 LMH/LMU) equipped with an energy dispersive X-ray spectroscopy (EDS) detector. SEM images were recorded in backscatter electron mode operating in a low vacuum of 0.5 Torr and 20 keV. The magnetism properties of the roasted samples were determined by a vibrating sample magnetometer (PPMS Dynacool, Quantum Design, America) at ambient temperature.

Results and Discussion

Phase Transformation

Magnetic roasting process was conducted under weak reduction atmospheres at various temperature ranges. Figure 2 illustrates the XRD patterns of the MnO_2 and Fe_2O_3 briquettes with molar ratio of 1:1 roasted in 10 vol% CO content at the temperature range of 800–1100 °C and in 4 vol% CO content at the temperature range of 1200–1300 °C for 30 min. Herein, CO content refers to the molar ratio of $CO/(CO + CO_2)$. At 800 °C, diffraction peaks of Fe_2O_3, Fe_3O_4 and MnO are observed. The formation of Fe_3O_4 and MnO were ascribed to the reduction of Fe_2O_3 and MnO_2, respectively. Moreover, the diffraction peaks of $Mn_xFe_{3-x}O_4$ appear over 900 °C. The intensity of the diffraction peaks of $Mn_xFe_{3-x}O_4$ increase with the rising of temperature while that of the Fe_2O_3 decreases. As the temperature goes up to 1200 °C, only $Mn_xFe_{3-x}O_4$ is found, indicating the almost complete chemical combination of MnO_2 and Fe_2O_3 with stoichiometric ratio of 1:1. In particular, it's also discovered by careful observation that the main diffraction peaks of $Mn_xFe_{3-x}O_4$ gradually shift from $2\theta = 35.482°$ (Fe_3O_4, PDF#75-0033) towards $2\theta = 34.918°$ ($MnFe_2O_4$, PDF#73-1964) with the increase of temperature, which indicates that the x value in the $Mn_xFe_{3-x}O_4$ increases to 1, and in other words, the $Mn_xFe_{3-x}O_4$ approaches to $MnFe_2O_4$.

Direct reduction roasting process was conducted under strong reduction atmospheres, and the reduction atmosphere (CO content) was gradually enhanced from the top to the bottom of the furnace. Figure 3 displays the XRD patterns of the MnO_2 and Fe_2O_3 briquettes roasted at 1000 °C in the CO content range of 2.5–80 vol% for 30 min. As observed, the diffraction peaks of $Mn_xFe_{3-x}O_4$ appear under 2.5 vol% CO content. The intensity of the diffraction peaks of $Mn_xFe_{3-x}O_4$ firstly increases with the rising of CO content to 10 vol% and then decreases as increasing the CO content over 10 vol%. The diffraction peaks of $(MnO)_{0.669}(FeO)_{0.331}$ occur under 25 vol% CO content, and disappear when the CO content is over 60 vol%. Further increasing the CO content to 80 vol%, the main phases are MnO and Fe. It's noteworthy that the diffraction peaks of MnO appear at

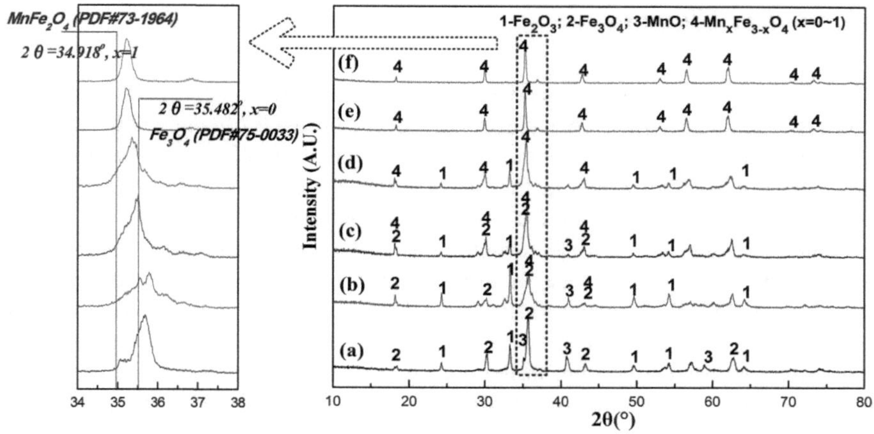

Fig. 2 XRD patterns of the MnO_2 + Fe_2O_3 briquettes roasted at different temperatures under 10 vol% CO content for 30 min. **a** 800 °C, **b** 900 °C, **c** 1000 °C, **d** 1100 °C, **e** 1200 °C, **f** 1300 °C (**a–d** 10 vol% CO content; **e, f** 4 vol% CO content)

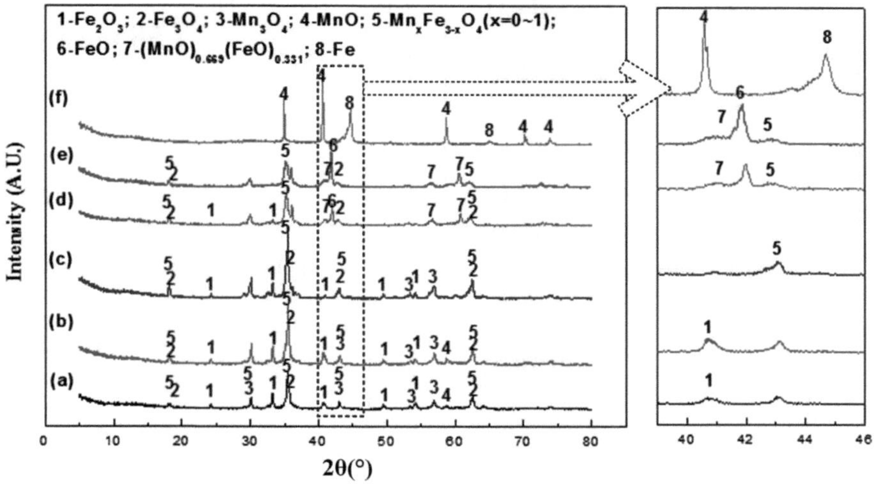

Fig. 3 XRD patterns of the MnO_2 + Fe_2O_3 briquettes roasted under different CO contents at 1000 °C for 30 min. **a** 2.5 vol% CO, **b** 5 vol% CO, **c** 10 vol% CO, **d** 25 vol% CO, **e** 60 vol% CO, **f** 80 vol% CO

the CO content range of 2.5–10 vol% and over 80 vol%. This indicates the MnO may be reacted with the iron oxides to form $Mn_xFe_{3-x}O_4$, and the $Mn_xFe_{3-x}O_4$ was then reduced by CO.

Magnetism Changes

Magnetic property of the roasted samples is crucial to the magnetic separation process. The magnetization hysteresis loops of the products obtained under different CO contents at 1000 °C for 30 min were investigated by VSM, and the results are drawn in Fig. 4. The results in Fig. 4 show that the maximum saturation magnetization (M_S) increases from 2.0 emu/g (raw materials) to 58 emu/g as the CO content increases to 10 vol%, which is ascribed to the formation of magnetic $Mn_xFe_{3-x}O_4$. With the increase of CO content to 60 vol%, the Ms of the products decreases to 28 emu/g owing to the formation of $(MnO)_y(FeO)_{1-y}$. When the CO content increases to 80 vol%, the Ms of the products increase to 70 emu/g due to the formation of metallic iron. In the reduction roasting process of high-Fe content manganese ores, the formation of $Mn_xFe_{3-x}O_4$ under weak reduction atmosphere and the formation of $(MnO)_y(FeO)_{1-y}$ under strong atmosphere are unfavorable to the magnetic separation of iron and manganese.

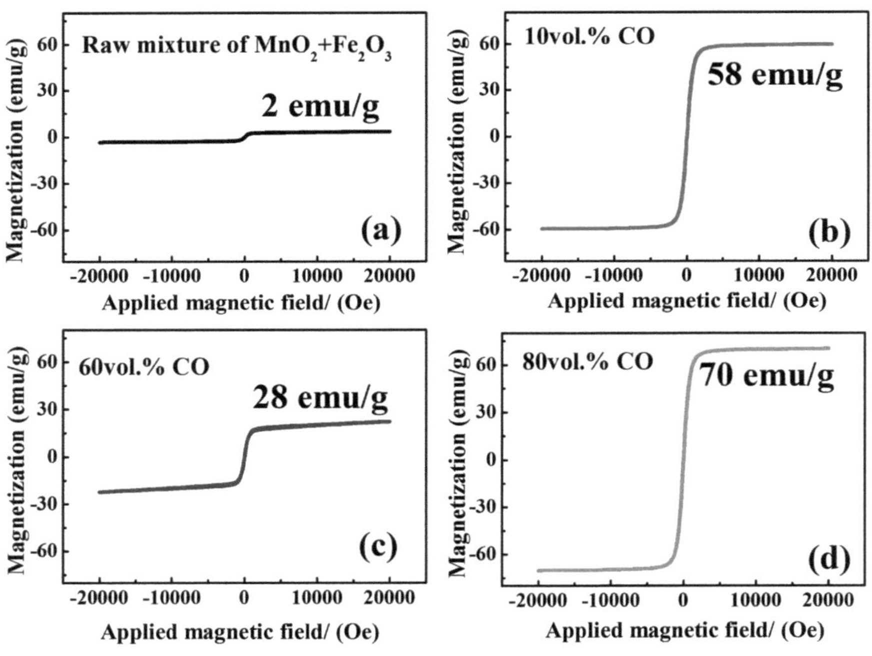

Fig. 4 Magnetic hysteresis loops of the samples roasted under different CO contents at 1000 °C for 30 min. **a** raw mixture of MnO_2 + Fe_2O_3, **b** 10 vol% CO, **c** 60 vol% CO, **d** 80 vol% CO

Discussion on the Phase Transformation

The formationPhase transformation of metallic iron shown in Fig. 3 may be from the reduction of iron oxides or the reduction of $Mn_xFe_{3-x}O_4$. Here, the $Mn_xFe_{3-x}O_4$ was first prepared under 4 vol% CO content at 1200 °C for 60 min, and then the as-prepared $Mn_xFe_{3-x}O_4$ product was reduced under 80 vol% CO content at 1000 °C for 30 min. The morphology analyses of $Mn_xFe_{3-x}O_4$ before and after reduced are presented in Fig. 5. As shown in Fig. 5a, the Fe, Mn and O elements are almost distributed uniformly in the $Mn_xFe_{3-x}O_4$ particles. When the $Mn_xFe_{3-x}O_4$ was reduced by CO, the Fe and Mn elements are separated and the $Mn_xFe_{3-x}O_4$ is reduced to metallic iron and MnO (Fig. 5b, c). This demonstrated that the $Mn_xFe_{3-x}O_4$ could be also reduced to metallic iron.

During the reduction process of MnO_2 and Fe_2O_3 briquettes, besides the stepwise reduction reaction of MnO_2 or Fe_2O_3, there exist some other reactions between MnO_2 and Fe_2O_3. From the analyses of Figs. 2, 3, 4, 5, we can confirm that the main condensed phases during the reduction roasting process were Fe_2O_3, Fe_3O_4, Fe, Mn_3O_4, $Mn_xFe_{3-x}O_4$ and $(MnO)_y(FeO)_{1-y}$ ($y = 0{-}1$). Note that, $(MnO)_y(FeO)_{1-y}$ refers to MnO or FeO when y value is equal to 1 or 0 respectively.

Table 1 lists the main phase and possible reactions between MnO_2 and Fe_2O_3 roasted under 10, 25, 60 and 80 vol% at 1000 °C for 30 min. As given in Table 1, under a lower CO content of 10 vol%, stepwise reductive reactions of MnO_2 to Mn_2O_3, to Mn_3O_4 and then to MnO occur while the Fe_2O_3 is reduced to Fe_3O_4. Then, MnO can react with Fe_2O_3 to generate $Mn_xFe_{3-x}O_4$ according the following reaction: $MnO + Fe_2O_3 = MnFe_2O_4$ ($\Delta G^\theta = -24655.52 + 4.07T$, calculated by Factsage 7.0). The ΔG^θ value of this reaction is very small, which indicates this reaction is readily to happen. When the CO content goes up to 25–60 vol%, Fe_3O_4 is reduced to FeO and the $Mn_xFe_{3-x}O_4$ is reduced to FeO and MnO. After that, the MnO and FeO reacted with each other to form $(MnO)_y(FeO)_{1-y}$. If the CO content increases to 80 vol%, the FeO is reduced to metallic iron, and both $(MnO)_y(FeO)_{1-y}$ and $Mn_xFe_{3-x}O_4$ can be reduced to MnO and metallic iron.

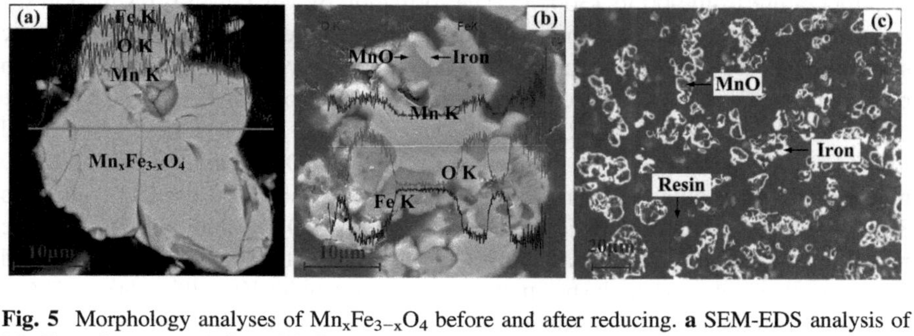

Fig. 5 Morphology analyses of $Mn_xFe_{3-x}O_4$ before and after reducing. **a** SEM-EDS analysis of $Mn_xFe_{3-x}O_4$ prepared from MnO_2 and Fe_2O_3 mixture roasted under 4 vol% CO content at 1200 °C for 60 min, **b** SEM-EDS analysis of as-prepared $Mn_xFe_{3-x}O_4$ reduced under 80 vol% CO content at 1000 °C for 30 min, **c** Optical microstructure of the reduced $Mn_xFe_{3-x}O_4$

Table 1 Phase evolution and main reactions between MnO_2 and Fe_2O_3

CO content/vol%	Main phase	Possible reactions
	Fe_2O_3, Fe_3O_4, Mn_3O_4, MnO, **$Mn_xFe_{3-x}O_4$**	$3Fe_2O_3 + CO = 2Fe_3O_4 + CO_2$; $2MnO_2 + CO = Mn_2O_3 + CO_2$; $3Mn_2O_3 + CO = 2Mn_3O_4 + CO_2$; $Mn_3O_4 + CO = 3MnO + CO_2$; $MnO + Fe_2O_3 = MnFe_2O_4$
	Fe_3O_4, FeO, MnO, **$(MnO)_y(FeO)_{1-y}$, $Mn_xFe_{3-x}O_4$**	$Fe_3O_4 + CO = 3FeO + CO_2$; $Mn_3O_4 + CO = 3MnO + CO_2$; $MnFe_2O_4 + CO = MnO + 2FeO + CO_2$ $yMnO + (1-y)FeO = (MnO)_y(FeO)_{1-y}$
	FeO, **$(MnO)_y(FeO)_{1-y}$, $Mn_xFe_{3-x}O_4$**	$MnFe_2O_4 + CO = MnO + 2FeO + CO_2$; $yMnO + (1-y)FeO = (MnO)_y(FeO)_{1-y}$
	MnO, Fe	$FeO + CO = Fe + CO_2$; $MnFe_2O_4 + CO = MnO + 2Fe + CO_2$; $(MnO)_y(FeO)_{1-y} + (1-y)CO = (1-y)Fe + yMnO + (1-y)CO_2$

Further study is being performed to determine the microstructure evolutions of $Mn_xFe_{3-x}O_4$ and $(MnO)_y(FeO)_{1-y}$ as well as elemental migration processes of Fe and Mn under $CO–CO_2$ atmospheres in order to understand the phase transformation of MnO_2 and Fe_2O_3 more clearly.

Conclusions

In summary, it's found that when the CO content was in the range of 0–10 vol%, the MnO_2 was easily reduced step by step from MnO_2 to Mn_2O_3, to Mn_3O_4 and then to MnO, while the Fe_2O_3 was only reduced to Fe_3O_4. Moreover, $Mn_xFe_{3-x}O_4$ ($x = 0-1$) phase was formed and the diffraction peaks of $Mn_xFe_{3-x}O_4$ enhanced with the increase of temperature from 900 to 1300 °C. The formation of $Mn_xFe_{3-x}O_4$ resulted in the inferior separation effect of manganese and iron by magnetization roasting-magnetic separation process. With the increase of CO content to over 10 vol%, the diffraction peaks of $Mn_xFe_{3-x}O_4$ decreased due to the reduction of $Mn_xFe_{3-x}O_4$. When the briquettes were reduced under 25–60 vol% CO content at 1000 °C, Fe_3O_4 is reduced to FeO and the $Mn_xFe_{3-x}O_4$ is reduced to FeO and MnO. Then, the MnO and FeO reacted with each other to form $(MnO)_y(FeO)_{1-y}$. As increased the CO content to 80 vol%, FeO is reduced to Fe, and both $(MnO)_y(FeO)_{1-y}$ and $Mn_xFe_{3-x}O_4$ are reduced to MnO and Fe. The maximum saturation magnetization (M_S) of the roasted products were higher when the briquettes was reduced under 10 vol% or 80 vol% CO content due to the plentiful formation of $Mn_xFe_{3-x}O_4$ or metallic iron, respectively, while the M_S of the product was lower when the briquettes were reduced under 60 vol% CO content owing to the vast formation of $(MnO)_y(FeO)_{1-y}$.

Acknowledgements The authors wish to express their thanks to the Co-Innovation Center for Clean and Efficient Utilization of Strategic Metal Mineral Resources, the Teachers' Research Fund of Central South University (No. 2013JSJJ028) and the Hunan Provincial Innovation Foundation for Postgraduate (No. CX2016B053).

References

1. Singh V et al (2011) Beneficiation and agglomeration process to utilize low-grade ferruginous manganese ore fines. Int J Miner Process 99(1):84–86
2. Wu Y et al (2014) Magnetic separation and magnetic properties of low-grade manganese carbonate ore. JOM 67(2):361–368
3. Tripathy SK, Banerjee PK, Suresh N (2015) Effect of desliming on the magnetic separation of low-grade ferruginous manganese ore. Int J Min Met Mater 22(7):661–673
4. Sharma T (1992) Physico-chemical processing of low grade manganese ore. Int J Miner Process 35(3–4):191–203
5. Gao YB et al (2012) Upgrading of low-grade manganese ore by selective reduction of iron oxide and magnetic separation. Metall Mater Trans B 43B:1465–1475
6. Mpho M, Samson B, Ayo A (2013) Evaluation of reduction roasting and magnetic separation for upgrading Mn/Fe ratio of fine ferromanganese. Int J Min Sci Technol 23(4):537–541
7. Huang ZC et al (2013) Reduction and separation of high iron content manganese ore and its mechanism. In: 4th international symposium on high-temperature metallurgical processing, TMS, San Antonio, pp 367–374
8. Hansen JS, Tress JE, Nafziger RH (1993) Carbothermic reduction of U.S. ferruginous manganese resources. JOM 45(4):59–63
9. Zhang YB et al (2016) Consolidation behavior of high-Fe manganese ore sinters with natural basicity. Min Proc Ext Met Rev. doi:10.1080/08827508.2016.1218870
10. Willard MA et al (1999) Magnetic properties of ordered and disordered spinel-phase ferrimagnets. J Am Ceram Soc 82:3342–3346
11. Ahmed YMZ (2010) Synthesis of manganese ferrite from non-standard raw materials using ceramic technique. Ceram Int 36:969–977
12. Wu XF et al (2016) Effect of polyacrylic acid addition on structure, magnetic and adsorption properties of manganese ferrite nanoparticles. Powder Technol 295:59–68

Contact Angle of Iron Ore Particles with Water: Measurements and Influencing Factors

Kai Tang, Senwei Xuan, Wei Lv, Xuewei Lv and Chenguang Bai

Abstract The contact angle between six iron ores and water were determined by the capillary pressure method with cyclohexane as the perfect wetting liquid. The factors influencing the contact angle were discussed, and a relationship between the content of oxy–hydroxides and the contact angle of iron ores was obtained. The results show that the ore F has the lowest contact angle but the ore D get the biggest, which are 32.09° and 78.16°, respectively. The volume fraction of oxy–hydroxides is linearly dependent on the contact angle. The contact angle decreased with increasing the volume fraction of $Fe_2O_3{\cdot}H_2O$.

Keywords Contact angle · Iron ore particles · Oxy–hydroxides

Introduction

With the decreasing of high grade iron ore, the granulation process, which is a typical method getting iron ore particles together to form pellets using water as the binder liquid before the iron ore sintering, has become more and more important in making a good permeability of the bed of granules in the sintering strand. In this process, wettability of iron ore particles, as well as the surface feature has a great effect on the granulation behavior.

It is believed that a decreasing contact angle makes the wettability of materials increased and easier to agglomerate, therefore, the granule strength can be enhanced and the amount of water needed in granulation would be reduced, because it increases the strength of the liquid bridges holding particles together [1–3].

K. Tang · S. Xuan · W. Lv · X. Lv (✉) · C. Bai
College of Materials Science and Engineering,
Chongqing University, Chongqing 400044, China
e-mail: lvxuewei@163.com

S. Ikhmayies et al. (eds.), *Characterization of Minerals, Metals, and Materials 2017*, The Minerals, Metals & Materials Series,
DOI 10.1007/978-3-319-51382-9_35

The measurement methods of contact angle can be divided into direct methods and indirect methods. Sessile drop method, captive bubble method, ESEM method [4–6] are the direct measurement methods, and the capillary rise methods [7–11] (include Washburn method, the height method, the pressure method and so on) are the indirect methods. For porous particles like natural iron ore particles, contact angles are greatly influenced by pores, so the indirect methods are recommended for contact angle measurements.

Contact angles are influenced by many factors. Iveson et al. [12, 13] found the value of contact angle is positively correlated to the ratio between oxides and oxy–hydroxides, and they give a relation about contact angle and the content of hematite. Works of other researchers confirmed Iveson's opinion, but instead, they considered that the content of goethite plays the key role. Surface hydroxyls were provided by oxy–hydroxides, and influence contact angles. Gentleman and Ruud [14] recently showed that dehydration of the metal oxide surface decreases the water contact angle of the surface due to the fact that metal–water interactions are stronger than hydrogen bonding between the surface hydroxyls and water.

In this paper, the capillary pressure method was used as the measurement method of contact angle, and the relation of three kinds of oxy–hydroxides and contact angles were obtained.

Experiments

Capillary Pressure Theory

In an ideal capillary tube, if neglecting inertial and gravity effects, the penetration of liquid can be described by Hagen-Poiseuille law [15]:

$$\frac{dh}{dt} = \frac{r^2 \Delta p}{8\eta h} \tag{1}$$

where r is the mean radius of capillary (m), Δp is the total pressure driving the penetration (Pa), η is the liquid (water) viscosity (Pa s), h is the height of the fluid column in the capillary (m) and t is the penetration time (s).

Join with the Young Laplace equation [16], the famous Washburn equation [7] is obtained:

$$h^2 = \frac{\gamma_{LG} \cdot r \cdot \cos\theta}{2\eta} t \tag{2}$$

where γ_{LG} is the surface tension of liquid-gas (N/m or J/m^2), θ is the contact angle (°). When the top of the column is enclosed, the pressure increment Δp is not equal to zero. The relationship between the pressure increment (Δp) and the penetration height (h) can be found in Eq. (3) [17]:

$$h = k\Delta p \tag{3}$$

the k is a pressure-height coefficient, and:

$$k = \frac{V_0}{P_0 \cdot A} \tag{4}$$

where A is the effective cross-sectional area, V_0 is the initial volume (m^3) and P_0 is the initial pressure (Pa) of the inner air. So, combining Eq. (2) with Eqs. (3) and (4), we obtain:

$$(\Delta p)^2 = \frac{c \cdot \gamma_{LG} \cos\theta}{2\eta} t \tag{5}$$

where c is a packing-related parameter ($N^2.m^{-5}$), which is determined by the packed status (r, A and V_0) and the initial air pressure P_0. In same sample, the parameter (c) is usually believed as a constant in these measurements. This equation can be used in the determination of contact angle, which is called the Washburn Osmotic Pressure method [12, 17, 18].

Compared with the original Washburn method, the capillary pressure method (Washburn Osmotic Pressure method) has a better way to measure the contact angle. For the Washburn method, the accurate height of liquid front is difficult to observe because the wetting status is different even in a cross section and the effect of the column edge.

But in practice, the value of c is difficult to obtain, which makes difficultly to calculate the value of contact angle with Eq. (5). This problem was settled by using a perfect wetting liquid or reference liquid which has a special contact angle equals zero to eliminate the constant c. Thus, the contact angle can be calculated by Eq. (6).

$$\cos\theta = \frac{k\eta/\gamma_{LG}}{k_0\eta_0/\gamma'_{LG}} \tag{6}$$

where k and k_0 are the slope of the $(\Delta p)^2 - t$ curve with reference liquid and water, respectively. η_0 is the viscosity of the reference liquid, γ_{LG} is the surface tension of water and gas, γ'_{LG} is the surface tension of reference liquid and gas.

Materials

The samples used came from some commercial brand. The size fraction of the six iron ore particles are of 240–400 meshes. The chemical composition of six test iron ore particles were shown in Table 1.

Table 1 Chemical composition (mass%) of six test iron ore particles

	TFe	FeO	SiO_2	CaO	Al_2O_3	MgO	TiO_2	S
A	58.27	0.22	5.55	0.04	1.37	0.08	0.05	0.009
B	62.30	0.22	4.40	0.03	2.43	0.07	0.08	0.015
C	61.26	0.32	3.69	0.03	2.25	0.06	0.08	0.022
D	62.24	0.23	4.40	0.04	2.31	0.08	0.06	0.009
E	59.14	0.21	4.38	0.05	1.50	0.08	0.06	0.005
F	61.53	0.29	3.05	0.02	2.06	0.16	0.07	0.018

Table 2 Physical properties of liquid

Liquid	Density (g/cm^3)	Viscosity (mPa s)	Surface tension (mN/m)
Water	1.000	1.002	72.1
Cyclohexane	0.779	0.908	25.0

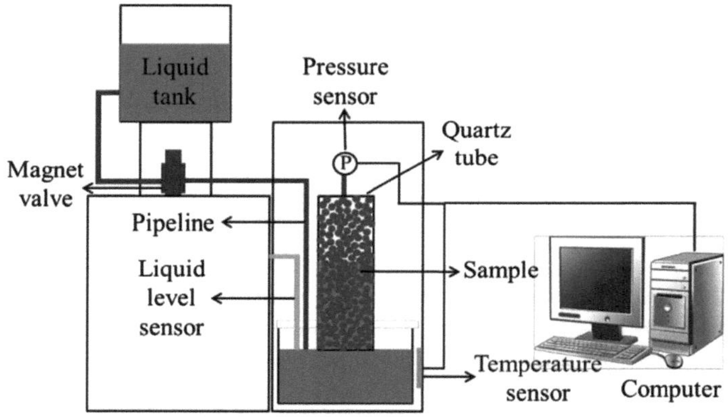

Fig. 1 Apparatus of contact angle measurement

The physical properties of test liquid and reference liquid are shown in Table 2. In this test, the test liquid is water and cyclohexane as the reference liquid.

Apparatus and Method

In this experiment, JF99A which was made by Shanghai Zhongchen Digital Technic Apparatus Co., LTD, was adopted in the contact angle measurements for these six iron ore particles. The schematic of this apparatus is shown in Fig. 1.

Iron ore particles were kept in an oven under 100 °C for 2 h, and then were put in a drying bottle and set aside. A filter paper about 10 mm diameter was laid horizontally on the bottom of the quartz tube, which can prevent the overflow of gas from the bottom when the upper pressure is large. Before the quartz tube was fixed in the apparatus, it was filled with dried iron ore particles and was shocked many times (about 300) to make the density of every packing bed was the same. The pressure sensor was placed on the top of the quartz tube to catch the top pressure and it's the same with the temperature sensor, connected with a computer.

The liquid was penetrated into the quartz tube when the liquid level reaches the bottom of it under the action of hydrostatic and capillary pressure. The changing pressure in the upper gas at the top tube is measured by the pressure sensor, and the data can be recorded in computer.

Results and Discussion

The fitting $(\Delta p)^2 - t$ curves of six iron ore particles with cyclohexane were obtained which are shown in Fig. 2, and the Fig. 3 shows the fitting $(\Delta p)^2 - t$ curves with water. The straight lines show that it fitting better with the data.

Using the Eq. (6) and the slope of $(\Delta p)^2 - t$ curves, the contact angle of six iron ore particles can be calculated, and are shown in Table 3. From the Table 3, it can be seen that the contact angles of the six iron ore particles are so different, for the biggest reaches at 78.16° and the smallest is 32.09°.

According to the previous literatures [13, 17], the factors influencing the contact angle are listed as follows:

- The surface morphology, pore volume and other physical properties
- The chemical compositions of iron ore particles
- The physical properties of test liquid

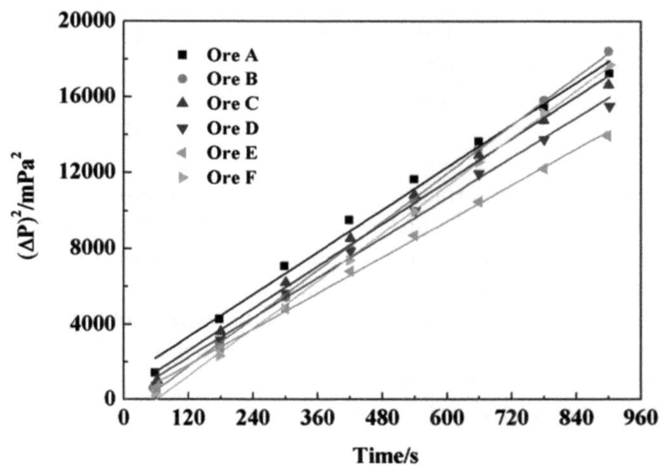

Fig. 2 Fitting $(\Delta p)^2 - t$ curves of six iron ore particles using cyclohexane

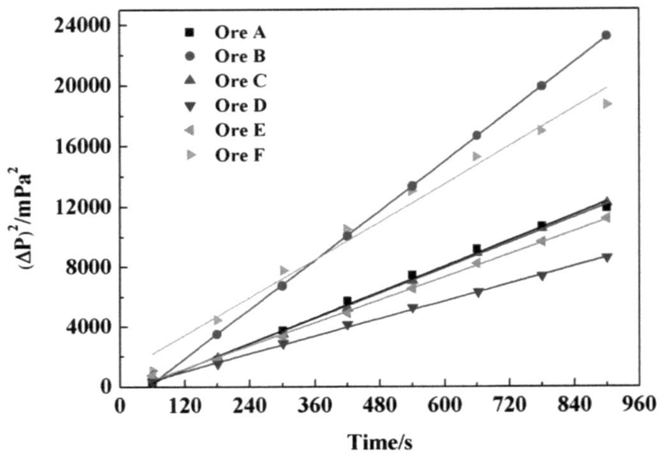

Fig. 3 Fitting $(\Delta p)^2$−t curves of six iron ore particles using water

Table 3 Contact angles of iron ore particles and water

Iron ore particle	A	B	C	D	E	F
Contact angle/°	75.78	52.40	73.69	78.16	71.98	32.09

Table 4 Bulk density (g/ml) and phase composition (vol%) for iron ore particles used

Iron ore particles	Bulk density	Phase composition				
		Fe_2O_3	Fe_3O_4	$Al_2O_3{\cdot}2SiO_2{\cdot}2H_2O$	$Fe_2O_3{\cdot}3H_2O$	$FeO{\cdot}H_2O$
A	1.984	56.65	0.60	0.00	42.75	0.00
B	2.293	74.07	0.60	25.33	0.00	0.00
C	1.98	66.23	1.07	11.00	21.71	0.00
D	1.889	79.79	0.73	10.78	0.00	8.71
E	1.805	55.88	0.59	0.00	43.54	0.00
F	1.935	56.28	0.97	10.14	32.61	0.00

The mechanisms of the surface morphology of these ore particles influenced on contact angles are still poorly understood [13] and the test liquid is water for measurements. In this experiment, it is focus on the influence of chemical compositions of these iron ore particles on contact angles.

The phase compositions of the six iron ore particles were obtained and shown in Table 4. The major ingredient of the six iron ore particles are Fe_2O_3, $Fe_2O_3{\cdot}3H_2O$ and some other crystalline hydrates.

Due to the finding of literature, we considered the influence of oxy–hydroxides on contact angles, and a formula was obtained by using Multiple Linear Regression.

$$\theta = -8.756_{Al_2O_3 \cdot 2SiO_2 \cdot 2H_2O} - 4.633_{Fe_2O_3 \cdot 3H_2O} - 11.515_{FeO \cdot H_2O} + 272.846\left(R^2 = 0.994\right) \quad (7)$$

where θ is the contact angle, $V_{Al_2O_3 \cdot 2SiO_2 \cdot 2H_2O}$ is the volume fraction of $Al_2O_3 \cdot 2SiO_2 \cdot 2H_2O$, $V_{Fe_2O_3 \cdot 3H_2O}$ is the volume fraction of $Fe_2O_3 \cdot 3H_2O$ and $V_{FeO \cdot H_2O}$ is the volume fraction of $FeO \cdot H_2O$.

The formula (7) indicates that a strong correlation between the oxy–hydroxides and the contact angle. Contact angle decreased by increasing the volume fraction of the three crystalline hydrates, and the volume fraction of $FeO \cdot H_2O$ has the most impact which is agree with the result of literature.

Conclusion

The measurement of contact angle between iron ore particles is experimentally demonstrated, and a relation was obtained: contact angle of the six iron ore particles decreases with increasing oxy–hydroxides content ($Al_2O_3 \cdot 2SiO_2 \cdot 2H_2O$, $Fe_2O_3 \cdot 3H_2O$, $FeO \cdot H_2O$) according to the formula:

$$\theta = -8.756 \cdot V_{Al_2O_3 \cdot 2SiO_2 \cdot 2H_2O} - 4.633 \cdot V_{Fe_2O_3 \cdot 3H_2O} - 11.515 \cdot V_{FeO \cdot H_2O} + 272.846$$

Acknowledgements The authors are especially grateful to The National Natural Science Foundation of China (No. 51234010).

References

1. Rumpf H (1962) International symposium on agglomeration, Interscience, London. pp 379–419
2. Aulton ME, Banks M (1979) Influence of the hydrophobicity of the powder mix on fluidized bed granulation. International Conference on Powder Technology in Pharmacy, Basel, Switzerland, Powder Advisory Centre
3. Maraglou A (1987) Fluidiser and granulation. PhD. Thesis, University of Birmingham, UK
4. Cieśliński Janusz T, Krygier Katarzyna A (2014) Sessile droplet contact angle of water–Al_2O_3, water–TiO_2 and water–Cu nanofluids. Exp Thermal Fluid Sci 59:258–263
5. Xue J, Shi P, Zhu L et al (2014) A modified captive bubble method for determining advancing and receding contact angles. Appl Surf Sci 296:133–139
6. Mather RR, Wei QF, Risnes OK et al (2003) Synthetic fibres studies by environmental scanning electron microscopy. Microsc Anal 9–12
7. Washburn EW (1921) The dynamics of capillary flow. Phys Rev 17(3):273
8. Alghunaim A, Kirdponpattara S, Newby BZ (2016) Techniques for determining contact angle and wettability of powders. Powder Technol 287:201–215
9. Kirdponpattara S, Phisalaphong M, Newby BMZ (2013) Applicability of washburn capillary rise for determining contact angles of powders/porous materials. Colloid Interface Sci 397:169–176

10. Delker T, Pengra DB, Wong P (1996) Interface pinning and the dynamics of capillary rise in porous media. Phys Rev Lett 76(16):2902
11. Lago M, Araujo M (2001) Capillary rise in porous media. Physica A 289(1):1–17
12. Iveson SM, Holt S, Biggs S (2000) Contact angle measurements of iron ore powders. Colloids Surf A 166(1):203–214
13. Iveson SM, Holt S, Biggs S (2004) Advancing contact angle of iron ores as a function of their hematite and goethite content: implications for pelletising and sintering. Int J Miner Process 74(1):281–287
14. Gentleman MM, Ruud JA (2009) Role of hydroxyls in oxide wettability. Langmuir Acs J Surf Colloids 26(3):1408–1411
15. Grosch CE, Salwen H (1968) The stability of steady and time-dependent plane Poiseuille flow. J Fluid Mech 34(01):177–205
16. Gaydos J (1998) The Laplace equation of capillarity. Interface Sci 6:1–59
17. Huang XB, Lv XW, Song JJ et al (2015) Contact angle of water on iron ore fines: measurement and analysis. J Min Metall Sect B 51(1):33
18. Huang XF, Gong FZ (2003) Measurement of contact angles of solid powder by Washburn osmotic pressure. Res Explor Lab 22(5):48–50

Important Factors to Consider in FIB Milling of Crystalline Materials

Jian Li and Pei Liu

Abstract In recent years FIB has become a powerful instrument to perform ion beam sectioning and site-specific TEM specimen preparation. However, there are still gaps in understanding FIB-induced artifacts. Gallium ion beam damage and specimen heating during FIB milling can produce unwanted artifacts during FIB work. Calculations of maximum specimen temperature increases have mostly been based on the assumption that ion beam implantation into substrate of infinite size. This deviates from finite specimen volume in FIB TEM specimen preparation that impedes specimen heat dissipation. In this paper, the authors will demonstrate a few commonly observed FIB artifacts, and discuss the concerns of specimen temperature rise during FIB milling process.

Keywords FIB · Artifacts · Temperature · Milling

Introduction

Since its development in the mid 1980s, focused ion beam (FIB) microscope has become a powerful instrument not only for TEM specimen preparation but also for ion beam cross sectioning and 3D EBSD work. With expanded applications and vast number of new FIB users, few have good understanding of FIB induced artifacts that could have detrimental effects in their analyses. The interaction between the primary gallium ion beam and substrate materials during FIB milling and imaging can lead to certain degree of specimen damage. Earlier generation of FIBs (e.g. Micrion) uses 50 and 30 kV in recent years to accelerate the Ga ions. These powerful Ga ion beams strikes the target not only remove materials (as the intended purpose), but also result in the remaining material (e.g. TEM foils) to lose

J. Li (✉) · P. Liu
CanmetMATERIALS, Natural Resources Canada, 183 Longwood Road South, Hamilton, ON, Canada
e-mail: jili@nrcan.gc.ca

S. Ikhmayies et al. (eds.), *Characterization of Minerals, Metals, and Materials 2017*, The Minerals, Metals & Materials Series,
DOI 10.1007/978-3-319-51382-9_36

long-range order when the density of point defects reaches certain level [1]. The damaged layers are typically on the order of a few tens of nanometers.

Rajsiri studied FIB induced amorphization and redeposition using Si. Using 30 kV and 1000 pA beam current, he reported the amorphized layer thickness on the FIB milled side wall to be 14 nm [2]. This is consistent with ion beam—specimen interaction theory. In addition to the amorphization, re-depositions are often common in FIB milling. However, they are usually difficult to differentiate from ion beam amorphization. Rajsiri suggested that re-deposition can be removed using gas-assisted etching (GAE) during FIB milling. More evidence is needed for this claim.

Mayer [3] reported damage layer thickness to Si to be 22, 2.5 and ~1 nm respectively when using ion beam energy of 30, 5 and 2 keV at 88° incident angle. It has been generally agreed that lowering beam acceleration voltage reduces ion beam damages, and it has been a common practice to reduce beam energy at the final step of TEM specimen preparation.

To date, most of the studies on FIB induced artifacts focus on beam induced amorphization and re-deposition. They are indeed very important in TEM specimen preparation as it is obvious that ~40 nm amorphized layers (20 nm on each side) on TEM foil (typically 100 nm thick) is a significant factor. However, as demonstrated by Li [4–6], FIB microscopes are much more than just fine milling machines to make site-specific TEM foils. Under circumstances when FIBs are used as an imaging tool, semiconductor circuit modifications and 3D EBSD studies, other FIB induced artifact can become a significant factor. This paper will briefly discuss FIB induced artifacts that we have encountered in our laboratories, and also discuss the concerns of specimen temperature increase during FIB milling.

Ga-Rich Phase Formation

During FIB milling, the incident Ga ions can react with the target material and form Ga-rich phases. The formation of these low temperature phases depends on the type of target materials and even their crystallographic orientation. This type of artifact is demonstrated by scanning a low-carbon steel in a FIB microscope. Prior to ion beam exposure, the sample is polished following a standard metallographic polishing routine. It was then exposed under the Ga ion beam in the FIB set at 30 kV, 800 pA beam current and 3 μs dwell time. A few scanning passes removed the thin surface oxide, and the microstructure of the steel start to appear as shown in Fig. 1a. Application of such FIB imaging technique was previously published [5]. After a few additional scan passes on the same area with increased beam current, small dark patches start to emerge as shown in Fig. 1b. With increased beam current and increased scan passes, these dark patches start to grow much larger (Fig. 1c). These dark islands are later confirmed to be Ga-rich phase.

The formation of Ga-rich phase depends on many factors that include beam energy, incident angle, dwell time and target materials. In some cases,

Fig. 1 Fe_xGa_y phase formation on low carbon steel during FIB imaging. **a** First image pass, **b** multiple imaging passes under 1.5 nA beam current, **c** multiple passes under 6 nA beam

crystallographic orientation can be a critical factor. Phaneuf [7] studied the effect of crystallographic orientation of single crystals Cu on the formation of Cu–Ga intermetallic phase. When Ga ion beam is perpendiculars to {111} plane of Cu target, Ga-rich phase does not form. Ga-rich phase forms almost instantly when the incident ion beam is perpendiculars to {110} plane. This results in four folds decrease in sputtering rate under the same beam condition. Similar effects are found when imaging Au and Ag.

Ion Beam Damage to HCP Materials

It is generally difficult to obtain crystallographic orientation contrast of HCP materials in FIB. Ion beam tends to cause more damage to this group of materials. Figure 2 show FIB images of galvanized Zn coating on steel substrate. After careful FIB sectioning, the first imaging pass with very small ion beam current of 32 pA show relatively weak orientation (or phase) contrast of galvannealed Zn coating. Ion beam has caused serious damage to the coating layer after just a few more imaging passes.

Similar problem have been found through our research at CanmetMATERIALS and FIBICS Inc. on other types of materials that include Mg alloys, Zr alloys and Ti alloys. At this point, we are not sure if such phenomenon is attributed by amorphization or redeposition. In some cases, we were able to eliminate the damage by milling with simultaneous flow of XeF_2 gas. At present, we still struggle to establish a set milling routine to cope with the problem.

Fig. 2 FIB images of a galvannealed Zn coating on steel. **a** One imaging pass with a 32 pA beam; **b** damaged coating after few additional imaging passes

Concerns of Specimen Temperature Rise

There have always been questions about potentials of target materials' temperature rise during FIB milling process. In modern FIBs with improved column design, the power density has increased significantly to obtain fast milling rate. For example the power density of FEI Helios NanoLab FIB can reach as high at $\sim 1.2E6$ kW/m^2. Some degree of beam heating can be expected. The energy of the primary Ga ion beam can transfer into secondary ions or neutrals, implant into substrate, or stored as defects in the specimen. A significant portion of energy that is deposited by nuclear stopping converts into heat. Potential energy stored in defects and energies of emitted particles are on the order of 10 eV per defect and sputtered particle. Many defects immediately recombine and release the stored energy again in the form of heat, and this contribute to specimen temperature rise. Specimen temperature rise mainly dependents on the beam energy, the nature of target material and substrate thermal conductivity.

There have been limited discussions on temperature increase during FIB millings in literature [8, 9], and most of them have been based on theoretical calculations with assumption of ion beam implantation into solid of infinite size. Carlslaw [10] made an assumption that when ion beam incident on a flat surface the transfer of heat away from incidence point is very effective. Using his heat dissipation model, we calculated the maximum specimen temperature rise in a FEI Helios NanoLab FIB at the highest beam density to be: aluminum alloys 6.6 K, stainless steel that has relatively poor thermal conductivity than aluminum to be 17.7 K, SiO_2 842 K, and polymer 3318 K. The calculated temperature rise should be regarded as the maximum temperature rise. In reality, a large percentage of the incident beam energy is carried away by sputtered atoms from the substrate material and backscattered Ga ions especially when the beam incident angle is close to 90° during TEM specimen preparation. The models and calculations are based on many assumptions, and there has been no experiment to measure the temperature rise.

In attempt to assess this commonly concerned subject, we used three types of materials with known melting points. Crystalbond™ 509 that has poor thermal conductivity and melts 121 °C; pure gallium that has a melting point of 29.76 °C, and Wood's Metal (50%Bi–27%Pb–13%Sn–10%Cd) with thermal conductivity of 20 W/mK that melts at 71 °C [11].

Crystalbond

Prior to FIB milling, the surface of the Crystalbond was covered with a thin layer of conductive graphite cement to reduce charging. It was then subjected to FIB trenching using a range of relatively low beam current of 2.5, 0.79, 0.43 nA and final milling was carried out with a fine beam of 0.23 nA at 30 kV. As shown in Fig. 3, the FIB milling did not seem to cause the sample to melt.

We were able to lift out a small sample and mount it onto a standard TEM grid. Upon further milling with 0.23 nA beam current, the specimen melted almost instantly (Fig. 3). The fact that no significant local melting when FIB milling the bulk specimen is partially attributed to the much enhanced heat dissipation from the conductive graphite cement layer.

The fact that this material maintained its structural integrity during FIB trenching suggests that the local temperature rise can be kept below 121 °C even the sample has relatively poor thermal conductivity. The thin foil mounted on copper grid can have much reduced thermal conductivity. This is due to the thin Pt deposition used to connect the foil to the copper grid limited heat dissipation efficiency [12].

Fig. 3 **a** FIB trenching into the bulk Crystalbond 509 sample; **b** lift out TEM sample melt upon further milling

Wood's Metal

When milling the Wood's Metal with the same milling routine, we observed no obvious melting. As shown in Fig. 4, the materials maintained its structural integrity during FIB trenching in the bulk materials and lift out TEM specimen preparation.

The fact that Wood's Metal does not melt under typical FIB milling conditions has shown that the bulk and local temperature of typical metal alloys during FIB milling should be less than 71°. This is important particularly for FIB-TEM preparation of metal alloys that has relatively low melting points. For example, aging of some aluminum alloys can occur at temperature as low as 150 °C. Precipitation of β phase in AM60 magnesium alloy can occur below 120 °C. It is of great interest to physical metallurgists to know, with confidence, that those features like nano precipitates or phase transformation identified through TEM work are not introduced by un-wanted sample anneal during FIB TEM specimen preparation process. Details of this study that includes discussion of increased temperature rise during FIB-TEM specimen preparation can be found elsewhere [12].

Gallium

Gallium melts at 29.76 °C. To avoid potential chamber contamination, a small chip of Ga (99.99%) was used in this study. The small chip of Ga was first placed on a standard aluminum sample stub, and then the bottom of the stub was heated by hot water. This melted the Ga chip and created good contact with the aluminum stub.

Room temperature was held constant at 24 °C during the course of the experiment. FIB milling was set on a 10 × 10 μm area with a dose of 1 nq/μm^2 using a range of beam currents of 0.79–21 nA under 30 kV. As shown in Fig. 5, even at the lowest beam energy used in this experiment, we were not able to create any FIB

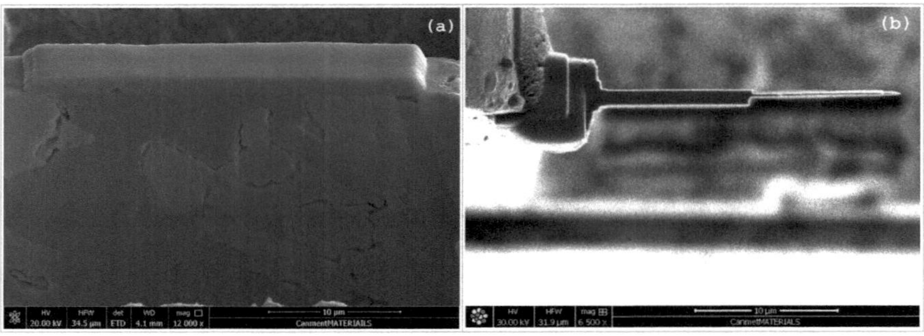

Fig. 4 FIB milling of Wood's metal. *Left* FIB trench in the bulk sample; *right* TEM foil

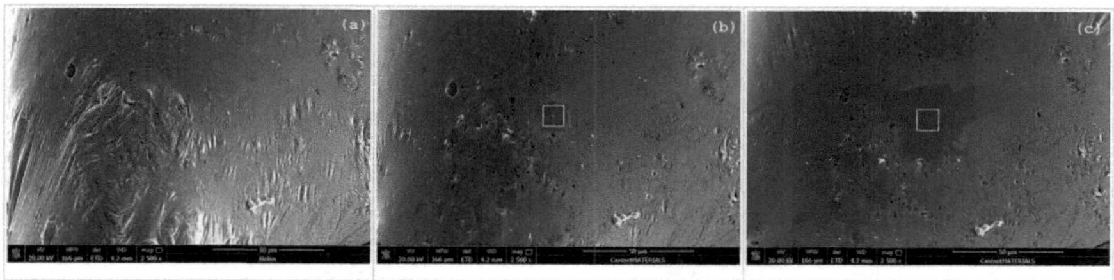

Fig. 5 FIB milling of pure Gallium. **a** Prior to milling; **b** after milling with 0.79 nA; **c** after milling with 21 nA. The square mark is the FIB milled area

trench. Local melting occurs immediately upon milling and creates flat melting zones. Size the melting zone enlarges with increased beam energy.

Temperature increase predictions based on Carlslaw [10] model suggest 6.8 °C at 0.79 nA and 34.6 °C at 21 nA (with our FEI Helio's NanoLab FIB). This is consistent with our experimental observations. In fact, under large beam current, even brief imaging with ion beam (much short dwell time of 100 nS) surface melting occurs with even one imaging pass. The melting has been limited to relatively small local area even when milling with large beam current (21 nA).

The current study suggests that FIB milling of bulk metal alloys result in temperature rise. The temperature rise is limited to small area around incident ion beam. Temperatures at beam incident location are generally more than 29.76 °C but less than 71 °C for most metal alloys. Without the conductive paint, insulators would melt instantly upon FIB milling.

Summary

FIB operator should be aware of limitation of FIB microscope and aware of commonly observed artifacts before performing any FIB work. Some degree of beam heating to metal alloys during FIB millings is inevitable. However the heating are limited to small local area, and temperature rises are typically more than 5 °C but less than 50 °C for most metal alloys.

References

1. Cerva H et al (1993) For testing and failure analysis. Inst Phys Conf Ser 134:133
2. Rajsiri S et al (2002) FIB damage in silicon: amorphization or redeposition? Microsc Microanal 8(Suppl 2):50–51
3. Mayer J et al (2007) TEM sample preparation and FIB-induced damage. MRS Bull 32:400–407

4. Li J (2008) The detection of local plastic strain in microscopic scale. Mater Lett 62(6–7):804–807
5. Li J (2008) Advances in materials engineering using state-of-the-art microstructural characterization tools. In: New material science research. Nova Science Publishers Inc. ISBN-13:978-1-60021-645-1
6. Li J (2008) Advanced techniques in TEM specimen preparation. In: Khan M (ed) The transmission electron microscope. ISBN: 979-953-307-311-7
7. Phaneuf MW, Li J, Casey JD Jr (2002) The focused-ion-beam microscope—more than a precision ion milling machine. Microsc Microanal 8(Suppl 2):52
8. Bassim ND et al (2012) Minimizing damage during FIB sample preparation of soft materials. J Microsc 245:288–301
9. Ishitani T, Kaga H (1995) Calculation of local temperature rise in focused-ion-beam sample preparation. Microsc 44:331–336
10. Carlslaw HS, Jaeger JC (1959) Conduction of heat in solids, 2nd edn. Oxford University Press, Oxford, p 164
11. Bulletin MRS (1996) Thermal conductivity of common solids
12. Li J (2015) Specimen temperature-increase considerations during FIB milling. Microsc 63(1):3–10

Part IX
Material Processing and Corrosion

Corrosion Behavior of Super-Ferritic Stainless Steels in NaCl Media

Natalia S. Zadorozne, Jorge D. Vier, Raúl B. Rebak and Alicia E. Ares

Abstract Recent worldwide demand has driven nickel and molybdenum prices to record high values. Alloys containing significant amounts of nickel and molybdenum, such as the austenitic and duplex grades, have experienced significant price increases and some spot shortages have resulted in some regions. With low or no nickel content and reasonable molybdenum content, super-ferritic stainless steels are now proving to be the most cost effective corrosion resistant materials. The electrochemical behavior APMT® (Fe–21Cr–5Al–3Mo–0.04C) and 4C54® (Fe–26.5Cr–0.8Mn–0.5Si) super-ferritic stainless steel in NaCl solutions has been studied using open circuit potential measurements, potentiodynamic polarization curves, and electrochemical impedance spectroscopy. The aim is to compare the behavior of these materials with austenitic stainless steel alloy 316 (Fe–18Cr–10Ni–2.5Mo). The results indicate that the APMT alloy presents a better performance than the AISI 316 stainless steel, while 4C54® does not, especially at anodic potentials. The good behavior of APMT® may be attributed to Al and Mo content in its composition.

Keywords Super-ferritic · NaCl · Corrosion behavior · APMT

N.S. Zadorozne (✉) · A.E. Ares
IMAM, UNaM, CONICET, FCEQyN, Calle Félix de Azara N° 1552,
N3300LQD Posadas, Misiones, Argentina
e-mail: nataliazadorozne@gmail.com

A.E. Ares
e-mail: aares@fceqyn.unam.edu.ar

N.S. Zadorozne · J.D. Vier · A.E. Ares
ProMyF—FCEQyN, Calle Félix de Azara N° 1552,
N3300LQD Posadas, Misiones, Argentina

R.B. Rebak
GE Global Research, Schenectady, NY 12309, USA
e-mail: rebak@ge.com

S. Ikhmayies et al. (eds.), *Characterization of Minerals, Metals, and Materials 2017*, The Minerals, Metals & Materials Series,
DOI 10.1007/978-3-319-51382-9_37

Introduction

AISI 316 stainless steel is perhaps the most commonly used material in industries that use chlorinated media, including desalination plants seawater [1–5]. However, despite the excellent properties that such an alloy steel exhibits, it begins to present problems as the temperature and the chloride concentration increase. In general, it is recommended that the chloride concentration does not exceed 5000 ppm Cl^- [6].

Even if the performance of type 316 SS has been very good, today's raw material prices have driven the price of the Ni and Mo containing alloys skyward. This has kindled the rediscovery of the super-ferritic stainless alloys. Alloys containing significant amounts of nickel and molybdenum, such as the austenitic and duplex grades, have experienced significant price increases and some spot shortages have resulted in some regions of the world. With low or no nickel content and reasonable molybdenum content, super-ferritic stainless steels are now proving to be the most cost effective [7, 8]. Since 2000, the use of super-ferritic stainless steel in other markets, such as the petrochemical industry and desalination and water recovery, has also grown significantly [8].

Recently, we have introduced two new alloys to the group of ferritic stainless steels. One is the APMT® (Fe–21Cr–5Al–3Mo–0.04C) alloy and the other is 4C54® alloy (Fe–26.5Cr–0.8M–0.5Si) [9–11]. Both are characterized by their high content of Cr. They exhibit excellent corrosion resistance at high temperature and have potential for use in nuclear applications and in the petrochemical industry [12–14].

The high performance stainless steels are commonly chosen for applications where high chloride, low pH, or high microbiological activity are present. Several alloying elements, such as Cr, Mo, and N, promote chloride resistance in this group of alloys [8].

Different parameters indicating the degree of susceptibility to localized corrosion of alloys, such as *PRE* (Pitting Resistance Equivalent), *CPT* (Critical Pitting Temperature) and *CCT* (Critical Crevice Temperature) have been established [5–11]. These parameters are functions of the alloy composition. Equation 1 corresponds to *PRE* parameter[15]. It was originally developed for stainless steels then modified to include nickel alloys [7, 11, 12]. Other expressions of *PRE* can be found in the literature [5].

$$PRE = \%\mathrm{Cr} + 3.3\%\mathrm{Mo} + 16\%\mathrm{N} \quad (1)$$

The electrochemical behavior APMT and 4C54® super-ferritic stainless steel in NaCl solutions has been studied using open circuit potential measurements, potentiodynamic polarization curves, and electrochemical impedance spectroscopy (EIS). The aim is to compare the behavior of the ferritic materials with austenitic stainless steel alloy 316.

Experimental Procedure

The chemical compositions of the tested alloys in weight percent are listed in Table 1. The specimens used for the electrochemical studies were in the form of parallelepipeds, a variation of the ASTM G 5 specimen, with approximate dimensions of 12 mm × 12 mm × 15 mm. Each specimen is screwed to a metal rod (used as external electrical contact) that was introduced in a glass tube and was isolated from the solution via a PTFE gasket. The specimen exposed area to the solution was approximately 10 cm^2. The specimens had a finish grinding of abrasive paper number 1200 and were degreased in acetone and washed in distilled water within the hour prior to testing.

All the electrochemical tests were conducted in a one-liter, three-electrode glass vessel. Nitrogen (N_2) was purged through the solution 1 h prior to testing and was continued throughout the entire test. The reference electrode was a saturated calomel electrode (SCE) ($V_{SCE} = V_{SHE} + 0.242V$). The reference electrode was connected to the solution through a water-cooled Luggin probe. The counter electrode consisted in a flag of platinum foil (total area 50 cm^2) spot-welded to a platinum wire. All the potentials in this paper are reported in the SCE scale. The open circuit potential measurements, cyclic potentiodynamic polarization curves, and electrochemical impedance spectroscopy were conducted in 1000 and 10,000 ppm Cl^- solution at room temperature.

Results and Discussion

Open Circuit

Figure 1 shows the open circuit potential (*OCP*) versus immersion time for the materials studied. The total registered time was 4 h. The potential initially increased rapidly and after 1-h immersion it tended to be stabilized. The increase of potential is due to the formation of a passive oxide film on the surface, probably chromium oxide. Considering the last recorded *OCP* value, for the highest concentration of chloride (10,000 ppm) the open circuit potential was 316 > 4C54® > APMT®. For the concentration of 1000 ppm Cl^-, alloys 316 and APMT® showed an *OCP* value close to −146 mV_{SCE}, while the 4C54® alloy recorded value of open circuit

Table 1 Chemical composition of used alloys (wt%)

Alloy	Fe	Cr	Mo	Si	Mn	Al	C	Ni	Other	PRE
316	Bal.	17	2.5	0.5	2	–	0.03	11	P, S	25
4C54®	Bal.	26.5		0.5	0.8	–	0.2		P, S	30
APMT®	Bal.	21	3	–	–	5	0.04			31

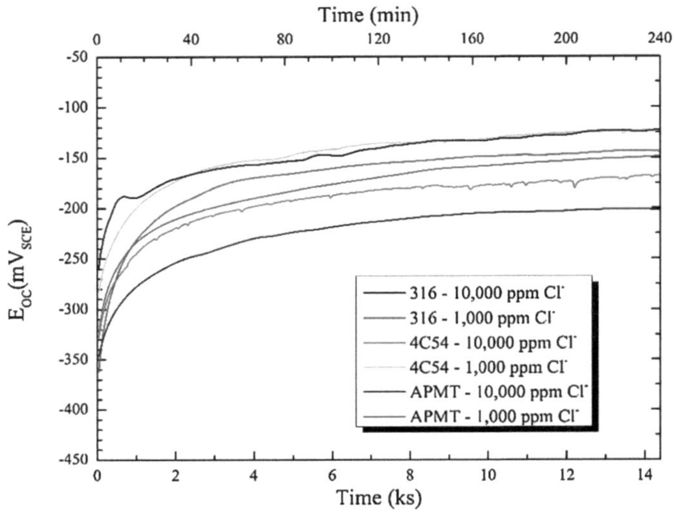

Fig. 1 Open circuit potential versus time for 316, 4C54® and APMT® alloys in deaerated Cl^- solution at room temperature

potential only 25 mV higher. From these results, it is observed that the influence of the alloying becomes important only in the more concentrated solution.

EIS (Electrochemical Impedance Spectroscopy)

The EIS measurements were immediately carried out after the stainless steel was immersed at *OCP*. The results of impedance measurements were performed in a wide frequency range from 10 kHz to 1 MHz, to obtain information related to the tested corrosion system and the electrochemical processes occurring on the surface of the electrode. EIS spectra obtained for the 316, 4C54® and APMT® alloys in 1000 and 10,000 ppm Cl^- solution at room temperature can be seen in Fig. 2. The interpretation of electrochemical impedance measurements involves the use of electrode equivalent circuits (EEC). The EEC with one R–Q which is shown as an inset in Fig. 2. The parameter *Rs* denotes the solution resistance; *Rt* represents charge transfer resistance; *Q* denotes capacitive behavior of the passive film. The constant phase angle element *Q* (CPE) is used to replace the pure capacitance parameter *C* to reflect the capacitance behavior, considering the dispersion effect caused by the roughness of the electrode interface. The electric circuit element *Q* could be formulated as Eq. (2).

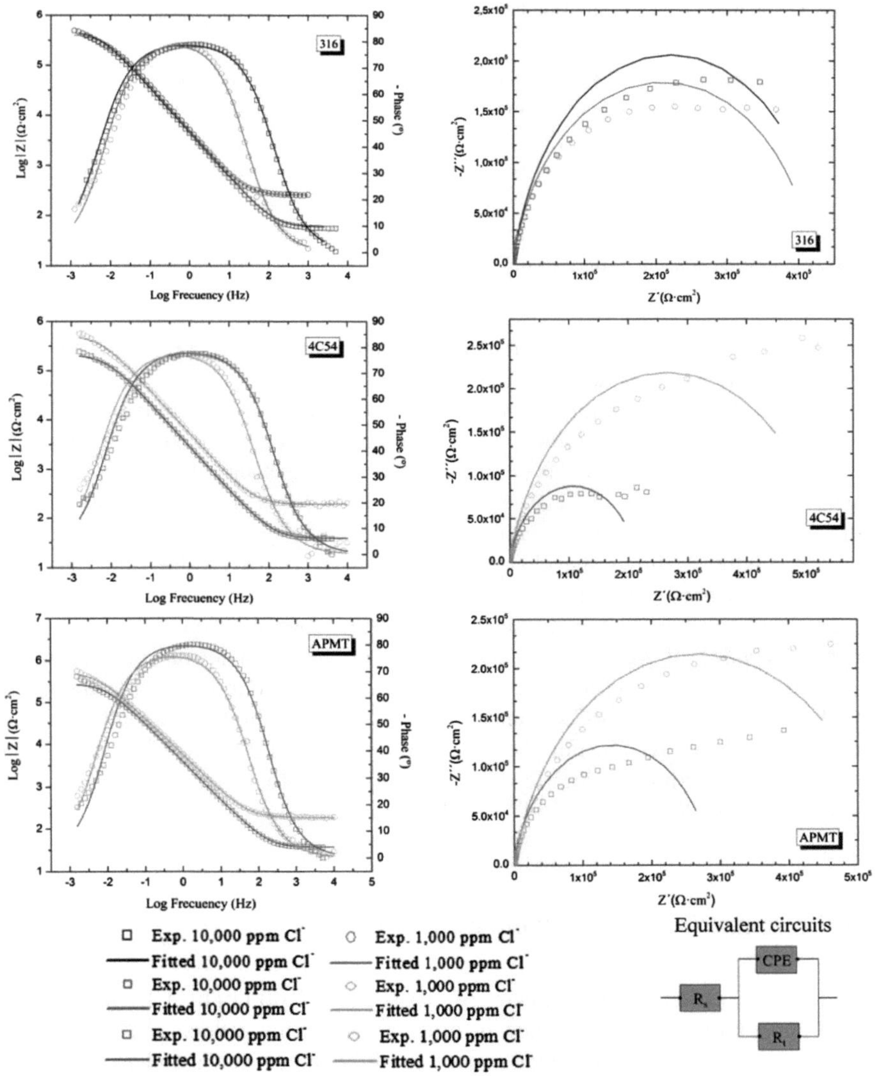

Fig. 2 Experimental (*symbols*) and fitted (*lines*) Bode and Nyquist diagram for, 316, 4C54® and APMT® alloys in 1000 and 10,000 ppm Cl^- solution at room temperature

$$Z(w) = Z_0 \cdot (iw)^{-n} \tag{2}$$

In which, Z_0 denotes the modulus of Q; w represents the angular frequency (in rad/s); $i^2 = -1$ is the imaginary number and n is the exponent of Q; $0 < n < 1$. As

Table 2 Parameters used in the simulation of impedance data

Alloy	Cl^- concentration	R_s (Ω cm^2)	R_t (Ω cm^2)	CPE (Ω^{-1} cm^{-2} s^{-n})	n
316	10,000 ppm	56.28 (2.65%)	463,404 (5.92%)	4.57E−05 (2.36%)	0.88 (0.72%)
4C54®		43.55 (3.80%)	212,774 (8.25%)	7.01E−05 (3.49%)	0.88 (1.09%)
APMT®		36.71 (4.21%)	284,829 (4.21%)	4.88E−05 (3.80%)	0.89 (1.07%)
316	1000 ppm	247.35 (1.7%)	419,390 (2.59%)	3.84E−05 (1.19%)	0.89 (0.44%)
4C54®		192.89 (1.69%)	531,560 (3.91%)	3.84E−05 (1.41%)	0.87 (0.50%)
APMT®		187.25 (1.2%)	534,679 (2.68%)	3.59E−05 (0.99%)	0.86 (0.35%)

(*) Data in parenthesis are the estimated error

$n = 0$, Q represents pure impedance resistance; as $n = 1$, Q represents pure capacitance; $n = 0.5$, Q denotes Warburg impedance.

Table 2 shows the fitting parameters corresponding to the equivalent electric circuit. High values of n obtained shows that our CPE acted like a capacitance. The charge transfer resistance, *Rt*, is commonly used as a measure of the resistance of a metal to the corrosion damage. A high value of *Rt* is associated with a high corrosion prevention capability; a low value of *Rt* indicates potential high corrosion activity. Figure 2 shows the charge transfer resistance (*Rt*) values for de tested specimens for both concentrations of chloride.

The results show that increasing the chloride concentration did not imply a change in value of *Rt* for Alloy 316. However the super-ferritic alloys show a different behavior for each concentration of chloride. For the concentration of 1000 ppm of Cl^-, alloys 4C54® and APMT® showed a similar and slightly higher value of *Rt* than 316 SS. For the concentration of 10,000 ppm Cl^-, 4C54® and APMT® alloys exhibited high corrosion activity due to lower values of *Rt*. The lowest value of *Rt* was for 4C54® (Table 2; Fig. 3).

Cyclic Potentiodynamic Polarization Curves

Cyclic polarization measurements in 1000 and 10,000 ppm Cl^- solution at room temperature were performed in order to assess the influence of alloy composition on the localized corrosion resistance of the three tested stainless steels. The cyclic potentiodynamic polarization was performed at a scan rate of 0.16 mV/s. The scans were reversed back to the starting potential when the specimens reached a current density value of 1 mA/cm^2. Cyclic polarization curves are shown in Fig. 4, and Table 3 shows the main parameters related to the corrosion processes that were extracted from graphical analysis of the potentiodynamic scans. The three tested materials showed similar behavior. That is, the potential breakdown (E_P) and the repassivation potential (E_{CO}) were lower for the higher chloride concentration.

All of them exhibited hysteresis anodic loops so that they were able to repassivate after the breakdown of the passive film. When the chloride concentration was higher, an apparent decrease in pitting potential was observed. After the tests, the

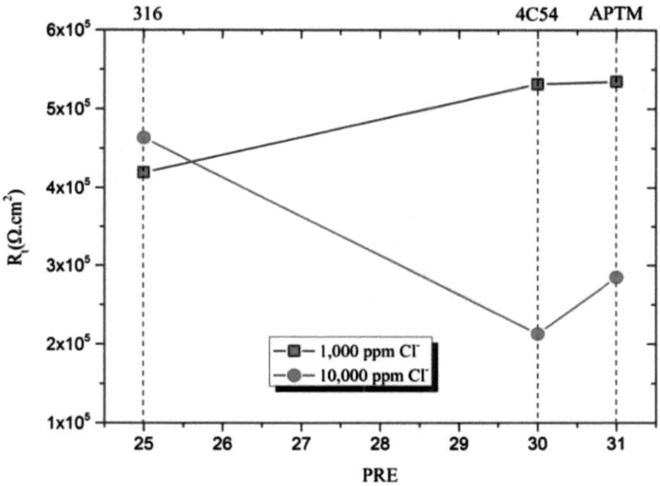

Fig. 3 Charge transfer resistance (*Rt*) for 316, 4C54® and APMT® alloys in 1000 and 10,000 ppm Cl^- solution at room temperature. The higher the *Rt* value the higher the resistance to corrosion

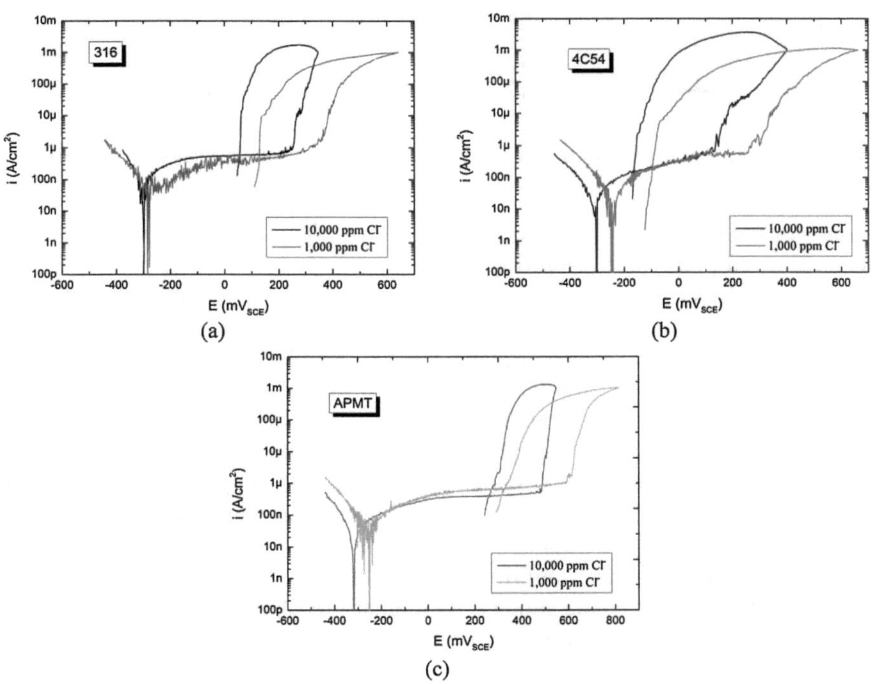

Fig. 4 Cyclic polarization curves for **a** 316, **b** 4C54® and **c** APMT® alloys in 1000 and 10,000 ppm Cl^- at room temperature

Table 3 Parameters obtained from the potentiodynamic polarization curves of 316, 4C54® and APMT® alloys in 1000 and 10,000 ppm Cl^- solution at room temperature

Alloy	Solution	E_{CORR}	E_P	E_{CO}
316	1000 ppm Cl^-	−277	252	132
4C54®		−300	256	−97
APMT®		−255	595	324
316	10,000 ppm Cl^-	−300	250	52
4C54®		−243	128	−163
APMT®		−322	475	261

E_{PIT} is the potential at which the current increased suddenly after the passive region and E_{CO} is the potential at which the reverse scan crosses over (CO) the forward scan, representing the repassivation potential

Fig. 5 Optical microscopy photos of a specimen surface after cyclic potentiodynamic tests

surface of the specimens was observed under an optical microscope. All the tested alloys showed pitting corrosion (Fig. 5).

Figure 6 shows the variation of the E_{CORR}, E_{PIT} and E_{CO} potentials obtained from the cyclic polarization curves as a function of the PRE. The ranking of the characteristic potentials (E_{CORR}, E_P and E_{CO}) are similar in both chloride concentrations. Of the two super-ferritic alloys studied in this work, only APMT® alloy showed a better performance against localized corrosion than alloy 316 at both concentrations of chloride. The alloy APMT® exhibited E_P and E_{CO} potentials values much higher than those obtained by the other two alloys. It is likely that this behavior is due to the presence of Mo and Al in its composition, even though containing less Cr that 4C45 alloy.

Cyclic polarization tests confirm the relationship between localized corrosion resistance and the *PRE* value (Eq. 1). Preliminary results show that the *PRE* relationship may be applicable to both austenitic and ferritic stainless steels.

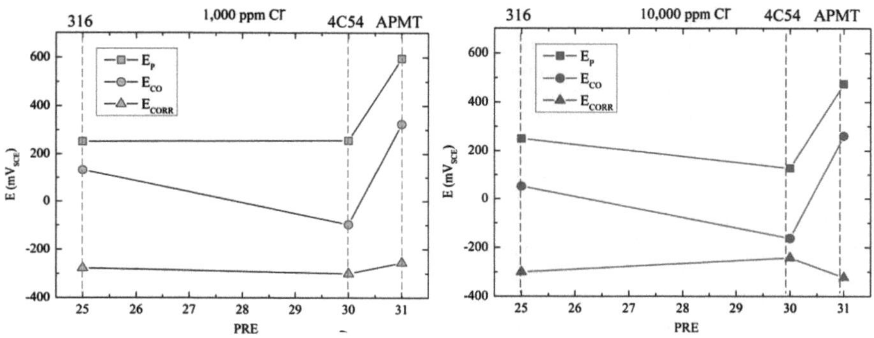

Fig. 6 Relationships between E_{CORR}, E_{PIT} and E_{CO} potentials obtained from the cyclic polarization curves with *PRE*

Conclusions

With the aim of finding a more economical substitute the alloy 316, the corrosion behavior was evaluated for two super-ferritic stainless alloys in solutions with 1000 and 10,000 ppm Cl^- concentrations. The results were compared with those obtained for 316 SS under the same conditions.

Current results obtained from EIS tests show that the ferritic alloys do not present a good performance in chloride solutions at room temperature compared to the alloy 316. For the concentration of 1000 ppm of Cl^-, alloys 4C54® and APMT® showed a similar and slightly higher value of *Rt* than 316 alloy. At a concentration of 10,000 ppm Cl^-, 4C54® and APMT® alloys exhibit higher corrosion activity due to the low values of measured *Rt*.

However, when evaluating the ferritic alloys behavior for localized corrosion resistance, it is clear that APMT® had a better localized corrosion resistance than type 316 SS. The APMT® alloy exhibit E_P and E_{CO} potentials values much higher than those obtained by the other two alloys. It is likely that this behavior is due to the presence of Mo and Al in its composition.

Acknowledgements This work was supported by PICT-2012-2952 of the National Agency for Promotion of Science and Technology. Thanks are due to the Argentinean Research Council (CONICET) for the financial support.

References

1. Moreira RR et al (2014) Electrochemical investigation of corrosion on AISI 316 stainless steel and AISI 1010 carbon steel: study of the behaviour of imidazole and benzimidazole as corrosion inhibitors. Adv Chem Eng Sci 503–514

2. Nawal Muhammad D (n.d.) Corrosion of stainless steel 316 L in various aqueous salt solutions. Collage Mater Eng Babylon Univ http://uobabylon.edu.iq/uobColeges/fileshare/articles/Corrosion%20of%20Stainless%20steel%20316%20L%20with%20Time%20in%20Various%20Aqueous%20Salt%20Solutions.pdf
3. Malik AU, Siddiqi NA, Ahmad S, Andijani IN (1995) The effect of dominant alloy additions on the corrosion behavior of some conventional and high alloy stainless steels in seawater. Corros Sci 37:1521–1535. doi:10.1016/0010-938X(95)00043-J
4. Jiménez-Come MJ, Turias IJ, Trujillo FJ (2014) An automatic pitting corrosion detection approach for 316L stainless steel. Mater Des 56:642–648. doi:10.1016/j.matdes.2013.11.045
5. Davison RM, Laurin TR, Redmond JD, Watanabe H, Semchyshen M (1986) A review of worldwide developments in stainless steels. Mater Des 7:111–119. doi:10.1016/0261-3069(86)90001-4
6. Grundfos (n.d.) Corrosion resistance chart—aggressive water. Www.us.grundfos.com
7. Handbook of stainless steel (2013) Outokumpu Oyj
8. Janikowski DS (2012) Super-ferritic stainless steels rediscovered. Stainless Steel World, KCI Publishing BV©
9. Kanthal, Kanthal APMT, 24 (2014). http://kanthal.com/en/products/material-datasheets/wire/na/kanthal-apmt/
10. Steel Sandvik, Seamless tube Sandvik 4C54 (1998)
11. Steel Sandvik, Sandvik 4C54 Datasheet (2015) 1–12
12. Gao W, Li Z (eds) (2008) Developments in high-temperature corrosion and protection of materials. Woodhead Publishing Limited, Abington Hall, Abington
13. Guria A, Charit I (2015) Observation of serrated flow in APMT™ steel. Mater Lett 160:55–57. doi:10.1016/j.matlet.2015.07.072
14. Engkvist J, Canovic S, Hellström K, Järdnäs A, Svensson JE, Johansson LG et al (2010) Alumina scale formation on a powder metallurgical FeCrAl alloy (Kanthal APMT) at 900–1,100 °C in dry O_2 and in $O_2 + H_2O$. Oxid Met 73:233–253. doi:10.1007/s11085-009-9177-7
15. Szklarska-Smialowska Z (2005) Pitting and crevice corrosion. NACE International, Houston

Effect of Bromide Ions on the Pitting Corrosion of Hafnium in Anhydrous *t*-Butanol and Acetonitrile

Wang Changhong, Yang Shenghai, Chen Yongming, Yang Xiyun, Wu Yanzeng, He Jing and Tang Chaobo

Abstract Pitting of Hf in Et_4NBr *t*-butanol and acetonitrile (ACN) solutions was studied by means of cyclic voltammetry, potentiodynamic anodic polarization, galvanostatic, potentiostatic and impedance techniques. The potentiodynamic anodic polarization curves did not exhibit an active dissolution region near corrosion potential due to the presence of an oxide film on the electrode surface, which was followed by pitting corrosion resulting from the passivity breakdown by the aggressive attack of bromide (Br^-) ion. The pitting potential (E_{pit}) increased with increasing potential scanning rate but decreased with increasing temperature and Br^- concentration. Cyclic voltammetry and galvanostatic measurements allowed the pitting potential (E_{pit}) and the repassivation potential (E_p) to be determined. Analysis of the potential/time transients revealed that the applied anodic current density had a significant influence on the values of E_{pit}. On the other hand, the E_p values were independent on the applied current density. The current/time transients indicated that the incubation time (t_i) for passivity breakdown decreased slightly with increasing potential and solution temperature. The impedance spectra showed that the resistance of passive layer decreased with increasing potential.

Keywords Hafnium · Polarization · EIS · Pitting corrosion · Passive films

Introduction

Hafnium is valve metal, and its electrochemical behavior has received little attention due to its high price and little application in the early 1990s. However, as the semiconductor industry ushered in nanometer era, hafnium as anode begins to be used to electrochemically synthesize hafnium alkoxides [1]. Hafnium alkoxide

W. Changhong · Y. Shenghai (✉) · C. Yongming
Y. Xiyun · W. Yanzeng · H. Jing · T. Chaobo
School of Metallurgy and Environment, Central South University, Changsha 410083, China
e-mail: Yangshcsu@163.com

S. Ikhmayies et al. (eds.), *Characterization of Minerals, Metals, and Materials 2017*, The Minerals, Metals & Materials Series,
DOI 10.1007/978-3-319-51382-9_38

($Hf(OR)_4$) is mainly used for the deposition of hafnium oxide(HfO_2) layers by atomic layer deposition (ALD), and deposited in this manner, the hafnium-based high-*k* dielectrics present much more stable electrical characteristics compared with the ones formed by chemical vapors or sputtering [2]. HfO_2 layers are among the most promising high-*k* candidates to meet the requirements for replacing the traditional SiO_2 gate oxide, which has reached its physical limits as a result of an increase in leakage currents due to tunneling effects in complementary metal–oxide–semiconductor devices [3–6].

At present, the most common method of preparing $Hf(OR)_4$ is based on the halide synthesis [7]. This method for producing $Hf(OR)_4$ has such disadvantages as a long process, poor working conditions, high cost, and low efficiency [8]. On the contrary, The electrochemical method has great promise for the direct conversion of the less electropositive metals to their alkoxides due to its simplicity, high productivity and its continuous and non-polluting character (with hydrogen as the major by-product) [7]. In our earlier works, we have prepared several tantalum and niobium alkoxides by electrochemical method [9–11]. In 1995, Hafnium *t*-butoxide was obtained for the first time by Turevskaya et al. [1] with the electrochemical method involving electrolysis of a *t*-butanol solution containing tetraethylammonium bromide with a platinum cathode and a hafnium anode. However, the electrochemical behaviors of this process have not been studied so far.

A program is in progress in our laboratory to study the electrochemical behaviour of Hf in anhydrous ethanol containing supporting electrolytes [12]. However, no results have been reported and discussed involving the corrosion behavior of hafnium in anhydrous *t*-butanol system. In the present work, to investigate the corrosion behavior of hafnium in anhydrous *t*-butanol and ACN solutions containing supporting electrolytes, measurements were performed using cyclic voltammetry, potentiodynamic anodic polarization, galvanostatic, potentiostatic and impedance techniques.

Experiment

Anhydrous *t*-butanol was supplied by Tianjing Damao Chemical Reagent Corporation. Anhydrous acetonitrile (ACN) and Et_4NBr were provided from Sinopharm Chemical Reagent Corporation Limited. The working electrode employed was made from a very pure hafnium rod (99.9%), which was supplied by the Northwest Institute for Non-ferrous Metal Research. The electrode used was cut as cylindrical rods and mounted into glass tubes using Araldite to offer an active flat disc shaped surface of 4.28 mm^2 geometric area to contact the test solution. Prior to each experiment, the electrode was successively polished with a series of emery papers from a coarse grade of 1200 to fine grade of 3500. The electrode was then successively rinsed with ethanol and *t*-butanol and finally dipped in the electrolytic cell.

The experiments were performed in a 250 mL volume glass electrolytic cell utilizing Pt foils of 1 cm × 1 cm and 2 cm × 2 cm geometric areas as auxiliary and reference electrodes, respectively. All potentials given in this paper were referred to this reference electrode. The experiments were carried out in anhydrous *t*-butanol and ACN solution containing Et_4NBr of various concentrations (0.01–0.05 M). All chemicals used were of analytical grade. For each experiment, a freshly prepared solution as well as a cleaned set of electrodes was used. Each run was carried out in aerated stagnant solutions at the required temperature (±1 °C), using water thermostat. The sample was immersed in the solution until the potential reached a stable value (about 3 h), and then each measurement was performed.

Electrochemical measurements were performed using a potentiostat/galvanostat (CHI660C Electrochemical Workstation provided by Shanghai CH Instrument Company, China) connected to a personal computer. Cyclic voltammetric measurements were carried out by sweeping the potential linearly from −2.5 V (Pt) more negative than the open circuit potential (E_{ocp}) in the positive direction at a scan rate of 5 mV s^{-1} up to 2 V (Pt), reversing with the same scan rate to the lowest potential, and finally returning to the starting potential to form one complete cycle. The potentiodynamic anodic current/potential curves were recorded from −2.5 V (Pt) more negative than the open circuit potential (E_{ocp}) up to 2 V (Pt) with the required scanning rate. In galvanostatic potential/time transients, a constant anodic current density is applied on the Hf electrode and the variation in potential was recorded as a function of time. The potentiostatic measurements were carried out at a given step anodic potential ($E_{s,a}$) at which the current transient was recorded. EIS measurements was measured using AC signals of amplitude 5 mV in the frequency range from 0.1 Hz up to 0.1 MHz. All impedance data were fit to appropriate equivalent circuits using computer program ZSimDemo 3.30d.

Results and Discussion

Cyclic Voltammetry Measurements

Figure 1 shows the cyclic voltammogram of Hf starting from −2.5 V (Pt) and reversed at 2 V (Pt) in anhydrous *t*-butanol (50%) and ACN (50%) solutions containing 0.04 M Et_4NBr at 30 °C and at a scanning rate of 5 mV s^{-1}. On the positive going scan, the cathodic current density drops gradually to a small value plateau and nearly reaches a zero value at the corrosion potential (E_{cor}). The anodic excursion scan does not exhibit an active dissolution region near E_{cor} as a result of the spontaneous passivation of Hf due to the presence of a stable HfO_2 film on the electrode surface. Subsequently, the passive current density (J_{pass}) increases slowly with the applied potential. This increase is assigned to the replacement of chemically dissolved HfO_2 by the aggressive attack of Br^- anion following its adsorption [13].

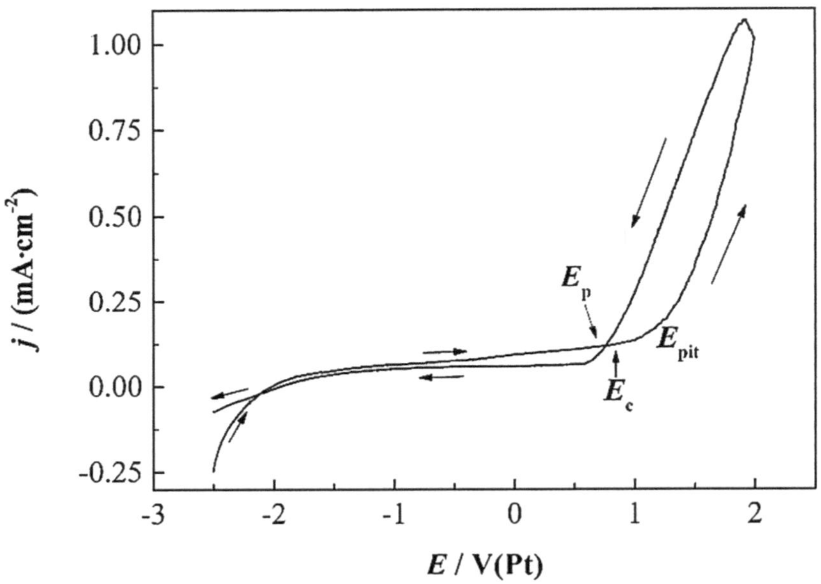

Fig. 1 Cyclic voltammogram of Hf in anhydrous *t*-butanol (50%) and ACN (50%) solutions containing 0.04 M Et_4NBr at 30 °C and at a scanning rate of 5 mV s^{-1}

However, when the passive region extends up to a certain potential, defined as the critical potential (E_c), the current observed begins to grow gradually until a pitting potential (E_{pit}) is reached. Afterwards, the current density goes up rapidly and accelerately, without any sign of gas evolution, which indicates the breakdown of the passive oxide film, initiation and propagation of pitting corrosion. Pitting corrosion may be attributed to the adsorption and incorporation of Br^- ions supported by a high electric field into the oxide film on Hf [13]. At flaws and defect sites, the oxide films can be totally removed and the base metal exposed to the electrolyte.

After the potential sweep reversal from 2 V (Pt), the current density rises still due to the autocatalytic character of pitting [14]. Subsequently, the current density shows to decline rapidly and linearly. This phenomenon suggests that an ohmic controlled process is taking place [15]. As a result, the flyback curve intersects the forward scanning curve at approximately 0.8 V (Pt), which defines the protection (repassivation) potential E_p for the Hf electrode under the test conditions. This denotes that the surface of the Hf returns again to the passivated state. Overall, a hysteresis loop is clearly observed, a crucial characteristic of the pitting corrosion phenomenon, which is similar to our previous study [12]. The existence of a hysteresis loop in a cyclic voltammogram suggests a delay in repassivation of an existing pit when the potential is scanned toward negative direction. The larger the hysteresis loop, the more difficult it becomes to repassivate the pit [16].

Potentiodynamic Measurements

Figure 2 shows the effect of the Et_4NBr concentration on the potentiodynamic polarization curves of Hf in anhydrous *t*-butanol (50%) and ACN (50%) solutions at a scan rate of 5 mV s^{-1} and at 30 °C. It is clear that enhancing Et_4NBr concentration shifts E_{pit} to more negative values, which corresponds to a decreased resistance towards pitting. This trend might be due to the weakening of the passive film resulting from the competition among its formation, the formation of the soluble intermediate and the increase in the number of aggressive Br^- ions that dissolve the passive layer [17]. In other words, the higher the bulk Br^- ion concentration, the higher will be the local Br^- ion concentration that is responsible for initiating pit formation at defect areas and facilitating the exposed metal dissolution.

Figure 3 shows the effect of the temperature on the potentiodynamic polarization curves of Hf in anhydrous *t*-butanol (50%) and ACN (50%) solutions with 0.04 M Et_4NBr at a scan rate of 5 mV s^{-1}. With an increase in solution temperature, E_{pit} shifts into a negative direction, indicating an increased susceptibility to pitting. These results may be interpreted on the basis that the higher temperature facilitates the solubility of oxide film. Moreover, the increasing temperature can accelerates the rates of diffusion of the reactant and product species into and from the pit cavity.

Figure 4 illustrates the effect of the potential scanning rate (υ) on the potentiodynamic polarization curves of Hf in anhydrous isopropanol (50%) and ACN

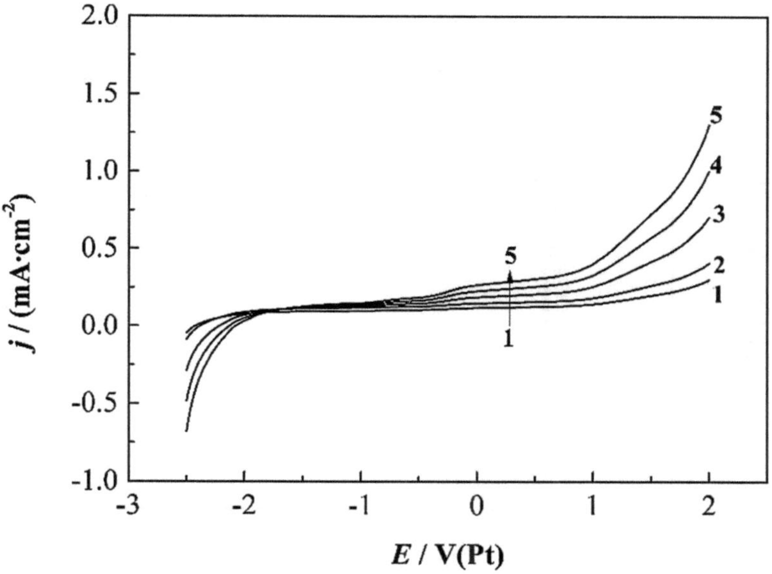

Fig. 2 Potentiodynamic polarization curves of Hf in anhydrous *t*-butanol (50%) and ACN (50%) solutions containing Et_4NBr of various concentrations at a scan rate of 5 mV s^{-1} and at 30 °C. (*1*) 0.01 M; (*2*) 0.02 M; (*3*) 0.03 M; (*4*) 0.04 M; (*5*) 0.05 M

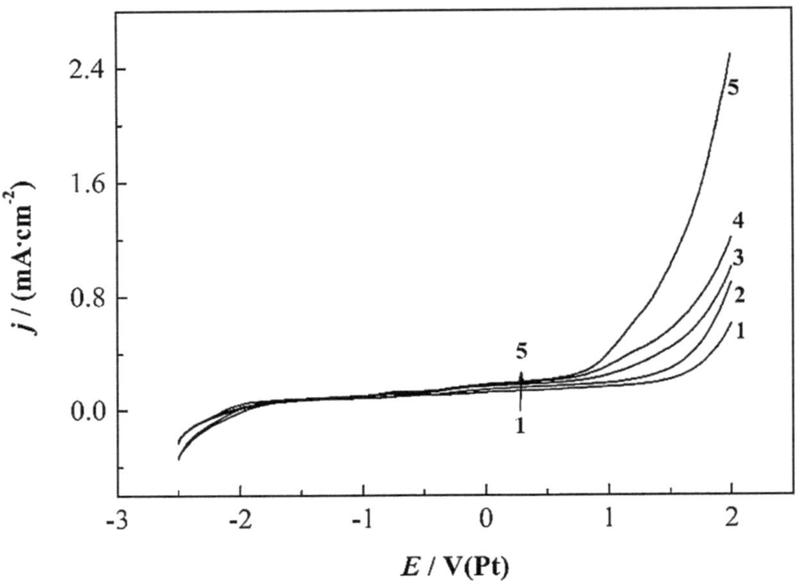

Fig. 3 Potentiodynamic polarization curves of Hf in anhydrous *t*-butanol (50%) and ACN (50%) solutions containing 0.04 M Et_4NBr at a scan rate of 5 mV s^{-1} and at various temperatures. (*1*) 20 °C; (*2*) 25 °C; (*3*) 30 °C; (*4*) 35 °C; (*5*) 40 °C

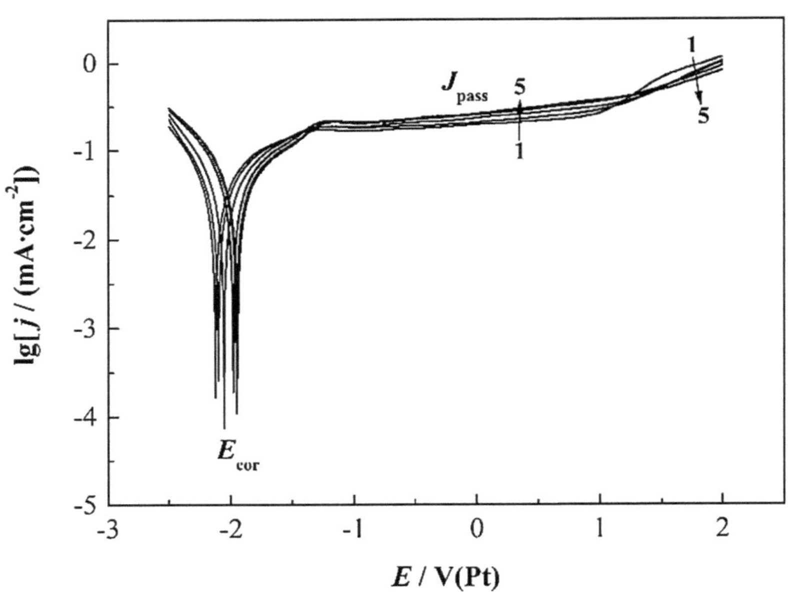

Fig. 4 Potentiodynamic polarization curves of Hf in anhydrous *t*-butanol (50%) and ACN (50%) solutions containing 0.04 M Et_4NBr at 30 °C and at various potential scan rates (v). (*1*) 5 mV s^{-1}; (*2*) 10 mV s^{-1}; (*3*) 20 mV s^{-1}; (*4*) 30 mV s^{-1}; (*5*) 50 mV s^{-1}

(50%) solutions with 0.04 M Et_4NBr at 30 °C. As the sweep rate is increased, E_{pit} slightly shifts to more noble values. This could be interpreted in terms of the incubation time [18] needed for Br^- to break down the oxide layer and make the base metal expose to the solution. Increasing the sweep rate corresponds to less time to form a soluble complex at a certain potential. As a result, the passive film is not rapidly dissolved, resulting in the formation of a thicker passive layer. Consequently, for a higher scanning rate, pitting initiation occurs only at more positive potentials, corresponding to a sufficiently short pit incubation time.

Potential/Time Transient Measurements

To investigate the growth kinetics of the oxide film formed on Hf surface and its breakdown, Hf electrode was anodized galvanostatically in 0.04 M Et_4NBr anhydrous *t*-butanol (50%) and ACN (50%) solutions at 30 °C and at different current densities (5–100 μA cm^{-2}), as illustrated in Fig. 5. The steep linear increase in potential with time is attributed to the initial formation and growth of the oxide film. A potential maximum appears at a certain potential. This maximum is thought to be the pitting potential (E_{pit}) caused due to the competition between two processes, namely further oxide film growth and its breakdown. After E_{pit}, the potential

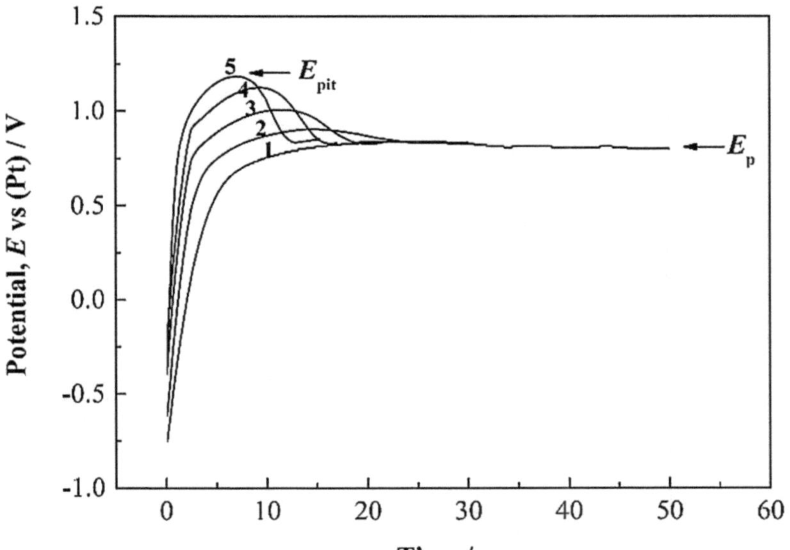

Fig. 5 Potential/time transients recorded for Hf in 0.04 M Et_4NBr anhydrous *t*-butanol (50%) and ACN (50%) solutions at 30 °C and at different current densities. (*1*) 5 μA cm^{-2}; (*2*) 25 μA cm^{-2}; (*3*) 50 μA cm^{-2}; (*4*) 75 μA cm^{-2}; (*5*) 100 μA cm^{-2}

declines to a steady value, corresponding to the repassivation potential (E_p). It is clearly observed in galvanostatic curves that E_p is the only potential which are nearly constant and independent of time and current [19]. The values of E_p obtained from cyclic voltammetry in Fig. 1 agree well with the ones from galvanostatic measurements in Fig. 5. Moreover, the values of maximum (E_{pit}) are dependent on the constant current density applied (see Fig. 5 again).

Current/Time Transient Measurements

Figure 6 displays the potentiostatic current-time transients for Hf in anhydrous *t*-butanol (50%) and ACN (50%) solution containing 0.04 M Et_4NBr at various potentials ($E_{s,a}$, before and after E_{pit}) and at 30 °C with a time of 200 s for the experiment. For $E_{s,a} \leq 1.0$ V (Pt) ($E_{s,a} < E_{pit}$, under these conditions), the current densities gradually decay to steady-state values, indicating that low potential is not enough to induce pitting corrosion. The fall in current density is attributed to the formation and growth of HfO_2 on the anode surface. However, For $E_{s,a} \geq 1.2$ V (Pt) ($E_{s,a} > E_{pit}$, under these conditions), the current density initially decreases to a minimum value at a certain incubation time (t_i) and then rises rapidly until a steady-state current is acquired, suggesting that pitting corrosion has taken place on the sample surface.

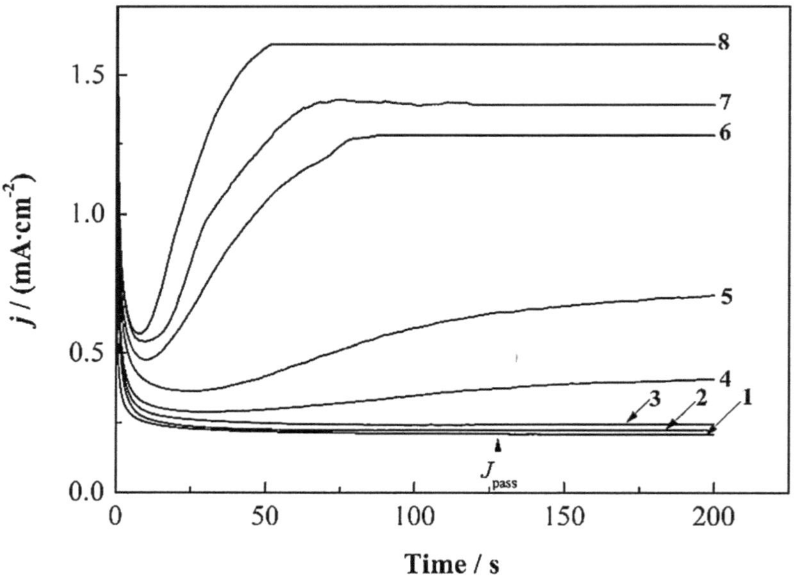

Fig. 6 Potentiostatic current/time transients recorded for Hf in *t*-butanol (50%) and ACN (50%) solutions with 0.04 M Et_4NBr at 30 °C and at various potentials ($E_{s,a}$). (*1*) 0.6 V; (*2*) 0.8 V; (*3*) 1.0 V; (*4*) 1.2 V; (*5*) 1.4 V; (*6*) 1.5 V; (*7*) 1.55 V; (*8*) 1.6 V

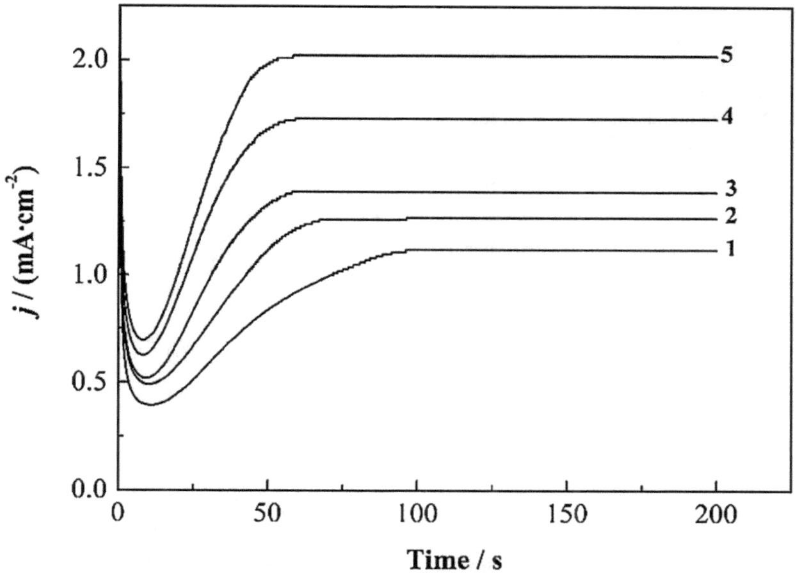

Fig. 7 Potentiostatic current/time transients recorded for Hf in *t*-butanol (50%) and ACN (50%) solutions with 0.04 M Et_4NBr at various temperatures and at 1.55 V (more positive than E_{pit}, under these conditions). (*1*) 20 °C; (*2*) 25 °C; (*3*) 30 °C; (*4*) 35 °C; (*5*) 40 °C

Figure 7 shows the potentiostatic current-time transients for Hf in anhydrous *t*-butanol (50%) and ACN (50%) solution containing 0.04 M Et_4NBr at 1.55 V (more positive than E_{pit}, under these conditions) and at different temperatures. The enhancing effect of solution temperature can be clearly observed. As the temperature is increased, incubation time (t_i) becomes shorter and the steady-state currents becomes higher.

EIS Measurements

Figure 8 presents the characteristic Nyquist and Bode diagrams of Hf in *t*-butanol (50%) and ACN (50%) with 0.04 M Et_4NBr at different potentials (Pt) and at 20 °C. Though different processes take place simultaneously, just only one semi-circle (i.e., time constant) in the Nyquist diagrams is observed. This phenomenon might either reflect that the time constants of the individual processes, suggested by Brett [20], strongly overlap, or that one process predominate and, thus, excludes the other processes in this frequency range. The other possible explanation for this capacitive time constant is the oxide film itself. The oxide film is thought as a parallel circuit of a resistor due to ionic conduction in the oxide, and a capacitance due to the dielectric properties of the oxide. It is clear that the diameter of the semicircle decreases with an increase in applied potential.

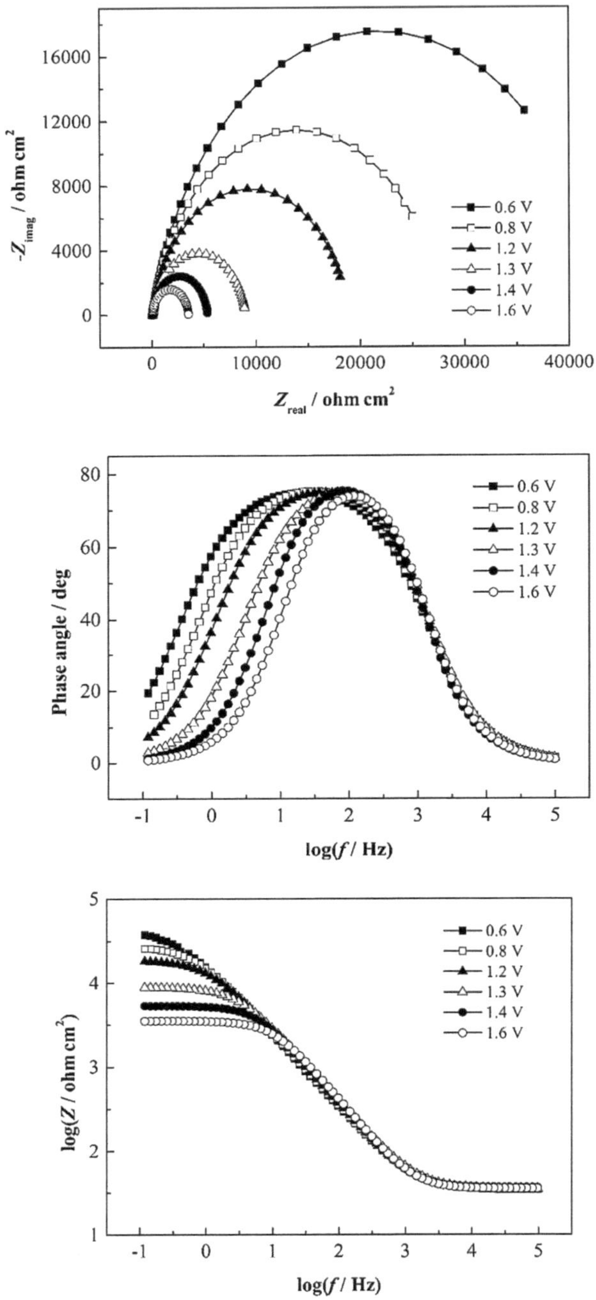

Fig. 8 Complex plane impedance plots recorded for Hf in 0.04 M Et_4NBr anhydrous *t*-butanol (50%) and ACN (50%) solution at different $E_{s,a}$ and at 30 °C

A more effective method, which better exhibits the frequency dependence of the impedance data is the so-called Bode diagram. The high frequency limit ($f > 10$ kHz) represents the solution resistance (R_s). The low frequency limit ($f < 0.2$ Hz) corresponds to the sum of R_s and the charge-transfer resistance R_{ct}, which is, in the first approximation, determined by both the electronic conductivity of the oxide film and the polarization resistance of the dissolution and repassivation processes. At medium frequencies capacitive phenomenon of the system is evident, determined by the dielectric properties of the oxide film. The phases shifting between current and voltage are approximately 75 Hz.

Conclusions

Cyclic voltammetry curve shows that a hysteresis loop is clearly formed, when the surface of Hf is subjected to a certain anodic potential. Its positive-sweep part does not exhibit an active dissolution region near corrosion potential caused due to the presence of an oxide film on the electrode surface, which is followed by pitting corrosion as a result of the passivity breakdown by the aggressive attack of Br^-. Potentiodynamic anodic polarization curves reveal that the pitting potential (E_{pit}) shifts to more positive values with increasing potential scanning rate, but shifts into a more negative direction with increasing solution temperature and Br^- concentration. Analysis of the potential/time transients reveals that the applied anodic current density has a significant influence on the values of E_{pit}. On the other hand, the E_p values are independent on the applied current density. The potentiostatic current-time transients show that the current density gradually decays to a steady value when the potential is low; however, when the potential is higher than a certain value, the current density initially decreases to a minimum value at a certain incubation time (t_i) and then rises to a steady-state value. As the solution temperature and the potential applied are increased, the incubation time (t_i) becomes shorter and the steady-state currents becomes higher. Impedance measurements show that the resistance of passive layer decreases with an increase in applied potential.

Acknowledgements We are grateful to the Natural Science Foundation of China (No. 51374254) for providing financial support.

References

1. Turevskaya EP et al (1995) The alkoxides of zirconium and hafnium: direct electrochemical synthesis and mass-spectral study. Do "M(OR)$_4$", where M=Zr, Hf, Sn, really exist. Russ Chem Bull 44(4):734–742
2. Bohr MT et al (2007) The high-k solution. IEEE Spectr 44:29–35
3. Usui T et al (2013) Approaching limits of dielectric breakdown for SiO_2 films deposited by plasma-enhanced atomic layer deposition. Acta Mater 61(20):1–11

4. Karmakov I et al (2009) Spectroscopic ellipsometry of very thin tantalum pentoxide on Si. Appl Surf Sci 255:9211–9216
5. Zhang F et al (2015) Solution-processed hafnium oxide dielectric thin films for thin-film transistors applications. Ceram Int 41:13218–13223
6. Hildebrandt EM (2012) Oxygen engineered hafnium oxide thin films grown by reactive molecular beam epitaxy, pp 11–17
7. Bradley DC et al (2001) Alkoxo and aryloxo derivatives of metals. Academic Press, San Diego
8. Shreider VA et al (1981) Direct electrochemical synthesis of metal alkoxides. Inorg Chim Acta 53:73–76
9. Yang SH et al (2008) Preparation of high-purity tantalum ethoxide by vacuum distillation. Trans Nonferrous Met Soc China 18(1):196–201
10. Yang SH et al (2009) Electrochemical synthesis and characterization of tantalum alkoxides. Trans Nonferrous Met Soc China 19(6):1504–1508
11. Cai YN et al (2011) Electrochemical synthesis, characterization and thermal properties of niobium ethoxide. J Cent South Univ Technol 18:73–77
12. Wang CH et al (2015) Effect of bromide ions on the corrosion behavior of hafnium in anhydrous ethanol. RSC Adv 5:34580–34587
13. Davydov AD (2001) Breakdown of valve metal passivity induced by aggressive anions. Electrochim Acta 46(24):3777–3781
14. Kaesche H (1988) Microtunnelling during selective alloy dissolution and during pitting. Mater Corros 39(4):153–161
15. Muñoz AG, Bessone JB (1999) Pitting of aluminium in non-aqueous chloride media. Corros Sci 4(41):1447–1463
16. Szklarska-Smialowska Z (1986) Pitting corrosion of metals. Electrochem Soc Interface 19 (1):33–38
17. Hassan HH, Fahmy K (2008) Pitting corrosion of tin by acetate anion in acidic media. Int J Electrochem Sci 3(1):29–43
18. Duc HD, Tissot P (1979) Anodic behaviour of tin in neutral phosphate solution. Corros Sci 19 (3):179–190
19. Szklarska-Smialowska Z (1986) Pitting corrosion of metals. Electrochem Soc Interface 19 (1):33–38
20. Brett CMA (1992) On the electrochemical behaviour of aluminium in acidic chloride solution. Corros Sci 33(2):203–210

Compression Behaviour of Semi-closed Die Forged AZ80 Extrusion

A. Gryguc, S.K. Shaha, S.B. Behravesh, H. Jahed, M. Wells and B. Williams

Abstract Warm forging to near final product geometry shows promising benefits in terms of microstructure refinement and texture modification. In the present study, Semi-closed die forging was performed on extruded AZ80 magnesium alloy at a forging temperature of 400 °C and various forging ram speeds. Compression tests were performed to investigate the influence of forging at various deformation rates on the performance of extruded AZ80 magnesium alloy. The microstructural analysis showed that the extruded AZ80 possesses a grain structure typical in wrought magnesium alloys, with sharp basal texture. In contrast, the forged samples had a recrystallized microstructures and a spatially varying textures and crystal orientations resulting from the complex flow behaviour within the closed die forging. Furthermore, once forged, a significant increase in the material isotropy, homogeneity as well as failure elongation were observed. An increase in UCS of 9% was observed in the forging direction, while the UCS in the extrusion direction was similar to that before forging. The changes in UCS can be attributed to both the refined grain microstructure and crystal reorientation. Successful microstructure recrystallization occurs for strain rates which traverse several orders of magnitude, and in general, UCS isotropy increases with increasing strain rate. It is believed that forging cause's grain refinement and texture modification which act to enhance the mechanical performance (UCS, fracture strain and superficial hardness) for the forging conditions investigated in this study.

A. Gryguc (✉) · S.K. Shaha · S.B. Behravesh · H. Jahed · M. Wells
Department of Mechanical & Mechatronics Engineering, University of Waterloo, 200 University Ave W, Waterloo, ON N2L 3G1, Canada
e-mail: agryguc@uwaterloo.ca

B. Williams
CanmetMATERIALS, Natural Resources Canada, 183 Longwood Road South, Hamilton, ON L8P 0A1, Canada

S. Ikhmayies et al. (eds.), *Characterization of Minerals, Metals, and Materials 2017*, The Minerals, Metals & Materials Series,
DOI 10.1007/978-3-319-51382-9_39

Keywords AZ80 · Forging · Material characterization · Wrought magnesium · Texture modification

Introduction

There has been a strong impetus towards increasing levels of fuel efficiency and decreasing emissions in the automotive industry throughout the past several decades [1]. Magnesium (Mg) and its alloys show a very promising future with widespread applicability in structural components since they are the lightest engineering metal and possess mechanical properties similar to that of the heavier materials which are currently used in industry [2]. The focus of this research project is to develop the technology to successfully forge magnesium front lower control arms to replace existing cast aluminum components in a high volume production passenger car. However, the strong crystallographic texture in wrought Mg alloys, which commonly forms during manufacturing processes, and the resultant anisotropic mechanical properties cause poor formability at room temperature which limits the application of wrought Mg alloys. Numerous studies [3, 4] have been performed with the aim of weakening the crystallographic texture which reduces the anisotropy in mechanical properties. The most commonly reported mechanism for texture weakening of Mg alloy is introducing rare earth materials. The addition of Y, Nb, Gd, Ce, Ng etc. in Mg alloys randomizes the texture during hot deformation processes such as extrusion, forging and rolling. This technique causes a reduction of crystallographic texture intensity and activates the basal slip system [4]. However, the cost associated with rare earth elements significantly limits their application. Recently, Sarker et al. [3] proposed multidirectional compression as a method for randomizing the texture and modifying the strain hardening behaviour in Mg.

Forging process has been used as a technique to improve the mechanical properties of Mg alloys [5]. A few studies [6–9] are available regarding the effect of forging process parameters on the mechanical properties of Mg are available in literature, however they are not numerous. Madaj et al. [6] investigated the effect which varying the forging parameters had on the Mg-Al-Zn group of Mg alloys in open-die forging at both hot and warm working temperatures. They determined the optimal temperature window for forging of cast-homogenized AZ31 Mg to be between 290 and 345 °C. Kobold et al. [7] performed both axial and radial open die forging of extruded AZ80-T5 at rates between 5 and 20 mm/s, and observed no significant difference on the anisotropy of the material flow regardless of the forging direction. Furthermore, they concluded that optimal isothermal forging temperature to be 350 °C, and also recommending that this optimum condition tends toward lower temperatures with increasing strain rate. Kurz et al. [8] investigated the behaviour of as-extruded AZ80-F Mg during closed die forging, and concluded that at higher deformation rates (300–400 mm/s) lower forging temperatures (240 °C) are desirable. Furthermore, they observed that increasing the forging temperature decreases the mechanical strength due to grain growth but enhances the elongation due to increased microstructure homogeneity. Miura et al. [10], investigated the effect of multi-directional multi-step forging on the mechanical properties of AZ61 alloy, and discovered that the effect that cumulative strain had on grain refinement was pronounced. Furthermore, they observed that with decreasing grain size came increasing levels of microstructure homogeneity and corresponding increases in strength which agree with Hall-Petch relation. Despite the previous works mentioned above, the study of the monotonic properties of forged Mg is very limited in the literature. In particular the level of anisotropy in compressive behaviour of extruded then forged Mg alloy has not been investigated in the alloys suitable for automotive structural applications. Specifically, it is not clear how the evolution of the microstructure and texture during forging of AZ80-F Mg alloy affect the level of anisotropy in the ultimate compressive strength (UCS).

Therefore, the aim of this study was to investigate the effect of forging conditions on compressive behaviour of AZ80 extruded Mg alloy which has been forged in a semi-closed die. The influence of forging on the microstructure and texture evolution is also discussed.

Material and Experiments

The material used in this investigation was commercially available AZ80 magnesium alloy. The material was received from Magnesium Elektron in the form of an extruded billet of diameter 63.5 mm in the as-fabricated condition. Then smaller diameter (20 mm) billets were extracted at a radial position of ~0.63R as according to Fig. 1a. The forging of the materials was conducted at CanmetMATERIALS using the small billets of AZ80 extruded which was cut into 45 mm lengths. All forging trials were carried out on a 110 Ton hydraulic press with an upper and lower platen (die) which had internal geometry as that found in Fig. 1b, to facilitate complex mixed LD/RD flow within the die. Graphite lubricant was used during the process on all test samples. The billet and tooling were heated separately to 400 °C (iso-thermal), this temperature was selected to ensure successful forging although lower temperatures (~300 °C) are also expected be feasible. The billet was held at 400 °C for ~1 h prior to forging to ensure all thermal gradients have decayed. Although the die temperature remained almost constant throughout the test, heat loss to the surrounding air during forging was expected, particularly for the slower forging rate condition. The forgings were then air cooled. The orientation of the billet to the press was such that the radial direction was along the direction of the press stroke (i.e. direction of forging was coincident to radial direction of the billet). Forging was carried out at four different ram speeds (0.04, 0.4, 4 and 40 mm/s). Please note that each condition will subsequently be referred to as S1, S2, S3, and S4 as shown in Table 1.

The as-extruded material was forged to a height of ~13 mm corresponding to 35% height reduction in the forging direction, and dimensional changes of 150% in

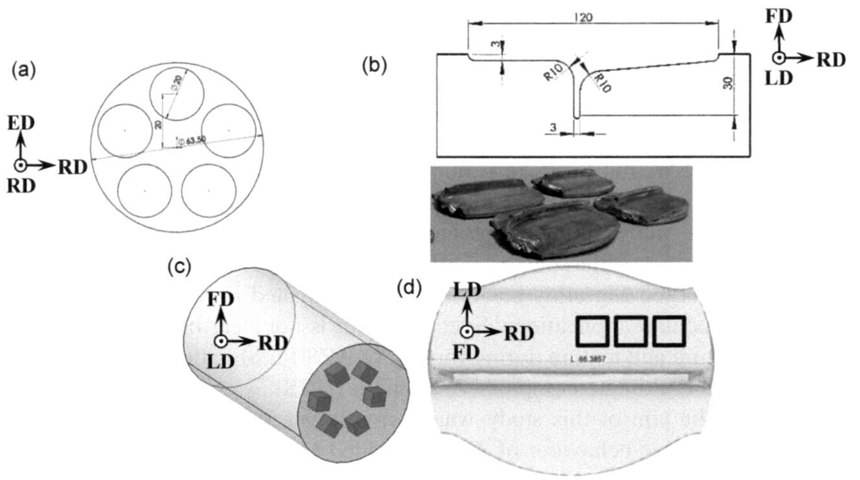

Fig. 1 **a** Scheme for machining small billets from as-extruded billet **b** forging die schematic and some forged samples at four different ram speeds, and schematic diagram showing the location of compressive sample extraction in (**c**) as-extruded billet (**d**) and semi-closed die forged sample

Table 1 The relationship between forging parameters, superficial hardness and UCS of AZ80 alloy forged at 400 °C at various rates

ID	Approx. forging rate		Temperature	Grain size	Rockwell hardness (30T)		UCS (MPa)	UCS (MPa)
	(mm/s)	1/s	(°C)	(μm)	(LD)	(FD)	(LD)	(FD)
Base	As-extruded		–	18.0 ± 0.7	35.3	–	412.1	349.8
S1	0.04	0.01	400	42.2 ± 7.4	33.9	25.7	346.4	317.3
S2	0.4	0.1		15.3 ± 0.4	32.2	25.6	393.1	382.8
S3	4	1		–	34.4	27.5	364.2	344.1
S4	40	10		–	35.0	27.5	353.2	346.4

the radial direction and ~47% in the longitudinal direction. Figure 1c, d show the orientation for which the metallographic, and monotonic compression specimens were extracted from both the extruded and forged billets. All specimens were extracted from a location which was ~50% of the radius of the as-extruded billet. Samples were extracted such that both light optical microscope (LOM) and monotonic compression tests could be performed in various material directions (LD/RD/FD) of the as-extruded and forged billets (LD, RD, and FD refer to longitudinal, radial, and forging directions, respectively).

The metallographic samples were prepared following standard metallographic techniques [11]. The microstructure was observed using a LOM and a scanning electron microscope (SEM), coupled with energy-dispersive X-ray spectroscopy (EDS). The texture measurements of the alloy before and after forging tests were performed using a Bruker D8 discover equipped with advanced 2D-detector of X-ray diffractometer on polished samples.

Compressive test samples of cuboid geometry with side lengths of ~7.6 mm were extracted from the as-extruded billet and forged samples as shown in Fig. 1c, d. The quasi-static compressive tests were performed using a MTS 810 Servo-Hydraulic test machine operating in displacement control mode. The displacement rate of the crosshead was 10^{-3} m/min for all tests, which corresponds to an approximate strain rate of 1.4E−3 s^{-1}. Strain measurement was accomplished using a GOM Aramis 3D 5MP DIC system.

Results and Discussion

Microstructure and Texture

The as-extruded sample (base) (Fig. 2) possesses moderately bimodal grain structure whose major axis is oriented along the extrusion direction, and surrounded by clusters of smaller grains. The calculated average grain size for the as-extruded material is 18.0 μm ± 0.7 with a significant bi-modal size distribution. The majority of precipitates have a discontinuous lamellar structure similar to those investigated by Lai et al. [12] and range in size from ~10 to 80 μm and always initiate at the grain boundaries. In the lower magnification images shown in Fig. 2, evidence of continuous precipitates where bands of β-phase ($Mg_{17}Al_{12}$) can be distinguished from the magnesium matrix Fig. 2a. Both of these precipitate morphologies are typical of as-fabricated AZ80-F extrusion where coarse precipitates are expected. In contrast, in materials which have undergone aging/heat-treatment more refined precipitate morphologies are anticipated [12].

The forged specimens all exhibited varying grain morphology, with the two investigated conditions (S1 and S2) both showing evidence of a recrystallized microstructure. Grain refinement of ~15% in size occurs at a forging rate of 0.4 mm/s (S2). This condition corresponds to the maximum UCS of all forged

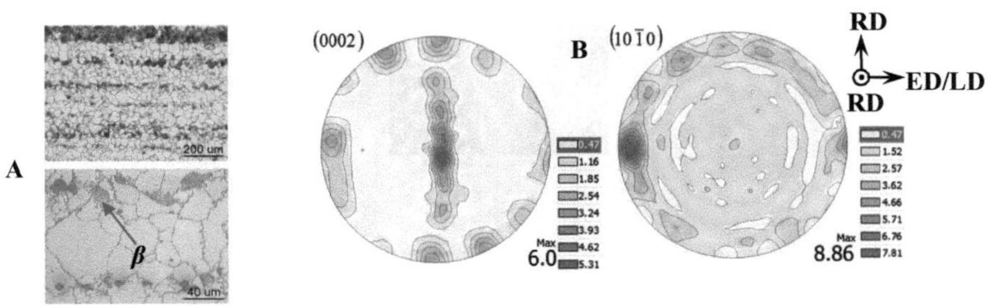

Fig. 2 Initial microstructure (**a**) and (0002) and (10$\bar{1}$0) pole figures (**b**) of the as-extruded AZ80-F Mg alloy

conditions and hence is considered to be optimal. It is expected that at higher forging rates (S3, S4) only marginal further grain refinement would occur, and microstructure homogeneity would increase which agrees with findings presented by Kurz et al. [8] in AZ80 die forgings. In contrast to the optimum forging condition S2, the slowest forging rate S1 shows significant evidence of a fully recrystallized microstructure as well as substantial grain growth (~135% increase in average grain size). This observation is supported by the softer superficial hardness and lower UCS in both directions (LD/FD) versus all other forging conditions. The grain morphology in the forged conditions (Figs. 3 and 4) appears to be much more "pancake" like in nature with obvious spatial variations in microstructures and grain orientations which depend upon the local deformation direction within the closed die. The precipitate morphology in the forged samples S1 and S2 is a discontinuous lamellar structure similar to that of the base as-extruded material, however the size of these precipitates is much reduced ~5 to 10 μm (i.e. the original precipitate structure of the base seems to have been broken

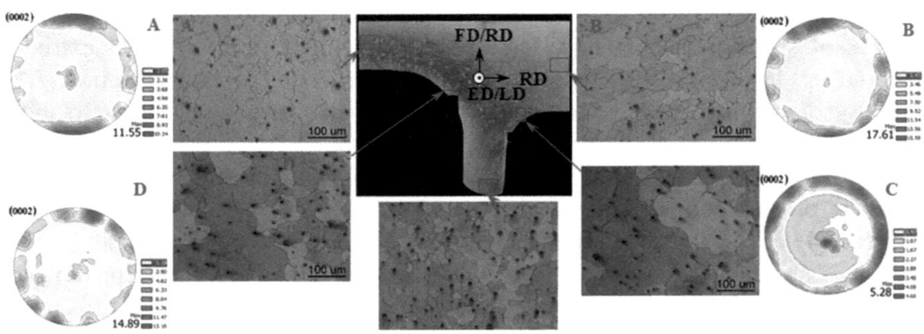

Fig. 3 Variation of microstructure and texture of AZ80 extruded Mg alloy forged at the temperature of 400 °C a ram speed of 0.04 mm/s in a semi-closed die

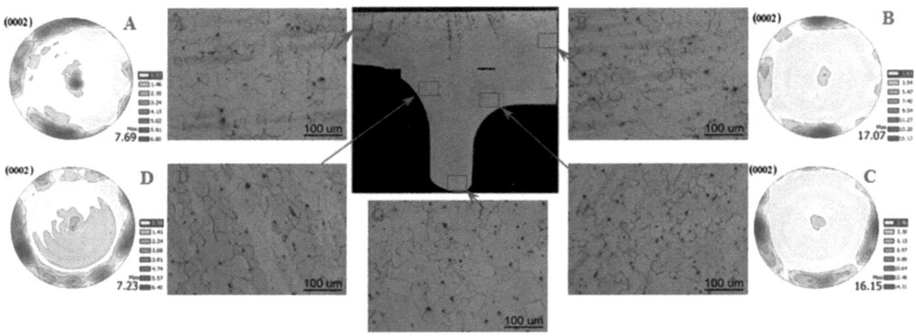

Fig. 4 Variation of microstructure and texture of AZ80 extruded Mg alloy forged at the temperature of 400 °C a ram speed of 0.4 mm/s in a semi-closed die

up due to the high temperature deformation). Further, the majority of precipitates seem to have nucleated at grain boundaries however there is a small proportion of intra-granular discontinuous precipitates in the forged material. The slowest forging condition S1 has the highest proportion of β-phase which remains out of solution from the matrix, whereas in S2, much of the β-phase seems to have dissolved into solution in the matrix. This significant difference in precipitation structure between S1 and S2 can be attributed to the higher plastic work rate and heat of deformation resulting higher forging rate which provides sufficient energy to dissolve some of the β-phase back into solution.

The calculated pole Figs for the basal (0002) and prismatic $(10\bar{1}0)$ planes obtained via XRD are shown for the as-extruded material in Fig. 2, and the two forged conditions S1 and S2 in Figs. 3 and 4, respectively. There is a strong texture in the as-extruded condition with the c-axis of the crystal orientated along the radial direction of the billet. The forged materials texture shows little texture sensitivity to forging rate, particularly in the prismatic plane, Furthermore, there appears to be significant spatial variation in the texture (as expected in a semi-closed die forging with complex flow behaviour) with the local c-axis orientation being approximately coincident with the local forging direction in the majority of cases.

Compressive Properties

Figure 5 shows the UCS for various forging rates (S1, S2, S3, S4) at a forging temperature of 400 °C for three different material directions (LD, FD, and RD). It can be seen that in the extruded material there is pronounced anisotropy in the UCS as expected based on the pole figures presented in Fig. 2. Once forged, the UCS becomes somewhat invariant of material direction, with increasing forging rate

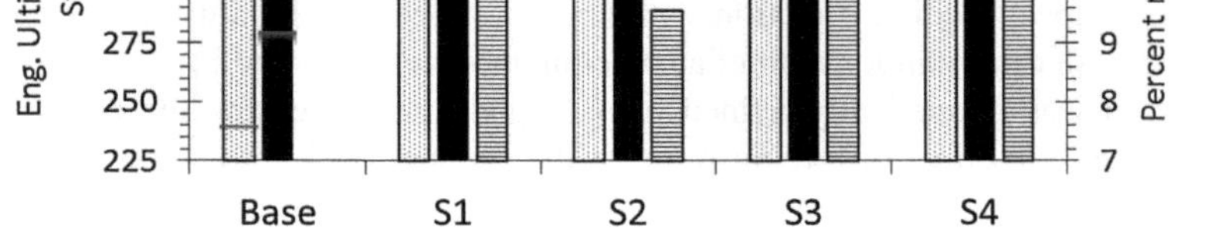

Fig. 5 UCS and percent reduction in height at failure for AZ80 forged at 400 °C and a variety of forging rates (S1: 0.04 mm/s, S2: 0.4 mm/s, S3: 4 mm/s and S4: 40 mm/s) in LD, FD, and RD directions

resulting in more isotropic behaviour. The optimum forging condition S2 exhibits a significant increase (+9%) in UCS in the forging direction with only a marginal decrease in the longitudinal direction. A similar increase in the mechanical properties in the forging direction were observed by Gryguc et al. for both the tensile [13] and compressive [14] strength of AZ31B after forging. This increase is attributed to the grain refinement and texture modification via the reorientation of the c-axis to the direction of forging, which locally, is coincident with the loading direction in compression, (please refer to sample extraction schematic in Fig. 1d). Figure 5 also illustrates the percent reduction in height until failure for samples loaded in monotonic compression. It can be observed that the reduction in height is greater in all forged conditions compared to the base material, which indicates an expected increase in ductility of the forged material in all directions.

As discussed previously, the mechanical properties of magnesium alloy are significantly influenced by the texture, especially the orientation of the crystallographic c-axis relative to the local loading direction of the material. It can be seen that the deformability in both the longitudinal and forging directions are superior in all forged conditions compared to the base material. Coupling this observation with the XRD results presented in Figs. 3 and 4 reinforces the conclusion that the improvement in mechanical properties can be attributed to the changes in both microstructure (grain refinement and homogeneity) and the modification of texture (from axisymmetric to a complex orientation whose crystal c-axis depends upon the local loading direction during forging).

Conclusions

The compression behaviour of as-extruded and semi-closed die forged AZ80-F magnesium alloy were investigated. On the basis of the microstructure, texture, and UCS characteristics, the following conclusions can be drawn:

1. Microstructural analysis shows that AZ80-F alloy in the as-extruded condition exhibited a significant bi-modal grain structure with average grain size of 18.0 μm while the sample forged at the optimum condition (400 °C and 0.4 mm/min) shows smaller, equiaxed grains with average grain size of about 15.3 μm. At the same time, the plastic deformation imparted to the sample via forging resulted in a texture modification and a rotation of the crystal to align with the local loading direction within the semi-closed die during forging.
2. The coarse discontinuous lamellar precipitate structure found in the as-extruded material was dramatically refined once forged, and increasing the forging ram speed had the effect of forcing more of the β-phase back into solution with the matrix.
3. The combination of the effect of grain refinement and precipitate morphology modification resulted in a variety of different UCS values among the forged conditions which were investigated. In general, higher strain rates improved

microstructure homogeneity and reduced the level of UCS anisotropy. Slower strain rates resulted in grain coarsening and lower strength.

4. Once forged, more compressive deformation was achievable before failure, regardless of the forging condition. This can be attributed to the recrystallized microstructure and modified texture.

Acknowledgements The financial support of the Natural Sciences and Engineering Research Council of Canada (NSERC) through the Automotive Partnership Canada (APC) under APCPJ 459269–13 grant with contributions from Multimatic Technical Centre, Ford Motor Company, and Centerline Windsor are acknowledged. The authors would also like to thank G. Yu from the University of Waterloo for design of the forging die and J. McKinley from CanmetMATERIALS, Hamilton for forging trials.

References

1. Skszek T, Zaluzec M, Conklin J, Wagner D (2015) MMLV: project overview. SAE International
2. Chen X, Conklin JL, Carpenter RM, Wallace J, Flanigan C, Wagner DA, Kiridena V, Betrancourt S, Logsdon J (2015) MMLV: chassis design and component testing. SAE International
3. Sarker D, Friedman J, Chen DL (2014) Influence of pre-strain on de-twinning activity in an extruded AM30 magnesium alloy. Mater Sci Eng, A 605:73–79
4. Pan H, Ren Y, Fu H, Zhao H, Wang L, Meng X, Qin G (2016) Recent developments in rare-earth free wrought magnesium alloys having high strength: a review. J Alloy Compd 663:321–331
5. Wang S, Xu D, Wang B, Han E, Dong C (2016) Effect of solution treatment on the fatigue behavior of an as-forged Mg-Zn-Y-Zr alloy. Sci Rep
6. Madaj M, Greger M, Karas V (2015) Magnesium-alloy die forgings for automotive applications. Mater Technol 49(2):267–273
7. Kobold D, Pepelnjak T, Gantar G, Kuzman K (2010) Analysis of deformation characteristics of magnesium AZ80 wrought alloy under hot conditions. J Mech Eng 56(12):823–832
8. Kurz G, Clauw B, Sillekens W, Ketzug D (2009) Die forging of the alloys AZ80 and ZK60. In: Magnesium technology—TMS 2009, San Francisco, California
9. Yoon J (2015) Process design of Warm-Forging with extruded Mg-8Al-0.5Zn alloy for differential case in automobile transmission. Int J Precis Eng Manuf 16(4):841–846
10. Miura H, Matsumoto K, Kobayashi M (2015) Multi-directional forging of AZ61 Mg alloy using die under decreasing temperature conditions. In: The 10th international conference on magnesium alloys and their application, Jeju, Korea
11. Roostaei AA, Jahed H (2016) Role of loading direction on cyclic behaviour characteristics of AM30 extrusion and its fatigue damage modellin. Mater Sci Eng, A 670:26–40
12. Lai W-J, Li Y-Y, Hsu Y-F, Trong S, Wang W-H (2009) Aging behaviour and precipitate morphologies in Mg–7.7Al–0.5Zn–0.3Mn. J Alloy Compd 476:118–124
13. Gryguc A, Shaha S, Jahed H, Wells M, Williams B, McKinley J (2016) Tensile and fatigue behaviour of as-forged AZ31B extrusion. Fract Struct Integrity (38):251–258. doi:10.3221
14. Gryguc A, Jahed H, Williams B, McKinley J (2015) MagForge—mechanical behaviour of forged AZ31B extruded. Mater Sci Forum 828–829:291–297

Nondestructive Characterization of Microstructures of Heat-Treated Steels by Magnetic Barkhausen Noise Technique

C. Hakan Gür

Abstract Optimization and control of the microstructure is important for improving the properties of steel components. Development of non-destructive techniques for microstructure characterization is a critical task. Magnetic Barkhausen Noise (MBN) technique is a promising and challenging non-destructive technique for automated evaluation of microstructures in steel components in a fast and reliable manner. This paper presents the results of MBN measurements for microstructure characterization of the quenched and tempered steels.

Keywords Nondestructive characterization · Magnetic Barkhausen noise · Steel · Heat treatment · Microstructure

Introduction

Microstructure and mechanical properties of steels can be modified by quenching and tempering procedures. If the primary requirement is hardness or strength the component is tempered at about 200 °C; if the principal property is toughness the part is then tempered above 400 °C. Besides, residual stresses are significantly relieved by tempering at about 500 °C.

Needs for improvements in the steel properties and quality control process enforce the industry to use nondestructive techniques in the manufacturing routes for a rapid and reliable product characterization. Magnetic Barkhausen Noise (MBN) technique is a promising alternative to the traditional methods for microstructure characterization of ferromagnetic materials.

C.H. Gür (✉)
Metallurgical and Materials Engineering Department,
Middle East Technical University, 06800 Ankara, Turkey
e-mail: chgur@metu.edu.tr

C.H. Gür
Welding Technology and Nondestructive Testing Center,
Middle East Technical University, 06800 Ankara, Turkey

S. Ikhmayies et al. (eds.), *Characterization of Minerals, Metals, and Materials 2017*, The Minerals, Metals & Materials Series,
DOI 10.1007/978-3-319-51382-9_40

Ferromagnetic materials retain a large spontaneous magnetic moment due to the cooperative alignment of unpaired electron spins which is called as domain. Oppositely magnetized domains are divided by domain walls. When a variable external magnetic field is applied, irreversible jumps of domain walls occur due to discontinuous domain wall motion, nucleation and annihilation of domains. Investigation of hysteresis loops at high resolution reveals discontinuous flux changes due to domain wall motion as external magnetic field strength is increased. If the field strength reaches the critical level, domain wall motion continues by Barkhausen jumps that can be detected as the voltage pulses induced in a pick-up coil positioned close to the surface.

Parameters derived from hysteresis loops (permeability, coercivity etc.) and MBN signals are strongly influenced by strain field (residual stress) and microstructure (grain boundary, phase, precipitate, dislocation etc.) [1–3]. Residual stresses influence the area of the domain walls whereas microstructure affects the pinning sites for domain walls. Investigations on steels have shown that the main microstructural parameters affecting the coercivity are the volume fractions and grain sizes of ferrite and pearlite, and domain walls are strongly pinned in cementite lamellae in pearlite grains [4]. The amplitude of the MBN signal decreases with decreasing grain size [5, 6]. It was reported that MBN in low carbon steels is dependent on the grain size and stress [7, 8]. The change of AISI/SAE 1060 microstructure from coarse lamellar carbides to uniformly dispersed spherical carbides in ferrite matrix due to spheroidization process caused an increase in the MBN activity [9].

The effect of tempering period on the MBN emission of the quenched low-C steel samples was explained based on the obstacles to the domain wall movement [10]. The MBN fingerprints and frequency spectra of a medium-C steel showed that MBN activity increases with increasing tempering temperature [11]. A correlation between hardness profile and MBN peak height was found for the quenched and tempered steels [12]. In this study, various magnetic parameters were used to characterize the quench and temper induced changes in the AISI/SAE 4140 steel by MBN method.

Experimental Studies

The AISI/SAE 4140 specimens with 7 mm thickness were austenitized at 860 °C for 30 min under controlled atmosphere, and then quenched in water. Some of the samples were tempered for 90 min at 200, 400, and 600 °C. Microstructures were investigated by SEM analyses. Average hardness values were determined by measurements at different locations on the sample surfaces. MBN signals were recorded and analysed using Rollscan-μscan 500-2 commercial system. A cyclic magnetic field with an excitation frequency of 125 Hz was induced in a small volume of the samples. The MBN signals were amplified (20 dB) and filtered (0.1–1000 kHz). The peak magnetizing voltage and sampling frequency was 10 V and 2 MHz.

Results and Discussion

SEM investigations (Fig. 1) show that typical martensitic and tempered martensitic structures were obtained. The microstructures are uniform along the information depth of MBN activity. Distributions of the phases and the grain sizes were uniform. The influence of retained austenite is negligible. Martensite with interstitial carbon atoms in solid solution and high dislocation density, that makes it the hardest of all specimens (Fig. 2).

The sum integral of rectified MBN bursts gives a local simulation of the hysteresis loop (Fig. 3). The remanence and coercivity values can be calculated, however, they are relative values because saturation magnetization is far beyond the level of local magnetization. The quenched structure contains large amounts of crystal imperfections, and therefore, elimination of them via atomic re-arrangement during temper causes an increase in the remanence and a reduction in the coercivity.

The MBN fingerprints of the samples are given in Fig. 4. Using a definite sampling frequency, a local MBN root-mean-square (rms) value is calculated and plotted against magnetic field strength. This value is averaged over several field cycles determined by the number of bursts. MBN peak height is directly dependent

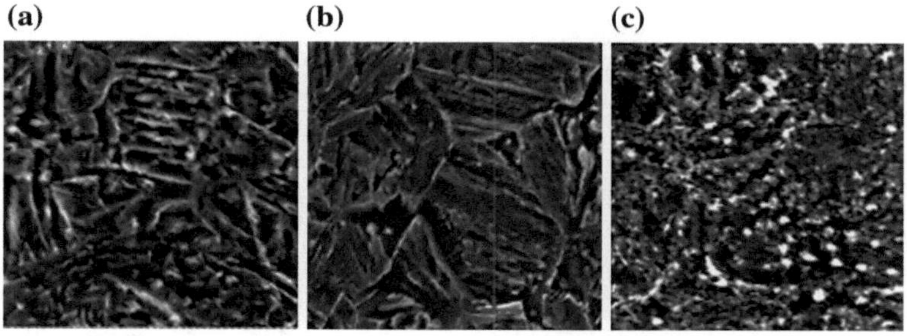

Fig. 1 SEM micrographs of the quenched and tempered samples (×5.000) SEM micrographs of the quenched and tempered samples (×5.000): **a** quenched, **b** tempered (200 °C), **c** tempered (600 °C)

Fig. 2 Effect of tempering on hardness

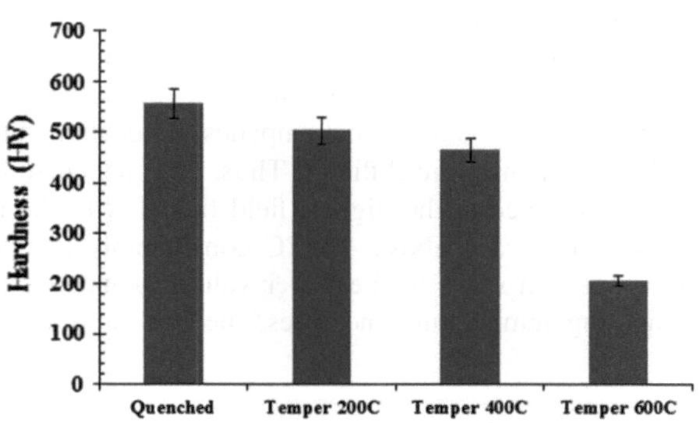

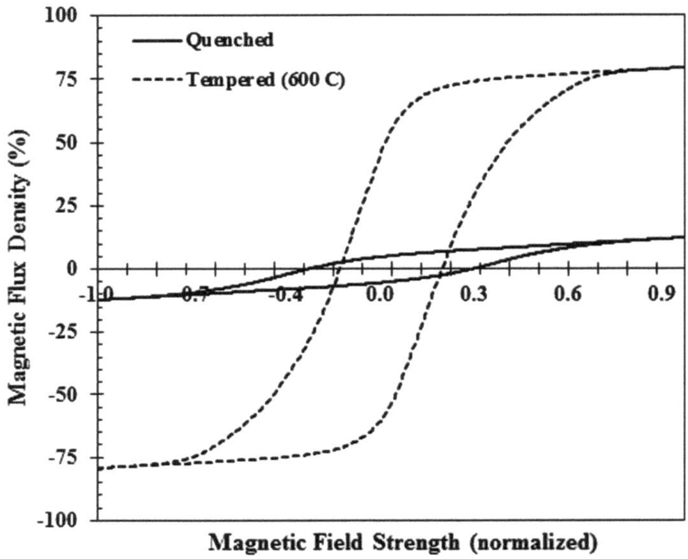

Fig. 3 Effect of quenching and tempering on the local magnetic hysteresis loop

Fig. 4 Effect of quenching and tempering on the MBN emission

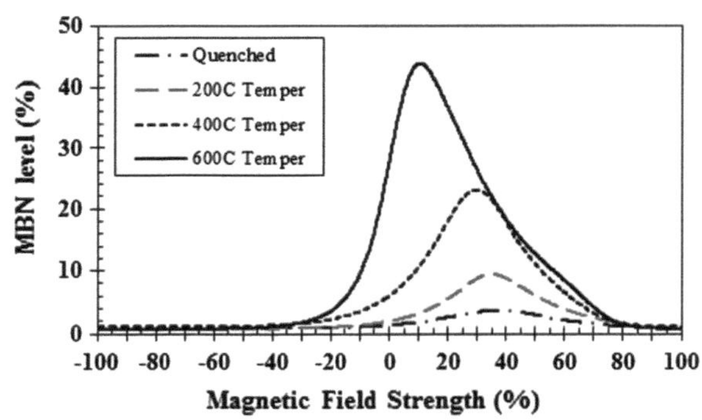

upon the number of domain walls moving and the mean free path of the domain wall displacement whereas the corresponding magnetic field represents the field strength required for the movement of the domain walls from the obstacles. The microstructural features that impedes dislocation movement, also makes domain wall movement more difficult. Thus, the quenched sample has the weakest MBN peak positioned at the highest field linked with the high coercivity of martensite. Tempering at and above 400 °C considerably increases the peak height, and the peak position shifts to the lower values of magnetic field. In other words, as the tempering temperature increases, the low amplitude broad peak of as-quenched martensite transforms into a high amplitude peak situated at low magnetic field. The results are in agreement with literature, however, some quantitative discrepancies exist due to different experimental set-ups.

In the quenched steel, small needles/lathes in martensite result in very small domains, i.e., a larger relative volume occupied by domain walls, and high dislocation density acts as obstacles to the domain wall motion. Moreover, difficulty in domain wall nucleation and small displacements of domain walls require a strong field for the reversal of magnetization. As tempering temperature increases, in contrast to the decrease in hardness, MBN-rms value increases. Tempering below 400 °C changes the microstructure slightly; although ε-carbides form, the microstructure is still needle shaped. Thus, the height and position of the MBN peak do not change significantly. Tempering at 600 °C, cementite replaces ε-carbides, carbides start to spheroidize, the crystal structure of martensite loses its tetragonality by recrystallization of ferrite, and dislocation density remarkably reduces. The average size of the domains increases due to coarsening of the microstructure. Thus, reverse domain nucleation and subsequent domain wall motions occur at the lower magnetic field strength. These microstructural variations make the domain wall motion easier which leads to a drastic increase in the MBN peak and a clear shift to lower external magnetic field in the peak position by reducing the resistance to the nucleation and movement of domains.

Conclusions

Microstructural variations of the quenched and tempered AISI/SAE 4140 specimens were investigated by the Magnetic Barkhausen Noise (MBN) method. MBN emission of martensite is very low, and the signal peak is at a higher magnetic field due to difficulty in domain nucleation and pinning of domain walls caused by excessive dislocation density and needle-like structure. Changes in the number of pinning sites and relative domain wall displacements upon tempering affect the MBN activity. As tempering temperature increases dislocation density decreases, microstructure coarsens and softens, and thus, MBN emission increases. MBN method can be utilized efficiently for microstructure characterization of the heat treated steels by establishing the quantitative relationships between magnetic parameters and microstructure.

Acknowledgements Author is thankful to Dr. Kemal Davut (Atılım University) and Dr. Ibrahim Cam (METU Central Lab.) who contributed to the experiments.

References

1. Williams HJ, Bozorth RM, Shockley W (1949) Magnetic domain patterns on single crystals of silicon iron. Phys Rev Lett 75:155–163
2. Montalenti G (1970) Barkhausen noise in ferromagnetic materials. Zeitschrift für angewandte Physik 28:295–300
3. Gaunt P (1987) Magnetic coercivity. J Phys 65:1194–1199
4. Hetherington MG et al (1987) High-voltage Lorentz electron-microscopy studies of domain-structures and magnetization processes in pearlitic steels. Philos Mag B 56:561–577
5. Gatelier-Rothea C et al (1998) Characterization of pure iron and (130 p.p.m.) carbon-iron binary alloy by Barkhausen noise measurements. Acta Mater 46:4873–4882
6. Anglada-Rivera J, Padovese LR, Capo-Sanchez J (2001) Magnetic Barkhausen noise and hysteresis loop in commercial carbon steel—influence of applied tensile stress and grain size. J Magn Magn Mater 231:299–306
7. Ng DHL et al (2003) Study of microstructure, mechanical properties, and magnetization process in low carbon steel bars by Barkhausen emission. Mater Sci Eng A 358:186–198
8. Blaow M, Evans JT, Shaw B (2004) Effect of deformation in bending on magnetic Barkhausen noise in low alloy steel. Mater Sci Eng A 386:74–80
9. Davut K, Gür CH (2010) Monitoring the microstructural evolution in spheroidized steels by magnetic Barkhausen noise measurements. J Nondestr Eval 29:241–247
10. Moorthy V et al (1997) Microstructural characterization of quenched and tempered 0.2% carbon steel using magnetic Barkhausen noise analysis. J Magn Magn Mater 171:179–189
11. Saquet O, Chicois J, Vincent A (1999) Barkhausen noise from plain carbon steels: analysis of the influence of microstructure. Mater Sci Eng A 269:73–82
12. Davut K, Gür CH (2007) Monitoring the microstructural changes during tempering of quenched SAE 5140 steel by magnetic Barkhausen noise. J Nondestr Eval 26:107–113

Part X
Method Development

Development of a New Recycling Process of PGM from Metal-Supported Catalyst Using Complex Oxide

Takashi Nagai, Hiroki Kumakura, Masahito Abe, Kotaro Seki and Daiki Noguchi

Abstract A large amount of platinum group metals (PGM) are used in automobile catalyst. It is important to recover the metals from wasted catalyst. PGM in the catalyst are supported by ceramics for four-wheel vehicle or metals such as stainless steel for two-wheel ones. The recycling process of PGM from the metal-supported catalyst has not been developed, while the process for ceramic-supported catalyst was developed, such as Rose process. We designed a noble recycling process of PGM from metal-supported catalyst using complex oxides between PbO and PGM oxide. In this study, complex oxides, $PbPt_2O_4$, $PbPdO_2$, $PbRh_2O_4$ and so on were synthesized at about 773–973 K. And the dissolution behavior of these compounds to hydrochloric acid was investigated. $PbPt_2O_4$ and $PbPdO_2$ were dissolved to hydrochloric acid easily, but $PbRh_2O_4$ was not.

Keywords PGM · Recycling · Automobile catalyst · Metal substrate · Separation

Introduction

Platinum (Pt), Palladium (Pd), and Rhodium (Rh) are used in automobile catalyst [1]. Resource of their metals is limited and mining of them has high environmental load. Therefore, the recycling of PGMs from wasted automobile catalysis is very important. The Rose process was developed as a process to recover PGM from wasted automobile catalysis [2]. In this process, the automobile catalyst is first crushed mechanically, and it was melted with extracting agent (copper, Cu), and flux at high temperature. The mixture separates into a metal phase, (copper (Cu) alloy containing PGM) and slag. The PGM is enriched by oxidized Cu to Cu oxide. The PGM are then recovered and separated to each element in a wet process. Generally, the catalysis with ceramics is treated in this process, that supported by

T. Nagai (✉) · H. Kumakura · M. Abe · K. Seki · D. Noguchi
Chiba Institute of Technology, 4-6-1 Tudanuma, Narashino 275-0016, Japan
e-mail: takashi.nagai@it-chiba.ac.jp

S. Ikhmayies et al. (eds.), *Characterization of Minerals, Metals, and Materials 2017*, The Minerals, Metals & Materials Series,
DOI 10.1007/978-3-319-51382-9_41

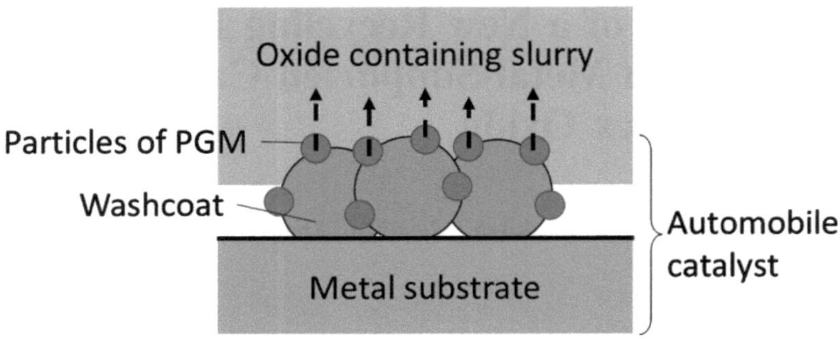

Fig. 1 Schematic configuration of a new recovery process of PGM from automobile catalyst supported by metal substrate

metals are not. Recycling of PGM from the catalyst supported by metals is limited. Therefore, it is needed to develop a recycling process.

In our laboratory, a new recycling process is designed. In the process, an extracting agent is coated to automobile catalyst, and then it was heated to synthesize complex oxide between PGM oxide and the extracting agent. Then the extracting agent is peeled off and PGMs are recovered from complex oxide (Fig. 1).

Lead oxide (PbO) is selected as an extracting agent. In this study, synthesization of the complex oxide between PGMs and PbO is investigated in this study. And the complex oxides' dissolution behavior to hydrochloric acid is also investigated.

Experimental Procedure

Platinum black, Palladium black or Rhodium black and PbO was mixed with 1:4 in molar ratio. The mixture was charged to Al_2O_3 crucible and heated at 773–973 K for 1–24 h in air atmosphere. The phase of specimen after experiment is identified by X ray diffraction analysis (XRD).

0.1–0.5 g of the specimen was dissolved to hydrochloric acid. Temperature was controlled by water bath at ca. 318 K, experimental time was 9 h. The concentration of PGM in the acid was analyzed by inductively coupled plasma atomic spectroscopy (ICP-AES). The dissolution rate was calculated by comparing the concentration in acid with the PGM content in complex oxide before dissolution.

Results and Discussion

Pt-PbO

By XRD pattern, it is indicated $PbPt_2O_4$ was synthesized. In order to synthesis stoichiometric $PbPt_2O_4$, the mixture of platinum black and PbO in 2:1 in molar ratio

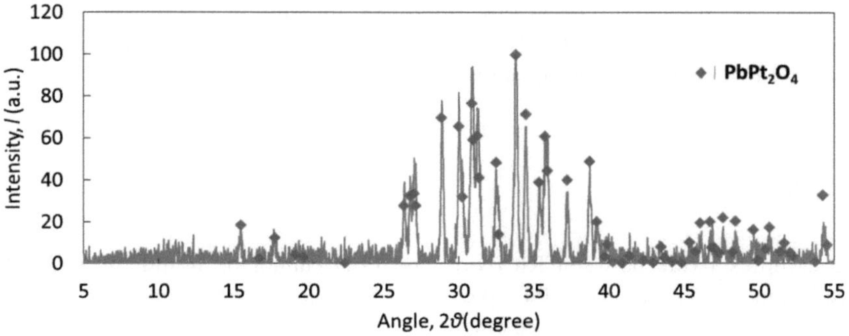

Fig. 2 XRD pattern of $PbPt_2O_4$

was heated at 973 K for 24 h. Figure 2 shows the XRD pattern of specimen after the heat treatment. $PbPt_2O_4$ was synthesized.

The dissolution rate of Pt from $PbPt_2O_4$ to concentrated hydrochloric acid was over 80%. Because that of pure platinum (heated platinum black) in same condition was 1–3%, it is indicated the dissolution rate of Pt is improved.

Pd-PbO

By XRD pattern, it is indicated $PbPdO_2$ was synthesized. In order to synthesis stoichiometric $PbPdO_2$, the mixture of platinum black and PbO in 1:1 in molar ratio was heated at 773 K for 1 h. Figure 3 shows the XRD pattern of specimen after the heat treatment. $PbPdO_2$ was synthesized.

The dissolution rate of Pd from $PbPdO_2$ to concentrated hydrochloric acid was 80–100%. Because that of pure platinum (heated platinum black) was ca. 30%, it is

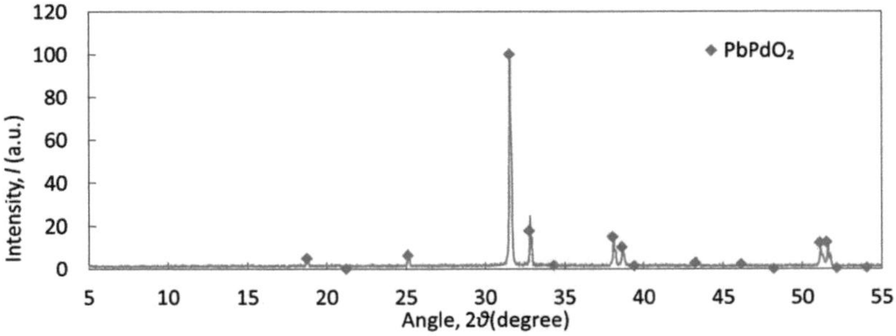

Fig. 3 XRD pattern of $PbPdO_2$

indicated the dissolution rate of Pt is improved. In addition, the dissolution rate to dilute hydrochloric acid (1, 0.1 N) was also 80–100%. This complex oxide can be dissolved by 0.1 N hydrotropic acid.

Rh-PbO

The complex oxide between Rh oxide and PbO cannot be found by XRD. The dissolution rate of Rh from the specimen after heating treatment was increased slightly from that without the treatment.

Conclusions

$PbPt_2O_4$, and $PbPdO_2$ were synthesized easily by heating with PbO at 773–973 K for 1–24 h in air atmosphere. Synthesized there complex oxides, dissolution rate of Pt or Pd is improved dramatically. However, in the case of Rh, complex oxide cannot be found by XRD.

References

1. Nomura K, Kageyama H (2013) Recycling technology for platinum group metals by using perovskite-type oxides. Materia Jpn 52(2):58–63
2. Matthey J (2013) Platinum 2013, UK

Part XI
Composites

High Thermal Conducting Composites Using Percolation Theory

Kenji Monden

Abstract A thermal solution composite is made by incorporating dense inorganic fillers with high thermal conductivity into a resin. Silica, alumina, aluminum nitride, silicon nitride, and boron nitride are used for fillers of thermal solution composite. The thermal conductivities of composites are measured and the effect of inorganic fillers' volume fraction is analyzed. The influence of percolation on thermal conductivity is considered. When percolation by random filling is assumed, the thermal conductivity of composite materials is expressed by the predictive expression using the effective thermal conductivity of filler. The relationship between the effective thermal conductivity of filler and its mean diameter is estimated by the equation. In the case of irregular AlN filler with average particle size 45 μm, the effective thermal conductivity of filler becomes 20 Wm^{-1} K^{-1}. When the volume fraction of the fillers is 0.75, the thermal conductivity of the composite has been reported to 8 Wm^{-1} K^{-1}. If the thickness of the interfacial region can be lowered, an effective larger thermal conductivity of the filler appears to set in and the thermal conductivity of the composite can be largely improved. The thermal conductivity of BN can approach 600 Wm^{-1} K^{-1}. The BN constitutes the matrix of composite by causing percolation without interfacial region; high thermal conductivity of 55–100 Wm^{-1} K^{-1} has been achieved.

Keywords Composite · Percolation · Thermal conductivity

Introduction

A thermal solution composite is made by incorporating dense inorganic fillers with high thermal conductivity into a resin. Because the composite materials have high thermal conductivities, they have been used in electrical products that generate excessive thermal energy during operation, such as inverters, amplifiers and motor

K. Monden (✉)
Denka Company Limited, Machida-shi, Tokyo, Japan
e-mail: kenji-monden@denka.co.jp

S. Ikhmayies et al. (eds.), *Characterization of Minerals, Metals, and Materials 2017*, The Minerals, Metals & Materials Series,
DOI 10.1007/978-3-319-51382-9_42

drivers. They function as electrically insulating and thermally conducting parts of the electric modules. As the uses of thermal solutions are expanded, composites of higher thermal conductivities will be required. This study was achieved by analyzing the effect of volume fraction of inorganic fillers and measuring the thermal conductivity. In addition, the influence of percolation on thermal conductivity is considered. The thermal conductivity of the composites is discussed by employing a predictive equation using literature data of thermal conductivities for both the filler and resin. As well as thermal conductivity measurements the effect of volume fraction of the filler was also explored.

Experimental

Specimen Preparation

The composite samples were prepared from silicone resin and aluminum nitride (AlN) fillers. The AlN fillers with an average particle size of 20 μm are shown in Fig. 1. A planetary centrifugal mixer was used for simultaneously dispersing the components uniformly and de-aerating the mixture. After mixing, these materials were placed in a silicone rubber case and cured at 150 °C for 1 h.

Measurement of Thermal Properties

Thermal diffusivity was measured using both the xenon flash (Flash) and temperature wave thermometric analysis (TWA) methods on sheets of the composite material. The material density is simply defined as mass divided by volume and a balance was used for measuring the mass. A gas pycnometer was used for measuring volume.

Fig. 1 Picture of aluminum nitride fillers

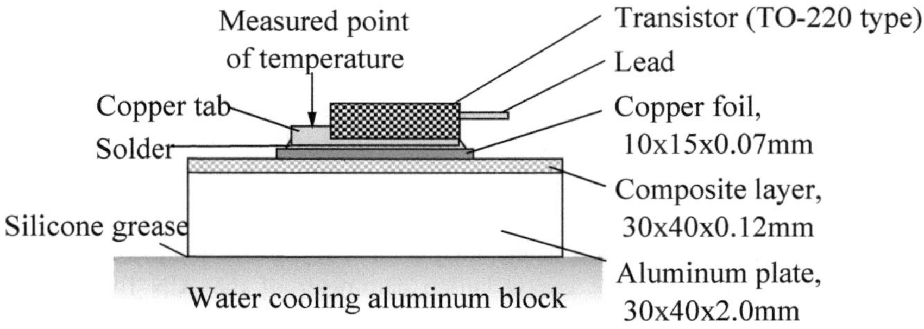

Fig. 2 Schematic view of the test assembly for thermal resistance measurement

Assembly Measurement Technique

A 0.12 mm-thick composite material layer is formed on a 30 × 40 × 2.0 mm-thick aluminum plate. After press laminating 0.07 mm-thick copper foil onto it, the resin is cured at 150 °C. The thermal conductivity of the composite layer is calculated from the measurement result of thermal resistance for the assembly set. The plastic molded TO-220 type transistor (2SC2233) that has the nickel plated copper tab is used to the measurement of thermal resistance. The transistor is attached on the copper pad by soldering. The temperature of the copper tab is measured by using a thermocouple. The schematic view of the assembly setup is shown in Fig. 2.

The temperature of a tab is increased by increasing the collector loss, power dissipation of a transistor. The thermal resistance is calculated from the relationship between temperature rise and power dissipation of the transistor. Some composite layers with different thickness were measured thermal resistance using the assembly set. The thermal conductivity of the composite Λ_1 is calculated by the following equation:

$$\Lambda_1 = \Delta x / (\Delta R \cdot A) \tag{1}$$

where Δx is the difference in thickness of composite layer, ΔR is the difference in thermal resistance, and A is the area of copper pad.

Results and Discussion

Thermal Conductivity of Fillers

Certain oxide binders are used for sintering AlN, the oxygen enters the AIN lattice and substitution occurs. Because the ionic radius of oxygen is smaller than that of

Table 1 Thermal conductivity of materials

Material	Thermal conductivity [W/(mK)]
AlN	159
Silicone resin	0.239

nitrogen, and because metal vacancies occur in the aluminum sites, the AlN lattice parameters should decrease as the oxygen content increases [1]. The c-axis lattice parameter of the AlN fillers was determined from X-ray diffraction pattern analysis. The relationship between the c-axis lattice parameter of the AlN filler and the oxygen impurities present is given as follows [2]:

$$\Delta c/\mathrm{c}_0 = -0.0516(\Delta n/\mathrm{n}_0). \tag{2}$$

Here, n_0 (=4.79 × 10^{28} m^{-3}) is the number of nitrogen atoms per unit volume in pure AlN, Δn is the number of nitrogen atoms that have been replaced by oxygen, c_0 (=4.9816 × 10^{-10} m) is the c-axis lattice parameter of pure AlN, and Δc is the displacement in c-axis lattice parameter caused by Δn. Because the decreases with increasing oxygen impurities, the difference in thermal conductivity between the AlN filler and pure AlN can be defined by the following equation [1]:

$$(1/\lambda_\mathrm{f}) - (1/\lambda_\mathrm{AlN}) = 0.9 \times 10^{-29}\Delta n. \tag{3}$$

Here, λ_AlN (=320 Wm^{-1} K^{-1}) is the thermal conductivity of pure AlN and λ_f is the thermal conductivity of the AlN filler. The predicted thermal conductivity of each material is shown in Table 1.

Thermal Conductivity of Composite Materials

Using the thermal diffusivity α measured by Flash or TWA, the density ρ and the specific heat c, the thermal conductivity λ can be defined by the following equation:

$$\lambda = \alpha\rho c. \tag{4}$$

The specific heat of the composite materials was evaluated by using those of AlN and silicone resin and the volume fraction of AlN fillers. The evaluated thermal conductivity of silicone resin is shown in Table 1. The relationship between the specific heat of the composite materials and volume fraction of AlN filler is shown in Fig. 3.

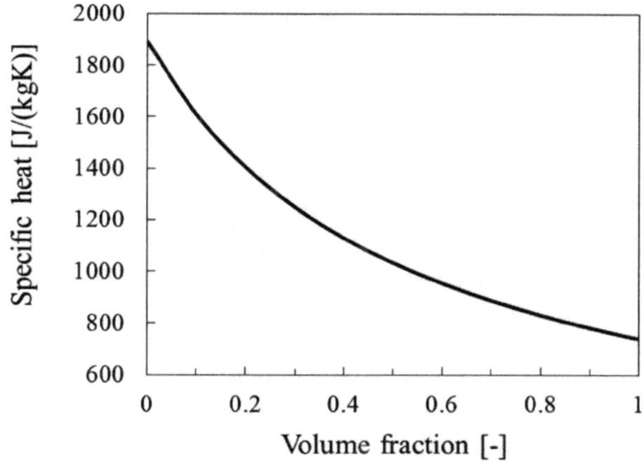

Fig. 3 Relationship between specific heat of composites and volume fraction of AlN filler

Effect of Volume Fraction of Fillers

Composite resin sheets with a volume fraction from 0.4 to 0.6 of AlN were prepared. The thermal diffusivity of these samples was measured by either Flash or TWA. The thermal conductivity was estimated by using Eq. (4) with measured density and estimated specific heat. The relationship between the thermal conductivity of composite materials and volume fraction of AlN filler is shown in Fig. 4. When the volume fraction of the fillers is 0.5 or less, there is very little difference between the thermal conductivity measured by TWA and that measured by Flash. For the 0.6 volume fraction samples, the values given by TWA are much larger than those given by the Flash method. It is thought that TWA is more sensitive than Flash to the percolation thermal conduction pass in these composite materials. The dependency of thermal conductivity of the composite material on

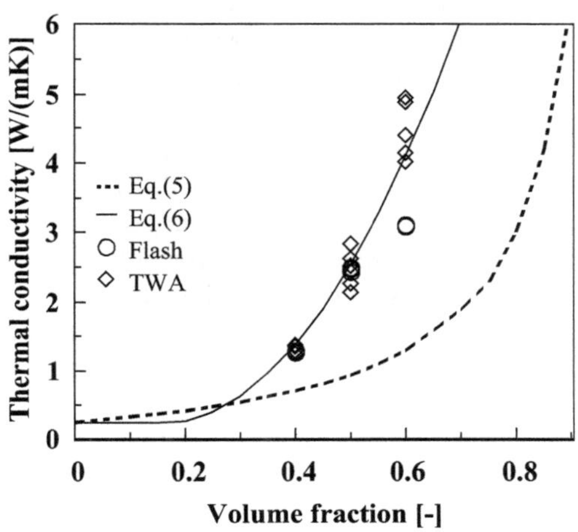

Fig. 4 Relationship between the thermal conductivity of the composite materials and volume fraction of AlN filler

volume fraction of filler is predicted in the effective medium model [3]. The effective medium model is expressed by the following formula:

$$\lambda_5 = \lambda_m + V_f/[1/(\lambda_f - \lambda_m) + (1 - V_f)/3\lambda_m]\,. \tag{5}$$

Here, V_f is the volume fraction of filler, λ_f is the thermal conductivity of filler and λ_m is the thermal conductivity of the matrix. In Fig. 4, the dashed line indicates the result of Eq. (5) using the thermal conductivities that are shown in Table 1. It exhibits considerably lower thermal conductivities than those measured experimentally because the influence of percolation is not considered in Eq. (5). However, the influence of the percolation cannot be ignored in composite materials with a high volume fraction of fillers. When percolation by random filling is assumed, the thermal conductivity of composite materials is expressed by the following formula [4]:

$$\lambda_6 = \lambda_m + \lambda_0\left(V_f - V_p\right)^k \quad \left(V_f > V_p\right). \tag{6}$$

Here, k is the critical index, V_p is the percolation threshold volume fraction and λ_0 is the effective thermal conductivity of filler. The relationship between the effective thermal conductivity of the filler and its mean diameter d_f is estimated by the following equation [5]:

$$\lambda_0 = \lambda_m + (1 + 2t_m/d_f)^{-3}/\{1/(\lambda_f - \lambda_m) + [1 - (1 + 2t_m)/d_f)^{-3}]/3\lambda_m\}. \tag{7}$$

Here, t_m is thickness of the interfacial region around filler. The thickness of the interfacial region of the spherical filler can be estimated as 150 nm [6]. In the case of the AlN (average particle size 20 μm) λ_0 becomes 20 $Wm^{-1}\,K^{-1}$. When $V_p = 0.16$ and $k = 2$ are used in Eq. (6), the predicted result for the AlN filler is shown as a solid line in Fig. 4. The predictive expression Eq. (6) accurately fits the experimental data measured by TWA.

The relationship between the thermal conductivity given by the assembly measurement technique (AMT) and the volume fraction of filler is shown in Fig. 5. These data are for the high volume fraction of AlN filler. These data fit the predictive expression Eq. (6) and the experimental data measured by TWA.

The thermal conductivities of resin composite materials with different volume fractions of AlN were evaluated. The thermal conductivity as measured by temperature wave thermometric analysis was larger than that measured by the xenon flash at high filler volume fractions. The thermal conductivities given by the assembly measurement technique fit the predictive expression and the experimental data measured by temperature wave thermometric analysis. Thermal conductivity behavior was successfully modeled using a predictive expression that took percolation into account.

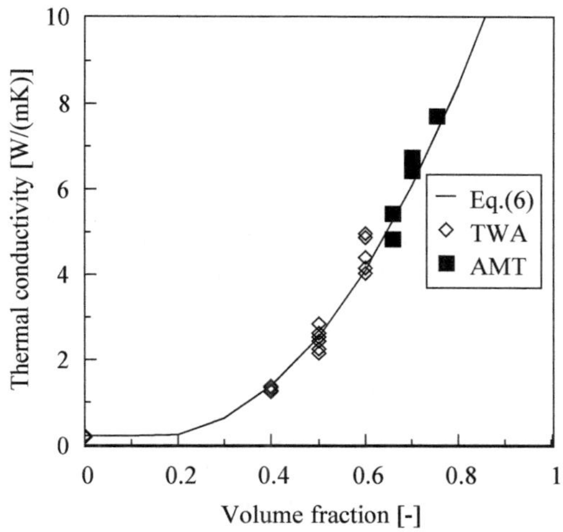

Fig. 5 Relationship between the thermal conductivities given by TWA and AMT and volume fraction of AlN filler

In the case of irregular AlN filler with average particle size 45 μm, λ_0 becomes 20 Wm^{-1} K^{-1}. When the volume fraction of the fillers is 0.75, the thermal conductivity of the composite has been reported to 8 Wm^{-1} K^{-1} [6]. If the thickness of the interfacial region can be lowered, an effective larger thermal conductivity of the filler appears to set in and the thermal conductivity of the composite can be largely improved. The thermal conductivity of BN can approach 10–600 Wm^{-1} K^{-1} [7]. The BN constitutes the matrix of composite by causing percolation without interfacial region; high thermal conductivity of 55–100 Wm^{-1} K^{-1} has been achieved [8].

Conclusions

The thermal conductivities of resin composite materials with different volume fractions of AlN were evaluated. The thermal conductivity as measured by temperature wave thermometric analysis was larger than that measured by the xenon flash at high filler volume fractions. The thermal conductivities given by the assembly measurement technique fit the predictive expression and the experimental data measured by temperature wave thermometric analysis. Thermal conductivity behavior was successfully modeled using a predictive expression that took percolation into account.

Acknowledgements The help of Mr. T. Asami, Denka Innovation Center, Denka Company Limited, in acquiring composite thermal properties is gratefully acknowledged.

References

1. Slack GA (1973) Nonmetallic crystals with high thermal conductivity. J Phys Chem Solid 34:321–335
2. Sato Y et al (1990) The relation between oxygen content of aluminum nitride and its thermal conductivity. In: Proceedings of annual meeting ceramic society of Japan, Kobe, p 10
3. Hashin Z (1983) Analysis of composite materials—a survey. J Appl Mech 50:481–505
4. Monden K (2008) Effect of percolation on the thermal conductivity in resin-matrix composites. In: Proceedings of 29th Japan symposium on thermophysical properties, Tokyo, pp 324–326
5. Monden K (2006) Equivalent thermal conductivity of insulating layer in insulated metal substrates. Adv Sci Technol 45:2664–2669
6. Monden K, Okajima Y, Yashima K (2011) Application of percolation theory and development of high conducting materials in a compound thermal component. J Jpn Inst Electron Packag 14:485–491
7. Duclaux L et al (1992) Structure and low-temperature thermal conductivity of pyrolytic boron nitride. Phys Rev B 46:3362–3367
8. Hirotsuru H, Terasaki R, Loutfy K (2015) Development of metal-ceramic composite materials to heat dissipation parts. Fine ceramics report 33:42–46

Sorption Characteristics of Low Density Polyethylene/Kola Nut Composite

G.C. Onuegbu, M.U. Obidiegwu and G.O. Onyedika

Abstract The effect of kolanut (KN) filler and compatibilizer (maleic anhydride) contents on the transport of toluene through low density polyethylene (LDPE) and low density polyethylene composite were studied at the temperatures 40, 60 and 80 °C.The kolanut powder used was ground to 25 μm particle size. In preparation of composite 0–5 wt% of kolanut powder and 0–2.5 wt% of the compatibilizer were mixed with 200 g of low density polyethylene in an injection moulding machine. The injected polyethylene and polyethylene/kolanut composites (LDPE/KN) were taken for sorption test. The sorption data obtained for low density polyethylene and kolanut powder filled low density polyethylene at the different temperature investigated were expressed as the molar percentage uptake (% Q_t) of toluene per gram of the kola nut composites. Results show that the uptake of toluene at the five filler content (0–5 wt%) generally increased with increase in sorption temperature but decreased with increase in compatibilizer and filler contents. This can be seen at 80 °C and 0, 1, 1.5 and 2.5 wt% of compatibilizer respectively. The molar percentage uptake of toluene at 1 wt% of filler yielded the following values; 8.6789, 8.4768, 8.3476 and 8.2679 respectively. At 2 wt% it yielded 8.6224, 8.1333, 7.9525 and 7.9042 respectively, at 3 wt% it gave 8.4336, 7.9450,7.8900 and 7.7823 respectively, at 4 wt% it yielded 8.0263,7.9614,7.8218 and 7.6800 respectively and at 5 wt% it gave 7.9635, 7.8630, 7.6256 and 7.5013 respectively.

Keywords Sorption · Kolanut · Low density · Polyethylene · Composite · Compatibilizer

G.C. Onuegbu (✉) · M.U. Obidiegwu
Department of Polymer and Textile Engineering,
Federal University of Technology, Owerri, Nigeria
e-mail: gc.onuegbu@yahoo.com

G.O. Onyedika
Department of Chemistry, Federal University of Technology, Owerri, Nigeria

S. Ikhmayies et al. (eds.), *Characterization of Minerals, Metals, and Materials 2017*, The Minerals, Metals & Materials Series,
DOI 10.1007/978-3-319-51382-9_43

Introduction

In order to overcome the draw backs of the addition of fillers, compatibilizer has to be added to reduce the repellence of the polymers and fillers respectively, thereby increasing the adhesive power between the polymer matrix and the filler which can enhance the properties, flexibility and solubility of the final mixture [1]. Such compatibilizers have to be on one hand compatible with the polymer, ideally they have to be the same chemistry as the polymer and on the other hand they have to interact as a glue to the filler. The most commonly used compatibilizers are titanates, silananes, zirconates and organic acid, chromium chloride coordination complex.

Numerous studies have used graft copolymers of synthetic polymers (examples are polypropylene and polyethylene) and maleic anhydride as coupling agents for thermoplastic composites (Kazayawoko 1999) [2]. In order to improve the similarity and adhesion between filler and thermoplastic matrices, several chemicals have been employed and Maleated coupling agents were found most suitable for organic fillers [3]. The coupling agent maleic anhydride grafted polypropylene and polyethylene can modify the interface by interacting with both the filler and the matrix, thus forming a link between the components [4]. Kokta and Raj [5] studied the use of silanes as coupling agents with different combination of solvent and additives in thermoplastic composites.

The effect of compatibilizer on the transport characteristics of polymers has been studied [6]. They studied jute fibre filled polypropylene and reported that the toluene absorption values of the composites increased with increase in the jute loadings and in the presence of maleic anhydride grafted polypropylene (compatibilizer). The amount of water absorbed was found to decrease with increases in compatibilizer contents. Similarly [7] filled wood sample at 175, 190 and 205 °C with different amounts of high density polyethylene and coupling agents. Toluene sorption thickness swelling and diffusion coefficients of composites were measured for a 4-week period. Immersion in water showed that composites with wood treatment at 190 and 205 °C had considerable high ware resistance. Addition of coupling agent reduced water sorption, thickness swelling and diffusion co-efficient. Diffusion co-efficient is more pronounced in composites with untreated wood. Treating wood at 190 °C resulted in good flexural properties and excellent water resistance. Andizeg et al. [8] reinforced polypropylene with wood fibre at different filler loadings in the presence of coupling agents (Maleic anhydride- grafted-polypropylene) and discovered that addition of coupling agent decreased the water sorption properties of the composites. Siti and Supri [9] investigated the effect of chemical modification on the mechanical properties, morphology and water absorption of low density polyethylene/egg shell powder composites. They concluded that the tensile strength and water sorption resistance of the modified low density polyethylene/Egg shell powder composites are higher than that of unmodified low density polyethylene/Egg shell powder composites.

This research work involves the use of kolanut in filling low density polyethylene due to the presence of N-nitroso compound [10]. The objective of this is to investigate the effect of kolanut filler contents, compatibilizer and temperature on the transport of toluene through low density polyethylene and kolanut filled low density polyethylene (the composite).

Materials and Methods

Materials

The low density polyethylene (LDPE) used in this study was bought from CEE PLAST Industry Limited, Osisioma Ngwa, Aba, Abia State, Nigeria. It is a product of INDORAMA group, a subsidiary of Eleme Petrochemical Company Limited, Rivers State, Nigeria. It has a melt flow index of 0.4 g/10 min at 190 °C and density of 0.922 g/cm ^{3}. The filler used in filling the low density polyethylene is kolanut. It was bought locally from Ezinihitte Mbaise, Imo State, Nigeria. This was sun dried, crushed to fine powder and sieved to 25 μm mesh size.

Compatibilizer: Maleic anhydride-grafted–polyethylene (MAPE) was used as compatibilizer.

Reagent: Toluene was obtained from stigma Aldrich Company, U.S.A.

Preparation of Polyethylene Composites

The LDPE composites of kola nut powder (KN) were prepared by thoroughly mixing 200 g of low density polyethylene with approximate filler quantities (0, 1, 2, 3, 4 and 5 wt%.). The low density polyethylene and kolanut/low density polyethylene were melted and homogenized with filler in an injection moulding machine at 173 °C and the resultant composites were produced as sheets.

Testings

In determination of sorption characteristics, 0.2 g of low density polyethylene and low density polyethylene composite samples were cut, weighed and place in sample bottles filled with twenty millilitres of toluene. The thermostatic water bath at 40, 60 and 80 °C was used for the sorption experiment. The samples were brought out after equilibration at the required temperature and time ranging from 30–510 min at 30 min intervals.

The wet samples were reweighed and the difference in mass between the dry and the wet samples were recorded. The molar percentage uptake of toluene per gram of the LDPE and LDPE/KN composites (Q_t) was expressed as:

$$Q_t = \frac{\Delta M}{M_0} \times 100 \quad (1)$$

where

ΔM Change in mass (difference between the dry and wet samples)
M_0 mass of dry sample (g)

Results and Discussion

Sorption Characteristics

Sorption Characteristics of LDPE and LDPE/KN at 0–5 wt% and 2 μm particle size were studied at the following temperatures 40, 60 and 80 °C. The sorption data obtained for the low density polyethylene and the composites at the different temperatures investigated were expressed as the molar percentage uptake (% Q_t) of toluene per gram of the kola nut composites. The molar percentage uptake (% Qt) at any particular temperature and filler contents of the compatibilized LDPE and LDPE/KN composites used were plotted against the square root of time and are shown in Figs. 1, 2, 3, 4, 5, 6, 7, 8, 9, 10, 11, 12, 13, 14 and 15.

Figure 1 shows the effect of compatibilizer and 1 wt% of kolanut filler on the molar percentage uptake of toluene in the low density polyethylene and low density polyethylene composite at 40 °C. This figure shows that the molar percentage

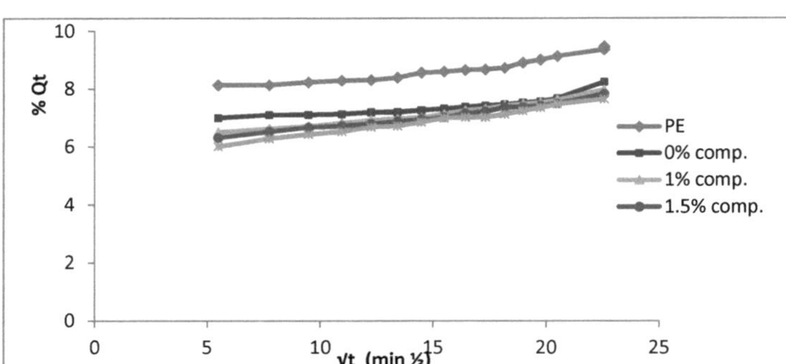

Fig. 1 Plot of % Q_t versus $\sqrt{t}$ (min ½) for kolanut powder (1 wt% of filler) filled low density polyethylene in toluene at different compatibilizer contents at 40 °C

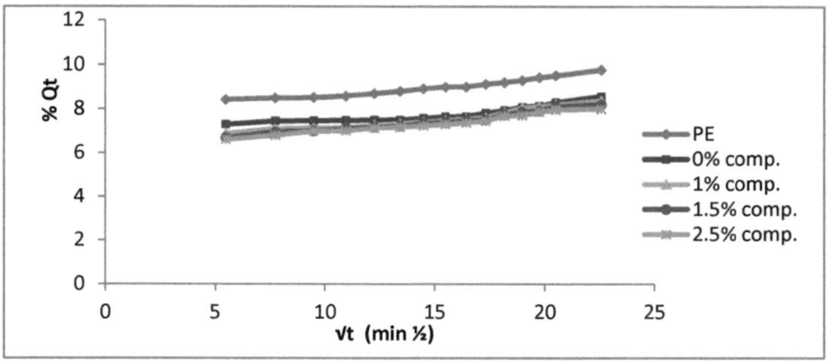

Fig. 2 Plot of % Q_t versus $\sqrt{t}$ (min ½) for kolanut powder (1 wt% of filler) filled low density polyethylene in toluene at different compatibilizer contents at 60 °C

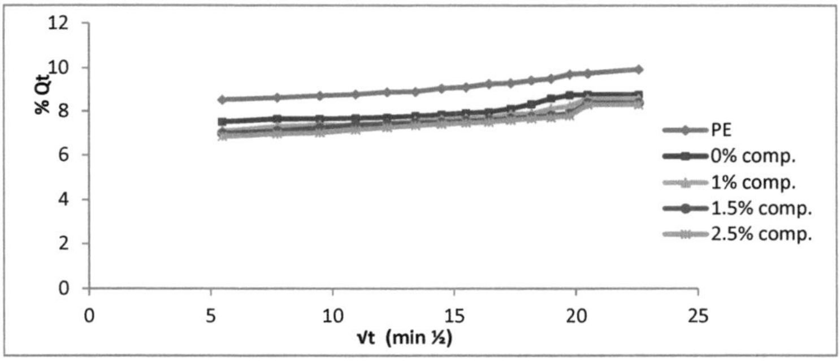

Fig. 3 Plot of % Q_t versus $\sqrt{t}$ (min ½) for kolanut powder (1 wt% of filler) filled low density polyethylene in toluene at different compatibilizer contents at 80 °C

uptake increased with time and temperature but decreased with increase in kolanut powder and compatibilizer contents.

Figure 2 shows the effect of compatibilizer and 1 wt% of kolanut filler on the molar percentage uptake of toluene in the low density polyethylene and low density polyethylene composite at 60 °C. This figure shows that the molar percentage uptake increased with time and temperature but decreased with increase in kolanut and compatibilizer contents

Figure 3 shows the effect of compatibilizer and 1 wt% of kolanut filler on the molar percentage uptake of toluene in the low density polyethylene and low density polyethylene composite at 80 °C. This figure shows that the molar percentage uptake increased with time and temperature but decreased with increase in kolanut and compatibilizer contents.

Figure 4 shows the effect of compatibilizer and 2wt% of kolanut filler on the molar percentage uptake of toluene in the low density polyethylene and low density polyethylene composite at 4 °C. This figure shows that the molar percentage uptake increased with time and temperature but decreased with increase in kolanut and compatibilizer contents.

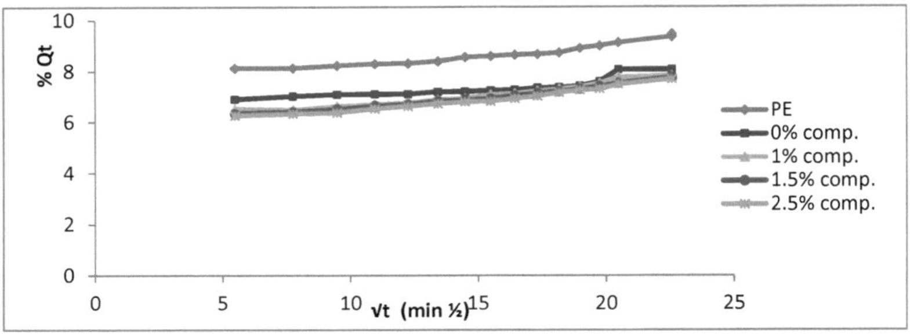

Fig. 4 Plot of % Q_t versus $\sqrt{t}$ (min ½) for Kolanut powder (2 wt% of filler) filled low density polyethylene in toluene at different compatibilizer contents at 40 °C

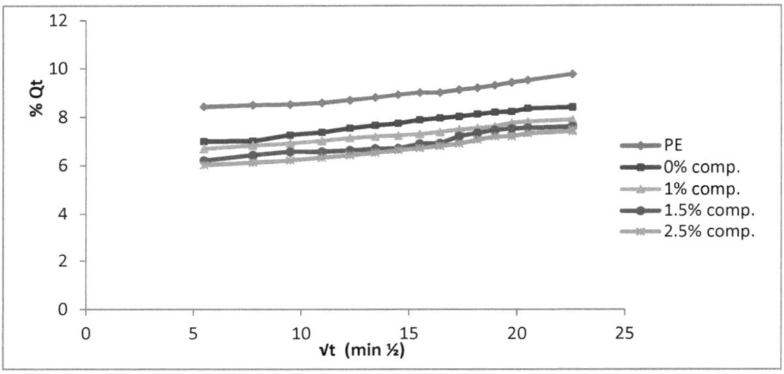

Fig. 5 Plot of % Q_t versus $\sqrt{t}$ (min ½) for kolanut powder (2 wt% of filler) filled low density polyethylene in toluene at different compatibilizer contents at 60 °C

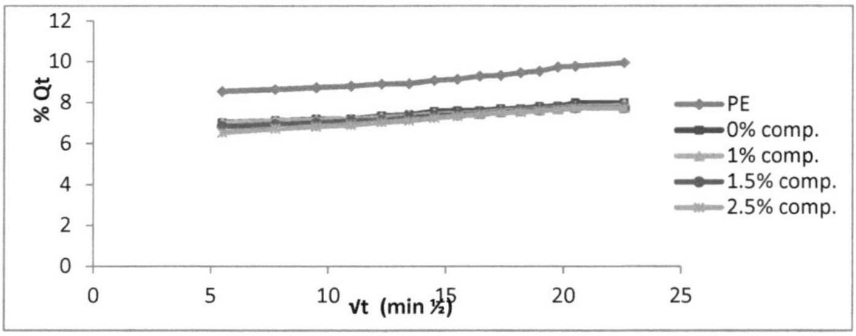

Fig. 6 Plot of % Q_t versus $\sqrt{t}$ (min ½) for kolanut powder (2 wt% of filler) filled low density polyethylene in toluene at different compatibilizer contents at 80 °C

Figure 5 shows the effect of compatibilizer and 2 wt% of kolanut filler on the molar percentage uptake of toluene in the low density polyethylene and low density polyethylene composite at 60 °C. The figure shows that the molar percentage

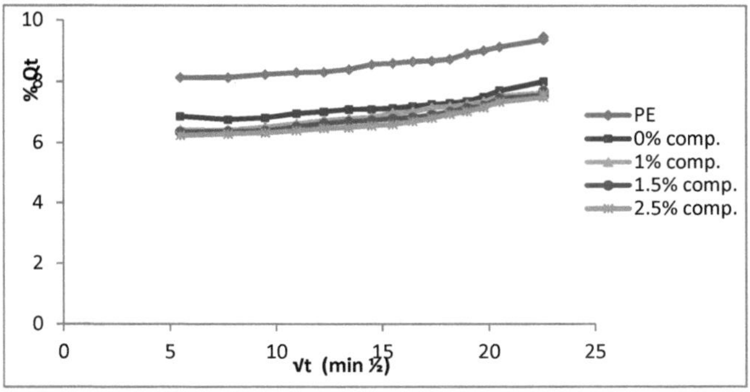

Fig. 7 Plot of % Q_t versus $\sqrt{t}$ (min ½) for kolanut powder (3 wt% of filler) filled low density polyethylene in toluene at different compatibilizer contents at 40 °C

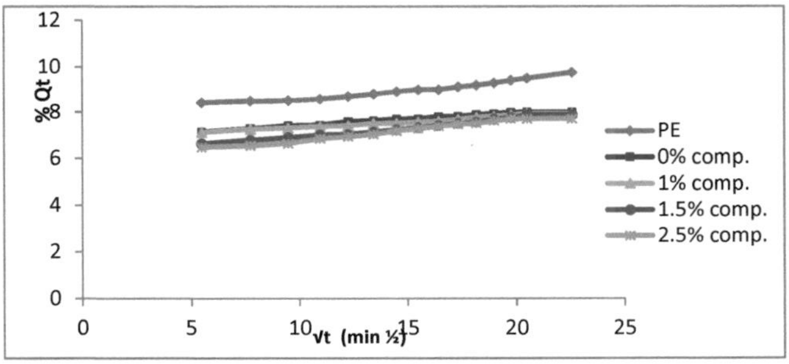

Fig. 8 Plot of % Q_t versus $\sqrt{t}$ (min ½) for kolanut powder (3 wt% of filler) filled low density polyethylene in toluene at different compatibilizer contents at 60 °C

uptake increased with time and temperature but decreased with increase in kolanut and compatibilizer contents.

Figure 6 shows the effect of compatibilizer and 2 wt% of kolanut filler on the molar percentage uptake of toluene in the low density polyethylene and low density polyethylene composite at 80 °C. This figure shows that the molar percentage uptake increased with time and temperature but decreased with increase in kolanut and compatibilizer contents.

Figure 7 shows the effect of compatibilizer and 3 wt% of kolanut filler on the molar percentage uptake of toluene in the low density polyethylene and low density polyethylene composite at 40 °C. This figure shows that the molar percentage uptake increased with time and temperature but decreased with increase in kolanut and compatibilizer contents.

Figure 8 shows the effect of compatibilizer and 3wt% of kolanut filler on the molar percentage uptake of toluene in the low density polyethylene and low density polyethylene composite at 60 °C. This figure shows that the molar percentage uptake increased with time and temperature but decreased with increase in kolanut and compatibilizer contents.

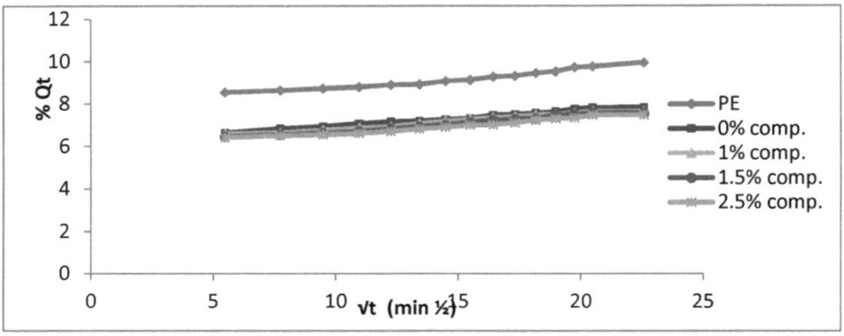

Fig. 9 Plot of % Q_t versus $\sqrt{t}$ (min ½) for Kolanut powder (3 wt% of filler) filled low density polyethylene in toluene at different compatibilizer contents at 80 °C

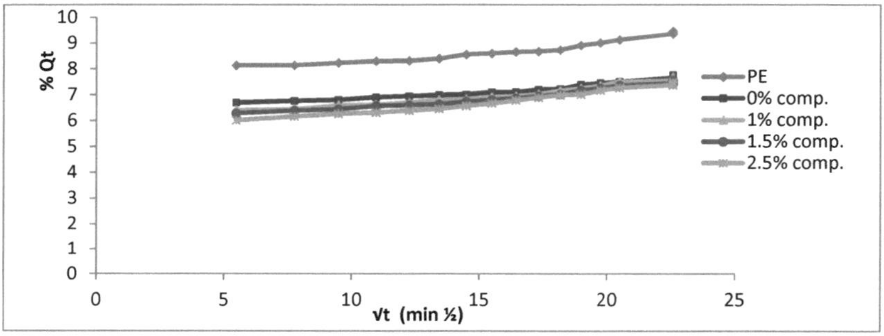

Fig. 10 Plot of % Q_t versus $\sqrt{t}$ (min ½) for olanut powder (4 wt% of filler) filled low density polyethylene in toluene at different compatibilizer contents at 40 °C

Figure 9 shows the effect of compatibilizer and 3 wt% of kolanut filler on the molar percentage uptake of toluene in the low density polyethylene and low density polyethylene composite at 80 °C. This figure shows that the molar percentage uptake increased with time and temperature but decreased with increase in kolanut and compatibilizer contents.

Figure 10 shows the effect of compatibilizer and 4 wt% of kolanut filler on the molar percentage uptake of toluene in the low density polyethylene and low density polyethylene composite at 40 °C. This figure shows that the molar percentage uptake increased with time and temperature but decreased with increase in kolanut and compatibilizer contents.

Figure 11 shows the effect of compatibilizer and 4 wt% of kolanut filler on the molar percentage uptake of toluene in the low density polyethylene and low density polyethylene composite at 60 °C. The figure shows that the molar percentage uptake increased with time and temperature but decreased with increase in kolanut and compatibilizer contents.

Figure 12 shows the effect of compatibilizer and 4 wt% of kolanut filler on the molar percentage uptake of toluene in the low density polyethylene and low density

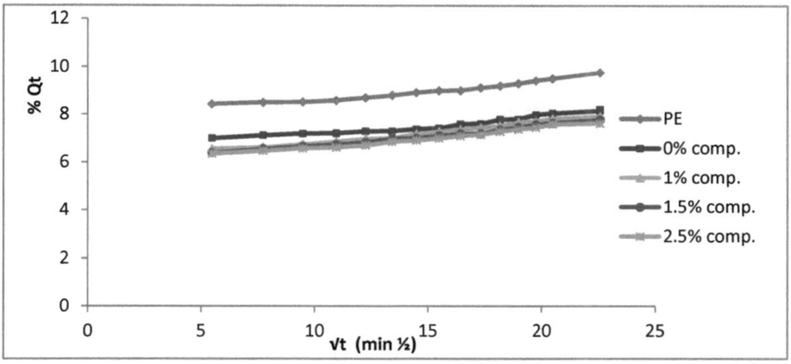

Fig. 11 Plot of % Q_t versus $\sqrt{t}$ (min ½) for kolanut powder (4 wt% of filler) filled low density polyethylene in toluene at different compatibilizer contents at 60 °C

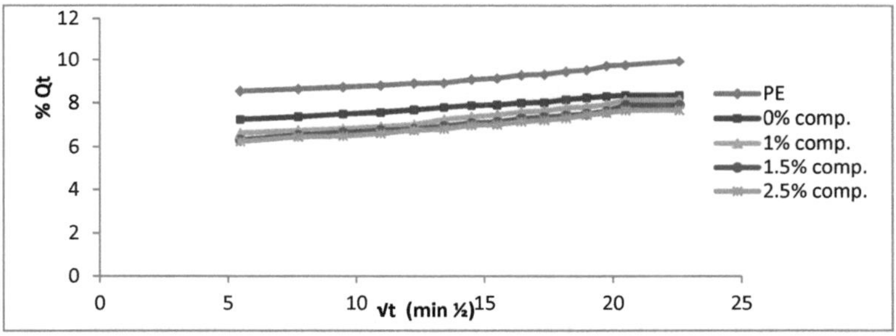

Fig. 12 Plot of % Q_t versus $\sqrt{t}$ (min ½) for kolanut powder (4 wt% of filler) filled low density polyethylene in toluene at different compatibilizer contents at 80 °C

polyethylene composite at 80 °C. This figure shows that the molar percentage uptake increased with time and temperature but decreased with increase in kolanut and compatibilizer contents.

Figure 13 shows the effect of compatibilizer and 5 wt% of kolanut filler on the molar percentage uptake of toluene in the low density polyethylene and low density polyethylene composite at 40 °C. This figure shows that the molar percentage uptake increased with time and temperature but decreased with increase in kolanut and compatibilizer contents.

Figure 14 shows the effect of compatibilizer and 5 wt% of kolanut filler on the molar percentage uptake of toluene in the low density polyethylene and low density polyethylene composite at 60 °C. This figure shows that the molar percentage uptake increased with time and temperature but decreased with increase in kolanut and compatibilizer contents.

Figures 1, 2, 3, 4, 5, 6, 7, 8, 9, 10, 11, 12, 13, 14 and 15 show initial increased in the mass of the toluene sobbed and the toluene uptake for LDPE and LDPE/KN at the five filler contents (0–5 wt%) are generally increased with increases in

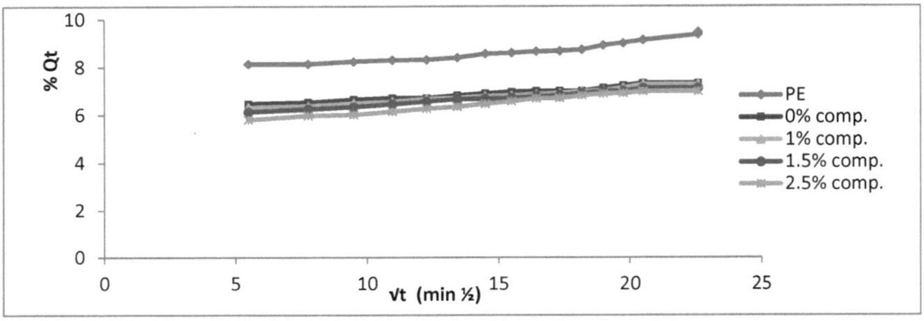

Fig. 13 Plot of % Q_t versus $\sqrt{t}$ (min½) for kolanut powder (5 wt% of filler) filled low density polyethylene in toluene at different compatibilizer contents at 40 °C

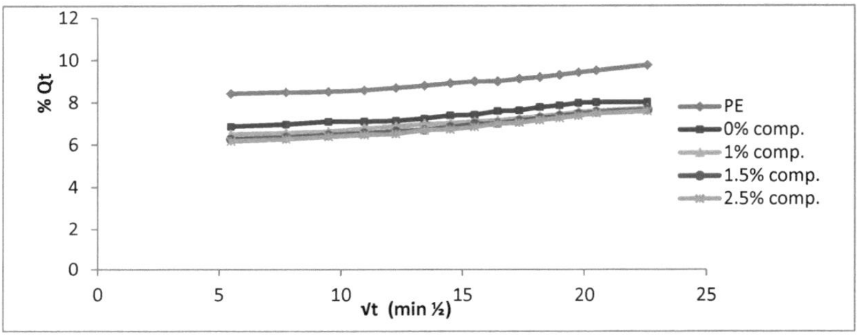

Fig. 14 Plot of % Q_t versus $\sqrt{t}$ (min ½) for kolanut powder (5 wt% of filler) filled low density polyethylene in toluene at different compatibilizer contents at 60 °C

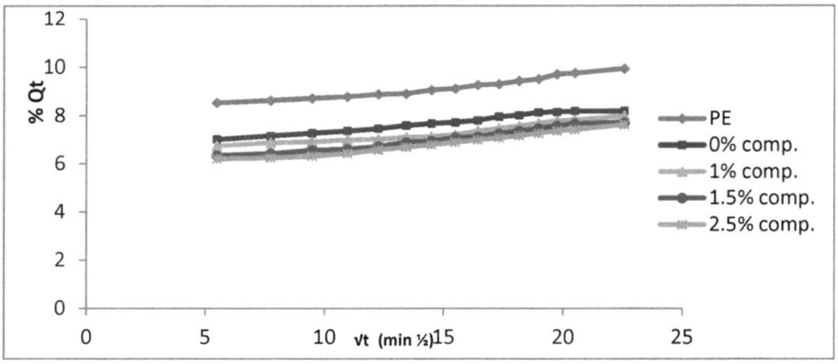

Fig. 15 Plot of % Q_t versus $\sqrt{t}$ (min ½) for Kolanut powder (5 wt% of filler) filled low density polyethylene in toluene at different compatibilizer contents at 80 °C

temperature, this is due to the increase in heat energy of the solvent and the increase in mobility of the polymer. This was observed by John and Thomas [11] who studied the effect of epoxidation on the transport behaviour and mechanical properties of natural rubber, he reported that the mole percentage uptake of hexane in epoxidized rubber showed increased with increases in temperature, Obasi et al. [12] who investigated the transport of toluene through linear low density polyethylene/natural rubber blend and Ismail and Suzaimah [13] who studied the swelling properties styrene butadiene rubber/epoxidized natural blends.

Figures 1, 2, 3, 4, 5, 6, 7, 8, 9, 10, 11, 12, 13, 14 and 15 also show that the sorption of the toluene into LDPE was higher than those of compatibilized LDPE/KN because LDPE is an amorphous polymer. It was found that molar percentage uptake of the composites for the toluene at filler contents (0–5 wt%) were generally decreased with increase in compatibility contents at any temperature. This is due to the strong interface between the filler and matrix which reduced the free volume of the composites. It is in agreement with the works of Alireza and Karl [7] who filled wood sample at 175, 190 and 205 °C with different amounts of high density polyethylene and coupling agents.

The figures also show that the sorption of LDPE/KN composites decreased with increase in kola nut filler contents due to increase in cross link density. This observation is in agreement with the work of Ahmed et al. [14] who investigated sorption properties of natural rubber/linear low density polyethylene blend in toluene.

Conclusion

The effect of compatibilizer on the sorption characteristics of kola nut powder filler low density polyethylene was studied and The molar percentage uptake (%Q) of toluene in the low density polyethylene and kolanut filled LDPE was found to show increase in the mass of the absorbed toluene with time at a given temperature and filler contents. Toluene uptake of LDPE/KN composites increased with increase in sorption temperature but decreased with an increase in compatibilizer and filler contents.

References

1. Ismail H (2010) Study on the effect of coupling agent on the mechanical properties of composite from kenaf and recycled polypropylene. J Mech Eng 6:201–206
2. Maldas D, Kokta BV (1994) An investigation of the interfacing adhesion between reclaimed newspaper and recycled polypropylene composites mechanical properties. J Adhes Sci Technol 8(12):1439–1451
3. Mengeloglu F, Kurt R, Gardnes DJO, Neille S (2007) Mechanical properties of extruded high density polyethylene (HDPE) and polypropylene (PP) wood flour decking bard. Iran Polym J 1007:477–487
4. Felix JM, Gatenholm P (1991) The nature of adhesion in composites of modified cellulose fibre and polypropylene. J Appl Polym Sci 42:609–620
5. Kokta BV, Raj RG (1991) The effect of processing conditions and binding material on the mechanical properties of bagasse fibre composites. J Polym Sci 35:204–216

6. Boggoera G, Dekanski A, Park V, Poleti D, Grozdanor A, Buzarovskam A, Avelle M, Gentile G (2006) Natural fibres in polymer composite materials. J Appl Polym Sci 42: 609–620
7. Alireza K, Karl RE (2010) Composite materials and engineering center. Washington State University Pullman, USA, 9, 9164–1806
8. Andizeg K, Bledzki RO, Omar F (2009) Effect of fibre and coupling agent on physical and mechanical properties of wood fibre reinforced polypropylene composite. J Appl Polym Sci 11:57–70
9. Siti S, Supri AC (2006) LDPE-isophthalic acid-modified egg shell powder composites. Polym Compos UNIMAP 5:25–32
10. Atawodi SE, Mendel PP, Funstein B, Pressmann R, Spiegelhalder B (1995) Nitrosatable & nitrosamide formation in natural stimulants, cola acuminate and garciccola. Food Chem Toxicol 33(8):625–630
11. John T, Thomas S (2000) The effect of epoxidation on the molecular transport behavior and mechanical properties of natural rubber. J Polym Sci 41(20):7511–7522
12. Obasi HC, Ogbobe O, Igwe IO (2009) Diffusion characteristics of toluene into natural rubber/linear low density polyethylene blends. Int J Polym Sci 23:132–142
13. Ismail H, Suzaimah S (2000) Styrene butadiene rubber/epoxidized natural blends, dynamic properties, curing characteristics and swelling studies. Polym Testing 19:879
14. Ahmed AM, Mohd DH, Abdullah H (2004) Mechanical properties of filled natural rubber/linear low density polymer. J Appl Polym Sci 13:31–73
15. Kazayawoko M, Balatinecz JJ, Wood Hans RT, Law S (2006) Effect of wood fibre surface chemistry on the mechanical properties of wood fibre-polypropylene composites. Int J Polym Mater 37:237–261
16. Smith MT (2010) Advances in understanding benzene health, effects and susceptibility. Ann Rev Pub Health 31:133–148
17. Zhang J (2011) Effect of coupling agent- keener et al maleated coupling agents for natural fibre composites. Eastman Chemical Company, Specialty Polymers Development and Technical Service. Langview, TX, U.S.A., pp 357–362

Residual Stress Analysis Within Steel Encapsulated Metal Matrix Composites Via Neutron Diffraction

Sean Fudger, Dimitry Sediako, Prashant Karandikar and Chaoying Ni

Abstract Neutron diffraction measurements were performed on steel encapsulated metal matrix composites (MMCs) in order to quantify bulk residual stresses. A coefficient of thermal expansion (CTE) mismatch induced residual compressive stress method is utilized as a means of improving the ductility of the MMCs and overall efficiency of several macro hybridized materials systems. Systems consisting of an A36, 304 stainless steel, or Nitronic® 50 stainless steel shell filled with an Al-SiC or Al-Al_2O_3 metal matrix composite are evaluated in this work. Upon cooling from processing temperatures residual strains are generated due to a CTE mismatch between each of the phases: steel, aluminum, and reinforcement. The analysis shows variation in the measured strain and stress results due to outer steel thickness, difference in CTE between materials, and relative position within the composite.

Keywords Encapsulation · Metal matrix composite · Neutron diffraction

S. Fudger (✉) · C. Ni
Department of Materials Science and Engineering, University of Delaware,
201 Du Pont Hall, Newark, DE 19716-1501, USA
e-mail: fudgersj@udel.edu

S. Fudger
Army Research Laboratory, Aberdeen Proving Ground, MD 21005, USA

D. Sediako
Canadian Nuclear Laboratories, Canadian Neutron Beam Centre,
Chalk River, ON K0J 1J0, Canada

P. Karandikar
M Cubed Technologies, Inc., 1 Tralee Industrial Park, Newark, DE 19711, USA

S. Ikhmayies et al. (eds.), *Characterization of Minerals, Metals, and Materials 2017*, The Minerals, Metals & Materials Series,
DOI 10.1007/978-3-319-51382-9_44

Introduction

The objective of this research is to develop a low cost/light weight macro composite materials system with increased specific stiffness, increased strength, increased damage tolerance, reduced density, and greater structural efficiency for structural applications. Since residual stresses present in the materials can result in improved mechanical properties, it is necessary to accurately quantify these stresses. Traditional laboratory X-Ray Diffraction for Residual Stress Analysis (XRD-RSA) was performed on several of the materials. As XRD is a superficial measurement method, neutron diffraction was performed as a means of measuring bulk material stresses within both the steel and MMC materials.

Three different macro composite material systems were evaluated via neutron diffraction. Each system consists of an identical MMC core consisting of an Al-10Si matrix alloy reinforced with 55 volume percent particulate silicon carbide (SiC). Two steels are evaluated in this work: A36 steel and 304 stainless steel, which have significantly different mechanical properties and CTE. By selecting a steel with a higher coefficient of thermal expansion than its MMC core, a residual compressive stress is introduced onto the core due to contraction of the steel upon cooling from processing temperatures. The ability to tailor a metal matrix composite's CTE by adjusting reinforcement percentage allows us to select a CTE mismatch which results in improved mechanical properties.

Experimental Procedure

The steel encapsulated metal matrix composites produced for this work are formed by casting or infiltrating a molten MMC into a steel shell. The casting/infiltration processes require no additional filler material, can be incorporated into complex geometries, and are produced at a relatively low cost [1]. Steel (A36, 304, and Nitronic® 50) bar with a diameter of greater than 1.0″ was bored to produce a tube with an inside diameter of 0.375″. The steel tubes were then processed to fill the internal void with various compositions of particulate MMCs, Al-SiC, Al-Al_2O_3, and Mg-B_4C. Finally, the outer diameter of the steel/MMC bars were turned to a diameter of 0.625″, 0.875″ or 1.00″. The various MMC core diameters remained constant at 0.375″ but with 3 different thicknesses of steel to provide for 3 different residual stress conditions [1–6].

XRD-RSA was performed with a Technology for Energy Corporation (TEC) Model 1610 X-Ray Stress Analysis System employing the $\sin^2\psi$ technique. Following Bragg's Law:

$$n\lambda = 2d\sin(\theta) \tag{1}$$

The incident x-ray beam penetrates the sample, interacting with some of the crystallites in its path. The beam diffracts to a characteristic 2θ angle based on the atomic spacing of the material. Here *d* is the spacing between diffracting planes, θ is the incident angle, *n* is any integer, and λ is the wavelength of the beam.

It was realized that additional variables, such as thermal and mechanical stresses resulting from the cutting method and polishing technique, might adversely affect the results of a surface measurement. Neutron diffraction was utilized as it provides a method capable of measuring through thickness strain profiles in each phase of the material, substantially larger gauge volumes, and can separate stress contributions from each phase within the hybrid composites. Neutron diffraction does however have limitations in the type of materials that it can measure, specifically B_4C. Boron has an extremely high neutron absorption cross section making neutron diffraction measurements nearly impossible in the MMCs reinforced with B_4C.

Similar to XRD-RSA, neutron diffraction follows Bragg's law, however the determination of strain within a material requires measurement of a stress-free "standard" for quantification. In Eq. (2), ε is strain, d is the measured atomic spacing of the sample, and d_0 is the measured atomic spacing of the stress-free "standard".

$$\varepsilon = \frac{d - d_0}{d_0} \tag{2}$$

The source of the observed strains varies greatly with the composite material constituents. Generally speaking, they can be divided into 3 categories based on the length scale over which the strain acts. Type I stresses, also called macro-stress and engineering stress, typically fluctuate on a length scale proportional to many grains (several millimeters in size). These stresses are often the result of fabrication including processing parameters, phase constituents and structural geometry. Type II stresses typically fluctuate on a length scale proportional to individual grains. These stresses result from inhomogeneity within a material or between phases in a composite, specifically a coefficient of thermal expansion (CTE) mismatch, elastic modulus mismatch, or discontinuity in plastic flow resulting from elastically deforming particles. Type III stresses fluctuate over the smallest distances down to atomic length scales. These stresses can result from variations in local stress fields around dislocations [7]. Type-I and Type-II stresses contribute to the shifts in the positions of diffraction lines and Type-II and Type-III stresses contribute to diffraction line-widths [8].

All neutron diffraction measurements were performed at the Canadian Neutron Beam Centre, (CNBC) Chalk River site utilizing the L3 diffractometer. (115) reflection of a single crystal Ge monochromator was used to filter the spectrum resulting in a beam of neutrons that possess a narrow energy band exhibiting a wavelength of 1.5509 Å.

Stress-free "standards" were fabricated in the form of rectangular prisms, approximately 2 × 2 × 20 mm in size, similar in appearance to match-sticks. In an effort to limit the amount of residual stress introduced during fabrication of these specimens, electrical discharge machining was utilized. Since there is no direct contact between the tool and sample, machining is performed without introducing additional mechanical stresses which results from traditional cutting methods. Samples were fabricated for each of the steels, the Al-10Si matrix, and the Al-SiC MMC. In addition, equivalent match-sticks were made from the MMC raw materials: an unreinforced Al-10Si alloy and powder SiC. The powder SiC was poured into a vanadium crucible so a similar gauge volume could be measured. Vanadium is typically used as a container since its nuclei hardly scatter neutrons, resulting in no effect on the SiC diffraction peaks. The average stress-free d_0 was measured from each of these samples with a gauge volume of 1 × 1 × 10 mm to be used for the corresponding strain calculations.

For each sample, limited by geometry, there are three measurement areas within the steel shell and four measurement areas within the MMC core. Since the steel thickness varied between the samples, step sizes were chosen such that they represent even spacing between measurements and span the thickness of the steel wall. The first measurement is ½ step size inside the outer diameter of the steel and subsequent measurements penetrate deeper into the material. This process is repeated for the MMC core except the final measurement is in the center of the MMC, see Fig. 1. Because the core is a composite consisting of Al-10Si and SiC, separate measurements were made at each MMC location since each of the constituent materials diffracts at different angles and will be subjected to different strains.

It is necessary to measure three separate orientations, axial, radial, and hoop for the cylindrical geometry, to properly map out stresses within the sample.

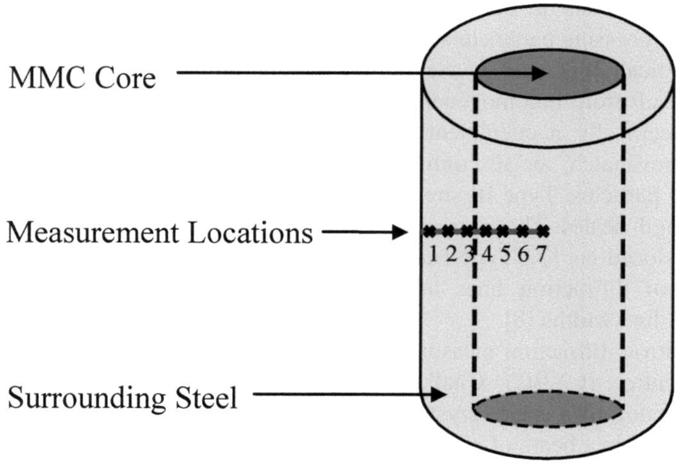

Fig. 1 Diagram of the measurement locations on a hybrid-composite; 1–3 within steel, 4–7 with MMC core

For example, Eq. (3) solves for stress, σ. E is Young's Modulus, ν is Poisson's ratio, and ε is strain as measured from Eq. (2). Subscripts A, R, and H represent axial, radial, and hoop directions respectively.

$$\sigma_A = \frac{E}{1+\nu}\left[\varepsilon_A + \frac{\nu}{1-2\nu}(\varepsilon_A + \varepsilon_R + \varepsilon_H)\right] \tag{3}$$

The measurement gauge volume was maximized to decease the required collection time and associated error. Radial and hoop measurements were performed with a 1 × 10 mm beam slit whereas the axial measurement was limited to a 1 × 1 mm beam slit due to geometry. Although measuring a larger area may be more representative of the bulk material condition, the smaller beam geometry is still capable of obtaining narrow peaks with a high signal to noise ratio, see Fig. 4, when the collection time is increased. A 32 line detector was used to capture the diffracted beam resulting in the 3.2° collection width observed in Fig. 2.

Throughout the set of experiments, each sample must be examined with exactly the same calibrated set-up, wavelength, collimators and slits as the reference standards. Furthermore, the alignment of the beam and position of the samples is also a crucial step in guaranteeing accurate results. First the sample is lined up optically to establish a datum, typically the edge of the cylindrical sample. This step is then followed by a second alignment, this time utilizing the count rate of the neutrons to more accurately determine the samples position relative to the path of the beam.

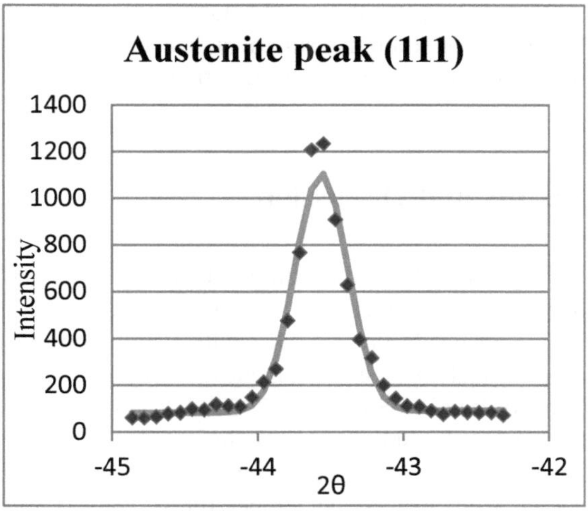

Fig. 2 Example of a typical Bragg peak (with Gaussian distribution) measured in 304 SS

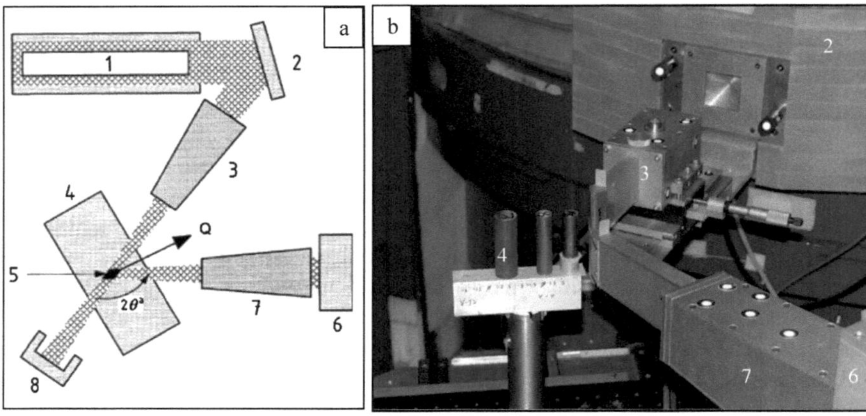

Fig. 3 **a** Schematic [9] **b** photo of steady-state neutron diffraction setup for strain measurement

1. Neutron beam source
2. Single crystal Germanium monochromator resulting in a 1.5509 Å wavelength.
3. Beam defining optics: Collimator and Cadmium slit to shape incident beam
4. Specimen(s)
5. Interaction volume
6. Detector
7. Beam defining optics: Collimator and Cadmium slit to shape diffracted beam
8. Beam stop: Lead shielding.

Figure 3 shows the steady-state neutron diffraction setup utilized at the Chalk River site for strain measurement. The current configuration illustrated in Fig. 3b is set up to measure the hoop direction on 3 macro-composite specimens.

Results and Discussion

During processing, the hollow steel rods are heated to a processing temperature between 600 and 750 °C. At these temperatures, the molten MMCs are cast or infiltrated inside the hollow cavity. Upon cooling the materials attempt to shrink proportional to their coefficient of thermal expansion. Evident from literature for 2009Al-25vol.% SiC_p [10], 2124Al-17vol.% SiC_p [11], and 6061Al-15vol.% SiC_w [12] thermal residual stresses due to CTE mismatch between the phases result in tension in the matrix and compression in the reinforcement when MMC's are cooled from processing temperatures. The ceramic particles have a substantially lower CTE compared to the aluminum matrix, 3 versus 23 ppm/K, see Table 1. The aluminum attempts to shrink around the SiC, upon doing this, the aluminum transfers a load onto the particle. Due to the high stiffness of the SiC, it won't break or yield but imparts a residual compressive stress on the particle. At the same time,

Table 1 Material properties

Material	Matrix	Vp	E (GPa)	ν	CTE (ppm/K)
A36	N/A	N/A	200	0.26	11.7
304 SS (annealed)	N/A	N/A	200	0.29	16.6
Al-10Si	N/A	N/A	66	0.35	22
Hot pressed SiC	N/A	N/A	453	0.19	3
Al/SiC	Al-10Si	55	202	0.25	11.8

E—Young's modulus; ν—Poisson's ratio; Vp—particle volume fraction, CTE—coefficient of thermal expansion (20–100 °C). *Sources* M Cubed Technologies (MCT) [13] or directly from manufacturer's websites or datasheets

the forces balance on the aluminum which can result in yielding and plastic deformation within the matrix, leading to residual tensile stress.

Neutron diffraction measurements were performed on 3 of the macro composite materials systems. In an effort to isolate the effect of the encapsulating steel/MMC CTE mismatch, each of the three steel systems with the 55 volume percent Al-SiC was measured. Figure 4 presents the results of residual stress calculations using

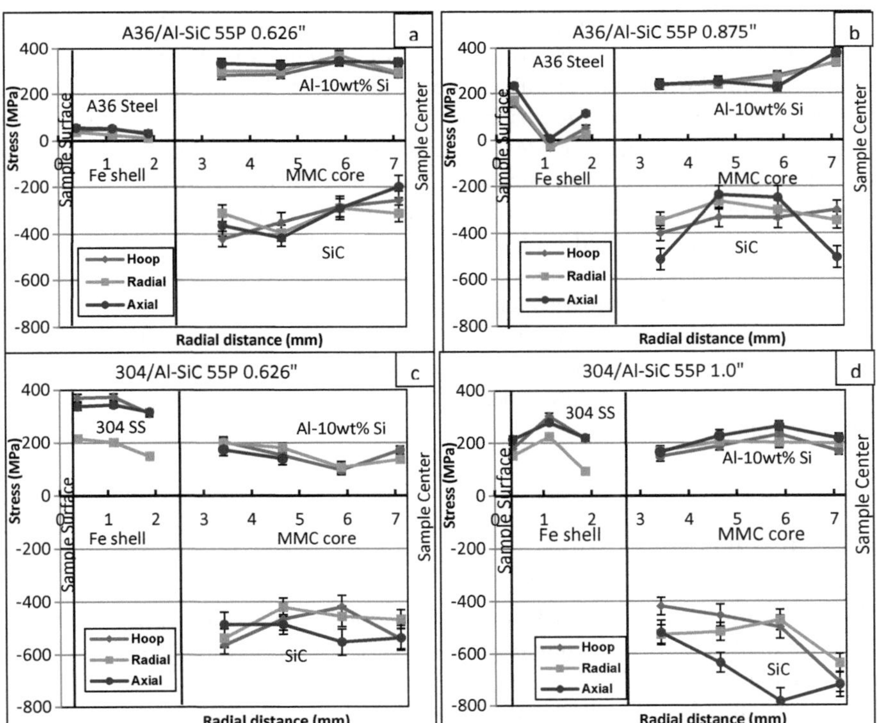

Fig. 4 Neutron diffraction residual stress measurements **a** A36/Al-SiC 55P 0.626″ steel outer diameter **b** A36/Al-SiC 55P 0.875″ **c** 304/Al-SiC 55P 0.626″ **d** 304/Al-SiC 55P 1.0″

Eq. (3) for the A36 and 304 stainless steels, each with the Al-SiC MMC and varying outer steel thicknesses.

It is clear that the type II stresses do not normally average to zero in each phase and thus, like the macro stress, cause a shift in the Bragg scattering angles which allows the average strain in each phase over the sample gauge volume to be measured.

Since the core is a composite of two materials: the matrix and the reinforcement, Al-10Si and SiC respectively, they were measured separately. The core diameter thickness is the same for all materials however the outer steel thickness varies according to the chart title. It can be seen that within the respective material systems, a thicker outer steel diameter results in a higher magnitude of residual compressive stress in the SiC reinforcement particles.

Furthermore, an additional strain is applied to the MMC core via the steel encapsulant. The surrounding steel will contract onto the MMC core affecting the residual stresses based again on the CTE mismatch. Considering the MMC as having one averaged value of CTE based on the volume percentage of its constituents the following trends can be observed. Since the CTE is comparable between the A36 steel and the 55P Al-SiC, the residual stresses observed on the MMC are dominated by the Al-SiC CTE mismatch. The weak confinement offered by the A36 steel has little effect on the observed stresses on the core. Conversely, the 304 SS with the 55P Al-SiC samples have a much higher CTE difference. The surrounding 304 SS contracts around the MMC compressing both the matrix and reinforcement. Lower tensile values are observed in the aluminum matrix and higher compressive values are observed in the SiC reinforcement. The effect of the steel can be isolated since the MMC core is identical between the samples and the elastic constants between the steels are comparable. This indicates the type II thermal misfit stresses are the dominant stress type present in these materials.

Uncertainty in strain and thus stress is a result of many factors relating to the measurement and reference values and include: uncertainty in the measurement of d and d_0, the incident beam wavelength, and the measurement position within the specimen. Furthermore, beam time was limited so shorter than ideal collection times were performed, to optimize the number of samples and measurements per sample, which resulted in lower peak to noise ratios and increased error in each measurement.

Conclusions

Neutron diffraction can be used as an accurate and reliable method for measuring bulk residual tensile and compressive stresses within steel encapsulated metal matrix composites.

Low magnitude residual stresses within the steel are observed in the A36/Al-SiC systems which exhibit a low CTE mismatch between outer steel and MMC core. Conversely, the steel in the 304SS/Al-SiC systems have significantly higher residual stresses due to a large CTE mismatch between adjacent materials.

The confinement of the 0.626″ 304SS on the Al-SiC MMC results in a 50% more compressive stress in both the matrix and the reinforcement as compared with the same system surrounded by A36.

The 1.0 inch thick surrounding steel results in a residual compressive stress in the MMC core approximately 15% greater than the 0.626 in. steel for the 304/Al-SiC system.

The results obtained provide reliable foundation for fine-tuning and verification of thermo-mechanical computer modeling created in, for example, ABAQUS. Furthermore the models can be used to extrapolate predictions on the B_4C reinforced system, as neutron diffraction measurements on this system are nearly impossible to accurately perform.

References

1. Aghajanian M et al (2013) Processing of hybrid structures consisting of Al-based metal matrix composites (MMCs) with metallic reinforcement of steel or titanium. In: TMS2013 annual meeting supplemental proceedings, Wiley, Hoboken, NJ, pp 629–638
2. Karandikar PG et al (2013) Al/Al_2O_3 MMCs and macrocomposites for armor applications. Ceram Eng Sci Proc 34(5):63–74
3. Karandikar PG et al (2013) Net shape Al/B_4C metal matrix composites (MMCs) for high specific stiffness and neutron absorption applications. In TMS 2013 annual meeting supplemental proceedings, Wiley, Hoboken, NJ, pp 683–689
4. Givens B et al (2009) Effect of particle loading on the properties of Al/SiC metal matrix composites. In: Yin et al (eds) Aluminum alloys: fabrication, characterization and applications II, TMS, Warrendale, PA, pp 197–202
5. Karandikar P et al Metal encapsulated MMC macrocomposites. Materials science & technology 2014 conference proceedings
6. Fudger S et al (2015) Mechanical properties of steel encapsulated metal matrix composites. Advanced composites for aerospace. Wiley, New York, pp 121–36 (Print)
7. Fitzpatrick ME, Hutchings MT, Withers PJ (1997) Separation of macroscopic, elastic mismatch and thermal expansion misfit stresses in metal matrix composite quenched plates from neutron diffraction measurements. Acta Mater 45(12):4867–4876
8. Holden TM (2013) Chapter 8 neutron diffraction. Practical residual stress measurement methods, Wiley (Print)
9. Non-destructive testing—stand test method for determining residual stresses by neutron diffraction. 1st edn. N.p.: ISO 21432, 205 (Print)
10. Bruno G, Girardin E, Giuliani A, Manescu A, Rustichello F, O'donnel B, McHugh PE (2002) Residual stress analysis in aerospace MMC materials by neutron diffraction. Appl Phys A Mater Sci Process 74(1):s1701–s1703
11. Levy-Tubiana R et al (2003) Relaxation of thermal mismatch stress due to plastic deformation in an Al/SiCp metal matrix composite. Mater Sci Eng A 341(1):74–86
12. Bruno G et al (2006) Gauge volume effects in residual stress determination by neutron diffraction: the strength differential effect in metal matrix composites. Mater Sci Eng A 437 (1):100–108
13. http://www.mmmt.com/materials/metal-matrix-composites.html

Tensile Behavior of Epoxy Matrix Composites Reinforced with Pure Ramie Fabric

Caroline G. de Oliveira, Janine F. de Deus, Ygor M. de Moraes, Marcos V.F. Fonseca, Djalma Souza, Frederico M. Margem, Luiz G.X. Borges and Sergio Monteiro

Abstract Recently, the natural fibers have been studied as alternative reinforcement material to the use of synthetic fibers. Aiming to improve the properties of natural fibers composites, it is investigated the use of natural fibers fabrics, since they provide multidirectional reinforcement and enables the introduction of a larger fiber volume fraction. In this work, it was investigated the tensile behavior of the epoxy matrix composites reinforced with different volume fractions of fabric made of Ramie, a highly resistant lignocellulosic fiber. The specimens were made by pressing the fabric with liquid resin to guarantee total impregnation, and then cutting the fabric in the mold format. Finally, the fabric layers were placed into the mold with the liquid resin and catalyst. They were cured for 24 h and tested. The results revealed an increase in the tensile strength and almost no effect in the elastic modulus with the increase of fabric volume fraction.

Keywords Natural fibers · Ramie fabric composite · Epoxy composites

Introduction

With the rapid development of technology over the past decades, many studies have been made in order to develop materials that meet the new demands. In this context, it is extremely important to consider not only the technical benefits of projects and researches, but also their environmental advantages aiming their long-term sustainability. One of the most important concerns is finding alternatives to non-renewable materials, which contribute to the air pollution, to resources shortage

C.G. de Oliveira (✉) · J.F. de Deus · Y.M. de Moraes · M.V.F. Fonseca · D. Souza
State University of Northern Rio de Janeiro, Campos dos Goytacazes, RJ, Brazil
e-mail: caroline.oliveyra@gmail.com

F.M. Margem · L.G.X. Borges
Faculdade Redentor, Itaperuna, RJ, Brazil

S. Monteiro
Military Institute of Engineering, Rio de Janeiro-RJ, Brazil

S. Ikhmayies et al. (eds.), *Characterization of Minerals, Metals, and Materials 2017*, The Minerals, Metals & Materials Series,
DOI 10.1007/978-3-319-51382-9_45

and to increasing the quantity of non-biodegradable residues. In this scenario, natural fibers emerge as an advantageous alternative [1–3].

Among the natural fibers, the lignocellulosic ones must be highlighted for having high mechanical resistance. The presence of lignine and cellulitis makes the fibers simultaneously flexible and resistant [4–6]. Moreover, they are abundantly cultivated all around the world and it offers the possibility of social development, once the cultivation can be a source of income for poor communities in many countries [1].

Ramie fiber must be particularly noted for being one of the most resistant, which can reach a tensile strength higher than 1000 MPa. Furthermore, epoxy matrix composites reinforced with Ramie fibers reached 102.26 MPa for 30% of fibers in tensile tests [7, 8]. Aiming to further improve the performance of the natural fiber composites, it is investigated the performance of composites reinforced with Ramie fabric (Fig. 1), instead of its aligned fibers. The fabric reinforces the matrix in more than one direction simultaneously and also enables the introduction of a higher fraction of reinforcement material. This new structure, regarding the use of Ramie fabric, has already been tested for bulletproof panels and jackets [9].

The objective of this work is to evaluate the tensile behavior of epoxy matrix composite reinforced with Ramie fabric.

Experimental Procedures

The fabric used for this work was obtained by the Chinese company Rose Natural Healthy Items Wholesale. The specimens with different fabric volume fractions were individually confectioned by incorporating 1, 2 and 3 fabric layers into the composite. For this, the fabric was initially cut in the mold format and then pressed

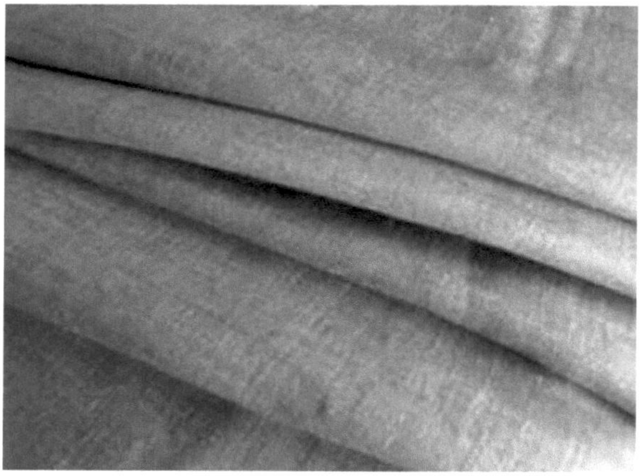

Fig. 1 Ramie fabric (*Source* Trade Korea)

with the liquid resin (Epoxy + DEGEBA/TETA) to guarantee a total impregnation. After this, enough amount of epoxy resin was poured into the mold and then the fabric layers were placed onto it alternately with new resin layers. The specimens were cured for 24 h and tested in an universal Instron machine, model 5582, at 25 °C.

The tensile specimens are shown in Fig. 2.

Results and Discussion

The typical load versus elongation curves for the pure epoxy (0% fabric) and different composites are exemplified in Fig. 3. These curves were recorded directly from the Instron software. After a straight elastic line, a sudden fracture occurs, indicating a brittle behavior for both the pure epoxy and the ramie fabric composites tensile specimens.

Figure 4 shows the macro aspect of tensile ruptured specimens corresponding to the different volume fraction of ramie fabric. In this figure, the fracture of all specimens tends to be transversal to the tensile axis. No macroscopic evidence of ramie fabric participation has been observed.

From the data observed in the curves shown in Fig. 3, it was possible to obtain the average tensile strength (maximum stress) and elastic modulus for different volume fractions of ramie fabric. This data is presented in Table 1. These results show that the presence of ramie fabric increases the composite tensile strength and has almost no influence in the stiffness.

Figure 5 shows the graphical results for the variation of tensile strength and elastic modulus with the volume fraction of ramie fabric. It can be seen that tensile

Fig. 2 Tensile specimens

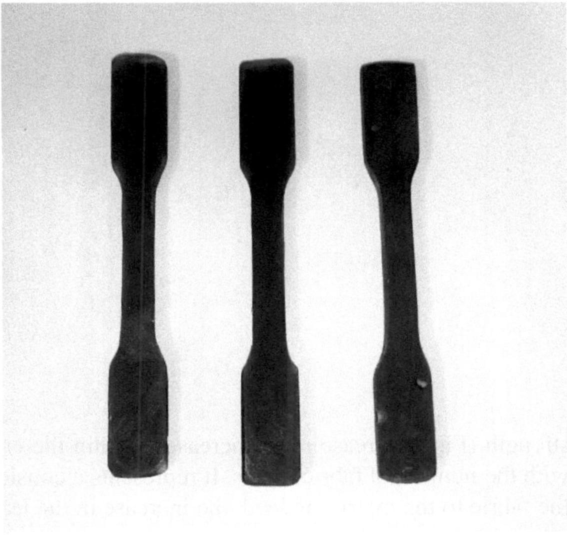

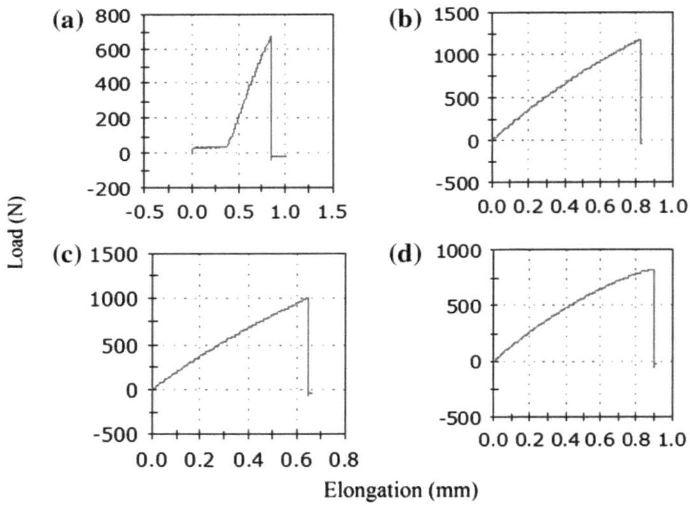

Fig. 3 Load versus elongation curves: **a** pure epoxy, **b** 1 fabric layer, **c** 2 fabric layers and **d** 3 fabric layers

Fig. 4 Ruptured tensile specimens

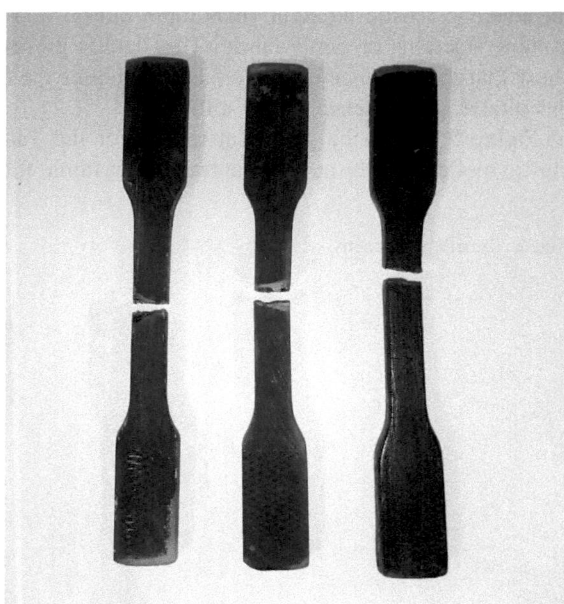

strength (Fig. 5a) reasonably increases, within the error bars, in almost linear way with the number of fabric layers. It represents a considerable reinforcement effect of the fabric to the matrix. Indeed, the increase in the tensile strength is approximately

Table 1 Tensile properties of epoxy composites reinforced with different volume fraction of ramie fabric

Number of fabric layers	Tensile strength (MPa)	Elastic modulus (GPa)
0	24.15 ± 3.88	1.851 ± 0.011
1	31.41 ± 3.57	1.844 ± 0.011
2	34.24 ± 3.09	1.830 ± 0.012
3	37.63 ± 2.06	1.827 ± 0.011

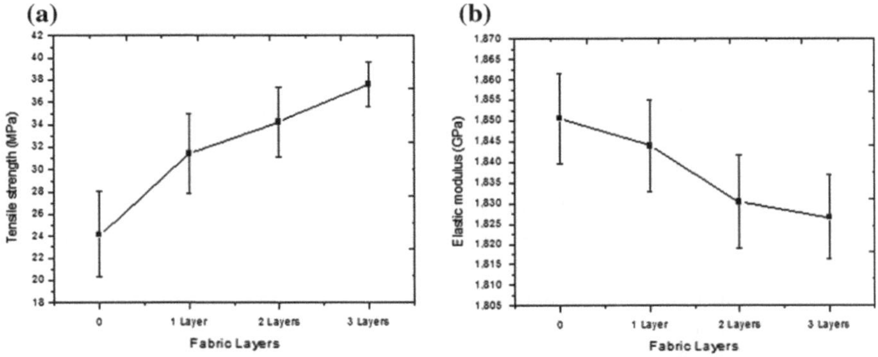

Fig. 5 Graphical results of (**a**) tensile strength (**b**) elastic modulus for composites with different ramie fabric volume fraction

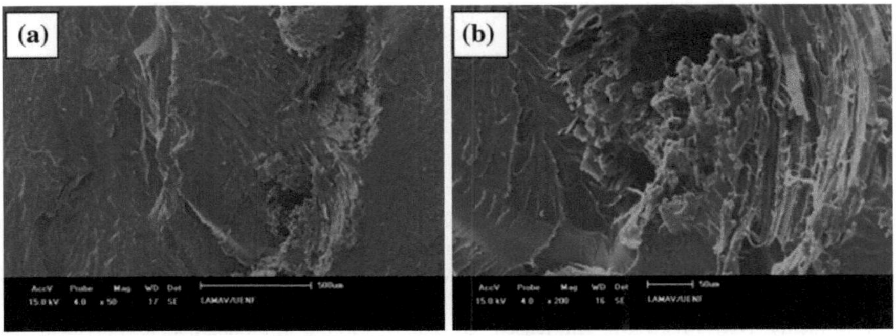

Fig. 6 SEM fractograph of a tensile ruptured epoxy composite specimen with 3 ramie fabric layers: **a** 50× and **b** 200×

55%. It is also possible to observe that the elastic modulus (Fig. 5b) is almost not affected by the presence of fabric, remaing the same, withing the error bars.

In Fig. 6 it is possible to see the SEM fractographs of the fracture surface. It is observed that the fabric slightly pulls out from the matrix. This is due to delamination effect, common in composites reinforced with laminate materials, such as natural fiber fabrics. However, the delamination effect is not highly intense in composites subjected to tensile stresses, thus, even with this effect, the fabric still retains a reasonable adhesion to the matrix. It can also be observed, based on the fracture marks on the epoxy matrix surface, that the fabric worked as a barrier to the crack propagation, increasing the composite tensile strength compared to the correspondent value of pure epoxy.

Conclusions

- Tensile tested epoxy composites reinforced with ramie fabric exhibit a reasonable increase in the tensile strength and almost no variation in the elastic modulus.
- With three fabric layers, the tensile strength increases approximately 55% while the elastic modulus remains the same, considering the error bars.
- Although the fabric provides multidirectional reinforcement and enables the introduction of higher volume fractions of reinforcement material, it is also subjected to delamination and debonding effects, which can mitigate the reinforcement effect.
- The SEM fractographs analysis reveals that, even with a slight delamination effect, the fabric has a reasonable adhesion to the matrix, working as barrier to crack propagation and reinforcing the matrix.

Acknowledgements The authors thank the support to this investigation by the Brazilian agencies: CNPq, CAPES, FAPERJ.

References

1. Mohanty AK, Misra M, Drzal LT (2002) Sustainable biocomposites from renewable resources: opportunities and challenges in the green material world. J Polym Environ 10:19–26
2. Crocker J (2008) Natural materials innovative natural composites. Mater Technol 2–3: 174–178
3. Mohanty AK, Misra M, Hinrichsen G (2000) Biofibers, biodegradable polymers and biocomposites: an overview. Macromol Mater Eng 276(277):1–24
4. Bledzki AK, Gassan J (1999) Composites reinforced with cellulose-based fibres. Prog Polym Sci 24:221–274
5. Satyanarayana KG, Guimarães JL, Wypych F (2007) Studies on lignocellulosic fibers of Brazil. Part I: source, production, morphology, properties and applications. Compos A 38:1694–1709
6. NabiSahed D, Jog JP (1999) Natural fiber polymer composites: a review. Adv Polym Technol 18:221–274
7. Bevitori AB et al (2013) Tensile behavior of epoxy composites reinforced with continuous and aligned ramie fibers. Charact Miner Met Mater, 465–471

8. Monteiro SN, Satyanarayana KG, Lopes FPD (2010) High strength natural fibers for improved polymer matrix composites. Mater Sci Forum 638–642:961–966
9. Milanezi TL (2015) Comportamento Balístico da Fibra de Rami em Blindagem Multicamadas (Major degree dissertation). Military Institute of Engineering

Hemp Fiber Density Using the Pycnometry Technique

Lázaro A. Rohen, Anna C.C. Neves, Dhyemila de P. Mantovani, F.V. Carlos Maurício, Janaina da Silva Vieira, Lucas de A. Pontes, Frederico M. Margem and Sergio Monteiro

Abstract The lower density of the lignocellulosic fibers such as the hemp fibers and the other natural ones is one of the characteristics that allow the scientists to obtain high levels of specific properties that when compared to the properties of ordinary reinforcements gives to the hemp fibers an advantage as a material with an expressive potential in the mechanical properties area. Then, the aim of this present work is to characterize the relative density of the hemp fiber related to the water using the pycnometry technique. This technique uses the pycnometer and it is divided in two steps. In the first step the correction factor and the moisture content of the fresh hemp fiber specimen and in the other step some weightings were done such as the dry specimen (g), empty pycnometer (g), pycnometer + water (g) and pycnometer + water + specimen (g) using a digital balance with a precision scale of 0.001 g of the bran INNERSCAN. Based on these measurement and calculations using some formulations already established in previous studies, the relative density of the hemp fiber found was around 1.393 g/cm^3 and the moisture content around 7.1%.

Keywords Relative density · Hemp fibers · Pycnometer

L.A. Rohen · A.C.C. Neves · D. de P. Mantovani ·
F.V. Carlos Maurício · J. da Silva Vieira · L. de A. Pontes
State University of the Northern Rio de Janeiro, UENF,
Advanced Materials Laboratory, LAMAV, Av. Alberto Lamego,
2000, Campos dos Goytacazes 28013-602, Brazil

S. Monteiro
Instituto Militar de Engenharia, IME, Praça Gen, Tibúrcio, nº 80 Urca,
Rio de Janeiro, RJ 22290-270, Brazil

F.M. Margem (✉)
Faculdade Redentor Rodovia, BR 356, nº 25, Itaperuna
RJ 28300-000, Brazil
e-mail: fmargem@gmail.com

S. Ikhmayies et al. (eds.), *Characterization of Minerals, Metals, and Materials 2017*, The Minerals, Metals & Materials Series,
DOI 10.1007/978-3-319-51382-9_46

Introduction

The interest in the composites reinforced with natural fibers is justified by the concern of the global population with the environmental and the use of renewable raw materials. This huge interest is focused in finding natural fibers that could replace the synthetic ones, such as the glass fiber [1]. Although, the use of natural fibers as reinforcement is a challenging task because of the big dispersion in shape and dimension allied to a lower mechanical performance presented in comparison to the synthetic fibers.

On the other hand, the principal justification of the use of natural fibers as reinforcement of composite materials is their low density that allows the obtainment of high levels of specific properties. These properties when compared to conventional reinforcements give to the hemp fibers an expressive potential to replace them with suitable mechanical properties, what favor the use of this renewable natural material, low toxicity and low density as reinforcement of composite materials [2].

Previous studies relate the characterization of the hemp fiber as its density using different methods; some scientists estimated the hemp fiber density and its moisture content. Analyzing their work the estimated density is from 0.86 to 1.51 g/cm^3 and the moisture content from 8 to 12%. The determination of the density of the hemp fiber is essential in a fiber analysis because this property is responsible for specific properties also related to mechanical properties that can be interesting for the use of these fibers as reinforcement of composite materials. Therefore, the aim of this present word is to evaluate the relative density of the hemp fiber using the pycnometry method, estimating the value of the relative density of the fiber according to the water and its moisture content [2–6].

Experimental Procedure

The determination of the relative density of the hemp fibers using the pycnometry method was performed in the Soil Laboratory of UENF (LSOL) using four pycnometers due to the test was done in four repetitions. The pycnometer consists in a glass flat-bottom flask with a glass stopper where a capillary channel is. The volume of the pycnometers vary from 25 to 250 mL or more, this volume depends on the application. The pycnometers used in this work are with 50 mL of volume as it's possible to see in the following Fig. 1.

The test was performed into steps. The first step consists in estimate the correcting factor of the masses and the moisture content of the specimen and the other step consists in the determination of the relative density of the hemp fiber related to the water using the pycnometry method.

To the determination of the correction factor (Cf) a fresh specimen was weighed in a digital balance with 0.001 g of precision and brand Innerscan, in the Laboratory of Soils at UENF (LSOL). The specimen was dried in the kiln at 105 °C

Fig. 1 Pycnometer: **a** model of a pycnometer; **b** empty pycnometer; **c** pycnometer with distilled water and the hemp fiber specimen

of temperature for 48 h, and could be found the relation between the fresh mass specimen and the dried mass one, in grams. This relation is the Cf that subtracted in one unit and multiplied by 100 determinates the moisture content of the specimen.

To the determination of the relative density of the hemp fiber, the liquid used as standard was water with the density around 0.9962 g/cm^3. The process was performed in a chronologic sequence of steps, as the following list:

1st weigh the fresh specimen (g)
2nd dry the fresh specimen at 105 °C for 48 h;
3rd weigh the clean and dried pycnometer;
4th fill the pycnometer with distilled water;
5th leave the pycnometer with water for 2 h in the vacuum desiccator to remove the air bubbles;
7th put the specimen in the empty pycnometer and fill with distilled water;
8th leave the pycnometer with water and the specimen for 2 h in the vacuum desiccator to remove the air bubbles;
9th weigh the pycnometer with distilled water and the specimen;
10th measure the temperature of the water;
11th calculate the relative density of the fiber using the following Eq. (1):

$$\rho_{fiber} = \frac{m_1/(m_3 - m_2) - (m_4 - m_2 - m_1)}{\rho_{water}} \tag{1}$$

m_1 mass of the clean and dried pycnometer (g);
m_2 mass of the pycnometer filled with water (g);
m_3 mass of the dry specimen (g);

m_4 mass of the pycnometer with water and the dry specimen (g);
$\rho_{H_2O}Z$ density of the water.

Results and Discussion

To determinate the density of the hemp fiber using the pycnometry method four pycnometers and four different specimens were used due to the test was performed in four repetitions aiming to decrease the margino of error of the results.

In the first moment, the test was performed to determinate the correction factor (Cf) that consists in the relation between the mass of the fresh specimen and the dried mass. This calculated gave the possibility to estimate the Cf of the sample which the founded value was 1.071. This value subtracted in one unit and multiplied by 100 allowed the estimative of the moisture content of the sample which the founded value was 7.1%.

Besides that, the test of density using the pycnometry method was performed. This is a practical method and it has a reasonable precision based on the reason between the absolute densities of two substances, using Eq. (2).

$$\rho_{1,2} = \frac{\rho_1}{\rho_2} \quad (2)$$

Generally ρ_2 is chose as standard and ρ_1 the density to determinate, in this case, the density of the hemp fiber. It's normal to consider the density of the water as standard because of its convenience and abundance, its absolute density is about 0.9962 g/cm^3 at room temperature (25 °C).

Then, the four specimens and the masses of the dry fiber (m_1), of the pycnometer (m_2), of the pycnometer + water (m_3) and of the pycnometer +water + specimen were weighed and reported in Table 1.

For each sample tested the density of the hemp fiber related to the water was calculated according to Eq. (1). Therefore, the densities were estimated as it's possible to observe in Fig. 2.

Table 1 Masses of the dried fiber, empty pycnometer, pycnometer + water and pycnometer + water + specimen, all of them in grams (g)

Specimens	Dried specimen (g) (m_1)	Empty pycnometer (g) (m_2)	Pycnometer + water (g) (m_3)	Pycnometer + water (g) (m_4)
1-a	0.784	28.065	80.333	80.604
1-b	0.683	37.977	86.472	86.653
1-c	0.825	28.065	80.333	80.604
1-d	0.854	37.977	86.472	86.752

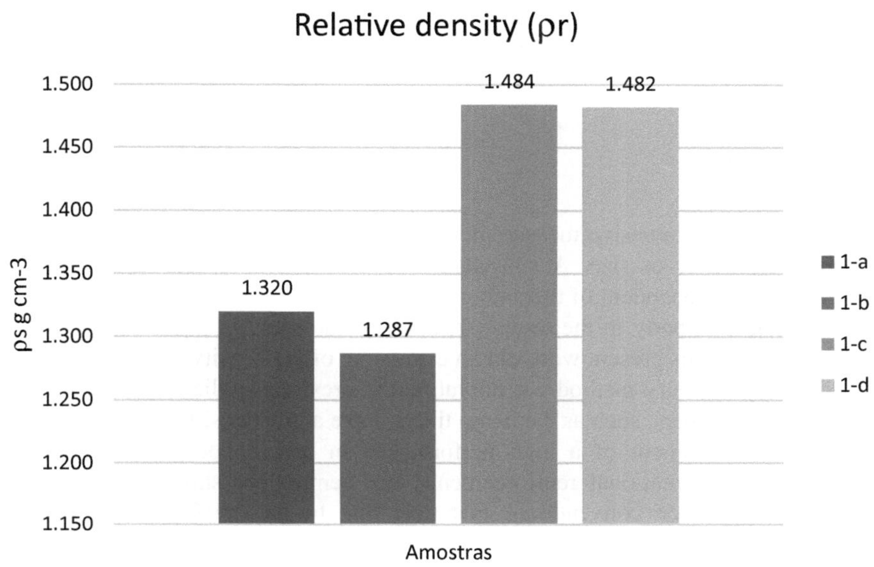

Fig. 2 Density of the tested samples

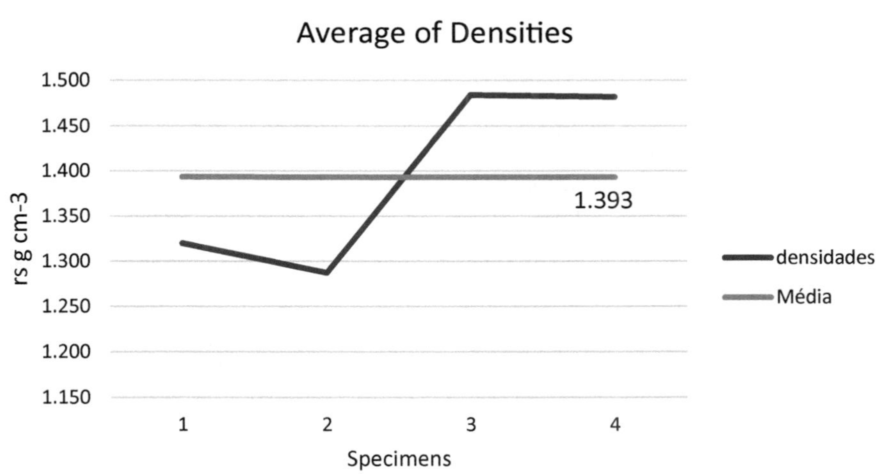

Fig. 3 Average of the densities of the specimens tested

An average of those densities resulted in a density of the hemp fiber estimated to 1.393 g/cm^3 as observed in Fig. 3.

This estimated value corroborates with previous studies that gives the density of the hemp fibers from 0.86 to 1.51 g/cm^3. Moreover the moisture content estimated to the sample of this present work differs with the previous studies due to it's not between 8 and 12% and the founded value was 7.1% [2–6].

Conclusion

The density is an intensive measurement, in other words, it doesn't depend on the quantity of matter or size. A substance will always exhibit the same intensive measurement independent of the amount that would be being examined, that's why this is a valid property in the characterization of samples.

Concluding, this present work did an estimative of the density of the hemp fibers using the pycnometry method corroborating the previous studies and also proving that the natural fibers, such as the hemp fibers, have a low density. This low density allows the obtainment of a high performance in specific properties and when compared to conventional reinforcements, the hemp fiber shows an expressive potential to replace conventional materials due to its satisfactory mechanical properties.

Acknowledgements The authors thank the support to this investigation by the Brazilian agencies: CNPq, CAPES, FAPERJ and TECNORTE/FENORTE.

References

1. Da Silva RV, Aquino EMF, Rodrigues LPS, Barros ARF (2008) Desenvolvimento de um compósito laminado híbrido com fibras natural e sintética. RevistaMatéria 13(1):154–161
2. Dhakal HN, Zhang ZY, Richardson MOW (2007) Effect of water absorption on the mechanical properties of hemp fibre reinforced unsaturated polyester composites. Compos Sci Technol 67:1674–1683
3. Koronis G, Silva A, Fontul M (2013) Green composites: a review of adequate materials for automotive applications. Compos B 44:120–127
4. Lu N, Oza S (2013) A comparative study of the mechanical properties of hemp fiber with virgin and recycled high density polyethylene matrix. Compos B 45:1651–1656
5. Mostefai N, Hamzaoui R, Guessasma S, Aw A, Nouri H (2015) Microstructure and mechanical performance of modified hemp fibre and shiv mortars: discovering the optimal formulation. Mater Des 84:359–371
6. Shahzad A (2011) Hemp fiber and its composites—a review. J Compos Mater 46(8):973–986
7. Brady JE, Humiston GE (1986) Química Geral. Rio de Janeiro: Livros Técnicos e Científicos 1 (2):14–17

Preparation and Characterization of Clay Exfoliation and Vegetal Fibre on Properties of Recycled Low Density Polyethylene

Amauche Cyprian Achusim-Udenko, Coida D.S. Renata, Francisco R. Valenzuela-Diaz, Gerald Okwuchi Onyedika, Moura E.B. Esperidiana, Martin Chidozie Ogwuegbu and Graca Valenzuela-Diaz

Abstract This work aims to incorporate Mgbee clay, vegetal fibre (*Costus afer*) in recycled low density polyethylene (rLDPE). The incorporation of clay and vegetal fibre in rLDPE was carried out using a twin screw extruder machine and the nano composites were obtained. The nano composites were characterized by the Young's modules test, tensile tests, bending test, XRD, SEM, DSC, and TG and other properties evaluated. The results of morphological and thermal properties showed a good interaction between the clay, and rLDPE as compared to neat rLDPE.

Keywords Clay · Recycled low density polyethylene · Costus afer

Introduction

Non-degradable synthetic fibres are associated with long-term pollution after use. Also large amount of energy is involved in their production, with consequent emission of toxic gases such as CO_2 and other "green house" gases. Such synthetic products cannot be incinerated in thermal power plants.

A.C. Achusim-Udenko (✉)
Department of Science Laboratory Technology, Imo State Polytechnic, Owerri, Nigeria
e-mail: achusimuac@gmail.com

C.D.S. Renata · F.R. Valenzuela-Diaz · G. Valenzuela-Diaz
Universidade de São Paulo Escola Politenica, Sao Paulo, Brazil

G.O. Onyedika · M.C. Ogwuegbu
Department of Chemistry, Federal University of Technology, Owerri, Nigeria

M.E.B. Esperidiana
Instituto de Pesquisas Energetics E Nucleares _IPEN-CNEN/SP, São Paulo, Brazil

S. Ikhmayies et al. (eds.), *Characterization of Minerals, Metals, and Materials 2017*, The Minerals, Metals & Materials Series,
DOI 10.1007/978-3-319-51382-9_47

According to Alexandre and Dubois [1], polymeric nanocomposites belong to a class of materials made of one reinforcing agent with at least one dimension on the nanometric scale, dispersed in a polymer matrix.

Nanocomposites have their application in areas such as automotive, packaging, aerospace, electronics, medicare and biotechnology [2–4]. The most improved composite system are based on polymer matrix and organphilic clays.

Some clay minerals (especially the montmorrilonite stand out as among the fillers incorporated to polymer matrix to form nanocomposites which are hybrid of organic and inorganic materials dispersed at the nano scale.

Due to interaction in this nano range, this kind of materials can be produced with a small amount of fillers compared to conventional composites and exhibit superior mechanical thermal, barrier and flow flame retardancy [5].

Several methods have been used to prepare nano composites of polymers and clay. Among them melt intercalation and exfoliation are the most famous.

In intercalated structures, single or more polymer chains are intercalated between the clay layers leading to an ordered multi layers structures of alternating polymeric and inorganic layers [6–8]. Exfoliated structures result from a situation whereby clay layers are completely and uniformly dispersed in the polymer matrix [9, 10].

The natural fibres are lignocellulosic in nature. Natural fibre thermoplastic composites are gaining importance due to their noncarcinogenic, biodegradable nature [11]. *Costus afer* belongs to the family of *zingiberaceae* and is a monocot relatively tall berbaceous, unbranched plant with creeping rhizome. It is commonly found in moist or shady forest of west and tropical Africa [12]. *Costus afer* is a perennial rhizomatous herb that attains a height up to 4 m. The leaves are arranged spirally, simple and entire [13]. *Costus afer* has about 70 species, which about 40 species are found in tropical Africa and about 5 in South East Asia [14]. *Costus afer* stem has been used in many African cultures to make table mats, baskets and hose construction. *Costus afer* stem has been successfully used to make paper. *Costus afer* is certainly under exploited as at today. Its numerous uses as medicinal plants and social-cultural values has informed this research into its fibre as reinforcement in composites. In this work, the evaluation of nanocomposite formed from modified Mgbee clay, *Costus afer* and recycled low density polyethylene (discarded water sachets that litter the environment in Nigeria) was carried out.

Materials and Methods

Materials

The materials used in this work were *costus afer* fiber, recycled low density polyethylene (rLDPE) from discarded pure water sachets, coffee colored clay from Mbee in Orlu Local Government Area of Imo state, Nigeria.

Preparation of Costus Afer Particles

The stems of the *costus after* plant were collected from Imo State Polytechnic within the Otamiri River area. The leaves and outer layer of the plant were removed and the parts containing the fiber were cut to an average diameter of 3.2 cm. The pieces were put in a laboratory oven at 60 °C for 4 days to dry. The dried samples were pulverized in a knife mill, then reduced to a pointer ball mill for another 48 h. The pulverized sample were graded into sizes using a nest of sieves. The size less than 75 μm were selected for composite formulation.

Clay Nanoparticle

The clay samples were ball milled and sized in nest of seizes and sizes less than 75 μm were selected.

Clay Organophilization

The clay was modified using 20% $NaCO_3$ and 50% ethanolic solution of quatering ammonium salt and underwent the processes of dispersion into water, stirring and heating for a determinate time, then was it filtered and dried in oven at 60 °C for 24 h for the disaggregation of one particle in another, and finally characterized.

Composite Preparation

The composite were prepared by melt extrusion using a twin-screw extruder Haake Rheomix with 16 mm and L/D = 25 rate from Thermo Scientific. The temperature profile was 160/170 185/175/180/185 °C. Screw speed was 40 rpm. The extrudates coming out of the extruder were cooled down for a better dimensional stability, pelletized by a pelletizer, dried again at 60 ± 2 °C for 24 h and fed into injection molding at 165/170/175 temperature profile and specimens test samples were obtained. The compositions of the blend and its composite are presented in Table 1.

Table 1 Compositions of the r-LDPE composite

Materials	Compositions A (wt%)	Compositions B (wt%)
r-LDPE	85	82
Clay	–	3
Fibre	15	15

X-ray Diffraction Analysis

The clay intercalation and exfoliation within the polymer matrix and dispersion at nanometric level were analyzed using X-ray Diffractometer (Philips XPERTMPD with Cuk radiation (Δ = 1.54 mm) operated at 40 kV and 40 MA with 2θ.

Thermal Analysis

The thermal behavior of the rLDPE and the composites were undertaken to determine thermal gravity (TG) using Differential Scanning Calorimeter (DSC). In this study, the TG analysis were carried out in a Mettler Toledo TGA module "TGA/SDTA851e" from 25 to 600 °C at a heating rate of 10 °C/min under nitrogen atmosphere (50 ml/min). DSC analyses were carried out using a Mettler Toledo DSC 822e from 25 to 250 °C at a heating rate of 10 °C/min under nitrogen atmosphere (50 ml/min). The samples were initially heated from 25 to 250 °C at a heating rate of 10 °C/min and the temperature was held at 250 °C for 5 min to eliminate the previous thermal history. Subsequently, the sample was cooled to room temperature at a cooling rate of 10 °C/min, and secondly heated to 250 °C at a heating rate of 10 °C/min rate to obtain heating thermograms. DSC were carried out to obtain melt temperature (Tm) and melting enthalpy (ΔHm).

Mechanical Test

The following tests were conducted on the rLDPE and the composites to assess their mechanical properties. They are; the Brinnel hardness, shear stress, tensile strength, elongation at break and Young modues. The tensile tests were carried out using an INSTRON Universal Testing Machine, model 5564 according to ASTM D 638 standard at 25 °C. Six replicates were made of each sample and the average values were reported.

Results and Discussion

X-ray Diffraction Analysis

By the XRD curves it can be observed that the Mgbee clay is composed basicaly of kaolin and quartz. Having major peaks of kaolin in 3.57 A and quartz in 1.82 A. The organophylic clay present a peak at 19 A which is peculiar to organoclays with organic compound existing at the surface of the clay mineral sheets Fig. 1.

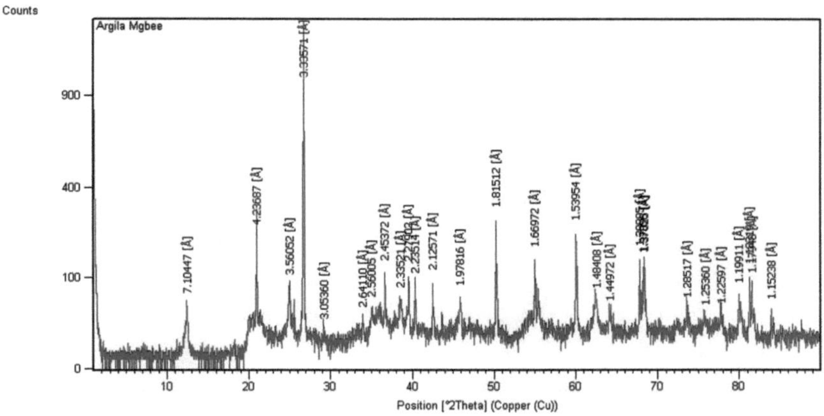

Fig. 1 XRD pattern for Mgbee clay

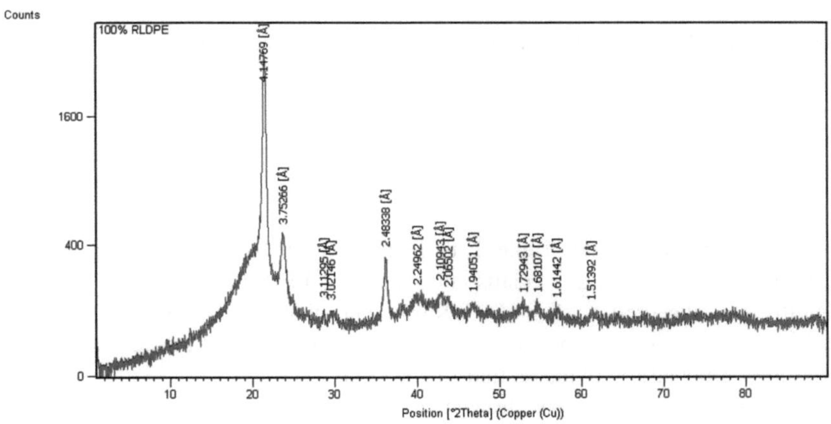

Fig. 2 XRD curves for recycled low density polyethylene (rLDPE)

Figures 2 and 3 shows the XRD pattern of the rDLPE and the composites (clay, costus afer and rLDPE).

Mechanical Test

Table 2 shows the Brinnel hardness test and shear stress properties of the rLDPE and the nanocomposite (clay, *costus afer* and rLDPE). The nanocomposite and rLDPE has high Brinnel hardness of 59.96 and 71.25 N/mm respectively which displays a harder material than the rLDPE alone. The shear strength of rLDPE. From the results, this difference in hardness is about 18.83% increase in the

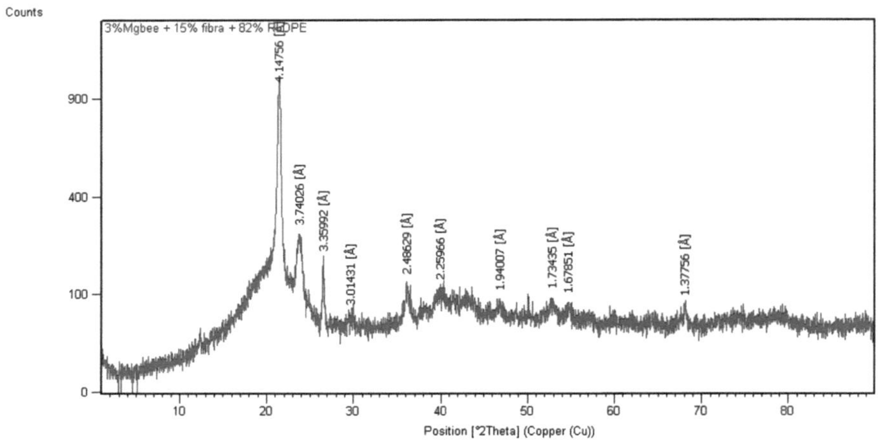

Fig. 3 XRD curves for 3% clay 15% costus afer fibre and rLDPE composite

Table 2 Brinnel hardness and Shear stress results of r-LDPE and its composites

Materials	Brinnel hardness (N/mm^2)	Shear stress (N/mm^2)	
r-LDPE	59.96 ± 5.0	5.75 ± 0.1	
rLDPE/COSTUS AFER/CLAY	71.25 ± 5.0	11.05 ± 0.1	

hardness of the composite. Table 2 also shows a very significant increase in the shear strength of the clay, costus afer and rLDPE nanocomposite. The shear strength was observed to be 11.05 N/mm and that of rLDPE was 5.75. This increase represent 92% in the shear strength. This implies that the stress required to pierce the composite is twice as that of the rLDPE.

The tensile tests results are showed in Table 3. It cam be seen in this table an important increase of about 85% in Young's modulus due to fibre addition and ca. 40% due to fibre and clay addition. On the other hand, the addition of fibre and clay lead to a reduction in the both, tensile stress at break and elongation at break properties of the rLDPE.

Table 3 Tensile tests results of r-LDPE and its composites

Materials	Tensile strength at break (MPa)	Elongation at break (%)	Young's modulus (MPa)
r-LDPE	1.48 ± 0.1	145.8 ± 22.3	6.04 ± 1.7
r-LDPE/COSTUS AFER FIBRE	1.31 ± 0.1	85.8 ± 19.8	11.16 ± 1.7
r-LDPE/MGBEE CLAY	1.45 ± 0.1	132.4 ± 10.5	7.15 ± 1.0
r-LDPE/COSTUS AFER FIBRE/MGBEE CLAY	1.40 ± 0.1	130.5 ± 10.5	8.25 ± 1.0

Thermogravicmetric Analysis

The results of the thermogravicmetric analyses of the rLDPE and its composites (clay, costus afer and clay) are as shown in Fig. 4. The TG shows that at T_{max} 378. 91 and 381.82 °C and weight loss of 98.52 and 91.89% for rLDPE and composite respectively.

It can be seen in Fig. 5 that when costus afer fibre is added in rLDPE, three-stage of weight loss could be observed. The first stage, a peak of around 222 °C characteristic of hemicelluloses, the second, one peak at 350 °C corresponded to the thermal degradation of cellulose and the third stage of degradation temperature occurred at 405 °C corresponds to the thermal degradation of composites [15].

The results of the DSC analyses of the rLDPE and its nanocomposites (clay, costus afer and clay) are as shown in Fig. 5.

The results of melting temperature and enthalpy of the neat rLDPE and its composite are showed in Table 4. From this Table, it could be inferred that melting temperature of the r–LDPE has not presented significant change due to vegetal fibre and clay addition. On the other hand the melting enthalpy of rLDE increase to about 10% because of fibre and clay addition.

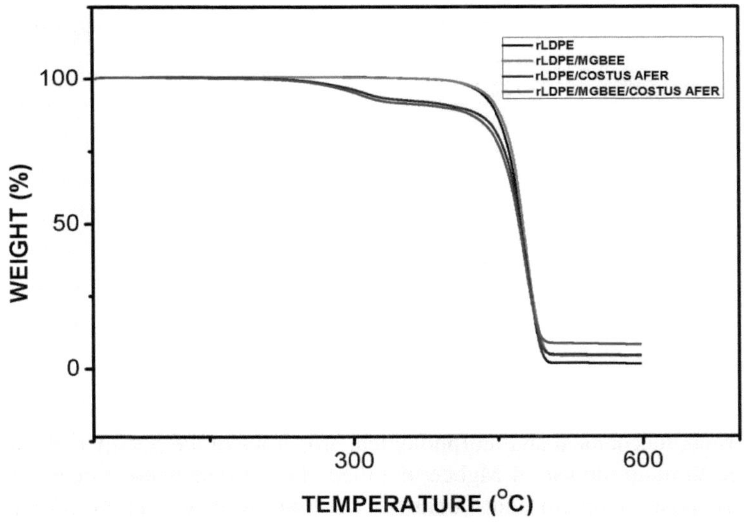

Fig. 4 Thermogravimetric analysis (TG) curves of neat rLDPE and its composites

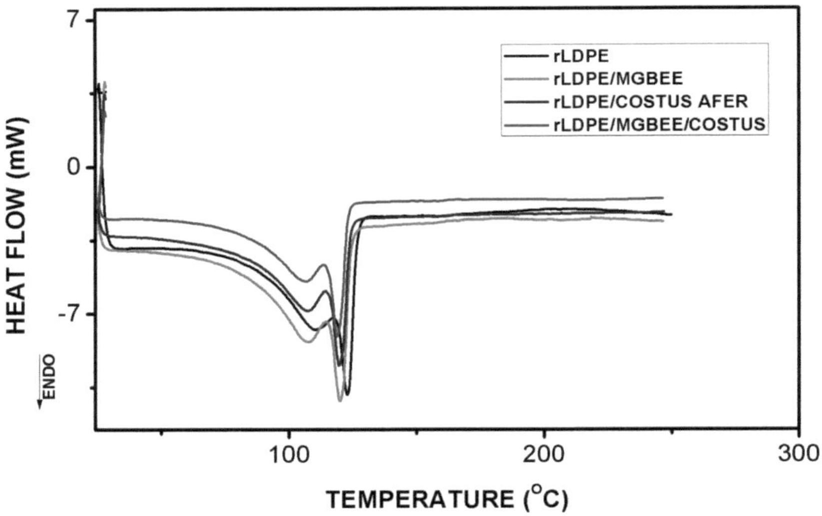

Fig. 5 DSC analysis results of the rLDPE and its composites

Table 4 Melting temperature and enthalpy results of rLDPE and its compositels

Materials	Melting temperature (T_m, °C)	Melting enthalpy (ΔH_m, Jg^{-1})
rLDPE[a]	122.4	89.9
rLDPE/Costus afer[b]	122.2	99.9
rLDPE/Mgbee clay[c]	122.8	98.4
rLDPE/Mgbee/Costus afer[d]	121.9	98.9

[a]recycled LDPE 100 wt%
[b]rLDPE/Costus afer 15 wt%
[c]rLDPE/Mgbee clay 3 wt%
[d]rLDPE/Mgbee/Costus afer (82/3/15 wt%)

Conclusion

The thermal, mechanical and morphological properties of the polymer (rLDPE) was enhanced through the use of Mgbee modified clay and the nanoparticles of Costus after. Composite strength and toughness was improved reasonably. The reuse of waste sachet water disposed indiscriminately in gutters, waterways as well as farmland thereby causing disastrous consequences to the environment can be drastically reduced. This also will encourage investment in this direction consequently wealth will be created for Nigeria.

References

1. Alexandre M, Dubois P (2000) Mater Sci Eng R 28:1–63
2. Meenakshi D, Kuksenok O, Balazs AC, Meenakshi D, Olga K, Balazs AC (2012) Designing tunable bio-nanostructured materials via self-assembly of amphiphilic lipids and functionalized nanotubes. Mater Res Soc 1464:12–1464
3. Nguyen D (2011) Active starvation responses mediate antibiotic tolerance in biofilms and nutrient- limited bacteria. Science 18:334(6058):982–6. doi:10.1126/science.1211037
4. Luo JJ, Daniel IM (2003) Characterization and modeling of mechanical behavior of polymer/clay nano composites. Compos Sci Technol 63:1607–1616
5. Beyer G (2006) Flame retardancy of nanocomposites based on organoclays and carbon nanotubes with aluminium trihydrate. Polym Adv Technol 17:218–225
6. Sharma SK, Nayak SK (2009) Surface Modified Clay/Polypropylene (PP) Nanocomposites: effect on physico-mechanical. Therm Morphol Prop Polym Degrad Stab 94(1):132–138
7. Picard E, Gauthier H, Gérard JF, Espuche E (2007) Influence of the intercalated cations on the surface energy of Montmorillonites: consequences for the morphology and gas barrier properties of polyethylene/montmorillonites nanocomposites. J Colloid Interface Sci 307:364–376. ISSN 0021-9797
8. Morales AR, de Paiva LB, Zattarelli D (2012) Morphology, structure and mechanical properties of polypropylene modified with organophilic montmorillonite. Polimeros 22(1): 54–60
9. Paiva LB, Morales AR, Díaz FRV (2008) Argilas Organofílicas: Características, Metodologias de preparação, compostos de intercalação e técnicas de caracterização. Cerâmica 54:213–226. ISSN 0366-6913
10. Ortiz AV, Teixeira JG, Gomes MG (2014) Preparation and characterization of electron beam treated HDPE Composites reinforced with rice husk ash and Brazilian Clay. Appl Surf Sci 310:331–335
11. D' Almeida JRM, Monteiro SN (2003) Composites reforcados por fibras de naturals-Opportunidades desatios. In: 580 Congressor Annul de associao Brasileira de metalurgia e materials, Rio de Janeiro. ABM, pp 1724–1434
12. Iwu MM (2009) Traditional igbo medicine applied science, vol 3(4). Institute of African Studies Publication. University of Nigeria Nsukka, pp 21–25
13. Omokhua GE (2011) Medicinal and socio-cultural importance of *Costus afar* (Ker Grawl) in Nigeria. Int Multi J Ethiop 5(5) 22:258–287
14. Soladoye MO, Oyesika OO (2008) A textbook of medicinal plants from Nigeria. University of Lagos Press, p 628
15. Rachini A, Le Troedec M, Peyratout C, Smith A (2009) Comparison of the thermal degradation of natural, alkali-treated and silane-treated hemp fibers under air and an inert atmosphere. J Appl Polym Sci 112:226–234

Part XII
Ferrous Metals

Estimation of Dislocation Density in Metals from Hardness Measurements

A.A.H. Ameri, N.N. Elewa, M. Ashraf, J.P. Escobedo-Diaz and P.J. Hazell

Abstract A novel methodology to estimate the dislocation density in cubic metals by using microhardness measurements has been established. The proposed methodology is based on the Indention Size Effect (ISE) phenomena and strengthening mechanisms of materials at the micro-level. The methodology was validated by using experimental data of lean duplex stainless steel 2101 alloy. The results are confirmed via X-ray diffraction measurements and they substantiate that the proposed approach can be used as a general method to estimate dislocation density in metals.

Keywords Hardness · Microstructure · Dislocations · Lean duplex

Introduction

Dislocations, the main type of linear defects, have a major influence on the mechanical properties of solid materials. The density of dislocation can be estimated directly by Transmission Electron Microscopy (TEM) [1]. However, the use of TEM is not only complex in terms of sample preparation, but also does not represent the bulk plastic deformation of the deformed metal due to the small size of a typical sample [2, 3]. Subsequently, alternative approaches, which are easier and provide more representative measurements of dislocation density are required.

During plastic deformation, two types of dislocations are generally present: the Statically Stored Dislocations (SSDs), which evolve during the uniform plastic deformation (uniform strain), and the Geometrically Necessary Dislocations

A.A.H. Ameri (✉) · M. Ashraf · J.P. Escobedo-Diaz · P.J. Hazell
School of Engineering and Information Technology,
The University of New South Wales, Canberra, ACT 2600, Australia
e-mail: Ali.Ameri@student.adfa.edu.au

N.N. Elewa
School of Physical, Environmental and Mathematical Sciences,
The University of New South Wales, Canberra, ACT 2600, Australia

S. Ikhmayies et al. (eds.), *Characterization of Minerals, Metals, and Materials 2017*, The Minerals, Metals & Materials Series,
DOI 10.1007/978-3-319-51382-9_48

(GNDs) that exist only to accommodate the strain gradient as in microhardness and nanoindentation tests [4, 5]. Many studies during the last few decades have shown that the hardness during microhardness and nanoindentation tests increase considerably by decreasing the depth of indentation (h) due to the influence of GNDs [6, 7], which is known as the Indentation Size Effect (ISE) [6]. Nix and Gao [6] developed a widely accepted numerical model (NG model) to represent the influence of the ISE by assuming that the total hardness (H) originates only from the hardness due to dislocations (H_d). They used Tabor relationship, Von Mises hypothesis and Taylor hardness model to link dislocation density (ρ) to hardness [6]. They suggested that H_d varies depending on the indentation depth. For macrohardness (H_o) with relatively large indentation depths, H_d results only from SSDs because of the absence of GNDs, whereas for microhardness (H_{mi}) with small indentation depths but larger than 200 nm [8], H_d originates from the superposition of GNDs and SSDs [6].

This paper proposes a general approach to estimate the density of SSDs for cubic metals by using microhardness experimental data with taking the advantage of ISE during microhardness test. The suitability of the suggested approach to determine the SSDs is examined by computing the SSDs for dual-phase Lean Duplex Stainless Steel alloy 2101 (LDSS2101), these results are also compared herein with X-ray diffraction (XRD) measurements.

Theoretical Approach

Material resistance against plastic deformation at the micro-level represents the summation of strengthening mechanisms: internal friction stress, solid solution strengthening, grain boundary strengthening and dislocation strengthening [9]. By correlating the flow stress with hardness by Tabor relationship, the total hardness (H) can be presented as the summation of hardness due to internal friction (H_{fric}), solid solutions (H_{SL}), dislocation density (H_d) and grain boundary (H_g) as shown in Eq. (1) [10, 11]:

$$H = H_{fric} + H_{SL} + H_d + H_g \tag{1}$$

The grain boundary effects on the hardness (H_g) can be neglected as long as the indentation occurs at the center of the grain, and the size of indentation is smaller than the grain size as there is no interaction between dislocations and grain boundaries (i.e. $H_g = 0$) [11].

In order to establish a general method valid for predicting the density of SSDs accurately, it is important to consider all the factors in Eq. (1). H_d can have two different quantities according to the indentation depth as mentioned in the previous section. For H_{fric}, this component of material strength represents a significant contribution in the material flow stress of BCC metals when compared with FCC metals [12] as BCC metals are not densely close-packed. Qiu et al. [13] showed that

H_{fric} is depth independent. Solid solutions show more complex behavior with solute atoms in non-pure metals by hindering the movement of dislocations during plastic deformation [7]. Durst et al. [7] studied the effects of solid solution in Ni-Fe alloys on hardness and they concluded that solid solution has two effects. Firstly, solute atoms interact with dislocations (this is the conventional influence of solid solution), and enhance the hardness by a constant value H_{SL} which depends on the amount of solid atoms. Secondly, they work as obstacles stopping the expansion of GNDs that leads to less stored volume for dislocations; as a result, they influence H_d. Hence, H_{SL} is also a constant value similar to H_{fric}. Therefore, Eq. (1) could be rewritten for macrohardness (H_o) and microhardness (H_{mi}), where H_o represents the hardness at the plateau region of the H-h curve and H_{mi} is the hardness within the ISE zone, as:

$$H_o = H_{fric} + H_{SL} + H_{SSD} \tag{2}$$

$$H_{mi} = H_{fric} + H_{SL} + H_{ISE} \tag{3}$$

where H_{SSD} is the hardness resistance resulting only from SSDs and H_{ISE} is the hardness resistance resulting from the superposition of SSDs and GNDs [10]. Nix and Gao [6] proposed that H_{SSD} and H_{ISE} can be calculated using Eqs. (4) and (5):

$$H_o = 3\sqrt{3}\,\alpha G b \sqrt{\rho_{SSDs}} \tag{4}$$

$$H_{ISE} = 3\sqrt{3}\,\alpha G b \left(\sqrt{\rho_{SSDs} + \rho_{GNDs}}\right) \tag{5}$$

where α is a constant that can be taken as 0.5, b is the Burgers vector magnitude, G is the shear modulus, ρ_{SSDs} is the density of SSDs and ρ_{GNDs} is the density of GNDs. Subtracting Eq. (2) from Eq. (3) and substituting Eq. (4) and Eq. (5) leads to:

$$\frac{H_{mi} - H_o}{3\sqrt{3}\,\alpha G b} = \sqrt{\rho_{SSDs} + \rho_{GND}} - \sqrt{\rho_{SSDs}} \tag{6}$$

$$\rho_{GND} = \frac{3\tan^2\theta}{2f^3 bh} \tag{7}$$

where θ is the angle between the surface and the indenter, f factor is defined to represent that the stored volume of GNDs is bigger than a hemispherical volume with a radius equals to the contact radius of indentation (a_c) as suggested by Nix and Gao [6]. This makes all parameters in Eq. (6) known, and then by using the Newton-Raphson method the ρ_{SSDs} for each indentation can be calculated. The value of ρ_{SSDs} should be depth independent; however, due to unavoidable sensitivity issues in hardness measurements and the approximations in the suggested approach, some minor discrepancies are typically inevitable. Therefore, the final value of the ρ_{SSDs} can be calculated by taking average of the estimated values of

ρ_{SSDs} for all indentations experiencing ISE. The ISE zone could be defined as the region that satisfies $\frac{H - H_o}{H_o} \times 100 > 10\%$ where H is the hardness value at each indentation depth.

Estimating the stored volume of GNDs accurately is essential to determine the contribution of GNDs in H_{ISE}, and subsequently defining the density of SSDs, which is the main goal in this study. *f* factor is defined by using the same technique suggested by Durst et al. [7] and considering the minimum value for $f = 1.00$ [12]. This approach based on using Eq. (4) to define the density of SSDs and Eq. (5) to define the most suitable value of *f* for the relevant experimental data. It is worth noting that the obtained value for *f* will be less than the actual as both Eq. (4) and Eq. (5) neglect the effects of H_{fric} and H_{SL}.

Experimental Procedure

Material and Sample Preparation

A 20 mm-thick lean duplex stainless steel (LDSS 2101) plate used in this study was obtained from Outokumpo®. The plate was hot rolled, heat treated at 1050 °C and then quenched in water. In the preparation for optical microscopy (OM), electron backscatter diffraction (EBSD), hardness and X-ray diffraction (XRD) measurements, the specimens were prepared following standard metallographic procedures up to vibratory polishing with 0.02 μm non-crystallizing colloidal silica suspension.

Micro-Hardness Experiment

Hardness tests were conducted by using Hysitron TI 950 TriboIndenter nanoindenter with the Berkovich indenter (this indenter has $\theta \approx 24°$) with loading range between 1 gm and 70 gm. After defining the area function of the indenter and performing the correction for thermal drift, the hardness was calculated according to the approach suggested by Oliver and Pharr [14]. For each load step, several indentations were performed and the average of them was used to represent the hardness of the material at the corresponding load.

X-ray Diffraction Measurement

The XRD measurements were performed by an Empyrean, PANalytical diffractometer with CuKα radiation operating at 45 kV, 40 mA and at a step size 0.013° and a count time 10.2 s./step. The measurement was obtained with angular range between 35° and 125°. Before analyzing the XRD profiles, the background of XRD profile was eliminated by using a polynomial function, and then followed by CuKα_2 stripping [2]. For the fitting process of XRD pattern, different functions were used: Voigt, Lorentzian, Gaussian and pseudo-Voigt fitting functions; the last function produced the best fitting when compared against other functions.

Results and Discussion

Microstructural Analysis

The full chemical composition of the investigated LDSS2101 listed in Table 1 was measured by Inductively Coupled Plasma Atomic Emission Spectroscopy, Spark Emission Spectroscopy and Leco methods. EBSD maps in Fig. 1 shows that the material has only two phases: α-ferrite (BCC) and γ-austenite (FCC) with volume fractions estimated at 42 and 58%, respectively.

Revealing grain boundaries of both phases was essential, as the indentations have to be performed far from the grain boundaries as well as phase boundaries to ignore the effects of H_g as explained in the theoretical approach section. Among the several etchants tried to reveal grain boundaries of both austenitic and ferritic phases, only two were suitable for this purpose: an acid etchant (10 ml HCl 32%, 10 ml HNO 63% and 10 ml $C_2H_4O_2$), and a glyceric etchant (30 ml HCl 32%, 25 ml $C_3H_8O_3$, 25 ml HNO 63%). The acid solution rendered the microstructure with a relatively better contrast as shown in Fig. 2, and thus it was used throughout this study.

Table 1 Full chemical composition (wt%) of the investigated LDSS 2101

Element	C	Mn	Si	S	P	Ni	Cr	Mo	Cu	N	Fe
Content	0.03	4.08	0.59	0.01	0.03	1.55	21.5	0.34	0.24	0.23	Bal.

Fig. 1 EBSD maps for both phases: austenite (γ) and ferrite (α) of as-received sample of LDSS 2101

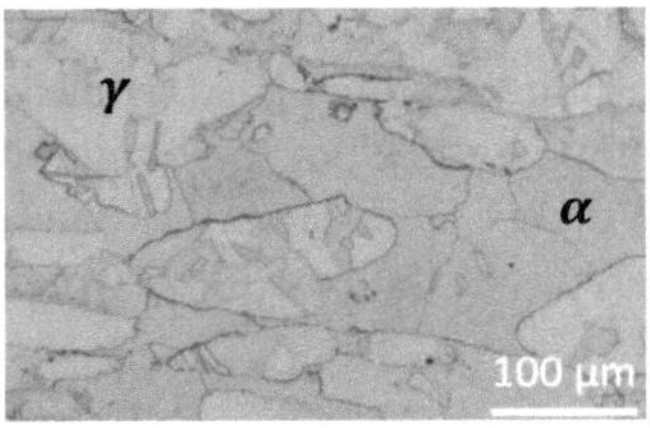

Fig. 2 The microstructure of LDSS 2101 by using the acid etchant

SSDs Calculations by Hardness Measurements

LDSS2101 was selected to test the suitability of the suggested method. All indentations were conducted within microhardness range (i.e. h > 200 nm) for both phases of LDSS2101. Indentations occurred far from phase boundaries and grain boundaries as illustrated in Fig. 3; subsequently each phase could be considered as a single material.

Figure 4 shows hardness versus indentation depth for ferritic phase. It is clear that this phase exhibits the ISE for indentation depths <3000 nm. From Fig. 4, a value of $H_o = 2.35$ GPa is obtained, and it is in consistent with the reported hardness for ferritic phase of LDSS2101 [15]. By applying the proposed approach in the theoretical approach section, *f* can be taken 1.0 as shown in Fig. 4. The density of SSDs can be estimated by using Eq. (6) with $H_o = 2.35$ GPa, $\alpha = 0.5$, b = 0.248 nm, G = 77 GPa and $\theta = 24°$. This results a value of 5.54×10^{14} m^{-2} for the density of SSDs.

For the austenitic phase, Fig. 5 illustrates the hardness versus indentation depth. These results indicate that austenite experiences similar behavior as ferrite in terms of generating the GNDs. The ISE for this phase starts at indentation depth <2000 nm and $H_o = 3.44$ GPa. This value of H_o agrees well with the reported values in

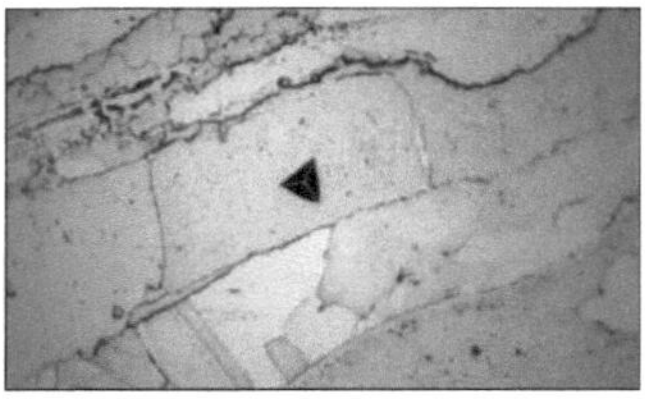

Fig. 3 A typical image of hardness test for ferritic phase (load = 30 gm); the *yellowish phase* is austenite and the *lighter phase* is ferrite (Color figure online)

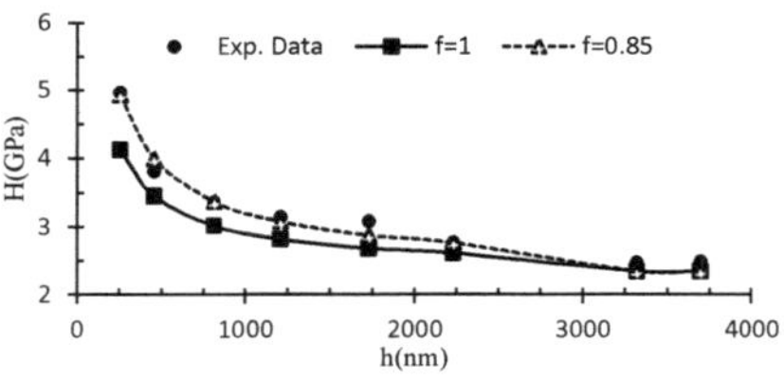

Fig. 4 Hardness versus indentation depth for ferrite phase of LDSS 2101

Fig. 5 Hardness versus indentation depth for austenitic phase of LDSS 2101

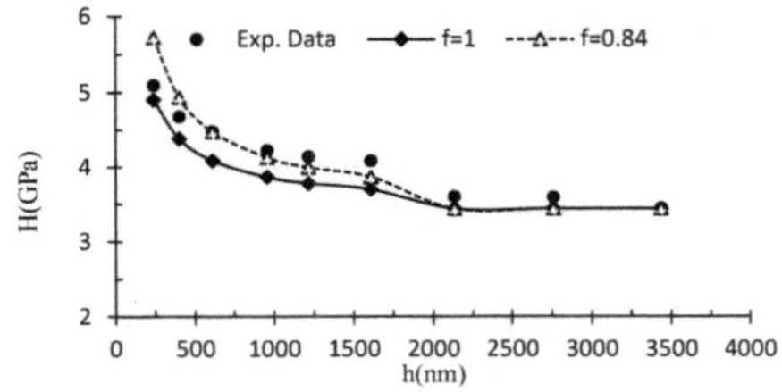

[15]. As shown in Fig. 5, *f* can be taken 1.0. Using Eq. (6) with same values for ferritic phase except $H_o = 3.44$ GPa and b = 0.25 nm gave the density of SSDs about 1.71×10^{15} m^{-2}.

SSDs Calculations by XRD

The broadening of diffraction lines in XRD experiment results mainly from dislocations and grain size effects [16]. The Modified Williamson-Hall method (MWH) and the Modified Warren-Averbach method (MWA), as modified by Ungar and Borbely [17] to consider strain anisotropy, were employed for analyzing the XRD profile broadening in the current study. The XRD profile of LDSS 2101 is shown in Fig. 6. For the ferritic phase (α) {110}, {200}, {211}, {220} and {310} were recorded and analyzed, whereas for austenitic phase (γ) {111}, {200}, {220}, {311}, {222} and {400} were noticed; however, only the first four reflections were used for the analysis because of the low intensities obtained for the others.

From XRD measurements, the densities of SSDs were estimated as 5.21×10^{14} m^{-2} and 3.29×10^{15} m^{-2} for ferritic and austenitic phases, respectively. In comparison, the results obtained from microhardness measurements in both phases, 5.54×10^{14} m^{-2} for ferritic and 1.71×10^{15} m^{-2} for austenitic phases, which are considerably close to those obtained by XRD method. This comparison shows that the suggested approach could be used to estimate the density of SSDs of cubic metals with indentation depth >200 nm. It is worth mentioning that this high dislocations density of such well annealed alloy results from the effect of mechanical polishing. As a result, the best way to define the dislocation density of a sample is to use electro-polishing for preparing the sample.

Fig. 6 XRD profile of LDSS 2101

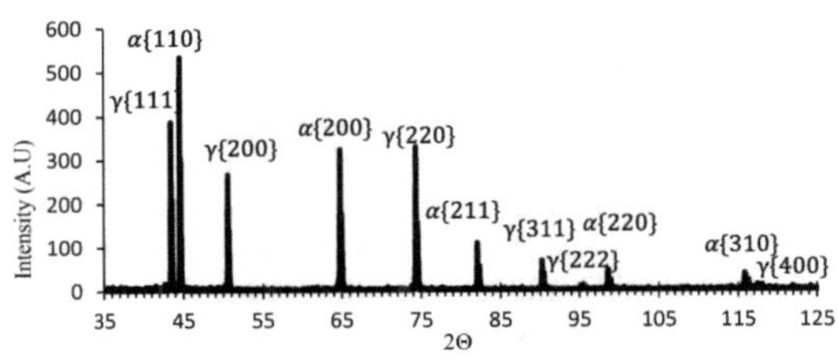

Conclusions

A novel approach to estimate the density of the SSDs by using microhardness test has been presented. This approach was validated for the calculated dislocation density for both phases of the Lean Duplex Stainless Steel LDSS 2101 complemented by XRD measurements. The following conclusions can be drawn from the presented study:

- Microhardness tests with indentation depths larger than 200 nm can be used as a simple and general technique to estimate the dislocation densities of cubic metals regardless of the type of crystal structures.
- The suggested approach has been validated for cubic crystalline metals, but further investigations are required to check the suitability of the suggested approach to estimate the dislocation density for HCP metals with different degrees of alloying.

Acknowledgements The authors would like to thank Prof. Sean Cadogan in the School of Physical, Environmental and Mathematical Sciences at The University of New South Wales Canberra for his support in performing X-ray diffraction experiments. Dr. Md. Quadir at Faculty of Science and Engineering, Curtin University is also acknowledged for his help with EBSD measurements. Moreover, the authors are grateful for the support of A/Prof. Jodie Bradby and Mr. Christopher Tanner in the Department of Electronic Materials Engineering in the Research School of Physics and Engineering at the Australian National University in conducting nanoindentation experiments.

References

1. Miyajima Y, Okubo S, Abe H, Okumura H, Fujii T, Onaka S, Kato M (2015) Dislocation density of pure copper processed by accumulative roll bonding and equal-channel angular pressing. Mater Charact 104:101–106. doi:10.1016/j.matchar.2015.04.009
2. Shintani T, Murata Y (2011) Evaluation of the dislocation density and dislocation character in cold rolled Type 304 steel determined by profile analysis of X-ray diffraction. Acta Mater 59:4314–4322. doi:10.1016/j.actamat.2011.03.055
3. Kishor R, Sahu L, Dutta K, Mondal AK (2014) Assessment of dislocation density in asymmetrically cyclic loaded non-conventional stainless steel using X-ray diffraction profile analysis. Mater Sci Eng A 598:299–303. doi:10.1016/j.msea.2014.01.043
4. Arsenlis A (1999) Crystallographic aspects of geometrically- necessary and statistically-stored dislocation density. Acta Metall 47:1597–1611. doi:10.1016/S1359-6454(99)00020-8
5. Rodríguez R, Gutierrez I (2003) Correlation between nanoindentation and tensile properties influence of the indentation size effect. Mater Sci Eng A 361:377–384. doi:10.1016/S0921-5093(03)00563-X
6. Nix WD, Gao H (1998) Indentation size effects in crystalline materials: a law for strain gradient plasticity. J Mech Phys Solid 46:411–425. doi:10.1016/S0022-5096(97)00086-0
7. Durst K, Franke O, Böhner A, Göken M (2007) Indentation size effect in Ni-Fe solid solutions. Acta Mater 55:6825–6833. doi:10.1016/j.actamat.2007.08.044
8. ISO, 14577 (2015) Metallic materials—instrumented indentation test for hardness and materials parameters

9. Totten GE, Xie L, Funatani K (2004) Handbook of mechanical alloy design. Marcel Dekker, 2004
10. Durst K, Backes B, Franke O, Göken M (2006) Indentation size effect in metallic materials: modeling strength from pop-into macroscopic hardness using geometrically necessary dislocations. Acta Mater 54:2547–2555. doi:10.1016/j.actamat.2006.01.036
11. Yang B, Vehoff H (2007) Dependence of nanohardness upon indentation size and grain size —a local examination of the interaction between dislocations and grain boundaries. Acta Mater 55:849–856. doi:10.1016/j.actamat.2006.09.004
12. Feng G, Nix WD (2004) Indentation size effect in MgO. Scr Mater 51:599–603. doi:10.1016/j.scriptamat.2004.05.034
13. Qiu X, Huang Y, Nix WD, Hwang KC, Gao H (2001) Effect of intrinsic lattice resistance in strain gradient plasticity. Acta Mater 49:3949–3958. doi:10.1016/S1359-6454(01)00299-3
14. Oliver WC, Pharr GM (1992) An improved technique for determining hardness and elastic modulus using load and displacement sensing indentation experiments. J Mater Res 7:1564–1583. doi:10.1557/JMR.1992.1564
15. Strubbia R, Hereñú S, Alvarez-Armas I, Krupp U (2014) Short fatigue cracks nucleation and growth in lean duplex stainless steel LDX 2101. Mater Sci Eng A 615:169–174. doi:10.1016/j.msea.2014.07.076
16. Ribárik G (2001) MWP-fit: a program for multiple whole profile fitting of diffraction profiles by ab-initio theoretical functions. J Appl Crystallogr 34:669–676. doi:10.1107/S0021889801011451
17. Ungár T, Borbély A (1996) The effect of dislocation contrast on x-ray line broadening: a new approach to line profile analysis. Appl Phys Lett 69:3173–3175. doi:10.1063/1.117951

Part XIII
Welding and Solidification

Interfacial Strength Characterization in a High-Modulus Low-Density Steel-Based Fe-TiB_2 Composite

Y.Z. Li and M.X. Huang

Abstract The titanium diboride (TiB_2)-reinforced steel composite, designed for automotive applications, exhibits a combination of high isotropic Young's modulus and low density as compared to existing advanced high strength steels (AHSS). The steel-based composite is produced by in situ precipitation of TiB_2 particles during eutectic solidification followed by hot rolling, and its microstructure displays a homogeneous distribution of both large primary TiB_2 and small eutectic TiB_2 particles in the ferrite matrix. Instead of interfacial debonding, particle cracking is the primary mode of damage, revealing very high interfacial strength. To investigate the intrinsic interfacial strength, a hybrid method combining both nanoindentation and finite element analysis (FEA) was used. A micron-sized sample containing a single crystal TiB_2 attached to the ferrite matrix was fabricated by focused iron beam (FIB), and was compressed using nanoindentation with a flat indentation tip. By combining the compression and FEA results, the interfacial strength was therefore determined.

Keywords Steel-matrix composites · Finite element method · Cohesive zone models · Interfacial shear strength

Introduction

The lightweight design in transportation systems has traditionally been achieved by gauge reduction using stronger steels, i.e., advanced high strength steels (AHSS) [1], as higher strength enables the use of thinner components. However, thickness reductions can be detrimental to the stiffness. When stiffness is considered to be the design criterion, the Young's modulus (E) of the materials has to be increased for further lightening. Steel-matrix composites (SMCs) reinforced with ceramic

---

Y.Z. Li · M.X. Huang (✉)
Department of Mechanical Engineering,
The University of Hong Kong, Hong Kong, China
e-mail: mxhuang@hku.hk

S. Ikhmayies et al. (eds.), *Characterization of Minerals, Metals, and Materials 2017*, The Minerals, Metals & Materials Series,
DOI 10.1007/978-3-319-51382-9_49

particles, which blend light and stiff phase with ductile matrix [2], exhibit increased specific modulus (i.e. E/ρ, where ρ is the density) and are now considered to be the potential materials in lightweight design. Among them, the Fe-TiB_2 system has received considerable attention during the past a few years. Bonnet and co-workers [3, 4] showed that a significant volume fraction of TiB_2 can be achieved by eutectic solidification in a conventional industrial process and can guarantee high productivity and low cost. Rana and Liu [2] found that TiB_2 particles (13 vol. %) lead to the increase of specific modulus by $\sim$26%.

A critical issue in the industrial application of the Fe-TiB_2 SMC is the behavior of particle/matrix interfaces during the macroscopic deformation of the bulk material, as the role of interface is crucial to the overall performance of the composite [5]. One accomplishment is the work done by Lartigue-Korinek and co-workers [6], who clarified the good coherency between TiB_2 and Fe phases in the atomic scale. However, no quantitative information on the strength and toughness of the interface has been provided, and the role of interfacial mechanical properties on the mechanical behaviors of the composite is still not clear. Therefore, it is necessary to perform mechanical tests on the TiB_2/ferrite interface.

The difficulties in accurately measuring the interfacial mechanical properties on micrometric scale can be solved by the nanoindentation technique [7]. The single fiber push-out technique, which has been intensively used for studying the interfacial shear behavior of composites, allows the use of indenter to push out a fiber perpendicularly aligned to the specimen surface and to measure the force and displacement with fine resolution. However, this technique is only suitable for the particles having cylindrical fibrous morphology, and is not applicable for the current Fe-TiB_2 system containing mainly irregular morphologies of reinforcements. For this reason, instead of using the traditional push-out tests, a modified nanoindentation method is adopted in the present work. A micron-sized sample, containing a wedge-shaped TiB_2 particle attaching to the ferrite matrix, was fabricated by focused iron beam (FIB), and was compressed using nanoindentation with a flat indentation tip. In order to further interpret the experimental results, simulation techniques were used. The experimental process was simulated by using the finite element method (FEM) and the cohesive zone model (CZM), in order to study the interfacial failure during the testing.

Experimental Procedure

Materials and Specimens

An ingot with a thickness of 80 mm was prepared by casting with in situ precipitation of TiB_2 particles during solidification, and was heated initially at 1473 K (1200 °C) and hot rolled in five rolling steps, with the finishing temperature of

1173 K (900 °C), to produce the sheet material with 3 mm thickness. Figure 1a shows the scanning electron microscopy (SEM) image of the cross-sectional microstructure of the sheet. The reinforcements consist of large primary TiB_2 particles surrounded by the small eutectic TiB_2 particles, as shown in Fig. 1a. The primary particles, 10–20 μm in size, are formed before eutectic solidification. The dominant morphology of primary particles is hexagonal prisms, as shown in Fig. 1b.

Several primary particles with the perfect hexagonal shape of cross sections were carefully chosen under the SEM, and were fabricated using FIB to form the wedge-shaped specimens, as schematically shown in Fig. 2.

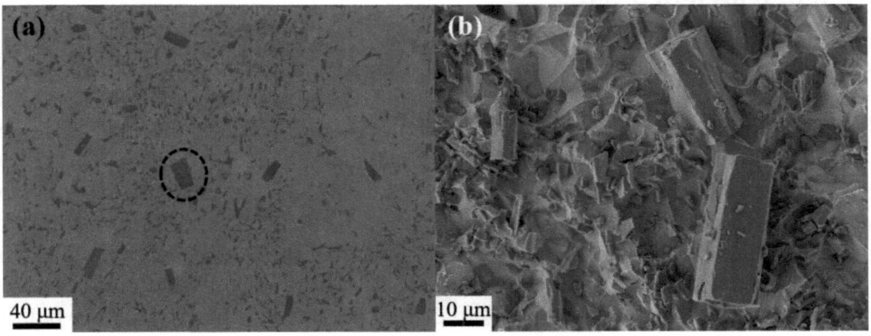

Fig. 1 **a** SEM micrograph showing the distribution of TiB_2 particles in ferrite matrix. The *dashed circle* indicates the primary particle; **b** SEM micrograph taken from the etched specimen, showing the morphologies of primary particles

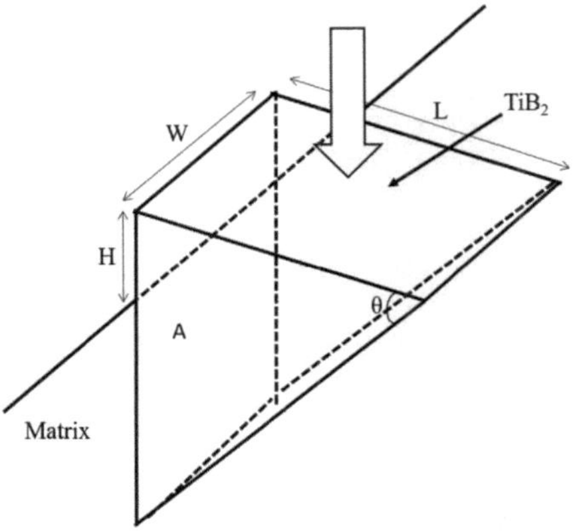

Fig. 2 Schematic representation of the wedge-shaped particle attaching to the matrix for the nanoindentation tests

Nanoindentation Tests

The specimen was loaded by a flat indenter tip. The flat tip can increase the contact area between the particle and the indenter, leading to the least particle damage. The nanoindentation tests were displacement controlled, and the maximum displacement is 2 μm.

Results and Discussions

Nanoindentation Test

A typical load-displacement curve of the nanoindentation test performed on the specimen shown in Fig. 3a is presented in Fig. 3b. When the compression load is first applied on the top surface of the wedge-shaped particle, the load increases linearly with the indenter displacement, indicating that the specimen deforms elastically at this stage. When the load reaches the value of around 30 mN, the slope starts to change, and when the applied load reaches P_{max}, the sudden transition of the indenter displacement occurs, which indicates the entire debonding of the particle/matrix interface. Once the particle debonds completely, the load will drop onto the bottom matrix, and that again causes the increase of load, as the bottom matrix can be plastically deformed and work hardened.

During the compression, the shear stresses at the interface are never homogeneous, and are likely to have the maximum value at the region near the top surface. The shear stresses, of course, increase with the applied load. This suggests that once the localized shear stress near the top surface reaches the critical value, the

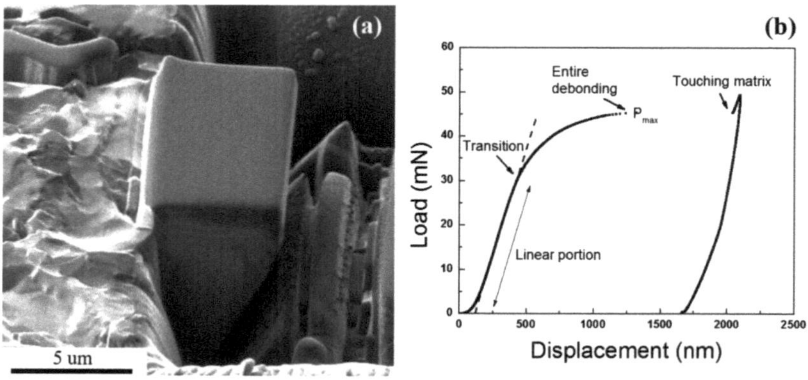

Fig. 3 **a** A sample fabricated using FIB, showing the wedge-shaped particle attaching to the matrix; **b** A typical load-displacement curve of the nanoindentation test performed on the specimen shown in (**a**)

debonding initiates, leading to the considerable reduction in the effectively bonded area of the particle/matrix interface. As the debonding proceeds, the location of maximum shear stress will continuously move as a response of the debonding process, until the entire particle gets debonded. In other words, once the applied load of the indenter reaches the critical value, the debonding initiates and proceeds until the entire debonding.

Simulation of Interfacial Behavior

In this work, the CZM in the bilinear form is used as the constitutive law to simulate the interfacial mechanical response during the nanoindentation test. As the interface mainly deforms in the shear mode, in the present work the normal traction at the interface is not considered. For the pure shearing, the relation between the tangential traction and the tangential separation is shown in Fig. 4.

In Fig. 4, τ_{max} is the interfacial shear strength, and Δ_{max} is the critical tangential separation at which the complete debonding is assumed. The two parameters are carefully selected to minimize the residuals between the experimental result and the simulation.

Finite Element Modeling

Figure 5 shows the FE simulation set-up, which is identical to the nanoindentation experiment. Both the matrix and the particle are modeled using the 8-node C3D8R nodes. The interface, which is assumed to be a thin cohesive layer connecting the matrix and the particle, is modeled by the 8-node COH3D8 nodes. The indenter is rigid in the FE simulation. To simulate the nanoindentation process, the elastic constitutive behavior is assumed for the TiB_2 particle, and the elastic-plastic behavior is assumed for the matrix. The Young's modulus E of particle and matrix

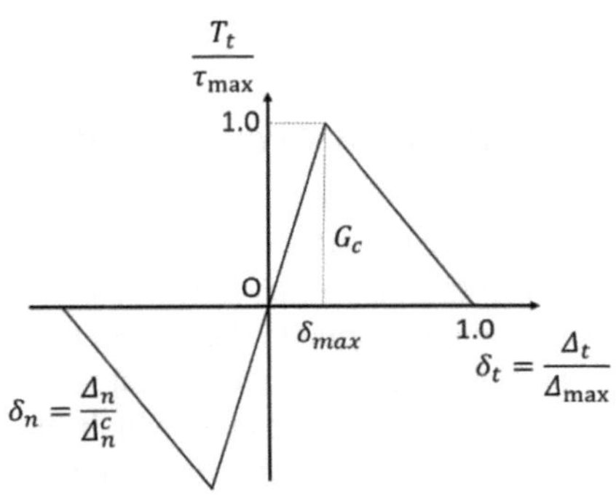

Fig. 4 Schematic diagram of the bilinear cohesive zone model representing the shear traction versus shear separation

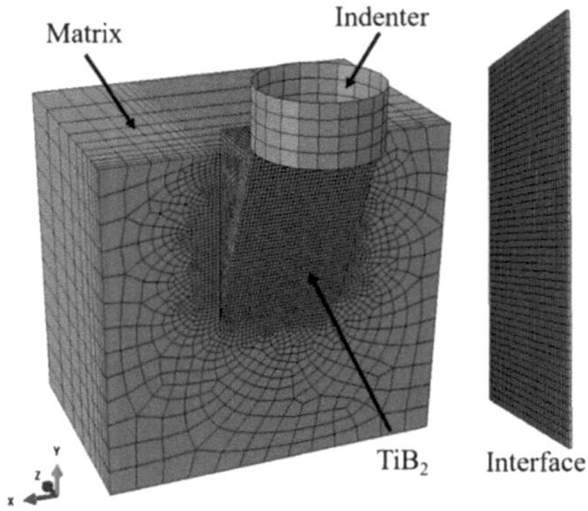

Fig. 5 Finite element model to simulate the experimental process of nanoindentation

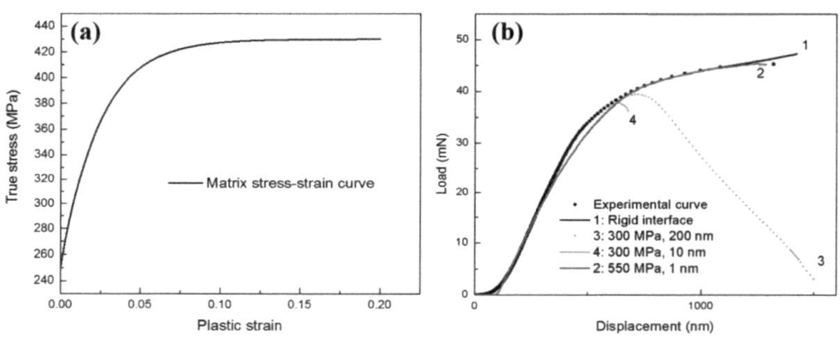

Fig. 6 **a** The true stress versus plastic strain curve, assigned as the plastic constitutive law of the matrix in the FE simulation; **b** Comparison of the nanoindentation tests: experiment and simulations

is assigned to be 565 and 210 Gpa, respectively, and the Poisson's ratio υ of particle and matrix is taken as 0.11 and 0.3, respectively. For the plastic constitutive behavior of the matrix, the true stress vs. plastic strain curve shown in Fig. 6a serves as the input for the matrix.

Comparison Between Simulation and Experimental Result

Figure 6b shows the various computational results together with the experimental load-displacement curve. It is shown by the simulation that the change in slope from linearity is due to the plastic flow of the matrix. It is also clear that the choices of both τ_{max} and Δ_{max} are crucial to reproduce the experimental curve. The value of τ_{max} determines the load at which the initial debonding takes place. By assuming that the interface is rigid (line 1 in Fig. 6b), the load will never decrease, suggesting that the drop of loading in the nanoindentation test is due to interfacial debonding. When the interfacial shear strength τ_{max} = 300 MPa, the load drops before reaching the experimental value of P_{max}, suggesting that the interfacial shear strength should be higher than 300 MPa.

In addition, the value of Δ_{max} determines the rate of the decrease of loading after debonding initiates. When τ_{max} is constant, the value of Δ_{max} is proportional to the fracture energy of the interface. In other words, the parameter Δ_{max} determines the fracture toughness of the interface, and its value should be carefully selected. With τ_{max} = 300 MPa, when the critical tangential separation distance is assumed to be 200 nm (line 3 in Fig. 6b), the load decreases gradually with the increase of displacement. The value of Δ_{max} is then altered to 10 nm (line 4 in Fig. 6b), and the entire debonding of interface takes place immediately after the initial debonding, matching well with experimental result.

The optimized values of τ_{max} and Δ_{max}, according to the best fitting of the experimental curve, are 550 MPa and 1 nm (line 2 in Fig. 6b), which indicate the considerably high strength but relatively low fracture toughness of the TiB_2/ferrite interface.

Summary and Conclusions

The present work can be summarized as follows:

- The hybrid method, combining the finite element simulation and cohesive zone model, provides one way to simulate the nanoindentation process and also to quantitatively interpret the interfacial mechanical properties.
- The slope change in the force-displacement curve of the nanoindentation test is due not to debonding, but to the plastic flow of matrix.
- The drop of the load is simulated to be caused by the initiation of interfacial debonding.
- The peak load of the indenter is determined by the interfacial shear strength τ_{max}, which is ~550 MPa based on the simulation.

References

1. Bouaziz O, Zurob H, Huang M (2013) Driving force and logic of development of advanced high strength steels for automotive applications. Steel Res Int 84:937–947
2. Rana R, Liu C (2014) Effects of ceramic particles and composition on elastic modulus of low density steels for automotive applications. Can Metall Q 53:300–316
3. Bonnet F, Daeschler V, Petitgand G (2014) High modulus steels: new requirement of automotive market. How to take up challenge? Can Metall Q 53:243–252
4. Hadjem-Hamouche Z, Chevalier JP, Cui Y, Bonnet F (2012) Deformation behavior and damage evaluation in a new titanium diboride (TiB_2) steel-based composite. Steel Res Int 83:538–545

5. Grande D, Mandell J, Hong K (1988) Fibre-matrix bond strength studies of glass, ceramic, and metal matrix composites. J Mater Sci 23:311–328
6. Lartigue-Korinek S et al (2015) Interfaces and defects in a successfully hot-rolled steel-based composite Fe–TiB_2. Acta Mater 98:297–305
7. Oliver WC, Pharr GM (1992) An improved technique for determining hardness and elastic modulus using load and displacement sensing indentation experiments. J Mater Res 7:1564–1583

Part XIV
Materials Extraction

Leaching of Copper–Cobalt Tailings from Democratic Republic of Congo

Y.R.S. Hara, S. Chama, D.M. Musowoya, G. Kaluba, J. Machona, P. Chishimba, Tina Chanda Phiri and S. Parirenyatwa

Abstract The investigation focuses on leaching of copper (2 wt%), cobalt (0.5 wt%) and iron (2.4 wt%) from complex tailings material from the Democratic Republic of Congo. The highest recovery of copper and cobalt into the leach solution were 78 and 10%, respectively, under direct leaching with sulphuric acid. Higher recoveries (90% for copper and 96% for cobalt) were attained when the material was pre-reduced with coal for 1.5 h in the temperature range of 700–950 °C. The increased recoveries were due to metallisation of all the copper and cobalt which is in the silicate matrix. The effects of pre-reduction temperature and sample to carbon weight ratios were analysed and found to be important in; (i) achieving high recoveries of copper and cobalt and (ii) decreasing the amount of iron in the leach solution. The samples were characterised by atomic absorption spectroscopy and scanning electron microscopy (SEM) techniques.

Keywords Leaching · Copper and cobalt · Tailings and silicate

Introduction

Copper ores from the Democratic Republic of Congo contain high amount of copper (2–20%) and small amounts of cobalt. The copper ores are commonly concentrated via froth flotation technique in order to achieve high grade concentrates. Nonetheless, a significant amount of copper is lost to the tailings during production of high grade (>25 wt%) copper concentrates.

Y.R.S. Hara (✉) · S. Chama · D.M. Musowoya · G. Kaluba · J. Machona · P. Chishimba · T.C. Phiri
Metallurgy Department, School of Mines and Mineral Sciences, The Copperbelt University, Kitwe Zambia, Zambia
e-mail: yotamuhara@gmail.com

S. Parirenyatwa
School of Chemical and Process Engineering, Institute of Materials Research, University of Leeds, Leeds LS2 9JT, UK

S. Ikhmayies et al. (eds.), *Characterization of Minerals, Metals, and Materials 2017*, The Minerals, Metals & Materials Series,
DOI 10.1007/978-3-319-51382-9_50

All cobalt that is associated with copper ore is lost to the tailings during flotation of copper concentrates. Most of the tailings dam from flotation plants in the Democratic Republic of Congo contain more than 0.3 wt% cobalt. According to London Metal Exchange (LME) in August, 2016, the copper and cobalt metals price per ton are $4800 and $26,000, respectively, such that it is very important to reclaim both copper and cobalt from tailings dam.

The cobalt and part of copper in the tailings dam, in the Democratic Republic of Congo is in the complex silicate form and hence it may not be recovered via flotation which is otherwise, the most economic route. As such, this work focuses on leaching of copper and cobalt from the flotation tailing dams. Leaching was considered because it is one of the most economic routes of treating low grade copper and cobalt materials [1, 2]. The effect of pre-treating the material prior to leaching was considered in order to improve the recovery of copper and cobalt into the leach solution.

Experimental

Materials

The material used in this experiment was obtained from the Democratic Republic of Congo. The sample was collected from fifty (50) different locations of the tailings dam in order to have a better representation of the sample. The samples were then mixed thoroughly and a representative portion was collected. Full chemical analysis of the sample is shown in Table 1 and it can be observed that the sample has 1.8 wt% Cu and 0.4 wt% Co. The main elemental constituents are Si, Mg and Al and these were found to be in the form of complex magnesium aluminium silicate. Iron, cobalt and part of copper were dissolved in this silicate matrix.

Leaching

For leaching experiments, the samples were ground to particle size of less than 106 μm. The weight of the sample in each leaching experiment was 100 g. Each sample was leached in sulphuric acid solution at a constant pH of 1.8 [3–5], under normal atmosphere. The pH was maintained by adding sulphuric acid. The total time for each experiment was 2 h. However, samples were collected at various

Table 1 Full elemental analysis of the sample (wt%)

Cu	Co	Fe	S	Si	Mg	Al	K	Ti	Ca	O
1.8	0.4	1.8	0.06	22.5	7.2	5.3	0.6	0.1	0.1	60.3

times such as 5, 10, 20 min in order to understand the progress of the reaction. The leach residue was washed with warm water (50 °C) to prevent crystallisation of copper.

Pre-treatment

100 g of the sample was reduced with 10 wt% coal for 1.5 h. The reduced sample was slightly sintered and hence it was ground for 5 min in a mortar and pestle. The ground sample was leached as described above.

Analysis and Characterisation of the Sample

The as-received and reacted samples were analysed and characterised by atomic absorption spectrometer (AAS) and scanning electron microscope (SEM). The weight of the sample for analysis was 0.5 g.

Results and Discussions

Direct Leaching

Direct leaching of the sample in sulphuric acid at a pH of 1.8 was carried out in order to dissolve copper and cobalt. The effect of time was studied by collecting the sample at various times and, analysing the leach residues via AAS. The copper and cobalt contents in the leach residues are shown in Table 2. It is evident from Table 2 that the grade of copper decreased from 1.8 to 0.9 wt% when the sample was leached for 5 min and this confirms that the reaction is extremely rapid. It can

Table 2 Analysis (wt%) of the leach residue after leaching the samples at various times

Duration (min)	Cu	Co
5	0.9	0.4
20	0.7	0.39
30	0.5	0.39
40	0.4	0.39
50	0.4	0.39
60	0.4	0.37
90	0.4	0.36
120	0.4	0.36

Table 3 Full analysis of the leach solution after leaching the samples at various times

Duration (min)	Cu	Co	Fe	S	Si	Mg	Al	Ca	O
30	21	1.2	0.4	11.8	2.7	1.3	0.8	1.9	58.2
90	23.1	1.3	0.4	12.9	0.6	1.4	0.5	1.5	57.8
120	22.4	1.2	0.2	13.3	0.8	1.1	0.5	3.78	56.3

be concluded from Table 2 that the leaching process completes in about 40 min as there was no further decrease in the grade of copper.

On the other hand, only about 10% cobalt was leached out even after 2 h and this is due to the fact that cobalt is in the complex silicate matrix. The copper and cobalt in the silicate matrix cannot be leached out, unless the silicate matrix is destroyed or leached out. Based on the results in Table 2, it can be concluded that the efficiency for direct leaching of copper and cobalt are 78 and 10%, respectively. Therefore, direct leaching should be accompanied by another method in order to recover cobalt and residual copper.

The leach solution was evaporated in order to fully analyse its elements. The overall elemental analysis of the leach solution is shown in Table 3 and it is evident from this table that the main constituent is copper (Cu) which is present in the form of copper sulphate ($CuSO_4$). The concentration of copper in the solution is very high, above 20 gpl. The main impurities in the leach solution are Mg and Ca which dissolved as magnesium and calcium sulphates, respectively. In general, the purity of the leach solution is higher than what is produced from most leaching plants.

Effect of Pre-treatment

As discussed above, cobalt and part of the copper are chemically dissolved in the silicate matrix. As such, the sample was pre-treated by reducing it with coal at various temperatures prior to leaching. The analysis for the leach residues are shown in Table 4. Three main observations can be made from the results in Table 4.

1. The grade of copper in the leach residue increases with increase in the pre-treatment temperature.

Table 4 Analysis of the leach residues, the samples were pre-treated with 10 wt% coal at various temperatures

Temperature (°C)	% weight		
	Cu	Co	Fe
700	0.59	0.05	1.64
800	0.36	0.02	1.24
850	0.49	0.02	1.71
900	1.78	0.02	0.36
950	1.69	0.04	0.58

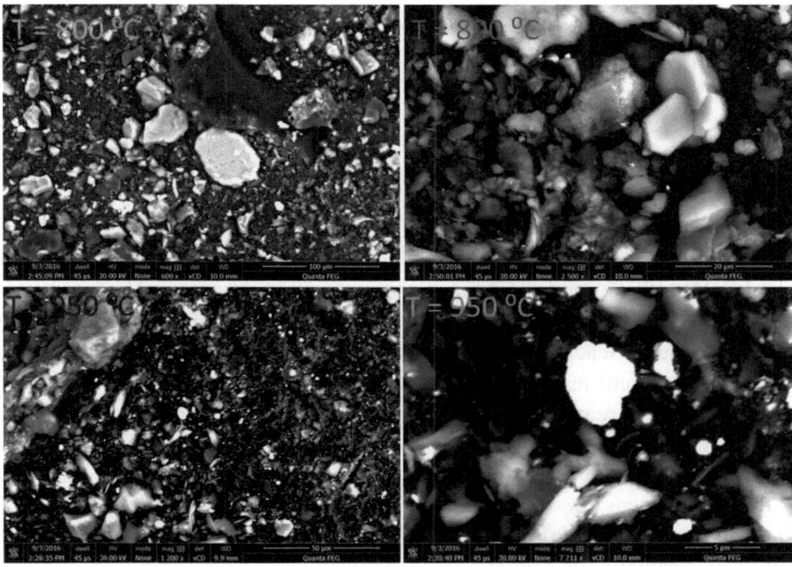

Fig. 1 Scanning electron microscopy (SEM) under backscattered electron imaging for the samples that was pre-treated and leached in sulphuric acid solution. The *bright particles* are metallic copper

2. The grade of iron increases with increase in pre-treatment temperature.
3. The grade of cobalt in the leach residue is very low.

From the above observations, it can be concluded that nearly all the cobalt was leached out after the samples were pre-treated via carbothermic reduction. The higher recovery of cobalt is due to the fact that, cobalt was reduced to metallic state during reduction and hence it was leached out. Iron was higher in the leach residue for the sample which was pre-reduced at low temperature (700 °C) and this is because, iron is in the magnetite (Fe_3O_4) form, which is less soluble in acid. Most of the iron was metallised and hence it was leached out as the pre-treatment temperature was increased. Conversely, most of the copper did not dissolved in acid when the samples were pre-treated at higher temperature which can be attributed to the fact that small amount of iron dissolved in copper, thereby making it less soluble in acid.

The scanning electron microscopy (SEM) images for the samples which were preheated at 800 and 950 °C and leached are compared in Fig. 1. The bright areas in the electron image are the un-leached copper particles. By comparison, there are more copper particles for the samples that were pre-treated at 950 °C than at 800 °C and this agrees with the analysis in Table 4. SEM-EDX analysis showed that the copper particles had about 0.3 wt% iron. The dissolved iron in copper prevented the dissolution of the later at pH of 1.8. The solubility of iron in copper increases with temperature and hence less copper was leached out at 900 and 950 °C.

Table 5 Analysis of the leach residues, the samples were pre-treated with 5 wt% coal at various temperatures

Pre-treatment temperature (°C)	% weight		
	Cu	Co	Fe
700	0.67	0.016	1.1
800	0.28	0.02	0.57
900	1.2	0.14	1.0

Since the leach residue for the sample which was pre-treated at 950 °C has high residual copper, it was leached again for 1 h at a pH of 1.3. The final leach residue only had 0.21 wt% Cu and the overall recovery of copper was calculated as 90%. Therefore most of the copper was leached out after double leaching. The main advantage of double leaching is that, cobalt and copper rich solutions are produced in the first and second leaching stages, respectively.

The effect of pre-treating temperature was also studied at coal addition of 5 wt% and the analysis of the leach residues are shown in Table 5. It can be observed from Table 5 that copper and cobalt decreased to 0.28 wt% and 0.02 wt%, respectively, when the sample was pre-treated at 800 °C. The recoveries of copper and cobalt into the solution for the sample which was pre-treated at 800 °C, were calculated as 86 and 96%, respectively.

As observed in Table 4, the recovery of copper decreased as the pre-treating temperature was increased to 900 °C. By comparison, more copper was leached out at coal addition of 5 wt% than at 10 wt% (see Table 4). Therefore, it can be concluded that the optimum coal addition is 5 wt%.

Conclusions

1. Copper decreased from 1.8 wt% to 0.4 wt% when the tailings material was leached for 40 min in sulphuric acid solution at pH of 1.8 (see Table 2).
2. About 90% of the cobalt in the material was not leached out at pH of 1.8. Decreasing the pH to 1.5 did not have any effect as the cobalt is chemically combined with iron, silica and magnesium.
3. Most of the cobalt was leached out when the sample was pre-treated via carbothermic reduction between 800 and 950 °C. However, the best pre-treatment temperature is 800 °C as the recoveries of copper and cobalt were 90 and 96%, respectively.
4. More iron dissolves together with cobalt as pre-treatment temperature is increased.
5. The best coal addition should be 5 wt% as more copper was leached (see Table 5).

References

1. Bodsworth C (1994) The extraction and refining of metals. In: Ralph B (ed) Materials science and technology. CRC Press, Boca Raton, Fl, p 346
2. Davenport WGL et al (2002) Extractive metallurgy of copper, 4th edn. Chemical, Petrochemical and Process. Elsevier, pp 1–452
3. Ntengwe FW (2010) The leaching of dolomitic-copper ore using sulphuric acid under controlled conditions. Open Mineral Process J 3:60–67
4. Stevanović Z et al (2009) Leach-sx-ew copper revalorization from overburden of abandoned copper mine cerovo, eastern serbia. J Min Metall 45B(1):45–57
5. Gharabagh M (2016) Acid leaching of malachite ore and separation of Cu(II) by di-(2-Ethylhexyl) phosphoric acid and tributyl phosphate in acetate buffer solution. Trans Indian Institute Metals doi:10.1007/s12666-016-0854-3

Optimum Operating Conditions and Characterisation of Lignin Extracted from Palm Fruit Bunch

E.I. Akpan, S.O. Adeosun and M.A. Usman

Abstract Preliminary studies on the extraction and characterisation of biodegradable lignin from palm oil fruit bunch (POFB) for carbon fibre production has been conducted. Central Composite Design was used to determine the optimum operating conditions for maximum yield and properties of the extracted lignin. Soda concentration (1–3 wt/vol%), temperature (50–90 °C) and reaction time (20–30 min) were analysed using R core software. Experimental results show that using reaction temperature of 90 °C, time of 30 min and 3 wt/vol% soda concentration yields 3.3 g of lignin from 30 g of the POFB. The path of steepest ascent from ridge analysis shows that with temperature of 130 °C, time of 40 min. and soda concentration of 3. wt/vol%, 8.7 g of lignin can be extracted from 30 g of POFB. Analysis of FTIR and SEM show that the lignin has the same structure in all cases with minimal impurities.

Keywords Palm fruit bunch · Lignin · Soda concentration · Reaction temperature · Reaction time

Introduction

Lightweight components designs are essential part of current technology geared at reducing CO_2 emission. In pursuit of this, attempts have been made to develop low cost, recyclable, biodegradable green composites from various renewable sources, which could substitute for existing metals and synthetic plastics in automotive,

E.I. Akpan (✉)
Institut fur Verbundwerkstoffe GmbH, 67663 Kaiserslautern, Germany
e-mail: emma_eia@yahoo.com

S.O. Adeosun
Department of Metallurgical and Materials Engineering,
University of Lagos, Lagos, Nigeria

M.A. Usman
Department of Chemical Engineering, University of Lagos, Lagos, Nigeria

S. Ikhmayies et al. (eds.), *Characterization of Minerals, Metals, and Materials 2017*, The Minerals, Metals & Materials Series,
DOI 10.1007/978-3-319-51382-9_51

building and packaging applications. Carbon fibres as reinforcement in polymers offer potentials for automotive applications and can compete with aluminium. These composites offer over 60% weight reduction without functional degradation [1].

However, the use of carbon fibre is limited because of the cost of conventional precursors such as Poly-Acrylic-Nitrile (PAN). On the other hand, most of the carbon fibres used is purely synthetic and not environmentally friendly. To satisfy this, there is need to develop renewable, biodegradable and cost-effective precursors in order to produce carbon fibres at a more competitive price for structural purposes. Polyethylene and lignin have been identified as alternative carbon fibre precursors with the highest potential for the automotive industry. Mainka et al. [1] reported that the use of lignin as carbon fibre precursor in place of PAN has the potential to reduce manufacturing cost by over 50%.

Lignin is one of the major components of the cell wall of higher plants. It is the second most abundant renewable bio-source next to cellulose [2]. Researchers have shown that a huge amount of technical lignin can be produced from plant and agricultural wastes sources [3–6]. An earlier study by Mussatto et al. [7] shows 81% recovery of lignin from Brewery spent grain black liquor. Sun et al. [8] also attempted isolation of lignin from the black liquor of oil palm empty fruit bunch and found that high quality lignin can be obtained through precipitation of pulp black liquor of the palm fruit bunch.

But, it is widely acknowledged that the diversity in structure of these sources prevents the use of single extraction, purification and processing methods for the low cost carbon fibres. Moreover, lignin purity, glass transition temperature, molecular weight and rheology depend on the source of biomass and extraction process used. The challenge therefore lies in the establishment of effective processing routes for each of the identified lignin source [3–6].

Having identified POFB as a source of lignin, which does not involve destruction of the ecosystem but utilization of waste materials from agro-based industries, there is need to ascertain; (i) how much of lignin can be extracted from POFB, (ii) what is the quality of the lignin from these sources (iii) what are the required modifications necessary for improve lignin properties. This preliminary study, therefore, focuses on the extraction of lignin from POFB, the optimization of the extraction process, the characterization and quantification of the morphological and physical properties of the extracted lignin as a source of carbon fibre for automotive applications.

Materials and Methods

Materials

POFB was obtained from an oil producing factory in Edo State, Nigeria. Prior to pulping process, the POFB was ground into 1 mm size particles and washed with distilled water thoroughly to remove impurities. The POFB was dried in the sun (average daily temperature of 32 °C) and later pre-treated with 75% hydrochloric acid. A mixture of 100 g acid per gram of dry matter, solid: liquid ratio of 1:8 g:g was treated at 100 °C for 25 min. After the reaction, a solid residue was obtained, separated, washed with distilled water until neutral pH and dried under sun (average daily temperature of 32 °C) until constant mass.

Experimental Design

A Central Composite Design (Response Surface Methodology) with three coded levels leading to 9 sets of experiments was used to evaluate the effect of three different variables: soda concentration (X_1), temperature (X_2) and pulping time (X_3) on the yield of lignin.

Soda Pulping

The pre-treated POFB solid residue was subjected to pulping reactions in a 200-ml water bath. Reactions were performed with POFB and the soda solution in a solid: liquid ratio of 1:20 g:g at different soda concentrations, temperatures and pulping times according to response surface methodology, central composite design (see Table 1). At the end of each reaction, the solutions were allowed to cool in still air to room temperature after which the black liquor was separated from the pulp by filtration in a 100% polyester cloth. The black liquor was stored in a beaker and the cellulose fibres obtained were exhaustively washed with distilled water until the pH becomes neutral. The fibres were dried (about 7 days) in the sun (average daily temperature of 32 °C) until a constant mass was obtained.

Lignin Recovery from Black Liquor

To sediment lignin from the black liquor 20 ml of the black liquor was treated with concentrated hydrochloric acid (98% w/w) until a solution pH of 2 was obtained. The modified liquor was centrifuged at 4000 rpm for 10 min and the precipitated lignin was separated, washed with distilled water until the supernatant was neutral.

Table 1 Experimental ranges and levels of the process independent variables

Independent variable	Symbol	Range (levels)		
		−1	0	+1
Soda concentration (% w/v)	X_1	1	2	3
Temperature (°C)	X_2	50	70	90
Pulping time (min)	X_3	20	25	30

The extracted lignin was dried in the sun (average daily temperature of 32 °C) until all the moisture was removed. The mass obtained per 100 ml of the black liquor was measured and recorded.

Analytics

The morphological characteristics of lignin and the cellulose fibrils were obtained by Scanning Electron Microscopy (SEM) using an ASPEX 3020 model variable pressure. The SEM is located in Covenant University Ota, Nigeria, and it operates with a 15 kV electron intensity beam equipped with Noran-Voyager energy dispersive spectroscope. The physico-chemical properties of the lignin and the cellulose fibrils were determined using Nicolet 6700 M Fourier Transform Infra-red spectrometer in transmission mode from Redeemers University, Ede, Nigeria in the range 400–4000 cm^{-1} at a resolution of 4 cm^{-1}.

Results and Discussion

Statistical Analysis of Process

In the optimization of the pulping conditions, there is need to identify the variables with the largest influence on the yield of the resulting lignin. The analysis of the effects of the variables used for POFB pulping was performed. A response surface model with two factor interactions gave the best regression coefficient with $R^2 > 0.91$. The high value of R^2 (>0.91) indicate suitability of the selected model for the process, as there is a close agreement between the experimental results and the theoretical values predicted by the model. The analysis shows that the effects of the factors are not independently significant but the interaction effects are important. Two and Three-dimensional response surfaces described by the model used were fitted to the experimental data point of the extracted lignin as shown in Figs. 1, 2 and 3.

Figure 1 shows the combined effect of temperature and concentration of the yield of lignin. It is obvious from Fig. 1a that increasing temperature and increase concentration will increase lignin yield. Within the experimental region a lignin content of 1.15 g/100 ml of black liquor is possible. The relationship between the factors and the yield is more or less linear. Figure 1b corroborates the observations in Fig. 1a but stresses that temperature has a major influence on the interaction.

The effect of interaction of temperature and time on the yield of lignin is shown in Fig. 2. Figure 2a is a two dimensional contour plot with a non-linear

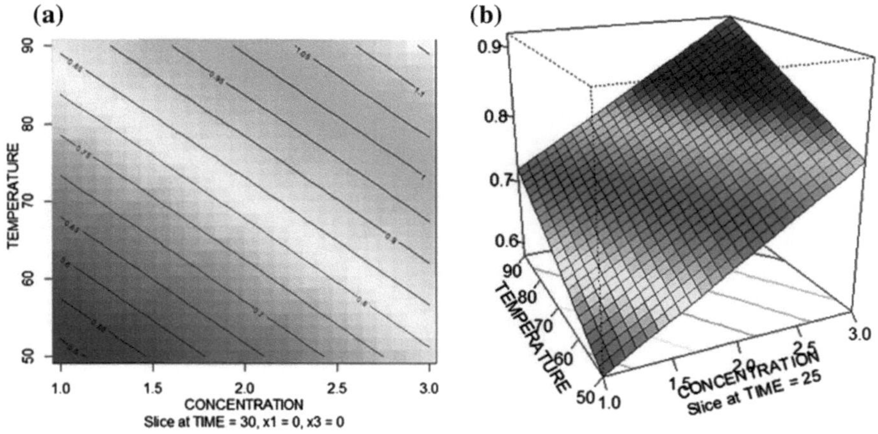

Fig. 1 Effect of change in temperature and concentration on yield of lignin

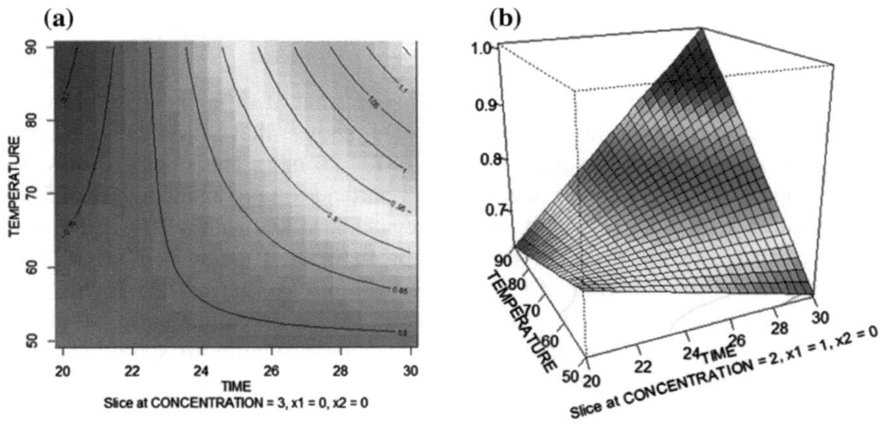

Fig. 2 Effect of change in temperature and time on yield of lignin

relationship. However, increase in both factors show increase yield of lignin. The possibility of obtaining up to 1.15 g/100 ml of the back liquor is presented within the experimental region. A similar feature is shown by the interaction between concentration and time as depicted in Fig. 3. Figure 3a is a non-linear relationship between concentration and reaction time with the same possibility as that shown by other interactions. However, the line of steepest ascent performed using the R-core suggest improvement yield with increase in temperature, time and concentration.

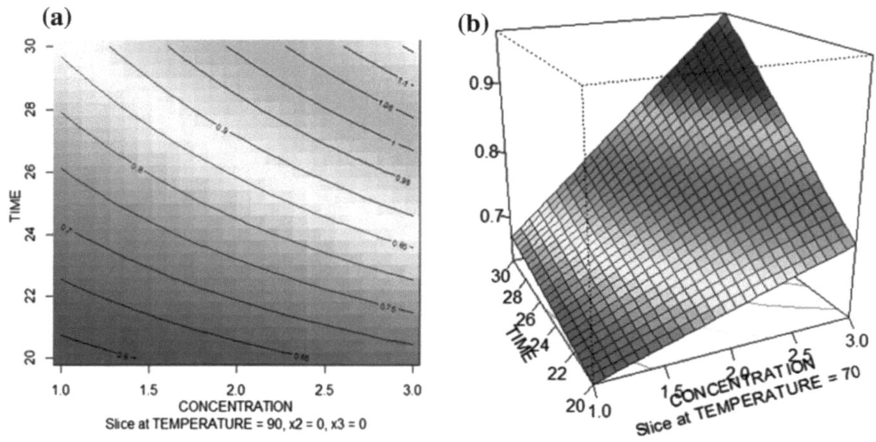

Fig. 3 Effect of change in time and concentration on yield of lignin

Morphology of Extracted Lignin and Cellulose Fibres

Figures 4, 5 and 6 are SEM/EDS of lignin extracted at different conditions and cellulose residue from the POFB. Figure 4a shows the SEM/EDS of lignin extracted with 3% NaOH solution at 90 °C for 20 min. The Figure shows that lignin appear as aggregation of many primary particles of average diameter of 5 μm with distinct size and shape while the aggregates size lies within 20–30 μm. This structure was recently reported by Kannangara et al. [9] and Sun et al. [10]. The EDS show the presence of Carbon (30.6%) and Oxygen (60.9%) with some sodium and chlorine impurities. Snowdon et al. [11] in an earlier study has shown carbon and oxygen content in this range for lignin. Figure 4b shows a highly ordered cellulose structure after delignifiation. This confirmed that lignin has been removed from the surface of the fibres.

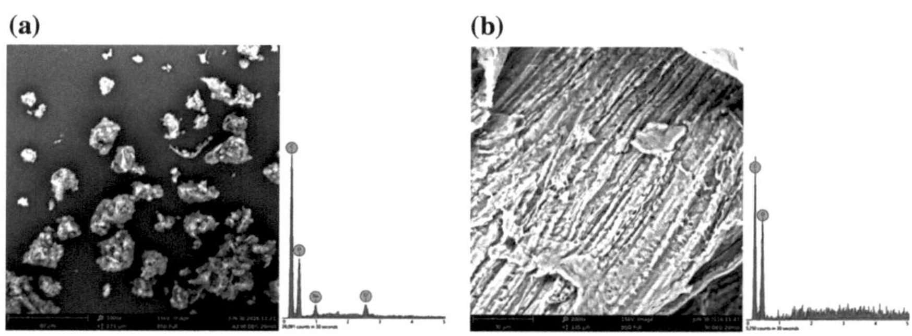

Fig. 4 SEM/EDS of **a** lignin and **b** cellulose fibres treated with 3% NaOH solution at 90 °C, and 20 min

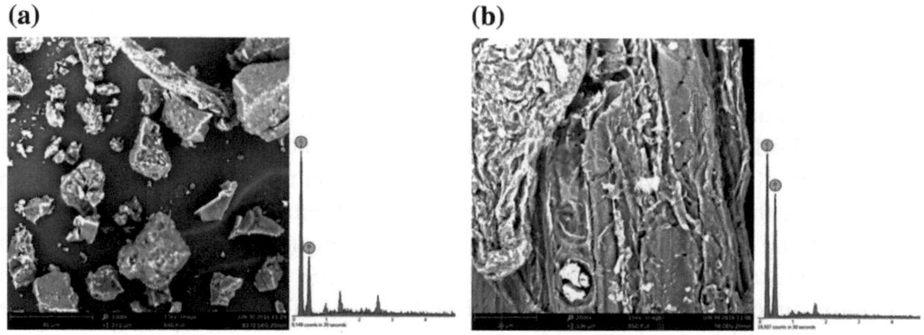

Fig. 5 SEM/EDS of **a** lignin and **b** cellulose fibres treated with 2% NaOH solution at 70 °C, and 25 min

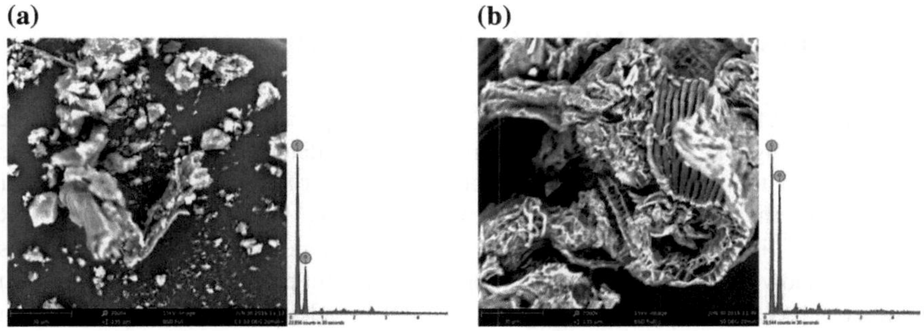

Fig. 6 SEM/EDS of **a** lignin and **b** cellulose fibres treated with 1% NaOH solution at 50 °C, and 20 min

SEM/EDS images of lignin and cellulose fibres extracted with 2% NaOH solution at 70 °C and 25 min reaction time are shown in Fig. 5. Figure 5a shows aggregated structure of lignin with large diameters compared to the one reported above. However, the aggregates here seem to be more compact. The EDS show the presence of 28.7% Carbon and 71.3% Oxygen. This indicates that lignin extracted with 3% NaOH solution at 90 °C for 20 min is richer in carbon than that extracted with 2% NaOH solution at 70 °C and 25 min. On the other hand Fig. 5b shows the network structure of cellulose fibres with small amounts of lignin on the surface. This indicates that lignin was not completely removed during the reaction. Ibrahim et al. [12] also observed similar structure of lignin and cellulse from selected biomas. Similar structures were presented but with different morphological features.

Figure 6 shows the SEM/EDS of lignin and cellulose fibres extracted with 1% NaOH solution at 50 °C, and 20 min. The SEM of the lignin show less compact aggregation of the lignin with very small diameter (see Fig. 6a) compared to those reported above. The EDS show the presence of 31.2% carbon and 68.8% oxygen. This is an indication that the lignin is rich in carbon and can be a source of carbon fibres for other applications. On the other hand, it is evident from Fig. 6b that lignin was to a large extent removed from the surface of the fibrils as the SEM did not reveal the presence of lignin. From the results shown here, it is clear that higher concentration and temperature favours the yield of lignin and the amount of carbon present in the lignin. The scenario is noted at low concentrations (1%) and temperatures (50 °C). However, at low concentration and temperature the yield of lignin is lower than that at higher concentrations (3%) and temperatures (90 °C). On the other hand, the use of medium concentrations and temperatures seem to give a high yield but low carbon content. This supports the position that the variables tested in this study directly affects both the yield and quality of lignin obtained from the pulping process. Lignin structure from this study is also significantly different from those reported in literature, which confirms the fact that lignin differs according to plant species and plant tissue types facing the external environment [13].

Structural Characteristics of Extracted Lignin

It is important to note that enhancing the industrial application of lignin will require both improving the lignin extraction processes and determination of the precise structure and the functional groups. The knowledge of this will help in further treatment of the extracted lignin for new applications. Figure 7 shows the FTIR spectrum of lignin extracted from the POFB. It is noted that all the samples extracted show the same functional groups. The spectra presented show absorption between 3200 and 3600 cm^{-1} assigned to (OH) broaden band of either hydrogen bonded or hydroxyl group in the phenolic and aliphatic compounds [14]. The absorption peak at 2936 cm^{-1} was due to (C–H) stretch band of methyl group [14, 15]. The stretching of (C=C) and (C–C) bands in the aromatic range were recorded at 1595 and 1462 cm^{-1} respectively [16]. The peak at 1124 cm^{-1} could be

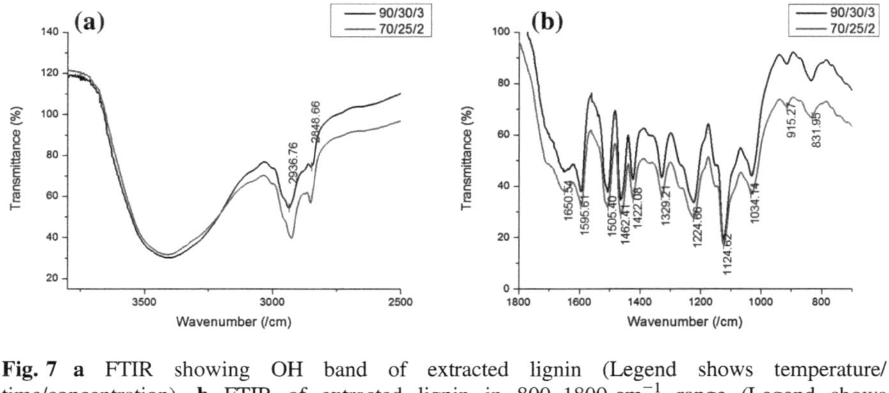

Fig. 7 **a** FTIR showing OH band of extracted lignin (Legend shows temperature/time/concentration). **b** FTIR of extracted lignin in 800–1800 cm^{-1} range (Legend shows temperature/time/concentration)

attributed to (C–O) bond in ethers. The band at 1224 cm^{-1} is attributed to syringyl and guaiacyl ring breathing with C–O stretching. The band at 1422 cm^{-1} is suggested to be attributed to aromatic ring vibrations, which are affected by the nature of the ring substituent and coupled to C–H in plane deformations. The band at 2848 cm^{-1} is assigned to C–H stretching in methyl, methylene and methoxyl groups.

The shoulder at 831 cm^{-1} is typical of aromatic C–H out of plane bending. The bands at 1329 and 1124 cm^{-1} are assigned to syringyl ring with C–O stretching and aromatic C–H in plane deformations respectively. The aromatic C–H in plane deformation for guaiacyl type lignin is shown in 1034 cm^{-1} band. It is suggested that the presence of syringyl (S) and guaiacyl (G) bands in the lignins show that the lignin extracted are similar to wood lignin that is normally HGS lignin [8, 17, 18]. The strong intensity band at 1505 cm^{-1} is assigned to the aromatic skeleton or the aromatic ring/aromatic methyl group vibration.

It is widely accepted that most important chemical functional groups in lignin include the hydroxyl, methoxyl, carbonyl and carboxyl groups. These are all present in the obtained spectra indicating that the lignin obtained is an important lignin. Ibrahim et al. [20] reported that the presence of guaiacyl-type (G) unit in lignin confirms that lignin has potential active site for polymerisation [18, 21]. This fact shades light on the possible post processing route to achieve the needed carbon fibres. The shoulder at 1650 cm^{-1} is attributed to the presence of conjugated carbonyl carboxyl stretching. The presence of this band may be an indication to the presence of more associated water or protein impurity. It is expected that the lignin contain considerable amounts of residual minerals, which may remain in the lignin fraction after pre-treatment process [22]. This may account for the presence of Na noted in Fig. 4a and suggest that the post processing route should involve the removal of mineral elements.

Conclusion

The extraction of lignin from Palm fruit oil bunch has been optimized using central composite designs. Palm fruit oil bunch has proven to be a good source of quality lignin for carbon fibre applications. Within the experimental design region the best combination of variables are reaction temperature of 90 °C, time of 30 min and 3 wt/vol% soda concentration, which yield 0.11 g/g of the POFB. Statistical analysis also show that the yield will double if the processing condition for extraction is temperature of 130 °C, time of 40 min. and concentration of 3.5 wt/vol%.

References

1. Mainka H (2015) Lignin—an alternative precursor for sustainable and cost-effective automotive carbon fiber. J Materials Res Technol 4:283–296
2. Lin J et al (2012) Chemical thermo-stabilisation for the preparation of carbon fibres from softwood lignin. BioResource 7:5634–5646
3. Sudo K, Shimizu K (1992) A new carbon fiber from lignin. J Appl Polym Sci 44:127–134
4. Baker FS, Griffith WL, Compere AL (2005) Low-cost carbon fibers from renewable resources. Automotive light weight materials. FY 2005 Process Report, 187–196
5. Kadla JF et al (2002) Lignin-based carbon fibers for composite fiber applications. Carbon 40:2913–2920
6. Eckert RC, Abdullah Z (2008) Carbon fibers from kraft softwood lignin. United States Patent US 0318043 A1, 25 Dec 2008
7. Mussatto SI, Fernandes M, Roberto IC (2007) Lignin recovery from brewer's spent grain black liquor. Carbohydr Polym 70:218–223
8. Sun RC et al (1999) Effects of precipitation pH on the physico-chemical properties of the lignins isolated from the black liquor of oil palm empty fruit bunch fibre pulping. Polym Degrad Stab 63:195–200
9. Kannangara M et al (2016) Effect of mixing hydrodynamics on the particle andfiltration properties of precipitated lignin. Chem Eng Res Des 105:94–106
10. Sun Y-C et al (2014) Dissolution, regeneration and characterisation of formic acid and Alcell lignin in ionic liquid-based systems. RSC Adv 4:2743–2755
11. Snowdon MR et al (2014) A study of carbonized lignin as an alternative to carbon black. ACS Sustain Chem Eng 2:1257–1263
12. Ibrahim MM et al (2010) Isolation and characterization of cellulose and lignin from steam. BioResources 5:397–416
13. Fox SC, McDonald AG (2010) Chemical and thermal characterization of three industrial lignins and their corresponding lignin esters. BioResources 5:990–1009
14. Ibrahim MNM et al (2010) Preparation and characterization of a newly water soluble lignin graft copolymer from oil palm lignocellulosic waste. Carbohydr Polym 80:1102–1110
15. Jin X-J et al (2005) Lignin characterization of triploid clones of populus tomentosa carr. For Stud China 7:52–56
16. Trinsoutrot I et al (2001) Assessment of the biochemical composition of oilseed rape (*Brassica napus L.*) 13C-labelled residues by global methods, FTIR and 13C NMR CP/MAS. Plant Soil 234:61–72
17. Anglès MN et al (2003) Characteristics of lignin obtained from steam-exploded softwood with soda/anthraquinone pulping. Wood Sci Technol 37:309–320
18. Tejado A et al (2007) Physico-chemical characterization of lignins from different sources for use in phenol-formaldehyde resin synthesis. Biores Technol 98:1655–1663
19. Sun R, Tomkinson J (2001) Fractional separation and physicochemical analysis of lignins from the black liquor of oil palm trunk fiber pulping. Sep Puri Technol 33:529–539
20. Ibrahim MNM et al (2011) Chemical and thermal properties of lignins from oil palm biomass as a substitute, for phenol in a phenol formaldehyde resin production. Carbohydr Polym 86:112–119
21. Hattalli S et al (2002) Adding value to Alfa grass (*Stipa tenacissima L.*) soda lignin as phenolic resins: lignin characterization. Polym Deg Stab 75:259–264
22. Abdelaziz OY, Hulteberg CP (2016) Physicochemical characterisation of technical lignins for their potential valorisation. Waste Biomass Valor. doi:10.1007/s12649-016-9643-9

Selection on the Process of Enriching Gold by Smelting from Refractory Gold Ores

Weifeng Liu, Xunbo Deng, Shuai Rao, Tianzu Yang, Lin Chen and Duchao Zhang

Abstract To overcome the shortcomings of traditional cyanidation process for refractory gold ores, the smelting processes have been proposed. These smelting processes could be divided into two kinds, namely, indirect smelting and direct smelting. In the indirect smelting process, the calcine of refractory gold ores are added in the copper or lead smelting processes during which gold can be enriched into copper matte or crude lead. Then gold is extracted from the anode slimes. In the direct smelting process, refractory gold ores are smelted directly with gold enriched into ferrous matte. Then gold is extracted from ferrous matte. Based on the direct smelting process, the metallurgy-beneficiation united process for refractory gold ores has been proposed. In this process, gold is firstly enriched into ferrous matte by direct smelting, then the grinding-flotation process is used to obtain gold-rich sulfurous iron ore, and finally gold is extracted via hydrometallurgical method. The metallurgy-beneficiation united process has advantages of high gold recovery ratio, environment-friendly and low cost, which provides a new idea for the extraction of gold from refractory gold ores.

Keywords Refractory gold ores · Pretreatment · Smelting · Enrichment

Introduction

Gold is a strategic precious metal which can be widely used in gold ornaments, currency reserves and high-tech industries. According to United States Geological Survey (USGS), the amount of identified gold is 89,000 tons in the world, while a third of which is contained in the refractory ores [1]. With the gradual depletion of the high grade resources, refractory gold ores will become more and more important. China produces 450 tons gold annually, which ranks 1st in the world.

W. Liu · X. Deng · S. Rao · T. Yang (✉) · L. Chen · D. Zhang
School of Metallurgy and Environment, Central South University, 410083 Changsha Shi, China
e-mail: tianzuyang@163.com

S. Ikhmayies et al. (eds.), *Characterization of Minerals, Metals, and Materials 2017*, The Minerals, Metals & Materials Series,
DOI 10.1007/978-3-319-51382-9_52

Refractory gold ores (RGO) stand for the minerals for which gold leaching ratio is less than 80% even after fine grinding. Especially for the arsenic containing refractory ore, with fine gold particles even in micron size encapsulated in pyrite or arsenopyrite, the gold leaching ratio can be as low as 50%. Therefore, refractory gold ores should be pretreated in order to improve gold leaching ratio.

There are many researches on the pretreatment methods of refractory gold ores. The most widely applied methods in industry are two stages roasting, pressure oxidation and bacterial oxidation [2]. During the two stages roasting process, arsenic and sulfur are oxidized and volatilized in the form of As_2O_3 and SO_2 while porous calcine is produced for leaching. This method has been widely applied due to its simple process and low cost. But Fe_2O_3 generated during the roasting process would cause a secondary encapsulation, resulting in gold leaching ratios vary between 75 and 92% and cyanide tailings with gold contents of 2.0–20.0 g/t. During pressure oxidation process, pyrite and arsenopyrite are decomposed in the acidic solution under high temperatures and oxygen pressures. This technology is efficient and less sensitive to other impurities. Moreover, high gold recovery (95–97%) can be obtained. However, the large investment, high processing cost and a large amount of waste water and residue constrained its wide industrial application. Bacterial oxidation is an advanced leaching technology. Pyrite and arsenopyrite can be oxidized with bacteria in an environment-friendly way. This method has a simple process, and high leaching rate of gold (92–95%) can be reached. The gold content of cyanide tailing is 2.0–5.0 g/t. The main disadvantages are long oxidation period and high investment, etc.

In the smelting process of copper, nickel, lead and other heavy metals, trace amounts of gold, silver and other precious metals will be enriched in the matte or metal phase. The precious metals can then be extracted from the corresponding anode slimes. It indicates that the matte or heavy metals can be used to collect precious metals [3]. Metallurgists have been trying to use heavy metals to enrich gold from refractory gold ores in the direct or indirect smelting process. Hence, we propose the metallurgy-beneficiation united process to treat refractory gold ores. The process has advantages of high recovery ratio, environment-friendly and low cost, which provides a new idea for extracting gold from refractory gold ores.

The Indirect Smelting Process of Refractory Gold Ores

Refractory gold ores contain a high content of arsenic as well as very low copper and lead, therefore, it is unsuitable to be added into heavy metals smelting process directly. It is necessary to remove sulfur and arsenic by oxidizing roasting before conducting heavy metals smelting process. And then gold can be enriched into copper matte or crude lead. Finally gold is extracted from anode slime.

The indirect smelting process has been successfully applied in some metallurgy enterprises [4–6]. Firstly, the As content was decreased to 0.5% by oxidizing roasting. And then the calcine was mixed with copper concentrate to conduct smelting process. After completing these processes, the contents of Cu, Au and Ag in flotation tailings were 0.50%, 0.50 g/t and 9.40 g/t respectively. Almost all precious metals were enriched into copper matte. Therefore, it can be inferred that the indirect smelting process is capable of extracting gold from refractory gold ores efficiently. The successful application of indirect smelting process provides a new development direction for the gold extraction from refractory gold ores. But adding the calcine not only causes huge energy consumption, but also brings some new problems:

(1) The calcine of refractory gold ores was used as the fusing agent during smelting process. However, the addition of calcine was relatively low due to its high Fe_2O_3 and SiO_2 content, constraining gold yield.
(2) An especially high amount of smelting slag causes copper and gold loss, decreasing gold recovery ratio.

To reduce the loss of gold in smelting slag, the indirect smelting process of enriching gold by crude lead smelting was proposed. The main reason is that it is more effective to enrich precious metals compared with copper smelting process. Gold can be enriched into crude lead while the calcine of refractory gold ores was added in the lead smelting process, and finally gold was recovered from anode slime. The small smelting slag production could reduce the loss of gold and improve the recovery of gold greatly. But the requirement of slag type in lead smelting process causes the low treatment capacity.

To solve the problem of low treatment capacity, the lead paste was also proposed to enrich gold [7]. Firstly, refractory gold ores (Au 18.05 g/t, As 1.62%, SiO_2 43.03%, S 20.76%, Fe 19.6%, CaO 7.42%, C 6.95%) and lead paste (Pb 69.86%, S 5.60%) were mixed to conduct smelting process, then gold was enriched in crude lead, and finally gold was extracted from lead anode slime. In this process, the gold recovery ratio was 97.02%, and the gold content of smelting slag was 0.39 g/t, while the gold content of crude lead was 49.56 g/t, which was 2.75 times than that of refractory gold ores. It can be concluded that the mixed smelting of refractory gold ores and lead paste not only increases the treatment capacity, but also improves the gold recovery.

In addition, the iron smelting process was studied to enrich gold, where the calcine of refractory gold ores were smelted in blast furnace after briquetted [8], then gold was enriched in the pig iron, and finally, gold was extracted from anode slime after electrolysis refining. In this study, the compositions of calcine of refractory gold ores were Fe 60.62%, SiO_2 9.38% and Au 9.82 g/t. When the raw materials were smelted at 1500 °C for 90 min, with the C/O of 0.93 and the binary basicity of slag of 1.0, the pig iron with Fe 97.27%, C 1.38% and Au 16.0 g/t was produced. In this process, the recovery ratios of iron and gold were 97.49 and 99.10% respectively, and the gold enrichment ratio of 1.63 was obtained. However, because the iron is selected as the agent to enrich gold, the process requires appropriate compositions of raw materials, and also has the shortcomings of high energy consumption and complicated technologies.

The Direct Smelting Process of Refractory Gold Ores

The direct smelting process of refractory gold ores is the other important method for extracting of gold from refractory gold ores. In general, it is suitable for antimony containing refractory gold ores. The principle is that gold can be enriched into smelting product directly.

The refractory gold ore with a high content of antimony (25.0–35.0% Sb, Au 30 g/t) is commonly used as an important raw material in the smelting of antimony [9]. The melting products are antimony oxide, crude antimony, antimony matte and slag. Crude antimony can be viewed as a significantly material to enrich gold. Above 90% of antimony can be volatilized and recovered in the form of antimony oxide in smelting process. The crude antimony contains 88.0–92.0% Sb, 3.0–7.0% Fe and 0.13% Au. The Fe, S, Sb and Au contents in antimony matte are 58–67%, 24–26%, 4–8% and 4–8 g/t, respectively. The ratios of gold among them are 6.2, 85.0, 6.8 and 6.2%, respectively.

During the electric furnace melting process, crude antimony can be produced by reduction smelting. And then crude antimony can be further processed to produce Au-Sb alloys [10]. The Melting products are crude antimony, antimony matte and slag. The crude antimony contains 92.1–97.4% Sb, 0.2–3.0% Fe, 0.04% Au. The contents of Fe, S, Sb and Au in antimony matte are 55–68%, 20–25%, 0.8–4.5% and 2–3 g/t, respectively. The proportions of gold among them are 95.4, 2.9 and 1.7% respectively.

Above mentioned two methods both make use of crude antimony to enrich gold. However, the gold content in the different antimony matte is absolutely various. The antimony and gold content obtained in the electric furnace melting process is lower compared with blast furnace smelting process, indicating that gold content in the antimony matte is strongly dependent of antimony content.

Studies of geochemistry show that gold is easy to migrate and enrich along with characteristic elements such as arsenic and antimony during the endogenous mineralization process of gold deposits. Therefore, arsenic and antimony are always regarded as indicator elements for gold prospecting. However, the coexistence of those elements makes it difficult to extract gold from refractory gold ores. The mineralization process in gold deposits suggests that it is better to enrich gold with arsenic and antimony rather than removing them from refractory gold ores. The smelting process also shows that there is a proportional relationship between the contents of antimony and gold in antimony-ferrous matte. Therefore, the direct smelting process of refractory gold ores was proposed, during which the refractory gold ores are mixed with fluxes and melted directly to enrich gold in ferrous matte, and then gold is further enriched by selective converting. Finally, gold is extracted from the products of converting process.

The antimonial refractory gold ore containing Au 48.0 g/t, Sb 4.5%, As 5.0% was directly smelted under following conditions: FeO/SiO_2 1.1, CaO/SiO_2 0.4,

1200 °C, 30 min. The antimony-ferrous matte and slag can be separated easily during the smelting process [11]. The contents of gold in antimony-ferrous matte and smelting slag were 143.58 and 0.75 g/t respectively. The recovery rate of Au reached 98.47%, which indicated that antimony-ferrous matte can be used as an effective carrier to enrich gold.

The smelting slag and antimony-ferrous matte were detected by digital automatic polarizing microscope and scanning electron microscope, and the results are shown in Fig. 1. It can be seen that the melting slag mainly contains FeO, SiO_2 and Al_2O_3, as well as a small number of As and Sb. Antimony-ferrous matte is mainly composed of FeS and FeO. And a small proportion of Sb and FeAs are present in it.

Refractory gold ore containing a small amount of copper is more efficient to enrich gold in direct smelting process. The refractory gold ore containing Au 72.07 g/t, Cu 3.0–5.0% and As 5.58% was melted directly under conditions: FeO/SiO_2 1.2, CaO/SiO_2 0.4, temperature 1250 °C, oxygen flux 0.8 L/min, ventilation time 10 min. The boundary between the copper-ferrous matte phase and slag phase is very obvious [12]. The contents of gold in copper-ferrous matte and smelting slag were 144.16 and 0.40 g/t. The recovery ratio of Au reached 98%, which indicated that copper-ferrous matte can be used as an effective carrier to enrich gold.

Based on above analysis, gold can be enriched effectively in antimony matte or copper matte. And gold can be further enriched by converting process. However, some researches have shown that it is difficult to separate the metal phase and slag phase during the converting process. Moreover, the extent of gold enrichment is relatively low.

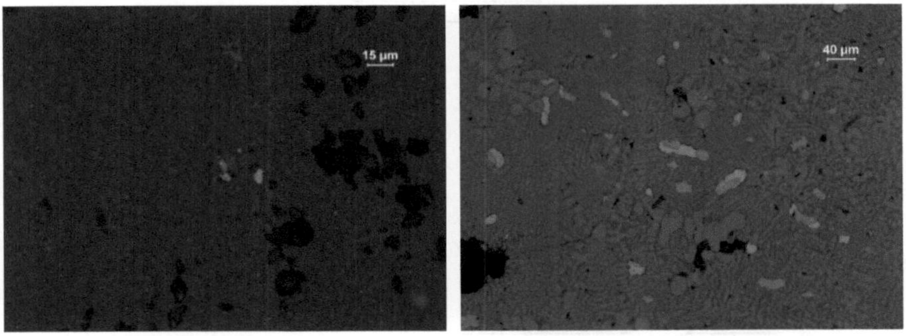

(a) smelting slag (b) antimony-ferrous matte

Fig. 1 Microscopic morphology of the melting products **a** smelting slag **b** antimony-ferrous matte

The Metallurgy-Beneficiation United Process of Refractory Gold Ores

Various pretreatment-cyanidation processes for extracting gold from refractory gold ores have some problems including low extraction of gold, high processing cost, long processing period and poor adaptability. The indirect smelting process for refractory gold ores requires not only an entire pretreatment system of roasting, but also a large smelting system of copper or lead. The enrichment of gold for refractory gold ores can be achieved by direct smelting process, but it is difficult to extract gold from matte phase in following converting processes. These disadvantages restrict the application for extracting gold from refractory gold ores by smelting process.

To solve the problems mentioned above, a novel metallurgy-beneficiation united process to recover gold from refractory gold ores is proposed. Firstly, the enriched gold matte is produced by smelting process. And then the enriched gold matte is cooled, grinded, floated to produce gold-rich pyrites. Last, gold is extracted by hydrometallurgical method. The united process has many advantages including high gold recovery, environment-friendly and low cost, which provides a new idea for extracting gold from refractory gold ores (Fig. 2).

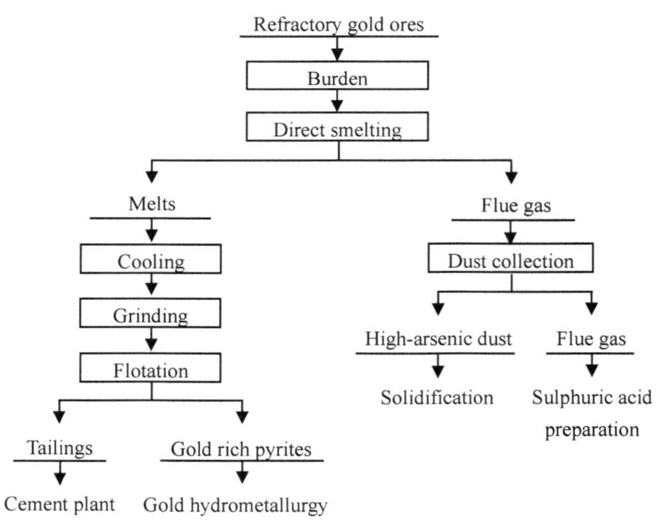

Fig. 2 Flowsheet of metallurgy-beneficiation united process for extracting gold from refractory gold ores

Conclusions

In order to overcome the shortcomings of traditional cyanidation process for refractory gold ores, the smelting process has been proposed to enrich gold from refractory gold ores. In the indirect smelting process, the calcines of refractory gold ores are added into the lead or copper pyrometallurgy system. And Gold can be recovered from lead and copper anode slimes. In the direct smelting process, gold can be enriched directly in the form of copper matte or antimony matte. The gold recovery is above 98.0% and the gold content in the slag is below 1 g/t, suggesting that matte is an effective carrier to enrich gold. Based on the view, the method on extraction of gold from refractory gold ores by metallurgy-beneficiation united process has been proposed. Firstly, refractory gold ores are used as materials to produce enriched gold matte by smelting process. Then the enriched gold matte is cooled, grinded, floated to produce gold-rich pyrites. Last, gold is extracted by hydrometallurgical method. The united process has many advantages including high gold recovery, environment-friendly and low cost, which provides a new idea for extracting gold from refractory gold ores.

Acknowledgements The financial support from National Science Foundation of China (No: 5140040395) is gratefully acknowledged.

References

1. George MW et al (2011) U.S. geological survey mineral yearbook. U.S. Geol Surv 5:1–18
2. Yang TZ (2005) Metallurgy and chemicals of precious metals. Central South University Press, Changsha, pp 210–215
3. Chen J (2007) Discussion on the micro-mechanism of precious metals trapped in pyro-metallurgical processes by base metals and matte phase. Eng Sci 9(5):11–16
4. Wang EX et al. Extraction method of multiple elements in high arsenic complex gold concentrate. Patent number:ZL200910020494.9
5. Cui ZX et al (2010) The theory and practice of copper smelting with oxygen enriched bottom blowing. China Nonferrous Metall 12(6):21–26
6. Qu SL et al (2012) Plant practice of and de bottom sign discussion on oxygen enriched blowing smelting. Nonferrous Metals (Extractive Metallurgy) 3:21–26
7. Luo XL et al (2014) Enriching gold from refractory gold concentrate by reducing sulfur-retention smelting. Chin J Nonferrous Metals 24(12):3129–3135
8. Wang Y et al (2013) Enriching gold by iron-making from roasting cinder of sulfur concentrate containing gold and arsenic. J Central South University (Sci Technol) 44(1):8–13
9. Liu WF et al (2008) Improvements on the process of gold-antimony concentrate smelting in China. Paper presented at TMS 2008 annual meeting, vol 8. New Orleans, Louisiana, 9 Mar 2008
10. Myzenkov FA, Taraov AV (2006) Development and creation of modern metallurgical technology for antimony production with gold recovery from domestic gold-antimony concentrates in Russia, vol 10. In: Paper presented at TMS 2006 annual meeting, San Diego, California, 27 Aug 2006

11. Liu ZL (2014) Research on the process of treating refractory gold ores containing antimony with bath smelting technology. Master Degree Thesis, Central South University, Changsha, Hunan
12. Xiao QK (2016) Enrichment of gold in copper-containing materials from gold concentrate via matte smelting method. Master Degree Thesis, Central South University, Changsha, Hunan

Selection on the Process for Removing and Recovering Antimony from Antimonial Refractory Gold Ores

Weifeng Liu, Xinxin Fu, Shuai Rao, Tianzu Yang, Duchao Zhang and Lin Chen

Abstract It is well known that the gold ores containing a high content of arsenic and antimony are very refractory. In this paper, the adverse effect of antimony on the extraction of gold by pretreatment-cyanide leaching was discussed in detail. Moreover, these processes of removing and recovering antimony from alkaline sodium sulfide or hydrochloric acid systems were introduced. Last but not the least, a novel cleaning technology for recovering antimony from refractory gold ores was proposed. Antimony was dissolved into the solution with sodium sulfide and sodium hydroxide as leaching agents. And then pressure oxidation technology was applied to prepare sodium pyroantimonate from sodium thioantimonite solutions. This process is more promising owing to its high removal ratio of antimony, low production cost and environment friendly, which provides a positive guidance for the extraction of antimony from antimonial refractory gold ores.

Keywords Refractory gold ores · Antimony · Leaching · Pressure oxidation

Introduction

Gold is a strategic precious metal which can be widely used in gold ornaments, currency reserves and high-tech industries. According to United States Geological Survey (USGS), the amount of identified gold is 89,000 tons in the world, while a third of which is contained in the refractory ores [1]. With the gradual depletion of the high grade resources, refractory gold ores will become more and more important. China produces 450 tons gold annually, which ranks 1st in the world.

Refractory gold ore (RGO) stands for the minerals for which gold leaching ratio is less than 80% even after fine grinding. Especially for the arsenic containing refractory ores, with fine gold particles even in micron size encapsulated in pyrite or

---

W. Liu · X. Fu · S. Rao · T. Yang (✉) · D. Zhang · L. Chen
School of Metallurgy and Environment, Central South University,
Changsha 410083, China
e-mail: tianzuyang@163.com

S. Ikhmayies et al. (eds.), *Characterization of Minerals, Metals, and Materials 2017*, The Minerals, Metals & Materials Series,
DOI 10.1007/978-3-319-51382-9_53

arsenopyrite, the gold leaching ratio can be as low as 50%. Therefore, refractory gold ore should be pretreated in order to improve gold leaching ratio.

There are many researches on the pretreatment methods of refractory gold ores. But the most widely applied methods in industry are two stages roasting, pressure oxidation and bacterial oxidation [2]. During the two stages roasting process, arsenic and sulfur are oxidized and volatilized in the form of As_2O_3 and SO_2 while porous calcine is produced for leaching. This method has been widely applied due to its simple process and low cost. But Fe_2O_3 generated during the roasting process would cause a secondary encapsulation, resulting in gold leaching ratios vary between 75–92% and cyanide tailing with gold contents of 2.0–20.0 g/t. During pressure oxidation process, pyrite and arsenopyrite are decomposed in the acidic solution under high temperatures and oxygen pressures. This technology is efficient and less sensitive to other impurities, moreover, high gold recovery (95–97%) can be obtained. However, the large investment, high processing cost and large amount of waste water and residue constrained its wide industrial application. Bacterial oxidation is an advanced leaching technology. Pyrite and arsenopyrite can be oxidized with bacteria in an environment-friendly way. This method has a simple process, high leaching ratio of gold (92–95%) can be reached. The gold content of cyanide tailing is 2.0–5.0 g/t. The main disadvantages are long oxidation period and high investment, etc.

The influence of antimony on refractory gold ores pretreatment and following leaching process was summarized in this paper. Moreover, the differences on removing and recovering antimony between acidic and alkaline systems were introduced. Finally, an environment friendly process of removing and recovering antimony from antimonial refractory gold ores was proposed. This process provides a positive guidance for the extraction of antimony from antimonial refractory gold ores owing to high removal ratio of antimony and low production cost.

Effect of Antimony on Cyanide Leaching and Pretreatment

Effect of Antimony on Cyanide Leaching

The main antimony phase of antimonial refractory gold ores is stibnite (Sb_2S_3). In the process of alkaline cyanide leaching, stibnite increases the consumption of OH^-, CN^- and O_2. The main reactions are listed as follows:

$$Sb_2S_3 + 6OH^- = SbO_3^{3-} + SbS_3^{3-} + 3H_2O \quad (1)$$

$$SbS_3^{3-} + 6OH^- = SbO_3^{3-} + 3S^{2-} + 3H_2O \quad (2)$$

$$4SbS_3^{3-} + 6CN^- + 6H_2O + 3O_2 = 2Sb_2S_3 \downarrow + 6CNS^- + 12OH^- \quad (3)$$

It can be seen from Eq. (3) that Sb_2S_3 is deposited on the surface of gold particles, forming a dense product layer and hindering the diffusion of CN^- and O_2 to the gold particles, leading to low leaching ratio [3, 4].

Effect of Antimony on Pretreatment Process

The melting point of antimony compounds, generated in the roasting process of antimonial refractory gold ores, is low(the melting point of Sb_2S_3 and Sb_2O_3 is 546 and 655 °C, respectively). It is easy to fuse with each other, which causes the gold particles wrapped again [5, 6], resulting in low gold leaching ratio [5, 6]. In the process of pressure oxidation pretreatment of refractory gold ores, the precipitates of antimony compounds were produced, covering with the surface of gold particles and hindering the further leaching reaction. In the process of biological oxidation pretreatment, the antimony minerals seriously affect the oxidation rate of gold bearing minerals [7].

Because of the adverse effects of antimony on pretreatment cyanidation process, it is very difficult for metallurgical enterprises to make full use of antimonial refractory gold ores. Therefore, it is necessary to remove antimony from antimonial refractory gold ores before conducting conventional pretreatment process. In general, removal antimony process can be divided into alkaline sodium sulfide system and hydrochloric acid system.

Recovering of Antimony from Sodium Sulfide Solution

Removing of Antimony from Antimonial Refractory Gold Ores

The main antimony phase of antimonial refractory gold ores is stibnite (Sb_2S_3). Sodium sulfide is proposed as leaching agent to remove antimony from antimonial refractory gold ores [8]. Meanwhile, sodium hydroxide is also used to prevent the hydrolysis of sodium sulfide. The main equations are expressed as follows:

$$Sb_2S_3 + 3Na_2S = 2Na_3SbS_3 \quad (4)$$

$$Na_2S + H_2O = NaOH + NaHS \quad (5)$$

Leaching of antimony in antimonial refractory gold ores with sodium sulfide alkaline solution has some advantages, such as high leaching ratio of antimony and limited dissolution of arsenic and gold.

Recovering of Antimony from Sodium Sulfide Solution

There are many researches on recovering antimony from sodium thioantimonite solution. The major methods are as below:

(1) Neutralization precipitation

The sodium thioantimonite solutions can be treated by neutralization precipitation method. The main neutralization agents are sulphuric acid, carbon dioxide or sulfur dioxide. And then antimony can be recovered in the form of antimony sulfide. The corresponding reaction equation can be expressed as follows:

$$2Na_3SbS_3 + 3H_2SO_4 = Sb_2S_3 \downarrow + 3Na_2SO_4 + 3H_2S \uparrow \quad (6)$$

$$2Na_3SbS_3 + 3CO_2 + 3H_2O = Sb_2S_3 \downarrow + 3Na_2CO_3 + 3H_2S \quad (7)$$

$$2Na_3SbS_3 + 3SO_2 + 3H_2O = Sb_2S_3 \downarrow + 3Na_2SO_3 + 3H_2S \quad (8)$$

Compared with other methods, although neutralization precipitation has the characteristics of short flow and low production cost, some shortcomings also are inevitable, including low-quality products and bad operation conditions.

(2) Displacement

The antimony in the sodium thioantimonite solutions can be displaced by metals powder as displacement agent, such as aluminium, zinc, copper and sodium amalgam. The main chemical reactions can be listed as follows:

$$2Na_3SbS_3 + 8NaOH + 2Al = 2Sb + 2NaAlO_2 + 4H_2O + 6Na_2S \quad (9)$$

$$2Na_3SbS_3 + 3Zn = 2Sb + 3ZnS + 3Na_2S \quad (10)$$

$$2Na_3SbS_3 + 3Cu = 2Sb + 3CuS + 3Na_2S \quad (11)$$

$$Na_3SbS_3 + 3Na(Hg) = Sb(Hg) + 3Na_2S \quad (12)$$

Although antimony can be recovered by displacement methods, the products are impure.

(3) Pressure hydrogen reduction

The antimony can be produced by pressure hydrogen reduction method with a high temperature. The main chemical reaction can be expressed as follows:

$$2Na_3SbS_3 + 3H_2 + 6NaOH = 2Sb + 6Na_2S + 6H_2O \quad (13)$$

Sodium sulfide can be recovered from the process, but slow reaction rate and extremely high production cost restricts its wide application.

(4) Electrodeposition

Electrodeposition is one of the most important methods to produce antimony. Moreover, sodium sulfide is also recovered [9]. The technology can be divided into diaphragm electrolysis and non-diaphragm electrolysis. For diaphragm electrolysis, sodium hydroxide and sodium thioantimonite are used as anode and cathode electrolytes, respectively. Sodium thioantimonite solution is regarded as electrolyte for non-diaphragm electrolysis. The main chemical reaction during electro-winning process is as follows:

$$4Na_3SbS_3 + 12NaOH = 4Sb + 12Na_2S + 6H_2O + 3O_2 \uparrow \quad (14)$$

The electro-winning process can obtain a qualified product, but heavy energy consumption and terrible working environment are the main problems.

(5) Air-oxidation

Sodium pyroantimonate and sodium thiosulfate can be recovered simultaneously by air-oxidation technology [10]. The main chemical reactions are described as follows:

$$2Na_2SbS_3 + 7O_2 + 2NaOH + 5H_2O = 2NaSb(OH)_6 + 3Na_2S_2O_3 \quad (15)$$

$$2Na_2S + 2O_2 + H_2O = Na_2S_2O_3 + 2NaOH \quad (16)$$

The precipitation ratio of antimony is up to 98.0% during air-oxidation process. This method has been successfully used on recovering antimony from jamesonite. High quality of products and low production cost can be obtained, but the oxidation rate is very slow.

The Comparison of Different Methods

When antimonial refractory gold ores are dealt in alkaline solution of sodium sulfide, the removing ratio of antimony is high. All these methods mentioned have their advantages and disadvantages for recovering antimony from sodium thioantimonite solutions. Therefore a satisfactory process must fully consider many factors such as the open circuit of sulfur, working environment, treatment cost and the stability of process.

Recovering Antimony from Hydrochloric Acid System

Leaching Antimony in Hydrochloric Acid System

Antimony is dissolved into the solution in the form of $SbCl_3$ in hydrochloric acid system with oxidizing agent added. The oxidizing agents include Cl_2, $FeCl_3$, $SbCl_5$, H_2O_2, ect. The mainly chemical reaction is as follows:

$$Sb_2S_3 + 3Cl_2 = 2SbCl_3 + 3S \tag{17}$$

$$Sb_2S_3 + 3SbCl_5 = 5SbCl_3 + 3S \tag{18}$$

$$Sb_2S_3 + 6FeCl_3 = 2SbCl_3 + 6FeCl_2 + 3S \tag{19}$$

$$Sb_2S_3 + 3H_2O_2 + 6HCl = 2SbCl_3 + 6H_2O + 3S \tag{20}$$

The high Cl^- concentration and acidity are required to prevent the hydrolysis of Sb^{3+}. This process has the advantage of high leaching ratio of antimony. But its disadvantages are also existent, such as high consumption of reagent, high loss of gold, serious corrosion to devices, and difficulty to recycle the leaching solution.

Recovering Antimony from Hydrochloric Acid System

The methods on the recovery of antimony from $SbCl_3$ acidic solutions are as below [11]:

(1) Hydrolysis

Sb^{3+} can be precipitated in the form of $SbCl_3$, SbOCl and Sb_2O_3 by adjusting pH of $SbCl_3$ solution by adding H_2O or NaOH. The reaction is as follows:

$$SbCl_3 + H_2O = SbOCl \downarrow + 2HCl \tag{21}$$

$$4SbOCl + H_2O = Sb_4O_5Cl_2 + 2HCl \tag{22}$$

$$2SbOCl + 2NH_4OH = Sb_2O_3 + 2NH_4Cl + H_2O \tag{23}$$

Hydrolysis process is a purify process in essence. During the formation process of $Sb_4O_5Cl_2$ precipitates, the acidity is about 0.5–0.6 mol/L. Main impurities are still present in the leaching solutions. Therefore, hydrolysis method for the recovery of antimony is widely used in industry.

(2) Distillation method

The main principle for distillation method is that with a temperature above $SbCl_3$ boiling point, $SbCl_3$, $AsCl_3$ and $SnCl_4$ can be separated from $CuCl_2$, $PbCl_2$ and

$ZnCl_2$ by distillation method due to the diffidence of boiling point. Then high purity $SbCl_3$ can be obtained by gradually cooling process. Chlorination-distillation process has been researched in depth. High energy consumption is the main problem for this method.

(3) Replacement method

Antimony can be recycled from leaching solutions with some metal powder added by replacement reactions. The most common reagent is iron power due to its low price. The reaction is as follows:

$$2SbCl_3 + 3Fe = 2Sb + 3FeCl_2 \quad (24)$$

Although antimony can be recycled effectively from leaching solution, it is difficult to recycle leaching solutions due to the presence of Fe^{2+}.

(4) Electrodeposition

Electrodeposition refers to a method that metal antimony is produced from leaching solutions by electrolysis. After completing the process, the residual electrolytes are returned to leaching process. The cathode and anode reactions are listed as follows:

$$\text{Cathode reaction: } Sb^{3+} + 3e = Sb \quad (25)$$

$$\text{Anode reaction: } Fe^{2+} - e = Fe^{3+} \quad (26)$$

High purity products can be obtained by electrodeposition, but high energy consumption and bad operation conditions constrain its widely application.

Slurry Electrolysis

Slurry electrolysis method is a novel hydrometallurgical technology. During this process, leaching and electrodeposition are conducted simultaneously in an electrolytic cell [12]. Metal is dissolved in anode by oxidation reaction and objective metal is deposited in cathode by reduction reaction. HCl–NaCl, HCl–$CaCl_2$ and HCl–NH_4Cl solutions are widely used in slurry electrolysis process. The reactions are as follows:

$$\text{Cathode reaction: } 2Sb^{3+} + 6e = 2Sb \quad (27)$$

$$\text{Anode reaction: } Sb_2S_3 - 6e = 2Sb^{3+} + 3S \quad (28)$$

Slurry electrolysis method has some disadvantages, such as high failure rate of devices, serious abrasion of anode plate, accumulation of ferrous iron and poor operating environment.

The Comparison of Different Acid Systems

The used recovery methods include hydrolysis, distillation and electrodeposition. Although these methods can extract antimony effectively, these problems such as high failure rate of device, poor operating environment and high production cost are inevitable. Moreover, Fe and As also are dissolved into solution, causing a high burden for following purification and electro-winning process. All these shortcomings restrict their widely application.

A Clean Process for Extracting Antimony

Process Flow

Based on the advantages and disadvantages of every process to extract antimony from antimonial refractory gold concentration, a clean process is proposed, and the process flow is presented in Fig. 1. Firstly, antimony will dissolve into the solution as sodium thioantimonite by leaching of antimonial refractory gold concentration with sodium sulfide. Then air oxidation technology is used to prepare sodium pyroantimonate from sodium thioantimonite solution, the oxidated solution are neutralized, purified, and concentrated to recover sodium thiosulfate. After pretreatment, the leaching residue can be raw material to extract gold by cyaniding technology.

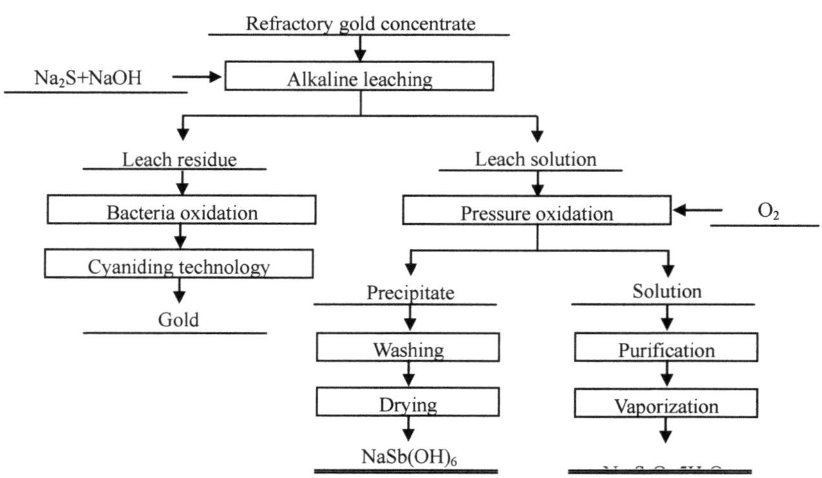

Fig. 1 A clean process for extracting antimony from antimonial refractory gold ores

Technics Characteristic

The antimonial refractory gold concentration (Sb 6.30%, As 5.50%, S 17.68%, Fe 15.74%, Au 58.8 g/L) is leaching in the solution containing sodium sulfide and caustic soda, the leaching efficiency of Sb can reach 97.08% under optimal conditions, only 0.20% Sb remains in the residue. The pressure oxidation technology is applied to prepare sodium pyroantimonate from sodium thioantimonite solution, and the Sb precipitation ratio can reach 99.80%. The product is composed of Sb_2O_3 (58.98%, by mass), Na_2O (12.55%, by mass) and Sb^{3+} (≤0.01%, by mass). The oxidated solution is neutralized with sulphuric acid, As (III) and Sb (III) in the solution will precipitate as As_2S_3 and Sb_2S_3 at pH 6.5, respectively. After liquid-solid separation, when the solution is concentrated till its density reaches more than 1.5 g/cm^3, sodium sulfite and sodium sulfate will crystallize together at 90–95 °C, then the heat solution is filtered again. Sodium thiosulfate will crystallize from the filtrate when it cools down to room temperature. The purity of sodium thiosulfate product can reach more than 95.0%.

Conclusion and Expectation

Considering antimony plays a negative role for extraction of gold from antimonial refractory gold ores, it is necessary to remove antimony before conducting pretreatment-leaching process to improve gold recovery. Antimony can be removed successfully from antimonial refractory gold ores with Na_2S–NaOH as leaching agents. However, many factors on the recovery of antimony from sodium thioantimonite solutions, including the open circuit of sulfur, working environment, treatment cost and the stability of process, must be considered adequately. Antimony also can be removed successfully from antimonial refractory gold ores in hydrochloric acid system with oxidizing agent added. However, some problems must be considered, such as high failure rate of devices, poor operating environment and high production cost.

In this paper, a clean process for extracting antimony from antimonial refractory gold ores is proposed. Antimony is dissolved into the solution with sodium sulfide as leaching agents. And then pressure oxidation technology is applied to prepare sodium pyroantimonate from sodium thioantimonite solution. Sodium thiosulfate is produced from oxidation residual solution by neutralization, purification and concentrated crystallization. The leaching residues obtained can be used to extract gold by pretreatment-cyanide leaching process.

Acknowledgements The financial support from Young Scientists Fund of National Natural Science Foundation of China (Grant No. 51404296) is gratefully acknowledged.

References

1. George MW et al (2011) U.S. geological survey mineral yearbook. U.S. Geol Survey 5:1–18
2. Yang TZ et al (2005) Metallurgy and chemicals of Piecious metals. Central South University Press, Changsha, NY, pp 210–215
3. Jiang T et al (1998) Chemical of extracting gold. Hunan Science and Technology Press, Changsha, NY, pp 8–15
4. Celep O, Alp I, Deveci H (2011) Improved gold and sliver extraction from a refractory antimony ore by pretreatment with alkaline sulphide leach. Hydrometallurgy 105:234–239
5. Jin SB et al (2009) Influence of antimony on refractory gold ore roasting and cyanide leaching. Gold 30(2):121–128
6. Yuan CX, Wang Y (2003) Roasting and cyanide leaching process study on As, Sb, and C content refractory gold concentrate. Nonferrous Metall 3:32–34
7. Cui RC et al (2011) Biooxidation-cyanidation leaching of gold concentrates with different arsenic types. Chin J Nonferrous Metals 21(3):694–699
8. Ubadini S et al (2000) Process of flow-sheet for gold and antimony recovery from stibnite. Hydrometallurgy 57(5–6):187–199
9. Sun LG et al (2015) Status and development of gold extraction from refractory gold ore. Nonferrous Mteals (Metallurgy part) 4:38–43
10. Yang TZ, Liu WF Jiang MX (2006) Production and processes of sodium pyroantimonate in China. In: Sohn international symposium advanced processing of metals and materials held during TMS 2006 Annual Meeting, vol 8, San Diego, CA, pp 27–31, pp 411–420
11. Zhao TC (1987) Metallurgy of antimony. Central South University of Technology Press, Changsha, NY, pp 358–462
12. Zhang YL et al (2014) Pant practice of slurry electrolysis of high arsenic gold-bearing stibnite concentrate. Nonferrous Metals (Metallurgy Part) 11:16–20

Upgrading of Copper and Cobalt from the Democratic Republic of Congo Tailings

Y.R.S. Hara, S. Chama, D.M. Musowoya, G. Kaluba, J. Machona, P.W. Chishimba, K. Nyirenda and S. Parirenyatwa

Abstract Our study focuses on upgrading of copper and cobalt from the Democratic Republic of Congo tailings, containing 2 wt% Cu and 0.5 wt% Co. Cobalt is in the complex iron silicate matrix such that its recovery by direct leaching is not possible. Two concentrating methods were employed in order to obtain copper and cobalt concentrates. Copper was concentrated via flotation and, at the optimal conditions its grade in the concentrate was 31.9 wt%. Cobalt was concentrated to 8.6 wt% by carbothermically reducing the flotation tailings with coal in a temperature range of 800–1050 °C and followed by magnetic separation. The reaction products were analysed by atomic absorption spectrometer (AAS) and scanning electron microscopy (SEM) techniques. The best carbothermic reduction temperature was found to be 950 °C due to maximum metallisation and less sintering between gangue and metallic/alloy phases.

Keywords Magnetic separation · Copper and cobalt · Flotation and concentration

Introduction

Tailings from the froth flotation process in the Democratic Republic of Congo contains significant quantities of copper (>1 wt%) and cobalt (>0.2 wt%). By comparison, the concentration of these metals in tailings is higher than what is found in the underground ore bodies, in most parts of the world. The main advantage of processing copper and cobalt from the tailing dam is that the mining cost is eliminated.

Y.R.S. Hara (✉) · S. Chama · D.M. Musowoya · G. Kaluba · J. Machona · P.W. Chishimba · K. Nyirenda · S. Parirenyatwa
School of Mines and Mineral Sciences, Metallurgy Department, The Copperbelt University, Kitwe, Zambia
e-mail: yotamuhara@gmail.com

S. Ikhmayies et al. (eds.), *Characterization of Minerals, Metals, and Materials 2017*, The Minerals, Metals & Materials Series,
DOI 10.1007/978-3-319-51382-9_54

Most froth flotation plants in the world aim at producing higher grade copper concentrates (>33 wt% Cu) but this results in higher loss of copper into the tailings. Therefore, the copper in the tailing dam may be recovered via froth flotation, by employing the right conditions such as reagent dosing, pH and particle size.

On the other hand, all cobalt and part of the copper from the tailings dams, in the Democratic Republic of Congo are dissolved in the complex silicate matrix. The copper and cobalt which is dissolved in the silicate matrix creates one main challenge during froth flotation. They float together with silica, thereby producing very low grade concentrates. As such, there is need of developing the right processing route for upgrading copper and cobalt that is in the complex silicate matrix.

Thermodynamic Analysis

Metallic cobalt is ferromagnetic and hence it may be upgraded or concentrated via magnetic separation. Consequently, reduction stage is necessary to convert cobalt to the metallic state [1, 2]. The Cu–Co–Fe–Ca–Si–O system at 950 °C is shown in Fig. 1 from which it can be observed that copper and cobalt may be selectively metallised at $\log_{10}(P(O_2))$atm > −15.7. Cu, Co and Fe are all metallised at lower partial pressure of O_2 gas. Consequently, the metallised cobalt and iron may be separated magnetically as they are ferromagnetic [1, 2].

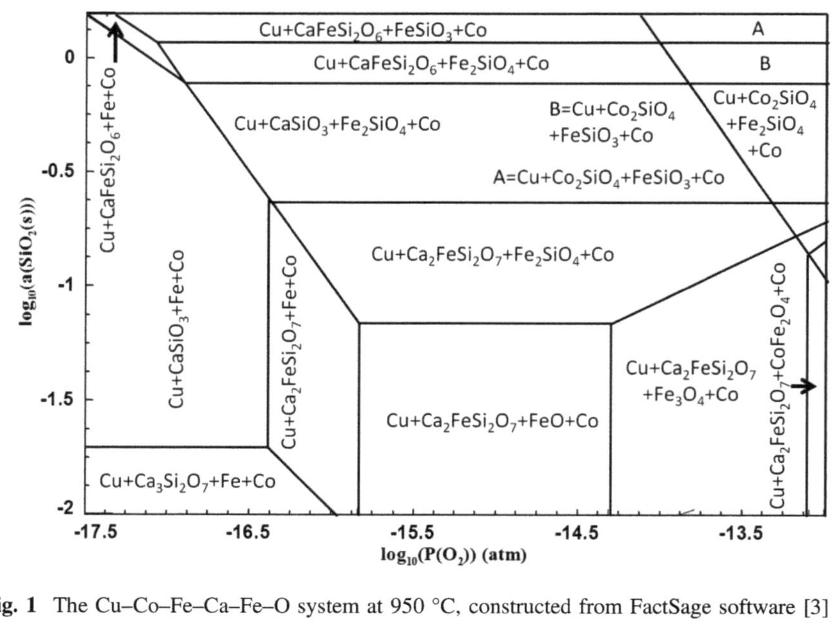

Fig. 1 The Cu–Co–Fe–Ca–Fe–O system at 950 °C, constructed from FactSage software [3]

Experimental

Materials

The material was obtained from the tailings dam in the Democratic Republic of Congo. The sample was collected from several different locations of the tailings dam in order to have a better representation of the sample. The sample which was collected from different sites of the tailing dam was then mixed thoroughly and a representative portion was obtained and analysed. Full elemental analysis of the sample is shown in Table 1 and it is clear that the sample has 1.8 wt% Cu and 0.4 wt% Co. The main elemental constituents in the sample are Si, Mg and Al and, these were found to be in the form of complex magnesium aluminium silicate. Iron, cobalt and part of the copper were dissolved in this silicate matrix.

Flotation

A Denver D-12 sub-aeration flotation cell of 2 l capacity was used for batch flotation studies. The sample was conditioned at 60% pulp density with the desired amounts of reagents for a predetermined time. In the flotation experiments, freshly prepared Sodium Butyl Xanthate (SBX) at strength of 1% was used as a collector for the rougher stages and, Potassium Amyl Xanthate (PAX) at strength of 1% was used as a collector for the scavenger stage. Beta Froth 65 and Pine oil were used as frothers. The pH was regulated by using quick lime.

The pulp density was further diluted to ~35% by adding water and further conditioning by addition of a frother. The flotation tests were carried out under the following, four different conditions in order to establish the best conditions:

1. The rougher stage was dosed with Sodium Butyl Xanthate at 50 g/t and sulphidiser at 60 g/t. Scavenger stage was dosed with Potassium Amyl Xanthate at 40 g/t and sulphidiser at 90 g/t. The impeller rotation speed was 1500 rev/min and pH was maintained at neutral (7).
2. The rougher stage was dosed with Sodium Butyl Xanthate at 50 g/t and sulphidiser (Na_2S) dosaging of 120 g/t. Scavenger stage was dosed with Potassium Amyl Xanthate at 40 g/t and the sulphidiser at 90 g/t. The impeller rotation speed was 1500 rev/min and pH was maintained at 10.

Table 1 Full elemental analysis of the as-received sample in weight %

Cu	Co	Fe	S	Si	Mg	Al	K	Ti	Ca	O
1.8	0.4	1.8	0.06	22.5	7.2	5.3	0.6	0.1	0.1	60.3

3. The rougher stage was dosed with Sodium Butyl Xanthate at 50 g/t and sulphidiser at 60 g/t. Scavenger stage was dosed with Potassium Amyl Xanthate at 40 g/t and sulphidiser at 90 g/t. The impeller rotation speed was 1200 rev/min and pH was maintained at 10.
4. The rougher stage was dosed with Sodium Butyl Xanthate at 50 g/t and sulphidiser at 60 g/t. Scavenger stage was dosed with Potassium Amyl Xanthate at 40 g/t and sulphidiser at 90 g/t. The impeller rotation speed was 1500 rev/min and pH was maintained at 9.

Reduction Experiments

Reduction experiments were carried out in order to metallise cobalt and residual copper in the flotation tailings. For the reduction experiments, 100 g of the flotation tailings sample was thorough mixed with 10 g coal and reduced at various temperatures (650–1050 °C). Each sample was reduced for 1.5 h in an elevating hearth furnace. The furnace was loaded with pieces of coal to prevent oxidation of the sample during the experiments. The reduced sample was removed from the furnace and cooled down to room temperature.

Characterisation of the Samples

The as-received and reacted samples were analysed and characterised by atomic absorption spectrometer (AAS) and scanning electron microscopy (SEM) techniques. Analysis of the samples was important in order to understand the recovery in the process. The % recovery for flotation was calculated according to reference [4].

Results and Discussion

The summary of the results which were obtained from the different flotation tests (conditions 1, 2, 3 and 4) are shown in Table 2. It is evident from Table 2 that copper was upgraded to as high as 31.9 wt% via froth flotation. The flotation result in Table 2 shows that the grade and recovery of copper was greatly influenced by pH modifiers and impeller speed.

Table 2 Results obtained from various froth flotation tests

Condition	% Cu in tailings	% Cu in concentrate	% Recovery	pH	Speed (Rev/min)
1	0.9	27.6	54.1	7.3	1500
4	0.84	28.7	57.2	9	1500
2	0.8	30.2	59.2	10	1500
3	0.58	31.9	70.6	10	1200

Effect of PH Modifiers

As stated above, pH was regulated by adding quick lime [4]. Quick lime alters the mineral surface by producing OH^{-} on the iron surface thereby generating hydrophilic ferric hydroxide film in the alkaline pulp which consequently depresses iron during the concentration process. When pH was increased from 7.3 to 9, the copper grade in the concentrate and recovery increased to 28.7 and 57.2%, respectively. A further increase in pH to 10, the copper grade in the concentrate and recovery increased to 30.2 and 59.2%, respectively. It can be argued that more iron was depressed with the increase in pH thereby resulting in increase in the grade of copper in the concentrate.

The sulphidiser (Na_2S) used in the test work also work as a pH modifier as it is highly alkaline. The higher the pH, the more stable the froth bed was which can also be attributed to improved performance of the circuit.

Impeller Speed

More particles attach themselves to the bubbles when the impeller speed is decreased and hence the grade of the concentrate and recovery increased owing to true hydrophobicity of the copper particles in the Democratic Republic of Congo tailings. A higher concentrate grade of 31.9 wt% and recovery of 70.6% were obtained after decreasing the speed of the impeller.

There is more entrainment at higher impeller speed which resulted in the recovery of both hydrophobic and hydrophilic particles. The recovery of hydrophilic particles results in decrease in the grade of copper in the concentrates.

The plot of % recovery against the grade of copper in the concentrate is shown in Fig. 2. It can be observed from Fig. 2 that; (i) concentrate grade is proportional to pH and (ii) the grade of the concentrate is proportional to the recovery. Observation (ii) agrees with findings in reference 4.

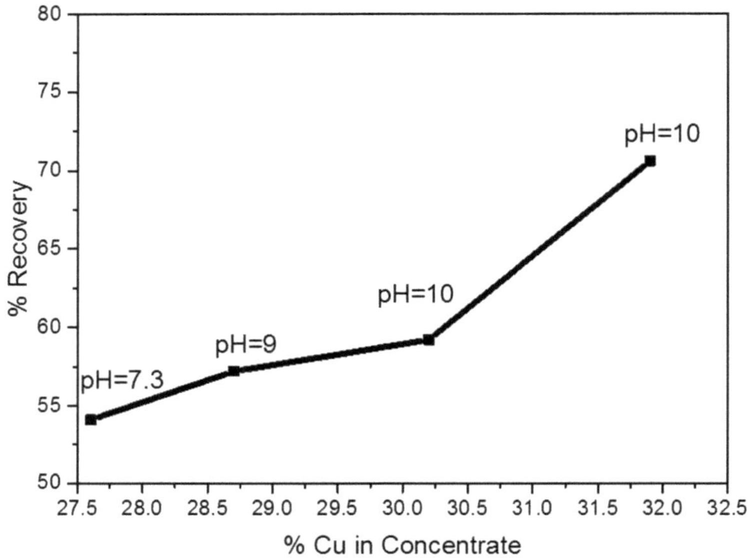

Fig. 2 Plot of % recovery against concentrate grade. The pH conditions have been inserted onto the graph

Magnetic Separation of the Tailings

The tailings from flotation contain most of the cobalt and hence it was reduced with coal and subjected to magnetic separation. The magnetic fraction was analysed for Cu, Co, Fe and Si and the results are shown in Table 3. It can be observed from Table 3 that copper and cobalt in the flotation tailing was upgraded to about 20 times. Copper is non-magnetic but it was separated via magnetic separation technique. SEM-EDX analysis revealed that the copper phase had about 0.2 wt% Fe and the presence of iron in copper makes it magnetic [5] which then enables magnetic separation. Three main observations can be made from Table 3:

i. The grade of copper and silicon in the magnetic fraction is increasing with increase in reduction temperature.
ii. The grade of iron in the magnetic fraction is decreasing with decrease in reduction temperature.
iii. The grade of cobalt in the magnetic fraction is highest at the reduction temperature of 950 °C.

Table 3 Analysis of Cu, Co, Fe and Si in wt% for the magnetic fraction

Reduction Temperature	Cu	Co	Fe	Si
850 °C	4.4	5.7	46.1	3.7
950 °C	6.8	8.6	41.4	5.1
1050 °C	7.6	7.5	33.6	10.5

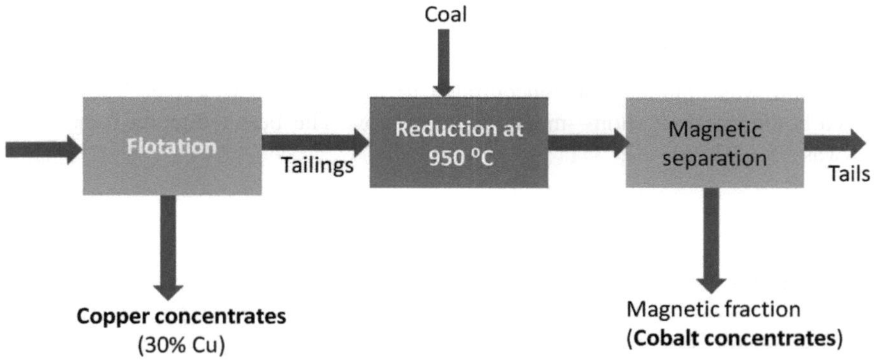

Fig. 3 Proposed flowsheet for treating the Congo tailings

The grade of copper was increasing with increase in the reduction temperature due to the increased amount of iron that dissolves in copper at higher temperatures. As stated above, the dissolution of iron in copper makes the later magnetic. On the other hand, the grade of iron decreases with increase in reduction temperature and this is due to sintering between iron and silica. From the results in Table 3, it can be concluded that the optimal reduction temperature is 950 °C.

The proposed flow sheet for upgrading or concentrating copper and cobalt in Democratic Republic of Congo tailings is shown in Fig. 3. Copper may be upgraded via froth flotation whereas cobalt may be upgraded from the flotation tailing via carbothermic reduction—magnetic separation. The copper and cobalt in the concentrates with greater than 25 wt% Cu and 8 wt% Co may be sold at about 60% of the London Metal Exchange (LME) price. Therefore, the copper concentrates (31.9 wt% Cu) and cobalt concentrates (8.6 wt%) produced from the present work may be sold at $900 and $1250, respectively, per ton.

Conclusions

1. Copper from the tailing dam was upgraded or concentrated via froth flotation as shown from the results in Table 2. The grade of copper in the concentrate and tailings was 31.9 and 0.58%, respectively, under optimal flotation conditions.
2. The best flotation conditions were determined from the experimental work as:
   a. Dosing the rougher stage with Sodium Butyl Xanthate (SBX) at 50 g/t and sulphidiser at 60 g/t.
   b. Dosing the scavenger stage with Potassium Amyl Xanthate at 40 g/t and sulphidiser at 90 g/t.

   c. Maintaining the pH of the pulp at 10 and impeller rotation speed at 1200 rev/min.

3. Cobalt was upgraded or concentrated to more than 8 wt% (see Table 3) via carbothermic reduction—magnetic separation. The best reduction temperature was 950 °C.

References

1. Anon (1960) Cobalt monograph. Centre d'Information du Cobalt, Brussels, Belgium, p 515
2. Betteridge W (1982) Cobalt and its alloys. In: Industrial metals, 1st ed. Ellis Horwood Limited, Chichester
3. Bale CW et al (2002) FactSage thermochemical software and databases. CALPHAD: Comput Coupling Phase Diagrams Thermochem 26(2):189–228
4. Wills B (2006) Mineral processing technology, an introduction to the practical aspects of ore treatment and mineral recovery, 6th edn. Butterworth-Heinemann
5. West EG (1982) Copper and its alloys. Ellis Horwood, Chichester
6. Habashi F (1969) Principles of extractive metallurgy. Gordon and Breach, London, pp 194–196
7. Davenport WGL et al (2002) Extractive metallurgy of copper, 4th edn. Chemical, Petrochemical and Process. Elsevier, pp 1–452
8. Bodsworth C (1994) The extraction and refining of metals. In: Ralph B (ed) Materials science and technology., CRC Press, Boca Raton: Fla, p 346

Characterization of Spent Printed Circuit Boards from Computers

Zhiwei Peng, Jiaxing Yan, Hongjin Zhang, Xiaolong Lin, Jiann-Yang Hwang, Yuanbo Zhang, Guanghui Li and Tao Jiang

Abstract The pollution caused by the spent printed circuit boards (PCBs) is currently a concern of researchers and environmental managers due to its large quantity and exceptionally complex properties. This study characterized printed circuit boards (PCBs) from discarded computers using a number of characterization techniques, including chemical analysis, X-ray diffraction (XRD), Fourier transform infrared spectroscopy (FTIR), granulometric analysis and scanning electron microscopy (SEM). The PCBs contain approximately 50 wt% metals, 25 wt% glass fibers and 25 wt% polymers. The principal metals include copper, calcium, tin, aluminum, zinc and lead. They are still wrapped by brominated epoxy resin after the PCBs are preliminarily crushed. As precious metals, such as gold and silver, are barely found in the PCBs, copper and tin are the main metallic elements of recycling value which warrants extra attention.

Keywords Spent printed circuit boards · Computers · Characterization techniques · Copper · Tin

Introduction

With the development of electronic industry, electrical production has been growing at a spectacular rate in recent years. Every year 20–50 million tons of electronic waste are generated worldwide [1]. Computer waste constitutes the main part of electronic waste and the spent PCBs account for about 6 wt% of the total

Z. Peng · J. Yan · H. Zhang · X. Lin · Y. Zhang · G. Li (✉) · T. Jiang
School of Minerals Processing and Bioengineering,
Central South University, Changsha 410083, Hunan, China
e-mail: liguangh@csu.edu.cn

J.-Y. Hwang
Department of Materials Science and Engineering,
Michigan Technological University, Houghton, MI 49931, USA

S. Ikhmayies et al. (eds.), *Characterization of Minerals, Metals, and Materials 2017*, The Minerals, Metals & Materials Series,
DOI 10.1007/978-3-319-51382-9_55

weight [2, 3]. The output of spent PCBs in China increases by 8.7% per year, and China will become the world's largest producer of spent PCBs in 2016 [4].

The solid electronic waste poses a great threat to both the environment and to the health of the human body in the world [5]. PCBs are composed of polymers, glass fibers and metals. They can be basically classified into two groups: monolayer and multilayer PCBs. The monolayer PCBs are made of glass fiber and phenolic resin with a copper layer, and the multilayer PCBs are constituted of epoxy resins, glass fibers and multiple copper layers [2, 3]. Many studies have revealed that the contents of metals, glass fibers and plastics could reach 40, 30 and 30 wt%, respectively [6, 7]. The spent PCBs can be recycled to obtain some valuable metals, such as copper, tin, aluminum, lead, gold and silver [5]. In the PCBs from spent personal computers, the main metal is copper originated from the electrical connections with content up to 33 wt% [8]. In addition, PCBs contain nonmetals, relatively high content of bromine and low concentration of chlorine and antimony, indicating the extensive use of brominated flame retardants (BFRs) for the manufacturing process of the PCBs [9].

In view of the potential economic and environmental benefits of spent PCBs, various approaches have been developed in the last decade to eliminate pollution and to recover useful components from spent PCBs [10, 11]. These processing methods can be classified into three categories, namely physical method, chemical processing and heat treatment [5]. The physical method has advantages of little secondary pollution and low cost, in spite of difficult separation of metals [5]. The chemical processing method is featured by low exhaust emission and easy handling of residue after the extraction of precious metals. However, it usually causes large consumption of chemical reagent and serious secondary pollution [12]. The heat treatment technique is simple for extraction of precious metals. Nevertheless, it has shortcomings of producing harmful gas, low metal recovery and costly investment [13]. Overall, these methods are useful in the treatment of spent PCBs but they all have technical demerits.

The limited development of recovery technologies for spent PCBs in the past decades was essentially due to lack of complete recognition of the characteristics of spent PCBs. Much work has focused on characterization of spent PCBs, but the composition of PCBs has been changing over the years. It was reported that in 1993 the PCBs contained 19 wt% copper and 1120 g t^{-1} gold and the main goal was to develop better strategies for recovery of gold [4]. Oliveira et al. [14] reported that in 2010 the PCBs contained 28 wt% copper and 29 g t^{-1} gold and recovering copper was the primary objective of the recycling process. The contents of precious metals, such as gold and silver, in PCBs have fallen in recent years. The main metals were copper, tin, lead and aluminum. The printed circuit boards from mobile phones (PCBs-MP) were constituted of 63 wt% metals, 24 wt% ceramics and 13 wt% polymers; and the PCBs from personal computers (PCBs-PC) were constituted of 45 wt% metals, 28 wt% ceramics and 27 wt% polymers. The PCBs-PC contained

20.19 wt% copper, and the PCB-MP contained 34.49 wt% copper, and thus the recovery of copper was the main goal [2].

In order to recycle spent PCBs effectively, further characterization of PCBs should be carried out. The aim of this work is to characterize PCBs from discarded computers by a number of characterization techniques, including chemical analysis, X-ray diffraction (XRD), Fourier transform infrared spectroscopy (FTIR), granulometry analysis and scanning electron microscopy (SEM).

Experimental

Material

Monolayer spent PCBs from discarded computers were collected for this study. They were crushed preliminarily to particle size less than 3 mm for further characterization.

Experimental Methods

The spent PCBs' chemical composition, phase composition, organic molecular structure, particle size distribution, density, specific surface area and physical microstructure were determined by various techniques. The composition of PCBs was determined by an X-ray fluorescence spectrometer (XRF, PANalytical, AxiosmAX). The polymeric fraction of the PCBs was determined by the loss-on-ignition method and the fraction of glass fibers in the PCBs was calculated by the mass difference [2]. The phase composition of the sample was determined by X-ray diffraction (XRD, West Germany Siemens Company, D500). The molecular structure of PCBs was determined by a Fourier transform infrared spectrometer (FTIR, Shimadu, IRAffinity-1). The samples were directly coated on a KBr tablet, and the wave number range examined was 400–5000 cm^{-1}. About 100 g crushed PCBs were used for particle size analysis using a sieve shaker. The results were expressed by cumulative weight percentage. The true density of the sample was measured using a pycnometer, while the bulk density, referring to the mass of the unit volume when the material is accumulated in natural state, was measured by using a bottle container of known volume. Both density measurements were repeated three times and their average values were used. The specific surface area of PCBs was measured by a full-automatic specific surface area analyzer (FBT-9) at 11 °C. The value was an average of three replicates. The physical microstructure of PCBs was determined by scanning electron microscopy (SEM, FEI Czech Republic s. r. o./Czech, Quanta 250 FEG).

Results and Discussion

Chemical Composition

The main chemical composition of the PCBs is presented in Table 1. According to Table 1 and loss-on-ignition data, the composition of the studied PCBs is constituted by approximately 50 wt% metals, 25 wt% glass fibers and 25 wt% polymers. The metals are principally copper, calcium, tin, aluminum, zinc and lead. Because of rare presence of precious metals, such as Ag, Au in the PCBs, copper and tin are the main valuable elements which warrant further attention. The high content of bromine (7.37 wt%) and low content of chlorine (0.11 wt%) in the PCBs indicate the presence of brominated flame retardants (BFRs). Hence, for the PCBs it is crucial to recover copper and tin with simultaneous reduction of bromine and chlorine pollution.

Phase Composition

Figure 1 shows the XRD pattern of powdered PCBs. Metallic copper and tin are found to be the main phase constituents. The other metals are present in complex forms, such as aluminosilicates. In view of the complex phase composition, extracting the metals by recycling PCBs via pyrometallurgical processes, such as microwave pyrolysis, seems promising [15].

Organic Molecular Structure

The FTIR analysis of powdered PCBs (organic substrate) is shown in Fig. 2. Based on the functional group analysis, the absorption bands around 3200 and 3500 cm^{-1} are associated with O–H stretching vibration. The absorption bands around 2800 and 3000 cm^{-1} are caused by C–H vibrations. In particular, the absorption bands

Table 1 Main chemical composition of PCBs

Element	wt%	Element	wt%	Element	wt%	Element	wt%
O	11.50	K	0.074	Ni	0.29	Sn	6.71
Mg	0.31	Ca	7.89	Cu	29.37	Ba	1.06
Al	4.37	Ti	0.23	Zn	1.58	W	0.65
Si	14.34	Cr	0.01	Br	7.37	Pb	1.53
P	0.022	Mn	0.007	Sr	0.055	Au	10 g/t
S	0.27	Fe	0.94	Zr	0.044		
Cl	0.11	Co	0.092	Ag	0.12		

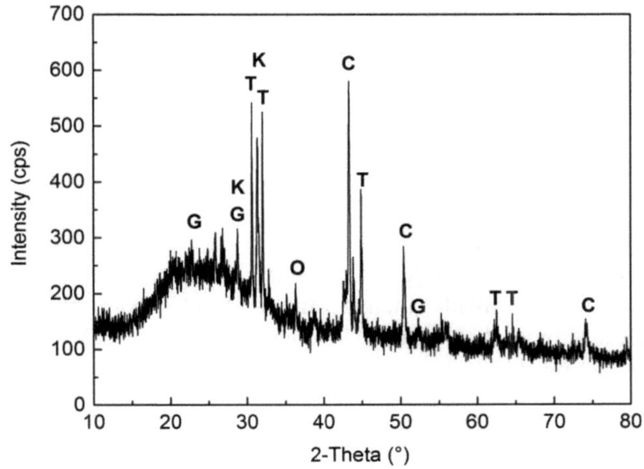

Fig. 1 XRD pattern of powdered PCBs. *C* Cu; *K* $Ca_2(Mg_{0.75}Al_{0.25}(Si_{1.75}Al_{0.25}O_7)$; *G* CaAl $(AlSiO_7)$; *O* CuO; *T* Sn

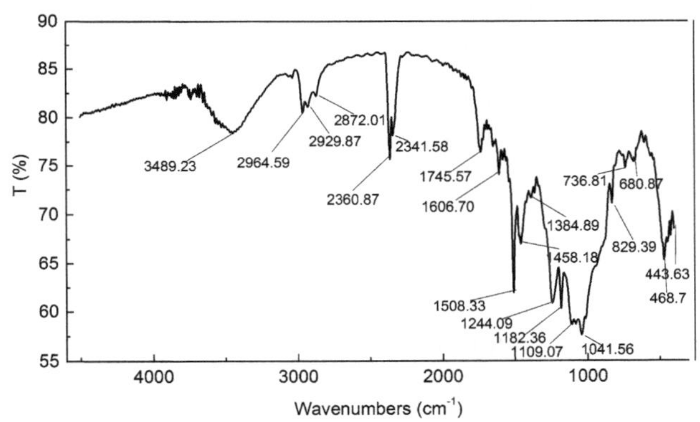

Fig. 2 FTIR spectrum of powdered PCBs

around 2964 and 2872 cm^{-1} are attributed to the $-CH_3$ methyl asymmetric vibration and symmetric vibration, respectively. The absorption bands around 2930 and 2850 cm^{-1} are the result of $-CH_2$ methylene asymmetric vibration and symmetric vibration, respectively, and the absorption bands between 2340 and 2360 cm^{-1} are assigned to the R–N≡N stretching vibration. The absorption band peaking at about 1745 cm^{-1} reveals the existence of C=O. The bands between 1450 and 1650 cm^{-1} can be assigned to benzene ring frame vibration. The absorption band near the 1385 cm^{-1} shows the presence of methyl and the absorption bands between 1000 and 1300 cm^{-1} are in association with C–O stretching vibration. The bands between 650 and 830 cm^{-1} show the existence of substituted benzene. The absorption bands between 500 and 750 cm^{-1} suggest the presence of halogen, caused by C–Br or C–Cl stretching vibration, especially C–Br [16, 17]. The main organic component of the PCBs is verified to be brominated epoxy resin.

Particle Size Distribution

The particle size distribution of powdered PCBs is shown in Fig. 3. The particle size of the sample is in the range up to 5 mm. It can be observed that more than 50% of the particles are smaller than 0.5 mm and only about 20% of the particles have a particle size more than 1 mm. The contents of 0–0.2, 0.2–0.5 and 0.8–1 mm size fractions size reach 33, 21 and 16%, respectively.

Density and Specific Surface Area

The true density of the PCBs is found to be 2.57 g cm^{-3}. The high true density is a result of the presence of approximately 50 wt% metals. The average value of bulk density is 0.82 g cm^{-3}. The large difference between true density and bulk density is attributed to the existence of large amounts of resin and glass fibers. The small bulk density is consistent with the large specific surface area of PCBs (1927.03 cm^2 g^{-1}). This feature is beneficial to subsequent processing for metal extraction.

Physical Microstructure

Figure 4 presents the microstructure of powdered PCBs at different magnifications. It shows that the PCBs powder particles accumulate together, with glass fibers and

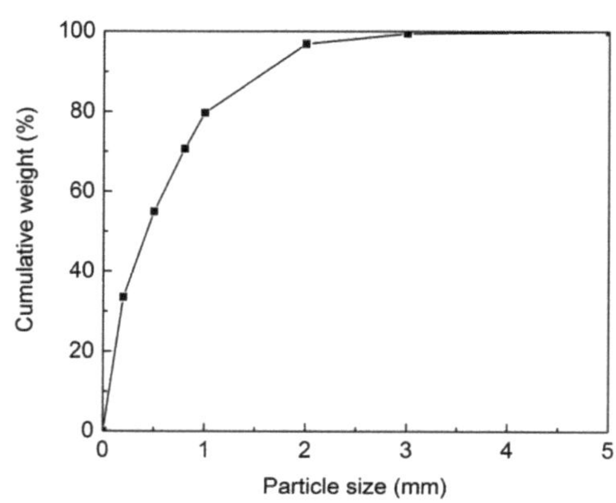

Fig. 3 Size distribution of powdered PCBs

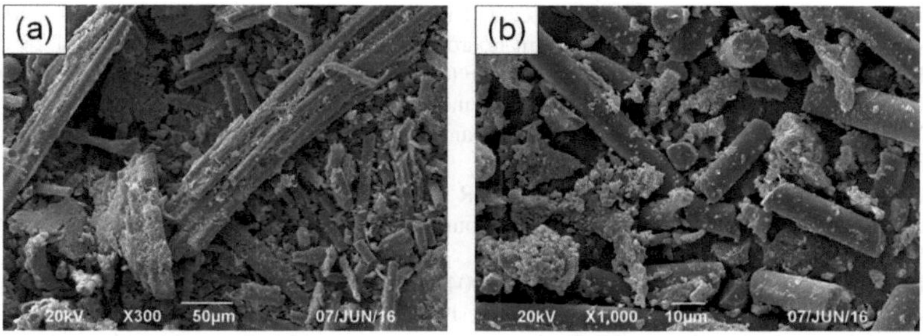

Fig. 4 SEM images of powdered PCBs

metals wrapped by resin. Obviously, the close association of the metals with other constituents will restrain extraction of valuable elements directly from PCBs via hydrometallurgical processes, such as leaching.

Conclusions

This study characterized PCBs using a number of techniques. The studied PCBs contain approximately 50 wt% metals, 25 wt% glass fibers and 25 wt% polymers. Because of the relatively high contents of copper, calcium, tin, aluminum, zinc and lead, the PCBs are deemed to be an important secondary resource for metal recovery. The true density of PCBs is 2.57 g cm^{-3} and the specific surface area is 1927.03 cm^2 g^{-1}. The main organic component of the PCBs is brominated epoxy resin, which may cause environmental problems in the recovery of PCBs. According to the characteristics of the PCBs from discarded computers, copper and tin are the main metallic elements of recycling value that warrant additional attention. It seems that pyrometallurgical processes will be useful for processing of these PCBs.

Acknowledgements This work was supported by the Science and Technology Major Project of Gansu Province, China under Grant 1602FKDC007, the National Undergraduate Training Program for Innovation and Entrepreneurship under Grant 201510533213, the Shenghua Lieying Program of Central South University under Grant 502035001, and the Innovation-Driven Program of Central South University under Grant 2016CXS021.

References

1. Estrada-Ruiz RH, Flores-Campos R, Gamez-Altamirano HA, Velarde-Sanchez EJ (2016) Separation of the metallic and non-metallic fraction from printed circuit boards employing green technology. J Hazard Mater 311:91–99
2. Yamane LH, Moraes VT, Espinosa DCR, Alberto J, Tenrio S (2011) Recycling of WEEE: characterization of spent printed circuit boards from mobile phones and computers. Waste Manage Res 31:2553–2558

3. Jha MK, Kumari A, Choubey PK, Lee J, Kumar V, Jeong J (2012) Leaching of lead from solder material of waste printed circuit boards (PCBs). Hydrometallurgy 121–124:28–34
4. Bizzo WA, Figueiredo RA, de Andrade VF (2014) Characterization of printed circuit boards for metal and energy recovery after milling and mechanical separation. Materials 7:4555–4566
5. Xu Y, Liu J (2015) Recent developments and perspective of the spent waste printed circuit boards. Waste Manage Res 33:392–400
6. Ogunniyi IO, Vermaak MKG, Groot DR (2009) Chemical composition and liberation characterization of printed circuit board comminution fines for beneficiation investigations. Waste Manage Res 29:2140–2146
7. Kim B, Lee J, Seo S, Park Y, Sohn H (2004) a process for extracting precious metals from spent printed boards and automobile catalysts. JOM 56:55–58
8. Veit HM, Diehl TR, Salami AP, Rodrigues JS, Bernardes AM, Tenrio JAS (2005) Utilization of magnetic and electrostatic separation in the recycling of printed circuit boards scrap. Waste Manage Res 25:67–74
9. Evangelopoulos P, Kantarelis E, Yang W (2015) Investigation of the thermal decomposition of printed circuit boards (PCBs) via thermogravimetric analysis (TGA) and analytical pyrolysis (Py–GC/MS). J Anal Appl Pyrol 115:337–343
10. Das A, Vidyadhar A, Mehrotra SP (2009) A novel flow sheet for the recovery of metal values from waste printed circuit boards. Resour Conserv Recycl 53:464–469
11. Veglio F, Quaresima R, Fornari P, Ubaldini S (2003) Recovery of valuable metals from electronic and galvanic industrial wastes by leaching and electrowinning. Waste Manage Res 23:245–252
12. Zhang Y, Liu S, Xie H (2012) Current status on leaching precious metals from waste printed circuit boards. Procedia Environ Sci 16:560–568
13. Fujita T, Ono H, Dodbiba G (2014) Evaluation of a recycling process for printed circuit board by physical separation and heat treatment. Waste Manage Res 34:1264–1273
14. Oliveira PC, Cabral M, Nogueira CA, Margarido F (2010) Printed circuit boards recycling: characterization of granulometric fractions from shredding process. Mater Sci Forum 636–637:1434–1439
15. Peng Z, Hwang JY (2015) Microwave-assisted metallurgy. Int Mater Rev 60:30–63
16. Guo X, Qin FGF, Yang X, Jiang R (2014) Study on low-temperature pyrolysis of large-size printed circuit boards. J Anal Appl Pyrol 105:151–156
17. Wu W, Qiu K (2015) Vacuum co-pyrolysis of Chinese fir sawdust and waste printed circuit boards. part II: influence of heating conditions. J Anal Appl Pyrol 111:216–223

Study of the Effect of the Initial Nucleation Mechanism of Lead Anode Oxidation Film on Internal Stress in Chromic Acid Electrolyte

Yunkai Wang and Jianzhong Li

Abstract Lead substrate anodes are widely used in the electroplating and electrolytic field. However, with the change of internal stress, the oxide films are gradually generated on the surface of lead anode. Their structure and stress properties were studied during the oxidation by means of Raman spectra, XRD, SEM. The result showed that the diffraction peaks of lead were mainly found at the initial oxidation stage. With increasing oxidation time, Pb diffraction peaks were weakened and the diffraction peaks of PbO/PbO_2 strengthened remarkably. When the oxidation time was above 10 min, monoclinic $PbCrO_4$ was also generated in the oxide film with CrO_4^{2-} ions solution as electrolyte. Raman spectra revealed that Raman peaks of PbO/PbO_2 were gained. In addition, Raman peaks of $PbCrO_4$ at 830–840 cm^{-1} were found using the CrO_4^{2-} ions as electrolyte. Based on the relationship of the internal stress of oxide film and Raman peaks shift, PbO/PbO_2 induced obvious stress generation in CrO_4^{2-} ion electrolyte. The nucleation style of oxidation film was 3D progressive nucleation at first. After the peak current, it turned to be 3D instantaneous nucleation. In the oxidation process, the internal stress of PbO/PbO_2.

Keywords Lead anode · Corrosion · Chromic acid · Internal stress · Nucleation mechanism

Introduction

Lead and its alloys possess good castability and mechanical property, low corrosion rate, high electrocatalytic efficiency and good conductivity of the contact layer. As a result, lead and its alloys have been widely used in industrial applications such as chromium plating [2], electrochemistry surface passivation [3], lead/acid batteries [4],

Y. Wang · J. Li (✉)
Physic-Chemistry of Metallurgy Department, School of Metallurgy,
Northeastern University, Shenyang 110819, Liaoning, China
e-mail: lijz@smm.neu.edu.cn

S. Ikhmayies et al. (eds.), *Characterization of Minerals, Metals, and Materials 2017*, The Minerals, Metals & Materials Series,
DOI 10.1007/978-3-319-51382-9_56

and hydrometallurgical processes [5]. In these applications, lead-alloy anodes undergo high current densities, in aggressive environments, such as sulphuric acid, CrO_3 solution or chlorine-included solution.

Although the new type anodes of Ti/TiO_2 [6–8], Pt alloy [9, 10] and cerement possess various outstanding properties than traditional lead matrix anode, lead anode still plays a vital role in manufacturing industry because of its characteristics of low cost and easy preparation. Thus, lead matrix anode still possess research potential. In the manufacturing process, A layer of lead oxide or dioxide is easily found on lead substrate anodes surface during use, which is essential because the PbO_2 layer is the electrochemically active component of the anode. However, after practical application the dioxide layer has a tendency to crack or break from the lead-alloy anode, exposing the underneath fresh lead surface to the electrolyte. When the thickness of oxide films is thin (about several hundred nm), the crystal growth is maintained to the same direction with the lead anode substrate. Due to prolonged oxidation, excessive growth of the oxide films promotes the lattice constant inconsistent with the initial oxide film, thereby generating internal stresses. As the oxidation proceeds, the structure change of oxide films on the surface is resulted in the lattice constant change accordingly, leading to the generated stress changing. So shear stresses on film interface are contributed whether a thin film stress is compressive stress or tensile stress. Meanwhile, the film interface lattice happens to mismatch, and a difference in coefficient of thermal expansion are increased via the evolution of the internal stress of the oxide films [11–13]. As a result, the oxide films could show micro-cracks and folds, or even fall off, when the stresses increase to overcome the shear stress. Thereby, the life of the anode is shortened. Till now, the corrosion mechanism of lead matrix anode still exists a series of problems. There are few studies focusing on the internal stress in oxide film and nucleation-growth mode of oxide film.

As a result, it presents theoretical and practical significance that internal stress formation and the nucleation-growth mode of lead matrix anode should be investigated in details. However, the alloying element in lead alloy anode will dramatically increases the study difficulty. As a result, the study object is selected to be pure lead anode.

In this study, the lead anode with oxide film was prepared using the different ion solution as the electrolyte. Its internal stress properties and structure change were then studied by means of Raman spectra, XRD and SEM.

Experimental

Preparation of Oxide Film

Using the commercial lead alloy anode material as research sample, the lead content of anode sample is 99.0%. Anode substrates with a size of 20 mm × 10 mm × 7 mm were used in research. The samples were mechanically polished with emery paper (up to 1200 grit) to ensure similar surface roughness, then thoroughly cleaned the lead anode surface in acetone and deionized water before drying, the samples were placed in sample dry container for later use.

In order to study internal stress formation and change of oxide film on lead substrate anode surface, the same set of electrolysis experiments is carried out over the oxidation time of 1 min, 5 min, 10 min and 20 min, respectively. The electrolyte solutions are used in these experiments via 1.5 mol/L^{-1} CrO_3 solution as deionized water solvent, to simulate plating solution under 40 °C. The lead alloy is used as the working electrode for preparing the oxide films. A low carbon steel is used as counter electrode for building the electrolysis system. During the above electrolysis, the current density is set at 20 A/dm^2. After washing in deionized water and then dried rapidly, lead anode samples with different oxide films are obtained under different conditions.

Properties Tests

ZEISS/EVO18 SEM characterized surface morphology and structure of oxide films. The phase purity and crystal structure of the synthesized samples were analyzed using Rigaku/UItima IV (K-alpha radiation source, $\lambda = 0.154056$ nm, 40 kV, 200 mA) in a 2θ range of 10° to 80°. The scanning rate and space length were 8°/min and 0.02°, respectively. Raman spectrum was obtained by a Jobin Yvon HR800 Raman Spectrometer system equipped with a 633 nm He–Ne laser. The power of the exciting laser measured at the sample position was 17mW. The scattered radiation from the sample passed through a SPEX triplemate monochromator and was detected by an OMAIII optical multi-channel analyzer with a photodiode array thermoelectrically cooled to −35°C.

The samples were pressed into self-supporting wafers. When a beam of photons strikes a molecule, the photons are scattered elastically and inelastically generating Stokes and anti-Stokes lines. Raman spectra are expressed in wave numbers which have units of inverse length. The Raman spectra of the samples were recorded at room temperature via long-focus lens for 50 times. Each sample was scanned at 3 times. When the peak position error is less than ±1 cm^{-1}, the figures cumulated by computer were considered in this research. Cyclic voltammetry and chrono amperometry tests were measured by Autolab PGSTAT302 N.

Results and Discussion

Figure 1 shows the images of oxide film with different electrolytic time in 150 g/L chromic acid. After electrolyzation for 1 min, there are a large amount of oxidation products generated on the lead matrix. A fraction of products have been grown into

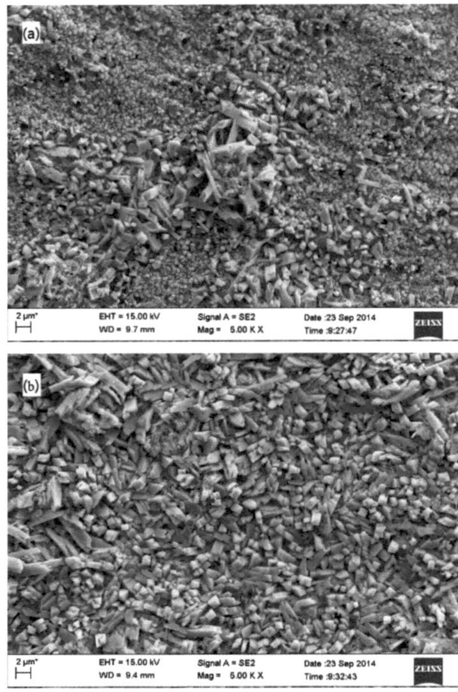

Fig. 1 SEM images of the oxide film with different electrolytic time in 150 g/L chromic acid. **a** 1 min **b** 5 min **c** 10 min **d** 20 min

rod-like grains, however, major grains are still small. When the electrolyzation time increases to 5 min, visible grains all turns into rod-like shape. When the electrolyzation time is 10 min, the rod-like grains continuously grow into larger scale grains. Meantime, the surface of some grains become irregular, which indicates the oxidation products change. With a electrolyzation time of 20 min, there are more irregular grains covered the surface of rod-like grains, which indicates further oxidation.

Figure 2 shows the X-ray spectra of lead anode surface with different electrolyzation time in 150 g/L chromic acid solution. When the electrolyzation time is 1 min, the anodized film mainly consists of matrix Pb, PbO and $PbCrO_4$. The intensity of matrix lead diffraction peak indicates that the anodized film is thin. When the electrolyzation time increases to 5 min, the composition change of anodized film is small, but the intensity of the matrix lead diffraction peak decreases and the intensity of the $PbCrO_4$ diffraction peak obviously increases. When the electrolyzation time is 10 min, the diffraction peaks of matrix lead disappear, which shows that with the increase of oxide film thickness, the matrix lead area is covered. With a electrolyzation time of 20 min, the diffraction peak of $PbO_x(x > 1)$ gradually increases and eventually becomes obvious. At the same time, the diffraction peak of PbO is enhanced evidently.

Compared with the SEM images, it asserts that in the initial stage of electrolyzation, the primary oxidation products are PbO on anode. With the chromic

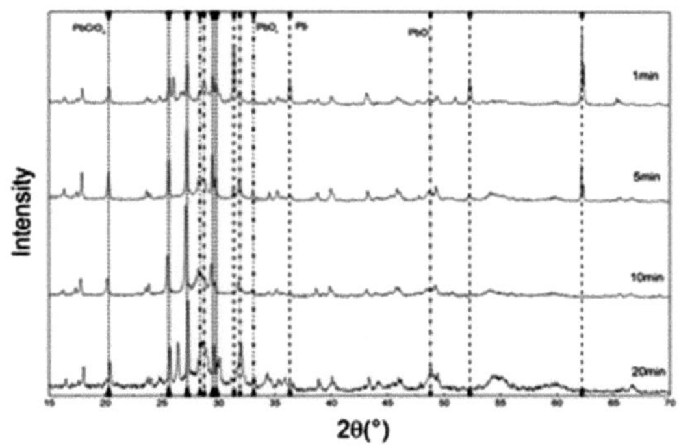

Fig. 2 X-ray spectra of lead anode surface with different electrolyzation time in 150 g/L chromic acid solution

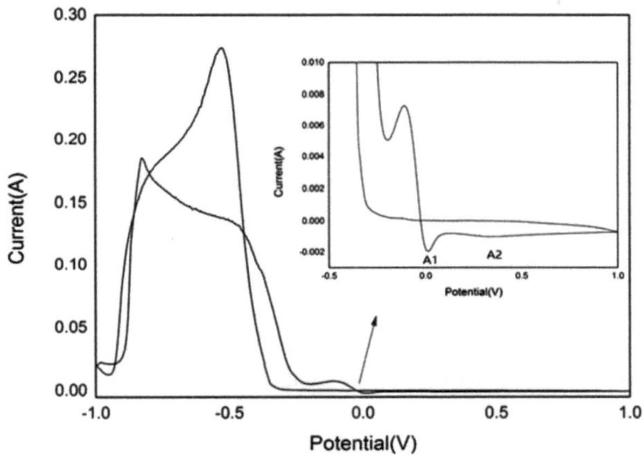

Fig. 3 Voltammogram for lead anode in 150 g/L chromic acid solution (Sweep rate, ν = 20 mv/s)

acid solution, PbO gradually turns into $PbCrO_4$, which is in a rod-like shape as shown in Fig. 1. Then as a result of the increasing thickness of anodized film, the reaction of matrix lead slows down, meanwhile, Pb^{2+} ions in oxidation products start to convert into higher valence states.

Figure 3 shows the cyclic voltammetry curves of the lead anode in 150 g/L chromic acid solution. According to Fig. 2, oxidation film is generated on Pb matrix. In the process of positive scan, the current is higher than that of negative scan and the curves intersect. The significant resistance current loop indicates that oxidation film is generated in “nucleation-growth” mode. There are two oxidation peaks. According to the analysis of XRD and SEM, peak A1 is the oxidation peak of Pb\PbO,

$$Pb + 2OH^- \rightarrow PbO + H_2O + e^- \quad (1)$$

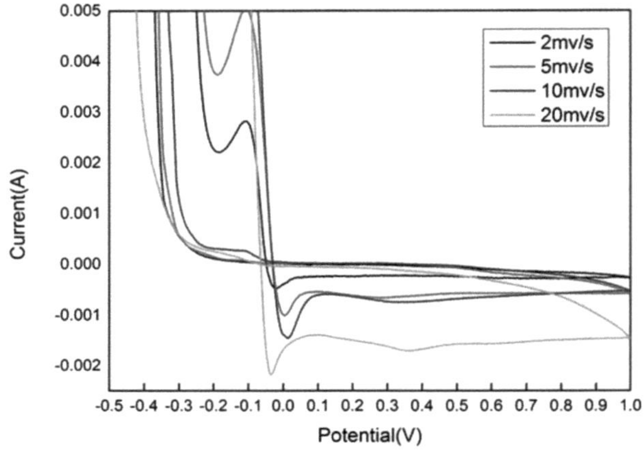

Fig. 4 Cyclic voltammogram of lead anode in 150 g/L chromic acid solution under different scan rate

Peak A2 is the oxidation peak of PbO\PbCrO$_4$,

$$2PbO + Cr_2O_7{}^{2-} + 2H^+ = PbCrO_4 + H_2O \tag{2}$$

Figure 4 shows the voltammogram of lead anode in 150 g/L chromic acid solution under different scan rate. It illustrates that with the increase of scan rate, the peak current of oxidation peak increases. The values of peak current are listed in Table 1.

Figure 5 shows the relation of oxidation peak current and root of scan rate. It illustrates the linear relation between peak current and root of scan rate, which indicates the reactions of peak A_1 and peak A_2 are irreversible reactions and controlled by diffusion [14–16]. According to the voltage value of peak A_1 and A_2 in Fig. 3, the transient current curve under different step voltage can be obtained.

Figure 6 shows the chronoamperometry curves of lead anode under different step voltages. It illustrates that after the peak, the current decreases with time and the

Table 1 Peak current value of cyclic voltammetry curve under different scan rate

Scan rate/mvs^{-1}	Peak current/A	
	Peak A_1	Peak A_2
2	−0.00047	−0.00039
5	−0.00101	−0.00065
10	−0.00145	−0.00101
20	−0.00218	−0.0017

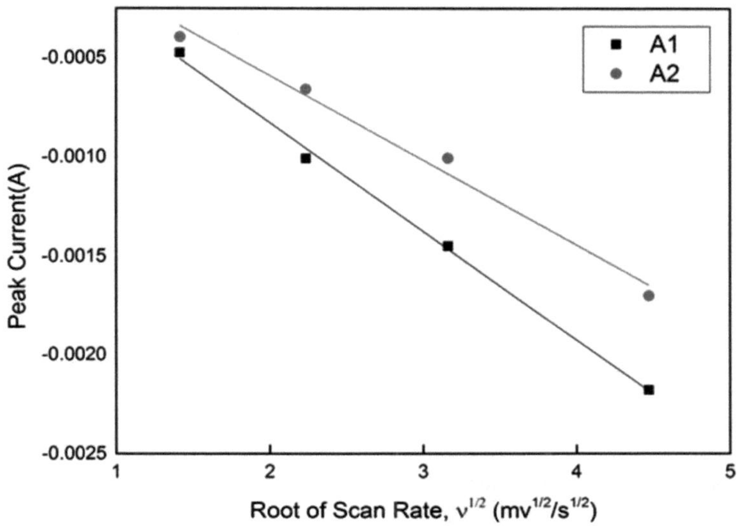

Fig. 5 The relation of oxidation peak current and root of scan rate

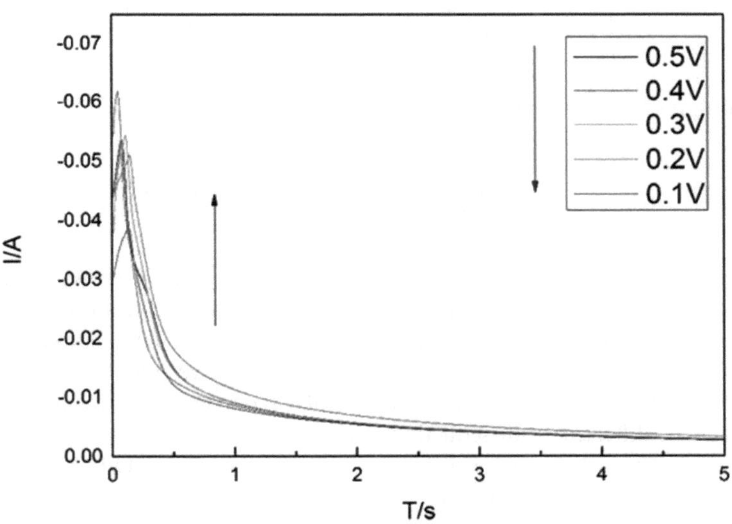

Fig. 6 Chronoamperometry curves of Pb lead anode under different step voltages

process is controlled by diffusion. To determine the growth mode, t_0 is corrected, and $i^2/i_m^2 - t'/t_m'$ is nondimensionalized. Compared with Scharifker-Hills 3D nucleation-growth mode [17] theoretical curve.

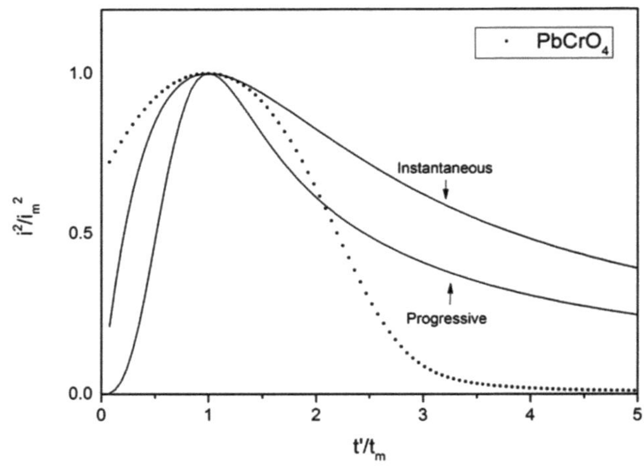

Fig. 7 Comparison of the adjust nucleation-growth curve and Scharifker-Hills mode theoretical curve

3D instantaneous nucleation,

$$\left(\frac{i}{i_m}\right)^2 = 1.9542\left(\frac{t}{t_m}\right)^{-1}\left\{1 - \exp\left[-1.2564\left(\frac{t}{t_m}\right)\right]\right\}^2 \tag{3}$$

3D progressive nucleation,

$$\left(\frac{i}{i_m}\right)^2 = 1.2254\left(\frac{t}{t_m}\right)^{-1}\left\{1 - \exp\left[-1.2564\left(\frac{t}{t_m}\right)^2\right]\right\}^2 \tag{4}$$

Figure 7 shows the comparison of the adjust nucleation-growth curve and Scharifker-Hills mode theoretical curve. It illustrates the nucleation style of oxidation film was 3D progressive nucleation at first. After the peak current, it turned to be 3D instantaneous nucleation. The deviation of curve may be because of the oxidation product $PbCrO_4$ on Pb matrix, which hampers the diffusion effect. On another hand, different from reduction deposition, the formation of oxide film is complicated. Thus, the transformation of PbO/$PbCrO_4$ and PbO/PbO_2 contributes to the decline of current. Besides, the crystal transition of α-PbO/β-PbO and α-PbO_2/β-PbO_2 may also leads to the decline. As a result, the nucleation-growth curve decreases dramatically.

Figure 8 shows the Raman spectrum of lead anode in 150 g/L chromic acid. It illustrates that the appearance of four Raman peaks under different conditions of electrolyzation time. Peak1 band is in the area of 60–70 cm^{-1}, peak2 band is in the area of 130–150 cm^{-1} and peak4 band is in the area of 820–840 cm^{-1}. Peak3 is a zigzagged peak in the area of 300–400 cm^{-1}, which is interacted by PbO, PbO_2 and $PbCrO_4$ and complicated to discuss. Combined with Table 2, when the oxidation

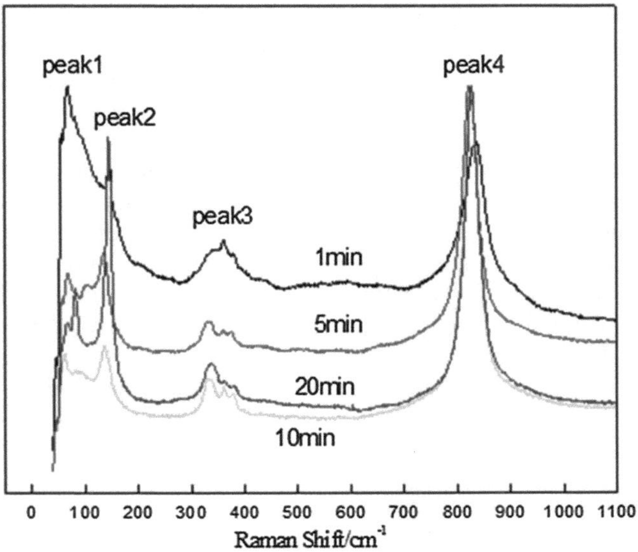

Fig. 8 Raman spectrum of lead anode in 150 g/L chromic acid

time increases from 1 min to 10 min, the Raman peak location of Peak1 changes obviously, which is from 67.8 to 62.8 cm^{-1}. The Raman peak location of Peak2 and Peak4 change slightly, which are from 139 to 136.2 cm^{-1} and from 837.6 to 825.6 cm^{-1}. When the oxidation time is 20 min, the location of Peak1, Peak2 and Peak4 change on the contrary of the former tendency. Based on the study of Kanai et al. [18], the Raman band of PbO corresponding to ~140.7 and ~84.9 cm^{-1}. Thus, Peak1 and Peak2 belongs to PbO. Based on the standard spectrum, the Raman band of $PbCrO_4$ lies around 839.2 cm^{-1}, so Peak4 belongs to $PbCrO_4$. Through the former analysis, in the initial stage of oxidation, there is a small amount of PbO_2 generated in the oxide film. As a result, the Raman band of PbO_2 should be covered by that of PbO and $PbCrO_4$. The values of Raman band of Peak1, Peak2 and Peak4 are listed in Table 2.

The inter stress of anodize film in the initial stage can be calculated by the equation as follow.

$$\sigma_{xx} = I_N * \Delta\omega(\text{GPa}) \tag{5}$$

Table 2 Raman peak positions of lead anode surface in150 g/LCrO_3 solution under different conditions

Oxidation time (min)	Raman band/cm^{-1}		
	Peak1	Peak2	Peak4
1	67.8	139.0	837.6
5	66.6	137.9	826.1
10	62.8	136.2	825.6
20	68.6	145.5	827.9

where I_N (GPa cm^{-1})is scale parameter, ω_0 is the first-order optical phonon characteristic peak wave number of Pb matrix, ω_s is the first-order optical phonon characteristic peak wave number of oxidation product. When the stress $\Delta\omega > 0$, σ_{xx} behaves as tensile stress, while the stress $\Delta\omega < 0$, σ_{xx} behaves as compressive stress.

Combined with Eq. 5, it is asserted that in the initial stage, after $PbCrO_4$ nucleation, the internal stress increase unconspicuously and behaves as compressive stress. The internal stress of PbO changes obviously and behaves as compressive stress. In the first 10 min, the change tendency of each oxidation product stay the same. It may be because of that after the maximum current, the oxide film starts to grow in 3D progressive mode. As the nucleation speed is low, the internal stress accumulates slightly. When the oxidation time increases to 20 min, the crystal type of PbO may start to change, which lead to the change tendency converting.

Conclusion

In this article, pure lead anode is taken as study object and the oxide film in the initial stage is studied. The results show that the initial oxidation products are mainly $PbCrO_4$, PbO and PbO_2, and $PbCrO_4$ takes the major part. When the oxidation process starts, $PbCrO_4$ generates on the matrix. As the reaction proceeds, $PbCrO_4$ layer continually increases, which hampers the reaction of lead matrix and chromic acid. However, with electrode reaction, lead matrix continues to be oxidized into PbO. As the oxidation time increases, $PbCrO_4$ and PbO start to turn into PbO_2.

In the initial oxidation stage, the nucleus of $PbCrO_4$ grows in 3D instantaneous nucleation type. After the maximum current, the oxide film grows in 3D progressive nucleation type. When the oxidation starts, the 3D instantaneous nucleation leads to an increase compressive stress in $PbCrO_4$ layer. After the maximum current, as 3D progressive nucleation, the internal stress changes little. The internal stress of PbO changes slightly until $PbCrO_4$ layer start to hamper the oxidation process. With the effect of $PbCrO_4$, PbO generates and the tensile stress increases. Finally, Pb^{2+} turns into Pb^{4+} and PbO_2 generates. As the stress continuously accumulates, the internal stress of oxidation film changes again.

References

1. Ivanov I, Stefanov Y, Noncheva Z, Dobrev TS, Mirkova L, Vermeersch R, Demaerel J-P (2000) Insoluble anodes used in hydrometallurgy: corrosion resistance of lead and lead alloy anodes. Hydrometallurgy 57:109–139
2. Perez-GonZalez FA, Camurri CG, Carrasco CA, Colas R (2012) Precipitation in a lead calcium tin anode. Mater Charact 64:62–68

3. Devilliers D, Dinh Thi MT, Mahe E, Xuan QL (2003) Cr(III) oxidation with lead dioxide-based anodes. Electrochim Acta 48:4301–4309
4. Li N, Zhang JS, Sencer BH, Koury D (2006) Surface treatment and history-dependent corrosion in lead. Nucl Instrum Methods Phys Res A 562:695–697
5. Egan DRP, Low CTJ, Walsh FC (2011) Electrodeposited nanostructured lead dioxide as a thin film electrode for a lightweight lead-acid battery. J Power Sources 196:5725–5730
6. Weng Z, Guo H, Liu X, Wu S, Yeung KWK, Chu PK (2013) Nanostructured TiO_2 for energy conversion and storage. RSC Adv 3:24758
7. Jiang C, Zhang J (2013) Nanoengineering titania for high rate lithium storage: a review. J Mater Sci Technol 29(2013):97
8. Chang YC, Lee CY, Chiu HT (2014) Porous inorganic materials from living porogens: channel-like TiO2 from yeast-assisted sol–gel process. ACS Appl Mater Inter 6(2014):31
9. Nassr ABAA, Sinev I, Pohl MM, Grünert W, Bron M (2014) Rapid microwave-assisted polyol reduction for the preparation of highly active PtNi/CNT electrocatalysts for methanol oxidation. ACS Catal 4:2449–2462
10. You H, Zhang F, Liu Z, Fang J (2014) Free-standing Pt-Au hollow nanourchins with enhanced activity and stability for catalytic methanol oxidation. ACS Catal 4(2014):2829–2835
11. Osorio WR, Peixoto LC, Garcia A (2009) electrochemical corrosion of Pb-1wt% Sn and Pb-2.5 wt% Sn alloys for lead-acid battery application. J Power Sources 194:1120–1127
12. Li Y, Jiang LX, Lv XJ, Lai YQ, Zhang HL, Li J, Liu YX (2011) Oxygen evolution and corrosion behaviors of co-deposited Pb/Pb-MnO_2 composite anode for electrowinning of nonferrous metals. Hydrometallurgy 109:525–527
13. Lafront AM, Zhang W, Ghali E, Houlachi G (2010) Electrochmical noise studies of the corrosion behavior of lead anodes during zinc electrowinning maintenance. Electrochim Acta 55:6665–6675
14. Remigiusz Kowalik, Piotr Zabinski, Krzysztof Fitzner (2008) Electrodeposition of ZnSe. J Electrochimica 53(21):6184–6190
15. Krzysztof Kowalik Remigiusz Fitzner (2009) Analysis of the mechanism for electrodeposition of the ZnSe phase on Cu substrate. J Electroanal Chem 633(1):78–84
16. Group Southampton Electrochemistry (1985) Instrumental methods in electrochemistry. Ellis Horwood, Chichester, pp 25–76
17. Mockute D, Bernotiene G (2000) The interaction of additives with cathode in a mixture of saccharin, 2-butyne-1, 4-diol and phthalimide during nickel electrodepositon in a Watts-type electrolyte. Surf Coat Technol 135:42–47
18. Kanai H, Ymashita Y, Kakihanab M, Yoshimurab M (1996) J Eur Ceram Soc 16:1157

Part XV
Poster Session

Addition of Cellulose Nanofibers in Reactive Powder Concrete

F.G.D. Machado, L.G. Pedroti, J.V.B. Lemes, G.E.S. Lima, L.A.F. Fioresi, W.E.H. Fernandes, R.C.S.S. Alvarenga and J. Alexandre

Abstract Cellulose fibers are extracted from renewable sources and have potential to be used in various fields. In this study, a type of Reactive Powder Concrete (RPC) was produced with the addition of Cellulose Nanofibers (CNF), a biopolymer extracted from lignocellulosic materials. Three different concrete compositions were produced, each with addition of 0.10, 0.20 and 0.25% of CNF in mass, proportionally to the cement's mass. During the analysis were performed Scanning Electron Microscopic (SEM) tests, compressive strength and Young's modulus tests after 1, 3, 7 and 28 days of production and shear and flexural strength after 28 days of production. The results showed that the addition of CNF in aqueous solution reduces the concrete's workability and negatively affects mechanical properties.

Keywords Cellulose nanofibers · CNF · RPC

Introduction

The study of materials from renewable sources and industrial waste has been increasing for the last years due to the growing concerns about the environment and economic factors. In this context, cellulose nanofiber (CNF) is an attractive and competitive material, not only because it is a renewable material, but also because of its low cost, low density, high strength, non-abrasive and non-toxic

F.G.D. Machado (✉) · L.G. Pedroti · J.V.B. Lemes · G.E.S. Lima
L.A.F. Fioresi · W.E.H. Fernandes · R.C.S.S.Alvarenga
DEC—Civil Engineering Departament, UFV—Federal University of Vicosa,
Av. PH Rolfs, Vicosa, Minas Gerais 36570-900, Brazil
e-mail: felipe.gabriel@ufv.br

J. Alexandre
LECIV—Civil Engineering Laboratory, UENF—Darcy Ribeiro State University
of Northern of Rio de Janeiro, Av. Alberto Lamego, 2000, Parque Califórnia,
Campos dos Goytacazes, Rio de Janeiro 28013-602, Brazil

S. Ikhmayies et al. (eds.), *Characterization of Minerals, Metals, and Materials 2017*, The Minerals, Metals & Materials Series,
DOI 10.1007/978-3-319-51382-9_57

behavior. In addition to this, it is widely found in nature and can be easily modified by chemical agents [1].

The most abundant component in nature is cellulose. It composes the plants' basic cell structure and approximately 40–45% of wood's mass, being located on the secondary cell wall [2].

The CNF is produced from mechanical disintegration of cellulose fibers, mostly its surfaces, causing the disruption of cell wall and subsequent exposure of the fibrils and microfibrils [3].

According to Eichhorn et al. [4], the main reason to utilize cellulose nanofibers on other materials' composition is the possibility of using the cellulose crystal's high stiffness for reinforcement.

The Reactive Powder Concrete (RPC) is a type of concrete made of materials with maximum diameter of 2 mm, such as quartz sand, Portland cement, crushed quartz, active silica, superplasticizer. It has a low water-cement ratio and is a fragile material. In order to tackle this brittleness, steel fibers are often added to the RPC mix.

With the goal of testing the addition of a natural fiber in the RPC mix, this study suggests the development of RPC replacing the steel fibers by CNF.

Materials and Methods

The CNF used in this study was provided by the Cellulose and Paper Laboratory from the Federal University of Vicosa's Forest Engineering Department. The material was provided in aqueous solution, with 2.5% of CNF in mass. With this concentration, the solution has the consistency of a gel, as shown in Fig. 1.

The material went through an Atomic Force Microscopy (AFM) test, an analysis technique that consists on scanning the surface of sample with a probe, in order to obtain a topographic image with atomic resolution. The test was performed by the Microcospy and Atomic Force Laboratory from the Federal University of Vicosa's Physics Department.

The following materials were used to produce RPC: Portalnd cement CPV ARI (high-initial resistance), quartz sand, crushed quartz, silica fume and superplasticizer. The composition used was the same as Vanderlei [5] in his Ph.D. thesis,

Fig. 1 CNF in aqueous solution

Table 1 Standard composition (in mass)

Série	w/c	Ciment (kg)	Quartz sand (kg)	Crushed quartz (kg)	Silica fume (kg)	Superplasticizer (kg)	Water (kg)
SR	0.23	1.000	1.101	0.235	0.246	0.04	0.23

changing only the water-cement ratio and superplasticizer mass. Table 1 shows the composition used.

Four different concrete compositions were produced, being the standard one without CNF addition (F0 series). The other compositions had addition of 0.25, 0.20 and 0.10% of CNF in mass proportionally to the cement's mass (F1, F2 and F3 series, respectively).

In each series, 12 specimens were prepared with 50 mm diameter and 100 mm height. They were tested for compressive strength and Young's modulus tests after 1, 3, 7 and 28 days of production. Regarding flexural strength, 3 specimens were prepared with dimensions of 40 × 40 × 160 mm and were tested after 28 days.

During RPC production, all materials (cement, sand, crushed quartz and active silica) were mixed inside a plastic bag until an homogeneus material was obtained. Then, it was placed in a laboratory mortar mixer. The superplasticizer was mixed with water and CNF in a beaker and the resulting solution was also placed in the mortar mixer. Mixing time changed according with the percentage of CNF used, ranging from 10 to 25 min.

The specimens were prepared according to the Brazilian Standard (ABNT NBR 5738—Concrete—Procedure for molding and curing concrete test specimens) [6]. Figure 2 shows the specimens produced for one composition.

After casting, the specimens were taken to a humidified curing chamber, where they stayed for 24 h. As the initial cure finished, they were demolded, properly named and put in a tank with saturated solution of calcium hydroxide, where they remained submerged until the testing day.

Fragments were removed from the specimens with CNF addition after the tests were performed. Scanning Electron Microscopic (SEM) tests were performed on the aforementioned fragments. The tested sample is shown in Fig. 3.

Fig. 2 Cylindrical and prismatic specimens

Fig. 3 Scanning electron microscopic (SEM)

Results

The CNF sample analyzed presented the AFM test results indicated in Figs. 4 and 5.

Both figures showed CNF in nanometric scale, which means that CNF is not going to harm the RPC homogeneity.

Figure 6 shows the result of SEM tests in the F1 series, which contains the highest proportion of CNF.

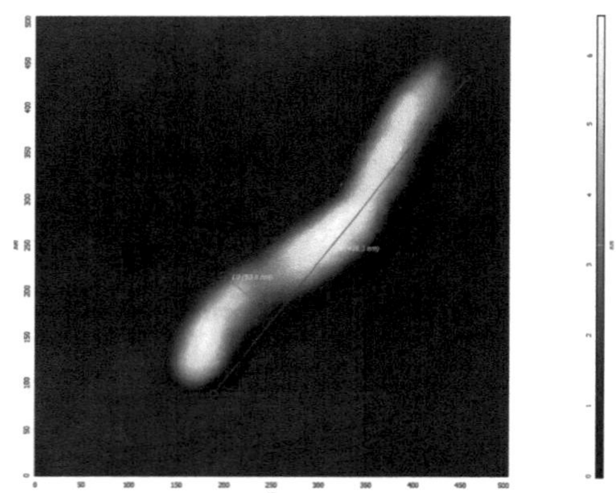

Fig. 4 View of CNF through microscope lens

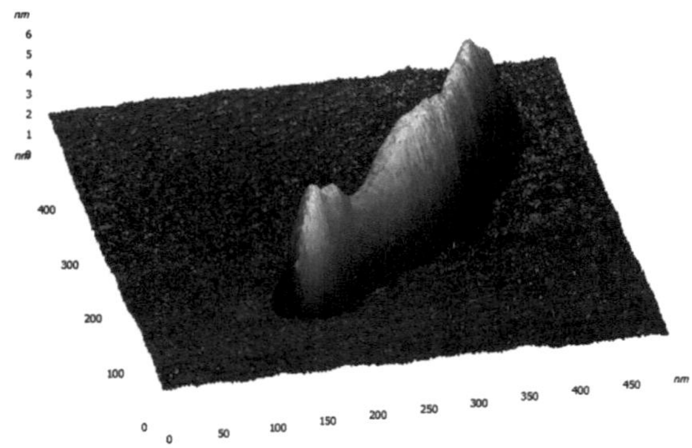

Fig. 5 Topography image of CNF

Fig. 6 SEM tests on specimen from F1 serie

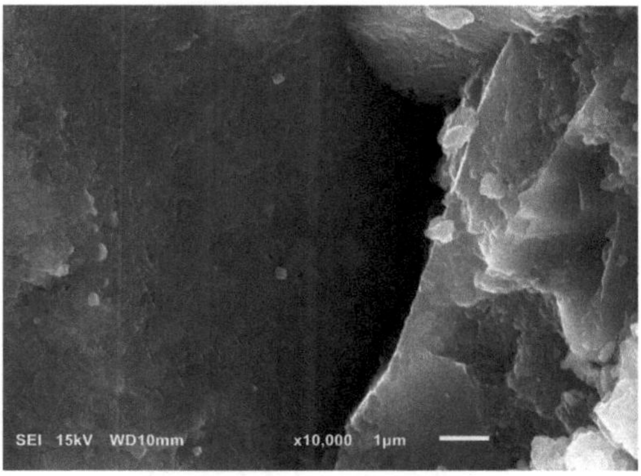

Figure 6 is an image of a specimen fragment with 10,000× zoom. CNF is not visible on it. It is also possible to attest the RPC homogeneity.

The results of compressive strength tests are shown in Fig. 7.

Compositions with CNF had 90% of its compression strength with 7 days of age. The increase in CNF proportion resulted in slow strength-gain in early ages. The standard composition had a noticeable strength-gain between 7 and 28 days, unlike the compositions with addition of CNF.

The Young's modulus tests presented similar results, between 43 and 47 GPa, as shown in Fig. 8.

With one day of age, RPC has 80% of the Young's modulus obtained after 28 days.

Figure 9 shows the flexural strength of RPC as a funcion of CNF proportion.

Figure 9 indicates that CNF addition did not cause any gain in flexural strength for RPC.

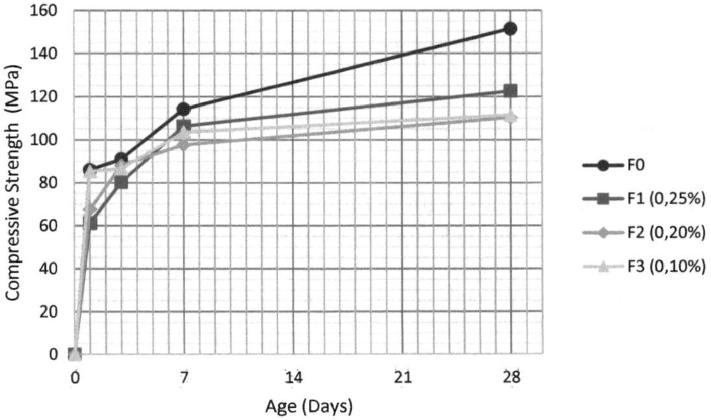

Fig. 7 Compressive strength of cylindrincal specimens as a function of age (time)

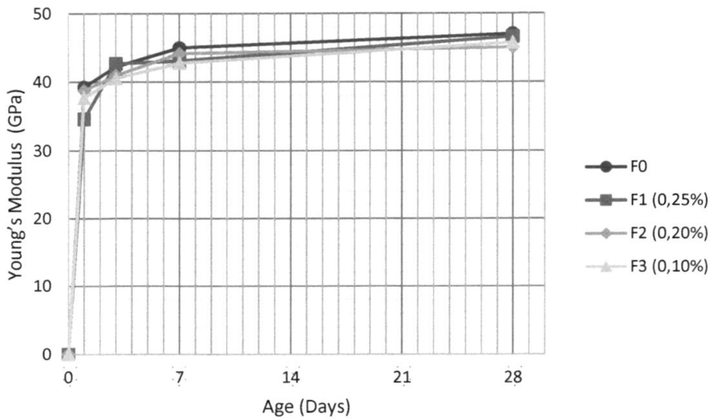

Fig. 8 Young's modulus of cylindricas specimens as a funtion of age (time)

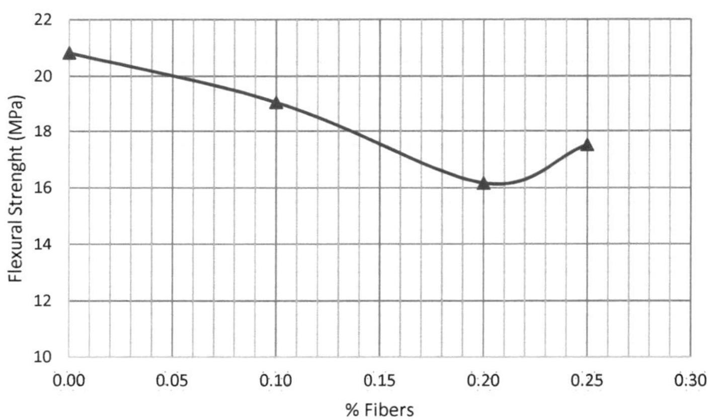

Fig. 9 Flexural strength of prismatic specimens as a function of CNF proporiton

Conclusions

After producing compositions with addition of CNF, it was noticeable that an increase in CNF proportion made the mixture drier, which diminished the RPC's workability. CNF has an elevated specific surface area and it could retain part of the added water. As a consequence of this, there is less water available for cement hydration.

With the RPC's workability harmed, the concrete compaction was also jeopardized. This could have influenced on the testing, because of the high dispersion found on the results. Additionally, it was perceived that humidity could directly influence workability and mixing time, being humid environments the more appropriate.

Results of compression strength tests were above the expected for standard concrete and in the range of high performance concretes (HPC) and not ultra-high performance concrete (UHPC). This was expected because there was no thermal cure, due to equipment restrictions. Thermal cure increases pozzolanic activity, which results in strength gains for concrete.

F0 series had an average compression strength of 151.40 MPa with 28 days of age, while F1, F2 and F3 series had, respectively, 122.62, 110.28, 111.26 MPa of average compression strength. According to an analysis made through Tukey's test, there was no significant difference between these values.

Tukey's test analysis also showed that F2 series had a significant reduction on flexural strength when compared to the standard series F0 (from 20.80 to 16.50 MPa). The other series had no significant difference.

Acknowledgements The authors thank the Brazilian agencies: CNPq, CAPES and FAPEMIG for the support provided to this investigation.

References

1. Miranda CS et al (2012) Thermal, mechanical and morphological properties of composites developed from glycerol and dicarboxylic acids reinforced with Piassava fiber. Macromol Symp 319:74–82
2. Sjöström E (1981) Wood chemistry fundamentals and applications. Academic Press Inc., San Diego, p 223
3. Syverud K et al (2010) A comparative study of *Eucalyptus* and *Pinus radiata* pulp fibres as raw materials for production of cellulose nanofibrils. Carbohydr Polym 84:1033–1038
4. Eichhorn SJ et al (2010) Review: current international research into cellulose nanofibres and nanocomposites. J Mater Sci 45:1–33
5. Vanderlei RD (2004) Análise Experimental do Concreto de Pós Reativos: Dosagem e Propriedades Mecânicas. Ph.D. thesis, Escola de Engenharia de São Carlos da Universidade de São Paulo, São Carlos, p 168
6. Brazilian Association of Technical Norms (2016) NBR 5738: concrete—procedure for molding and curing concrete test specimens. Rio de Janeiro, p 9

Advanced Ion Column Solution for Low Ion Damage Characterization and Ultra-Fine Process

Sang Hoon Lee, Mostafa Maazouz, Liang Zhang, Mauricio Gordillo, Micah Ledoux and Jeff Blackwood

Abstract To meet rising requirements for the low-damage ultra-fine process with the focused ion beam (FIB) in the materials science and semiconductor industry, the advanced ion column and corresponding advanced applications as a solution are developed and demonstrated. Novel high-speed high kV milling, low kV cleaning and imaging capabilities have enabled the creation of impeccable ultra-thin structures such as TEM samples and nanometer scale structures. Minimizing Ga+ induced sample damage is achieved through the optimized final cleaning at 500 eV ion landing energy with the new ion optics, which utilizes close control of the ion dose, beam incidence angle, spot size, pattern algorithm and imaging performance. Newly introduced process solutions achieve the improved process throughput. Initial demonstration is performed on the latest 14 and 10 nm scale FinFET device. Validation of the thinnest <0.8 nm thick ion damage layer on the <7 nm thick lamella structure was confirmed through TEM analysis.

Keywords Ion damage · Ion column · Low kv · 500 V · FIB · Focused ion beam · Thin · Lamella · TEM sample · Acceleration voltage · Gallium · Phoenix

Introduction

It is obvious that most of the small-scale components and devices in the research and applications development field of industries, especially semiconductor, biomechanical and metals and materials science drives increasingly to the atomic level characterization minimizing any artifact which comes from the design of experiments and of analysis [1]. The focused ion beam (FIB) is nowadays preferred in most research fields as one of the major methodologies and transmission electron microscopy (TEM) is accompanied by FIB for the most angstrom-scale research [2].

S.H. Lee (✉) · M. Maazouz · L. Zhang · M. Gordillo · M. Ledoux · J. Blackwood
FEI, Thermo Fisher Scientific, 5350 NE Dawson Creek Drive,
Hillsboro, OR 97124, USA
e-mail: sanghoon.lee@fei.com

S. Ikhmayies et al. (eds.), *Characterization of Minerals, Metals, and Materials 2017*, The Minerals, Metals & Materials Series,
DOI 10.1007/978-3-319-51382-9_58

Until recently, ion-induced damage-layer thickness was estimated varying from 5 nm to 40 nm. Another artifact, called "curtaining," gets worse as the ion acceleration voltage drops from a few kilovolts to hundreds of volts. Both ion and this curtain effect are drawbacks to using the FIB, especially as the sizes of devices and samples continue to shrink [3]. To meet and exceed the current demands, the invention of a novel concept ion column and its advanced application is immediately followed by diverse approaches and various characterizations of ion beam behavior. Ion induced surface damage is minimized to the sub nanometer scale on samples that are finely processed. The new ion column's increased scanning resolution based on the 20% finer spot size in the ion acceleration range from 500 V to 30 kV, along with the enhanced scanning algorithm enables the fine ion imaging and ultra-fine ion milling and deposition with gas injection systems (GIS). Ion column lens structure, ion dose and ion incident angle on target are also optimized for the best performance. All hardware and software components are scrutinized to demonstrate the best known method (BKM).

Experiment

The fundamental concept of the new ion column solution, called the Phoenix column, aims at accomplishing three key targets: minimizing damage presence of the ion beam treated surface, producing finer milling performance and minimizing the application process time in any processing of materials. The ion beam is multi-functional and can be used for imaging, milling and deposition simultaneously at each use-case of the beam current and beam acceleration voltage. The limiting factor to many FIB processes (such as TEM sample preparation) is the low-energy ion beam performance, in particular, the poor ion beam spot size in previous generation columns at <1 kV makes it difficult to accurately perform the final sample cleaning step. Therefore, considerable focus was put on improving this aspect of the Phoenix column performance. The result was improved image resolution at low acceleration ion voltage, specifically from 500 V to 2 kV. Figures 1 and 2 demonstrate the

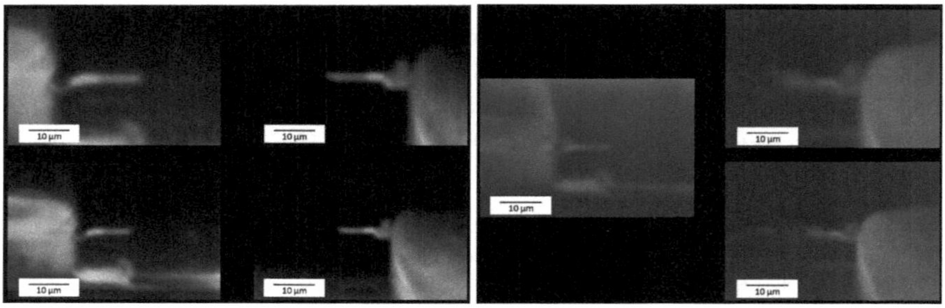

Fig. 1 500 V ion beam scanning images on the 10 μm wide 7 nm thick lamellae with TLD detector *(left four images)* and with ICE detector (*right three images*)

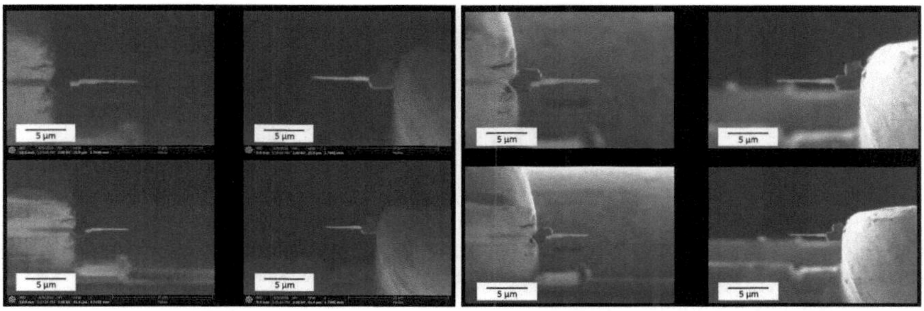

Fig. 2 2 kV ion beam scanning images on the 10 μm wide 7 nm thick lamellae with TLD detector (*left four images*) and with ICE detector (*right four images*)

imaging performance of 500 V and 2 kV using ion beam with Phoenix column. TEM samples that are 10 μm wide and 7 nm thick are well defined due to the enhanced image quality using two different imaging detectors, TLD and ICE.

The newly designed ion column produces a <20% reduced beam spot size compared to the equivalent ion current of previous generation ion columns and also achieves a >90% projected spot roundness. The imaging resolution range is between 3 nm and 10 nm when utilizing a beam current in the 1–100 pA range at 30 kV respectively, which correlates to 300 nm at 500 V. Two types of design of experiments (DOE) are applied to quantify the sophisticated milling competence for many applications. Figure 3 depicts the first DOE, which evaluates the amount of beam-induced erosion of the cross-section face and also determine the profile of the cross-section figure. The experiment begins by the cross-sectioned surface is created and is covered by Pt, W and TEOS (Tetraethyl orthosilicate) through a 1 keV electron-scanning deposition process. Following this, it is then cross-sectioned to measure the profile erosion by counting the pixels (Yellow colored, top right area) at each ion beam process condition.

The second DOE is described in Fig. 4. It evaluates the cap layer erosion rate, the correlation between the cap layer erosion and the remaining structure thickness after the thinning-down structure process. Most of the ion beam scanning methods generate a beam tail effect, and this affects the ultra-fine application process in the picometer to nanometer scale. The DOE is used to predict how precisely the ion

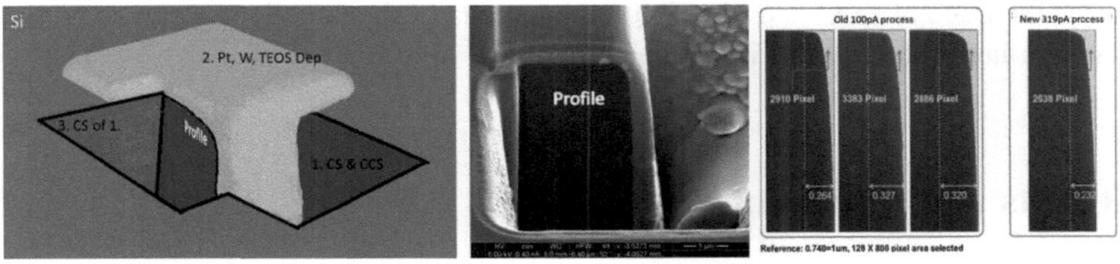

Fig. 3 From the *left* A simulated 3D image of the cross-section of cross-section. A reference SEM image of the cross-sectioned profile. The pixel count quantification with post-imaging process with FEI's image analyisis software. The higher pixel count accounts for a higher erosion rate

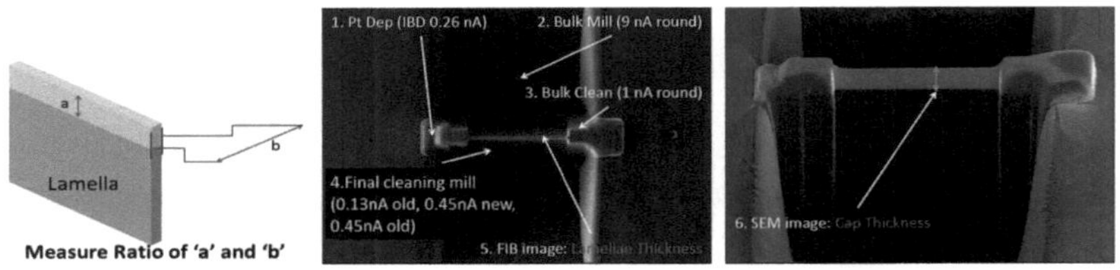

Fig. 4 Measuring ratio for "a" and "b" of the structure that is milled with various ion beam conditions; 0.13 nA previous scanning algorithm, 0.42 nA new scanning algorhythm, 0.45 nA previous scanning algorithm

beam creates a fine structure with minimized side effects, such as erosion of the cap layer. All DOEs are run by the fully automated process design software developed by FEI, iFast, which is part of the ion column solution system. Automated software ensures consistent and reliable data collection, eliminating human error.

One of the most critical requisites of a focused ion beam process is the amount of ion damage on the target surface [4]. The novel beam scanning algorithm integrated with the beam overlap ratio in X, Y and control of Z, the depth of focus, multi-scanning direction, scanning frequency optimization, and new design of column power supply, new electro-lens tables are developed and optimized and correspond to the new column design. The thickness of damage layer, not only for the ion projection but also for the actual subsequent milling process, can be expressed as an equation.

$$D_{\text{layer}} = \alpha \cdot F(\text{glance angle}, \text{time(dose)}) \tag{1}$$

In the Eq. (1), D_{layer} is ion bombardment damage layer thickness. α is assumed to be a constant, proportionate to the ion acceleration voltage. The glance angle is the ion trajectory angle perpendicular to the target surface. The dose is represented as pC/um^2 and is the process time based on the unit surface.

Two DOE sets for the damage layer analysis correspond to the various beam conditions to be determined as the best optimized condition of the new ion column. Both the lamella of the cross-section and the lamella of the lamella method are verified by TEM, EDS and SEM measurement. To minimize artifacts on the surface, 1 kV electron beam deposition is made prior to the 30 kV ion beam deposition on both the lamella of the cross-section and the lamella of the lamella. TEM-EDS scans the trace of C, O and Si ratio on the lamella of the lamella samples to confirm damage layer boundary (Figs. 5 and 6).

Results and Discussion

The erosion rate analysis result, as shown in Figs. 7 and 8, show that the new design performance is equivalent to one third of the beam current of the old ion column design. This implies that the new column solution can triple the beam

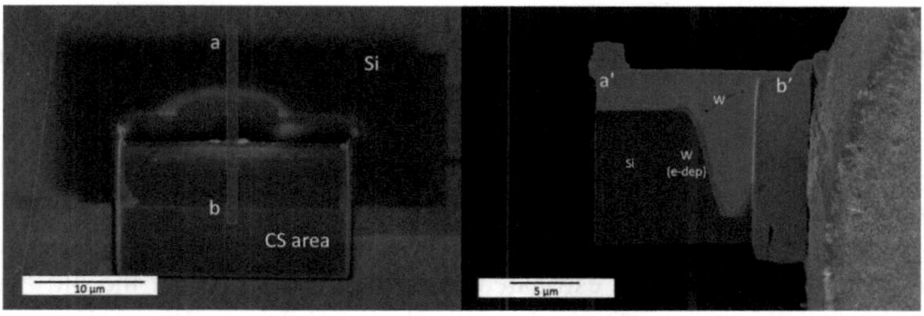

Fig. 5 Top view of the cross-section (CS) is shown in the *left* SEM image. "a" and "b" are the position and direction of the lamella presented in the *right* SEM image. Each cross-section of each beam condition is generated and collected as damage layer analysis data

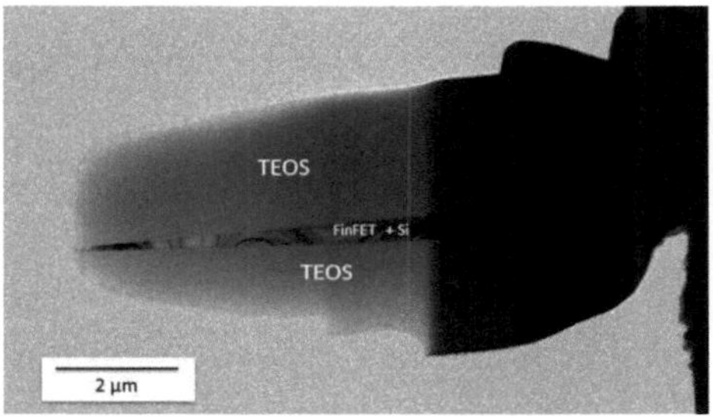

Fig. 6 The lamella of the lamella method is demonstrated. A thin TEM sample (lamella) is cross-sectioned and protected by TEOS through 1 keV electron beam deposition to minimize beam damage. It is thinned down to 30 nm. The TEM-EDS scanned the interface between TEOS and the lamella (Si)

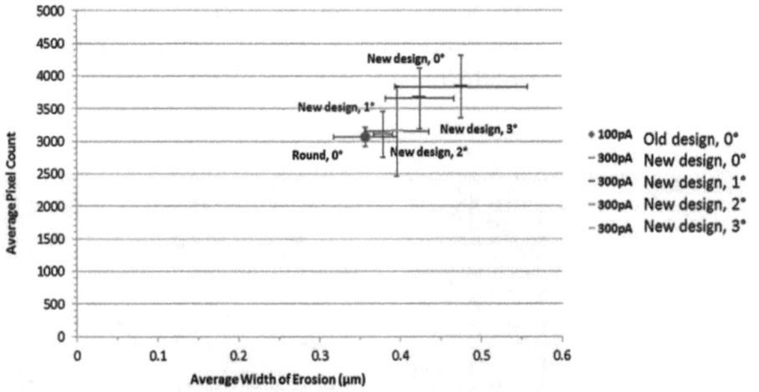

	100pA (0°)		300pA (0°)		300pA (1°)		300pA (2°)		300pA (3°)	
	Pixels	Width(µm)	Pixels	Width(µm)	Pixels	Width(µm)	Pixels	Width(µm)	Pixels	Width(µm)
Average	3069	0.36	3832	0.48	3655	0.42	3107	0.38	3149	0.40
STDEV	142	0.04	475	0.08	470	0.04	689	0.04	349	0.01

Fig. 7 Conventional 100 pA beam versus newly desinged 300 pA beam performance comparison. Lower pixel count in Fig. 3 means less erosion during the milling process. Rotational Tolerance, 0° to 3°, is one of the scanning procedures in the new algorithm

current and achieve similar fine structures in a third of the current process time using the previous design solutions. It also implies that there is improved beam spot position selectivity using the new column.

According to Fig. 8, the remaining cap thickness percentage at 100 nm thick lamella using previous 0.13 nA is 81 nm. Comparatively, at 0.42 nA the new design has a 77.8 nm thickness, while at 0.45 nA the old design only has 57.1 nm. If the 0.13 nA thickness result of the old design is normalized as 100%, then the 0.42 nA of the new design is estimated as 96% and 0.45 nA of the old design is 70.5%. As for the process time, the new design only requires 31% of the old designs 0.13 nA process time using the same milling dimensions (Fig. 10).

Lower-acceleration energy ions suffer from larger aberrations in the column which causes them to produce larger beam spot sizes. This prevents the ions from precisely milling the target and produces poor image quality. With the new column

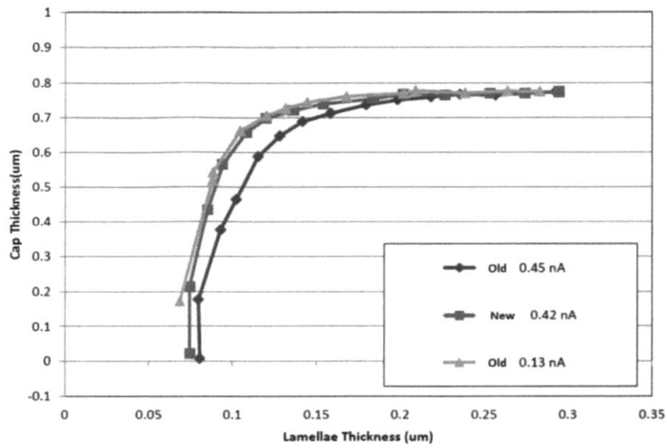

Fig. 8 The result of the second DOE as shown in Fig. 4. Representative 0.42 nA beam current of the new scanning algorithm has finer milling character and better region of interest (ROI) edge-protective behavior compared with a similar beam current of the former design. Simultaneously, it traces the same performance as the 0.13 nA of the old design

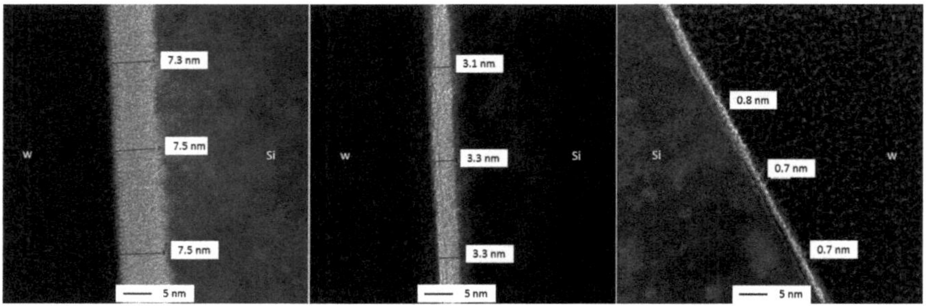

Fig. 9 Three TEM images show the thickness measurement of the ion damage layer on the Si target. From *left to right*; 7 nm thickness corresponds to 5 kV final process, 3 nm layer thickness corresponds to 2 kV final process, and 0.7 nm layer thickness corresponds to 500 V final process

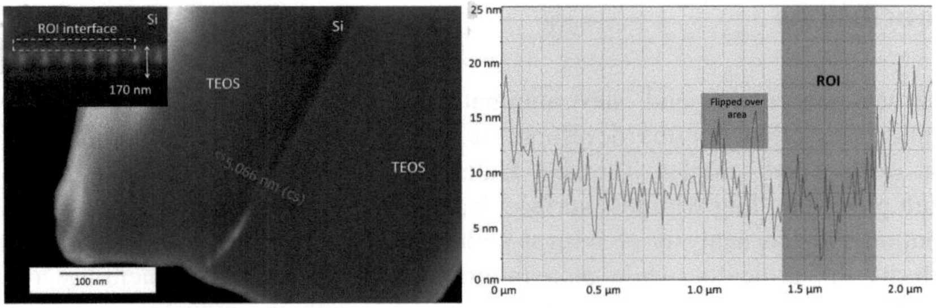

Fig. 10 The finest structure created by ion column solution is analyzed. Both the SEM cross-section image (*left*) and the TEM EELS thickness measurement (*right*) validate <7 nm thick lamella in the structure of the 10 nm node FinFET device

Fig. 11 A TEM example image demonstrates a <7 nm thick sample on the 14 nm node FinFET device. The clean Si boundary and ion damage-free device structures at the exact end-point are also evident in the image

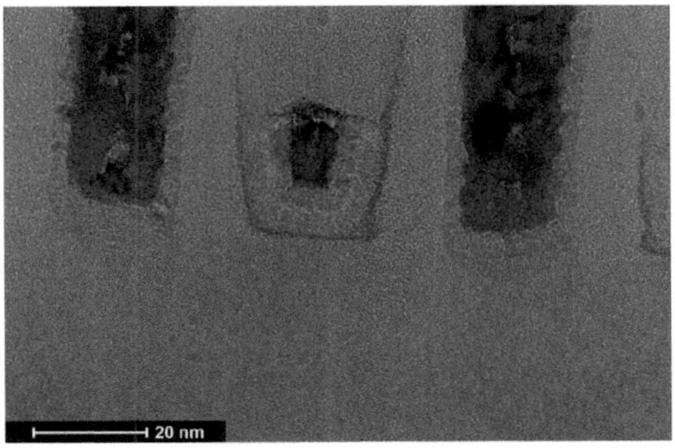

design these issues are mitigated to allow the use of target-oriented low damage processes on any region of interest (ROI) materials in the sub nanometer scale. With the highly increased ion scan resolution in the low kV energy range, especially from 500 V to 5 kV, it embodies both the high quality imaging and minimized ion damage process for sub nanometer targets simultaneously.

Figure 9 shows the cross-sectional TEM micrographs of the amorphized Silicon layer. The bright interlayer between Tungsten and Silicon target is seen as ion damaged amorphous Si. The thickness of its layer is measured at each beam acceleration voltage; 5 kV, 2 kV and 500 V.

The new solution demonstrates the capability of producing the finest structures, the world's thinnest lamella, at the commercial FinFET device on the Si substrate. In Fig. 10, the lamella structure is covered with TEOS using an electron scanning process and is cross-sectioned to measure the thickness of the lamella. At the bottom of the image, the white needle-like structure is the FinFET lamella generated by a 500 V ultra-fine focused ion milling process (Fig. 11).

Conclusions

The new ion column and latest automation software solutions paired with novel applications allow for the development of sub-nanometer scale modification on real-world TEM samples. The new Phoenix FIB column solution system delivers more precise milling with less surface damage and higher resolution imaging. New solution-assisted analysis of critical structures that are 7 nm and smaller greatly reduces the difficulty while also improving the speed and reducibility of the materials preparation process in any scientific research and industrial development fields.

References

1. Moon M-W, Lee SH, Sun J-Y, Oh KH, Vaziri A, Hutchinson JW (2006) Wrinkled hard skins on polymers created by focused ion beam. Proc Natl Acad Sci U S A 104:1130–1133
2. Lee SH (2011) Ion induced fabrication and analysis of wrinkle formation. PhD thesis, University of Texas at Dallas, pp 13–61
3. Alvis R, Blackwood J, Lee SH, Bray M (2012) High-throughput, site-specific sample preparation of sub-20 nm thick lamella. Int Symp Test Failure Anal 38:391–398
4. Giannuzzia LA, Stevieb FA (1999) A review of focused ion beam milling techniques for TEM specimen preparation. Micron 30:197–204

Application of Membrane Separation Technology in Wastewater Treatment of Iron and Steel Enterprise

Lei Zhang, Shining Chen, Lina Wang, Pu Liu, Benquan Fu and Jiannyang Hwang

Abstract The present status of wastewater advanced treatment and reuse in iron and steel enterprise is discussed, and the membrane technology used in wastewater advanced treatment and reuse from iron and steel enterprise was investigated. Its advantages, disadvantages and the development tendency were discussed and some advice and prospects of membrane technology for the reuse of wastewater treatment in iron and steel enterprise were pointed out.

Keywords Membrane separation technology · Wastewater treatment · Reuse · Iron and steel enterprise

Introduction

Large amount of water is used in the iron and steel industry, and wastewater discharged from the production process is mainly from the production process wastewater, equipment and product cooling water, equipment and site cleaning water, etc. More than 70% of the wastewater is from the cooling water, and small part of the wastewater discharged is from production process. Most of the pollutants in wastewater are chemical oxygen demand (COD), suspended solids, oil, ammonia nitrogen and cyanide, etc. So far, wastewater reclamation and reuse level in China is not high on the whole, whereas the potentiality of wastewater reuse is huge. Compared with famous enterprises such as Nippon Steel & Sumitomo Metal (Japan), POSCO (South Korea), there is a big gap in the water consumption, water reuse rate for Chinese iron and steel enterprises. Therefore, discharged wastewater

L. Zhang (✉) · L. Wang · P. Liu · B. Fu
R&D Center of WISCO, Wuhan 430080, China
e-mail: leizhang21c@163.com

S. Chen
Wuhan Institute of Shipbuilding Technology, Wuhan 430050, China

J. Hwang
Michigan Technological University, Michigan, MI 49931, USA

S. Ikhmayies et al. (eds.), *Characterization of Minerals, Metals, and Materials 2017*, The Minerals, Metals & Materials Series,
DOI 10.1007/978-3-319-51382-9_59

should be reasonably reused for Chinese iron and steel enterprises, as well as saving water and water conservation policy, which not only can reduce the pollution of water environment, but also reduce the quantity of water per ton steel and cost of production, it is an effective way to keep healthy and sustainable development for the iron and steel industry.

Membrane separation technology has gradually become an important part in the area of water treatment in recent years, especially in the area of wastewater treatment and reuse, which has been successfully applied in municipal/domestic wastewater, metallurgical wastewater, food wastewater, petrochemical wastewater, pharmaceutical wastewater, textile printing and dyeing wastewater treatment [1]. The objectives of this paper are to compressively analyze the current status of wastewater reclamation and reuse in Chinese steel industry and to summarize several kinds of membrane separation technology used in iron and steel wastewater treatment and reuse, and then to put forth recommends for future wastewater reclamation and reuse in Chinese steel industry.

Status Analysis of Wastewater Treatment and Reuse in Steel Industry

The production process of steel industry is very complex, and wastewater quality is also different. According to the production process, there are mining wastewater, coking wastewater, sintering wastewater, iron smelting wastewater, steel making wastewater and steel rolling wastewater, etc. There are five types of pollutes in wastewater from steel industry, such as suspended matter, heavy metals, oil, acid and alkali wastewater and organic pollutants, and so on [2]. Now Chinese iron and steel enterprises have taken various measures to improve water utilization efficiency and water recycling, such as developing a reasonable water standard and improving the circulation of water supply facilities to increase water recycling rate; improving quality of water, and strengthen the series of water and multi-purpose water; finding a new source of water (reclaimed water) and strengthen the development of water-saving technology and equipment [2]. Due to the complex composition of iron and steel wastewater and large amount of water, source of wastewater and reuse should be considered when we selected appropriate integrated technology/process. Most of the comprehensive wastewater reuse treatment process used in Chinese iron and steel enterprises are the “coagulation sedimentation tank + multi medium filter technology”, “high density sedimentation tank + V type filter technology”, and “high density sedimentation tank + V type filter + double membrane desalination recycling process”, respectively. Wastewater is discharged from units of steel industry, municipal wastewater and reclaimed water and so on. The practical results showed that the membrane integrated process was more effective than other treatments, which could guarantee the stability of the effluent water quality [3, 4].

Application of Membrane Separation Technology in Wastewater Treatment of Steel Industry

The membrane separation technology combined with pretreatment steps like physicochemical treatment was shown to be more effective for removal of pollutes than traditional physicochemical alone. Among these advantages, good water quality, physical separation, low energy consumption, high efficiency, energy saving, without secondary pollution, convenient operation, small occupied area will promote faster commercialization of membrane technology. In recent years, great progress was made in the research and application of recycled water reuse and industrial water recycling in membrane separation technology. The principle of membrane separation is the use of selective membrane as the separation medium, and we can achieve the purpose of separation and purification through the external driving force (such as concentration difference, pressure difference, potential difference, etc.) under the action of material components in the selective membrane [5]. Microfiltration (MF), ultrafiltration (UF), nanofiltration (NF), reverse osmosis (RO) and their combination process are widely used in industrial wastewater reuse. External pressure as the driving force, separation of the solute and solvent solution, grading or purification can be achieved, the applicable scope of various membrane units are listed in Table 1.

Table 1 The suitability of several membrane units

Membrane unit	Filtering accuracy (μm)	MWCO (Daltons)	Function	Application
MF	0.1 ~ 10	>1 × 10^5	Removal of suspended solids, particles, microorganisms, and large scale colloids	Removal of turbidity of drinking water; water reuse, nanofiltration or reverse osmosis pretreatment
UF	0.002 ~ 0.1	(1–10) × 10^4	Removal of colloids, proteins, microorganisms and organic compounds of macromolecules	Drinking water purification, water reuse, nanofiltration or reverse osmosis pretreatment
NF	(1–3) × 10^{-3}	200–1000	Removal of 200–1000 Daltons in organic polyvalent ions and molecular weight	Removal of hardness, chroma and dissolved salts
RO	(1–6) × 10^{-4}	>100	Remove of dissolved salts and organic compounds of >100 Daltons	Water desalination, boiler water supply, industrial pure water preparation

Application of MF and UF in Steel Industry

MF, UF can effectively remove the suspended solids in the wastewater, colloid and microorganisms, and has good turbidity removal effect. MF or UF can be used to replace the secondary settling tank for the suspended solids, colloid and oil removal discharged from steel wastewater. Cold rolling wastewater is produced in cold rolled steel before and after rolling, containing various contaminants, such as acid, alkali, oil and chromium. Emulsion wastewater has higher oil concentration, good chemical stability, which is one of the most difficult to treat in the steel plants. Membrane separation technology can effectively remove oil, suspended solids in the wastewater and reduce the load of the lower wastewater treatment. Inorganic ceramic ultrafiltration membrane was used in China Panzhihua Iron and Steel Corporation to treat the emulsification wastewater. The results showed that oil removal rate could reached above 95%, and the effluent oil content was lower than 10 mg/L. Comparing with the original the oily wastewater treatment technology, inorganic ceramic ultrafiltration membrane could effectively achieve the separation of oil and water, eliminate secondary pollution, and reuse the waste oil [6]. Suction PVDF hollow fiber ultrafiltration membranes were used in China Handan Iron and Steel Corporation for treatment of comprehensive wastewater from Handan iron and steel sewage station. Ultrafiltration separation as pretreatment unit of reverse osmosis membrane separation, the results showed that the effluent quality can meet the requirements of reverse osmosis membrane, the turbidity removal rate reaches was above 95%, the SDI value of water was kept below 3 [7].

Application of RO in Steel Industry

RO is mainly used to remove dissolved inorganic salts, which is used for the advanced treatment and reuse of wastewater. At present, China Taiyuan Iron and Steel Corporation has built the largest reverse osmosis membrane industrial wastewater reuse project, and the desalinated water from the first stage reverse osmosis device is used for cold rolling system. The maximum yield of 1200 m^3/h is designed, and the recovery rate of the reverse osmosis system is over 75%, and the salt removal rate is above 97%. Almost 28 thousand cubic meter per day reclaimed water is produced, and treatment cost is about 3 CNY/t [8]. China Hebei Tangshan Iron and Steel Corporation has adopted the reverse osmosis system to treat high salt containing industrial wastewater, and 40000 m^3 fresh water per day can be saved from the RO system [9, 10].

Application of Membrane Integrated Process in Steel Industry

Membrane integrated process refers to the combination process of membrane separation technology with the traditional physical, chemical and biological treatment methods. Membrane integrated process has been widely used for industrial wastewater. There are two typical treatment process, such as UF + RO membrane separation technology (double membrane process), membrane separation technology + biological treatment process (Membrane bioreactor). Wu et al. [4] adopted the conventional physicochemical treatment + biochemical treatment + dual membrane technology for desalination of comprehensive wastewater discharged from iron and steel company. The operation results showed turbidity was less than 0.1 NTU, conductivity less than 20 μS/cm, oil content is zero and the desalting rate is >98% Fujian Iron and Steel company in China adopted the coagulation + sand filtration + UF and RO process for advanced treatment of high salt containing wastewater treatment. The design scale of wastewater reuses was 30 m^3/h, and reuse rate was above 60%, the effluent water quality indicators were as follows, such as SS <1 mg/L, COD <10 mg/L, total hardness <1.5 mg/L and conductivity <5 uS/cm, respectively. The effluent could meet the requirements of effluent reuse standard of steel industry, and total operation cost is 16.11 CNY/m^3 [11].

MBR is considered as an integration of conventional activated sludge system and advanced membrane separation. It has many advantages as fellows: high membrane retention ability, low production of excess sludge, the independent control of sludge retention time and hydraulic retention time, retaining a high concentration of sludge biomass in the reactors. The effluent discharged from MBR has little suspended solids and turbidity, which can be directly recycled [12, 13].

Rolling oily wastewater contains mixtures of hydrocarbons and surfactants. Currently available conventional biological wastewater treatment techniques are mostly incapable of complete elimination of hydrocarbons in stable emulsions. These compounds are only poorly biodegradable under such conditions, while MBR has superior performances in treating oily wastewater. Wu et al. [14] developed a combined process of MBR in treating cold rolling oily wastewater treatment, and long-term and stable membrane flux, high treatment efficiency, low operation cost were obtained. The emission effluent could meet the requirements of China national discharge standard (GB8978-1996). China Laiwu Steel and Iron Corporation adopted electrochemical oxidation, MBR and RO membrane integrated process for advanced treatment and reuse of comprehensive waster from cold rolling plant. The practical results showed that water quality index was better than these of fresh water supply, and treatment cost is 3.47 CNY/m^3 [12]. In general, MBR technology shows specific advantages over conventional bioreactor system. The high quality of the obtained reclaimed water opens the possibility of recycling the effluent for reuse in the steel industrial process which can significantly reduce the demand for fresh water.

Conclusions

Membrane separation technology has made great progress in the areas of wastewater treatment and the same time, membrane technology is currently facing some research and development challenges. If these challenges, such as membrane fouling, membrane cleaning, membrane lifespan, capital costs and novel membrane modules to enhance their hydrodynamic conditions, etc., are well solved, it could lead to a more competitive and mature market for application in steel industry. Therefore, we can use the advantages of membrane technology to further expand the application of membrane integrated process with physical, chemical, biological process, which not only solves the requirements of fresh water for steel enterprises, but also reduce the pollution of water and improve the water environment. To a great extent, these measures will undoubtedly push forward the commercial application of membrane technology in areas of steel industry.

References

1. Liu M et al (2005) Application of membrane technology in wastewater treatment and reuse. Beijing Chemical Industry Press, Beijing, pp 36–45
2. Wang S et al (2008) Reuse technology and application of wastewater in iron and steel industry. Beijing Metallurgical Industry Press, Beijing, pp 12–16
3. Sui Z et al (2013) Zero discharge engineering example of the factory integrated wastewater of Tianjin Iron & Steel Group. Ind Water Treat 33(11):85–89 (in Chinese)
4. Wu Y et al (2013) Application of double membrane process in advanced treatment of integrated wastewater from Iron and Steel Industry. China Water Wastewater 29(22):102–109 (in Chinese)
5. John AH (2004) Future of membranes and membrane reactors in green technologies and for water reuse. Desalination 162(2):1–11
6. Zhang M et al (2006) Application of inorganic ceramic ultra-filtering film technology in disposal of cold rolling waste water. Metall Power 5:64–66 (in Chinese)
7. Gao S et al (2007) Pilot study on reclamation of sewage from a steel mill by UF. Chin J Environ Eng 1(2):43–46 (in Chinese)
8. Jin Y et al (2015) Application of RO technology in the big project of steel plant wastewater reuse. Membr Sci Technol 35(3):102–106 (in Chinese)
9. Zhong W et al (2013) Application of RO technology to the Project of Steel Plant Wastewater Reuse. Ind Water Treat 33(8):83–85 (in Chinese)
10. Yao L et al (2011) Study on the RO pretreatment at iron and steel enterprises. Ind Water Treat 31(2):21–24 (in Chinese)
11. Pei G (2016) Design and operation of high-salinity wastewater reuse in a steel plant in Fujian. Water Wastewater Eng 42(6):73–75 (in Chinese)
12. Yang W et al (2006) State-of-art of membrane bioreactor: worldwide research and commercial application in North America. J Membr Sci 270(1–2):201–211
13. Wang Z et al (2008) Research and applications of membrane bioreactors in china: progress and prospect. Sep Purif Technol 62(2):249–263

14. Wu Z et al (2014) Example project of cold rolling wastewater treatment. China Water Wastewater 30(10):64–68 (in Chinese)
15. Zhang J et al (2012) Advanced treatment of cold-rolling wastewater from Laiwu Steel Group and Reuse Engineering. China Water Wastewater 28(10):70–74 (in Chinese)

Brillouin Scattering Study on Elastic Properties of Bulk hcp ZnO Single Crystal

Ping-Ping Fan and Yong-Quan Wu

Abstract The polarized Brillouin scattering spectra were systematically measured to study the elastic properties of bulk hexagonal close-packed (hcp) ZnO single crystal in this paper. The peak polarizations of all modes in spectra were analyzed firstly. The elastic properties, such as the whole set of Elastic constants were accordingly calculated and compared with the previous literatures. Also investigated were the effects of the rotation of the sample in the (0001) plane perpendicular to the z axis on different acoustic modes. The results show that the quasi-longitudinal and quasi-transverse modes perpendicular to the (0001) plane act as the expected trigonometric function of the rotation angle, while the longitudinal and transverse modes on the (0001) plane remain constant.

Keywords Brillouin scattering · ZnO · Elastic constants

Introduction

ZnO as a new type of wide bandgap semiconductor material, has its unique characteristics such as large photoelectric coupling coefficient, low temperature coefficient, small dielectric constant and high light transmittance. All the fundamental physical advantages make ZnO one of the key materials in surface acoustic wave devices, piezoelectric and pyroelectric devices, and gas sensors. It is also a promising material for visible to ultraviolet light emitters and transparent field effect transistors [1]. For efficient design of these devices, it is important to determine the fundamental material parameters.

Up to now, there are lots of ways to investigate elastic properties, such as Mechanical stress method [2–4], Ultrasonic pulse and Ultrasonic resonance [1, 5, 6], as well as Brillouin scattering [7]. Recently, Brillouin scattering has proved to be

P.-P. Fan · Y.-Q. Wu (✉)
State Key Laboratory of Advanced Special Steels, Shanghai University, Shanghai 200072, People's Republic of China
e-mail: yqwu@shu.edu.cn

S. Ikhmayies et al. (eds.), *Characterization of Minerals, Metals, and Materials 2017*, The Minerals, Metals & Materials Series,
DOI 10.1007/978-3-319-51382-9_60

one of the most powerful means to study elastic properties of materials without any mechanical contact [1, 8]. Sound velocities and thus elastic constants can be determined with high precision [9–11].

In this paper, we report a back scattering configuration to measure elastic constants of bulk ZnO single crystal in different polarization states, and the results are compared with those reported previously. Moreover, we have investigated the effects of the rotation of the sample in the (0001) plane on different acoustic modes.

Experimental Work

Material and Method

The sample used here was a colorless bulk ZnO single crystal. It was a (0001) substrate with both sides polished, and it was $10 \times 10 \times 0.5\ mm^3$. Brillouin scattering measurements were performed at room temperature using a 532 nm line of an Diode-Pumped laser as a light source and a 180° back scattering geometry for collection. The scattering light was analysed by a Sandercock-type six-pass tandem Fabry-Pérot spectrometer [12], detected by a photon counting photomultiplier. Further details of the experimental setup can be found elsewhere [13].

The relationship between sound velocity V and the Brillouin scattering frequency shift ω can be determined by back scattering geometry sketched in Fig. 1, where the angle θ between incident light and normal of surface plane is 45°, and ki, ks, kr are wave vectors of incident, scattered and reflected light outside the sample, respectively.

The equation between sound velocity V and frequency shift ω can be described as:

$$V = \omega/q \tag{1}$$

$$q = 4n\pi/\lambda \tag{2}$$

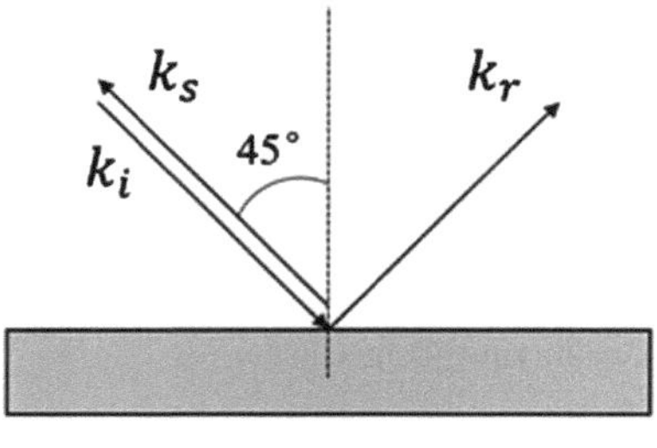

Fig. 1 Scattering geometry used in the present study. ki, ks, kr are wave vectors of incident, scattered and reflected light outside the sample, respectively

where q is the wave vector of phonons, n is the refractive index of the crystal, and λ is the wavelength of laser. Then based on the sound propagation equations and the measured velocity data, a full set of elastic constants can be determined.

Results and Discussion

Polarization Selection Rules

For the ZnO crystal with a wurtzite structure, there are five independent elastic constants, C_{ij} $(i\,j = 11, 13, 33, 44, \text{and } 66)$, and $C_{12} = C_{11} - 2C_{66}$. In order to obtain those parameters successfully, different polarization states are used to distinguish each kinds of acoustic phonon modes firstly. Herein, the polarization perpendicular (parallel) to the scattering plane is defined as *V*(*H*) polarization. For example, *VH* polarization denotes that incident light is *V* polarized and scattered light is *H* polarized. Acoustic phonon modes, their sound velocities, and polarization selection rules are tabulated in Table 1 for the back scattering geometry. Here, ρ is the crystal mass density and we use $\rho = 5.605$ g/cm^3 calculated from the lattice constants ($a = 3.325$ Å and $c = 5.213$ Å) and atomic masses of Zn and O.

From Table 1, two elastic constants, C_{11} and C_{66} can be determined directly by measuring LA and TA phonon velocities along (0001) plane of the ZnO in back scattering. In order to determine the remaining independent constants, we need to determine the velocities from quasi-longitudinal and quasi-transverse waves perpendicular to the (0001) plane for an scattering angle of 45°. And then the elastic constants of ZnO were calculated by combining algebraic computations with a computer-based nonlinear least-squares fitting method.

Table 1 Acoustic phonon modes, sound velocities, and polarization selection rules for ZnO

Propagation direction	Mode	Sound velocity	Selection rule	
			VV	VH
q//[100]	Longitudinal	$\sqrt{\frac{C_{11}}{\rho}}$	Allowed	Forbidden
	Transverse	$\sqrt{\frac{C_{44}}{\rho}}$	Forbidden	Forbidden
	Transverse	$\sqrt{\frac{C_{66}}{\rho}}$	Forbidden	Allowed
q//[sinθ, cosθ]	Quasilongitudinal	V_{QLA}^{a}	Allowed	Forbidden
	Quasitransverse	V_{QTA}^{b}	Allowed	Forbidden
	Transverse	$\sqrt{\frac{C_{66}\sin^2\theta + C_{44}\cos^2\theta}{\rho}}$	Forbidden	Forbidden

[a] $\rho V_{QLA}^2 = \frac{C_{11}\sin^2\theta + C_{33}\cos^2\theta + C_{44}}{2} + \sqrt{\frac{\left[(C_{11}-C_{44})\sin^2\theta - (C_{33}-C_{44})\cos^2\theta\right]^2}{4} + (C_{13}+C_{44})^2\sin^2\theta\cos^2\theta}$

[b] $\rho V_{QTA}^2 = \frac{C_{11}\sin^2\theta + C_{33}\cos^2\theta + C_{44}}{2} - \sqrt{\frac{\left[(C_{11}-C_{44})\sin^2\theta - (C_{33}-C_{44})\cos^2\theta\right]^2}{4} + (C_{13}+C_{44})^2\sin^2\theta\cos^2\theta}$

The Peak Polarizations of All Modes

Brillouin scattering spectroscopy of hcp ZnO single crystal with the back scattering angle of 45° is exhibited in Fig. 2. There are four high intensity peaks in both the Stokes and Anti-Stokes sides of elastic peak, and the frequency shifts are 7.48, 16.24, 23.32, 47.94 GHz respectively.

Figure 3 shows the polarized Brillouin scattering spectra with Fig. 3a for V-V polarization and Fig. 3b for V-H polarization. According to polarization selection rules, the peaks at 7.48 and 16.24 GHz are assigned to modes corresponding to C_{66} and C_{11}, respectively. Strictly speaking, these two peaks are assigned to phonons excited by the incident light reflected from the backside of the sample, the peaks satisfy the polarization selection rules for platelet geometry and their frequency shifts are almost the same as those in Fig. 3, thus they can be choosed as transverse and longitudinal modes that propagating along the (0001) plane.

Moreover, the peaks at 23.32 and 47.94 GHz are respectively due to quasi-transverse and quasi-longitudinal modes, which can be used to determine the remaining constants. According to the characteristics of the quasi modes, which mainly affected by the bulk acoustic modes, the relationship between the sound velocity of quasi modes and the angle of the bulk wave vector to the z axis must be investigated.

Calculation of Elastic Constants

Table 2 shows the velocity of quasi-longitudinal and quasi-transverse changing with the rotation angle of the sample in the (0001) plane perpendicular to the z axis from 0° to 110°. Herein, we assume the original angle of the sample stage to be 0°.

Sound velocities from Table 2 were fitted by trigonometric function, and the results were listed in Fig. 4. It's shown that calculated sound velocities fit the trigonometric curves fairly well, which revealed the sound velocities of QTA and QLA act as the expected trigonometric function of the rotation angle. Furthermore,

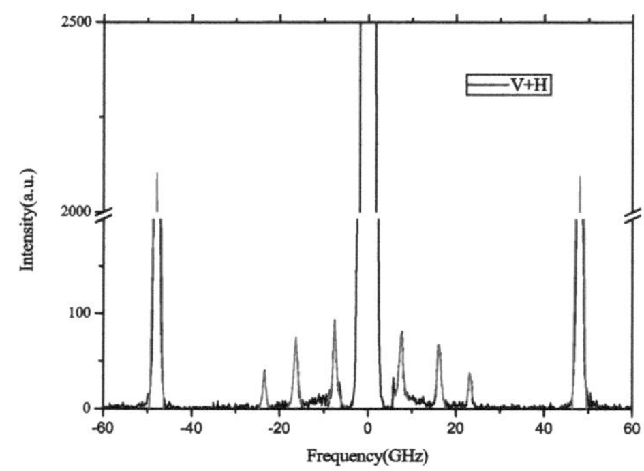

Fig. 2 Brillouin scattering spectroscopy of hcp ZnO single crystal, with the back scattering angle of 45°

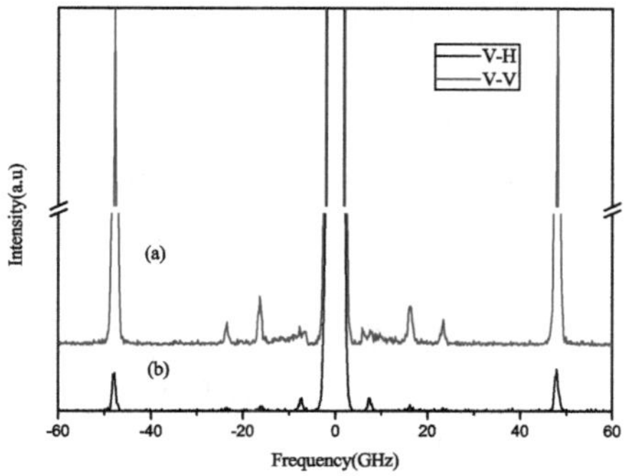

Fig. 3 Brillouin scattering spectra in back scattering geometry for V-H polarization

Table 2 The sound velocities of QTA and QLA with different rotation angles

Angle (deg)	V_{QLA} (m/s)	V_{QTA} (m/s)
0	6675.035	3243.152
10	6672.267	3245.74
20	6669.199	3251.011
30	6669.48	3247.304
40	6673.903	3240.769
50	6674.388	3244.92
60	6672.972	3243.298
70	6668.246	3242.923
80	6666.952	3250.84
90	6667.987	3244.754
100	6671.683	3240.45
110	6672.128	3244.361

the maximum and minimum of the curves undergo an unique variation with the rotation angle, which the maximum of QLA meets the minimum of QTA in 20° while the converse results occurs in 50°. It indicats that the effects of rotation of the sample in the (0001) plane on QLA and QTA act as the expected trigonometric function of the rotation angle with the period of 60°, which is in great agreement on the reciprocal velocity curves of ZnO single crystal. This is aroused from the intrinsic anisotropy and symmetry of the hexagonal crystal structure of ZnO.

The propagation direction of sound velocities must be determined, since the intrinsic anisotropy for single crystal ZnO. Herein, we need to know the crystallographic orientation angle first. To fix this problem, nonlinear least-square fitting method is used, the crystallographic orientation angle is calculated to be 70.43° and the whole set of elastic constants of ZnO are listed in Table 3. The results is also compared with previous literatures. As we can see, some discrepancy can be found clearly in Table 2. As mentioned before, crystal density and refractive indices are used to calculate elastic constants from Brillouin shifts. Thus we can assume that because of the discrepancy in crystal density and refractive indices of each sample that result from the difference in $c_{i\,j}$, rather than the material parameters. Moreover,

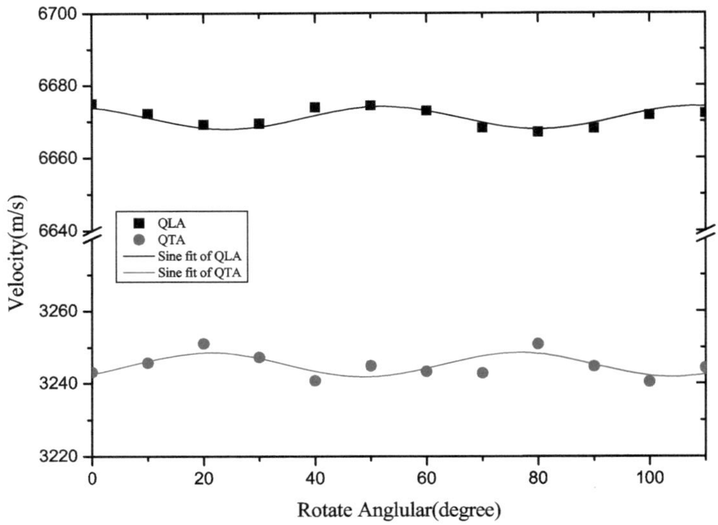

Fig. 4 The sound velocity anisotropy of single crystal ZnO

the cause of the difference might be the sample quality, the accuracy of each experimental method, and phonon dispersion.

Figure 5 shows the effects of rotation of the sample in the (0001) plane on LM and TM. Unlike QLA and QTA, the velocities of LM and TM in the (0001) plane remain constant.

Table 3 Elastic constants of ZnO in units of GHz

Sample	C_{11}	C_{12}	C_{13}	C_{33}	C_{44}	C_{66}
Single crystal[a]	20.97	12.11	10.51	21.09	4.25	4.43
Single crystal[b]	20.70	11.77	10.61	20.95	4.48	4.46
[0001]-oriented polycrystal on sapphire[c]	15.70	8.90	8.30	20.80	3.80	3.40
[0001]-oriented polycrystal on silicon[d]	20.60	11.70	11.80	21.10	4.43	4.46
Single crystal[e]	19.40	10.20	9.40	21.70	–	4.60
Present result	16.31	12.30	8.64	22.38	3.29	2.01

[a]Ultrasonic pulse technique [14]
[b]Ultrasonic resonance method [15]
[c]Acoustic investigation technique [16]
[d]Brillouin scattering spectroscopy [17]
[e]Ultrasonic resonance method [18]

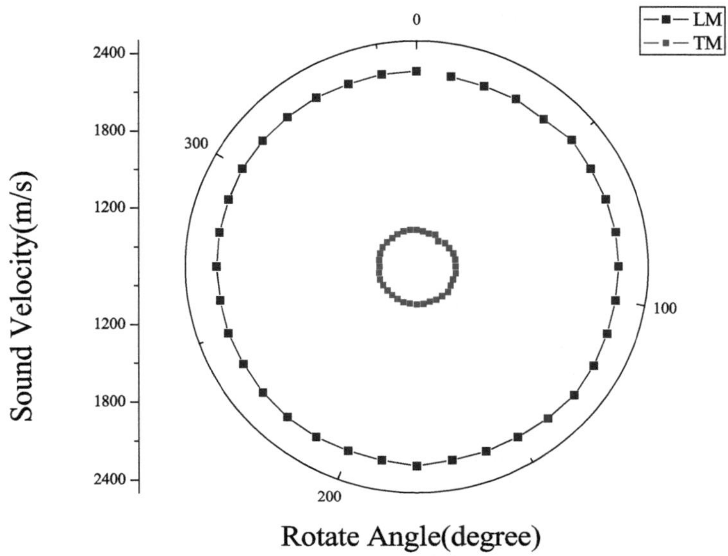

Fig. 5 Sound velocities of LM and TM in the (0001) plane for different rotation angles

Conclusions

In this work, polarized Brillouin scattering were systematically measured to study the elastic properties of bulk hexagonal close-packed (hcp) ZnO single crystal. From the Brillouin shifts observed, the whole set of elastic constants was determined. Moreover, the effects of rotation of the sample in the (0001) plane perpendicular to the *z* axis on QLA and QTA act as the expected trigonometric function with the period of 60°, while the LA and TA tend to remain constant.

Acknowledgements This work is supported by National Natural Science Foundation of China (Nos. 50504010, 50974083, 51174131, 51374141), Joint Funds of the National Natural Science Foundation of China (No. 50774112), National Basic Research Program of China (No. 2012CB722805).

References

1. Azuhata T et al (2003) Brillouin scattering study of ZnO. J Appl Phys 94(2):968–972
2. Sundqvist J et al. (2015) Identifying residual stresses in laser welds by fatigue crack growth acceleration measurement. J Laser Appl 27(4):042002-1–042002-8
3. Taskonak B et al (2008) Analysis of subcritical crack growth in dental ceramics using fracture mechanics and fractography. Dent Mater 24(5):700–707
4. Shakhvorostov D, Pöhlmann K, Scherge M (2006) Structure and mechanical properties of tribologically induced nanolayers. Wear 260(4–5):433–437

5. Cottam RI, Saunders GA (1974) An investigation of the validity of ultrasonic pulse-superposition velocity measurements using thin-film transducers. J Phys D Appl Phys 7(7):216–225
6. Akiyama I et al (2012) Development of multiple-frequency ultrasonic imaging system using multiple resonance piezoelectric transducer. Jpn J Appl Phys 51(7):1514–1522
7. Wrobel BM, Time RW (2011) Improved pulsed broadband ultrasonic spectroscopy for analysis of liquid-particle flow. Appl Acoust 72(6):324–335
8. Carlotti G et al (1991) Brillouin scattering by surface acoustic modes for elastic characterization of ZnO films. IEEE Trans Ultrason Ferroelectr Freq Control 38(1):56–61
9. Kotane LM et al (2011) Surface Brillouin scattering measurement of the elastic constants of single crystal $InAs_{0.91}Sb_{0.09}$. J Phys Conf Ser 278(1):68–72
10. Crowhurst JC et al (1999) Surface Brillouin scattering at high pressure: application to a thin supported gold film. Phys Rev B 601(601):R14990–R14993
11. Comins JD et al (2000) Surface Brillouin scattering of opaque solids and thin supported films. Ultrasonics 38(8):450–458
12. Every AG, Kotane LM, Comins JD (2010) Characteristic wave speeds in the surface Brillouin scattering measurement of elastic constants of crystals. Phys Rev B 81(81):1601–1614
13. Sandercock JR (1982) Trends in Brillouin scattering. Studies of opaque materials, supported films, and central modes. In Light scattering in solids III. Recent results, vol 51(2), pp 173–206
14. Sinogeikin SV, Katsura T, Bass JD (1998) Sound velocities and elastic properties of Fe-bearing wadsleyite and ringwoodite. J Geophys Res 103(B9):20819–20825
15. Bateman TB (1962) Elastic moduli of single-crystal Zinc oxide. J Appl Phys 33(11):3309–3312
16. Kobiakov IB (1980) Elastic, piezoelectric and dielectric properties of ZnO and CdS single crystals in a wide range of temperatures. Solid State Commun 35(3):305–310
17. Carlotti G et al (1987) Acoustic investigation of the elastic properties of ZnO films. Appl Phys Lett 51(23):1889–1891
18. Fioretto D et al (1995) Brillouin-scattering determination of the elastic constants of epitaxial fcc C_{60} film. Phys Rev B Condens Matter 52(52):R8707–R8710
19. Solbrig C (1965) Piezoelektrische Messungen an Zinkoxydkristallen. Zeitschrift Für Physik 184(3):293–298

Characteristics of Stamp Charging Coke and Top Charging Coke

Gao Bing, Xiao Hong and Zhang Wenqiang

Abstract The characteristics of top charging coke (coke 1) and stamp charging coke (coke 2) were investigated. The weight loss ratios of two kinds of cokes with and without alkalis were measured after reacting with CO_2 under the temperature of 900, 1000, 1100 and 1200 °C respectively. The microstructures of original cokes were observed by SEM. The experiment results show that, with no addition of alkali, the weight loss ratios of coke 1 and coke 2 were nearly the same under the temperature of 900 and 1000 °C. While the weight loss ratio of coke 2 was much larger than coke 1 under the temperature of 1200 °C. With alkali-rich, the weight loss ratio of the two cokes was almost the same at 1100–1200 °C.

Keywords Top charging coke · Stamp charging coke · Weight loss ratio · Microscopic structure

Introduction

In recent years, the technology of stamping charge coking developed quickly which can expand coking coal source and ultimate poorly coals [1–3]. Generally speaking, the cold strength of stamping cokes is better than that of top charging cokes. This is because by stamping process, coal particles contact dense and the resistance to escape of the gas increases too. The structures of stamping coke are more compact which help to improve the broken strength and abrasion strength of the coke. However, the studies have shown that the hot performance of stamping cokes was not surely better than that of top charging cokes in blast furnace, and even some were poorer than the top charging cokes [4–6]. The hot performance of cokes has a significant impact on the operation of the blast furnace, so it is necessary to study

G. Bing · X. Hong · Z. Wenqiang (✉)
HeSteel Group TangSteel Co. Ltd, 9 Binhe Road, Lubei District, Tangshan 063000, Hebei, People's Republic of China
e-mail: sjzgaobing@163.com

S. Ikhmayies et al. (eds.), *Characterization of Minerals, Metals, and Materials 2017*, The Minerals, Metals & Materials Series,
DOI 10.1007/978-3-319-51382-9_61

the degradation of stamping cokes in blast furnace, and make a right evaluation for the quality of stamping coke so as to guide the actual production.

Experimental

Samples

Two cokes were selected for this study, coke 1 is top charging coke, coke 2 is stamping coke, both cokes are derived from domestic steel mills. Their ultimate analysis and proximate analysis are provided as shown in Table 1. The CRI and CSR value of two cokes are shown in Table 2.

Experimental Procedure of Coke Gasification

Firstly, coke samples were dried at 105 °C for more than 2 h before the experiment of coke gasification. Then, the cokes were weighed 100 ± 0.5 g and put into the reaction vessel. The particle size ranges from 23 to 25 mm. Cokes were heated up to specified temperatures at 10 °C/min with flowing N_2 gas at 0.8 L/min. The coke gasification reaction could start with flowing CO_2 gas at 3 L/min when the temperatures reached at specified temperatures. The reaction time is 1 h. Then close CO_2 gas and open N_2 gas to cool the reaction vessel to room temperature. The reaction specified temperature of coke was 900, 1000, 1100, 1200 °C respectively. The test samples were coke 1 and coke 2 with no addition of alkali and alkali-rich (K_2O = 2 %). And the reaction characteristics of cokes under different temperature and conditions were investigated. Take the reaction temperature of 1000 °C for example, the programs of temperature and atmosphere in the reactor were shown in Fig. 1. The others were similar to it.

Table 1 Proximate analysis and ultimate analysis of cokes (wt%)

Samples	Ultimate analysis					Proximate analysis		
	C	H	O	N	S	A_{ad}	V_{ad}	F_C
Coke 1	86.69	0.25	0.08	0.42	0.6	12.28	2.33	85.39
Coke 2	86.12	0.31	0.64	0.37	0.63	13.21	1.71	85. 08

V_{ad}, volatile matters; A_{ad}, ash; Fc, fixed carbon on dry basis

Table 2 CRI and CSR value of cokes (wt%)

Samples	CRI	CSR
Coke 1	23.89	67.89
Coke 2	25.36	52.37

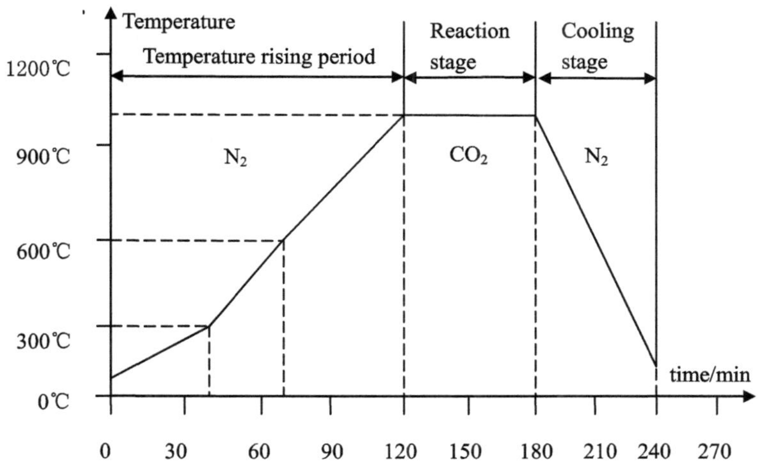

Fig. 1 The temperature program of coke gasification

Results and Discussion

The experiment results of weight loss ratio of cokes under different reaction conditions are shown in Table 3. And Figs. 2 and 3 show the results of gasification reaction of cokes with and without alkalis.

As shown in Figs. 2 and 3, it can be concluded that with no addition of alkali, the weight loss ratios of coke 1 and coke 2 are similar after reacting under 900 °C, the two kinds of cokes have low weight loss ratio at the low temperature. And the weight loss ratio of coke 1 increases slower than coke 2 under 1000 °C. When reaction temperature reaches 1100 and 1200 °C, the weight loss ratio of coke 1 increases relatively moderate, while the coke 2 increases dramatically. The reason may be that coke 2 contains more closed pores, which can reach up 20–30% and has a directly influence on coke gasification rate. When closed pores open and increase the reaction surface area, it will lead to the internal diffusion of CO_2 to coke. So with the temperature increasing, the weight loss ratio of coke 2 is larger than coke 1.

Table 3 Weight loss ratio of cokes under different conditions

Sample	Alkali-rich rate%	Weight loss ratio/%			
		900 °C	1000 °C	1100 °C	1200 °C
Coke 1	0	3.4	7.7	13.4	16.1
	2.0	15.5	28.0	33.9	47.3
Coke 2	0	0.7	6.8	17.8	29.2
	1.7	15.4	24.7	36.3	46.0

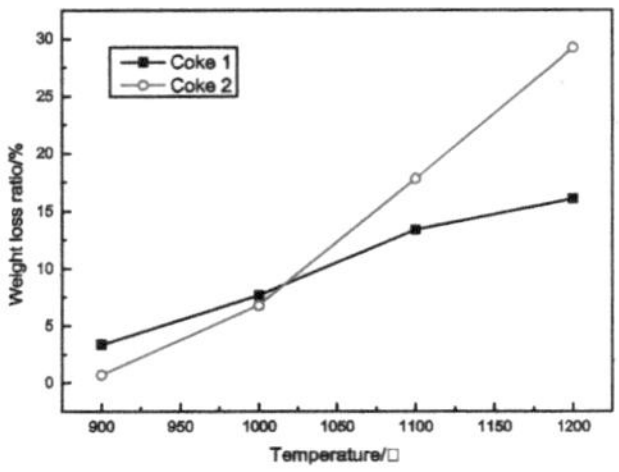

Fig. 2 Weight loss ratio of cokes without alkalis

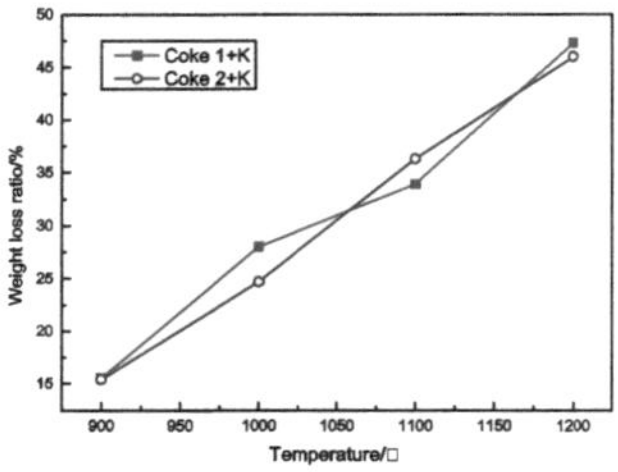

Fig. 3 Weight loss ratio of cokes with alkalis

With alkali-rich (K_2O), the two kinds of cokes weight loss ratio both increase obviously. This is because that the presence of alkalis has effects on the solution loss reaction, including that increasing the reactivity, reducing the apparent activation energy and initial reaction temperature. At the lower temperature (<1000 °C), the potassium adsorbed on the surface of coke has nearly a similar catalyst effect on the solution loss reaction of two kinds of coke. While the temperature is higher, the catalysis of potassium to coke 1 is stronger than to coke 2. This is possibly because the micro-pores of coke 2 are much smaller and potassium is easy to enrich in them, which hinders the CO_2 gas internal diffusion and the opening of closed pores. So the weight loss ratio of the coke 2 increases little, while the increasing amount of weight loss ratio of coke 1 is much larger than coke 2 with alkali-rich under high temperature.

Microstructure of Cokes

Coke 1 and coke 2 which were cut from center were examined by FEI Quanta250 scanning electron microscope (SEM). Figures 4 and 5 show the morphology of original coke 1 and coke 2.

As shown in Figs. 4 and 5, the amount of pores of coke 2 is obviously more than coke 1, but the large pores number of both cokes is almost the same. Compared to coke 1, the coke 2 has even more micro-pores and internal closed pores, so the porosity of coke 2 is much larger than coke 1. The ideal structure of coke contains

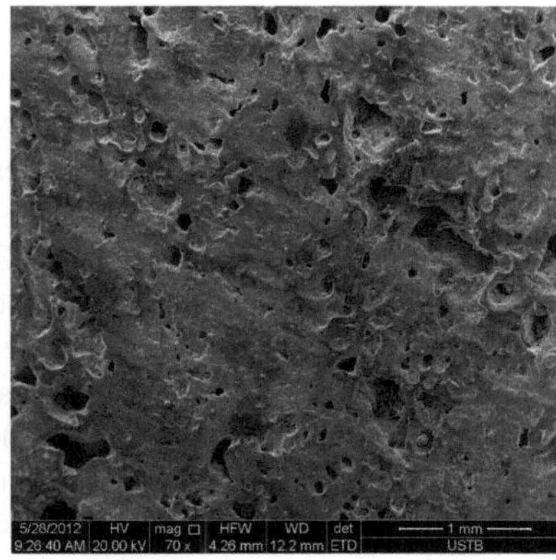

Fig. 4 The morphology of original coke 1

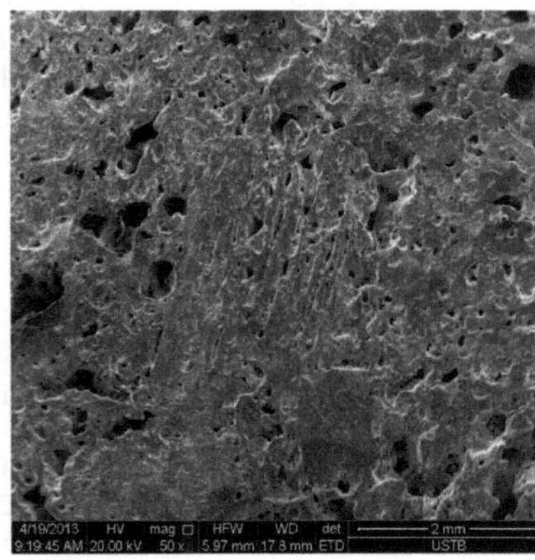

Fig. 5 The morphology of original coke 2

homogeneous pore diameter, much thick pore wall, few cracks. The pore size of coke distributes more uniform, its strength will be better. Because the stamping coke may ultimate more high volatile coals during coking, its pore size shows an uneven distribution. While the good quality coals are used more in top charging coking process, so its pore size distributes more uniform. The morphology and amount of micro-pore in stamping coke have an important role on its strength. On the one hand, because the micro-pore of stamping coke is long and flat, when it subjected to load, it is easy to produce micro-cracks on the wall of pore's radial direction, which has an influence on the structure of pore. And on the other hand, much micro-pores wall are reacted totally, so these micro-pores may be connected to form through holes, which increases the amount of large pores of stamping coke. During the gasification reaction, with the thinning of the pore wall, the strength of stamping coke reduces.

Conclusion

1. With no addition of alkali, the weight loss ratios of stamping coke and top charging coke nearly had no difference at the lower temperature(900–1000 °C). As the temperature reaches up to 1100 and 1200 °C, the stamping coke closed pores was eroded seriously by CO_2, which caused the closed pores opened and increased the reaction area, so the weight loss ratio of stamping coke is much larger than the top charging coke.
2. With alkali-rich, the catalysis of alkali to stamping coke and top charging coke was almost the same at 900–1000 °C. While the weight loss ratio of the two cokes was almost the same at 1100–1200 °C.
3. Through examining the two cokes by SEM, we can know that the pore ratio and closed pore ratio of stamping coke was larger apparently than top charging coke. The stamping coke had less large pores, more middle pores and micro-pores. Formation and distribution of large pores, thinner pore walls and the strength of the coke matrix may lead to the declining of stamping coke strength after reaction with CO_2 under high temperature.

References

1. Krishnan SH, Dash PS, Guha M et al (2004) Application of binder in stamp charge coke making. ISIJ Int 44(7):1150–1156
2. Sun C, Zhou Y (2005) Discussion on application of incoherent low-ash coal in stamping coking production. Gas Heat 65(5):1225–1235
3. Shi S, Lei Y, Cao S et al (2011) Influence of bulk density of coal charge on the performance of stamp-coking coke. J Wuhan Univ Sci Technol 34(4):285–288
4. Wang Y (2011) Problems existing in ramming coking and improvement measures. Technol BX Steel 5:33–37
5. Zhong Y (2011) Several worth attention issues of cokemaking with stamping coke oven. Fuel Chem Process 42(1):1–5
6. Han G (2009) Study of blending technology for stamping coke-making. Fuel Chem Process 40 (5):21–26

Characterization and Leaching Proposal of Ag(I) from a Zn Concentrate in an $S_2O_3^{2-}$-O_2 Medium

Teja R. Aislinn Michelle, Juárez T. Julio Cesar, Hernández C. Leticia, Reyes P. Martín, Flores G. Mizraim Uriel, Reyes D. Iván Alejandro and Mendez R. Eliecer

Abstract The present study shows the characterization of a zinc concentrate collected at Hidalgo, Mexico, in order to identify its mineralogical species and to propose a leaching methodology for silver recovery. The results obtained by X-Ray Diffraction (XRD) show Sphalerite (ZnS) as the mineralogical matrix and silver into the specie proustite (Ag_3AsS_3). The techniques of Scanning Electron Microscopy-Energy Dispersive Microanalysis (SEM-EDS) and Polarized Optical Microscopy (POM) identified the presence of sulfides of Fe, Cu, Pb and Zn, in addition to silicates and feldspar. The proposed dissolution method of Ag(I) was carried out with $[S_2O_3^{2-}] = 0.5$ M and $[NaOH] = 0.1$ M to regulate the medium at pH = 9 for 12 h at 298 K. Oxidation of the concentrate was favored by injection of O_2 to a partial pressure of 1 atm. Monitoring of the reaction was carried out by reading the [Ag] and [As] by AAS.

Keywords Silver · Arsenic · Thiosulfate · Zinc and leaching

Introduction

The volume of mining-metallurgical production in Mexico, has declined in recent decades because of the massive exploitation that has taken place in mining districts as it is the case of Zimapán, Hidalgo, where mining represents the socioeconomic

T.R.A. Michelle · JuárezT.J. Cesar (✉) · HernándezC. Leticia · R.P. Martín · M.R. Eliecer
Área Académica de Ciencias de La Tierra Y Materiales, Universidad Autónoma Del Estado de Hidalgo, Carretera Pachuca-Tulancingo Kilómetro 4.5, 42180 Pachuca, HGO, Mexico
e-mail: jcjuarez@uaeh.edu.mx

F.G.M. Uriel
Área de Electromecánica Industrial, Universidad Tecnológica de Tulancingo, Camino a Ahuehuetitla 301 Col. Las Presas, 43642 Tulancingo, HGO, Mexico

R.D.IvánAlejandro
Instituto de Metalurgia, Universidad Autónoma de San Luis Potosí, Av. Sierra Leona no. 550, Lomas 2da Sección, 78210 San Luis Potosí, SLP, Mexico

S. Ikhmayies et al. (eds.), *Characterization of Minerals, Metals, and Materials 2017*, The Minerals, Metals & Materials Series,
DOI 10.1007/978-3-319-51382-9_62

region livelihoods for more than 400 years [1, 2]. Silver production by 2015 nationally decreased 4.39% in comparison with the previous year and the positive variation in the mining of interest metals such as Zn and Cu [3]. The typical zoning of the Hidalgo deposits consists of veins of galena-sphalerite, chalcopyrite, silver antimonides and an abundant presence of silica and calcium as gangue mineral. These skarns are characterized by fluids with different properties that determine the species that compose them, indicating an overall decrease of temperature, with respect to the formation of back areas [4]. Metallic elements of higher economic value occurring in the mineralogy of Zimapán are Pb, Zn, Cu and Fe, said elements are sold as concentrates, which, even after processing, generally include significant concentrations of Mo, Cd, Sb, Ag and As that could recover through different leaching methods [5]. In the case of arsenic, this element mainly occurring as arsenopyrite, however, also is involved in the forming of lead, copper and silver sulfosalts, in addition to iron arsenides and arsenates [6, 7]. The recovery of metals from secondary sources represents an opportunity area in the hydrometallurgy field. Previous studies have reported the reactivity of sulfosalts such as pyrargyrite (Ag_3SbS_3) and proustite (Ag_3AsS_3) with respect to the complexing ability of CN^-, reporting a significant deficiency in the recovery of Ag, due to the complex nature of the mineral species containing it [8], which hinders the viability of the cyanidation to make necessary pre-treatments such as crushing, calcining or oxidation that allow the direct contact of metal with leaching solution [9], for this reason is that the study of alternative leaching processes have exponentially increased its importance in recent years [10], proposing the use of reagents such as thiourea, thiocyanate and thiosulfate [11, 12]. The thiosulfate leaching system represents one of the best alternatives for recovery of metals contained in sulfide minerals in less toxic conditions that cyanidation [13], reaching higher dissolutions of precious metal also decreasing the leaching time, due to its greater selectivity [14, 15].

The present study reports the chemical and mineralogical characterization of a concentrate of Zn from Zimapán, Hidalgo in order to know the elements that interfere with the dissolution reaction of Ag, in $S_2O_3^{2-}$-O_2 system.

Experimentation

Prior to dissolution testing, mineral zinc concentrate was dried, homogenized and quartered, with the purpose of obtaining a representative all-in-one sample of 0.2 kg. Concentrate powders were embedded in a test-tube of epoxy resin, which is roughed and polished in order to carry out the analysis of the sample surface and know the morphology particles using techniques of Polarizing Light Microscopy (PLM), Scanning Electron Microscopy with Energy Dispersion X-Ray Spectroscopy (MEB-EDS) and X-Ray Diffraction (XRD). For chemical analysis, the powders of concentrate of Zn at 44, 53 and 74 μm were measured. 1 g of each sample was digested with aqua-regia (3 parts of HCl per 1 part of HNO_3) on a heating mesh at 338 K to achieve higher dissolution of solids, finally, the insoluble remainder was filtered and the remaining solution was diluted at 0.1 L.

The general procedure for leaching experiments was carried out under the following conditions: 40 GL^{-1} of concentrate of Zn, stirring speed = 670 min^{-1}, pH = 9 adjusted with addition of 0.1 M NaOH, PpO_2 = 1 atm, temperature = 298 K, solution volume = 0.5 L, particle size = 74 μm, $[Na_2S_2O_3]$ = 0.5 M and reaction time of 720 min. The monitoring of dissolution reaction of Ag(I) was carried out by consecutive extraction of aliquots of 0.01 L at different time ranges, which were analyzed by technique of Atomic Absorption Spectrophotometry (AAS).

Experimental Methods

To carry out the mineralogical characterization were used equipment of X-ray Diffraction INEL model EQUINOX 2000 with a radiation of Co-Kal 81.789010 Å), Polarizing Light Microscope model BH41 with different magnifications and Scanning Electron Microscope JEOL model JSM 6701F using a voltage of 25 kV and a field depth of 16.5 mm at different magnifications using backscattered and secondary electrons.

To carry out the silver dissolution was used a glass reactor with capacity of 0.5 L on a heating mesh Thermo Scientific Super Nuova equipped with magnetic stirring, oxygen was injected through a diffuser and was regulated by a flow meter. The chemical analysis of samples and the monitoring of leaching reaction were carried out using the EAA technique, using a spectrophotometer Perkin Elmer-Analyst 200.

Results and Discussion

The results obtained from chemical and mineralogical characterization of concentrated Zn and the leaching of Ag and As in the proposed system are presented in this section.

X-ray Diffraction (XRD)

The results obtained by X-Ray Diffraction (XRD) reported that concentrate matrix corresponds to the sphalerite species (ZnS) as shown in Fig. 1, as well as the presence of copper sulfides, composition of which is shown in Table 1. Ag containing species was identified as proustite (Ag_3AsS_3), a typical sulfosalt of the mining district where it comes from the concentrate of Zn [5].

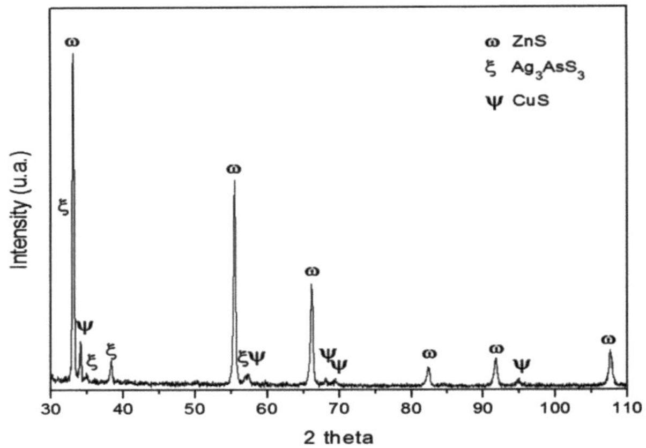

Fig. 1 X-ray diffractogram of the powder of pull size concentrate of Zn

Table 1 Mineral species present in the concentrate of Zn

No. PDF	Composition	Mineral phases
[96-900-9992]	Ag_3AsS_3	Proustite
[96-900-0108]	ZnS	Sphalerite
[96-900-9724]	CuS	Covellite

Polarizing Light Microscopy (PLM)

Using a polarizing light microscope, was identified that grain size of mineralization of concentrate of Zn is mostly formed by fine grains and was confirmed that matrix correspond to sphalerite species (ZnS). Also was identified the presence of metal sulphides such as pyrite and covellite as observed in Fig. 2. The galena occurs as embedding in a sphalerite matrix, this species is generally related to the silver content in Hidalgo deposits where the typical average zoning is characterized by the presence of abundant sulphides of Zn-Pb [4, 16].

Scanning Electron Microscopy with Energy Dispersion X-ray Spectroscopy (MEB-EDS)

A varied morphology, particles of which display flat angles and semi-spherical shapes obtained with secondary electrons is observed in Fig. 3a. The dispersive energy spectrum shown in Fig. 3b reports the presence of Fe, S, Cu, Zn, Si, Ca and O. The contents of C, Pd and Au, are related to epoxy resin and coating which submitted the Zn concentrate sample to analysis by SEM-EDS.

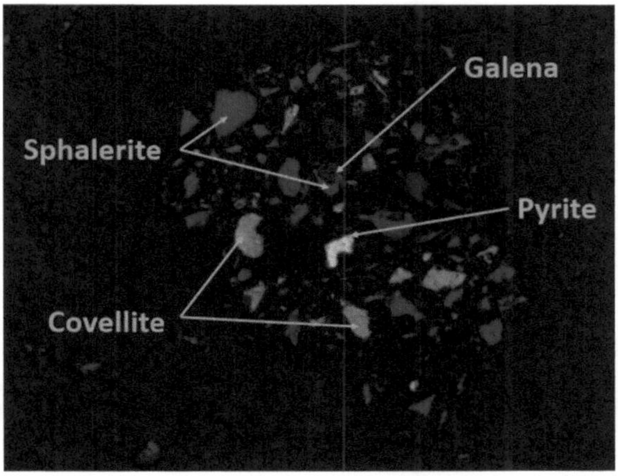

Fig. 2 Micrograph of a powder sample in a Zn concentrate obtained at 10×

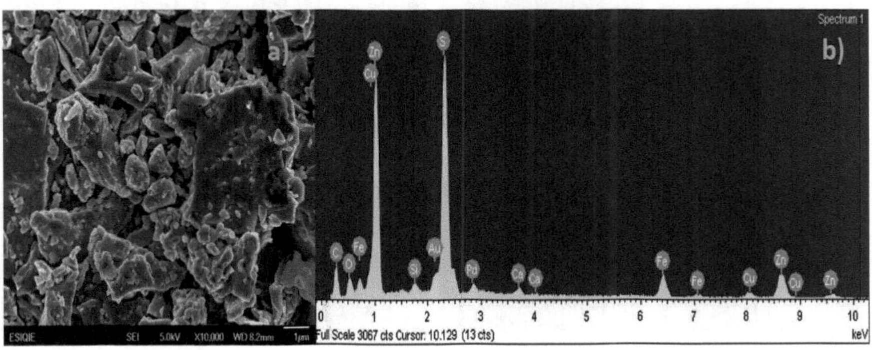

Fig. 3 **a** Micrograph of a powder sample in a concentrate of Zn obtained by secondary electrons. **b** Energy-dispersive X-ray spectrum performed to pull size powder sample of concentrate of Zn

The semi-quantitative analysis obtained from general dispersive energy microanalysis, shows significant concentrations of Zn and S that confirm the elemental composition of the concentrate matrix, in addition to abundant amounts of Fe and Cu. The As presence in concentrations below 0.62% is observed. In the concentrate of Zn, the content of non-metal elements, which are shown in Fig. 4, corresponds to silicates and feldspar, typical of mineral deposits of Zimapán [17].

Mapping performed to fine particles of concentrate of Zn allowed to relate the contents of S with metal elements such as Zn, Fe and Cu, as shown in Fig. 5. The presence of As and Ag is observed homogeneously distributed in the concentrate sample, which confirms that the mineral body where it was obtained the concentrate of Zn, includes massive base metal sulphides [16, 17], arsenates and other species of economic value carriers of mineral phases of Ag [6]. The ratio between elements

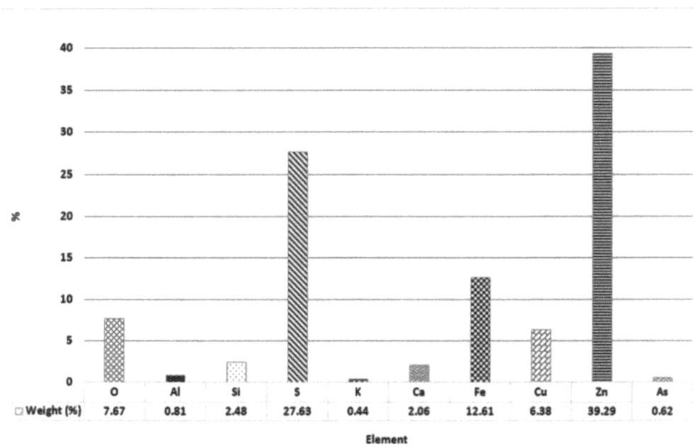

Fig. 4 Semi-quantitative analysis of the concentrate of Zn

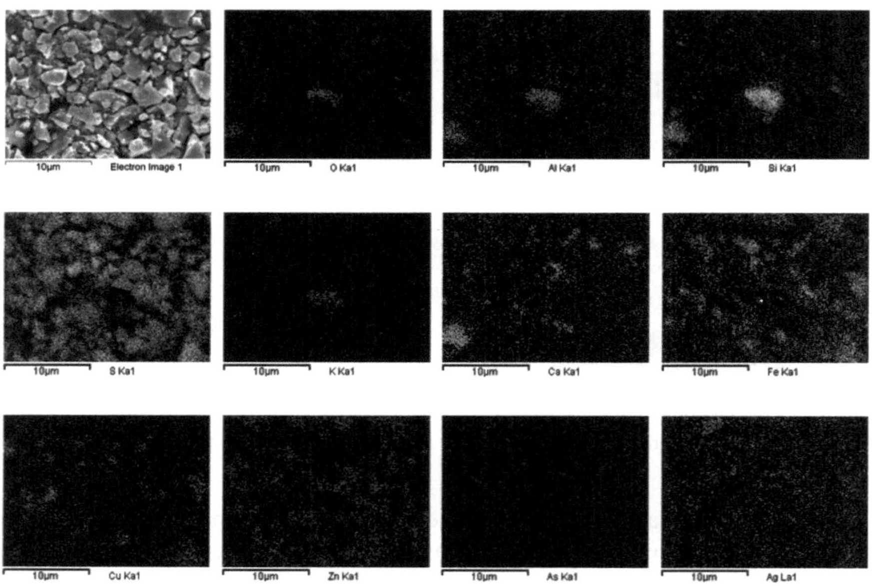

Fig. 5 Mapping of polished particles of concentrate of Zn

Ca, O, Al, and Si indicated the formation of typical gangue species such as orthoclase ($KAlSi_3O_8$) and anorthite ($CaAl_2Si_2O_8$) [5, 18].

At item 1 in Fig. 6a. obtained with backscattered electrons, is observed the presence of a polymetallic particle as demonstrated by the microanalysis of Fig. 6b. The above suggests that said particle corresponds to silver species identified by XRD.

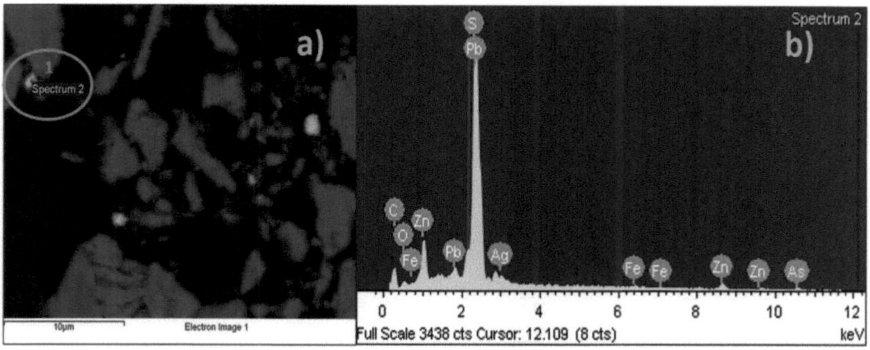

Fig. 6 **a** Backscattered electron micrograph showing the presence of metal species in the concentrate. **b** Scattered energy analysis performed to a metal particle

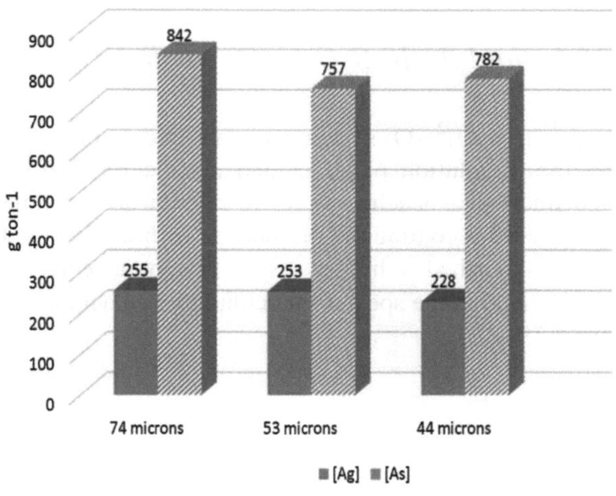

Fig. 7 Chemical analysis of the concentrate of Zn by mesh size

Chemical Analysis by Atomic Absorption Spectrometry (AAS)

The chemical analysis carried out by Atomic Absorption Spectrophotometry (AAS) to powders of concentrate of Zn measured at 44, 53 and 74 μm, reported the highest concentration of Ag and As in the 200 mesh (74 μm) of the Tyler series, therefore, this particle size for leaching tests was used, however, in Fig. 7, is observed the homogeneous distribution of Ag and As being constant with what was observed in the mapping performed to all-in-one powder sample.

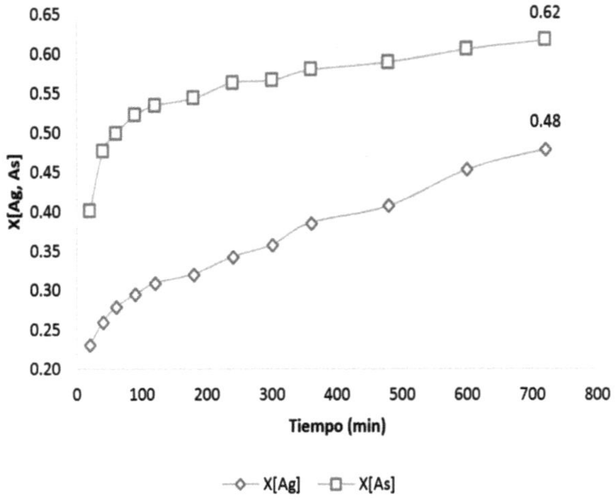

Fig. 8 Dissolution of Ag and As in $S_2O_3^{2-}$-O_2 system

Dissolution of Ag and As in $S_2O_3^{2-}$-O_2 Medium

The leaching reaction in $S_2O_3^{2-}$-O_2 system was carried out at room temperature and was followed by the dissolution of silver and arsenic for 720 min. The results obtained from the fraction of leached silver in relation to the time are shown in Fig. 8, where is observed a continuous dissolution of both metals, however, for a same reaction time is reached a higher dissolution of arsenic, due to the high concentration of this metal in the species that contains it, reaching maximum values of 48% of Ag and 62% of As.

Conclusions

According to the results of the characterization of zinc concentrate, is confirmed the presence of metal sulphides of Fe, Cu, Pb and Zn, latter being the concentrate matrix (ZnS), while the coexistence of gangue species such as orthoclase ($KAlSi_3O_8$) and anorthite ($CaAl_2Si_2O_8$) is allocated with the alkaline nature of concentrate pulp.

Despite the refractory nature of the sulfosalt containing Ag, leach process with thiosulfates and the addition of an oxidizing agent, is considered viable for the recovery of Ag, since has been reached 48% of precious metal dissolution in an alkaline medium and at room temperature.

References

1. Labastida I, Armienta MA, Lara-Castro RH, Aguayo A, Cruz O, Ceniceros N (2013) Treatment of mining acidic leachates with indigenous limestone, Zimapán Mexico. J Hazard Mater 262:1187–1195
2. Díaz L (1995) A simultaneous view of history, the creation of a hypermedia database. Leonardo, Hidalgo, pp 257–264
3. INEGI (2016) Estadística de la Industria Minero-metalúrgica. Instituto Nacional de Estadística y Geografía 1:1–4
4. González-Partida E, Carrillo-Chávez A, Levresse G, Tritlla J, Camprubi A (2003) Genetic implications of fluid inclusions in skarn chimney ore, Las animas Zn–Pb–Ag (–F) deposit, Zimapán, Mexico. Ore Geol Rev 23:91–96
5. Moreno RT, Téllez JT, Monroy J, Fernández MG (2012) Influencia de los minerales de los jales en la bioaccesibilidad de arsénico, plomo, zinc y cadmio en el distrito minero Zimapán. Revista internacional de Contaminación Ambiental. 28:203–218
6. Ongley LK, Armienta MA, Heggeman K, Lathrop AS, Mango H, Miller W, Pickelner S (2001) Arsenic removal from contaminated water by the Soyatal Formation, Zimapán Mining District, Mexico and a potential low-cost low-tech remediation system. Geochem Explor Environ 26:23–31
7. Armienta MA, Villaseñor G, Rodríguez R, Ongley LK, Mango H (2001) The role of arsenic-bearing rocks in groundwater pollution at Zimapán Valley, México. Environ Geol 40:571–581
8. Meléndez AM, Arroyo R, González I (2010) On the reactivity of sulfosalts in cyanide aqueous media: structural, bonding and electronic aspects. ChemPhysChem 11:2879–2886
9. Jeffrey MI, Breuer PL, Choo WL (2002) A kinetic study that compares the leaching of gold in the cyanide, thiosulfate and chloride system. Metall Mater Trans 32:979–986
10. Muir DM, Aylmore MG (2005) Thiosulfate as an alternative to cyanide for gold processing-issues and impediments. Miner Process 15:113–116
11. Schmitz PA, Duyvesteyn S, Johnson WP, Enloe L, McMullen J (2001) Ammoniacal thiosulfate and sodium cyanide leaching of preg-robbing gold strike ore carbonaceous matter. Hydrometallurgy 60:25–40
12. Breuer PL, Jeffrey MI (2000) Thiosulfate leaching kinetics of gold in the presence of copper and ammonia. Hydrometallurgy 52:442–452
13. Lewis RJ (1997) Hazardous chemicals desk reference. International Thomson Publishing Company, New York, pp 123–137
14. Luna RM, Lapidus GT (2000) Cyanidation kinetics of silver sulfide. Hidrometallurgy 56:171–188
15. Arima H, Fujita T, Yen WT (2003) SME annual meeting and exhibit, phoenix. Miner Metall Process 20:81–92
16. Villaseñor CG, Gómez C, Avedaño S, Reyes-Salas M, Islas-Cortéz V (1995) Minerales argentiferos de Zimapán, Hidalgo, México. Actas INAGEQ 1:123–127
17. Villaseñor CG, Peterson E, Avedaño S, Gómez-Caballero A, Sousa J, Reyes-Salas M (1996) Minerales del grupo de la tetrahedrita en las minas lomo de toro y las Ánimas, Zimapán Hidalgo. Actas INAGEQ 2:124–129
18. Moreno RT, Monroy MG, Castañeda EP (2012) Evaluación Geoquímica de residuos mineros (Jales o colas) de mineralización de tipo epitermal. Revista internacional de Contaminación Ambiental 14:133–143

Mechanical Properties and Behavior of Additive Manufactured Stainless Steel 316L

M.A. Bevan, A.A.H. Ameri, D. East, D.C. Austin, A.D. Brown, P.J. Hazell and J.P. Escobedo-Diaz

Abstract Additive Manufacturing (AM) has become increasingly popular due to the ability to produce components with complex geometries and properties. The efficiency of this manufacturing method is investigated in this paper by examining the mechanical performance of AM austenitic stainless steel 316L under different strain rates. Dynamic compression experiments were conducted by using a Split Hopkinson Pressure Bar (SHPB). Microstructural observations were performed by utilizing optical microscopy. The behavior of this material is compared against stainless steel 316L manufactured by conventional methods. The results show that the AM 316L used in this study display a considerable difference in the mechanical properties compared with conventional stainless steel alloys.

Keywords Stainless steel · 316L · Compression · Deposition direction · Additive manufacture

Introduction

316L is an austenitic stainless steel alloy has an attractive mechanical properties make it suitable for different engineering applications. 316L has a face centered cubic crystalline structure that boasts high toughness from room temperature down to low levels of temperatures [1], this make it has a high tensile strength at a significant strain (40% elongation in tension [2–4]). Furthermore, its corrosion resistance makes it suitable for marine applications compared with other types of

---

M.A. Bevan · A.A.H. Ameri · D.C. Austin
A.D. Brown · P.J. Hazell · J.P. Escobedo-Diaz (✉)
School of Engineering and Information Technology, UNSW Australia
at the Australian Defence Force Academy, Canberra, ACT 2600, Australia
e-mail: j.escobedo-diaz@adfa.edu.au

D. East
Commonwealth Scientific and Industrial Research Organisation,
Clayton, VIC 3168, Australia

S. Ikhmayies et al. (eds.), *Characterization of Minerals, Metals, and Materials 2017*, The Minerals, Metals & Materials Series,
DOI 10.1007/978-3-319-51382-9_63

austenitic stainless steel as 304 [5]. The 'L' variant of this steel is a low carbon alternative being desirable for large welding applications due to its immunity to grain boundary carbide precipitation. Due to the demand for this material, founding a suitable manufacturing process which can produce products with different geometries is required.

Laser Engineered Net Shaping, known also as Laser Powder Forming, is a powder feed, laser cladding process that works by metallic powder being borne through a nozzle onto the build surface which is melted by a laser to create a single layer of the powder into the desired shape, the process being repeated layer by layer until the 3D part is completed [6]. The method is specialized toward building parts from Computer Aided Design files that the system software processes and creates a "map" that the nozzle and laser focusing lens follow, creating a weld pool that the powder is deposited into. The metal is deposited orthogonally to the Z plane with the nozzle and lens moving upwards, taken as the positive z direction, all this occurs within a controlled argon atmosphere that contains less than 10 ppm oxygen. The melting process will generate a heat affected zone (HAZ) which is of significance as the printed part will constantly be affected by this zone given that the AM process takes hours to complete, this leaves the material in an annealed state upon build completion [7–9].

This paper presents our results on the characterization of the microstructure and the mechanical response additive manufactured stainless steel 316L.

Materials and Experimental Methods

The results of the comprehensive characterisation of our additive manufactured (AM) 316L material is provided in Fig. 1. The SEM image (Fig. 1a) shows long grains being stacked on top of another, with the stacking along the deposition direction (DD). The orientation map (Fig. 1b) shows that the microstructure has grain in almost random orientations. The texture pole figure however shows a slight tendency for a preferred 111 texture (Fig. 1c), this is further supported by the, X-ray diffraction (XRD) measurements (Fig. 1d), this is expected for austenite steels [10]. The average grain size was estimated from the EBSD map to be about 90 μm.

One of the major side effects of AM processing is porosity [10].

Figure 2a shows a representative image of the as-received material along DD. It shows rampant porosity throughout the cross section. Figure 2b is negative image of the area shown by the red square in Fig. 2a. The negative is created using ImageJ and better shows both the smaller and larger pores present in this plane. The size of larger pores found to be of up to 4000 μm^2.

Optical micrographs show that one of the transverse directions (TD-1) has the greatest amount of pores, however in terms of surface area its pores covered the least amount of the total area with 0.05%. The DD contained pores of the greatest size yielding that 0.15% of its surface area was covered in pores. The second transverse direction TD-2 had 0.10% of its surface area being porous and had the

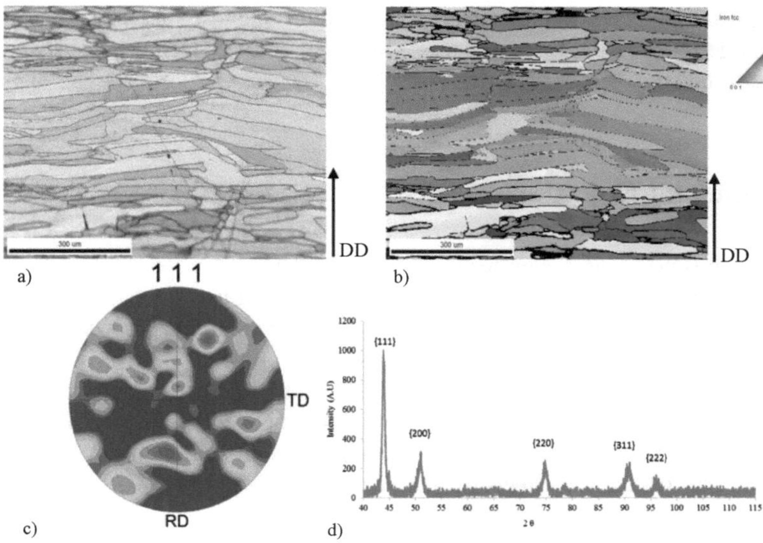

Fig. 1 **a** SEM image of DD for reference sample. **b** EBSD of DD. **c** 111 pole figure. **d** XRD of powder from reference sample

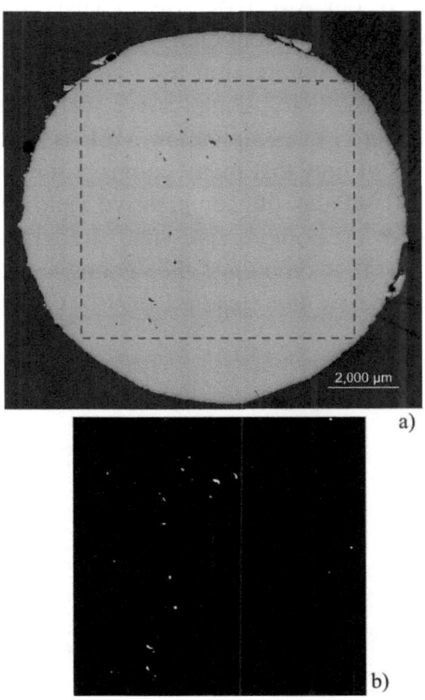

Fig. 2 **a** 5× magnification of DD for reference sample. **b** Negative image of area used for porosity analysis

least amount of pores overall. These observation suggest that the AM process leaves large pores in the layer it places, with smaller pores forming in between the deposition layers.

Experimental Procedure

Quasi-static (QS) compression tests at strain rate 10^{-3} s^{-1} were conducted using a Shimadzu (100 kN capacity) testing machine, Rockwell hardness and compression experiments using the Spilt Hopkinson Pressure Bar (SHPB) for dynamic characterization (intended strain rates of 5×10^2 s^{-1}, 10^3 s^{-1} and 2×10^3 s^{-1}) [11–13]. QS compression samples were 13 mm × 19.5 mm (diam × length) for an L/D ratio of 1.5 to reduce the effect of friction as recommended by the ASTM standards [12]. Test specimens for the SHPB were of equal length and diameter, 7.5 mm, for a ratio of 1 to reduce the effects of inertia as well as friction [14].

Results and Discussion

The measured hardness for our AM 316L was 11 HRC, this is ~30% lower than the reported 16 HRC value for conventional 316L.

Results of two quasistatic compression tests are provided in Fig. 3. Measured values for the modulus of elasticity, compressive yield stress, hardening rate and strain are reported in Table 1. In compression 316L is reported as having a yield strength of 350 MPa, it is of note that many common sources incorrectly quote the tensile yield stress of 170 MPa as the compressive yield strength [15].

The obtained values for the compressive yield strength were comparable to the data for conventional 316L. Comparing the Young's Modulus shows that, on average, AM 316L falls below the standard of 193 GPa [16]. Samples 1 and 3

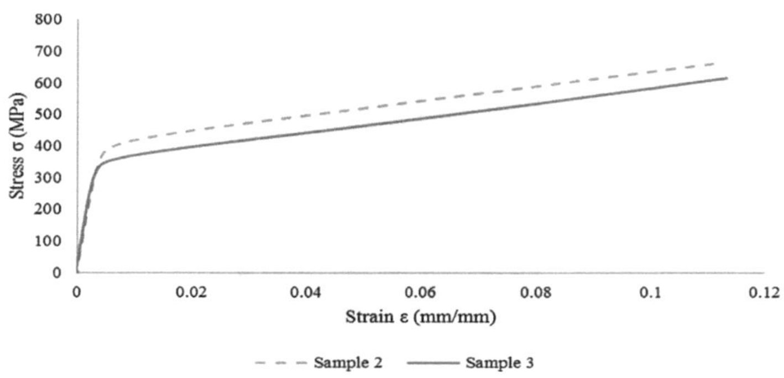

Fig. 3 Results of QS compression testing

Table 1 Mechanical properties found from compression testing of AM 316L

Sample	Modulus of elasticity (GPa)	Yield stress (MPa)	Hardening rate (GPa)	Strain (mm/mm)
AM 316L 1	166	332	2.294	0.13
AM 316L 2	206	376	2.358	0.11
AM 316L 3	151	343	2.342	0.13

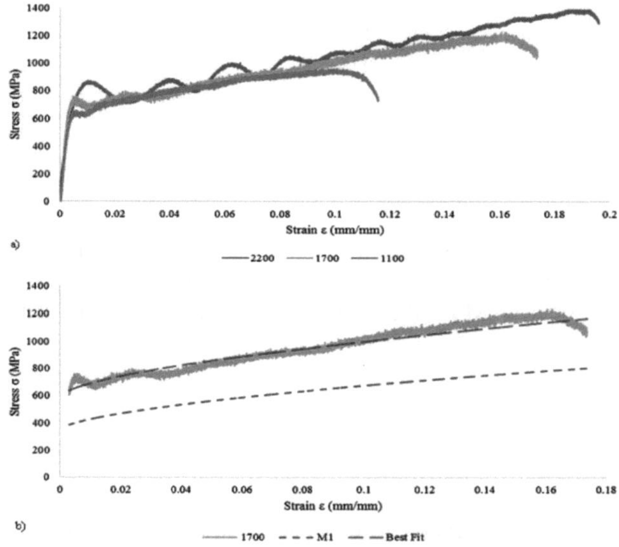

Fig. 4 **a** Results of SHPB compression testing at varying strain rate. **b** Johnson-Cook model with best fit from found parameters

values were similar, however, Sample 2 performed better than both the other test samples and the obtained data for conventional 316L. Again using ImageJ, the porosity of the tested specimen was recorded with both Sample 2 having a porosity percentage of 0.10% and Sample 1 and 3 being 0.11% porous. While a marginal difference, this can be an indication that an AM metal that is less porous does preform to a greater level than that of its conventional counterpart.

Figure 4a shows the results from compression testing at high strain rates (1100, 1700, 2000 s^{-1}) using the SHPB. Strain hardening and strain both increased as the strain rate increased as well as the yield stress, as shown in Table 2.

These results are expected and follow the same trend as conventional 316L, with some deviation in the Young's modulus [17]. A fit using the Johnson-Cook model is reported in Fig. 4b. The parameters used are reported in Table 3, they were taken from literature on conventional 316L.

While none of the prescribed models would perfectly fit the response of our AM 316L, the material does fall within the spread of these differing models. As such an

Table 2 Mechanical properties found from test with SHPB

Strain rate	Modulus of elasticity (GPa)	Yield stress (MPa)	Hardening rate (GPa)	Strain
1100	190	632	3.032	0.11
1700	245	756	3.177	0.16
2200	209	863	3.731	0.19

Table 3 Johnson-Cook parameters used to construct models for AM 316L

Reference	A	B	C	n
M1	305	1161	0.01	0.61
M3	301	1472	0.09	0.807
M5	514	514	0.042	0.533
Best fit	514	1472	0.01	0.61

accurate model is likely to be able to be created with different parameters. It was found that the best fit to our data was by using the yield stress from M5, the hardening modulus from M3 and the strain rate sensitivity and hardening coefficient from M1.

Conclusions

The microstructure and mechanical response of an additive manufactured stainless steel 316L has been characterized and these are our main findings:

- AM 316L displays a lower Rockwell hardness at 11 HRC compared to the conventional literature data of 16 HR C, a ~40% difference
- In quasi-static compression the AM metal displayed an average Young's Modulus of 174 GPa, 10% less than normal processed steel at 193 GPa
- The average compressive yield stress was 350 MPa, matching the reported value of 350 MPa for conventional 316L

References

1. Frazier WE (2014) Metal additive manufacturing: a review. J Mater Eng Perform 23(6): 1917–1928
2. Sultan A, Sharif S, Kurniawan D (2015) Chip formation when drilling AISI 316L stainless steel using carbide twist drill. Procedia Manufact 2:224–229. doi:10.1016/j.promfg.2015.07.039
3. Lind J, Hanninen J, Kotila J, Nyrhila O, Syvanen T (2002) Rapid manufacturing with direct metal laser sintering. In: MRS proceedings, p 758. doi:10.1557/proc-758-ll1.2

4. Gu D, Meiners W, Wissenbach K, Poprawe R (2012) Laser additive manufacturing of metallic components: materials, processes and mechanisms. Int Mater Rev 57(3):133–164. doi:10.1179/1743280411y.0000000014
5. Balla V, DeVasConCellos P, Xue W, Bose S, Bandyopadhyay A (2009) Fabrication of compositionally and structurally graded Ti–TiO_2 structures using laser engineered net shaping (LENS). Acta Biomater 5(5):1831–1837. doi:10.1016/j.actbio.2009.01.011
6. Wu D, Liang X, Li Q, Jiang L (2011) Fabrication of SS316L/Ni25 Functionally gradient materials using laser engineered net shaping. Mater Sci Forum 675–677:803–806. doi:10.4028/www.scientific.net/msf.675-677.803
7. Riza S, Masood S, Wen C, Ruan D, Xu S (2014) Dynamic behaviour of high strength steel parts developed through laser assisted direct metal deposition. Mater Des 64:650–659. doi:10.1016/j.matdes.2014.08.026
8. Ahmed AA, Mhaede M, Wollmann M, Wagner L (2014) Effect of surface and bulk plastic deformations on the corrosion resistance and corrosion fatigue performance of AISI 316L. Surf Coat Technol 259:448–455
9. Lindholm U (1964) Some experiments with the split hopkinson pressure bar*. J Mech Phys Solids 12(5):317–335. doi:10.1016/0022-5096(64)90028-6
10. Ellwood S, Griffiths L, Parry D (1982) Materials testing at high constant strain rates. J Phys E Sci Instrum 15(3):280–282. doi:10.1088/0022-3735/15/3/009
11. Song B, Chen W, Antoun B, Frew D (2007) Determination of early flow stress for ductile specimens at high strain rates by using a SHPB. Exp Mech 47(5):671–679. doi:10.1007/s11340-007-9048-6
12. Standard ASTM (2014) Standard test methods of compression testing of metallic materials at room temperature. In: 2014 annual book of ASTM standards, ASTM, West Conshohocken, PA
13. Sencer B, Maloy S, Gray G (2005) The influence of shock-pulse shape on the structure/property behavior of copper and 316L austenitic stainless steel. Acta Mater 53 (11):3293–3303. doi:10.1016/j.actamat.2005.03.037
14. ASM Material Data Sheet (2016) Asm.matweb.com. Retrieved 9 Apr 2016, from http://asm.matweb.com/search/SpecificMaterial.asp?bassnum=MQ316Q
15. Lee WS, Chen TH, Lin CF, Luo WZ (2011) Dynamic mechanical response of biomedical 316L stainless steel as function of strain rate and temperature. Bioinorg Chem Appl
16. Olleak AA, Nasr MN, El-Hofy H (2015) The influence of Johnson-Cook parameters on SPH modelling of orthogonal cutting of AISI 316L
17. Umbrello D, M'saoubi R, Outeiro JC (2007) The influence of Johnson–Cook material constants on finite element simulation of machining of AISI 316L steel. Int J Mach Tools Manuf 47(3):462–470

Characterization of Mercury Jarosite

Sayra Ordoñez, Francisco Patiño, Mizraim Uriel Flores, Iván Alejandro Reyes, Elia Guadalupe Palacios, Víctor Hugo Flores, Martín Reyes, Ister Mireles and Hernán Islas

Abstract This paper presents the characterization of a sample of mercury jarosite that was synthesized by slowly adding $Hg(NO_3)_2{\cdot}H_2O$ into a $Fe_2(SO_4)_3{\cdot}nH_2O$ solution. X-ray diffraction (XRD) analysis performed on the sample confirmed that the compound presents a crystal phase corresponding to mercury jarosite. Analysis through Scanning Electron Microscopy with Energy Dispersive X-ray Spectroscopy (SEM-EDS) determined that the precipitate is composed of mercury, sulfur, iron and oxygen. It is made up of spherical particles with an average size of 38 μm; these particles are made of rhombohedral micro-crystals with sizes ranging between 1 and 2 μm. Chemical characterization was performed through dichromatometry analysis, gravimetry and atomic absorption spectroscopy (AAS),

S. Ordoñez (✉) · MartínReyes · HernánIslas
Área Académica de Ciencias de la Tierra y Materiales, Universidad Autónoma del Estado de Hidalgo, Carretera Pachuca-Tulancingo Km. 4.5, 42184 Mineral de la Reforma, HGO, Mexico
e-mail: sayraoh@hotmail.com

F. Patiño
Ingeniería en Energía, Universidad Politécnica Metropolitana de Hidalgo, Boulevard Acceso a Tolcayuca no. 1009, Ex Hacienda San Javier, 43860 Tolcayuca, HGO, Mexico

M.U. Flores · I. Mireles
Área de Electromecánica Industrial, Universidad Tecnológica de Tulancingo, Camino a Ahuehuetitlan no. 301 Col. Las Presas, 43642 Tulancingo, HGO, Mexico

I.A. Reyes
Instituto de Metalurgia, Universidad Autónoma de San Luis Potosí, Av. Sierra Leona no. 550, Lomas 2da. Sección, 78210 San Luis Potosí, SLP, Mexico

E.G. Palacios
Departamento de Ingeniería en Metalurgia y Materiales ESIQIE, UPALM, Instituto Politécnico Nacional, 07738 México, DF, Mexico

V.H. Flores
Escuela Superior de Zimapán, Universidad Autónoma Del Estado de Hidalgo, Av. Jorge Preisser S/N Col. Llano Norte, 42330 Zimapán, HGO, Mexico

S. Ikhmayies et al. (eds.), *Characterization of Minerals, Metals, and Materials 2017*, The Minerals, Metals & Materials Series,
DOI 10.1007/978-3-319-51382-9_64

thus obtaining the following composition: 27% Fe^{3+}, 37% SO_4^{2-} 14% Hg^{2+}, 22% OH^- + H_3O^+ + H_2O. The following approximate formula of the compound was determined from chemical analysis: $[Hg_{0.39}(H_3O)_{0.22}]Fe_{2.71}(SO_4)_{2.17}(OH)_{4.79}(H_2O)_{2.09}$.

Keywords Mercury · Jarosite · Characterization

Introduction

The widespread distribution of mercury coming from man-related activities and natural sources has become a threat to the conservation of resources as well as to human health around the world. This is due to its high bioavailability, trophic biomagnification and extreme toxicity [1]. It affects the liver, kidneys, central nervous system and central respiratory system [2–4], as the exposition to mercury occurs through inhalation (elemental Hg) and food ingestion (organic Hg for the most part), or through contaminated water (mainly inorganic Hg).

In the last decades, developing countries such as Brazil, China, India and Mexico have been particularly vulnerable to mercury and other effects of environmental pollution due to a fast-paced industrialization and urbanization, together with a lack of proper waste control [5]. This has led to the establishing of guidelines to control the mercury emissions through the report of the US Environmental Protection Agency; the maximum contaminant level of mercury in drinking water has been set at 2 $\mu g\ L^{-1}$ [6]. Considering the adverse environmental and biological effect of water contamination by mercury, a considerable number of techniques has been developed to remove this element, such as: ion exchange, adsorption, membrane filtration, and electrochemical treatment. Although these methods are highly efficient in reducing the amount of mercury, they usually require an excess of costly adsorbents and chemical products, thus generating high economic expenses for the functioning of the devices and required materials [7, 8]. Therefore, their application in a wide range of industries remains limited [9].

An alternative for the reduction of mercury would be the precipitation of jarosite-type compounds as an efficient method to concentrate metals in solution for the removal of toxic elements such as As(VI), Cr^{3+}, Tl^+, ½ Pb^{2+} [10–13]. Jarosites occur in nature near volcanic vents, in acid saline lakes [14], and during aerobic processes in a pH range between 1 and 3 [15]. They also form in acid mine drainage, where iron and sulfates are typical, and where the geochemical conditions necessary for its formation are present [16]. It is a low-cost technology that has been widely used in hydrometallurgical plants in the zinc industry as a means to remove iron and unwanted impurities from the leaching liquids in order to make filtering easy [17–19]. Jarosite type compounds are represented by the general formula $MFe_3(SO_4)_2(OH)_6$, where M can be occupied by H_3O^+, Na^+, K^+, Rb^+, Ag^+, NH_4^+, Tl^+, ½ Pb^{2+} and ½ Hg^{2+} [20]. Some of these elements are considered of environmental importance, and can be incorporated into the jarosite structure, thus reducing their environmental impact [21, 22].

Mercury jarosite ($Hg_{0.5}Fe_3(SO_4)_2(OH)_6$) was synthesized by Dutrizac and Kaiman in 1976 by adding a solution of mercuric nitrate monohydrate ($Hg(NO_3)_2{\cdot}H_2O$) slightly acidified with nitric acid (HNO_3) into a hot ferric sulfate solution. They incorporated 14.62% weight of mercury [23]. Later in 1981, Dutrizac and Chen determined the effect of adding different amounts of seeds; they synthesized mercury jarosite samples, achieving incorporations ranging between 7.95 and 8.01% weight of mercury. The concentration of this element was relatively low because the amounts of added mercury were stoichiometrically lower [24].

The aim of this study is to conduct the characterization of synthetic mercury jarosite in order to determine its morphology, structure and mercury incorporation in its structure.

Experimental Methodology

For this study, we characterized the mercury jarosite from synthesis number 4 reported by Ordoñez et al. [25]. We used the slow addition technique: 10 mL h^{-1} of $Hg(NO_3)_2{\cdot}H_2O$ 0.58 mol L^{-1} slightly acidified with concentrated hydrochloric acid were added for 10 h into a 750 mL solution of $Fe_2(SO_4)_3{\cdot}nH_2O$ 0.27 mol L^{-1} with an initial pH of 1.3, contained in a 1 L (Pyrex) glass reactor immersed in oil (Quaker State SAE 25W-60). It was placed on a heating plate with automatic temperature control (Super-Nuova/Barnstead-Thermoline) coupled to a coil condenser to recirculate water, and a mechanical stirrer (IKA RW16 basic) with a 3-blade propeller. The solution was kept at a stirring rate of 400 rpm with a temperature of 93 °C (366 K) for 24 h. The obtained precipitate was filtered and washed with distilled water at 70 °C (343 K); it was then set to dry on a heating stove for 1 h at 65 °C (338 K). In order to confirm the formation of the mercury jarosite compound, the solids were analyzed by X-ray diffraction (XRD) using a Bruker D8 Focus diffractometer with a radiation of Cu kα1 (λ = 1.5406 Å) (this piece of equipment does not have a monochromator). Morphology and structure of the precipitate were observed through a JEOL JSM-6701F Scanning Electron Microscope (SEM) equipped with an INCA energy-dispersive X-ray detector (EDS). Particle size distribution was measured by using Tyler meshes (USA Standard Testing Sieve, ASTM-11 specifications) with the following openings: 120, 170, 200, 270, 325, 400, 500 and −500. The density of the mercury jarosite was determined with a 10 mL Pyrex pycnometer using water as immersion medium. The chemical composition of the precipitate was determined by quantifying the percentage of mercury through atomic absorption spectroscopy, AAS (PerkinElmer Analyst 200). Iron (Fe^{3+}) and sulfate (SO_4^{2-}) contents were determined through chemical analysis techniques such as dichromatometry and gravimetry respectively [26].

Results

The synthesized product leads to the formation of a precipitate of a yellowish color, which is typical of jarosite-type compounds; it has a density of 3.23 g cm^{-3}. It is similar to the density reported by J.E. Dutrizac and S. Kaiman, which was 3.40 g cm^{-3} [23].

Characterization of the solids by X-ray diffraction confirmed that the obtained compound corresponds to the crystallographic planes of the diffraction pattern ICDD-PDF 00-030-0837 of the International Center for Diffraction Data-Powder Diffraction Files of the mercury jarosite structure, as seen in Fig. 1. This compound is considered a pure phase, because only the previously mentioned reagents were used for its formation during the synthesis. The peaks that do not match the diffraction pattern are iron and sulfur compounds, and are considered residues of the reaction.

The energy-dispersive X-ray microanalysis spectrum presented on Fig. 2a shows the distribution of the chemical elements that make up the structure of the mercury jarosite, such as sulfur, iron, mercury and oxygen. The morphology of the particles of the obtained precipitate was observed through SEM, and it was determined that they present a spheroidal geometry, as seen on the general micrograph shown in Fig. 2b. The particle size is of approximately 38 μm, as observed in Fig. 2c. These particles are made up of rhombohedral crystals with sizes ranging from 1 to 2 μm (Fig. 2d). This is consistent with the data obtained in the particle size distribution analysis (shown in Table 1), where it can be observed that the predominant particle diameter is 38 ± 2 μm, which corresponds to 400 mesh.

Based on chemical analysis, the following composition was determined: 27 wt% Fe^{3+} by dichromatometry, 37 wt% SO_4^{2-} by gravimetric analysis, 14 wt% Hg^{2+} by AAS, and 22 wt% OH^- + H_3O^+ + H_2O obtained by difference. With these data, and taking the theoretical formula indicated by J.E. Dutrizac and S. Kaiman on the mercury jarosite ($Hg_{0.5}Fe_3(SO_4)_2(OH)_6$) [23] as a reference, we calculated the

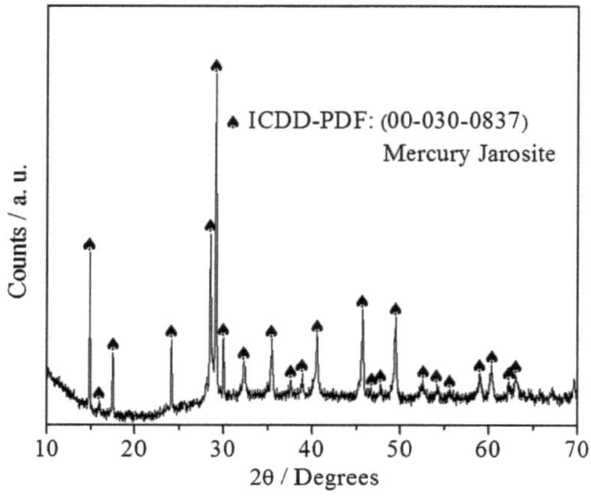

Fig. 1 X-ray diffraction pattern of the synthesized mercury jarosite compared to the ICDD-PDF file 00-030-0837

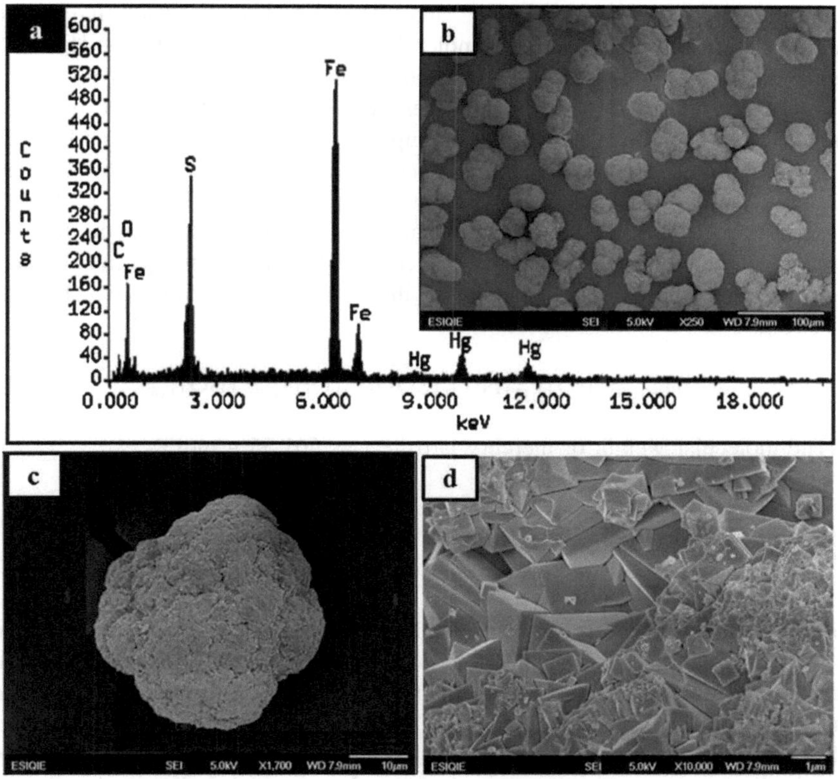

Fig. 2 **a** EDS analysis of mercury jarosite. **b** General micrograph of the synthesized precipitate. **c** Micrograph of a spheroidal particle. **d** Micrograph of the rhombohedral crystals that make up a spheroidal particle

Table 1 Particle size distribution (Tyler mesh) of the synthesized mercury jarosite

Mesh size	Diameters (μm)	Weight (g)	Retained weight %
+120	125	0.033	0.34
−120 + 170	90	0.016	0.17
−170 + 200	75	0.026	0.28
−200 + 270	53	0.369	3.86
−270 + 325	44	1.627	16.97
−325 + 400	38	5.962	62.17
−400 + 500	25	1.468	15.31
−500	<25	0.086	0.90
Total		9.59	100

weight percent of Hg^{2+}, Fe^{3+}, SO_4^{2-} and OH^-. We also performed the stoichiometric relationships to obtain the moles by considering that $Hg^{2+} + H_3O^+ = 1$, $Fe^{3+} = 3$, $SO_4^{2-} = 2$ and $OH^- + H_2O = 6$, assuming that site M can be partially substituted by the ion H_3O^+ [18]; the presence of water in the jarosite type compounds for compensating the six OH^- groups was also taken into account [23]. The H_2O moles were calculated by subtraction of the percentage obtained by difference (22 wt%) and by the sum of the weight percent of $OH^- + H_3O^+ = 15.22$ wt%; subsequently, a stoichiometric relationship was performed to convert the H_2O percentage into moles. Finally, we carried out the stoichiometric calculation to determine the approximate formula of the synthesized product, which is the following:

$$\left[Hg_{0.39}(H_3O)_{0.22}\right]Fe_{2.71}(SO_4)_{2.17}(OH)_{4.79}(H_2O)_{2.09} \quad (1)$$

The molecular weight of mercury jarosite was 561.55 g mol^{-1}.

Conclusions

The synthetic mercury jarosite compound incorporated 14 wt% of mercury within its structure. The particles of this compound have a tendency towards sphericity; they are made up of rhombohedral microcrystals ranging between 1 and 2 μm. These sizes allowed us to conduct studies in order to determine the kinetic model in a heterogeneous reaction. The particle size analysis determined that the average particle diameter was of approximately 38 ± 2 μm, which indicates that the precipitate is made up of relatively fine particles. The energy-dispersive X-ray microanalysis spectrum shows the distinctive elements of this jarosite. X-ray diffraction analysis confirmed that the obtained precipitate was mercury jarosite with an approximate formula $[Hg_{0.39}(H_3O)_{0.22}]Fe_{2.71}(SO_4)_{2.17}(OH)_{4.79}(H_2O)_{2.09}$, molecular weight of 561.55 g mol^{-1} and density of 3.23 g cm^{-3}. This study demonstrates that jarosites are capable of incorporating considerable amounts of elements of environmental importance.

References

1. Eagles-Smith CA et al (2016) Mercury in western North America: a synthesis of environmental contamination, fluxes, bioaccumulation, and risk to fish and wildlife. Sci Total Environ. doi:10.1016/j.scitotenv.2016.05.094
2. Rizzetti DA, Fernandez F, Moreno S, Uranga JA, Peçanha FM, Vera G, Vassallo DV, Castro MM, Wiggers GA (2016) Egg white hydrolysate promotes neuroprotection for neuropathic disorders induced by chronic exposure to low concentrations of mercury. Brain Res 1646:482–489

3. Black P et al (2016) Assessing occupational mercury exposures and behaviours of artisanal and small-scale gold miners in Burkina Faso using passive mercury vapour badges. Environ Res. doi:10.1016/j.envres.2016.06.004
4. Zhou C, Sun L, Zhang A, Ma C, Wang B, Yu J, Su S, Hu S, Xiang J (2015) Elemental mercury (HgO) removal from containing SO_2/NO flue gas by magnetically separable $Fe_{2.45}Ti_{0.55}O_4/H_2O_2$ advanced oxidation processes. Chem Eng J 273:381–389
5. Oliveira CS, Oliveira VA, Costa LM, Pedroso TF, Fonseca MM, Bernardi JS, Fiuza TL, Pereira ME (2016) Inorganic mercury exposure in drinking water alters essential metalhomeostasis in pregnant rats without altering rat pup behavior. Reprod Toxicol 65:18–23
6. U.S. Environmental Protection Agency (2007) Treatment technologies for mercury in soil, waste, and water. Report EPA 542-R-07-003, USEPA: Office of Superfund Remediation and Technology Innovation, Washington, DC
7. Zhang R, Shen B, Li C, Zheng C, Hou X (2016) Integrating photochemical vapor generation with photo-oxidation trapping for effective mercury removal from polluted water and its on-line monitoring. Microchem J 129:98–103
8. Heidel B, Rogge T, Scheffknecht G (2016) Controlled desorption of mercury in wet FGD waste water treatment. Appl Energy 162:1211–1217
9. Johari K, Saman N, Song ST, Cheu SC, Kong H, Mat H (2016) Development of coconut pith chars towards high elemental mercury adsorption performance–effect of pyrolysis temperatures. Chemosphere 156:56–68
10. Patiño F, Flores MU, Reyes IA, Reyes M, Hernández J, Rivera I, Juárez JC (2013) Alkaline decomposition of synthetic jarosite with Arsenic. Geochem Trans 14:2
11. Flores MU et al (2012) Kinetic modeling of the alkaline decomposition of potassium arsenojarosite. J Braz Chem Soc 23(6):1018–1023
12. Mireles I et al (2016) Kinetic analysis of the decomposition of the $KFe_3(SO_4)_{2-x}(CrO_4)_x(OH)_6$ Jarosite Solid Solution in $Ca(OH)_2$ Medium. J Braz Chem Soc 26(7): 1014–1025
13. Islas H, Patiño F, Roca A, Reyes IA, Flores MU, Palacios EG, Reyes M (2015) Decomposition nature in alkaline medium of a thallium jarosite. In: 24th international conference on metallurgy and materials, metal 2015, Brno, Czech Republic, EU
14. Elwood ME, Madden AS, Rimstidt JD, Zahrai S, Kendall MR, Miller MA (2012) Jarosite dissolution rates and nanoscale mineralogy. Geochim Cosmochim Acta 91:306–321
15. Johnston SG, Burton ED, Keene AF, Planer-Friedrich B, Voegelin A, Blackford MG, Lumpkin GR (2012) Arsenic mobilization and iron transformations during sulfidization of As (V)-bearing jarosite. Chem Geol 334:9–24
16. Norlund KLI, Baron C, Warren LA (2010) Jarosite formation by an AMD sulphide-oxidizing environmental enrichment: implications for biomarkers on Mars. Chem Geol 275:235–242
17. Alonso M, López A, López FA (1998) A kinetic study of the thermal decomposition of Ammoniojarosite. J Mater Sci 33:5821–5825
18. Ntumba E, Mulaba AF, Nheta W (2015) Alkaline leaching of nickel bearing ammonium jarosite precipitate using KOH, NaOH and NH_4OH in the presence of EDTA and Na_2S. Hydrometallurgy 155:69–78
19. Patiño F et al (2011) Decomposition kinetics of argentian lead jarosite in NaOH media. J Mex Chem Soc 55(4):208–213
20. Patiño F et al (2010) Kinetic modeling of the alkaline decomposition and cyanidation of argentojarosite. J Mex Chem Soc 54(4):216–222
21. Reyes IA, Mireles I, Patiño F, Pandiyan T, Flores MU, Palacios EG, Gutiérrez EJ, Reyes M (2016) A study on the dissolution rates of K-Cr(VI)-jarosites: kinetic analysis and implications. Geochem Trans 17:3
22. Patiño F, Reyes IA, Flores MU, Pandiyan T, Roca A, Reyes M, Hernández J (2013) Kinetic modeling and experimental design of the sodium arsenojarosite decomposition in alkaline media: implications. Hydrometallurgy 137:115–125

23. Dutrizac JE, Kaiman S (1976) Synthesis and properties of jarosite-type compounds. Can Mineral 14:151–158
24. Dutrizac JE, Chen TT (1981) The synthesis of mercury jarosite and the mercury concentration in jarosite-family minerals. Can Mineral 19:559–569
25. Ordoñez S, Patiño F, Reyes IA, Flores MU, Flores VH, Palacios EG, Reyes M (2015) Synthesis and topology of the reaction of mercury jarosite in NaOH medium. In: European metallurgical conference, vol 1, pp 537–552
26. Arthur I (1960) Vogel, Química analítica cuantitativa. Kapelusz S.A, Argentina, pp 317–636

Characterization of Steel Production Dust and Their Use in Structural Ceramics

A.T. Machado, J.R. Matos, F.M.S. Carvalho, A.A.S. Araujo and M.G. Silva-Valenzuela

Abstract Electric arc furnace steel dust is a by-product of the steelmaking process and contains high amounts of iron and zinc and significant amounts of lead, chromium, and cadmium and others chemical compounds in minor proportion. However, the material recycling is very complex, mainly due to their mineralogical composition. In this study the characterization and application of this material is presented. Ceramics were produced with clay and variable amounts of steel dust. The bulk material was fired between 800 and 1000 °C. The influence to insertion of this material on the ceramic composition on the mechanical strength, linear shrinkage, water absorption, apparent density and bending strength of the ceramic samples was investigated. A blend of clay with 5% dust presents potential for use in structural ceramics production.

Keywords Structural ceramics · Electric-arc furnace dust · Incorporation

A.T. Machado (✉) · J.R. Matos · F.M.S. Carvalho
Universidade de São Paulo, São Paulo, Brazil
e-mail: machadoat@globo.com

J.R. Matos
e-mail: jdrmatos@gmail.com

F.M.S. Carvalho
e-mail: flavioms@usp.br

A.A.S. Araujo
Universidade Federal de Sergipe, Aracaju, Brazil
e-mail: adriasa2001@yahoo.com

M.G. Silva-Valenzuela
Universidade Federal do ABC, Santo Andre, SP, Brazil
e-mail: gracavalenzuela@gmail.com

S. Ikhmayies et al. (eds.), *Characterization of Minerals, Metals, and Materials 2017*, The Minerals, Metals & Materials Series,
DOI 10.1007/978-3-319-51382-9_65

Introduction

Electric-arc furnace dust (EAFD) is one of the major by-products generated by the steel industry, about 15–20 kg/ton of steel [1]. EAFD is a complex powder with fine particle diameter and high-density. It contains high amounts of zinc and iron and significant amounts of calcium, manganese, magnesium, lead and chromium depending on the steelmaking process. A small amount of EAFD is recycled for zinc recovery and a large amount is disposed in landfills despite the fact that EAFD cannot be disposed in regular landfills due to the leaching of potentially toxic elements that contaminate the environment [1, 2].

Powders of electric arc furnace (PAE) are solid waste obtained from the steel-making process in electric arc furnaces (EAF). It was found that one of the largest sources of air pollution in the steel production process is due to solid particles that are detached from the process, carbon dioxide, nitrogen oxides and the volatilization of organic components.

Modern industries have installed on their process systems for retention of this material—both of volatiles that come off in the process, as the post of fog remain suspended. This material retained in this system is called electric arc furnace dust (EAFD). For containment and fixation of this material is used a sleeve filter system. In this dusting retained system are generated two types of waste, a very fine powder that is taken further dried by vacuum ducts to the courtyard of the plant and a slurry that is from the washing of filters. In both cases, the material is stored in the yards of the plant in the open. Due to the significant presence of heavy metals in their chemical composition and, in some cases, be rated as class I (dangerous), this material is difficult to recycle [3, 4].

Heavy metals present in the PAE are derived mainly from the scrap metal used in the casting of steel along with iron ore or sponge iron. Thus, the heavy metal content in the SAP, mainly Pb, Zn, Cr, Cd and Ni, is closely related to the percentage of scrap used in steel mill and the type of raw materials used in the furnace or converter.

It is estimated that about 20 kg of powder are generated by ton of steel produced, which, in 1996, representing approximately about 5 million tons produced worldwide. Between 1998 and 2002, in the US, the generation of EAF dust reached the amount of about 1 million tons, and only in Canada, about 77,000 t/ano [3]. Currently, it is estimated to be generated in the world about 2.0–7.0 million tons of electric arc furnace dust (between 1.0 and 2.0% of steel production in electric arc furnaces).

In Brazil, this number is in the order of 81.5 t/year. Recent information collected in steel in Bahia confirm a generation of EAF dust of about 500 t/month, which contain, on average, 4% Zn and 2–3% lead. At low zinc content—by virtue of the raw material used in the process—this residue is little sought in Bahia to be used as fertilizer. For lack of technology that enables other applications, it is being disposed in dumpsites [3].

Typically, the EAF dust contains high levels of iron and zinc (those with potential for recycling) and low levels of heavy metals such as lead, cadmium, hexavalent chromium and sometimes selenium. The water solubility of these elements is high, which leads to the classification of as hazardous waste in accordance with the contents of metals that are evaluated in the leaching test [3].

Currently, some companies have invested in the development of techniques for PAE reuse in own steel production process. However, as this is a new technology, this application does not consume all the material produced, and its even greater thrust of the industry patios.

The chemical, physical and mineralogical characterization of the dust was reported in several studies [1, 5, 6]. In a typical EAFD, the particle size ranges from 0.1 to 100 μm with a d50 of about 0.5 μm [5]. The major element contents of a typical EAFD are Fe: 30–42%, Zn: 5–13%, Ca: 2–10%, Cr: 1–11%, Mn: 1.7–1.9%, Si: 1.3–2.5%, Na: 1–2.7%, and Pb: 1.0–1.5%, depending on the raw materials [1, 5, 6].

The recycling of zinc and iron of EAFD was addressed in several studies [5, 7]. The use of pyrometallurgical and hydrometallurgical processes for metal recovery depends on the physical and chemical characteristics of the EAFD and the production process. When the zinc content of EAFD is about 15–20% and the amount of EAFD processed is about 50,000 tons/year, pyrometallurgical processes may be used to recycle the metals profitably [5]. However, finding a cost-effective and environmental-friendly process to recycle the metal content of the EADF is still a major challenge, and for this reason most recycling processes have never exceeded the pilot scale [5].

The stabilization of the metals in the EAFD is the common option to treat this industrial residue [2, 6, 8]. In this study, EAFD was stabilized by mixing it with clay for the production of ceramics.

Experiments

Samples of EAFD were collected at a carbon steel company in Bahia, Brazil (Usina Siderurgica da Bahia, USIBA) which processes iron ores from mines located in Minas Gerais State (Brazil) and scrap iron. About 20 kg of the EAFD were collected. The clay was collected from a factory of roofing tiles and hollow bricks (Cerâmica do Nordeste Ltda.) whose raw material comes from a clay mine located in the region of Feira de Santana, Bahia, Brazil.

The clay was blended with 0, 5, 50, and 90% of EAFD. Samples were prepared with a total mass of 12 g of the blend in a prismatic mold of 6.0 × 2.0 × 0.5 cm. The powder was compressed in the mold at 204 kPa using a Forney Model QC-200 DR compression tester. A total of 35 samples were produced for each EAFD content. Most samples were dried at room temperature for five days. Five samples were dried at 110 °C. Thirty dried samples were fired in an electric furnace at 800 and 1000 °C (ten samples at each temperature). In this case, the heating rate was

10 °C/min and the firing time was 2 h. The cooling rate was not automatically controlled.

The raw materials and the ceramic samples were analyzed at the University of São Paulo (Brazil). Scanning electronic microscopy (SEM) images were taken with the Philips XL–30 with an EDAX. X-ray diffraction patterns were obtained with CuKα radiation using the Philips X'Pert MPD and the software EVA. The chemical analysis of the clay and the EAFD were performed with the ICP Spectro Co Ciros CCD and X-ray fluorescence Philips PW 2404.

Five ceramic samples of each firing temperature were used to measure linear shrinkages, water absorption and apparent density. The other five samples were used to measure the bending strengths. These tests were performed according to the Brazilian Standards [9].

Results and Discussion

Characterization

The chemical analysis showed a large amount of Si, Al and Fe in the clay (SiO_2 69.08%, Al_2O_3 14.82%, Fe_2O_3 5.19%, TiO_2 1.47%, K_2O 1.37%, MgO 1.13%, CaO 0.43%, P_2O_5 0.07%, MnO 0.07%, and Na_2O 0.05%); the total amount of these oxides was 93.4%. EAFD contains large amounts of Fe and Zn (Fe_2O_3 60.72%, ZnO 11.59%, SiO_2 4.31%, CaO 3.68%, Al_2O_3 2.08%, MgO 2.03%, K_2O 1.91%, MnO 1.35%, Na_2O 1.30%, PbO 1.23%, P_2O_5 0.71%, and TiO_2 0.06%); total of 91.0%. Among the major and trace elements of the EAFD, the Pb and Cd content (194 mg/kg) potential toxicity.

Water Absorption and Apparent Porosity

The amount of water absorbed decreased with increasing firing temperatures but was not strongly affected by the EAFD content (Table 1). In addition, the porosity of the fired material decreased with increasing firing temperature and EAFD content (Table 1) in agreement with other studies [10, 11] (Fig. 1).

The SEM image of the samples with 90% EADF fired at 800 °C (Fig. 2a) shows that the EAFD and clay particles ware separated demonstrating incomplete sinterization. The SEM image of the sample with 90% EADF fired at 1100 °C (Fig. 2b) indicating the higher degree of sintering and the reduced porosity. The EAFD changed the shape during sintering making the ceramics less porous (Table 1) [6, 8]. Some elements such as Fe in EAFD retarded sintering blends with an EAFD content N 50% fired at N 950 °C revealed increased apparent porosity and water absorption with increasing EAFD content [3].

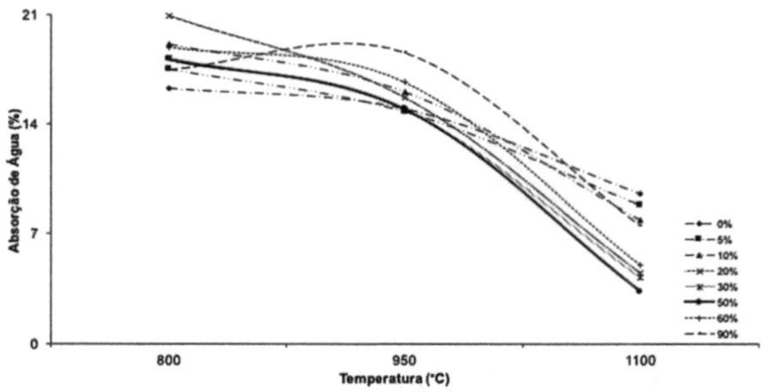

Fig. 1 Water absorption × temperature

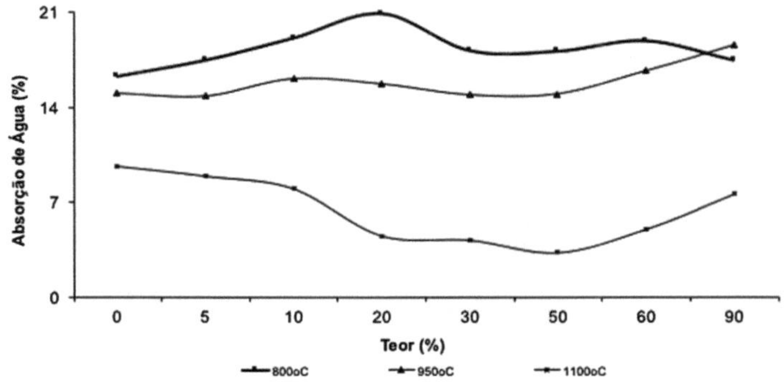

Fig. 2 Waster absorption × EAFD ratio

Water absorption can restrict the use of ceramics as building material [3]. Ceramics used as flooring should have a water absorption of 1%, as roofing tiles of 20%, and as hollow bricks about 25%. The materials described in this study can be used as roofing tiles, except for the samples with 20% EAFD and hollow bricks fired at 800 °C did not show an adequate water absorption for the use as road surface. Moreover, the samples fired at 800 and 950 °C presented apparent porosity values within the range expected for Brazilian clays (5–45%) [3, 10, 11].

Linear Shrinkage

At 800 °C the linear shrinkage increased with the firing temperatures. Samples dried at 110 °C and fired at 800 °C showed low linear shrinkage and the influence of EAFD addition was not significant. These samples presented linear shrinkage values closer to traditional ceramics [12] (Figs. 3 and 4).

The low linear shrinkage presented by the samples dried at 110 °C and fired at 800 °C is advantageous for traditional ceramics manufacturing because it reduces

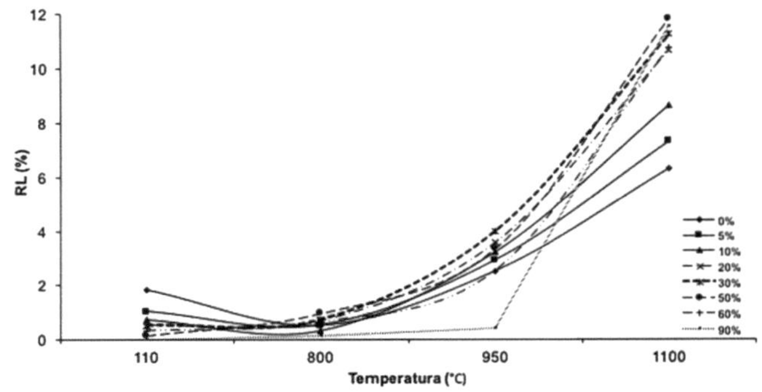

Fig. 3 Linear shrinkage × temperature

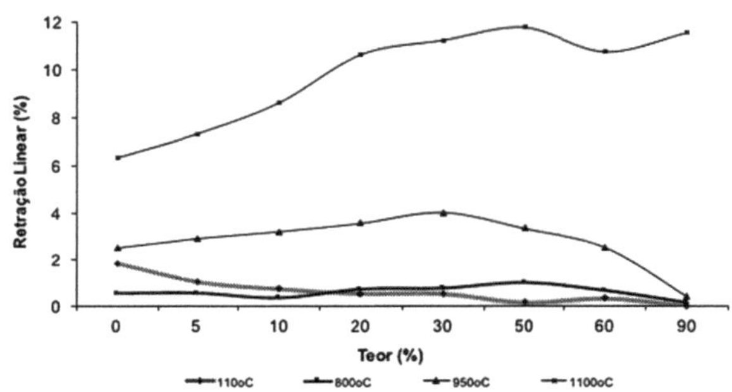

Fig. 4 Linear shrinkage × EAFD ratio

cracking and volume changes during firing. The linear shrinkage of the EAFD containing samples fired at 800 or 950 °C waste 5.0%, is similar to usual ceramics [13]. However, EAFD samples containing samples fired at 1000 °C showed linear shrinkage values 7.0%, inappropriate for ceramics manufacture.

Mechanical Strength

The bending strength increased with the firing temperature (Table 1) due to the decreased porosity. The addition of EAFD decreased the rupture tension of the sample fired at 800 and 950 °C. This is caused by the increased porosity after the EAFD addition. When the samples were dried at 110 °C, EAFD addition decreased the mechanical strength. This is probably related to the different morphology of the particles of EAFD and the clay. When the samples contained N 10% of EAFD and ware are fired at 1000 °C, the mechanical strength increased with increasing EAFD addition.

The recommended bending rupture tension for bricks dried at 110 °C is 15.3 kPa and fired at 800 °C is 20.4 kPa; for hallow bricks dried at 110 °C is

Fig. 5 Mechanical strength

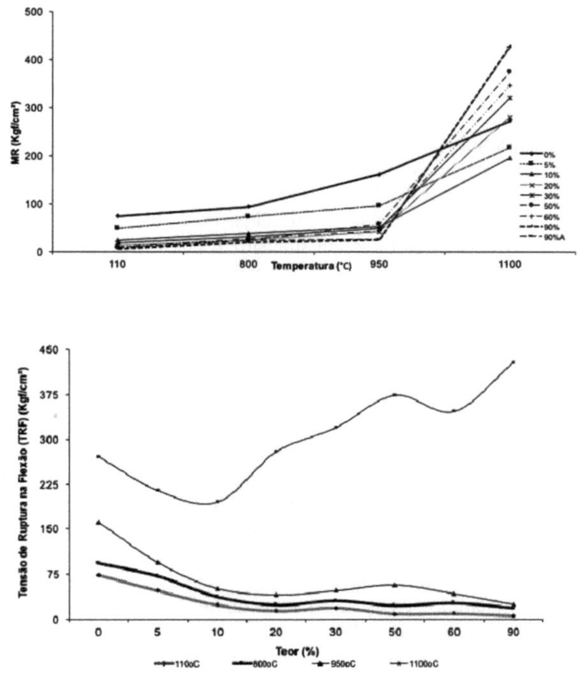

25.5 kPa and fired at 800 °C is 56.1 kPa; roof tiles dried at 110 °C is 30.6 kPa and fired at 800 °C is 66.3 kPa [9]. As shown in Table 1, mixes with 5 and 10% EAFD can be used in bricks, but only the blend containing about 5% EAFD is appropriate for the use of hollow bricks and roof tiles (Fig. 5).

The samples fired at 800 and 950 °C can be used to produce bricks. The sample containing 90% EAFD and fired at 800 °C is the only exception. Only the blend with about 10% EAFD can be used for hollow bricks and roof tile production. All blends fired at 1100 °C have adequate mechanical strength to be used as bricks, hollow bricks and roof tiles.

Conclusions

The samples fired at 800 and 950 °C can be used to produce bricks. The sample containing 90% EAFD and fired at 800 °C is the only exception. Only the blend with about 10% EAFD can be used for hollow bricks and roof tile production. All blends fired at 1100 °C have adequate mechanical strength to be used as bricks, hollow bricks and roof tiles.

References

1. Hagni AM, Hagni RD, Demars C (1991) Mineralogical characteristics of electric arc furnace dust. J Miner Met Mater Soc (JOM) 43:28–30
2. Domínguez EA, Ullmann R (1996) Ecological bricks made with clays and steel dust pollutants. Appl Clay Sci 11(1996):237–249
3. Machado AT (2005) Estudo da Incorporação e Estabilização do Pó de Aciaria Elétrica em Matrizes de Cerâmica Vermelha. Ph.D. thesis, São Paulo University (in Portuguese)
4. NBR (1987a) Brazilian standard 10004: solid waste-classification
5. Dutra AJB, Paiva PRP, Tavares LM (2006) Alkaline leaching of zinc from electric arc furnace steel dust. Miner Eng 19(2006):478–485
6. Pelino M, Karamanov A, Pisciella P, Crisucci S, Zonetti D (2002) Vitrification of electric arc furnace dusts. Waste Manag 22:945–949
7. Nyirenda RL (1991) The processing of steelmaking flue-dust: a review. Miner Eng 4:1003–1025
8. Mikhail SA, Turcotte AM, Aota J (1996) Thermoanalytical study of EAF dust and its vitrification product. Thermochim Acta 287:71–79
9. NBR (1987b) Brazilian standard 10005: "leaching test-procedures"
10. Melo MAF, Neto SG, Melo DMA, Carvalho LP, Galdino JN, Silva SAG (2002) Color and mechanical properties of some clays found in Rio Grande do Norte State for whiteware. Cerâmica 48:183–186
11. Moravia WG, Oliveira CAS, Gumieri AG, Vasconcelos WL (2006) Microstructural evaluation of expanded clay to be used as lightweight aggregate in structural concrete. Cerâmica 52:193–199
12. Dutra RPS, Varela ML, Nascimento RM, Gomes UU, Martinelli AE, Paskocimas CA (2009) Comparative study of fast firing with traditional firing on the properties of ceramic clay raw materials. Cerâmica 55:100–105
13. Bergaya F, Theng BKG, Lagaly G (ed) (2006) Handbook of clay science. Elsevier, Amsterdam

Charpy Toughness Behavior of Jute Fabric Reinforced Polyester Matrix Composites

Foluke S. de Assis, Artur C. Pereira, Fábio O. Braga and Sergio Monteiro

Abstract The jute fiber is one of the most investigated natural lignocellulosic fibers due to its easy processing and extensive occurrence. However, the studies with some varieties of jute fiber such as the *Corchorus capsularis* are rare, so this work investigated the toughness behavior of polyester matrix composites reinforced with up to 30% in volume of a fabric made of jute fiber by means of Charpy impact tests. It was found that the addition of jute fabric results in a marked increase in the absorbed energy by the composites. Macroscopic observation of the post-impact specimens and SEM fracture analysis showed that transversal rupture through the jute fabric interface with the polyester matrix is the main mechanism for the remarkable toughness of these composites.

Keywords Jute fabric · Composite · Polyester matrix · Charpy impact tests

Introduction

A significant change in the transportation sectors mainly the aerospace and automotive, occurred last century with the introduction of stronger and lighter materials. The most relevant example was the polymer composite reinforced with synthetic fiber. They are nowadays replacing conventional materials such as the aluminum alloys in airplane body owing to superior specific strength. In recent years, however, environmental problems in a global scale are imposing a reversion in the use of synthetic fiber specially the glass fiber. In spite of its higher specific strength, this fiber and its polymer composites can neither be recycled nor incinerated in a thermal plant to generate electricity [1]. Moreover, the fabrication, processing and transportation of a glass fiber composite are associated with CO_2 emission, the major responsible for global warming [2].

F.S. de Assis · A.C. Pereira · F.O. Braga (✉) · S. Monteiro
Military Institute of Engineering, IME, Praça Gen. Tibúrcio, nº80 Urca, Rio de Janeiro-RJ 22290-270, Brazil
e-mail: fabio_obraga@yahoo.com.br

S. Ikhmayies et al. (eds.), *Characterization of Minerals, Metals, and Materials 2017*, The Minerals, Metals & Materials Series,
DOI 10.1007/978-3-319-51382-9_66

A possible substitute for glass fiber without the above-mentioned environmental problems, could be a natural fiber. In fact, natural fibers obtained from cellulose-rich plants, also called lignocellulosic fiber, are already being used in several engineering fields, particularly the automobile industry [3–6].

In recent years a growing number of publications [7–12] have been dedicated to lignocellulosic fibers and their polymer composites. The results justify the current application as new engineering materials with advantages not only in terms of environmental issues but also economical, societal and even technical [10–12]. Moreover, recent studies [13, 14] show that the polymeric composites reinforced with lignocellulosic fibers can be applied in ballistic armor as intermediate layer in multilayered armor.

The major challenge faced by the lignocellulosic fiber and their composites is to shorten the distance to the performance of the corresponding synthetic ones. Here, drawbacks such as non-uniform properties and limited dimension as well as weak adhesion to polymeric matrix, are still impeding technical factors to the expansion on the use of lignocellulosic fibers [10].

The jute fiber (*Corchorus capsularis*) has been investigated in many studies, principally polymeric composites reinforced with jute fiber. These composites can replace the products of wood, and also they can be utilized as materials of construction or auto parts [15]. Then the jute fiber present very interesting potential improvement as seen before in literature. This work evaluated the Charpy impact resistance of polyester composites reinforced with jute fabric.

Experimental Procedure

The jute fabric (Fig. 1) investigated in this work were commercially supplied by Companhia Têxtil Castanhal do Pará.

After cleaning at room temperature, the jute fabric was mixed in amounts of 0, 10, 20 and 30% in volume with unsaturated polyester resin to prepare the

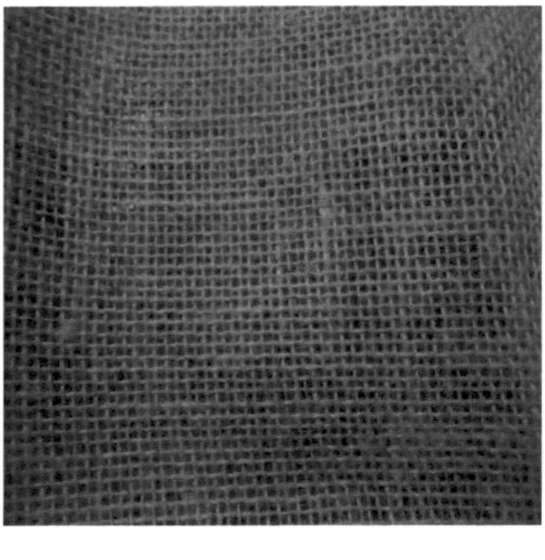

Fig. 1 Jute fabric

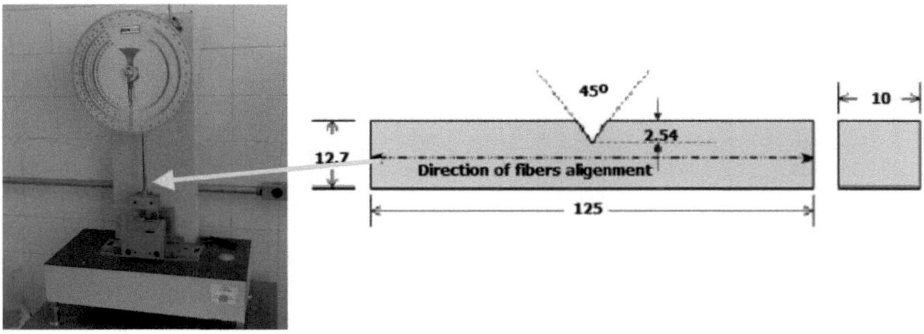

Fig. 2 Charpy equipment and standard specimen schematic

composites. Plates of the composites with 13 mm thickness were fabricated in a rectangular steel mold with dimensions of 152 × 125 mm.

The still liquid polyester resin, together with 0.5% catalyst based on methyl ethyl ketone, was poured into the jute fabric inside the mold. The composite thus formed was allowed to cure for 24 h at room temperature. The plate of each different composite was then cut in bars measuring 10 × 125 × 12.7 mm. These bars were used for preparation of samples for Charpy impact test, according to ASTM D256 [16].

Figure 2 illustrates the Charpy impact pendulum and the schematic specimen with standard dimensions. The notch with 2.54 mm in depth, angle of 45° and a tip curvature radius of 0.25 mm was machined with a DIN 847 milling tool. For each volume fraction of fiber 10 specimens were tested to assure a statistical validation, those were impact tested in a PANTEC hammer pendulum.

The impact fracture surface of the specimens was gold sputtered and analyzed by scanning electron microscopy, SEM, in a model SSX-500 Shimadzu microscope with secondary electrons imaging at an accelerating voltage of 15 kV.

Results and Discussion

Table 1 presents the results of Charpy Impact Tests of polyester matrix composites reinforced with different volume fractions of jute fabric.

Based on the results shown in Table 1, the variation of the Charpy Impact Energy with the amount of jute fabric in the polyester matrix is shown in Fig. 3.

Table 1 Charpy impact energy for polyester composites reinforced with jute fabric

Amount of jute fabric (%)	Energy (J/m)
0	16.1 ± 0.6
10	41.9 ± 6.7
20	92.4 ± 7.6
30	186.8 ± 17.6

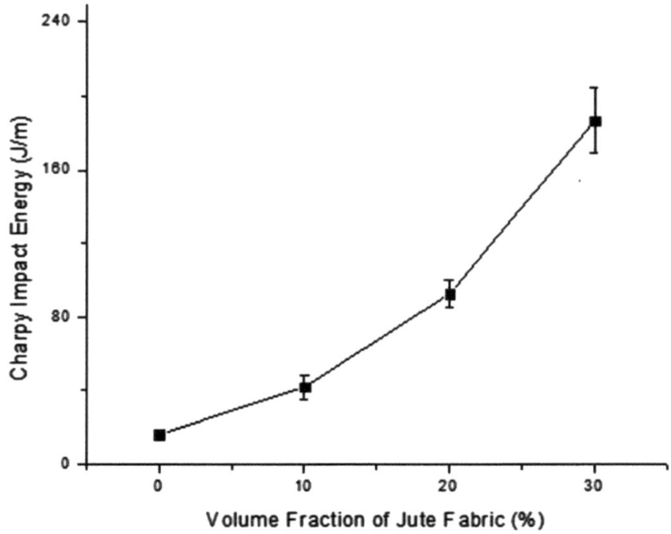

Fig. 3 Charpy impact energy as a function of amount jute fabric

In this figure it should be noticed that the jute fabric incorporation into the polyester matrix significantly improves the impact toughness of the composite. Within the standard deviation, this improvement can be considered almost a linear function with respect to the amount fiber up to 30%. The values shown in this figure are consistent with results reported in the literature. The reinforcement of a polymeric matrix with both synthetic [17] and natural [18–20] fibers increases the impact toughness of the composite.

The relatively low interface strength between a hydrophilic natural fiber and a hydrophobic polymeric matrix contributes to an ineffective load transfer from the matrix to jute fabric. This results in relatively greater fracture surface and consequently, higher impact energy is spent for the rupture [21].

The SEM analysis of the microstructure of the fractured region resulting from Charpy impact, Fig. 4 as seen in, permitted a better understanding of the mechanisms responsible for the toughness of jute fabric reinforced composites as compared to the pure resin. Regarding the pure resin, Fig. 4a, the fracture appearance is typical of polyester rupture. In this figure, the darker sideband at the right corresponds to the groove made by the cutter. The left part, smoother and clearer, corresponds to fracture by transverse impact.

The general SEM aspect of the surface ruptured by Charpy Impact of a specimen with 10% by volume of reinforcing jute fabric shows fibers sticking out of the polyester matrix. Some fibers were pulled out from the matrix by crack propagation through the surface due to the relatively low interfacial tension. A longitudinal crack propagation results in two aspects favorable to increase the energy absorbed during impact. First, the longitudinally separated area is larger than the section

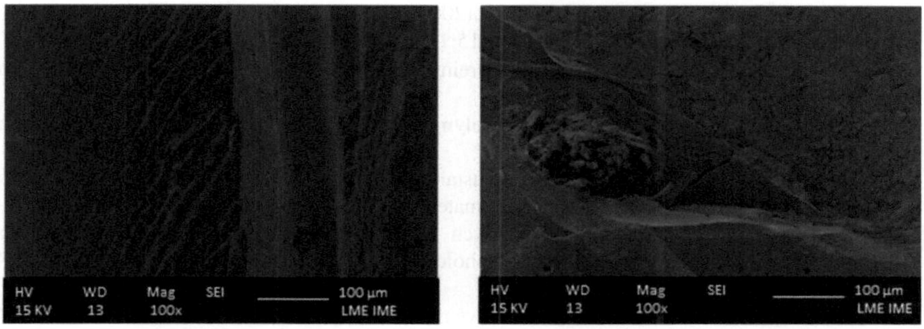

Fig. 4 Impact fractured surface of (**a**) the specimen of pure polyester resin and (**b**) of the composite with 10% in volume of jute fabric

through the matrix. This is associated with a higher energy and, consequently, higher impact resistance [21]. Second, jute fabric, being kept intact, hinders the separation of the specimen in two parts.

Conclusions

Composites made of jute fabric used as reinforcement of polyester matrix showed an almost linear increase in notch toughness, measured by Charpy impact tests, as compared to the pure polyester resin.

The incorporation of 30% jute fabric in the polyester matrix significantly improves the toughness of the composite, 187 J/m, as compared to pure polyester resin, with 16 J/m.

Most of this increase in toughness is apparently due to the low jute fabric/polyester matrix interfacial shear stress. This results in a higher absorbed energy as a consequence of the longitudinal propagation of the cracks throughout the interface, which generates larger rupture areas.

References

1. Wambua P, Ivens I, Verpoest I (2003) Natural fibers: can they replace glass and fibre reinforced plastics? Compos Sci Technol 63:1259–1264
2. Gore A (2006) An inconvenient truth. The planetary emergency of global warming and what we can do about it. Rodale Press, Emmaus, Pennsylvania, USA
3. Hill S (1997) Cars that grow on trees. New Sci 153(2067):36–39
4. Larbig H, Scherzer H, Dahlke B, Poltrock R (1998) Natural fiber reinforced foams based on renewable resources for automotive interior applications. J Cell Plast 34:361–379
5. Marsh G (2003) Next step for automotive materials. Mater Today 6(4):36–43

6. Zah R, Hischier R, Leão AL, Brown I (2007) Curaua fibers in automobile industry—a sustainability assessment. J Cleaner Prod 15:1032–1040
7. Bledzki AK, Gassan J (1999) Composites reinforced with cellulose-based fibers. Prog Polym Sci 4:221–274
8. Sahed DN, Jog JP (1999) Natural fiber polymer composites: a review. Adv Polym Technol 18:221–274
9. Mohanty AK, Misra M, Drzal LT (2002) Sustainable biocomposites from renewable resources: opportunities and challenges in the green material world. J Polym Environ 10:19–26
10. Satyanarayana KG, Guimarães JL, Wypych F (2007) Studies on lignocellulosic fibers of Brazil. Part I: source, production, morphology, properties and applications. Compos A 38:1694–1709
11. Crocker J (2008) Natural materials innovative natural composites. Mater Technol 2–3:174–178
12. Monteiro SN, Lopes FPD, Ferreira AS, Nascimento DCO (2009) Natural fiber polymer matrix composites: cheaper, tougher and environmentally friendly. JOM 61:17–22
13. Monteiro SN, Lima EP, Louro LHL, Silva LC, Drelich JW (2014) Unlocking function of aramid fibers in multilayered ballistic armor. Metall Mater Trans A
14. Monteiro SN, Candido VS, Braga FO, Bolzan LT, Weber RP, Drelich JW (2016) Sugarcane bagasse waste in composites for multilayered armor. Eur Polym J
15. Wambua P, Vangrimde B, Lomov S, Verpoest I (2007) The response of natural fiber composites to ballistic impact by fragment simulating projectiles. Compos Struct 77:232–240
16. ASTM D 256–84 (1984) Standard test method for impact resistance of plastic and electrical insulating materials. ASTM, USA, pp 81–102
17. Fu SY, Lauke B, Mäder E, Hu X, Yue CY (1999) Fracture resistance of short-glass-fiber-reinforced and short-carbon-fiber-reinforced poly-propylene under Charpy impact load and dependence on processing. J Mater Process Technol 89(90):501–507
18. Monteiro SN, Margem FM, Santos Jr LFL (2009) Impact tests in epoxy matrix composites reinforced with ramie fibers'. In: Proceedings of the 64th international congress of the Brazilian Association for metallurgy and materials, July 2009. Belo Horizonte, Brazil, pp 1–9 (in Portuguese)
19. Monteiro SN, Margem FM, Santos Jr LFL (2008) Impact tests in polyester matrix composites reinforced with ramie fibers'. In: Proceedings of the 63rd international congress of the Brazilian Association for metallurgy and materials, July 2008. Santos, Brazil, pp 1–12 (in Portuguese)
20. Monteiro SN, Costa LL, Santafé HPG (2008) Characterization of the Charpy impact resistance of coir fiber reinforced epoxy matrix composites. In: Proceedings of the 18th Brazilian congress on materials science and engineering—CBECIMAT, November 2008. Porto de Galinhas, PE, Brazil, pp 1–12 (in Portuguese)
21. Yue CY, Looi HC, Quek MY (1995) Assessment of fiber-matrix adhesion and interfacial properties using the pullout test. Int J Adhes Adhes 15:73–80

Chemical and Mineralogical Characterization of a Mixed Sulphide Ore at Zimapan, Hidalgo, Mexico

Laura Angeles, Martin Reyes, Miguel Perez, Elia Palacios, Francisco Patiño, Ivan Reyes and Mizraim Flores

Abstract Chemical and mineralogical characterization of a sulphide ore from the mining district of Zimapan, Hidalgo, Mexico, was performed in order to obtain accurate information on the composition of the ore to optimize the subsequent flotation process. A combination of techniques was used: Scanning Electron Microscopy (SEM) with Energy Dispersive Spectrometer (EDS), X-ray diffractometry (XRD), Inductively Coupled Plasma (ICP) and Fourier transform infrared spectroscopy (FTIR). The SEM mapping confirmed the presence of the elements: Ca, Si, Fe, K, S, Al, Zn, Mg and Mn. The XRD analysis suggests that the ore matrix is calcite; additional species found were wollastonite, johannsenite, and pyrite, among other impurities. The FTIR analysis performed for the aqua regia insoluble showed the presence of calcium oxide, copper oxide and silicon oxide. The results for the ICP analysis were:13.31% S, 2.55% Fe, 0.46% Zn, 0.03% Cu, 0.06% Pb, 0.07% Mn, 0.09% As, 0.01% Sb and 0.08 grton-1 Ag.

Keywords Characterization · Sulphide ore · Flotation · Zimapan

L. Angeles (✉) · M. Reyes · M. Perez
Autonomous University of Hidalgo State, Academic Area of Earth Sciences and Materials, Abasolo 600 C.P., 42000, Pachuca, Hidalgo, Mexico
e-mail: laura_angeles_palazuelos@yahoo.com

E. Palacios
School of Chemical Engineering and Extractive Industries, National Polytechnic Institute, Adolfo Lopez Mateos C.P., 07738, Mexico city, Mexico

F. Patiño
Energy Engineering, Metropolitan Polytechnic University of Hidalgo, Boulevard Access to Tolcayuca 1009 C.P., 43860 Tolcayuca, Hgo, Mexico

I. Reyes
Institute of Metallurgy, Autonomous University of San Luis Potosi, Av. Sierra Leona 550,Lomas 2ª. Sección C.P., 78210, San Luis Potosí, S.L.P, Mexico

M. Flores
Electromechanics Industrial Area, Technological University of Tulancingo, Ahuehuetitla #301, 43642 Tulancingo, Hidalgo, Mexico

S. Ikhmayies et al. (eds.), *Characterization of Minerals, Metals, and Materials 2017*, The Minerals, Metals & Materials Series,
DOI 10.1007/978-3-319-51382-9_67

Introduction

Mining activity has been of great importance for the economic development of Mexico. Particularly in the mining district of Zimapan, region of the state of Hidalgo, it is known that the extraction of stones and metals has been practiced since the beginning of our era by Teotihuacan, Otomi and Toltec, who had worked and polished obsidian [1]. Formally, the mining history of the district begins at 1632, with the discovery of oxide ores at "El Carrizal" [2]. Initially, the ore was processed by casting, then from mid-twentieth century, ore processing is performed by the flotation process [3].

Currently, the municipality of Zimapan has an annual production about 1322 tons of copper; 25,906 kg of silver; 2553 tons of lead and 7120 tons of zinc [4]. Geological studies of the ore body [5, 6] classifies it as a calcium skarn of Zn-Pb-Cu paragenesis, with a complex mineralogy, including metal sulphides, a number of small minerals in very low quantities, as well as non-metallic minerals.

The sulphide minerals are the most diverse, important and rich in terms of their chemical, physical and structural properties [7]; its diversity relies on possessing a more complex crystal structure compared to other materials, also they present a variety of oxidation states, coordination numbers, symmetry, stoichiometry, density and acid-base surface properties, that make them very important compounds [8], furthermore sulphide minerals are the primary source of most base metals [9].

Complex sulphide deposits present different difficulties along the various stages of benefit and metal extraction, because of the peculiar mineralogical associations and structure of the host rock [10]. Despite of this, sulphide minerals are processed at industrial scale by the method of froth flotation, obtaining significant results [11].

Froth flotation process, takes advantage of the surface properties of the minerals in order to separate valuable materials (hydrophobic) from gangue minerals (hydrophilic) and its efficiency is measured in terms of good recovery and high grade [12]. There are several features involved in the success of the process such as the addition of reagents, chemical pulp and adequate knowledge of the mineral composition. Thus it becomes evident that the identification and characterization of minerals is of fundamental importance for the selective flotation process [13, 14].

For this reason, the objective of the present work is to make a mineralogical and chemical characterization from a sample of sulphide ore, in order to optimize the recovery of lead, copper and zinc concentrates, by means of the flotation process.

Experimentation

It was used an ore sample from "El Carrizal", a mining area located at the district of Zimapan, Hidalgo. Firstly, the mineral was dried, homogenized, broken and pulverized. Then, a representative sample of 500 g was taken and subsequently a wet sieving was conducted to classify the mineral by particle size. These classified samples were assembled in epoxy resin and mirror polished for SEM analysis.

Characterization Methods

Scanning Electron Microscopy (SEM)

Morphological and semi-quantitative analysis were performed for the samples embedded in epoxy resin, which were carbon coated to make mineral surface conductive. An scanning electron microscope with high resolution model JEOL JSM 6701 was used. Chemical analysis of the elements of the sample was determined by EDS.

X-Ray Powder Diffraction (XRD)

XRD analysis was performed to identify the major phases within the mineral, using the powder method. A diffractometer INEL EQUINOX 2000, with a Cu-Ka1 radiation of 1.540598 Å was employed. This method was accomplished for bulk mineral, a sample of mesh 400 from Tyler's series, as well as for hand-selected particles, according to their color: metallic, black and white.

Inductive Coupled Plasma (ICP)

Chemical analysis to determine the minor elements, was performed in triplicate, for different particle sizes, by the following process [15]: It was weighed 1 g of mineral and digested with aqua regia (3HCl-1HNO3, v/v) and heated until dry, then another small amount of aqua regia was added. The solution was filtered to get the insoluble. This solution was transferred to a 100 ml volumetric flask and diluted with deionized water. For this chemical analysis it was used a Perkin Elmer 8300 spectrometer.

Fourier Transform Infrared Spectroscopy (FTIR)

This technique was conducted in order to identify the main functional groups present in the insoluble material remaining after acid digestion of mineral; for this purpose, an infrared spectrometer Perkin Elmer System 2000 with Fourier transform was utilized.

Results and Discussion

Scanning Electron Microscopy (SEM)

SEM micrographs obtained for different particle sizes are shown in Fig. 1. As can be seen, the morphology is angled and irregular and it does not change with the particle size analyzed. It is also noted that there are no crystalline clusters; generally it has a good degree of particle release.

Meanwhile, microanalysis by EDS indicated that major elements present are *Ca*, *Si*, *Fe*, *K*, *S*, *Al*, *Zn*, *Mg* and *Mn*. Figure 2 illustrates the results for the punctual analysis performed on the light particle marked with letter A, corresponding to mesh size 38 μm. The most abundant elements are *Si and Ca*, suggesting the presence of calcite ($CaCO_3$) and wollastonite ($CaSiO_3$).

Figure 3 shows the mapping for the same set of particles of mesh size 38 μm. It is appreciated that the element *Ca* is scattered along all particles, thus arises the matrix mineral is calcite. The element Silicon is widely found, indicating the presence or wollastonite. It is possible to observe an association between *S* and *Fe*, demonstrating the presence of pyrite (FeS_2). Likewise it is present an association between *S* and *Zn* manifesting the presence of sphalerite (*ZnS*).

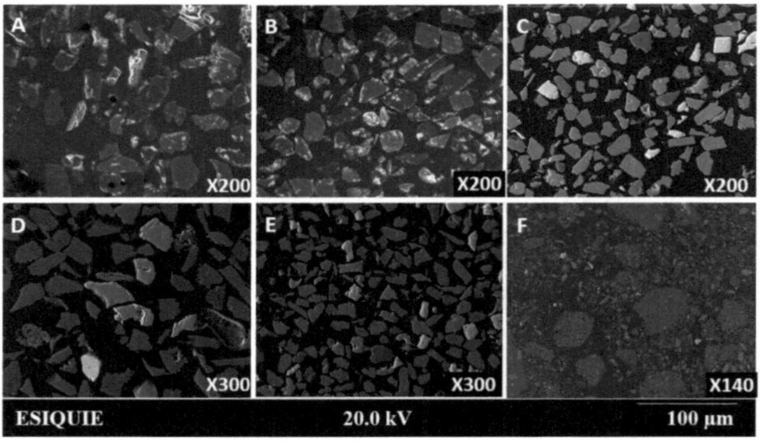

Fig. 1 SEM micrographs of the ore from different particle sizes. **a** Mesh size 74 μm, **b** Mesh size 53 μm, **c** Mesh size 45 μm, **d** Mesh size 38 μm, **e** Mesh size 25 μm and **f** Mesh size < 25 μm

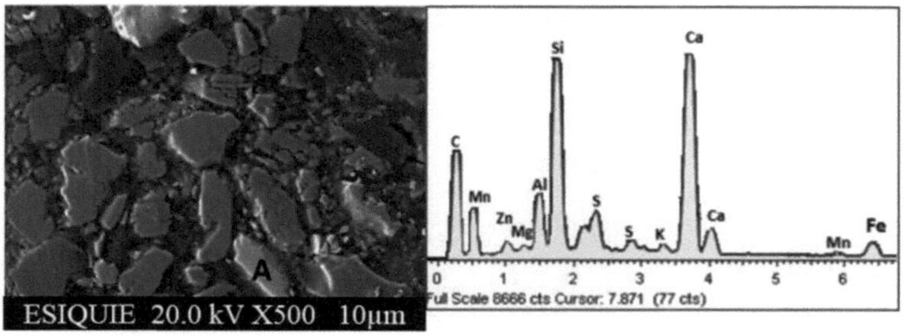

Fig. 2 EDS results for punctual analysis of particles mesh size of 38 μm

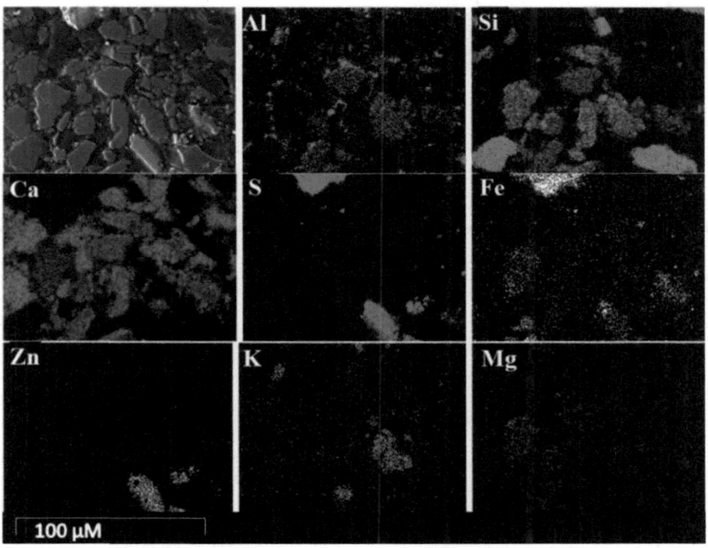

Fig. 3 Mapping for punctual analysis of particles mesh size of 38 μm

X-Ray Powder Diffraction (XRD)

In Fig. 4 the pattern of X-ray diffraction of the bulk sample is observed.

The most significant peaks correspond to Calcite ($CaCO_3$), which is the mineral matrix. Another species that are found in lots are johannesita ($CaMnO_6Si_2$) and wollastonite ($CaSiO_3$). Additionally, Fig. 5 shows the diffractogram for the particles of metallic color, in this case, the major species is pyrite ($As_{0.026}FeS_{1.974}$), followed by merwinite ($Ca_3MgO_8Si_2$) and quartz (SiO_2). For this metallic sample there are no presence of calcite and wollastonite.

X-ray analysis also was performed for particle size of 38 μm, black and white particles. In all these cases calcite was observed as mineral matrix and abundant

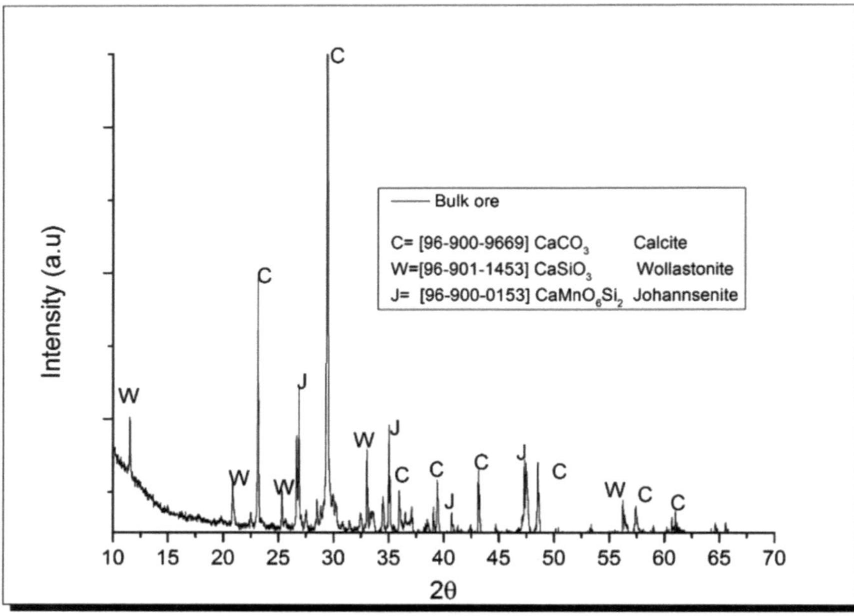

Fig. 4 X-ray diffractogram for bulk ore

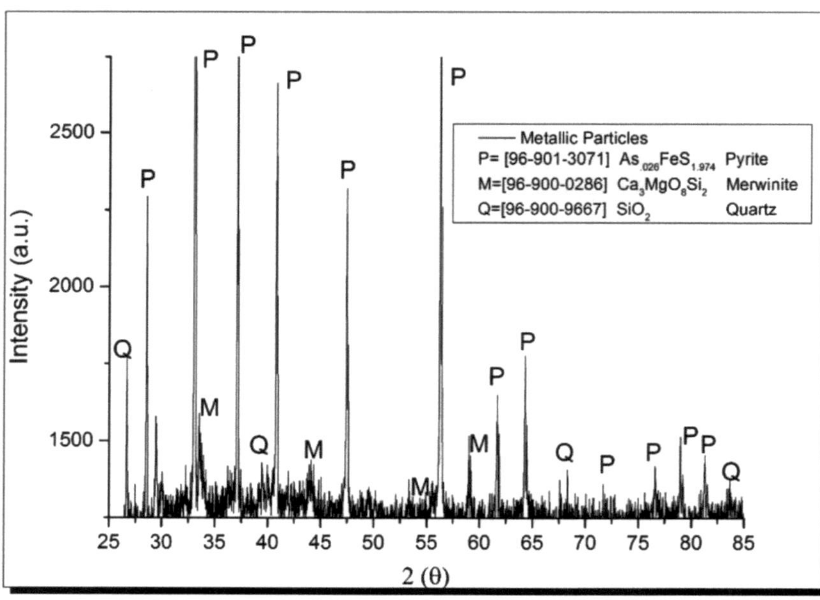

Fig. 5 X-ray diffractogram for metallic color particles

presence of wollastonite was detected; in the case of 38 μm. particle size, the sinnerita species ($As_4Cu_6S_9$) was also found and in the white particles, orthoclase species ($KAlSi_3O_8$) was observed. As can be seen, the diffractograms performed only found ore gangue but they could not show metal sulfides, except pyrite.

Inductive Coupled Plasma (ICP)

The results of the elemental analysis via ICP, for different mesh sizes, are shown in Table 1. As can be appreciated, sulfur (*S*) on average, is by far the most abundant element with 13.31%, followed by (Fe) with 2.55%. The remaining elements were found in lesser percentage: *Zn*, *Mg*, *As*, *Mn*, *Pb*. *Cu*, *Sb* and *Ag*. The high sulfur content suggests the presence of a variety of metal sulfides, mainly pyrite.

It can be appreciated that concentrations behave very homogeneous throughout the different particle sizes, so it is not possible to establish that there is an optimum size.

Fourier Transform Infrared Spectroscopy (FTIR)

This analysis is performed for the resulting insoluble material after acid digestion of the mineral. On average, this material represents a percentage higher than 50% of the sample. Figure 6 shows the infrared spectra for particle sizes 37 and 88 μm.

Table 1 Results of chemical composition by ICP

Particle size (μm)	Element percentage									
	Cu	Fe	Zn	Pb	Sb	Mg	Mn	S	Ag gr/ton	As
Bulk	0.03	2.64	0.36	0.06	0.01	0.21	0.09	12.18	0.09	0.08
<25	0.03	2.95	0.22	0.06	0.01	0.20	0.12	3.23	0.07	0.04
25	0.02	2.76	0.33	0.05	0.01	0.22	0.07	11.93	0.05	0.08
37	0.04	2.62	0.64	0.05	0.01	0.10	0.07	13.18	0.07	0.09
44	0.05	2.41	0.81	0.05	0.01	0.13	0.06	13.73	0.07	0.10
53	0.03	2.44	0.64	0.05	0.01	0.09	0.06	17.20	0.07	0.11
74	0.04	2.79	0.36	0.06	0.01	0.09	0.06	17.87	0.11	0.12
88	0.03	2.31	0.48	0.05	0.01	0.09	0.06	15.96	0.07	0.10
149	0.03	2.72	0.34	0.09	0.01	0.12	0.07	12.29	0.07	0.09
173	0.05	1.87	0.37	0.06	0.01	0.10	0.06	15.55	0.11	0.11
Average percentage	0.03	2.55	0.46	0.06	0.01	0.13	0.07	13.31	0.08	0.09

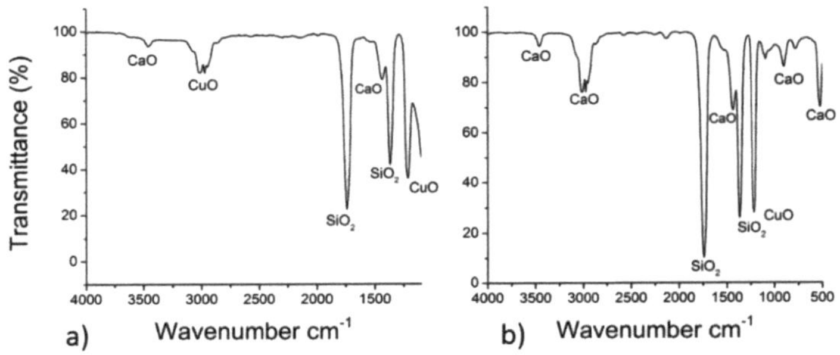

Fig. 6 Infrared Spectrum of insoluble material for size particle of: **a** 37 μm and **b** 88 μm

This analysis showed for all particle sizes, that most abundant compound in insoluble material after acid digestion is CaO, followed by SiO_2 and by lesser extent by CuO. These results are consistent, because it is known that the mineral matrix is calcite and which also has a high content of wollastonite.

Conclusions

Through the combination of different methods of characterization, it was possible to obtain a good overview of the content of the ore from the area of "El Carrizal", which has calcite as matrix and wollastonite in high proportion, the latter species are constituents of gangue material. However through the combined results of EDS, XRD and chemical analysis, it can be inferred that there is indeed presence of sulfides, of which the most abundant is pyrite. Since the sulfides are the minority species in the mineral composition, it is confirmed that the deposit is complex and low-grade, so that selective flotation is the only viable method to benefit the mineral. After having the knowledge of the species of the mineral, suitable reagents may be proposed to increase the recovery of *Pb*, *Cu* and *Zn* and decrease the presence of pyrite, arsenopyrite, calcite and other contaminants in the concentrates.

References

1. Canudas E (2005) Las venas de plata en la historia de México. México, Utopía
2. Servicio Geológico Mexicano, Panorama Minero del Estado de Hidalgo (México: 2014)
3. Espinosa E et al (2009) Geochemical distribution of arsenic, cadmium, lead and zinc in river sediments affected by tailings in Zimapan, a historical polymetalic mining zone of Mexico. Environ Geol 58:1467–1477
4. Servicio Geológico Mexicano, Anuario estadístico de la minería mexicana (México: 2012)
5. Villaseñor MG, Gómez JA Origen de la zonación en granates de la chimenea las Ánimas, Zimapán, Hidalgo, GEOS, 20(2):89–97

6. Moreno R, Téllez J, Monroy M (2012) Influencia de los minerales de los jales en la bioaccesibilidad de arsénico, plomo, zinc y cadmio en el distrito minero Zimapán, México. Rev. Int. Contam. Ambie, 203–218
7. Tossel J, Vaughan D (1992) Theoretical geochemistry: application of quantum mechanics in the earth and mineral sciences. Oxford University Press, New York
8. Hanumantha Rao K, Chernyshova I (2011) Challenges in sulphide mineral processing. Open Min Process J 4:7–13
9. Ralston J (1991) Eh and its consequences un sulphide mineral flotation. Min Eng 4(7):859–878
10. Holl C, Bromley A (1988) Application of mineralogy to the beneficiation of a tin-sulphide ore. Min Eng 1(1):19–30
11. Rao G (2004) Problems associated with complex sulphide ores and their beneficiation practices, Mineral characterization and processing. Allied Publishers, India, pp 140–142
12. Lynch A, Johnson N, McKee D, Thorne G. The behavior of minerals in sulphide flotation processes, with reference to simulation and control. J S Afr Inst Min Metall, 349–362
13. Hope G, Woods R, Munce C (2001) RAMAN microprobe mineral identification. Min Eng 14 (12): 1565–1577
14. Cook N (2000) Mineral characterization of industrial mineral deposits at the geological survey of Norway: a short introduction. Nor Geol 436:189–192
15. Chen M, Ma LQ (2001) Comparison of three aqua regia digestion methods for twenty florida soils. Soil Sci Soc Am J 65(2)

Contribution to the β Relaxation Study of the HDPE, LDPE and LLDPE

Washington Luiz Oliani, Duclerc Fernandes Parra, Luis Filipe Carvalho Pedroso Lima, Harumi Otaguro, Hélio Fernando Rodrigues Ferreto and Ademar Benevolo Lugao

Abstract The activation of three polyethylenes, the low density, the high density and the linear low density polyethylene, was determined by means of dynamic mechanical analysis (DMA). Storage modulus and tan δ spectra were obtained as a function of the temperature from −60 to 50 °C with a heating rate of 2° min^{-1} at frequencies of 1, 2 and 5 Hz and measurement strain amplitude of 10^{-5}. In the tan δ spectra, a small peak in the β region of the three polyethylenes was found and substantiated by subtracting the growing exponential background. This procedure was based on a model that associates the dislocation motion on metals under cyclic stress to vibrating strings and considering the same for the polymeric chains. For the determination of the activation energy we have utilized the frequency versus the reciprocal of the absolute temperature peak of the tan δ. The values obtained for the three polyethylenes, the low density, the high density and the linear low density were, 362 ± 12, 266 ± 9 and 197 ± 6 kJ mol^{-1}, respectively, which were compared with those presented in the literature.

Keywords Polyethylene · β relaxation · Activation energy

Introduction

To use materials adequately, it is essential to be well informed about its properties. This applies to polymers, and in particular, polyethylene (PE). In this case, various factors need to be taken into consideration to obtain information about its properties. Microstructural features of PE resulting from its synthesis and subsequent

W.L. Oliani (✉) · D.F. Parra · L.F.C.P. Lima · H.F.R. Ferreto · A.B. Lugao
Nuclear and Energy Research Institute—IPEN—CNEN/SP, Av. Professor Lineu Prestes, 2242, São Paulo, SP 05508—000, Brazil
e-mail: washoliani@usp.br

H. Otaguro
Universidade Federal de Uberlândia, Instituto de Química, Av. João Naves de Ávila, 2121, Uberlândia, MG, Brazil

S. Ikhmayies et al. (eds.), *Characterization of Minerals, Metals, and Materials 2017*, The Minerals, Metals & Materials Series,
DOI 10.1007/978-3-319-51382-9_68

processing, such as variation in the number of lateral chains of different lengths as well as differences in crystallinity in the three types of polyethylene—high density (HDPE), the low density (LDPE) and low density linear (LLDPE)—make it difficult to interpret the isochronal dynamic—mechanical analysis spectrum obtained as a function of temperature. This technique is appropriate to study phenomena associated with molecular movement (or relaxation) in these polymers. The spectrum obtained by this technique reveals three peaks or relaxations as a function of increasing temperature, namely, γ, β and α, respectively [1, 2].

Among these, the one that has been studied most, albeit least understood, is the β relaxation and this is often observed in LDPE, and to a much lesser extent or almost imperceptibly, in HDPE [3]. This fact and other relaxation processes that affect the spectrum, such as α relaxation located at higher temperatures, make it almost impossible to carry out a systematic study of parameters like activation energy associated with β relaxation [4].

The aim of this study was to infer the activation energy of the β relaxation of three kinds of polyethylene substantiating this relaxation by means of the subtraction of a growing exponential background from the tanδ. This contribution to the tanδ is due to the movement of polymeric chains considered as vibrating strings under cyclic stress [5, 6].

Experimental Section

Sample Preparation

Rods of three kinds of polyethylene were used as specimens in this experiment. In order to prepare the rods, pellets of HDPE, LDPE and LLDPE supplied by Braskem were heated to 190 °C for 10 min then the molding pressure was increased to 8 MPa and maintained for 5 min. Finally the mould and polymer were plunged into a bath filled with water at room temperature. The rod dimensions were $50 \times 8 \times 2$ mm^3.

Measurements

Viscoelastic Measurements

The melt flow index was measured by using a plastometer Ceast, at 190 °C with charge of 2.16 kg for the LDPE and LLDPE, and 5.00 kg for the HDPE (ASTM D1238).

The dynamic storage modulus and damping factor (tanδ) were carried out at frequencies of 1, 2 and 5 Hz over the temperature range from—60 to 50 °C with heating rate of 2° min^{-1} by using a dynamical mechanical analyzer (DMA 242) from Netzsch [7]. The length of the specimen between the support points was 40 mm and the dynamic amplitude 60 μm (strain ≈ 5×10^{-5}) (ASTM D4065).

Results

Melt Flow Index Measurements

The values of melt flow index obtained for HDPE, LDPE and LLDPE and their typical mean molecular weight are presented in Table 1. It can be noted that the smallest melt flow index 0.33, corresponds to the LDPE, which has the higher mean molecular weight and degree of branching. On the other hand, the polyethylene linear, HDPE, showed the higher melt flow index as expected.

Dynamic Mechanical Measurements

The results of the dynamic mechanical measurements for the three types of polyethylene are given in Figs. 1 and 2. Figure 1 shows the storage elastic modulus as a

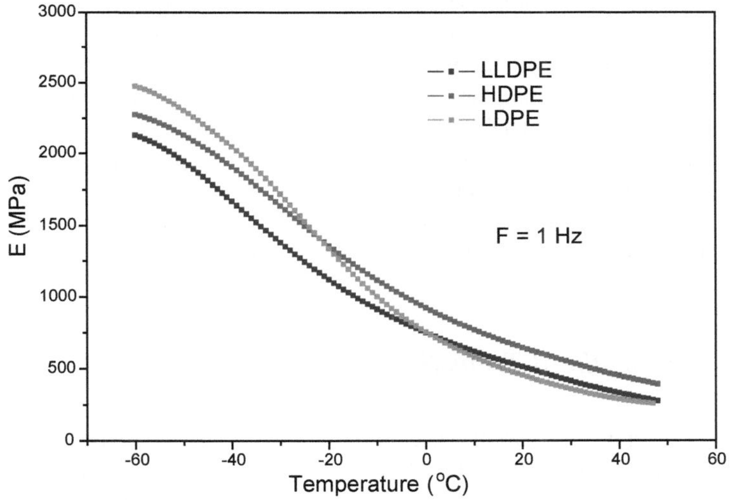

Fig. 1 Storage modulus as a function of temperature for the HDPE, LDPE and LLDPE

Table 1 Typical mean molecular weight of HDPE, LDPE, LLDPE, and respective melt flow index obtained in this work

Sample	Mean molecular weight (g mol^{-1}) [8]	Melt flow index (dg min^{-1})
HDPE	200,000	1.76
LDPE	500,000	0.33
LLDPE	150,000	0.69

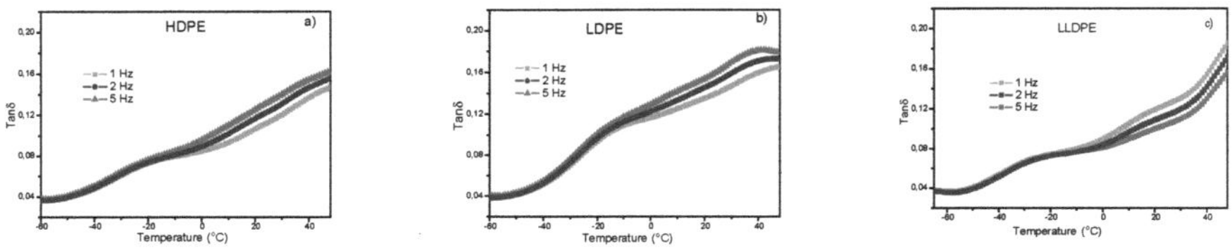

Fig. 2 Tan δ as a function of temperature for the three polyethylene's

Table 2 Storage modulus at −60 °C and at room temperature, storage modulus variation due to the β relaxation and β relaxation temperature for the three polyethylene's

Sample	E(MPa) (−60 °C)	E(MPa) (25 °C)	ΔE_β (MPa)	T_β (°C)
HDPE	2291	597	1694	−23 ± 1
LDPE	2481	409	2072	−16 ± 1
LLDPE	2127	466	1661	−28 ± 1

function of the temperature for HDPE, LDPE and LLDPE at the frequency of 1 Hz. Values of the storage modulus obtained at—60 °C and at room temperature (25 °C) and their variation due to the β relaxation are given in Table 2. These results indicate an important modulus variation of the LDPE that denotes a higher contribution of the amorphous phase to the relaxation process. On the other hand the modulus change of the HDPE is much lower but its elastic modulus at room temperature is the highest due to the crystallinity of this linear polyethylene.

Values of damping factor (tanδ) versus temperature for the PEAD, PEBD and LLDPE are given in Fig. 2a–c, respectively. As can be seen in these Figures the peak localized in the range of the β relaxation is not well defined.

Obtainment of the β-Transition Peaks

In order to display this peak the background considered as having an exponential growing behavior with temperature has been subtracted from the tanδ vs. temperature curve as shown in Fig. 3. In this figure are plotted the experimental curve for the HDPE obtained at the frequency of 1 Hz, the fitting background curve and β relaxation peak. The same procedure has been employed to obtain the β peak of the three polyethylene's at the frequencies of 1, 2 and 5 Hz.

The curves depicted in Fig. 4a–c, reflect the damping behavior due to the β relaxation for the HDPE, LDPE and LLDPE, respectively. The values of the β peak temperature are also shown in Table 3.

Fig. 3 Tanδ, exponential background and tanδ—exponential background as a function of temperature for the HDPE at 1 Hz

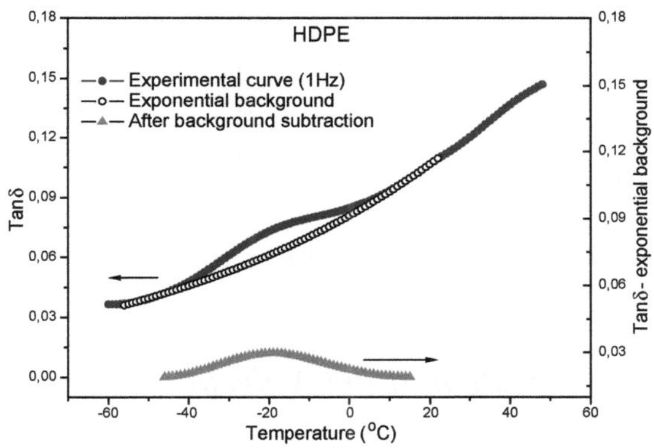

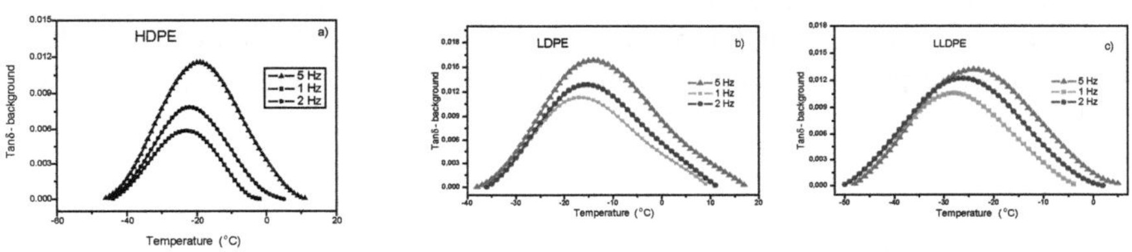

Fig. 4 Tanδ—exponential background as a function of temperature for the three polyethylene's: **a** HDPE, **b** LDPE and **c** LLDPE

Table 3 Activation energy (E_a) for polyethylene's and an isotactic polypropylene

Sample	E_a (kJ mol^{-1})	Frequency (Hz)	Reference
SCPE	235 ± 31^a	3.5;11 and 110	Mandelkern et al. [10]
LDBPE	200	0.01;0.1 and 1	Alberola et al. [11]
LPE	90	10	Nitta and Tanaka [12]
SCBPE	250	10	Nitta and Tanaka [12]
UHMWPE	77^b, 95^c	1; 3; 10 and 30	Zamfirova et al. [13]
BPE	114–115	10	Matsuo et al. [14]
UHMWPE	116	0.1–21.5	Goschel and Ulrich [15]
LDPE	185^d	10^3–10^6	Suljovrujic [17]
LLDPE	210^d	10^3–10^6	Suljovrujic [18]
HDPE	280^d	10^3–10^6	Suljovrujic et al. [19]
HDPE	266 ± 9	1; 2 and 5	This work
LDPE	362 ± 12	1; 2 and 5	This work
LLDPE	197 ± 6	1; 2 and 5	This work
iPP	380	0.01; 0.1 and 1	Jourdan et al. [20]

SCPE Short chain polyethylene, *LDBPE* low density branched polyethylene, *LPE* linear polyethylene, *SCBPE* short chain branched polyethylene, *UHMWPE* ultra high molecular weight polyethylene, *BPE* branched polyethylene, *iPP* isotactic polypropylene

[a]Calculated from the data presented for the D quenched sample on Table 2

[b]Calculated from E″ spectra

[c]Calculated from tanδ spectra

[d]Calculated from dielectric loss (tanδ) analysis

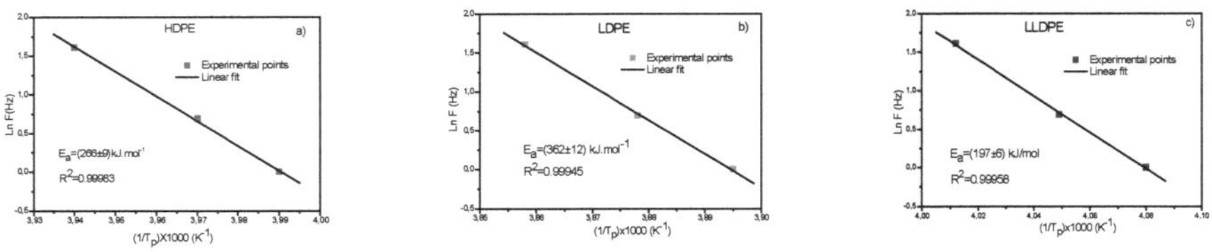

Fig. 5 Ln F as a function of the reciprocal of peak temperature for the three polyethylene's: **a** HDPE, **b** LDPE and **c** LLDPE

Determination of the Activation Energy

From the curves presented in Fig. 4, it has been possible to plot ln F as a function of the reciprocal absolute peak temperature for the HDPE, LDPE and LLDPE, shown in Fig. 5a–c, respectively, and determine their activation energy (E_a) presented in Table 3.

Comparison of the Three Polyethylene Spectra

In order to compare the β relaxation behavior of the three polymers the Fig. 6 shows the β peaks of these samples at the frequency of 1 Hz.

Discussion

Dynamic Mechanical Spectra

The presence of a varying number of side branches having different lengths and varying levels of crystallinity in the three polyethylenes is responsible partly for the differences of the tanδ presented in Fig. 6. The crosslinking of the amorphous phase by the crystallites, acting as physical ties, induces a decrease in the magnitude of the tanδ for the HDPE, which has the higher crystallinity level. On the other hand, the other two polyethylenes, LDPE and LLDPE, due to their branching chains that lower the crystallinity and improve the amorphous phase, show a higher β relaxation [3]. These two polymers due to their specific type of branching presented β relaxation temperatures very different (see Fig. 6 and Table 2). In LLDPE the short side branches in the backbone are responsible for the volume increase near the main chain which facilitates its motion and lowers the peak temperature [9] and activation energy. On the other hand, the higher T_β value of LDPE is due to the increase in size of the jumping unit of their molecular chain, which has short and

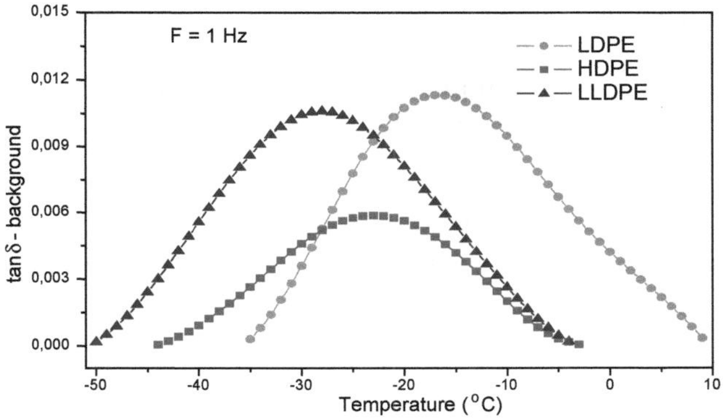

Fig. 6 Tanδ—exponential background as a function of temperature for the three polyethylene's at the frequency of 1 Hz

long side branches. These results agree with those of Khanna et al. [3] that observed that T_β increases with increasing molecular weight.

Activation Energy

Literature

Despite the extensive work that has been reported on polyethylene, few authors have determined the activation energy (E_a) of the β relaxation by dynamic mechanical methods. Table 3 shows the activation energies collected from the literature for various kinds of polyethylenes, for an isotactic polypropylene and from this work. The values for the polyethylene vary from 77 to 362 kJ mol^{-1} and the main factors responsible for this difference are the amounts and kind of branching. Moreover, slight changes could show up, depending on how spectra were analyzed. Nitta and Tanaka [12] and Matsuo et al. [14] obtained activation energy by the Arrhenius plots of logarithm of shift factor versus the reciprocal of the absolute temperature whereas the other authors and ourselves utilized the frequency versus the reciprocal of the absolute temperature peak of E″ or tan δ spectra [10, 11, 13, 15]. Another variable that affect the activation energy are the level or degree of crystallinity and the crystallite thickness. Zamfirova et al. [13] and Goschel and Ulrich [15] obtained 77 and 116 kJ mol^{-1} for UHMWPE with 60 and 52.2% of crystallinity, respectively. Nitta and Tanaka [12] obtained activation energy of 90 kJ mol^{-1} for the LPE sample with 55% of crystallinity and Matsuo et al. [14] with BPE films of low molecular weight with approximately 45.6 and 45.9% of cristallinity presented activation energies of 114 and 115 kJ mol^{-1}, respectively. These last results agree with the report of Kawai et al. [16] where the energy of β relaxation for linear polyethylene films became lower as crystallinity increase and the activation energy became less than 100 kJ mol^{-1}. The quenched samples of SCBPE analyzed by Mandelkern et al. [10] and by Nitta and Tanaka [12] that presented $E_a \approx 250$ kJ mol^{-1} can be compared because they have $M_w \approx 10^5$, narrow molecular weight distribution 1.31 and 1.79, and 36 and 34% of crystallinity, respectively.

The results obtained by the Suljovrujic group that studied the dielectric behaviour of different polyethylene's (LDPE, LLDPE and HDPE) through dielectric loss (tanδ) analysis are also presented and will be compared with our results in next section.

Our Results

Concerning the mechanical relaxation the values of the activation energy obtained in this work are higher than those presented in the literature with exception of the LLDPE value of 197 ± 6 kJ mol^{-1} that can be compared with (235 ± 31) kJ mol^{-1} [10] and 250 kJ mol^{-1} [12] of the SBCPE samples because the LLDPE has short chain branches too, all the samples have been quenched and their molecular weight is approximately 10^5 g mol^{-1}. These values are lower than the HDPE value of (266 ± 9) kJ mol^{-1} due to the fact that the short branches increase the volume and facilitates the backbone motion as pointed out in "*Discussion: Dynamic mechanical spectra*". On the other hand, the higher value of the activation energy of LDPE, (362 ± 12) kJ mol^{-1}, is due to the size of the jumping unit of their molecular chains, which have short and long branches. Comparing the activation energy of 380 kJ mol^{-1} calculated from isochronal measurements for the isotactic polypropylene by Jourdan et al. [20] and associated by them to the glass transition with our values, we are tempted to regard the β relaxation of the polyethylene as its glass transition, as suggested by Khanna et al. [3]. In that case the values of the peak temperature of the β relaxation of the three polyethylenes presented in Table 2 will be their glass transition temperatures.

In respect to the results obtained with LDPE, LLDPE and HDPE by Suljovrujic et al. summarized in [19] we can conclude that with the exception of LDPE the other two activation energies agree well to our results. The value $E_a = 185$ kJ mol^{-1} for the LDPE is much lower than the value $E_a = 362 \pm 12$ kJ mol^{-1} obtained in this work. In the work concerned with this polyethylene [17] the author observed that drawing and irradiation processes increased the crystallinity from 27 to 34% (our sample 33%) and influenced strongly the activation energy that attained values higher than 300 kJ mol^{-1}.

Conclusion

In this work we have taken advantage of a model that associates the dislocation motion under cyclic stress in metals to vibrating strings. By analogy, considering the polymeric chains as strings we could substantiate β relaxation in polyethylenes and then determine their activation energy. These values have been compared among themselves and with those of the literature demonstrating an activation energy variation from 77 to 362 kJ mol^{-1}. According to the values of the activation energy the samples can be divided in two groups. The first one presented values approximate to 100 kJ mol^{-1} and, the second, values higher than 200 kJ mol^{-1}. The samples of the former group are linear in its majority, presented high degree of crystallinity and mean molecular weight higher than 10^6 g mol^{-1}. On the other hand, the samples of the latter group with exception of HDPE are branched, with short branches (SCBCP, LLDPE) and short and long branches (LDPE), low crystallinity and mean molecular weight lower than 10^6 g mol^{-1}.

The samples analyzed by dielectric studies showed activation energies with values close to ours results for the HDPE and LLDPE, but not to the LDPE, which supports the fact that the activation energy in this polyethylene is strongly dependent upon changes in microstructure of the amorphous phase.

The samples presented in this work showed a β relaxation whose intensity and temperature increase with increase in amorphous volume by branching, and increasing molecular weight, respectively. The activation energies of this work compared with the activation energy of β relaxation of isotactic polypropylene (380 kJ mol^{-1}), consider as its glass transition, and suggest that the β relaxation could be related to the glass transition of the polyethylene.

Acknowledgements The authors are grateful to FAPESP process nº 1995/5630-2 and CAPES-Process: CSF-PVE's—88887.115684/2016-00.

References

1. Woodward AE, Sauer JA (1958) The dynamic mechanical properties of high polymers at low temperatures. Adv Polym Sci 1, Bd.1.S:114—158
2. McCrum NG, Read BE, Williams G (1967) Anelastic and dielectric effects in polymeric solids. Wiley, New York
3. Khanna YP, Turi EA, Taylor TJ, Vickroy VV, Abbott RF (1985) Dynamic mechanical relaxations in polyethylene. Macromolecules 18(6):1302–1309
4. Fried JR (2007) Physical properties of polymers handbook, (Edited by James E. Mark. New York: Springer; 2007), p 217
5. Granato A, Lucke K (1956) Thory of mechanical damping due to dislocations. J Appl Phys 27 (6):583–593
6. Granato A, Lucke K (1956) Application of dislocation theory to internal friction phenomena at high frequencies. J Appl Phys 27(7):789–805
7. Netzsch—Dynamic Mechanical Analyser (1998) DMA 242, Instruction manual
8. Schramm G (1998) A practical approach to rheology and rheometry. Gebrueder Haake GmbH, Karlsruhe, p 134
9. Plazek DJ, Ngai KL (2007) Physical properties of polymers handbook, (Edited by James E. Mark. New York: Springer, 2007), p 187
10. Mandelkern L, Glotin M, Popli R (1981) The influence of crystalline morphology (supermolecular structure) on the dynamic mechanical properties of polyethylene. J Polym Sci: Polym Lett Ed 19(9):435–441

11. Alberola N, Cavaille JY, Perez J (1992) Mechanical γ and β relaxations in polyethylene—I. Glass transitions of polyethylene. Eur Polym J 28(8):935–948
12. Nitta K, Tanaka A (2002) Dynamic mechanical properties of metallocene catalyzed linear polyethylene's. Polymer 42(3):1219–1226
13. Zamfirova G, Pereña JM, Benavente R, Perez E, Cerrada ML, Nedkov E (2002) Mechanical properties of ultra high molecular weight polyethylene obtained with different cocatalyst systems. Polym J 34:125–131
14. Matsuo M, Bin Y, Xu C, Ma L, Nakaoki T, Suzuki T (2003) Relaxation mechanism in several kinds of polyethylene estimated by dynamic mechanical measurements, pósitron annihilation, X-ray and ^{13}C solid-state NMR. Polymer 44(15):4325–4340
15. Goschel U, Ulrich C (2009) Mechanical relaxation of medical grade UHMWPE of different crosslink density as prepared by electron beam irradiation. J Appl Polym Sci 113(1):49–59
16. Kawai H, Suehiro S, Kyu T, Shimomura A (1983) Rheo—optical properties of spherulitic polyethylene's in relation to the alpha and beta mechanical dispersions. Polym Eng Rev 3 (2–4):109–196
17. Suljovrujic E (2002) Dielectric studies of molecular β-relaxation in low density polyethylene: the influence of drawing and ionizing radiation. Polymer 43:5969–5978
18. Suljovrujic E (2005) Some aspects of structural electrophysics of irradiated oriented LLDPE. Nucl Instrum Methods Phys Res B 232:399–406
19. Suljovrujic E, Micic M, Milicevic D (2013) Structural changes and dielectric relaxation behavior of uniaxially oriented high density polyethylene. J Eng Fibers Fabr 8(3):131–143
20. Jourdan C, Cavaille JY, Perez J (1989) Mechanical relaxations in polypropylene: a new experimental and theoretical approach. J Polym Sci, Part B: Polym Phys 27(11):2361–2384

Determination of Ten Impurity Elements in Tin Concentrate and Smelting Products by ICP-AES

Yunke Wang, Ping Long, Jian Wu, Wenli Zhang, Peipei Liu, Xinlin Ren and Bin Yang

Abstract In this research, ICP-AES was used for determining ten impurity elements (Fe, Pb, Ca, Mg, Cu, Zn, Al, As, Sb, and Bi) in tin concentrate and smelting products, such as crude tin and slag. Samples were dissolved by hydrochloric acid and nitric acid. Determination conditions were investigated in detail. Influences of matrix and coexisting elements were eliminated by selecting proper analysis spectral lines. Under the optimum conditions, the correlation coefficient of calibration curves were higher than 0.999. The detection limits of these ten impurities were in the range between 0.0009 to 0.0621 μg/ml. Relative standard deviation (RSD) values were lower than 6.0%, and the recoveries of standard addition were between 85 and 109%. The proposed method was demonstrated to be reliable by comparing with standard analysis methods and certified values of standard sample. It could be applied for determining impurity elements in tin concentrate and smelting products successfully.

Keywords Inductively coupled plasma atomic emission spectrometry · Tin concentrate · Crude tin · Slag · Impurity elements

Y. Wang (✉) · P. Long (✉) · J. Wu · W. Zhang · P. Liu · B. Yang
National Engineering Laboratory for Vacuum Metallurgy,
Kunming University of Science and Technology, Kunming 650093,
Yunnan, China
e-mail: 2463047321@qq.com

P. Long
e-mail: pkulongping@sohu.com

X. Ren
College of Chemical Engineering, Kunming University of Science and Tecnology,
Kunming 650500, Yunnan, China

P. Long · J. Wu · W. Zhang · B. Yang
State Key Laboratory of Complex Nonferrous Metal Resources Clean Utilization,
Kunming 650093, Yunnan, China

S. Ikhmayies et al. (eds.), *Characterization of Minerals, Metals, and Materials 2017*, The Minerals, Metals & Materials Series,
DOI 10.1007/978-3-319-51382-9_69

Introduction

Tin has been applied continuously since the third century B.C [1], because it can be melted at a low temperature. In modern times, it is used in many alloys or tin soft solders, which contain 60% tin or more. Tin is also largely applied in the technology of corrosion-resistant tin-plated steel [2]. Tin-plated metal is used as tin cans in food packaging field for its low toxicity. It is also known as "industrial monosodium glutamate" for its great appearance and ductility [3]. Nowadays, the shortage of tin ore resources leads to the decrease of tin grade and the increase of complicate composition in tin ore [4, 5]. Lots of impurity elements, such as Fe, Pb, Ca, Mg, Cu, Zn, Al, As, Bi, and Sb, are brought to tin smelting products through smelting process. Therefore, fast and accurate determination methods of impurity elements in tin concentrate and its smelting products are important to improve and monitor the production process of tin.

According to current Chinese National Standards or traditional methods of determining the impurities in tin products, Fe is determined by ceric sulfate method, Pb is determined by flame atomic absorption spectrometry (AAS) and EDTA titration method, Cu is determined by flame atomic absorption spectroscopy method, Al_2O_3 is determined by chromium azure S spectrophotometric method, As is determined by arseno-antimono-molybdenum blue spectrophotometry and distillation iodometric method [6–10], respectively. In these methods, different elements are determined from different process, leading to a tedious and inefficient operation process.

In recent years, lots of determination methods, such as energy dispersive X-ray fluorescence spectrometry (ED-XRF) [11], hydride generation atomic fluorescence spectrometry (HG-AFS) [12], and inductively coupled plasma atomic emission spectrometry (ICP-AES), have been reported to determine multi-element in the tin concentrate or tin ore simultaneously. However, ED-XRF method has the disadvantage of high detection limit [13]. The technique HG-AFS can only be applied to determine a small part of elements, and the linear range is narrow. ICP-AES have lots of advantages, such as high sensitivity, low detection limit, less interference, and wide linear range. For these reasons, simultaneous determination of multi-element in the tin concentrate or tin ore by ICP-AES has become one of the important detection means in the modern analytical techniques. Tuo Li and J. H Yang studied the detection of Sn, Cu, Pb, Fe in tin concentrate by ICP-AES method, which obtained the great recoveries of 92.0–101.6% and the low detection limits of 0.006 μg/ml [14]. H. L Yang et al. have simultaneously determined Sn, W, Mo, Cu, Pb and Zn in tin ores by ICP-AES [15]. The result of relative standard deviation (RSD) was less than 5.0%, which showed that the method was of good precision. A method of determining impurity elements, including Pb, As, Cu, Zn, Bi, by ICP-AES was established by Y.X Chen et al. When they used this method to analyze the standard sample and actual sample, the results were in accordance with the certified values, as well as the values obtained by other determination methods within the allowable error range [16]. However, few researches on the simultaneous determination of ten elements (Fe, Pb, Ca, Mg, Cu, Zn, Al, Sb, As, Bi) in tin concentrates and its smelting products (coarse tin, slag) were reported.

In this paper, ten impurity elements in tin concentrates and their smelting products have been determined simultaneously by inductively coupled plasma-atomic emission spectrometer (ICP-AES) method. The dissolving processes of tin samples were investigated firstly. Afterwards, the operating conditions of the instrument, such as analysis of the line selection, spectral interference, etc., were studied in detail. Precision and accuracy of this method were discussed finally. This analytical method of impurities in tin products is fast, simple and reliable, and is expected to be applied as the common used method in smelting process of complex tin ore greatly.

Experimental

Instruments and Operating Conditions

Ten elements (Fe, Pb, Ca, Mg, Cu, Zn, Al, As, Bi, and Sb) in the samples were determined by ICP-AES (OPTIMA 8000, PerkinElmer Corporation, America). The operating parameters of the ICP-AES are given in Table 1. Atomic Absorption Spectrophotometer (AAS) (WFX-320, Beijing Rayleigh Analytical Instrument Co., Ltd) was also used in this research. Milli-Q water purification system (Millipore Corp., France) was used to prepare ultrapure water (18.2 M cm).

Experimental Material and Reagents

Tin concentrate and its smelting products (coarse tin and slag) were collected from YUNNAN TIN GROUP. All the reagents used in dissolving process are shown as follow: nitric acid, hydrochloric acid, and ammonium fluoride (guarantee reagent, ChengDu KeLong Chemical Co., Ltd). Mixed standard solution (100 μg/ml) of the ten elements was purchased from NCS Testing Technology Co., Ltd. A series of standard working solutions were prepared from the mixed standard solution mentioned above and were used to plot the standard curve.

Table 1 ICP-AES operating conditions

Parameters	Parameters value
Power	1300 W
Plasma gas-flow	15 L/min
Aux. flow	0.55 L/min
Nebulizer	0.2 L/min
Viewing height	15 mm
Pump rate	1.5 ml/min

Experimental Methods

0.3 g (accurate to 0.0001 g) sample was taken into 250 ml Teflon beaker and 15 mL hydrochloric acid was also added. The beaker with sample was heated on the electric furnace at low temperature (about 300 °C) for 5 min. Next, 5 mL nitric acid was added into the beaker. Then the beaker was heated continuously until the sample was dissolved completely. Took it off, and cooled it in the air for a few minutes, later, 10 mL HCL (1 + 1) was added into the beaker, after that the solution was heated continuously until it was nearly dry. After cooling down to room temperature, the solution was transferred into 100 mL volumetric flask and diluted with ultra pure water to volume. Meanwhile, blank test was carried out in the same process.

Results and Discussion

Investigation of Sample Dissolution Method

Tin concentrate and smelting product (coarse tin, slag) samples were dissolved by aqua regia [V (HNO_3):V (HCL) = 3:1)]. In order to investigate the effect of ammonium fluoride, ammonium fluoride was added as masking agent in the sample dissolving process. It is found that the calcium concentration decreased. The formation of calcium fluoride might be the reason of this abnormality. Therefore, ammonium fluoride could not be added in the dissolving process of samples. When the sample solution was nearly dry, the solution of coarse tin and slag became transparent, which indicated that coarse tin and slag in the sample were dissolved completely. Although a few undissolved residues were observed in the solutions of tin concentrate standard samples and tin concentrate, the test results showed that the residues had no effect on the determination of target elements. Therefore, this dissolution method by ICP-AES is reliable for determining of ten impurity elements in tin concentrate and smelting products.

Selection of Spectral Line and Interference of Coexisting Elements

The main components of the tin concentrate, coarse tin, and slag samples are Sn, Fe, Si, Ca, Pb. Interferences of non-determination elements should be avoided in selection process of spectral line. Lots of spectral lines are provided for each determination element in instrument system. Considering various facts, such as noise-signal ratio (SNR), the intensity of signal, background equivalent concentration (BEC) and detection limit, the analytical spectral lines of ten elements

Table 2 Analytical wavelengths

Elements	Fe	Pb	Ca	Mg	Cu	Zn	Al	As	Sb	Bi
Wavelengths/(nm)	238.2	220.3	422.7	285.2	317.9	206.2	396.1	193.7	206.8	223.0

(Fe, Pb, Ca, Mg, Cu, Zn, Al, As, Bi, and Sb) with high sensitivity and low spectral interference were selected. Analytical wavelengths of these elements are given in Table 2.

The mass fractions of the main component elements (Sn, Fe, Si, Ca, Pb) were 20–70%, 16–23%, 0.1–9%, 0.2–15%, 0.3–5%, respectively. In order to examine the interference effects of coexisting elements, single element standard solutions (1000 μg/ml) of Sn, Fe, Si, Ca, Pb were tested under the wavelength of the under-determined element. The results showed that contents of these elements were less than 0.2 μg/ml, which indicated that there was little interference. This kind of interference belongs to background interference, and which can be eliminated by the high resolution of the instrument and the appropriate background deduction.

Regression Equation, Correlation Coefficients and Detection Limits

Under the optimum working conditions of the instrument, standard curves of each element were plotted. In order to calculate standard deviation of each element, blank sample solution was determined for 11 times. Three times value of standard deviation were used as detection limit. The linear equation, linear correlation coefficients, and detection limits of these elements are shown in Table 3.

Table 3 Regression equation, correlation coefficients and detection limits

Elements	Linear regression equation (k = 1000 cps)	Correlation coefficient	Detection limit (μg/ml)
Fe	$y = 57.29\ kx - 6.9175\ k$	0.9998	0.0036
Pb	$y = 23.36\ kx - 0.0108\ k$	0.9992	0.0336
Ca	$y = 95.85\ kx - 3.8236\ k$	0.9998	0.0621
Mg	$y = 1238\ kx - 444.4\ k$	0.9999	0.0012
Cu	$y = 398.2\ kx + 39.82\ k$	0.9999	0.0009
Zn	$y = 80.44\ kx - 0.046\ k$	0.9999	0.0009
Al	$y = 524.5\ kx - 108.59\ k$	0.9999	0.0123
As	$y = 2.968\ kx + 0.5123\ k$	0.9998	0.0579
Sb	$y = 7.487\ kx - 0.181\ k$	0.9999	0.0090
Bi	$y = 19.42\ kx - 0.5102\ k$	0.9990	0.0039

Accuracy and Precision

In order to test the accuracy and precision of this determination method, relevant experiments were conducted. The standard solutions with certified concentration values were added into the sample solutions to obtain the recovery rates and test accuracy. It can be seen from Table 4 that recovery rates are between 85 and 109%. Relative standard deviations(RSD) of ten elements (Fe, Pb, Ca, Mg, Cu, Zn, Al, As, Bi, and Sb) in tin concentrate raw materials, crude tin and slag are lower than 6%. These values indicate the great accuracy and precision of this method. Meanwhile, parts of these elements in tin standard sample were determined by this method. The results are partly shown as certified values rank in Table 4. The other elements in experimental samples were also determined by other feasible determination methods for comparison. The results are given as reference values column in Table 4. It should be noted that the certified values column and reference values column are put together for limited space. The notes on these two groups of values are stated in detail at the end of Table 4. These values coincide with each other greatly, which prove that this determination method is accurate and reliable.

Table 4 Test results for recovery, precision and analytical results of samples

Elements	Sample	Certified/reference# value (w/%)	Sample found (w/%)	Added (w/%)	Total found (w/%)	Recovery (w/%)	RSD (n = 6) (w/%)
Fe	Tin concentrate standard samples	9.53	9.48	–	–	–	–
	Tin concentrate	13.15	13.20	13.20	26.65	102	1.2
	Coarse tin	20.37#	20.10	13.38	33.40	99	0.93
	Slag	21.75#	21.40	13.31	33.57	91	0.56
Pb	Tin concentrate standard samples	1.62	1.61	–	–	–	–
	Tin concentrate	3.46#	3.40	3.31	6.38	90	2.7
	Coarse tin	4.61#	4.45	3.34	7.61	95	2.2
	Slag	0.30#	0.27	0.33	0.55	85	6.0
Ca	Tin concentrate standard samples	–		–	–	–	–

(continued)

Table 4 (continued)

Elements	Sample	Certified/reference[#] value (w/%)	Sample found (w/%)	Added (w/%)	Total found (w/%)	Recovery (w/%)	RSD (n = 6) (w/%)
	Tin concentrate	0.93[#]	0.96	0.66	1.58	94	3.1
	Coarse tin	0.20[#]	0.21	0.33	0.55	103	2.5
	Slag	16.61[#]	16.90	13.30	28.90	90	1.3
Mg	Tin concentrate standard samples	–		–	–	–	–
	Tin concentrate	0.31[#]	0.33	0.66	0.93	91	5.2
	Coarse tin	0.032[#]	0.034	0.067	0.10	99	4.9
	Slag	2.75[#]	2.56	1.66	4.16	96	4.5
Cu	Tin concentrate standard samples	0.043	0.043	–	–	–	–
	Tin concentrate	0.10[#]	0.11	0.066	0.18	106	5.6
	Coarse tin	0.20[#]	0.21	0.18	0.39	100	3.5
	Slag	0.035[#]	0.037	0.033	0.073	109	4.2
Zn	Tin concentrate standard samples	0.12	0.14	–	–	–	–
	Tin concentrate	0.35[#]	0.34	0.33	0.63	88	3.8
	Coarse tin	0.00090[#]	0.00083	0.0033	0.0043	105	2.3
	Slag	0.0038[#]	0.0035	0.0033	0.0069	103	2.3
Al	Tin concentrate standard samples	–		–	–	–	–
	Tin concentrate	–	0.15	0.17	0.31	94	4.0
	Coarse tin	–	0.032	0.067	0.10	101	5.2
	Slag	–	1.52	1.66	3.25	104	2.3
As	Tin concentrate standard samples	0.31	0.33	–	–	–	–
	Tin concentrate	–	0.16	0.17	0.33	100	3.2

(continued)

Table 4 (continued)

Elements	Sample	Certified/reference# value (w/%)	Sample found (w/%)	Added (w/%)	Total found (w/%)	Recovery (w/%)	RSD (n = 6) (w/%)
	Coarse tin	–	0.17	0.067	0.24	104	1.9
	Slag	–	0.030	0.033	0.065	106	5.8
Sb	Tin concentrate standard samples	0.016	0.013	–	–	–	–
	Tin concentrate	0.0045#	0.0046	0.0063	0.010	86	2.2
	Coarse tin	0.031#	0.030	0.067	0.087	85	2.5
	Slag	0.0085#	0.0080	0.017	0.023	88	4.7
Bi	Tin concentrate standard samples	0.02	0.018	–	–	–	–
	Tin concentrate	0.0092#	0.0090	0.033	0.037	85	3.0
	Coarse tin	0.013#	0.012	0.033	0.040	85	2.8
	Slag	–	–	0.033	0.032	97	–

Note (1) The meaning of "#" is a reference value. Among of them, iron was determined by cerium sulfate titrimetric method, and Pb, Ca, Mg, Cu, Zn, Sb, Bi was determined by AAS method; (2) The series number of tin concentrate standard samples is GBW 7232

Conclusions

In this research, ICP-AES method is used to determine the contents of Fe, Pb, Ca, Mg, Cu, Zn, Al, As, Bi, and Sb in tin concentrate and its smelting products (coarse tin and slag). A kind of feasible sample dissolving method is obtained in the pretreatment process. The detection limits of these ten impurities are in the range between 0.0009 to 0.0621 μg/ml. Relative standard deviation (RSD) values are lower than 6.0%, and the recoveries of standard addition are between 85 and 109%. A fast, reliable, and precise determination method has been established, and which is expected to be applied as the common used determination method in smelting process of complex tin ore.

Acknowledgements This work was subsidized by the National Basic Research Program of China (No. 2012CB722803).

References

1. Cierny J, Weisgerber G (2003) Bronze Age tin mines in Central Asia. The Problem of Early Tin, 23–31
2. Hua YiXin (1986) Introduction to nonferrous metallurgy. Metallurgical Industry Press, Bei Jing, BJ, pp 124–129
3. Integrated branch of Ganzhou Municipal Commission of industry and information technology, "Guidance on the Development of Tin Industry in Ganzhou" (Report B30040-0202-2011-0008, Huichang County People's Government, 2011)
4. Izard CF, Muller DB (2010) Tracking the devil's metal: historical global and contemporary US tin cycles. Resour Conserv Recycl 10:1436–1441
5. Ronguo C, Liu SC (2010) Analysis of tin resources and international competitiveness in China. Land Resour Inf, 29–33
6. GB/T1819.3-2004, Methods for chemical analysis of tin concentrate
7. GB/T1819.4-2004, Methods for chemical analysis of tin concentrate
8. GB/T1819.14-2004, Methods for chemical analysis of tin concentrate
9. GB/T1819.11-2004, Methods for chemical analysis of tin concentrate
10. GB/T1819.5-2004, Methods for chemical analysis of tin concentrate
11. Tan ZM, Huang XM, Huang X (2009) Energy dispersion X-ray fluorescence spectrometry determination of tin concentrate arsenic, zinc, iron and copper. Res Appl Mat, 3(4):287–289
12. Yuan YH, Yin CH, Yuan ZH, Xu H (2016) Simultaneous determination of arsenic and antimony by hydride generation atomic fluorescence spectrometry in tin ore. Metall Anal 36 (3):39–43
13. Zhang LY, Dai T (2008) Present status of energy dispersive X ray fluorescence spectrometer. Modern Instrum 14(5):50–53
14. Yang JH, Li T (2012) Determination of tin, copper, lead and iron in tin concentrates by ICP–AES. Chem Anal Meterage 6:73–75
15. Yang HL, Hui X, Du TJ, Bai L, Qin JH,, Tie JL (2013) Simultaneous determination of tin, tungsten, molybdenum, copper, lead, zinc in tin ore by inductively coupled plasma atomic emission spectrometry. Rock Miner Anal, 12:887–892
16. Chen YX, Lv ZE, Liu SQ, Liu GW (2008) Inductively coupled plasma atomic emission spectrometric determination of impurity elements in tin concentrate. Metall Anal, 4:20–22

Effects of Wet Grinding on the Structure and Granularity of Biological Origin Aragonite and Its Polymorphic Transformation into Calcite

Yunhui Tang and Mingsheng He

Abstract The structure and granularity of biological origin aragonite from abalone shell during wet grinding in a ball mill were studied with X-ray diffractometer, with morphology studied by scanning electron microscope (SEM). It is shown from the results that the particle sizes of aragonite decreased with grinding time, so did the X-ray diffraction (XRD) peak intensities, while the full widths at half maximum increased. A calcite XRD peak with low intensity and great width was observed in nearly all the samples with various grinding time, whereas the intensity of the peak increased obviously after grinding time 300 min. It is shown from the SEM results that the size of aragonite dropped with grinding time until it reached 20 nm and then kept constant, while that of calcite dropped first and then rose at the minimum about 10 nm until to the constant size as in aragonite. It was observed that the mineral content obtained from XRD show an obviously change that the calcite increases while aragonite decreases, which indicated the occurring of a polymorphic transformation of aragonite into calcite.

Keywords Wet grinding · Aragonite · Calcite · Structure · Granularity · Phase transformation

Introduction

It is well-known that mechanical effects on mineral particles during grinding may result in physical changes on both morphology and structure of crystal, and even chemical reactions in the raw materials [1]. Therefore, the study of changes on

Y. Tang (✉)
Materials Sciences and Engineering College, Beijing University of Technology, Beijing 100124, People's Republic of China
e-mail: tangyh@bjut.edu.cn

M. He
R&D Center of WISCO, Wuhan 430080, China

S. Ikhmayies et al. (eds.), *Characterization of Minerals, Metals, and Materials 2017*, The Minerals, Metals & Materials Series,
DOI 10.1007/978-3-319-51382-9_70

mineral particles during grinding has practical significance in fields such as food processing, pharmaceutical preparation, cosmetics manufacturing, etc.

Aragonite and calcite are the most common crystals studied in researches on grinding of calcium carbonate salts. Calcite is considered as the most stable phase of calcium carbonate in the nature and aragonite can spontaneously change into calcite in higher temperature. Burns et al. found calcite could change into aragonite during mechanical grinding first in 1956 [2]. Subsequently Momota proved that during mechanical grinding aragonite could change into calcite [3]. Since then on, more and more researchers verified the alteration between aragonite and calcite at room temperature [3–6]. It was observed that the time for calcite transforming into aragonite (e.g., several hundreds hours) is much longer than that for aragonite into calcite (several minutes to several hours). However, there are fewer credible opinions or detailed data deeply into the mechanism of the phase transformation by far [7]. Comparing with inorganic-origin calcite, both of the biological-origin calcium carbonates show more thermal stability as embedding in organic materials [8, 9]. Recently Wan et al. [10] found a reversible aragonite-calcite transformation during vibration mill grinding, but with no enough data being described in detail.

In this paper, wet grinding by ball mill was chose for biological-origin aragonite from abalone shell, aiming to study the temporal changes of structure, granularity and phase transform during the wet grinding process. The results might be useful for practical utilization of ultra-fine calcium carbonate as biomimetic materials.

Materials and Methods

The samples we used are abalone shells with white luster or rainbow sheen, purchased from Dalian market, P.R. China. It is well-known that the abalone shell is composed of the inner nacreous layer (aragonite), the prismatic layer (calcite) and outside organic layer. A two-step preprocessing treatment was applied. The first step is to get rid of muscle tissues and then cleaning, dried up in cool environment. The second step is to get rid of the brown organic addition and calcite prismatic layer on the surface by grinding wheel so as to obtain the inner nacreous layer for experiments. After the two-step treatment, the samples are almost pure aragonite composition, and ready for selective X-ray diffraction check.

The samples were ground by using a SFM-3 type high-vibration ball-mill with a motor operating at a voltage of 380 V, a power of 1100 W; the rotational speed was 1400 r/min. The ball mill had two 25-ml stainless steel pots with some 15 mm-diameter stainless steel balls. At regular intervals, all powders removed from the grinding pot were dried up in the shade and passed through a 200 mesh sieve. And then 0.2 g powder was taken for test, and the residual parts were put back into the pot and continue to grind with alcohol for cooling.

The powder X-ray diffraction spectrums (XRD) were collected with a Shimadzu XRD-7000 type X-ray diffractometer at Beijing University of Technology, with Cu_k X-ray of 1.5406 Å wavelength and scanning speed of 2°/min from 20° to 80° (2θ) at 40 kV and 30 mA in a 0.02° step. The data treatment was done with Jade 6.0 program from Materials Data Inc., including phase identification, quantity calculation, crystallite size calculation and cell parameter calculation. Imaging of particles was collected with A Hitachi S-3400N scanning electron microscope (SEM) at Beijing University of technology. The samples were coated with gold before imaging.

Results and Discussions

Particle Morphology

The particle morphology and average particle size of the powder materials changed obviously during the initial stage of grinding, as shown in SEM image (Fig. 1). At about 1 min., there appear irregular fragments nearly only by mechanical broken rather than grinding, which is composed of coarse fractions and fine particles fractions. By grinding time about 30 min., it is shown that the main components are fine fragments and granular particles with broken edge. By time 300 min., there are finer particles evenly distributed with some sphere-like aggregate. In addition, by time 600 min., numerous ultra-fine particles merge together forming sphere-like aggregate, indicating aggregation formation of ultra-fine particles in the grinding process. The breakage and aggregation of particles are the two opposing processes during grinding. Aggregation spontaneously happened during grinding, so as to reduce the surface energy consumption and keep the system stable. During the aggregation stage, particles bond together via van der Waals force and form agglomerates, which can be dispersed by dispersing agents.

Changing of Aragonite Structure

It is shown from the XRD spectrums that there are aragonite's characteristic peaks with sharp shapes, representing the crystal plane (111), (021), (002), (012), (200), (031), (112), (022), (211), (122), (221), (202), (132), (113), (023), indicating well crystalline (Fig. 2). However, compared with the starting material, the diffraction peaks are obviously weakened and broadened during grinding.

The three strongest diffraction peaks, (111), (021) and (012), indicated similar fluctuation of their absolute strength values for both crystal breakage and structure damage (Fig. 3a). For samples from the first 300 min of grinding, the diffraction peak strength fluctuated first and then decreased sharply, which may be due to oriented crystal plane's uneven distribution caused by breakage and structure damage during breakage. In the subsequent time, the strengths decreased evenly

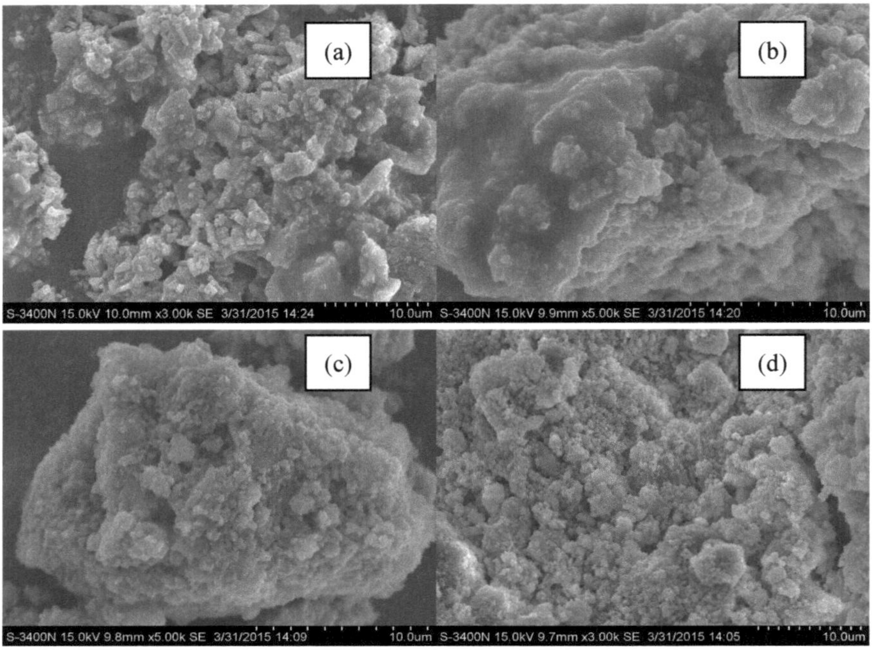

Fig. 1 Scanning electrical microscope images of abalone shells during wet grinding. The images of **a** irregular fragments after broken at 1 min of grinding; **b** fragments and particles with broken edge at 30 min of grinding; aggregations of fine particles after ground **c** 300 min and **d** 600 min

and seemed to reach a balance, indicating the distortion in aragonite structure which resulting in the reduction of diffraction intensity.

The intensity ratio to that of (111) indicates the crystal shape change in crystalline direction during grinding (Fig. 3b). For samples during the first 15 min, the ratio I_{012}/I_{111} decrease rapidly while I_{021}/I_{111} decreased evenly, indicating that (012) is the easiest crystal plane by abrasion and (021) is the more stable one. For samples in the subsequent time, both the ratio I_{012}/I_{111} and I_{021}/I_{111} decreased evenly, although with a little fluctuation, indicating the nearly isotropic-length granular crystal shapes after grinding.

The full width at half maximum (FWHM) of the diffraction peak can represent crystalline degree of a mineral, i.e., the larger the value of the FWHM, the lower the crystalline degree and the smaller crystal size. The FWHM value of (111), (012) and (021) increased quickly for the samples during the first 300 min. of grinding, indicating the crystal grains beginning to minifying due to breakage(see Fig. 3c). For samples after 300 min., the FWHM value reached to a steady value, indicating a stable crystal size, which might be related to the formation of aggregates. So it can be apparently concluded from the results that the first 300 min. of grinding was the breakage stage, during which the aragonite crystals were mainly broken into fine particles with some crystal structure damaged; and the later

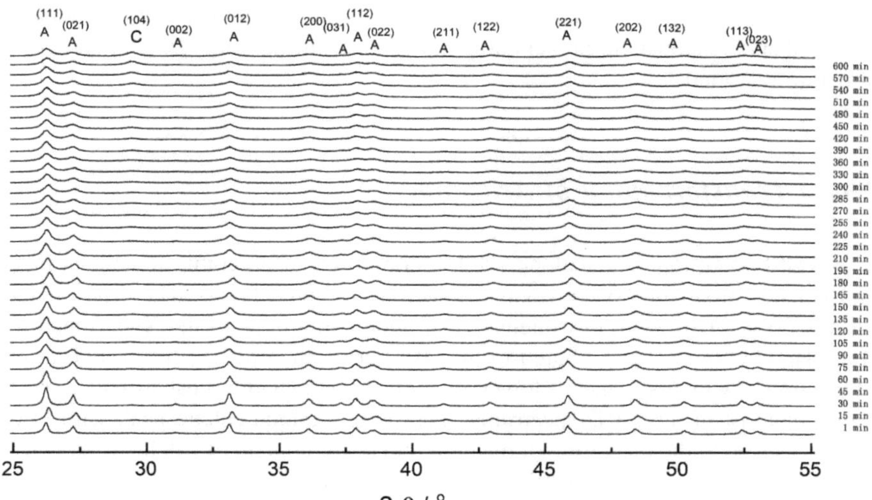

Fig. 2 X-ray diffraction spectrums of abalone shells during wet grinding. Most of spectrums indicated almost pure aragonite (*A*) and trace calcite (*C*) with a weak peak exposed

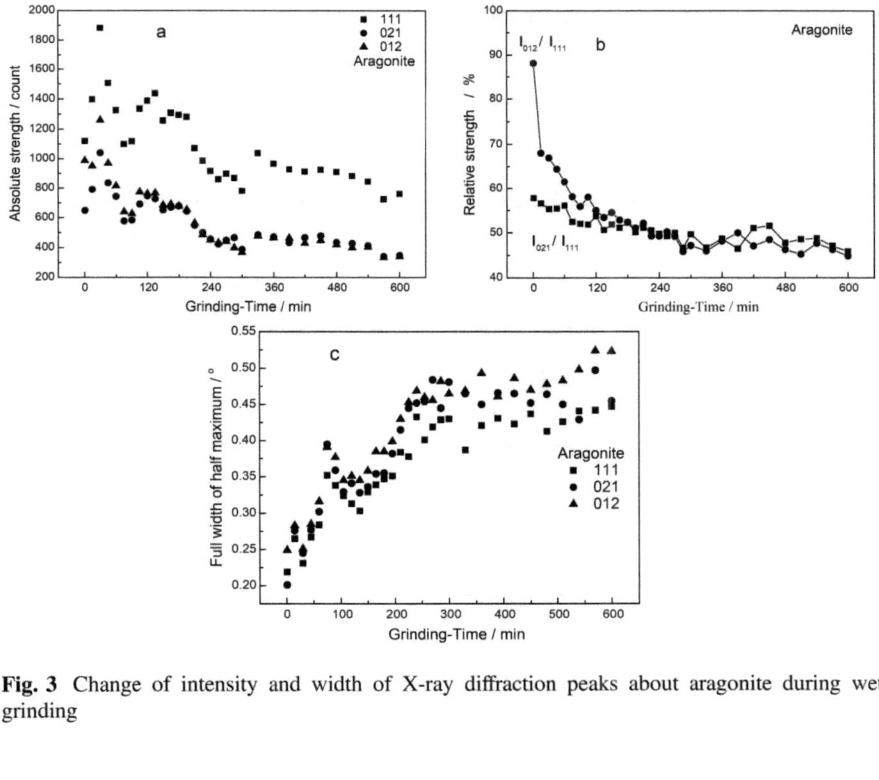

Fig. 3 Change of intensity and width of X-ray diffraction peaks about aragonite during wet grinding

300 min was the aggregation process stage, during which particles aggregation occurred spontaneously to prevent aragonite crystal further minifying and aragonite structure distorted.

Phase Transformation of Aragonite

It is well-known that the abalone shell is composed of aragonite (nacreous layer) and calcite (prismatic layer), organic materials (organic layer). There is a calcite peak at $2\theta = 29.46°$ with low intensity and wide width in all of the XRD spectrums (Fig. 4), which may be attributed to the (104) crystalline plane, the strongest peak of calcite. Low intensity means lower content and wide width means smaller crystal size. For most of the XRD spectrums, there is the unique calcite peak with lower intensity close to the background, meaning the lower content of calcite. Meanwhile, after 300 min. of grinding, the intensity of the peak (104) began to rise up. However, after 600 min. grinding, there appears some diffraction peaks at 23.06°, 39.62°, 43.04°, which might be attributed to (012), $(1\ 1\ \bar{3})$, (202)crystalline plane of calcite, respectively. The initial (104) peak might be ascribed to the impurity packed between aragonite crystals during abalone shell growth, but the sharply rising of (104) peak intensity and the appearance of other calcite peaks indicate new calcite crystals formation during grinding.

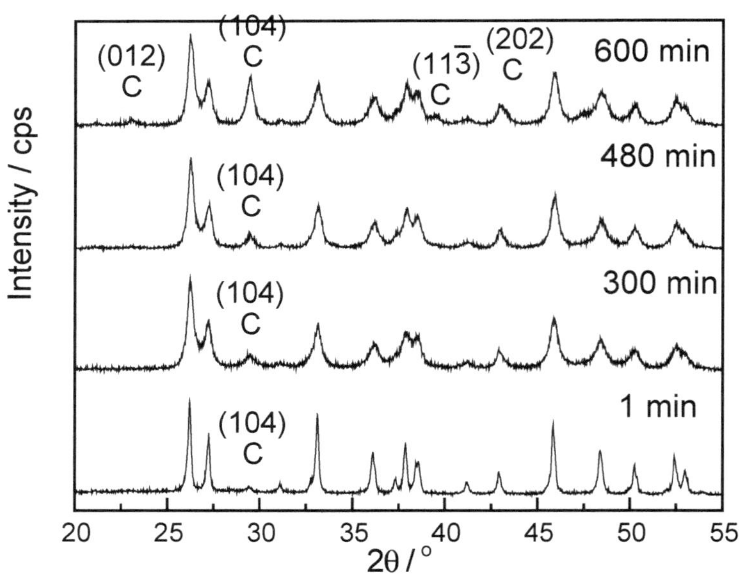

Fig. 4 Peak height (104) rising and some new peaks appearing of calcite (*C*), reflected calcite contents raising during wet grinding

By quantity analysis (Fig. 5a), it was disclosed that during the 600 min. grinding, there are two stages. For the first stage, the material contents keep constant, meaning a crystalline stable stage. While after grinding time of 480 min, begins the second stage as the content of calcite rising from 6 to 17% and that of aragonite dropping from 94 to 83%, implying a phase transformation procession. Unfortunately, the calcite content didn't reached maximum due to less of time in this experiment. Former researchers suggested that there were a process of first rising and then falling about calcite contents, with 33% as maximum of vibrating milling and 50% as maximum of planetary mill; and the aragonite contents were complementary with those calcite contents [10, 11]. The complementary content change of aragonite and calcite disclosed a phase transformation possibility from aragonite to calcite.

The average size of aragonite crystal was calculated by Scherrer equation (Fig. 5b). It is shown from the calculation that the average size of aragonite crystal dropped greatly from 70 to 20 nm and that of calcite dropped from 20 to 10 nm during the first 300 min. However, for the later 300 min, the size of calcite rose surprisingly while that of aragonite kept around 20 nm. It seemed that calcite crystals have a tendency to reach the same size with aragonite. At the end of grinding (600 min) both sizes of aragonite and calcite arrived to the same value of 18 nm. This might be an indirect evidence of calcite transformation from aragonite. The more the aragonite grains transformed into calcite, the more the calcite grains appearing with the size of aragonite, so the average size of calcite in the powder material should be more close to or equal to the size of aragonite.

The mechanism for the aragonite-calcite phase transformation in room temperature is still under argument by far. Some researchers ascribed it either to the application of extreme pressures at point-point grain contacts or to accumulation of strain energy in the crystal lattice [4, 6, 12]. It is well-known that the lattice spacing in calcite is much more similar to that of aragonite, allowing a degree of matching across the calcite–aragonite interface. A characteristic of these relationships is the continuity of the Ca-atom positions across the interface, because in aragonite Ca distribution approximately hexagonally close-packing, whereas in calcite it is

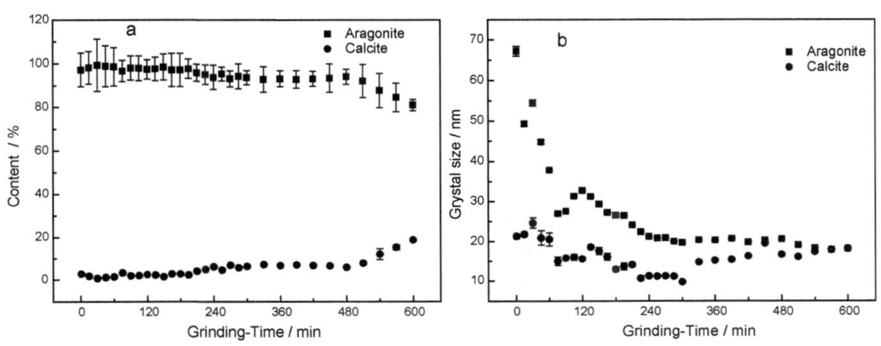

Fig. 5 Change of phase contents and crystal sizes about aragonite and calcite during wet grinding

Table 1 Change of cell parameter of aragonite and calcite during grinding

Cell parameter		a (nm)	b (nm)	c (nm)	Vol. (nm^3)
Calcite	Average value	0.500	0.500	1.708	0.370
	Increase or decrease rate (per min)	$\downarrow 7.72 \cdot 10^{-7}$	$\downarrow 7.72 \cdot 10^{-7}$	$\downarrow 5.12 \cdot 10^{-7}$	$\downarrow 1.23 \cdot 10^{-4}$
Aragonite	Average value	0.496	0.795	0.575	0.227
	Increase or decrease rate (per min)	$\downarrow 0.33 \cdot 10^{-4}$	$\downarrow 1.33 \cdot 10^{-4}$	$\downarrow 9.39 \cdot 10^{-4}$	$\downarrow 2.63 \cdot 10^{-3}$

Notice The increasing/decreasing rates were obtained by linear fitting of unit cell parameters in OriginLab software

approximates cubic close-packing [12]. Therefore, a dislocation glide mechanism may provide solutions for the phase transition of aragonite-calcite, i.e. from the hexagonal close-packing array of aragonite to the cubic close-packing array of calcite or from the cubic close-packing array to the hexagonal close-packing array [12]. Probably during grinding, the compressive and shearing forces applied on the powder materials can stretch or dislocate the aragonite crystals, producing a calcite phases.

During grinding, the main compressive and shearing forces are applied on the treated materials [13]. As shown in crystal cell parameters in Table 1, a slight and almost equivalent shrinkage of calcite happened in three axial directions during grinding, which might be due to compressive mechanic force. A great expansion of aragonite at c- and b-axel direction seemed to be due to crystal structure's dislocation glide under shearing force than for compressive force. From the changing rate of cell parameters in this experiment, it seemed to need much more time from bigger calcite particles into smaller aragonite ones by compressive force than the time from aragonite into calcite by dislocation glide force. It might be a proper explanation for transformation from aragonite to calcite and transformation from calcite to aragonite.

Conclusion

During wet grinding, aragonite's structure and size was changed in certain degree. At the beginning of grinding, aragonite crystals were broken into finer granular fragments. When the grain size decreased to a certain value, aggregation happens spontaneously. Since then, the aragonite size was kept stable for protection of particles aggregation. In the aggregation period effects of the mechanic forces varying from crystal minifying to structure's distorted and reconstructed; when accumulation of strain energy gets to a certain value, phase transformation happens.

Subsequently the phase transformation from aragonite into calcite became main process during grinding, which caused the calcite content rising and sizes of both calcite and aragonite turning to the same. In this experiment crystal structure's dislocation glide might be the main reason for phase transformation from aragonite to calcite rather than compressive force.

Protection by particle aggregation, there is a lower limit of average grain size of powder materials after a long period of grinding. The contents of newly-formed phase also may have an upper limit, which are decided by mechanic forces during grinding. In preparation of ultrafine powder materials, phase categories, as well as their contents and sizes might affect the performance of powder materials, which would have important results for pharmaceutical preparation, cosmetics manufacturing and so on. The specific grinding progress should be paid need more attentions by scientific researchers when grinding samples in the laboratory.

References

1. Garcia F, Bolay NL, Frances C (2002) Changes of surface and volume properties of calcite during a batch wet grinding processs. Chem Eng J 85:177–187
2. Burns JH, Bredig MA (1956) Transformation of calcite to aragonite by grinding. The J Chem Phys 25(96):1281
3. Momota H, Senna M, Takagi M (1986) Effects of wet vibro-milling on the polymorphic conversion of aragonite into calcite. J Chem Soc Faraday Trans 76:790–796
4. Gammage RB, Glasson DR (1976) The effect of grinding on the polymorphs of calcium carbonate. J Colloid Interface Sci 55:396–401
5. Lin IJ, Somasundaran P (1972) Alterations in properties of samples during their preparation by grinding. Powder Technol 6:171–179
6. Li T, Sui F, Li F, Cai Y, Jin Z (2014) Effect of dry grinding on the structure and granulatity of calcite and its polymorphic transformation into aragonite. Powder Technol 254:338–343
7. Teke E, Yekeler M, Ulusoy U, Canbazoglu M (2002) Kinetics of dry grinding of industrial minerals: calcite and barite. Int J Miner Process 67:29–42
8. Zhang G, Ding S (2005) Effect of grinding on IR spectra of aragonite. Mine Resour Geol 19 (4):129–131
9. Zhang G, Guo Y, Ao J, Yang J, Lv G, Shih K (2013) Thermally induced phase transformation of pearl powder. Mater Sci Eng, C 33:2046–2049
10. Wan G, Zhang G (2014) Reversible transformation between aragonite and calcite induced by mechanical milling at low temperature. J Mine Petrol 34(2):8–12
11. Zorkovska A, Balaz P (2011) High energy milling induced calcite to aragonite transformation in eggshell. Mater Struct, pp 188–190
12. Putnis A (1992) Transformation processes in minerals, Introduction to mineral sciences. University of Cambridge, Great Britain
13. Al-Wakeel MJ (2005) Effect of mechanical treatment on the mineralogical constituents of Abu-Tartour phosphate ore, Egypt. Int Miner Process 75:101–112

Evaluation of Ballistic Armor Behavior with Epoxy Composite Reinforced with Malva Fibers

Lucio Fabio Cassiano Nascimento, Luane Isquerdo Ferreira Holanda, Luis Henrique Leme Louro, Sérgio Neves Monteiro, Alaelson Vieira Gomes and Édio Pereira Lima Júnior

Abstract In the present study, we used natural fibers malva (Urena lobata, Linn) in 0, 15 and 30 vol.% as reinforcement of epoxy matrix composites to ballistic application. Ballistic efficiency of these composites was assessed by measuring the energy absorbed and residual velocity after impact 9 mm FMJ ammunition projectile, aiming to compare with traditional materials, such as: aramid fabric Kevlar™ used in vests for personal protection. The results showed by visual analysis and scanning electron microscopy (SEM) that the main active failure mechanisms in composites were delamination layers and fiber rupture. In test groups can be seen that the absorbed energy and the residual velocity varies with the fiber percentage used. Compared to aramid there is both a gain in weight and cost of the ballistic armor, which makes the fiber malva a promising material applications involving personal protection against 9 mm caliber ammunition.

Keywords Ballistic test · Malva fibers · Composites

Introduction

Ballistic armors were basically composed by metals until World War II [1]. Weaponry development has been increasing the demand for lighter and more efficient ballistic armor, especially for Armed Forces. In military applications, the basic requirements are directly linked to mobility and impact perforation resistance. At a first moment, increasing the projectile's potential for perforation leads to heavier armors, reducing mobility [2]. For this reason, it's important to seek for

L.F.C. Nascimento (✉) · L.I.F. Holanda (✉) · L.H.L. Louro · S.N. Monteiro · A.V. Gomes · É.P. Lima Júnior
Militar Institute of Engineering, IME, Rio de Janeiro, Brazil
e-mail: lucio_coppe@yahoo.com.br

L.I.F. Holanda
e-mail: luaneferreira12030@gmail.com

S. Ikhmayies et al. (eds.), *Characterization of Minerals, Metals, and Materials 2017*, The Minerals, Metals & Materials Series,
DOI 10.1007/978-3-319-51382-9_71

Table 1 Properties of aramid fiber, malva fiber and epoxy resin [5, 6]

Materials	Density (g/cm^3)	Tensile resistance (MPa)	Young's modules (GPa)	Specific resistance (MPa)	Specific modules (GPa)
Aramid	1.4	3000–3150	63–67	2143–2250	45–48
Malva	1.4	160	17.4	116	13
Epoxy	1.1–1.3	60–80	2–4	46–73	1.5–3.6

solutions different from simply incrementing thickness of layers that constitute ballistic armor.

Nowadays it's interesting to use less synthetic materials—mainly aramid fiber in typical armor systems—and more natural materials like lignocellulosic fibers. Those vegetal fibers are both biodegradable and renewable sources so they may contribute to social and economic country's development [1]. According to CONAB ("National Supply Company"), Brazil has been the world's fourth largest producer of jute and malva since 2005 [3]. Those products get social importance because they contribute to financial support of many Amazon rural families, principally indigenous peoples in Amazonas and Pará [4].

Another point is that aramid fabrics are imported and have limited lifetime [2]. On the contrary, natural fiber composites are more attractive options because of their components' price (resin and fiber), much smaller than those of synthetic materials. For instance, aramid costs US$ 63.60/kg whereas epoxy resin and malva fiber cost US$ 2.80/kg and US$ 0.30/kg respectively [2]. Besides the malva fiber's density is similar to aramid's, which is about 1.4 g/cm^3. Other properties can be found in Table 1.

In the meantime, it is necessary to consider some issues related to the use of natural fibers. It is known that load distribution happens throughout the interface between matrix and fiber. On the one hand, increasing fiber percentage makes the impact resistance increases as well. On the other hand, when fiber percentage is too high the wettability of the resin on the fibers becomes less effective, causing the resistance to decrease. Another point is that natural fibers are hydrophilic, while matrix is generally hydrophobic, so adhesion degree between them diminishes. In both cases the effective interface area between fiber and matrix become smaller, leading to worse load distribution.

The goal of this study is to analyze the ballistic efficiency of epoxy matrix composites in different percentage of malva fibers (0, 15 and 30 vol.%). The ballistic test consists of determining the residual velocity of the bullet after its impact on the target (specimen). This velocity is related to the energy that is absorbed by the composite during the impact—the greater that energy, the more efficient the composite will be. After, a visual analysis is made in scanning electron microscopy (SEM) to observe fracture mechanisms in the composites.

Materials and Methods

The malva fiber has been used as reinforcement in the composite. The fibers were provided by State University of Northern Rio de Janeiro (UENF) and commercially supplied by the Brazilian firm Castanhal Textile Company. Firstly, the fibers were cleaned, aligned and cut in length of 15 cm. After, they were maintained in stove for 24 h in order to reduce the absorbed moisture. The as received bundle of fibers and the aligned cut fibers can be seen in Fig. 1.

The composite matrix were diglycidyl ether of the bisphenol A (DGEBA), which is a commercial epoxy resin mixed with trietylene tetramine (TETA) hardener in proportion of 13 parts of hardener per hundred parts of resin in weight. Both components are fabricated by the Brazilian firm Dow Chemical and commercially supplied by the distributor RESINPOXY Ltda.

Moreover, it was used auxiliary equipment (Fig. 2), such as:

- Hydraulic press for loading until 30T (SKAY);
- Rectangular metallic mold with dimensions of 150 × 120 mm;
- Accurate balance (GEHAKA, BK300 version);
- Stove (ELKA); and
- Exhausting system.

Composites with 0, 15 and 30% in volume of malva fibers were manufactured, being 4 unites of each concentration. For malva fibers it was used as initial reference density of 1.40 g/cm^3 [7] and for the epoxy resin (DGEBA-TETA) value of 1.11 g/cm^3 [1]. A pressure of 5 MPa was applied and the composite plate cured for 24 h [1, 2].

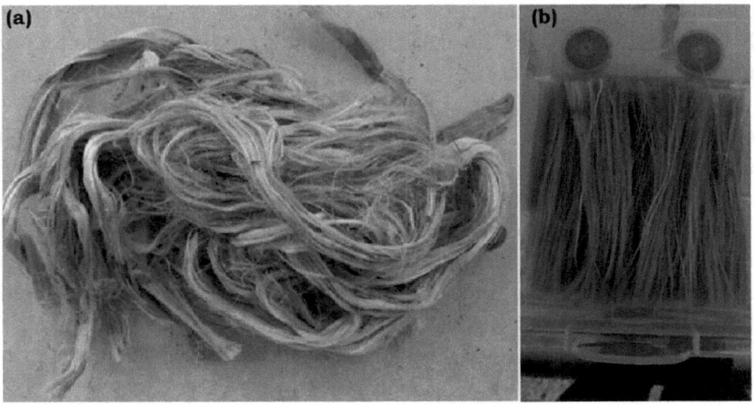

Fig. 1 Malva fibers: **a** as received bundle of fibers; **b** aligned cut

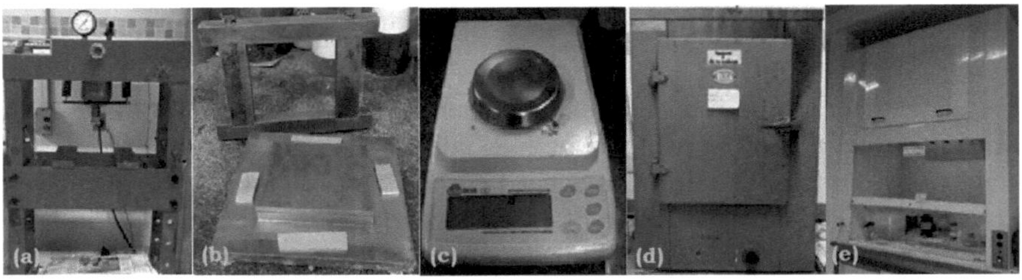

Fig. 2 Auxiliary equipment **a** hydraulic press; **b** metallic mold; **c** accurate balance; **d** stove; **e** exhausting system

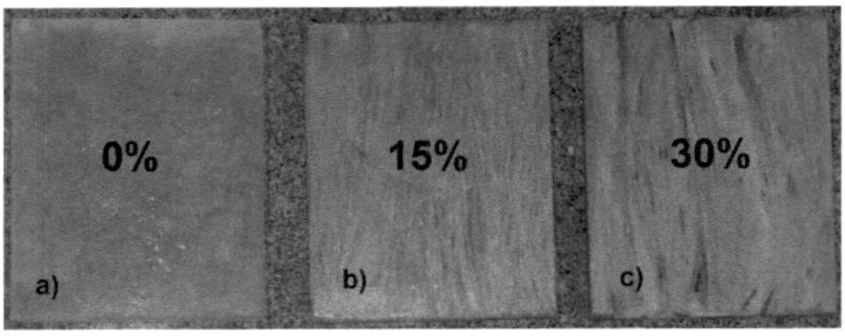

Fig. 3 Composite plates of epoxy reinforced with 0, 15 and 30 vol.% of malva fibers

The aramid fabric plies were supplied by the Brazilian firm LFJ Blindagem Com. Serv. S.A (São Paulo, Brazil). They've been used 4 sheets of aramid, manufactured with average dimensions 150 × 150 × 453 mm. Figure 3 show composite plates of epoxy reinforced with different percentages of malva fibers. The approximate dimensions of the composite plates were 150 × 120 × 11.5 mm.

The ballistic tests were conducted at the Brazilian Army shooting range facility, CAEX, in the Marambaia peninsula, Rio de Janeiro. All tests, 4 for each type of percentage, were carried out according to the NIJ 0101.04 standards using 9 mm FMJ military ammunition (m = 8 g—Armor level III-A). Figure 4 shows, schematically, the exploded view of the ballistic test setup. A dashed straight line indicates the projectile trajectory. A steel frame was used to position the target, which was held in place by spring clips (Fig. 5). The gun, located 5 m from the target, was sighted on its center with a laser beam. The exact velocity of the projectile at three moments: leaving the gun, immediately before impacting (V_i) in the plate and after outcoming projectile (residual velocity—V_r) was measured by an optical barrier and a model SL-52 OP Weibel fixed-head Doppler radar system (Fig. 6). The Kinetic energy ΔE_d, dissipated inside the target, could then be estimated by Eq. (1):

$$\Delta E_d = \frac{1}{2} m\left(V_i^2 - V_r^2\right) \tag{1}$$

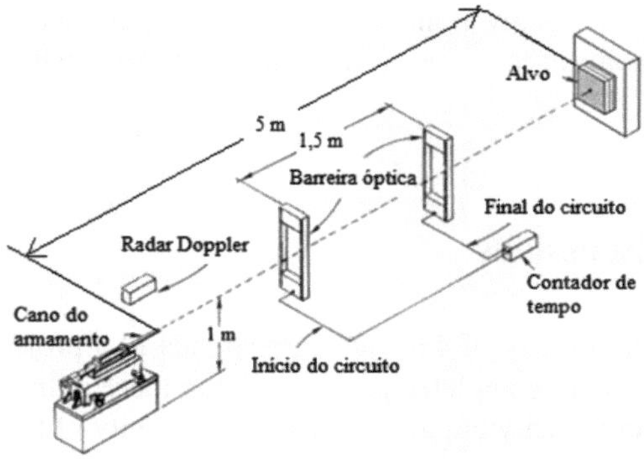

Fig. 4 Schematic exploded view of the ballistic experimental setup [1]

Fig. 5 Specimens fixed in the metal bracket with spring clips: composite plate (*left*) and aramid plate (*right*)

Fig. 6 Apparatus of CAEX for ballistic testing with ammunition 9 mm (*left*); Doppler radar

The system response is given in the form of a spectrum depicting the velocity of the projectile from the time it leaves the arms to a point where the system can't detect.

Results and Discussions

The average impact velocity of 436 m/s corresponds to a projectile kinetic energy of 760 J. All targets were completed perforated after these individual tests. Not only the projectile impact velocity but also the residual velocity of the fragments passing through the target were measured by the Doppler radar system. Table 2 presents the impact and residual velocities as well as the internally dissipated energy, Eq. (1), from ballistic tests of individual plates. In this table, it is important to realize that the energy absorption increases with the fiber percentage in the tested plates. Tables 2 and 3 shows the dissipated energy by composites reinforced with malva fibers and

Table 2 Impact and residual velocities together with internally dissipated energy in ballistic tested to composites reinforced with malva fibers

0% (v/v) fiber			15% (v/v) fiber			30% (v/v) fiber		
CP 1	V_i (m/s)	429.93	CP 5	V_i (m/s)	426.87	CP 9	V_i (m/s)	425.69
	V_r (m/s)	335.59		V_r (m/s)	367.21		V_r (m/s)	366.37
	ΔE (J)	288.87		ΔE (J)	189.50		ΔE (J)	187.93
CP 2	V_i (m/s)	428.18	CP 6	V_i (m/s)	432.01	CP 10	V_i (m/s)	425.99
	V_r (m/s)	329.78		V_r (m/s)	367.65		V_r (m/s)	357.61
	ΔE (J)	298.33		ΔE (J)	205.87		ΔE (J)	214.34
CP 2	V_i (m/s)	425.68	CP 7	V_i (m/s)	420.43	CP 11	V_i (m/s)	426.11
	V_r (m/s)	328.20		V_r (m/s)	342.65		V_r (m/s)	360.92
	ΔE (J)	293.95		ΔE (J)	237.41		ΔE (J)	205.24
CP 4	V_i (m/s)	426.26	CP 8	V_i (m/s)	426.03	CP 12	V_i (m/s)	423.57
	V_r (m/s)	332.65		V_r (m/s)	361.24		V_r (m/s)	371.07
	ΔE (J)	284.16		ΔE (J)	204.05		ΔE (J)	166.90
ΔE average 0% (J)		291.33 ± 6.15	ΔE average 15% (J)		209.21 ± 20.18	ΔE average 30% (J)		193.60 ± 20.90

Table 3 Impact and residual velocities together with internally dissipated energy in ballistic tested to aramid fabric

Aramid plate	V_i (m/s)	V_r (m/s)	ΔE (J)
Aramid 1	427.346	347.150	248.446
Aramid 2	434.711	377.519	185.812
Aramid 3	425.949	355.664	219.743
Aramid 4	428.515	354.878	230,747
		ΔE average (J)	221.187 ± 26.381

Table 4 Evaluation of weight and cost of the different armor components

Armor component	Volume (cm^3)	Density (g/cm^3)	Weight (kgf)	Price per kg (U.S. dollars)	Component cost (U.S. dollars)
Aramid fabric	225	1.44	0.324	63.60	20.61
Composite 0%	Epoxy (100%)—225	1.11	0.250	2.80	0.70
	Malva (0%)—0	1.40	0	0.30	
Composite 15%	Epoxy (85%)—191.25	1.11	0.212	2.80	0.61
	Malva (15%)—33.75	1.40	0.047	0.30	
Composite 30%	Epoxy (70%)—157.5	1.11	0.175	2.80	0.52
	Malva (30%)—67.5	1.40	0.095	0.30	
Total weight with aramid fabric (kgf)—0.324			Total cost with aramid fabric (US$)—20.61		
Total weight with malva composite (kgf)		CP 0%—0.250	Total cost with malva composites (US$)—0.52		CP 0%—0.70
		CP 15%—0.259			CP 15%—0.61
		CP 30%—0.267			CP 30%—0.52
Percentage decrease of weight (%)		CP 0%—22.84	Percentual decrease of costs (%)		CP 0%—96.60
		CP 15%—20.06			CP 15%—97.04
		CP 30%—17.60			CP 30%—97.48

aramid fabric respectively. The results show the great potential of composites reinforced with malva fibers, compared to aramid fabric, especially those with a percentage of 30 vol.% fibers.

This shows that aramid fabrics such as ballistic protection, are more efficient in terms of energy absorbed, which composite boards reinforced with malva fibers (84, 163 and 187% respectively).

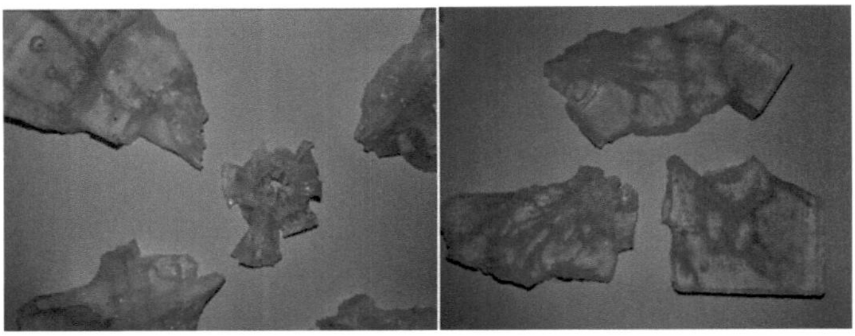

Fig. 7 Shatterproof body with 0% fiber after impact

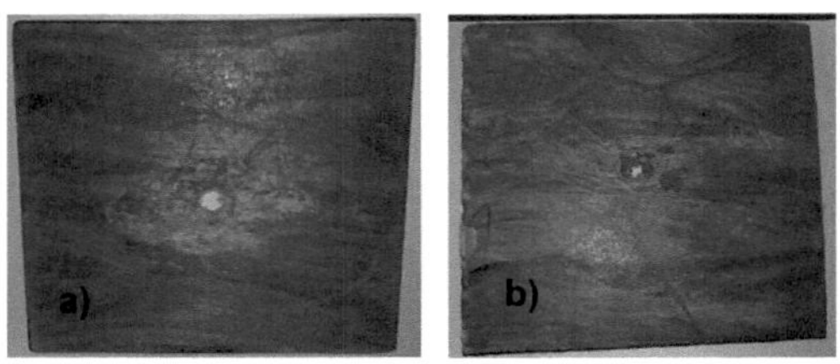

Fig. 8 Epoxy composites reinforced with malva fibers after ballistic test 9 mm ammunition. **a** 15% v/v fibers and **b** 30% v/v fibers

Moreover, it should be recalled that the aramid to be imported and a synthetic product has a high cost related to the acquisition and processing. In addition, it has a limited lifespan, causing environmental damage due to the degradation time after disposal. Malva fibers, in turn, being natural, are biodegradable and have much lower cost. Table 4 shows an estimate of costs between the composites reinforced with malva fibers which showed weight reduction of around 20% and cost around 97% in favor of composites reinforced with malva fibers.

Figure 7 shows the plate with 0% malva fibers completely destroyed after the ballistic test and Fig. 8 shows the composite board to 30% reinforcing fibers with the integrity largely preserved, enabling the occurrence of new impacts on the same plate.

Conclusions

The aramid fabric absorbed more energy than the composites reinforced with epoxy malva fibers against 9 mm ammunition. Around 84, 163 and 187% respectively for the samples containing 0, 15 and 30% of fibers.

Although absorb lower energy impact test specimens reinforced with 15 and 30% of malva fibers were more intact compared the samples produced with 100% epoxy, after ballistic tests.

The aramid cloth was more effective in ballistic testing against 9 mm ammunition, than the composites reinforced with fibers mauve. However, the use of these natural fibers may represent weight reduction of around 20% and cost reduction around 97%.

Acknowledgements The authors acknowledge the support to this investigation by the Brazilian agencies CNPq, CAPES, FAPERJ and UENF for supplying the malva fibers.

References

1. da Silva LC (2014) Ballistic behavior of natural fiber epoxy composite multilayer armor. Military Engineering Institute, Rio de Janeiro (in Portuguese)
2. Luz FS (2014) Ballistic behavior evaluation multilayered shield with epoxy composite reinforced with jute fiber. Military Engineering Institute, Rio de Janeiro (in Portuguese)
3. National Supply Company (CONAB) (2015) Juta Malva, Conab, 2015. [Online]. Available: http://www.conab.gov.br/OlalaCMS/uploads/arquivos/15_04_14_17_05_50_juta_malva. Accessed 18 May 2016 (in Portuguese)
4. Veríssimo A, Pereira D (2014) Production forest in the Amazon: characteristics, challenges and opportunities. Strateg Partnerships 19:13–44 (in Portuguese)
5. Teixeira AMJ (2007) Collapsible bridge composite material fiberglass. Federal University of Rio de Janeiro, Rio de Janeiro (in Portuguese)
6. Margem JI (2013) Analysis of structural characteristics and properties of polymer composites reinforced with malva fibers. State University of Norte Fluminense (in Portuguese)
7. Oliveira JTS (1998) Characterization of eucalyptus wood for construction. Polytechnic School of the University of São Paulo, São Paulo (in Portuguese)

Evaluation of the Pozzolanic Activity of Residue From the Paper Industry

A.R.G. Azevedo, J. Alexandre, L.J.T. Petrucci, E.B. Zanelato and T.F. Oliveira

Abstract The paper and cellulose industry is responsible for generating large amounts of solid waste, which are produced during various stages of the production process. Currently this material goes to places like landfills and other generating costs for the company. Furthermore, cement, mortar and concrete represent a significant economic and environmental cost. This study aims to determine the pozzolanic activity index in the residue studied to verify its viability in the incorporation of the same in cement. It will be used Luxan method for this determination. The data found show that the material has the possibility of being incorporated in the mortar due to its pozzolanic activity index.

Keywords Mortar · Paper waste · Pozzolanic activity

Introduction

Environmental problems caused by improper discharge of solid waste on the environment has become increasingly important to study on the appropriate destination of these materials. In Brazil, the environmental legislation is becoming increasingly demanding in respect, especially with the institution of the National Solid Waste Policy.

The paper industry is highly diversified and of great importance due to its strong coverage in all over the country. According to the Brazilian Association of Pulp and

A.R.G. Azevedo (✉) · J. Alexandre · E.B. Zanelato · T.F. Oliveira
LECIV—Civil Engineering Laboratory, UENF—State University of the Northern Rio de Janeiro, Av. Alberto Lamego, 2000, Campos dos Goytacazes, Rio de Janeiro 28013-602, Brazil
e-mail: afonso.garcez91@gmail.com

A.R.G. Azevedo · L.J.T. Petrucci · T.F. Oliveira
DIRINF—Direction of Infrastructure Rectory, IFF—Federal Institute Fluminense, St. Cel. Valter Kramer, 357—Parque Vera Cruz, Campos dos Goytacazes, Rio de Janeiro 28080-565, Brazil

S. Ikhmayies et al. (eds.), *Characterization of Minerals, Metals, and Materials 2017*, The Minerals, Metals & Materials Series,
DOI 10.1007/978-3-319-51382-9_72

Table 1 World production of cellulose and paper

Position	Cellulose		Paper	
	Country	A thousand tons	Country	A thousand tons
1st	USA	50,351	China	102,500
2nd	China	18,198	USA	74,375
3rd	Canada	17,073	Japan	26,083
4th	*Brazil*	*13,977*	Germany	22,630
5th	Sweden	11,672	Sweden	11,417
6th	Finland	10,237	South Korea	11,333
7th	Japan	8642	Canada	10,751
8th	Russian	7519	Finland	10,694
9th	Indonesia	6710	*Brazil*	*10,260*

Paper—BRACELPA [1] Brazil is among the ten largest producers in the world and contributes to the export of this product, as shown in Table 1. One of the explanations is the large domestic consumer market as well as the availability of large areas for timber planting, product required to the extraction of cellulose. The paper industry also has its drawbacks, such as extraction of natural resources such as wood, for the extraction of cellulose, release gases during the production process as well as large amount of solid waste at different stages of the production process.

The incorporation of a residue in a building material is something that should be thought. It must investigate the behavior of the material, this must be within the array, particularly in composite materials, such as mortars. The pozzolans substances are constituted of silica and alumina in presence of water combine with calcium and didroxido different cement components forming stable compounds to water and binders properties [2].

This work has as main objective the evaluation of the pozzolanic activity, using the method of Lúxan et al. [3], the residue from the cellulose and paper industry enabling the choice of the manner in which the residue can be incorporated in materials composite the cement base, such as mortar and concrete.

Materials and Methods

The residues analyzed in this work comes from various stages of the papermaking process and pulp an industry located in Fluminense northwest, municipality of Santo Antonio de Padua—RJ, that after generated is taken to an effluent treatment plant, Fig. 1, where it is brought to their subsequent destination, which are generally landfills.

The compound gets wet and is dried in an oven for 24 h at 110 ± 5 °C and cooled at room temperature. The grinding of the waste is held in a ball mill for 30 revolutions per minute of the drum Grinder load 12 steel balls (about 48 mm in

Fig. 1 Residue on wastewater treatment plant

diameter and mass between 390 g and 445 g) which is normally used to meet NBR NM 51 2001—Coarse Aggregates—Abrasion test "Los Angeles" [4]. After this process is expected to be in the residue incorporation of conditions composite materials such as mortars, according Fig. 2.

Among the different techniques for studying the pozzolan, it was held tests of electrical conductivity in the residue, a method that measures the evolution of pozzolanic activity [5]. The method monitors variation in electrical conductivity within 2 min after addition of residue in the mixture.

According Lúxan et al. [3] a rapid method for assessing a pozzolanic activity by measuring conductivity may be applied to many natural product. In this method, it

Fig. 2 Waste after process

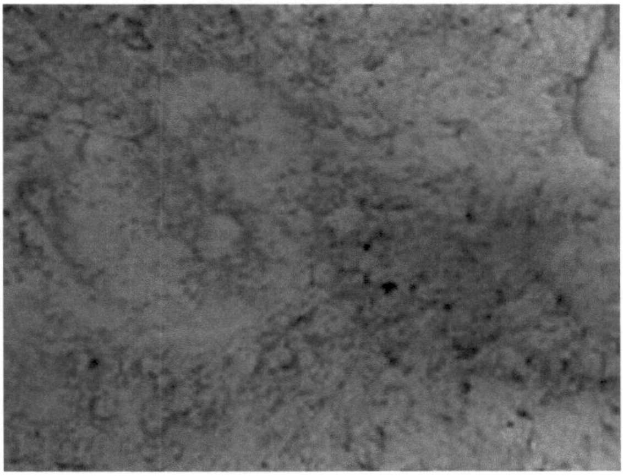

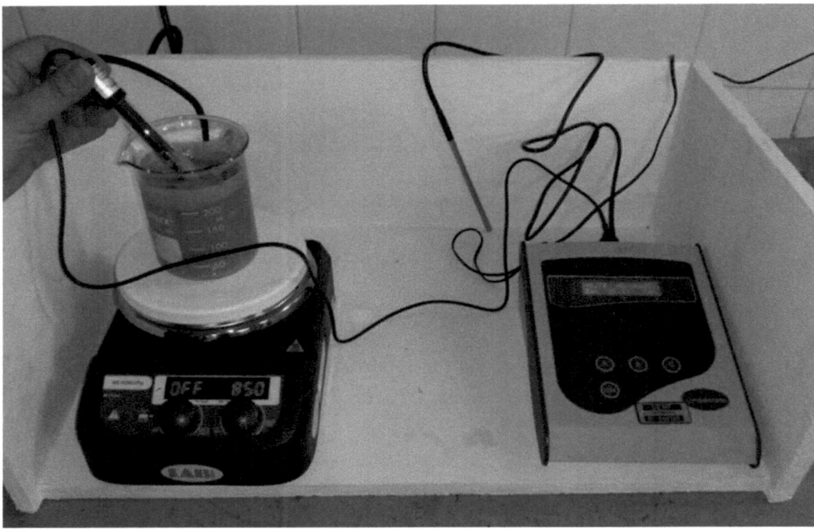

Fig. 3 Pozzolanic activity method

defines a pozzolanic activity index based on the variation of electrical conductivity of a material in a saturated solution of calcium hydroxide at a given time interval.

The evaluation of the conductivity measuring pozzolanic material can be classified as non-pozzolan, medium or good pozzolanic according to Fig. 3.

The procedure used was: Put water in a beaker; Put 70 ml of distilled water in another beaker and heat to 40 °C; Add 98.7 mg of $Ca(OH)_2$ and homogenizing magnet in lower beaker; Place the speed 4 and equilibrate for 20 min; Set the thermostat and conductivity at the same time; Connect the conductivity (first stop: 0–20 mS); Add 1.75 g of material (possible pozzolan).

For the execution of the test were selected three samples using a sample separation methodology [6], each sample was properly numbered and identified for the implementation of the test.

Results

Table 2 shows the results of electrical conductivity of residue of paper and cellulose measured during 20 min to classify the residue as a pozzolan.

The tested residues showed a conductivity variation of two minutes, which classifies a moderate Pozzolanic activity waste, according to classification proposed by Lúxan et al. [3], according Table 3.

It was regarded as the final result the mean of three samples, since the lower range of values is not necessary to dispose values. Thus the value of 0.94 mS/cm

Table 2 Result of the electrical conductivity of residue of paper and cellulose

Pozzolanicity			
Sample	Conductivity (mS/cm)		Difference Conductivity
	Initial	End	
1	6.95	5.95	1.00
2	6.82	5.98	0.84
3	6.57	5.58	0.99

Table 3 Leaderboard pozzolanicity the material proposed by Lúxan et al. [3]

Material classification	Variation conductivity (ΔC)
Not pozzolanic	$\Delta C < 0.40$
Moderate pozzolanicity	$0.40 < \Delta C < 1.20$
Good pozzolanic	$\Delta C > 1.20$

shows average pozzolanic. In cases where the values found in the results are discrepant, i.e. there is great variation, needs to be made a greater number of tests in order to validate the results and make them viable.

Conclusion

After analysis of the results found on the pozzolanicity of studied residue can infer that:

- The average pozzolanicity shows that the studied material has application potential as a replacement within a cementitious matrix in proportion of any binder.
- The replacement binders, in the case of mortars as cement or lime, is shown to be extremely efficient [7] regarding mechanical strength of these composites.
- Others Properties of composites analyzed was reinforce the pozzolanic behavior of material [8], the determination of the capillary mortars coefficient showing decrease capacity of the mortar voids thereby increasing strength and improving durability thereof.
- It is important to analyze the interaction of the residue with the other constituents, such as fine aggregates in mortar, and as the hydration reactions occur, so he will need further study to a full guarantee this pozzolanicity.
- The stage in which the residue is extracted from paper and cellulose production can influence the final result, then it is recommended run tests at different stages of the production process, to investigate whether cause or not real impact.

Acknowledgements The authors thank the Brazilian agencies: CNPq, CAPES and FAPERJ for the support provided to this investigation.

References

1. BRACELPA (2015) Brazilian association of paper and cellulose. Evolution of the Brazilian production of paper. Available at: http://bracelpa.org.br/bra2/?q=node/140 (in Portuguese)
2. Hossaina MM, Karimb MR, Hasanb M, Hossainb MK, Zaina MFM (2016) Durability of mortar and concrete made up of pozzolans as a partial replacement of cement: a review. Constr Build Mater 116:128–140
3. Lúxan MP, Madruga F, Saavedra J (1989) Rapid evaluation of pozzolanic activity of natural products by conductivity measurement. Cem Concr Res 19:63–68
4. Brazilian Association of Technical Norms—ABNT NBR NM 51:2001 (2001) Small-size coarse aggregate—test method for resistance to degradation by Los Angeles machine (in Portuguese)
5. Madhkhana M, Azizkhania R, Torki ME (2012) Harcheganib. Effects of pozzolans together with steel and polypropylene fibers on mechanical properties of RCC pavements. Constr Build Mater 26:102–112
6. Vana VTA, Röblerb C, Buia DD, Ludwig HM (2014) Pozzolanic reactivity of mesoporous amorphous rice husk ash in portlandite solution. Constr Build Mater 59:111–119
7. Azevedo ARG, Alexandre J, Xavier GC, França FCC, Silva FA, Monteiro SN (2015) Addition of paper sludge waste into lime for mortar production. Mater Sci Forum 820:609–614
8. Gao JM, Qian CX, Wang B, Morino K (2002) Experimental study on properties of polymer-modified cement mortars with silica fume. Cem Concr Res 31:41–45

Evaluation of the Properties of the Adhesive Mortar in the Fresh State with Addition of Glass Waste

D.P. Santos, A.R.G. Azevedo, J. Alexandre, S.N. Monteiro, G.C. Xavier, B.C. Mendes and L.G. Pedroti

Abstract The glass cutting industry is responsible for generating large quantities of waste, during the stages of cutting and polishing. The disposal of this material generates a cost to the company, thus looking for ways to repackage it, so there is the possibility of incorporating it into the adhesive mortar. The objective of this work is the characterization of waste and assessment of the properties of these mortars, fresh, and check the possibility of their use. As for the characterization, were carried out particle size testing, density and pozzolanic activity, for the fresh state were performed tests of air content, water retention consistency and entrained particular index. Were incorporated 0, 10, 15 and 20% of the amount of sand. The results showed that the material on display has potential for use in adhesive mortar, with values within the limits stipulated by Brazilian standards.

Keywords Mortar · Glass waste · Fresh state

D.P. Santos · A.R.G. Azevedo (✉) · J. Alexandre · G.C. Xavier
LECIV—Civil Engineering Laboratory, UENF—State
University of the Northern Rio de Janeiro, Av. Alberto Lamego,
2000, Campos dos Goytacazes, Rio de Janeiro 28013-602, Brazil
e-mail: afonso.garcez91@gmail.com

A.R.G. Azevedo
DIRINF—Directorate of Infrastructure Rectory, IFF—Federal
Institute Fluminese, St. Cel. Valter Kramer, 357—Parque Vera Cruz,
Campos dos Goytacazes, Rio de Janeiro 28080-565, Brazil

S.N. Monteiro
Department of Materials Science, IME—Military Institute of Engineering,
Square General Tibúrcio, 80, Rio de Janeiro 22290-270, Brazil

B.C. Mendes · L.G. Pedroti
Federal University of Viçosa, UFV, Av. Peter Henry Rolfs, s/n—Campus
Universty, Viçosa, MG 36570-000, Brazil

S. Ikhmayies et al. (eds.), *Characterization of Minerals, Metals, and Materials 2017*, The Minerals, Metals & Materials Series,
DOI 10.1007/978-3-319-51382-9_73

Introduction

The great technological development significantly increased the production of industrial goods and the volume of exploited mineral resources. Associated with this great productive expansion there was a large increase in the amount of waste produced, causing millions of tons of waste be generated every day around the world. These wastes are defined as products not used for human, social and industrial activity.

Among all types of waste, solid waste (SW) are worth mentioning, as they represent a significant portion of all the waste generated and, when mismanaged, can become a sanitary, environmental and social problem.

In recent years several studies were carried out about a partial replacement of conventional materials by waste from several industries. This fact has prompted the industry to develop alternative technologies, the level of processing and even the raw material consumption in order to save resources for future generations and preserve the environment. In some cases, the opportunity arose to develop new materials and products with similar mechanical and physical characteristics or even higher when compared to conventional materials and products [1].

In the process of manufacturing glass articles, during polishing is produced a solid residue consisting essentially of glass powder. This residue, unlike the pieces, cannot be recycled in a conventional manner, for entails among other issues the presence of bubbles. The recycling of this waste is a concern of the glass industry today, to preserve the environment and the accumulation of waste that require space to be stored [2].

Research in the use of crushed glass as a partial substitute for aggregation was done for many decades, but were not obtained good results due to the size of the grains, because the cement alkali Portland and the glass silica are reacted in the presence of humidity resulting in a chemical reaction called alkali-silica, resulting in a lower strength and overexpansion [3].

The glass production in Brazil has increased over the last years, as shown in Fig. 1 was, however, there was a significant decrease in the production of household glass in the year 2008, different from the flat glass. However, between 2007 and 2011, production increased by 27% [4].

The Viminas Industry, processing company and transformation of the glass in the municipality of Serra in the state of the Espírito Santo, Brazil, produces, for example, 150,000 m^2 of glass per month which generates 100 tons of waste per month from cutting, drilling and glazing of glass [5].

The objective of this work is the characterization of waste from glass cutting, as granulometry tests, density and index of pozzolanic activity. The influence of this residue on properties in fresh mortar state also will be evaluated (water retention, entrained air content and consistency index).

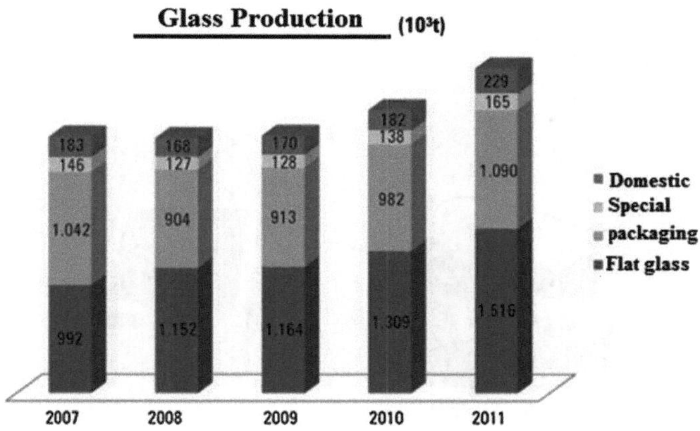

Fig. 1 Glass production progress in Brazil. *Source* ABIVIDRO

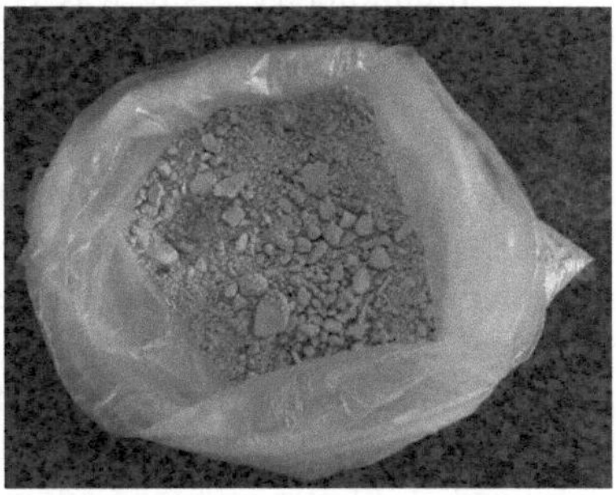

Fig. 2 Dried and ground residue. *Source* Own Author

Materials and Methods

The residue used was collected in ETE installed in Viminas, processing glass industry located in the city of Serra, metropolitan region of Vitoria, Espirito Santo state, Brazil. To implement this study, the residue was dried and ground.

Despite the waste, the result of a process of polishing that generates very fine particles, it underwent a harrowing process because after passing the press in ETE, the residue is grinded after drying. Two grindings were carried out with the harrowing duration of 1 h each, being used in each of these 4 kg of dry waste glass in an oven at 100 °C to constant weight. At the end of waste grinded it was screened for the first time, to match the grain size of the sand.

After this, it was found that the grains themselves had a still larger than desired (Fig. 2). The milling process uses the ball mill so that after grinding, the fineness residue should match with cement fineness. However, prior to the grinding process, the residue was homogenised and packed in plastic bags.

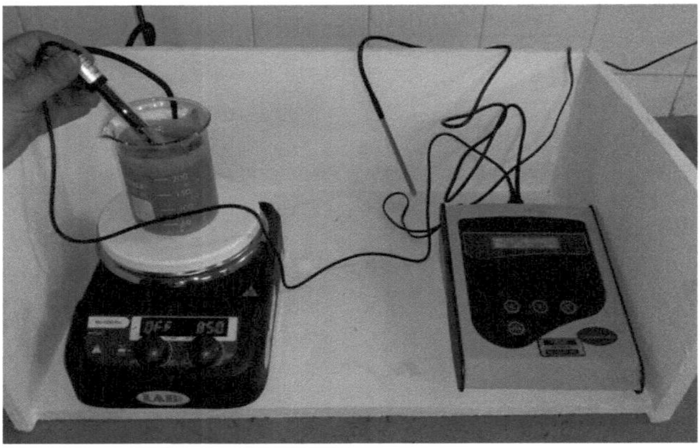

Fig. 3 Pozzolanic activity method. *Source* Own Author

The granulometry assay was performed by sieving and sedimentation residue. The determination of the density was performed according to NBR NM 23 [6] and the specific surface was determined by the Blaine method, standardized by NBR 16372 [7].

Among the different techniques for the study of pozzolanicity, it was held in the residue, testing electrical conductivity, a method that measures the evolution of pozzolanicity proposed by Lúxan et al. [8]. The method monitors variation in the electrical conductivity within 2 min after addition of the residue to the mixture. The assay is the addition of 98.7 mg of calcium hydroxide [$Ca(OH)_2$] in 70 g of distilled water at 40 ± 1 °C. With the aid of a magnetic stirrer, the solution was stabilized at 20 min and then adding 1.75 g of powdered glass residue [9], as seen in Fig. 3.

Assays for determining consistency were made by the truncated cone of the spreading table, performed as NBR 13276 [10]. This test allows to evaluate the workability of the mortar, and therefore carried out with the mortar in fresh state. In the water retention test is evaluated water holding capacity using a suction proceeding, and determined the amount of water (later expressed as a percentage) retained in the mortar after the proceeding (Fig. 4). Water retention in mortar is evaluated by the method proposed by NBR 13277 [11].

The control of the entrained air content is critical to the control of the mortar quality, whether to check maximum and minimum limits of desirable entrained air, or to identify gaps in levels of air. Entrained air content determination tests were performed according to NBR 13278 [12].

Fig. 4 Water retention test. *Source* Own Author

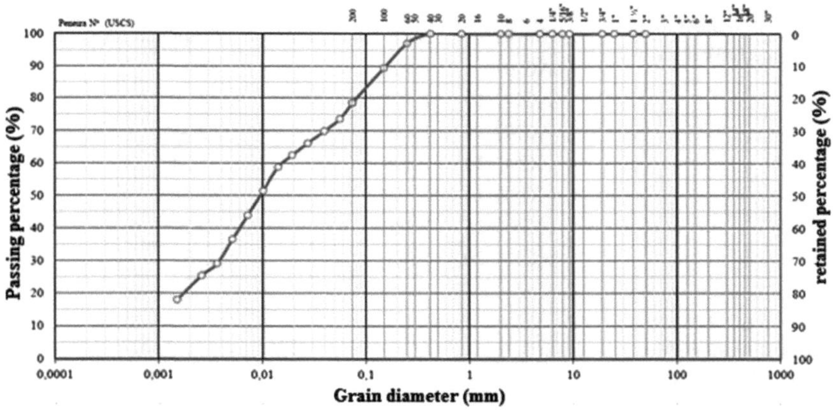

Fig. 5 Particle size distribution of waste

Results

The particle size distribution can be seen in Fig. 5.

The value of the density obtained was 2488 g/cm^3 because the specific surface area was 3685.70 cm^2/g, a slightly thicker than the cement used which has a specific surface of 4450 cm^2/g. In Table 1 shows the results of electrical conductivity of residue of pulp and paper measured during 20 min to classify the residue as a pozzolan.

The tested residues showed a conductivity variation of two minutes, which classifies a moderate pozzolanic activity waste, according to classification proposed by Lúxan et al. [8], according Table 2.

All results about pozzolanicity show that the glass residue shows a good pozzolanic activity. Table 3 shows the results for the entrained air content and mass

Table 1 Result of the electrical conductivity of glass waste

Pozzolanicity			
Sample	Conductivity (mS/cm)		Difference Conductivity
	Initial	End	
1	8.12	6.88	1.24
2	8.82	7.26	1.56
3	8.50	7.23	1.27

Table 2 Leaderboard pozzolanicity the material proposed by Lúxan et al. [8]

Material classification	Variation conductivity (ΔC)
Not pozzolanic	$\Delta C < 0.40$
Moderate pozzolanicity	$0.40 < \Delta C < 1.20$
Good pozzolanic	$\Delta C > 1.20$

Table 3 Entrained air content and bulk density fresh

Traces	Bulk density (g/cm^3)	Entrained air content (%)
AC 0	1.469	28.4
ACSC 10	1.451	28.99
ACSC 15	1.439	29.45
ACSC 20	1.398	31.34
ACSA 10	1.389	31.68
ACSA 15	1.386	32.2
ACSA 20	1.381	32.37

Table 4 Entrained air content and bulk density fresh

Traces	Water retention (%)
AC 0	99.28
ACSC 10	97.21
ACSC 15	99.75
ACSC 20	99.06
ACSA 10	98.89
ACSA 15	99.61
ACSA 20	98.43

density fresh. The traces presented on without replacement none (AC) in place with cement (ACSC) and sand in place (ACSA) in different ratios ranging 10, 15 and 20%. This same methodology was used in the water retention test.

Table 4 shows the results for the different water retention characteristics.

Conclusion

Regarding the effect of the addition of the residue in mortars can be stated that:

- The characterization of glass waste allowed a better evaluation of the pozzolanic properties of this material. From the tests it was possible equate the size distribution of the waste with the cement, the residue with 70% of particles under 40 μm having a specific surface lower than the cement. Thus, the results of the specific surface area obtained for the residue may influence the rate of hydration and workability of the prepared mortar with cement replacement at residue.
- With the addition of sand substitution at residue, the surface area of the mixture components increases, so that to obtain higher water retention.
- The entrained air content is a parameter which greatly influences on workability, so the mortar containing residues is lighter, has higher workability and the time it remains workable. This makes it possible to reduce stress during application, resulting in higher productivity. However, it is worth remembering that the entrained air, which corresponds to the voids formed by the air inlet inside the mortar, while improving their workability, impairs the mechanical strength. Soon the values found admits to its use.

Acknowledgements The authors thank the Brazilian agencies: CNPq, CAPES and FAPERJ for the support provided to this investigation.

References

1. Pereira FR (2006) Industrial waste recovery as mineral alternative source: ceramics and cimentíceas composition. Engineering Department of Ceramics and Glass of the Federal University of Rio de Janeiro. [S.l.] (in portuguese)
2. Pereira-de-Oliveira LA, Castro-Gomes JP, Santos PMS (2012) The potential pozzolanic activity of glass and red-clay ceramic waste as cement mortars components. Constr Build Mater, pp 197–203
3. Shao Y, Lefort T, Moras S, Rodriguez D (2000) Studies on concrete containing ground waste glass. Cem Concr Res, pp 91–100
4. Abividro (2015) Brazilian association of technical glass automatic industries. Available at: http://www.abividro.org.br/ (in Portuguese)
5. Santos DP, Azevedo ARG, Hespanhol RL, Alexandre J (2016) Characterization of incorporation the glass waste in adhesive mortar. Characterization Mine Met Mater
6. Brazilian Association of Technical Norms—ABNT NBR NM 23 (2001) Portland cement and other powdered material—Determination of density (in Portuguese), Rio de Janeiro, Brazil
7. Brazilian Association of Technical Norms—ABNT NBR 16372 (2015) Portland cement and other powdered materials—Determination of fineness by the air permeability method (Blaine method) (in Portuguese), Rio de Janeiro, Brazil
8. Lúxan MP, Madruga F, Saavedra J (1989) Rapid evaluation of pozzolanic activity of natural products by conductivity measurement. Cem Concr Res, pp 63–68

9. Barroso TR (2011) Study of the pozzolanic activity and application in concrete ashes of sugarcane bagasse with different physical and chemical characteristics, Dissertation. State University of Norte Fluminense (in Portuguese)
10. Brazilian Association for Technical Norms—ABNT NBR-13276 (2005) Mortars applied on walls and ceilings—Preparation of mortar for unit masonry and rendering with standard consistence index, Rio de Janeiro, Brazil (in Portuguese)
11. Brazilian Association of Technical Norms—ABNT NBR-13277 (2005) Mortars applied on walls and ceilings—Determination of the water retentivity, Rio de Janeiro, Brazil (in Portuguese)
12. Brazilian Association of Technical Norms—ABNT NBR-13278 (2005) Mortars applied on walls and ceilings—Determination of the specific gravity and the air entrained content in the fresh stage, Rio de Janeiro, Brazil (in Portuguese)

Experimental Evaluation of the Influence of Mortar's Mechanical Properties on the Behavior of Clay Masonry

Rita de C.S.S. Alvarenga, Gustavo H. Nalon, Lucas A.F. Fioresi, Monica C. Pinto, Leonardo G. Pedroti and José C.L. Ribeiro

Abstract This study aims to evaluate the influence of mortar's mechanical properties on the behavior of clay masonry. Experimental studies were performed using prisms of clay blocks under axial compression. The clay units were produced with a clay from Tres Rios, Rio de Janeiro, Brazil and had compressive strength of 4.9 MPa. Two mortar compositions were used, which had compressive strength of 94.7 and 50.4% of the clay blocks' net area compressive strength. It was noticed a slight increase on the prism's compressive strength with the increase of the mortars' compressive strength and stiffness. Prisms made with flexible mortars had a ductile failure and collapsed under lower compressive stresses, while prisms made with stiff mortars resisted higher compressive stresses, but had a brittle failure, crushing the blocks. A very strong mortar is not recommended. Using more flexible mortars does not compromise the strength and stiffness of the masonry and prevents an excessively brittle failure of the structure.

Keywords Mortar properties · Clay blocks · Masonry structures

---

R. de C.S.S. Alvarenga (✉) · G.H. Nalon · L.A.F. Fioresi · M.C. Pinto · L.G. Pedroti · J.C.L. Ribeiro
DEC—Civil Engineering Department, UFV—Federal University of Vicosa, Av. PH Rolfs, 36570-900 Vicosa, Minas Gerais, Brazil
e-mail: rcassia.alvarenga@gmail.com

G.H. Nalon
e-mail: gustavohnalon@gmail.com

L.A.F. Fioresi
e-mail: lucasaleksander@gmail.com

M.C. Pinto
e-mail: monicapinto36@gmail.com

L.G. Pedroti
e-mail: lpedroti@gmail.com

J.C.L. Ribeiro
e-mail: jcarlos.ribeiro@ufv.br

S. Ikhmayies et al. (eds.), *Characterization of Minerals, Metals, and Materials 2017*, The Minerals, Metals & Materials Series,
DOI 10.1007/978-3-319-51382-9_74

Introduction

As a result of technological advances, the design of masonry structures has evolved from empirical methods into rational ones. The study of each component of masonry structures (e.g. block unit, mortar and grout) and the interaction between them has become common practice in the construction industry.

More specifically, the mortar is used to provide uniform support and adherence between blocks. Mechanical resistance along with endurance to environmental changes are key requisites to hardened mortars. In plastic conditions, the mortar should allow bricklaying and compensate dimensional variations between block units [1].

Traditionally in Brazil, laying mortar is composed of a mix between Portland cement, hydraulic lime, sand and water. The Brazilian Standard [2] specifies that mortar compressive strength should be at least 1.5 MPa and not more than 70% of the clay blocks' net area compressive strength. On the other hand, the ASTM C 1329 [3] and C 270 [4] categorize mortars in different types (e.g. type M, S, N, O), which have different applications and physical properties.

The analysis of masonry's mechanical properties can be made using prisms. ASTM C 1314 [5] defines a masonry prism as the assemblage of masonry units and mortar constructed to serve as a test specimen for determining properties of masonry assemblages. In this study, compressive strength and Young's modulus tests were performed in masonry prisms along with an analysis of failure type.

Materials and Methods

The fired clay hollow blocks shown in Fig. 1 were produced in Tres Rios, Rio de Janeiro, Brazil, with a nominal size of (140 × 190 × 390) mm. Experiments were performed to determine actual dimensions, absorption, net area, and compressive

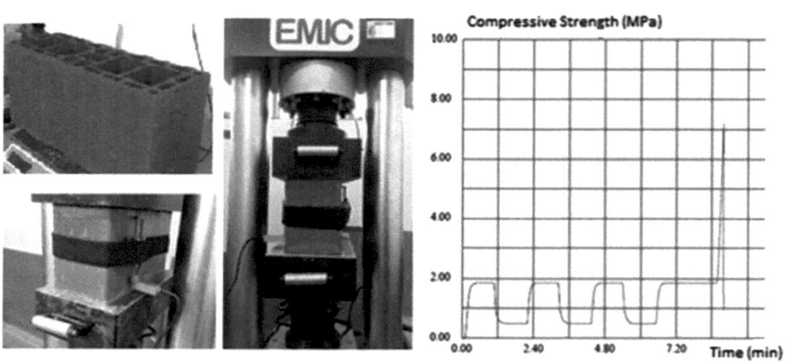

Fig. 1 Determination of clay blocks' modulus of elasticity

strength of the units, according to the Brazilian standard ABNT NBR 15270-3:2005 [6]. The tensile strength was determined according to the ASTM C1006 – 07 [7]. To determine the blocks' elasticity modulus, a test was performed on the DL 60000 EMIC testing machine, placing the blocks between (200 × 420) mm plates. Loadings and unloadings were conducted with the velocity of 0.05 MPa/s. The applied load was recorded by the machine's load cells. Longitudinal strain was measured by two strain gauges placed on the two larger faces of the block as shown in Fig. 1. The blocks' elasticity modulus was calculated as the slope of the straight line between the points of compressive stress of 0.5 MPa and the 30% of the block's estimated ultimate strength.

Mortar was produced with natural and quartzite sand from Porto Firme, Minas Gerais, Brazil; Brazilian Portland cement CP II E-32, produced by Tupi S.A.; and hydrated lime CHIII Supercal, produced by Ical, from Sao Jose da Lapa, Minas Gerais, Brazil.

The two mortar mixtures produced were designated as T50 and T100, which had respectively compressive strength of 50 and 100% of the clay blocks' net area compressive strength.

After mixing the mortar, six (40 × 40 × 160) mm prims and four 50 mm cylinders were cast. Specimens were left inside the environmental room during 28 days. After that, flexure tests and compression tests on prism halves were conducted as shown in Fig. 2, according to ABNT NBR 15270-3:2005 [6]. The cylinder specimens were used to determine the mortars' elasticity modulus, which was calculated as the slope of the straight line between the points of compressive stress of 0.5 MPa and the 30% of the mortar's estimated ultimate strength.

Six prisms were built with two blocks and each mortar composition. After 28 days of curing, their compressive strength and modulus of elasticity were determined in the DL 60000 EMIC testing machine, according to the loading procedures shown in Fig. 3.

A velocity of 0.05 MPa/s was used until the level of 50% of the failure load. Thereafter, a velocity of 0.15 MPa/s was used until the failure. The applied load was also measured by load cells. DT-100A Kyowa displacement transducers with 100 mm of rated capacity were calibrated and placed on the sides of the prism, as

Fig. 2 Flexure, compression, and elasticity modulus of mortar tests

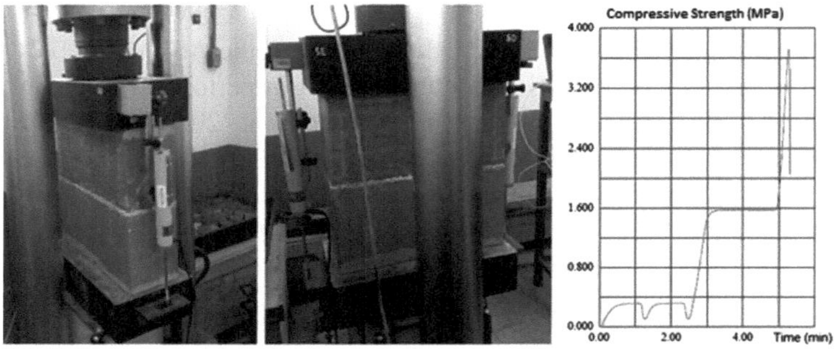

Fig. 3 Placement of Kyowa transducers and loading of the prisms tests

shown in Fig. 3. Fixed by magnetic bases on the top plate, they measured the displacement of the lower plate in the vertical direction. The prism deformations in the direction of the load axis were determined from the average of the displacements measured by each transducer.

The displacements measured by the transducers were obtained at a 5 Hz rate by the Data Acquisition System (DAQ) QuantumX and Catman®AP software from HBM. The transducers were connected to channels 1 and 2 of the MX840A universal amplifier. The modulus of elasticity of each prism was calculated in the interval corresponding to the secant straight line between 5 and 30% of the prism's actual ultimate strength.

Results and Discussion

Some properties obtained from the clay unit tests are summarized in Table 1. It also indicates the number of specimens used in each test and the coefficient of variation associated with the measurements. The compressive strength was calculated using the procedure described in ABNT NBR: 15270-2:2005 [8], which assume a normal distribution of the specimens' strength and determines the strength that more than 95% of the specimens can take.

The tensile strength of the clay units is about 9% of its gross area compressive strength, which is close to other sources' data. Brooks [9] presents a general trend for the modulus of elasticity of ceramic masonry, based on data of different sources. It indicates that the modulus of elasticity of clay units, in GPa, is about 40% of its compressive strength, in MPa, but there is a large scatter of data. The modulus of elasticity calculated for the blocks on this research is close to Brooks' prediction.

Table 2 presents for each mortar mixture the volumetric ratios of cement, lime, and sand; the water/cement ratio; the flow measured according to ABNT NBR 13276:2005 [10]; and some environmental conditions during its production process.

Table 1 Properties of clay units

Property	Value	Number of specimens	CV (%)
Average width (mm)	141	15	0.33
Average height (mm)	388	15	0.52
Average length (mm)	189	15	0.34
Water absorption (%)	19.5	6	0.93
Average net area A_n (mm^2)	20,535	13	0.99
Average gross area A_g (mm^2)	55,039	13	0.75
Ratio of A_n/A_g (%)	37.31	–	–
Gross area compressive strength $f_{bk\ \mathrm{Gross\ Area}}$ (MPa)	4.9	13	12.66
Net area compressive strength $f_{bk\ \mathrm{Net\ Area}}$ (MPa)	13.1	–	–
Tensile strength f_{bt} (MPa)	0.44	5	27.87
Modulus of elasticity E_b (MPa)	2,538	5	11.63

CV: coefficient of variation associated with the measurements

Table 2 Mortar compositions and environmental conditions during their production

Mortar	PC : lime : sand (volume)	Water/cement ratio	Flow (mm)	Temperature (°C)	Humidity (%)
M100	1:1:5	3.08	265	18.4	62
M50	1:1.5:6.75	3.66	235	19.3	66

Table 3 lists the properties obtained from the mortar tests, performed after 28 days curing. As expected, the compressive strength of the mortar M100 and M50 was 94.7 and 50.4% of the clay blocks' net area compressive strength, respectively.

The stress-strain curve of the prisms P50 and P100 are plotted in Figs. 4 and 5, ranging between the beginning of testing and 50% of the prism's estimated ultimate strength, when the displacement transducers were removed in order to protect equipment from a sudden specimen's failure. The modulus of elasticity of each prism was calculated from its stress-strain curve, considering the slope of a straight line between the points of 5 and 30% of the prism's actual ultimate strength.

Table 4 shows the mechanical properties obtained from the prism tests. The values found for the prism's compressive strength are pretty close to the values found in the literature, which range from 0.30 to 0.60 of the clay block's compressive strength, according to Parsekian et al. [1]. After analysis of experimental data from different sources, these authors found a general trend for the modulus of elasticity of ceramic masonry, which is between 390 and 444 f_{pk}. The modulus of elasticity calculated for P100 and P50 are within that range.

Table 3 Mechanical properties of mortar

Property	M100		M50		Number of specimens
	Mean	CV (%)	Mean	CV (%)	
Average compressive strength f_m (MPa)	12.4	5.36	6.6	6.69	12 (prism halves)
Average tensile strength f_{tm} (MPa)	3.5	4.35	2.1	10.47	6 (prisms)
Modulus of elasticity E_m (MPa)	13,155	5.89	8,733	2.61	3 (cylinders)

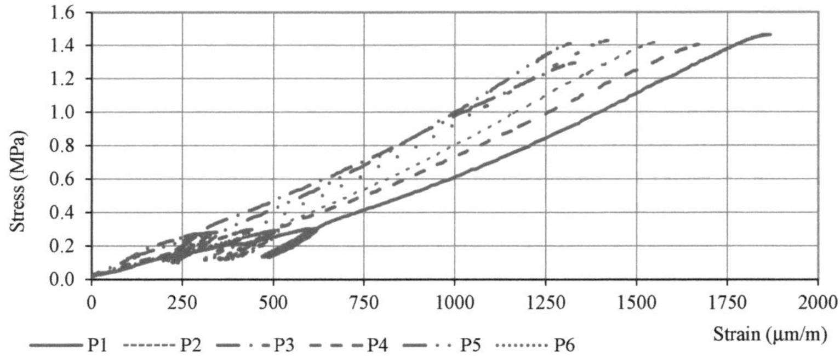

Fig. 4 Stress-strain curves—prisms P50

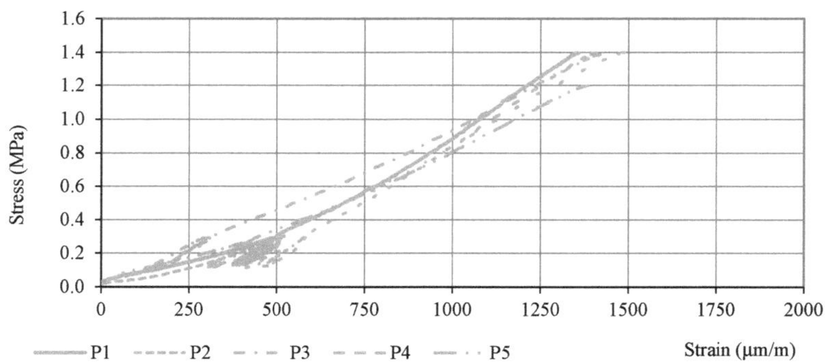

Fig. 5 Stress-strain curves—prisms P100

It was noticed an increase of 7.4% on the prism's compressive strength and an increase of 9.7% on the prism's modulus of elasticity with the 87.8% increase of the mortars' compressive strength.

The prisms P50 were produced with the weakest and most flexible mortar. The modulus of elasticity of the mortar M50 (8,733 MPa) is about 33.6% lower than

Table 4 Mechanical properties of prisms

	Compressive strength of prisms			Modulus of elasticity of prisms		Ratio of f_{pk}/f_{bk} gross area	Number of specimens
	Average compressive strength f_{pm} (MPa)	CV (%)	Characteristic compressive strength f_{pk} (MPa)	Average modulus of elasticity E_p (MPa)	CV (%)		
P100	3.4	21.34	2.9	1167	13.95	0.59	6
P50	3.2	7.70	2.7	1064	9.18	0.55	6

Fig. 6 Failure mechanism—prisms P50

Fig. 7 Failure mechanism—prisms P100

the modulus of elasticity of the mortar M100 (13,155 MPa), while its compressive strength (6.6 MPa) is about 46.8% lower than the compressive strength of the mortar M100 (12.4 MPa). While the uniaxial load is applied during the prism's compression test, the mortar and the blocks tend to expand laterally. The block/mortar adherence tries to prevent these lateral expansions, which leads to lateral compressive/tensile stresses in the prisms components. When used the weakest mortar (M50), which is much more flexible than the other, lateral tensile stresses occurred in the block when the mortar tried to expand. Consequently, the prism failed when the lateral tensile stresses reached the tensile strength of the blocks, occurring vertical fissures on the block's walls, as shown in Fig. 6. When used the strongest mortar (M100), which is much stiffer than the other, lateral compressive stresses occurred in the blocks, since the mortar deformation is no longer so substantial. In this case, the prism presented a very brittle failure, by crushing the blocks, as shown in Fig. 7.

Conclusions

Mechanical properties of mortar and clay units has significant influence on the behavior of structural clay masonry. The mechanical properties of clay blocks, mortar mixtures and prisms determined by this experiment were close to the data from literature.

Strength and stiffness of the mortar does not affect so significantly the compressive strength and stiffness of masonry. On the other hand, the mortar's strength and stiffness greatly affected the masonry failure mechanisms. Differential movements between its components result in very distinct ruptures of the structure. In order to obtain a more ductile and less explosive failure mechanism for the clay masonry, it is recommended to use weaker and less stiff mortar. When using the weakest mortar, the prisms did not break so suddenly. Vertical fissures on the block's walls were previous visual warnings before fracture. When using the strongest mortar, masonry failed abruptly and explosively, since the very stiff mortar could not effectively accommodate differential movements.

In conclusion, it is preferable not to use a very strong mortar, which does not compromise the strength and stiffness of the masonry and prevents an excessively brittle failure of the structure.

Acknowledgements The authors thank the Brazilian agencies: National Counsel of Technological and Scientific Development (CNPq), Coordination for the Improvement of Higher Education Personnel (CAPES) and Research Support Foundation of Minas Gerais (FAPEMIG) for the support provided to this study.

References

1. Parsekian, GA, Hamid AA, Drysdale RG (2012) Comportamento e dimensionamento de alvenaria estrutural. Edufscar, São Carlos, p 625
2. Brazilian Association of Technical Norms (2010) NBR 15812-1: structural masonry—clay blocks part 1: project. Rio de Janeiro, p 41
3. ASTM C1329/C1329 M-16 (2016) Standard specification for mortar cement. ASTM International, West Conshohocken, PA. www.astm.org
4. ASTM C270-14a (2014) Standard specification for mortar for unit masonry. ASTM International, West Conshohocken, PA. www.astm.org
5. ASTM C1314-16 (2016) Standard test method for compressive strength of masonry prisms. ASTM International, West Conshohocken, PA. www.astm.org
6. Brazilian Association of Technical Norms (2005) NBR 15270-3: ceramic components part 3: structural and non-structural ceramic blocks—test methods. Rio de Janeiro, p 27
7. ASTM C1006-07 (2013), Standard test method for splitting tensile strength of masonry units. ASTM International, West Conshohocken, PA. www.astm.org

8. Brazilian Association of Technical Norms (2005) NBR 15270-2: ceramic components part 2: structural ceramic block, perforated block, load-bearing masonry—terminology and requirements. Rio de Janeiro, p 11
9. Brooks JJ (2015) Concrete and masonry movements. Elsevier, Waltham, p 599
10. Brazilian Association of Technical Norms. NBR 13276: Mortars applied on walls and ceilings—Preparation of mortar for unit masonry and rendering with standard consistence index. Rio de Janeiro, 2005, p 3

Influence of Operation Conditions on Normal Stress and Flow Pattern of Burden Materials in Blast Furnace Based on Discrete Element Method

Wenxuan Xu, Shusen Cheng and Guolei Zhao

Abstract Nowadays, the flow pattern and stress distribution of burden materials in blast furnace has been widely calculated by discrete element method. In this study, the influence of operation conditions such as ore batch, the root position of cohesive zone and the slope of cohesive zone on the flow pattern and stress distribution of burden materials in blast furnace were investigated. The normal stress of burden materials in deadman increases with the increase of ore batch, but the normal stress of burden materials in deadman decreases with the increase of the root position of cohesive zone and the slope of cohesive zone. The average descending velocity of burden materials increases with the increase of ore batch, but the average descending velocity of burden materials decreases with the increase of the root position of cohesive zone and the slope of cohesive zone.

Keywords Blast furnace operation · Normal stress · Flow pattern · Discrete element method

Introduction

Blast furnace is a complicated and closed multi-phase reaction box, which includes solid phase, liquid phase, gas phase and powder phase. The coke and ore are charged into blast furnace layer by layer, with the consumption of coke in raceway and the periodic discharge of slag and iron, the burden materials in blast furnace descends slowly. However, the stable operation of blast furnace is closely to the stable descending of burden materials, some abnormal phenomena appear when the burden materials descend unsteadily. Therefore, the motion of burden materials in blast furnace has been the research focus of scholars at home and abroad.

W. Xu (✉) · S. Cheng · G. Zhao
School of Metallurgical and Ecological Engineering, University of Science and Technology Beijing, Beijing 100083, China
e-mail: xuwenxuan_ustb@126.com

S. Ikhmayies et al. (eds.), *Characterization of Minerals, Metals, and Materials 2017*, The Minerals, Metals & Materials Series,
DOI 10.1007/978-3-319-51382-9_75

Most researchers are investigated the motion of burden materials by experimental study of physical model [1–3]; calculation of continuous quasi fluid model based on particle dynamics [4, 5] and numerical calculation of discrete element method [5–9]. Takahashi et al. [10] obtained the flow pattern of burden materials in raceway and deadman by cold model, three characteristic regions were determined which are dominant flow region, quasi-stagnant region and stagnant region; Natsui et al. [11] investigated the influence of number of tuyeres on the motions of burden materials and stress field based on Discrete Element Method; Zhou et al. [12] also obtained that the motion of burden materials in blast furnace could be divided into three regions, rapid flow region, slow flow region and stagnant region based on Discrete Element Method; Yang et al. [13] investigated the influence of geometry boundary condition and simulation boundary condition on motion behaviors of burden materials in blast furnace. However, the effects of operating parameters on motions behaviors of burden materials in blast furnace are not clear. Therefore, in this study, the influence of operation conditions such as ore batch, the root position of cohesive zone (CZ) and the slope of cohesive zone on the flow pattern and stress distribution of burden materials in blast furnace were investigated.

Model Description

Mathematical Model

In DEM calculation, every individual particle is tracked and the motion is calculated by Newton's second law of motion, and the inter-particle contact model is schematically depicted in Fig. 1, which is composed of normal spring, normal damping, slider, tangential spring and tangential damping between two particles, the governing equations for particle *i* contact with another particle *j* are as follow:

$$m_i \frac{d\boldsymbol{u}_i}{dt} = \sum_{j=1}^{N} \left(\boldsymbol{F}_{cn,ij} + \boldsymbol{F}_{dn,ij} + \boldsymbol{F}_{ct,ij} + \boldsymbol{F}_{dt,ij} \right) + m_i g \quad (1)$$

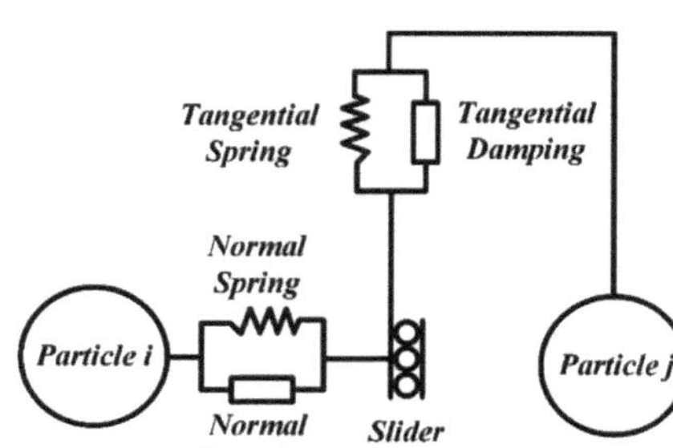

Fig. 1 Schematic of contact model between two particles, i and j, in DEM

$$\boldsymbol{I}_i \frac{d\omega_i}{dt} = \sum_{j=1}^{N} \left(\boldsymbol{T}_{t,ij} + \boldsymbol{T}_{r,ij}\right) \tag{2}$$

where, m_i, $\boldsymbol{u}_i$, $\boldsymbol{\omega}_i$, $\boldsymbol{I}_i$ are the mass, velocity, angular velocity, the moment of inertia of particle i, respectively; g is the acceleration of gravity; $\boldsymbol{F}_{cn,ij}$, $\boldsymbol{F}_{dn,ij}$, $\boldsymbol{F}_{ct,ij}$, $\boldsymbol{F}_{dt,ij}$ are the normal contact force, normal damping force, tangential contact force, tangential damping force, respectively; The torque acting on particle i includes a component from tangential force, $\boldsymbol{T}_{t,ij}$ and another from the rolling friction, $\boldsymbol{T}_{r,ij}$; N is total number of other particles contact with particle i.

Physical Model and Boundary Conditions

The dimension of blast furnace is shown in Fig. 2. The ore particles start to melt at the cohesive zone, and the coke particles consume in raceway and hearth at a certain rate. The length and height of raceway are 2.5 and 1 m, respectively. The diameter of coke and ore in simulation are 300 and 150 mm, respectively. The value of ore batch, root position of cohesive zone and slope of cohesive zone in different case are shown in Table 1. The physical parameters of coke and ore materials are shown in Table 2.

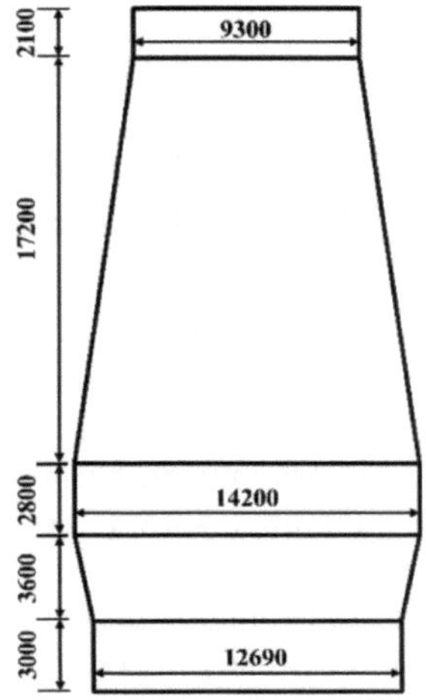

Fig. 2 The dimension of blast furnace (*unit* mm)

Table 1 The value of ore batch, root position of cohesive zone and slope of cohesive zone in different case

Case	Ore batch/kg	Root position of CZ/mm	Slope of CZ/°
Base case	**28,000**	**6600**	**60**
Case 1	**22,400**	6600	60
Case 2	**16,800**	6600	60
Case 3	28,000	**3000**	60
Case 4	28,000	**9400**	60
Case 5	28,000	3600	**45**
Case 6	28,000	3600	**30**

Table 2 The physical parameters of burden materials

Item	Density/kg m^{-3}	Shear modulus/Pa	Poisson's ratio/-	Coefficient of static friction/-	Coefficient of rolling friction/-	Coefficient of restitution/-
Coke	1050	2.2×10^9	0.22	–	–	–
Ore	3750	1.0×10^{10}	0.25	–	–	–
Wall	4500	1.0×10^{11}	0.30	–	–	–
C–C	–	–	–	0.56	0.15	0.18
C–O	–	–	–	0.66	0.10	0.17
C–W	–	–	–	0.41	0.09	0.20
O–O	–	–	–	0.67	0.15	0.18
O–W	–	–	–	0.35	0.10	0.20

Notes *C* represents coke; *O* represents ore; *W* represents wall

Results and Discussion

Influence of Ore Batch on Flow Pattern and Stress Field of Burden Materials

Figure 3 shows the influence of ore batch on the layer structure of burden materials, the batch of ore in base case, case 1 and case 2 are 28, 22.4 and 16.8 t, respectively. As shown in figure, with the decrease of ore batch, the thickness of ore layers becomes thinner, but the layer structure of burden material almost unchanged.

Figure 4 shows the influence of ore batch on the normal stress of burden materials. With the decrease of ore batch, the normal stress of burden materials in deadman region decreases. The reason is that the density of ore is about four times of the density of coke, therefore, the weight of burden materials in the region above the cohesive zone decreases.

Figure 5 shows the influence of ore batch on the velocity distribution of burden materials. With the decrease of ore batch, the velocity distribution of burden materials in the upper part of blast furnace becomes smaller. The reason is that the

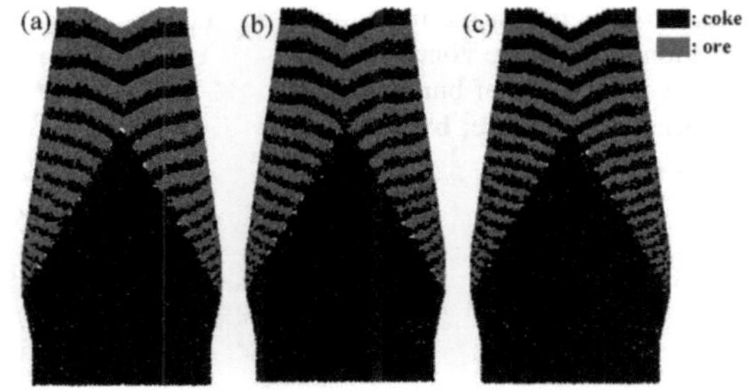

Fig. 3 Influence of ore batch on the layer structure of burden materials **a** base case; **b** case 1; **c** case 2

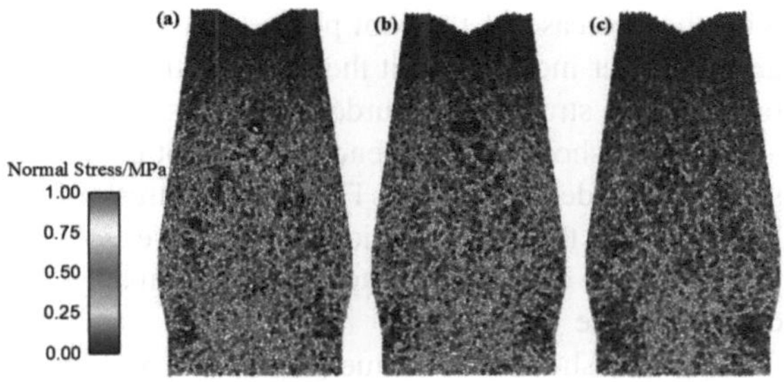

Fig. 4 Influence of ore batch on the normal stress of burden materials **a** base case; **b** case 1; **c** case 2

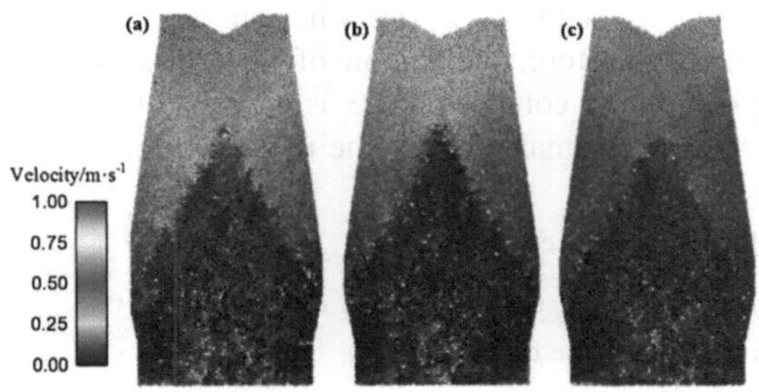

Fig. 5 Influence of ore batch on the velocity distribution of burden materials **a** base case; **b** case 1; **c** case 2

density of ore is larger than coke, with the decrease of the number of ore particles, the gravity of burden materials in the upper part of blast furnace decreases, therefore, the descending velocity of burden materials decreases.

Influence of the Root Position of Cohesive Zone on Flow Pattern and Stress Field of Burden Materials

The influence of the root position of cohesive zone on the layer structure of burden materials is shown in Fig. 6, the thickness of coke layer and ore layer are same in base case, case 3 and case 4. In base case, the root position of cohesive zone at the bottom of belly; the root position of cohesive zone at the top of hearth in case 3; and the root position of cohesive zone at the top of belly in case 4. As shown in Fig. 6,

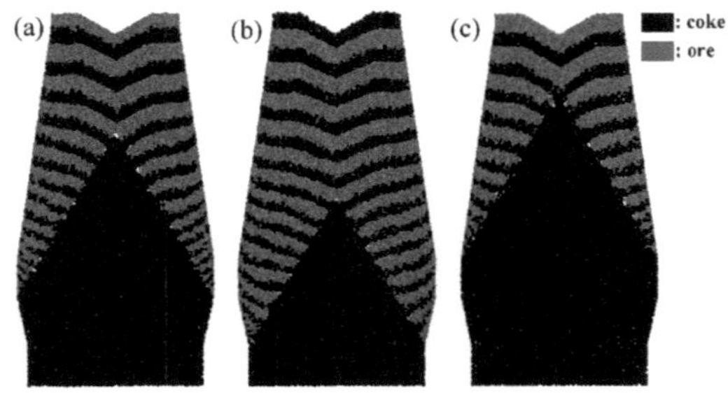

Fig. 6 Influence of the root position of cohesive zone on the layer structure of burden materials **a** base case; **b** case 3; **c** case 4

with the decrease of the root position of cohesive zone, the number of coke layer and ore layer increases, but the root position of cohesive zone has a little influence on the layer structure of burden materials.

Figure 7 shows the influence of the root position of cohesive zone on the normal stress of burden materials. The normal stress of burden materials in deadman is higher when the root position of cohesive zone is higher. On the contrary, the normal stress of burden materials in deadman is lower when the root position of cohesive zone is lower.

Figure 8 shows the influence of the root position of cohesive zone on the velocity distribution of burden materials. The descending velocity of burden materials becomes slower when the burden materials descend below the cohesive zone, therefore, the region of high descending velocity is larger when the root position of cohesive zone is lower. Otherwise, the region of high descending velocity is smaller when the root position of cohesive zone is higher.

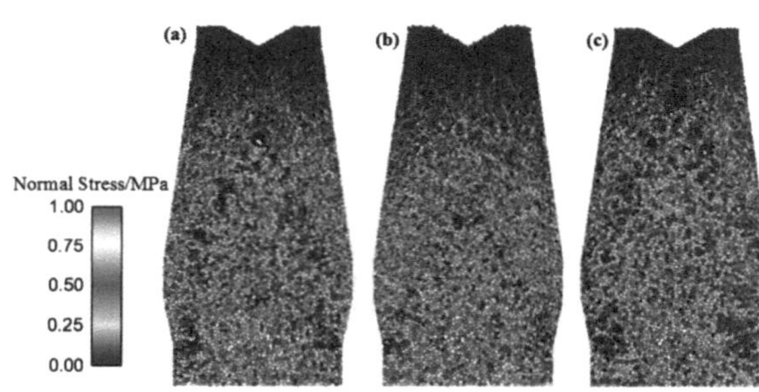

Fig. 7 Influence of the root position of cohesive zone on the normal stress of burden materials **a** base case; **b** case 3; **c** case 4

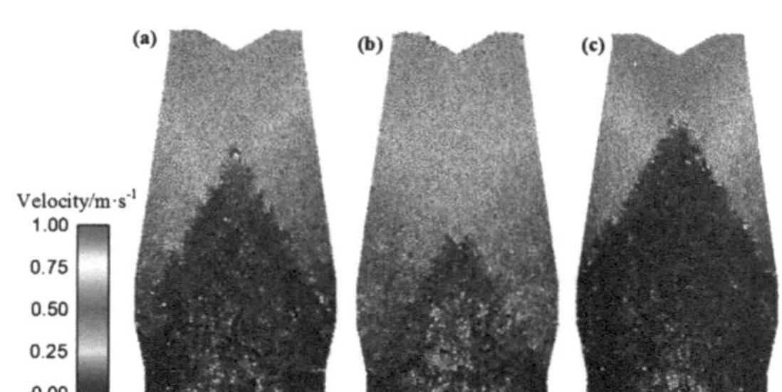

Fig. 8 Influence of the root position of cohesive zone on the velocity distribution of burden materials **a** base case; **b** case 3; **c** case 4

Influence of the Slope of Cohesive Zone on Flow Pattern and Stress Field of Burden Materials

Figure 9 shows the influence of the slope of cohesive zone on the layer structure of burden materials. Although the slope of cohesive zone is different, the layer structure of burden materials is almost same in different case, but the weight of ore above the cohesive zone increases with the decrease of slope of cohesive zone.

Figure 10 shows the influence of the slope of cohesive zone on the normal stress of burden materials. Owing to the weight of ore above the cohesive zone is larger when the slope of cohesive zone is smaller, the normal stress of burden materials in deadman becomes larger. It could be obtained that the normal stress of burden materials in deadman becomes larger with the decrease of the slope of cohesive zone.

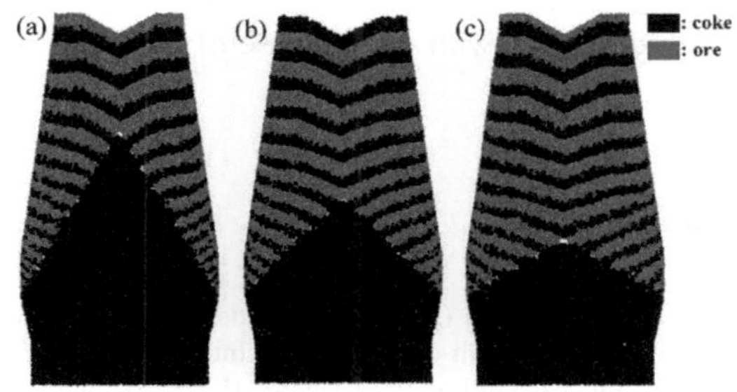

Fig. 9 Influence of the slope of cohesive zone on the layer structure of burden materials **a** base case; **b** case 5; **c** case 6

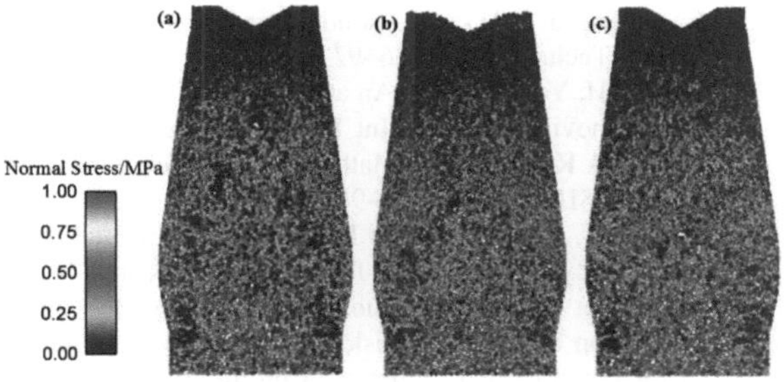

Fig. 10 Influence of the slope of cohesive zone on the normal stress of burden materials **a** base case; **b** case 5; **c** case 6

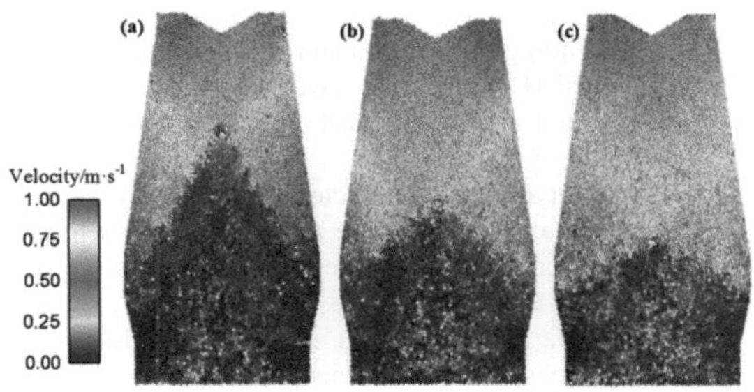

Fig. 11 Influence of the slope of cohesive zone on the velocity distribution of burden materials **a** base case; **b** case 5; **c** case 6

Figure 11 shows the influence of the slope of cohesive zone on the velocity distribution of burden materials. The descending velocity of burden materials becomes smaller when the burden materials descend below the cohesive zone, therefore, the region of high descending velocity is larger when the root position of cohesive zone is lower. It could be concluded that the cohesive zone is an important boundary of high descending velocity region and low descending velocity region.

Conclusions

(1) The normal stress and average descending velocity of burden materials increases with the increase of ore batch.
(2) The normal stress and average descending velocity of burden materials decreases with the increase of the root position of cohesive zone.
(3) The normal stress and average descending velocity of burden materials decreases with the increase of the slope of cohesive zone.

References

1. Nogami H et al (2004) Cold-model experiments on deadman renewal rate due to sink-float motion of hearth coke bed. ISIJ Int 12(44):2127–2133
2. Zaïmi SA, Guillot JB, Biausser H (2013) "A solid flow model for blast furnace based on hypo-plasticity theory". Ironmaking Steelmaking 6(30):475–482
3. Wright B et al (2011) Gas–solid flow in an ironmaking blast furnace — I: Physical modelling. Powder Technol 1(208):86–97
4. Nogami H, Yagi J (2004) An application of bingham model to viscous fluid modeling of solid flow in moving bed. ISIJ Int 11(24):1826–1834
5. Safavinick R et al (2013) Mathematical model of solid flow behavior in a real dimension blast furnace. ISIJ Int 6(53):979–987
6. Nouchia T, Yu AB, Takeda K (2003) Experimental and numerical investigation of the effect of buoyancy force on solid flow. Powder Technol 1–2(134):98–107
7. Fan ZY et al (2010) Influence of blast furnace inner volume on solid flow and stress distribution by three dimensional discrete element method. ISIJ Int 10(50):1406–1412
8. Zhang J et al (2011) Influence of profile of blast furnace on motion and stress of burden by 3D-DEM. J Iron Steel Res Int 11(18):01–06
9. Ueda S et al (2010) "Influences of physical properties of particle in discrete element method on descending phenomena and stress distribution in blast furnace". ISIJ Int 7SI (50):981–986
10. Takahashi H et al (2005) Two dimensional cold model study on unstable solid descending motion and control in blast furnace operation with low reducing agent rate. ISIJ Int 10 (45):1386–1395
11. Natsui S et al (2010) Characteristics of solid flow and stress distribution including asymmetric phenomena in blast furnace analyzed by discrete element method. ISIJ Int 2(50):207–214
12. Zhou ZY et al (2005) Discrete particle simulation of solid flow in a model blast furnace. ISIJ Int 12(45):1828–1837
13. Yang WJ, Zhou ZY, Yu AB (2015) Discrete particle simulation of solid flow in a three-dimensional blast furnace sector model. Chem Eng J 278:339–352

Polymer Blend Based on Recycled Polyethylene and Ethylene Vinyl Acetate Copolymers Reinforced with Natural Fibers from Agricultural Wastes

Renata D.S. Coiado, Gisele D. Lazo, Rene R. Oliveira, Rita C.L.B. Rodrigues and Esperidiana A.B. Moura

Abstract Agricultural residues, which are produced with large quantities annually throughout the world, may be used as reinforcement plastic to replace the wood and produce particleboard for application in the development of low cost construction elements and reduced environmental impact. The main aims of this study was to investigate the effects of agricultural wastes and glass residues addition on the properties of recycled-HDPE/EVA blend for use in particleboard manufacture. The recycled-HDPE/EVA blend reinforced with corncob fiber (15 wt%), coffee parchment (15 wt%) and glass residues (1 wt%) were processed by melt extrusion, using a twin screw extruder and injection molding machine to obtain specimen test samples. The samples were characterized by mechanical test, XRD, TG, DSC, and FE-SEM analysis to understand the nature of interaction between the fillers reinforcement and recycled-HDPE/EVA blend matrix and their properties were discussed.

Keywords Recycled-HDPE · EVA · Agricultural wastes · Glass residues · Mechanical test

Introduction

Agricultural residues such as cereal straw, cornstalks, corn cob, sugarcane bagasse, rice husk, cotton stalks, banana stalks, coconut coir, bamboo, among others, which are produced with large quantities annually throughout the world, represent a potential source of fiber that may be used as reinforcement plastic. These agricultural

R.D.S. Coiado · G.D. Lazo · R.R. Oliveira · E.A.B. Moura (✉)
Instituto de Pesquisas Energeticas e Nucleares—IPEN-CNEN/SP, Av. Prof. L. Prestes, 2242, Sao Paulo-SP 05508-000, Brazil
e-mail: eabmoura@ipen.br

R.C.L.B. Rodrigues
Escola de Engenharia de Lorena, Departamento de Biotecnologia, Universidade de Sao Paulo, Lorena-SP, Brazil

S. Ikhmayies et al. (eds.), *Characterization of Minerals, Metals, and Materials 2017*, The Minerals, Metals & Materials Series,
DOI 10.1007/978-3-319-51382-9_76

residues cam be used to replace the wood polymer composite and produce particleboard, medium-density fiberboard (MDF) and hardboard for application in the development of low-cost construction elements, furniture industry, thermal and acoustic isolation [1–3]. It has been reported the use of agricultural residues as a potential alternative material to replace wood to make composite materials, particularly for particleboards [2]. Use of the agricultural residues as reinforcement polymer composite materials contribute to reduction of environmental impact, once promote the removal of residues unutilized in the field, preventing their inadequate deposition that can lead to accidental burning or burned which are caused by the man of the field for the purpose of land clearing, growing crops or training pastures. The fire in these residues spreads with great speed and can quickly destroy kilometers of green areas that become unproductive; increase the release carbon dioxide, a major cause of global warming, cause the destruction of natural habitats, and the soil erosion [4–6].

Over the years the consumption of plastic products made of conventional polymers (petroleum resources and non-biodegradable) has produced a large amount of waste which has led to disposal problems in the worldwide. This waste has high degradation resistance taking long years to decompose and represent a large part of the municipal solid waste landfills in many parts of the world. The inappropriate disposal of this waste contribute to pollution and can lead to contamination of soil, groundwater sources and generate negative environmental impacts. Among the alternatives to minimize these problems are the reuse and recycling practices that are increasing over time [7–9]. In the last decade, polyolefin, like polyethylene, polypropylene and ethylene vinyl acetate due to their low-cost, good mechanical properties, and light weight are widely used to produce plastic products. The large increase of use these polymers has also increased concerns about of final disposal of these residues [8–10]. Thereof, the recycling of polyolefin and the use of recycled to produce new materials reinforced with agricultural residues for application in the development of low-cost construction elements, it can be an alternative to reduce the waste whose traditional disposal practice include incineration and discarding landfill.

The need to add value to wastes with the purpose of promoting increased sustainability has led to the opening towards the use of different wastes as reinforcements of polymer materials, in replacement to synthetic fiber. Among these waste, glass residues are an important candidate. The use of glass powder as reinforcement in polymer materials can be advantageous with respect to hardened and shine properties. Glass, a highly transparent material has proven its importance in our lives through manufactured products such as sheet glass, bottles, glassware, and several others. However, it non-biodegradable characteristics have been creating a serious problem for solid waste disposal. Then, the use of glass residues as reinforcement in polymeric materials besides to improve the polymer materials properties can be also a worthy solution to the environmental problem caused by this solid waste.

The aim of this work was to investigate the effects of agricultural wastes and glass residues addition on the properties of recycled-HDPE/EVA blend for use in particleboard manufacture.

Experimental

Materials:

In this work were used the following materials: recycled-HDPE (high density polyethylene), recycled-EVA (ethylene vinyl acetate), corncob fiber residues, coffee parchment residues and glass residues.

Preparation of Agricultural Residues:

The corncob fiber and coffee parchment residues were dried at 80 ± 2 °C for 24 h in an air-circulating oven, then pulverized into 250 μm powders with a ball mills and dried again at 80 ± 2 °C for 24 h in an air-circulating oven to reduce its moisture content to less than 2%.

Preparation of Glass Residues Powder:

The glass residues used in this study were obtained from of disposal of food packaging glass pots. The waste glass were washed, kept in distilled water for 4 h and dried at 110 ± 2 °C for 24 h in an air-circulating oven. Then ground the sample into fine powder, with particle sizes equal to or smaller than 30 μm by use the ball mills and then dried again at 110 ± 2 °C for 24 h to reduce the moisture content to less than 2%.

Preparation of blend and composite:

The recycled-HDPE/EVA blend (r-HDPE/EVA) (70/30 wt%) and its Composite were processed by melt extrusion using a twin-screw extruder Haake Rheomix with 16 mm and L/D = 25 rate from Thermo Scientific. The temperature profile was 170/175/185/190/200/200 °C. Screw speed was 40 rpm. The extrudates coming out of the extruder were cooled down for a better dimensional stability, pelletized by a pelletizer, dried again at 60 ± 2 °C for 24 h and fed into injection molding at 170/175/180 temperature profile and specimens test samples were obtained. The Compositions of the blend and its composite are presented in Table 1.

Characterization:

Tensile tests: Tensile tests were carried out using an INSTRON Universal Testing Machine, model 5564 according to ASTM D 638 standard at 25 °C. Six replicates were made of each sample and the average values were reported.

Table 1 Compositions of the r-HDPE/EVA blend and its composite

Materials	Compositions (wt%)
r-HDPE/EVA blend[a]	70/30
r-HDPE/EVA blend and corncob/coffee/glass[b]	70 and 15/15/1

[a]Recycled HDPE/EVA

[b]Recycled HDPE/EVA/corncob fiber residues/coffee parchment residues/glass residues

X-Rays Diffraction (XRD): XRD patterns were recorded on a Siemens—D5000 diffractometer operated at 40 kV and 40 mA, with CuKα radiation (λ = 15.4 Å).

Thermogravimetric analysis (TG): In this study, the tests were made in a Mettler Toledo TGA module "TGA/SDTA851e" from 25 to 500 °C at a heating rate of 10 °C/min under nitrogen atmosphere (50 ml/min).

Differential Scanning Calorimetry (DSC): (DSC) analyses were carried out using a Mettler Toledo DSC 822e from 25 to 250 °C at a heating rate of 10 °C/min under nitrogen atmosphere (50 ml/min). The samples were initially heated from 25 to 250 °C at a heating rate of 10 °C/min and the temperature was held at 250 °C for 5 min to eliminate the previous thermal history. Subsequently, the sample was cooled to room temperature at a cooling rate of 10 °C/min, and secondly heated to 250 °C at a heating rate of 10 °C/min rate to obtain heating thermograms. DSC were carried out to obtain melt temperature (Tm) and melting enthalpy (ΔHm), and crystallinity percentage variation X_C (%) of r-HDPE/EVA blend and its composite. The X_C is directly related to melting enthalpy (ΔHm) and was determined from the following equation:

$$X_C = (\Delta Hm/\Delta Hm^0(1 - wf)) \times 100\ \% \quad (1)$$

where:

ΔHm = melting enthalpy of sample.
ΔHm^0 = melting enthalpy of HDPE assuming 100 % crystallinity = 293 J/g; The crystallinity of EVA was calculated using the enthalpy of fusion of neat polyethylene, assuming that only the polyethylene sequences in EVA crystallize [11, 12].
wf = mass fraction of the GO in the flexible films.

Topography of samples: FE-SEM of cryofractured samples under liquid nitrogen were carried out using a JEOL-JSM-6701 F, microscope with an accelerating voltage of 1–30 kV, using EDS Thermo-Scientific mod. Noran System Six software, in carbon sputtered samples.

Results and Discussion

X-ray Diffraction (XRD): Fig. 1 shows the XRD pattern for the r-HDPE/EVA Blend and its composite. It can be seen that XRD spectrum of r-HDPE/EVA Blend presented a large reduction of the intensity due to agricultural and glass residues addition.

Thermogravimetric analysis (TG) results: Fig. 2 shows the TG thermograms of r-HDPE/EVA Blend and its composites. As can be seen in Fig. 2, the r-HDPE/EVA Blend showed a great reduction in the onset temperature due to agricultural and glass fiber residues addition.

The decomposition temperatures and weight loss of the r-HDPE/EVA blend and its composite are presents in Table 2. In the case of composite, the TG thermograms

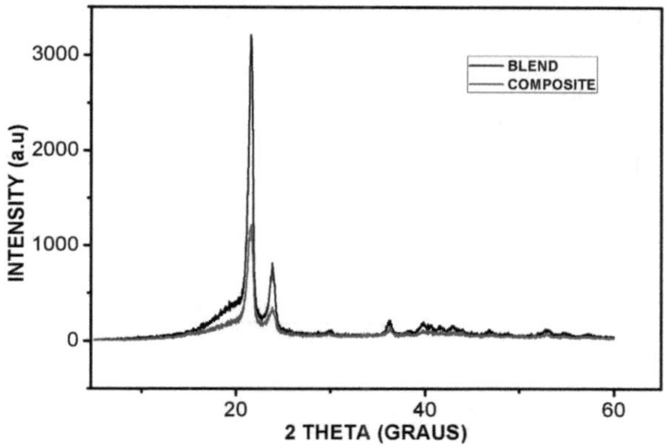

Fig. 1 X-ray diffraction patterns for the r-HDPE/EVA blend and its composite

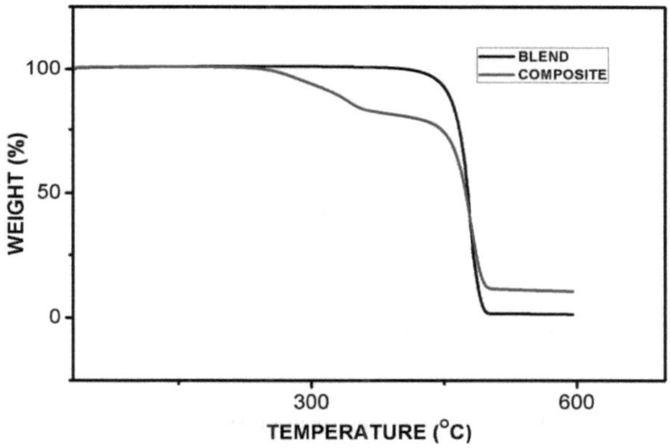

Fig. 2 TG thermograms of the r-HDPE/EVA blend and its composite

showed three-stage loss of weight. The first, a decomposition shoulder peak of low intensity centered of around 222 °C, characteristic of low molecular weight fiber constituents, such as hemicelluloses and pectin, as reported by [13, 14]. The second one, appearing at higher temperature, a centered peak at 350 °C corresponded to the thermal degradation of cellulose [13, 15]. This second stage loss of mass may be explained also by the contribution of lignin content in these fiber residues, whereas that of lignin decomposition range is between 280 and 500 °C. The third stage of degradation temperature occurred at 405 °C corresponds to the thermal degradation of composite.

Differential Scanning Calorimetry (DSC) analysis results: Fig. 3 shows the DSC analysis results of r-HDPE/EVA Blend and its composites.

The results of DSC are presented in Table 3 and they show no difference in the melting temperature of the neat blend when compared with its composite.

Table 2 Decomposition temperatures and weight loss of the r-HDPE/EVA blend and its composite

Materials	T_{onset} (°C)	T_{max} (°C)	Weight loss (%)
r-HDPE/EVA blend[a]	404	498	98.5
r-HDPE/EVA blend and corncob/coffee/glass[b]	222/350/405	503	89.3

[a]Recycled HDPE/EVA
[b]Recycled HDPE/EVA/corncob fiber residues/coffee parchment residues/glass residues

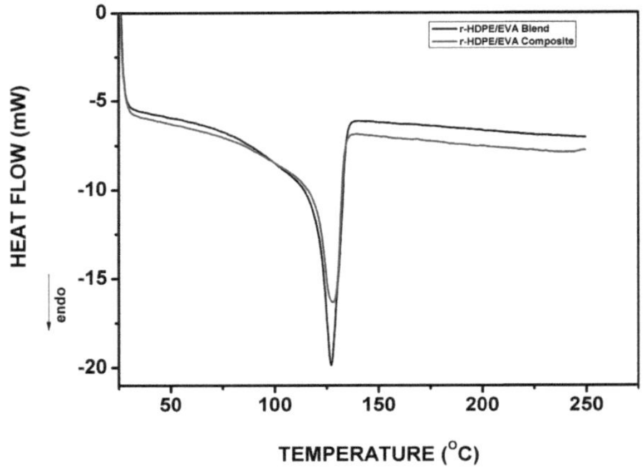

Fig. 3 DSC analysis results of the r-HDPE/EVA blend and its composite

Table 3 DSC analysis results of the r-HDPE/EVA blend and its composite

Materials	Melting temperature (T_m, °C)	Melting enthalpy (ΔH_m, J g^{-1})	Crystallinity percentage (Xc, %)
r-HDPE/EVA blend[a]	127.1	149.3	51.0
r-HDPE/EVA blend and corncob/coffee/glass[b]	127.9	111.3	54.3

[a]Recycled HDPE/EVA
[b]Recycled HDPE/EVA/corncob fiber residues/coffee parchment residues/glass residues

From Table 3, it could be inferred that compared with the r-HDPE/EVA blend the endothermic melting enthalpy of its composite decreased considerably due to agricultural residues addition. The decreases in the melting enthalpy can be attributed to the increase in crystallinity of r-HDPE/EVA composite.

Tensile tests results: Table 4 presents the results of tensile tests of the r-HDPE/EVA Blend and its composites. The results presented shows the average values calculated from the data obtained in tests for five test specimens.

From Table 4 it can be seen increase of the tensile strength at break and Young modulus properties of blend due to agricultural and glass residues addition. On the

Table 4 Tensile test results of the r-HDPE/EVA blend and its composite

Materials	Tensile strength at break (MPa)	Elongation at break (%)	Young's modulus (MPa)
r-HDPE/EVA blend[a]	10.2 ± 0.4	161.4 ± 15	619 ± 56
r-HDPE/EVA blend and corncob/coffee/glass[b]	13.2 ± 0.4	31.9 ± 2.9	723 ± 44

[a]Recycled HDPE/EVA
[b]Recycled HDPE/EVA/corncob fiber residues/coffee parchment residues/glass residues

other hand, it can be seen observed a large reduction in elongation at break of composites when compared with blend.

Topography of samples analysis results: FE-SEM micrographs of cryofractured surfaces of the r-HDPE/EVA Blend and its composite specimens were studied to understand the failure mechanisms and also study possible interaction between different components. FE-SEM micrographs of the r-HDPE/EVA Blend and its composites in different magnifications are showed in Fig. 4. Figure 4a is an overview of r-HDPE/EVA blend showing a uniform dispersion of r-EVA in the r-HDPE matrix. Micrographs surface of r-HDPE/EVA/Corncob Fiber

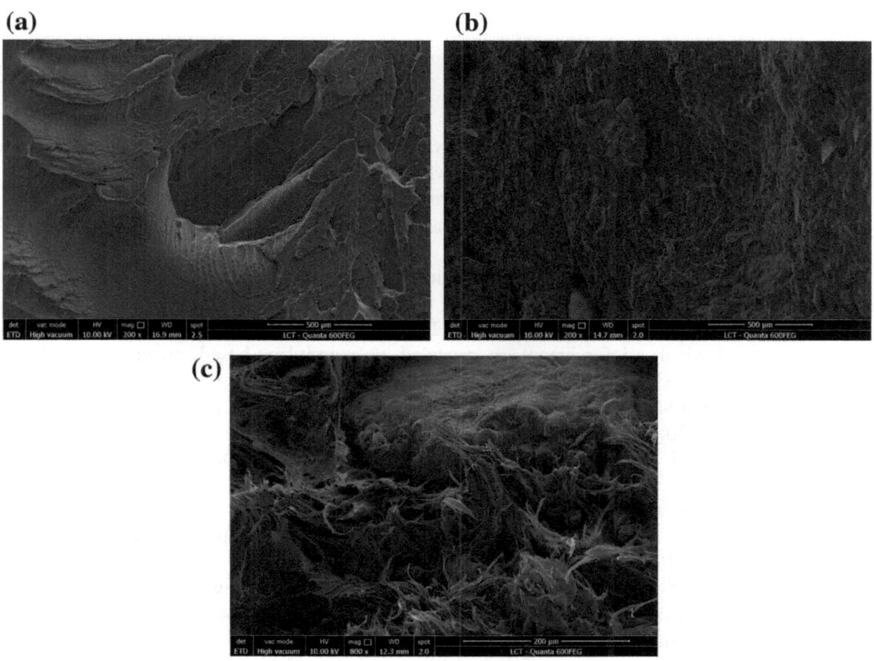

Fig. 4 FE-SEM micrographs of cryofractured surfaces of the r-HDPE/EVA Blend and its composite, **a** r-HDPE/EVA blend (200X), **b** r-HDPE/EVA/corncob fiber residues/coffee parchment residues/glass residues (200X), **c** r-HDPE/EVA/corncob fiber residues/coffee parchment residues/glass residues (800X)

Residues/Coffee Parchment Residues/Glass Residues with 200X and 800X of magnifications are showed in Fig. 4b, c, respectively. It can be seen in these figures a homogeneous distribution of agricultural fiber residues particles, but with some holes r-HDPE/EVA Blend due to agricultural fiber residues pull out.

Conclusion

The results showed gains in tensile strength at break, elongation at break and Young's modulus properties of blend due to agricultural residues addition. Compared with the r-HDPE/EVA blend the endothermic melting enthalpy of its composite decreased considerably due to agricultural residues addition. The decreases in the melting enthalpy can be attributed to the increase in crystallinity of r-HDPE/EVA composite. XRD spectrum of r-HDPE/EVA Blend presented a large reduction of the intensity due to agricultural and glass residues addition. FE-SEM showed homogeneous distribution of agricultural fiber residues particles, but with some holes in r-HDPE/EVA Blend matrix of composites due to agricultural fiber residues particle pull out.

References

1. Madurwar MV, Ralegaonkar RV, Mandavgane SA (2013) Application of agro-waste for sustainable construction materials: a review. Constr Build Mater 38:872–878
2. Panthapulakkal S, Sain M (2007) Agro-residue reinforced high-density polyethylene composites: fiber characterization and analysis of composite properties. Compos Part A 38:1445–1454
3. Faustino J et al (2012) Impact sound insulation technique using corn cob particleboard. Constr Build Mater 37:153–159
4. Ardanuy M, Antunes M, Velasco JI (2012) Vegetable fibres from agricultural residues as thermo-mechanical reinforcement in recycled polypropylene-based green foams. Waste Manag 32:256–263
5. Sahari J, Sapuan SM (2011) Natural fibre reinforced biodegradable polymer composites. Rev Adv Mater Sci 30:166–174
6. Yan L, Chouw N, Jayaraman K (2014) Flax fibre and its composites—a review. Compos B 56:296–317
7. Scott G (2000) Green' polymers. Polym Degrad Stab 68:1–7
8. Najafi SK (2013) Use of recycled plastics in wood plastic composites—a review. Waste Manag 33:1898–1905
9. Azwa ZN, Yousif BF, Manalo AC, Karunasena W (2013) A review on the degradability of polymeric composites based on natural fibres. Mater Des 47:424–442
10. Ammala A, Bateman S, Deana K, Petinakis E, Sangwan P, Wong S, Yuan Q, Yu L, Patrick C, Leong KH (2011) An overview of degradable and biodegradable polyolefins. Prog Polym Sci 36:1015–1049
11. Bistac S, Vallat MF, Schultz J (1999) Prog Org Coat 37:49
12. Kodjie SL, LI L, LI B, Cai W, LI CY, Keating M (2006) Morphology and crystallization behavior of HDPE/CNT nanocomposite. J Macromol Sci Part B Phys 45:231–245

13. Rachini A, Le Troedec M, Peyratout C, Smith A (2009) Comparison of the thermal degradation of natural, alkali-treated and silane-treated hemp fibers under air and an inert atmosphere. J Appl Polym Sci 112:226–234
14. Einhorn-Stolla U, Kunzek H, Dongowsk G (2007) Thermal analysis of chemically and mechanically modified pectins. Food Hydrocolloids 21:1101–1112
15. Monteiro SN, Calado V, Rodriguez RJ, Margem FM (2012) Thermogravimetric stability of polymer composites reinforced with less common lignocellulosic fibers—an overview. J Mater Res Technol 1(2):117–126

Preliminary Study of the Effect of Stirring Rate, Temperature and Oxygen Pressure on the Leach Rate of Copper Powder, Generated by Grinding of Printed Circuit Boards of Computer

M.A. Mesinas Romero, I. Rivera Landero, M.I. Reyes Valderrama, E. Salinas Rodríguez, J. Hernández Ávila, E. Cerecedo Sáenz and E.G. Palacios Beas

Abstract For this study, a sample of 5 g of copper powders obtained by grinding printed circuit boards of computers fixe to 50 mesh (270 μm) according Tyler® series was used. This sample was wholly characterized by XRD, SEM–EDS and AAS. The leaching medium used was $O_2-S_2O_3^{2-}$. From the evaluated variables it was obtained an apparent order of the reaction of n = 2.34 s^{-1}, whereas according the effect of partial oxygen pressure it was obtained an apparent order of reaction of n = 0.9516 and the activation energy of system was of 6.5 kJ mol^{-1}, the results so achieved suggest a transport control to the solid–liquid interface. Under certain work conditions such as: $S_2O_3^{2-} = 0.5$ M, 500 mL, 750 min^{-1}, P/PO_2 = 1 atm, 338 K, 5 g of sample and 9000 s of reaction time, it is possible get up to 97.35% of leached copper.

Keywords E-waste · Thiosulfate · Leaching · PCB · Copper

Introduction

An electronic waste (or e-waste) could be those devices both electrical and electronic that have fulfilled their useful life of have ceased to be valuable, and whose service time is market by both technological and scientific development which

M.A. Mesinas Romero (✉) · I. Rivera Landero · M.I.R. Valderrama · E. Salinas Rodríguez · J. Hernández Ávila · E. Cerecedo Sáenz
Área Académica de Ciencias de la Tierra y Materiales, Universidad Autónoma del Estado de Hidalgo, Carr. Pachuca-Tulancingo Km 4.5, 42184 Mineral de La Reforma, Hidalgo, Mexico
e-mail: me132370@uaeh.edu.mx

E.G. Palacios Beas
Instituto Politécnico Nacional, Escuela Superior de Ingeniería Química e Industrias Extractivas, Avenida Instituto Politécnico Nacional s/n, 07738 Gustavo A. Madero, Lindavista, Mexico

S. Ikhmayies et al. (eds.), *Characterization of Minerals, Metals, and Materials 2017*, The Minerals, Metals & Materials Series,
DOI 10.1007/978-3-319-51382-9_77

promote a continuous process use and disposal, thus producing large quantities of electronic wastes that nowadays represent an environmental problem and they must be reduced or discarded [1–5].

In recent years, have been proposed solutions of recovery and recycling of e-waste, focused on the development of a circular economy proposed by the exploitation of resources from the e-waste [6], which constitute one of the fastest growing fractions, representing 8% or more of all municipal solid wastes [7, 8], of which 3% corresponds to the printed circuit boards of computers, of total e-waste [9, 10].

Printed circuit boards of computers are considered the basis of the electrical/electronics industry and can contain up to 60 elements, some of which are valuable and other toxic; can classify these elements in metals, non-metals and organic materials [10–12]. These printed circuit boards are complex and heterogeneous structures whose composition varies according to the type of printed circuit board, the device type, year of manufacture and other factors [12, 13].

Generally the printed circuit boards are constituted approximately by; 16–20% Cu, 4% Sn, 3% Fe, 2% Ni, 1% Zn, 0.03% Au, 0.05% Ag, 0.01% Pd [14–17], whose purity is up to 10 times higher than that found in economically exploitable mineral deposits, which can be considered as an "urban mineral resource" [12, 13].

Some authors have conducted many studies to leach copper present in the printed circuit boards, finding that this metal can be leached effectively by acid or ammoniacal via [18]. On the other hand, some researchers have also evaluated the use of thiosulfate as a complexing reagent for precious metals such as Au and Ag, adding oxidants such as Cu^{2+} or oxygen, besides of stabilizing reagents such ammonia; observing in their results that the use of any of these systems allows to get dissolution rates quite similar to those obtained with the use of cyanide [19–21]; however, has observed the formation of complexes not only of precious metals but also of ammonia and copper [22, 23], so taking advantage of this feature, copper leaching was carried out in this work from printed circuit boards of computer using a basic medium, $O_2-S_2O_3^{2-}$.

Methodology

The material used in this work was given by the Mexican company "CVR, S.A de C.V." dedicated to recycling electronic devices, and the experimental procedure used in experimental part of work was that used by Isauro [24] and proposed by Mesinas et al. [25].

Characterization

The characterization of sample was done using an Scanning Electron Microscope (SEM–EDS), Jeol JSM–6300, and an X–ray Diffractometer (XRD) Equinox 2000 using the MATCH 1.1 software to results found which contains the PDF files. Finally, to analyze the copper content in sample, 0.1 g of sample were dissolved in concentrate solution of HNO_3, the obtained solution was decanted and then adjusted to a volume of 200 mL with distilled water, dilutions and standards were prepared with identical matrix and then analyzed using an Atomic Absorption Spectrophotometer (AAS) Perkin Elmer Precisely model Aanalyst 200.

Experimental Procedure and Materials

All leaching experiments were carried out in a 500 mL reactor, using distilled water and 5 g of sample pre - sieved and limited to mesh 50 (270 μm) according sieve Tyler series, treated with a 0.5 M pentahydrate sodium thiosulfate solution (brand *J.T. Baker*), 1 atm of partial oxygen atmosphere injected to solution, 750 min^{-1} of stirring rate, pH = 10, t = 10800 s and a temperature of 318 K; maintaining constant the pH by adding a 2 M NaOH solution. Aliquots were taken at predetermined time intervals, and the pH of the solution was constantely measured with a pH meter *Dual Star PH/ISE* with electrode and *ATC* probe, brand *Thermo Orion*.

Results and Discussion

Characterization

Characterization carried out by SEM–EDS and XRD show that the majority element present in sample is copper. Figure 1 shows a general image of particle size distribution found in mesh 50 (270 μm), where can be seen particles with irregular morphology which is the principal characteristic that shows these kind of material after grinding, also an EDS analysis is presented showing the typical composition

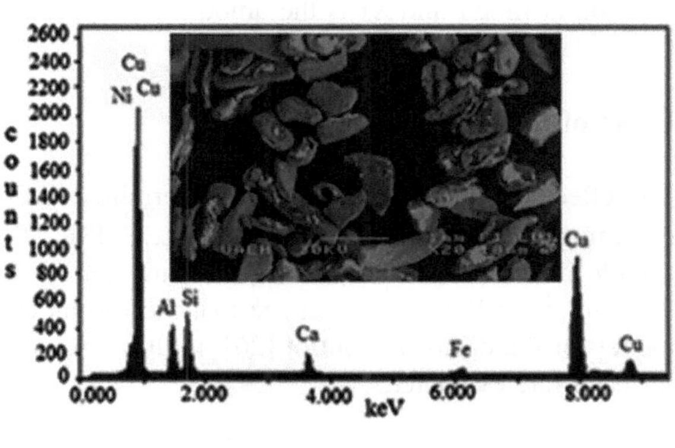

Fig. 1 General image of particles of copper powder (SEM-EDS)

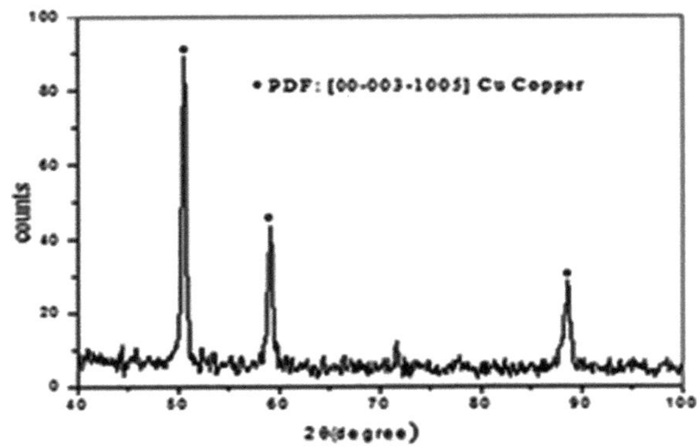

Fig. 2 X-ray diffraction spectrum of copper powders

of these particles [9]. Figure 2, shows the X–ray Diffractogram in which three characteristic peaks of copper are observed and were indexed with the PDF file [00-003-1005] that corresponds to copper. Chemical analysis done to sample by AAS reveals a 71.96% of copper present in sample, that is the majority phase.

Preliminary Kinetic Study

To choice a kinetic model and determine so the controlling stage of copper disolution process, were carried out some experiments, and and the results found are according to the model of spherical particles of constant size with an unreacted core and by diffusive control $X = 1 - 3(1 - X_{Cu})^{2/3} + 2(1 - X_{Cu})$.

The most useful variable that describes the evolution of solid during conversion, is X, that is an adimensional number that corresponds to the reacted fraction (Eq. 1):

$$X = \frac{Xt}{X\tau} \tag{1}$$

where X is the copper fraction that has already reacted, Xt is the amount of copper in solution at time t and $X\tau$ is the amount of copper at end of reaction.

Effect of the Stirring Rate

The effect of the stirring rate was determined under the following experimental conditions: $[S_2O_3^{2-}] = 0.5$ M, 500 mL, 318 K, $P/PO_2 = 1$ atm, 338 K, 5 g of sample and 10800 s of reaction, varying the stirring rate in the range of 350–900 min^{-1}. In Fig. 3, we can see that the results conform to the above–described model for the diffusion control [26], in this case, showing the relevant K_{exp} in s^{-1}, for each rate evaluated.

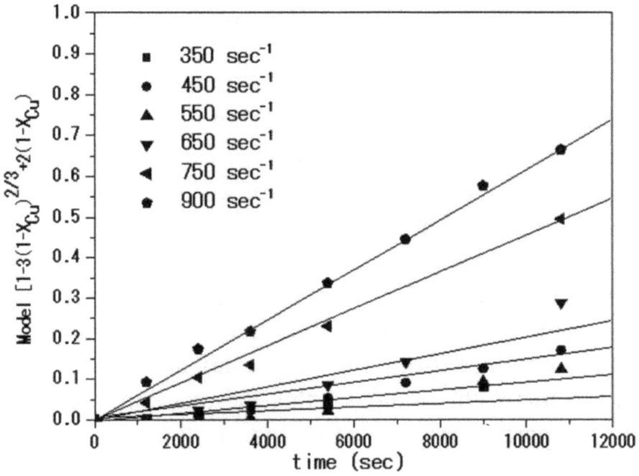

Fig. 3 Leaching of copper: effect of stirring rate

In Fig. 4, is shown the rate range here studied, and by linear regression of the obtained slope it was possible to calculate the order of the reaction for this variable, getting so an apparent order of reaction of $n = 2.34\ s^{-1}$.

The obtained results of this work show that stirring rate has an important effect over leaching rat of copper for this system, presenting so an increasing in the K_{exp} as stirring rate also increases, which is indicative that to a higher stirring rate can be reduced the thickness of the liquid–liquid film, or Nernst´s layer. Which could indicate that there is a concentration gradient of the reactants or products, between the bulk solution and the metal surface, so that is there a decreasing of the thickness of the liquid film, and will appear an increasing in the dissolution rate of copper, due to the existence of a minor impediment in transferring oxygen to system [24].

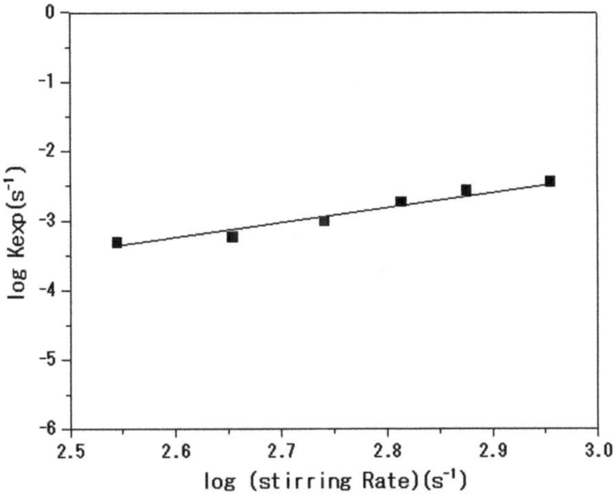

Fig. 4 Leaching of copper: effect of stirring rate. The apparent order of the reaction is $n = 2.34\ s^{-1}$

Effect of Oxygen Partial Pressure

Two experiments were executed according to the working conditions previously describes in experimental procedure section, only varying the oxygen partial pressure. The first experiment was done to a $P/PO_2 = 1$ atm and the second to a $P/PO_2 = 0.2$ atm. Figure 5 shows the results obtained for both experiments according copper leaching; where can be seen that results fix to the model of spherical particles of constant size and decreasing diffusion control. Figure 6 shows the dependence of the rate of copper leaching with respect to the oxygen partial pressure to the range studied here. When performing the linear regression of the slope obtained from Fig. 6, the apparent order of reaction is obtained, which is n = 0.9516, this indicates that the reaction rate for copper leaching is proportional to the partial pressure of oxygen in the system.

Effect of Temperature

Figure 7 shows the experimental results obtained during evaluation of the effect of temperature in the range of 289–338 K for copper leaching, as in the above effects, the results also fit well for the model of spherical particles of constant size and unreacted core for diffusion control. According the overall results, it has been found that is possible to leach up to 97.35% of copper under the following experimental conditions: $[S_2O_3^{2-}] = 0.5$ M, 500 mL, 750 min^{-1}, $P/PO_2 = 1$ atm, 338 K, 5 g of sample and 9000 s. Figure 8 shows the slope obtained by plotting the natural logarithm of the leach rate of copper versus 1000/T, from which the activation energy of system studied is calculated, giving a value of 6.5 kJ mol^{-1}, indicating that the temperature has little effect on the rate of copper leaching and confirming the diffusive leaching process control.

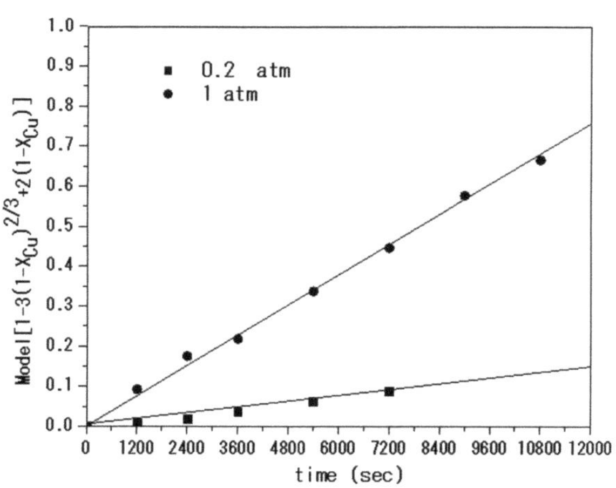

Fig. 5 Leaching of copper: effect of oxygen partial pressure

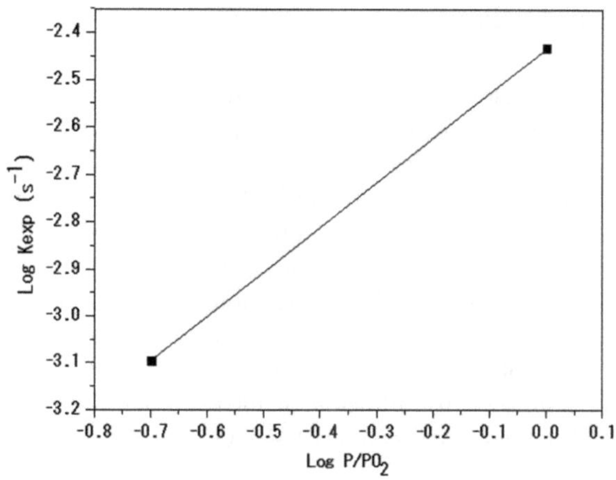

Fig. 6 Leaching of copper: effect of oxygen partial pressure. The apparent order of the reaction is n = 0.9516

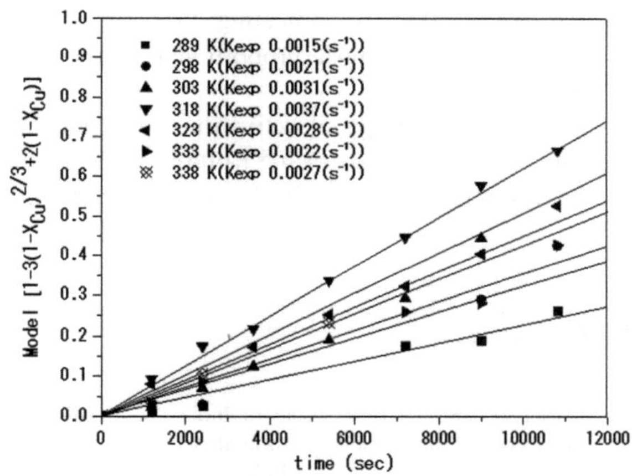

Fig. 7 Leaching of copper: effect of temperature

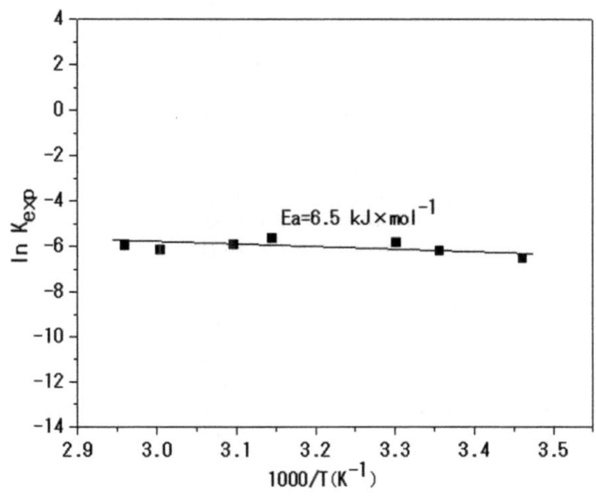

Fig. 8 Leaching of copper: determination of the activation energy

Conclusions

- Characterization done to grinded printed circuits boards of computes by SEM–EDS and XRD, shows that copper is present as majority element in this material. Chemical analysis done by AAS reveals a content of 71.96% of copper in the sample, from which can be leached up to 97.35% under the following experimental conditions: $[S_2O_3^{2-}] = 0.5$ M, 500 mL, 750 min^{-1}, $P/PO_2 = 1$ atm, 338 K, 5 g of sample and 9000 s.
- The apparent order of reaction with respect to the stirring rate effect was of $n = 2.34\ s^{-1}$, which is indicative of a strong dependence of this hydrodynamic variable with the rate constants obtained, indicating so that the reaction is controlled by diffusion; it means, controlling stage of reaction rate of copper leaching in the system $O_2-S_2O_3^{2-}$ is the transport of oxygen to the solid–liquid interphase.
- For the effect of the partial pressure of oxygen, the apparent order of the reaction was of $n = 0.9516$, indicating that rate of copper leaching in the system $O_2-S_2O_3^{2-}$ is almost proportional to the oxygen dissolved in the proper system.
- The apparent energy of activation calculated in the studied range of temperatures 289–338 K was of 6.5 $kJ \cdot mol^{-1}$, which suggests a control by transport of the system $O_2-S_2O_3^{2-}$ in the copper leaching.

References

1. Garlapati VK (2016) E-waste in India and developed countries: management, recycling, business and biotechnological initiatives. Renew Sustain Energy Rev 10:874–881
2. Needhidasan S, Samuel M, Chidambaram R (2014) Electronic waste—an emerging threat to the environment of urban India. J Environ Health Sci Eng 12(36):1–9
3. Orlins S, Guan D (2016) China´s toxic informal e-waste recycling: local approaches to a global environmental problem. Cleaner Prod 114:71–80
4. Abdollahi Y et al (2015) Enhacement of electronic protection to reduce e-waste. J Ind Eng Chem 29:400–407
5. Wei L, Liu Y (2012) Present status of e-waste disposal and recycling in China. Proc Environ Sci 16:506–514
6. Cucchiella F et al (2015) Recycling of WEEEs: an economic assessment of present and future e-waste streams. Renew Sustain Energy Rev 51:263–272
7. Widmer R et al (2005) Global perspectives on e-waste. Environ Impact Assess Rev 25:436–458
8. Sthiannopkao S, Wong MH (2013) Handling e-waste in developed and developing countries: initiatives, practices, and consequences. Sci Total Environ 463–464:1147–1153
9. Flandinet L et al (2012) Metal recovering from waste printed circuit boards (WPCBs) using molten salts. J Hazard Mater 213–214:485–490
10. Zhou Y, Qiu K (2010) A new technology for recycling materials from waste printed circuit boards. J Hazard Mater 175:823–828
11. Szałatkiewicz J (2014) Metals content in printed circuit board waste, Pollution J Environ Stud 123(6):2365–2369

12. Wu GQ, Zhang ZK (2010) Recycling of waste printed circuit boards. Circuit World 36 (4):35–39
13. Huang K, Guo J, Xu Z (2009) Recycling of waste printed circuit boards: a review of current technologies and treatment status in China. J Hazard Mater 164:399–408
14. Havlik T et al (2011) Hydrometallurgical treatment of used printed circuit boards after thermal treatment. Waste Manage 31:1542–1546
15. Eswaraiah C et al (2008) Classification of metals and plastics from printed circuit boards (PCB) using air classifier. Chem Eng Process 47:565–576
16. Neto I et al (2016) A simple and nearly-closed cycle process for recycling copper with high purity from end life printed circuit boards. Sep Purif Technol 164:19–27
17. Lu Y, Xu Z (2016) Precious metals recovery from waste printed circuit boards: a review for current status and perspective. Resour Conserv Recycl 113:28–39
18. Ghosha B et al (2015) Waste printed circuit boards recycling: an extensive assessment of current status. J Clean Prod 94:5–19
19. Aylmore M, Muir D (2001) Thiosulfate leaching of gold—a review. Miner Eng 14(2):134–174
20. Rivera I et al (2015) Kinetics of metallic silver leaching in the O_2-thiosulfate system. Hydrometallurgy 156:63–70
21. Salinas E et al (2016) Leaching of silver contained in mining tailings, using sodium thiosulfate: a kinetic study. Hydrometallurgy 160:6–11
22. Jeffrey MI, Brunt SD (2007) The quantification of thiosulfate and polythionates in gold leach solutions and on anion exchange resins. Hydrometallurgy 89:52–60
23. Puente-Siller DM, Fuentes-Aceituno JC, Nava-Alonso F (2013) A kinetic-thermodynamic study of silver leaching in thiosulfate-copper-ammonia-EDTA solutions. Hydrometallurgy 134–135:124–131
24. Isauro RL (2003) Estudio Cinético de la Precipitación/Lixiviación de Plata en el Sistema O_2–$S_2O_3{}^{2-}$–$S_2O_4{}^{2-}$. Aplicación en Efluentes Industriales. Ph.D. thesis, Universitat de Barcelona
25. Mesinas M et al (2015) Efecto de la temperatura sobre la velocidad de lixiviación de polvo de cobre contenido en circuitos impresos de computadoras en medio oxígeno–tiosulfato (O_2–$S_2O_3^{2-}$). Tópicos de Investigación en Ciencias de la Tierra y Materiales 2:30–39
26. Levenspiel Octave (2013) Ingeniería de las Reacciones Químicas. Limusa Wiley, México, pp 565–581

Preparation and Characterization of Polyethylene Nanocomposites with Clay and Silver Nanoparticles

Washington Luiz Oliani, Danilo Marin Fermino, Luiz Gustavo Hiroki Komatsu, Ademar Benevolo Lugao, Vijaya Kumar Rangari, Nilton Lincopan and Duclerc Fernandes Parra

Abstract Polyethylene is widely used for packing applications. Different grades of polyethylene, LLDPE and LDPE play a major role in the film blowing industry. The use of nanoclay in polymer matrices (non-polar) requires the use of a compatibilizer agent, ethylene graft maleic anhydride copolymer (PE-g-MA), which acts as a bridge for nanoclay-matrix-interaction. In our study, blown films of LLDPE, LDPE, and their blends were produced using a twin screw extruder. The combination of PEs as matrix, montmorillonite (MMT) and silver nanoparticles (AgNPs) intended to improve mechanical and thermal properties. Nanocomposites films with 1 wt% of montmorillonite and 0.5–1.0 wt% silver nanoparticles were prepared by blow extrusion and evaluated. Nanoclay intercalation/exfoliation degree was investigated by X-ray diffraction spectroscopy (XRD) and scanning

W.L. Oliani (✉) · L.G.H. Komatsu · A.B. Lugao · D.F. Parra
Nuclear and Energy Research Institute, IPEN-CNEN/SP,
Av. Prof. Lineu Prestes, 2242, Cidade Universitária, São Paulo,
SP CEP 05508-000, Brazil
e-mail: washoliani@usp.br

D.M. Fermino
Department of Metallurgical and Materials Engineering,
EPUSP (Escola Politécnica, Universidade de São Paulo),
Av. Prof. Mello Moraes, 2463, Universitária, São Paulo, SP, Brazil

V.K. Rangari
Department of Materials Science and Engineering, Tuskegee University,
Tuskegee, AL, USA

N. Lincopan
Department of Microbiology, Institute of Biomedical Sciences,
University of São Paulo, São Paulo, Brazil

N. Lincopan
Department of Clinical Analysis, School of Pharmacy,
University of São Paulo, São Paulo, SP, Brazil

S. Ikhmayies et al. (eds.), *Characterization of Minerals, Metals, and Materials 2017*, The Minerals, Metals & Materials Series,
DOI 10.1007/978-3-319-51382-9_78

electron microscopy (SEM). Mechanical and thermal properties were evaluated. The results showed optimum dispersion at intercalation/exfoliation levels for polyethylene nanocomposite (PENC), slight increasing in mechanical and especially improvement in thermal properties.

Keywords Polyethylene · Montmorillonite · Film blowing · Silver nanoparticles

Introduction

Polyethylene (PE) is a general purpose thermoplastic resin with good processing and mechanical properties as well as relative low price [1]. The rheology of linear low density polyethylene (LLDPE), low density polyethylene (LDPE) and blends is relevant to the film blowing process. The poor performance of LLDPE in the blowing phase of film manufacture is attributed to the low elongational viscosity of LLDPE owing to macromolecules linear structure. In contrast, LDPE performs differently in such processes due to its high elongational viscosity of long chain branched macromolecules [2–4].

Thermal processing such as melt blending, extrusion, and injection molding has been applied for incorporating antimicrobials into polymers, but it should be considered the heat stability and chemical compatibility of antimicrobials in polymer matrix, in order to evenly distribute antimicrobial chemicals [5].

In recent work [6] the low-density polyethylene (LDPE) nanocomposite films containing different concentrations of silver nanoparticles (AgNPs) (0.5 and 1% of polymer weight, w/w) were manufactured via extrusion. The results indicated that LDPE nanocomposite films containing AgNPs could potentially be used as antimicrobial packaging for food applications.

The ultrasound radiation was used to Ag nanoparticles chemical reduction for addition in Polyethylene fibers. The effect of reducing reagent, power of ultrasound radiation, reaction time and temperature in growth of the AgNP were studied. Results show a decrease in the particles size as increasing power of ultrasound irradiation [7].

In other work [8], montmorillonite (MMT) was first ion-exchanged with silver ions (Ag + MMT) then organically modified. Active carrier MMT-layers were then compounded into low density polyethylene (LDPE) polymer using a high shear force co-rotating twin screw extruder. Antibacterial activity of the Ag + OMMT − LDPE nanocomposite films was then investigated against powerful gram negative *Escherichia coli* (*E. coli*) bacteria. It seems to have strong synergetic effect between silver ions and OMMT nanoparticles as evidenced by the bacterial inhibition properties of thin blow molded Ag + OMMT − LDPE nanocomposite samples using liquid surface bacterial testing at a much lower active component concentration.

Interesting research [9] describes the synthesis and antimicrobial activity of silver–clay nanohybrid structure as combination of accelerated and diffusion-controlled antimicrobial activity, with long term impact. The antimicrobial activity was assessed in terms of interaction with *Escherichia coli*, where the constituents of the nanohybrid structure play a synergistic role. Clay provided a physically stable surface for nucleation of silver nanoparticles. Additionally, the parallel stacked layered structure of clay facilitated diffusion-controlled antimicrobial activity of in situ precipitated silver.

Other important research work was published by [10], where the silver incorporation in MMT plays an important role due to the possibilities of creating an antibacterial material with slow release of silver species in order to obtain a prolonged antibacterial activity. However, additional studies must be done for understanding the chemical nature of those silver species, along with its biological activity and the mechanisms of silver release over time. Its potential for application depends on how to overcome the aforementioned issues and how the antibacterial activity could be effectively transferred to complex matrices like polymers.

The processing by extrusion of polyolefin (polyethylene and polypropylene) nanocomposites with acquisition of properties and biocide activity should be an economically viable method and is being researched [11–13].

In the present research work, LDPE/LLDPE has been employed as matrix for films nanocomposites. The compounds have been altered by the addition montmorillonite (MMT) and silver nanoparticle and with the use of ultrasound. Such materials can be very valuable in films for biocide activity applications.

Materials and Methods

Materials

LDPE with a melt flow index—MFI (190/2.16) of 0.27 dg min^{-1} and density: 0.922 g mL^{-1}, LLDPE with a melt flow index—MFI (190/2.16) of 0.80 dg min^{-1} and density: 0.920 g mL^{-1}, Braskem (Brazil) was provided in the form of pellets. The LLDPE pellets were placed in a nylon container with nitrogen. The irradiation process of the plastic container was carried out at room temperature and at a dose rate of 5 kGy h^{-1}, using a multipurpose ^{60}Co gamma irradiator. The dose used was 20 kGy monitored by a Harwell Red Perspex 4034 dosimeter. The AgNPs were purchased from Sigma Aldrich, and poly(N-vinyl-2-pyrrolidone) (PVP) (average molecular weight = 1,300,000 g mol^{-1}), which acted as a surfactant for the AgNPs, was Plasdone. Silver nanoparticles (AgNPs) were purchased from Sigma Aldrich. The commercial clay Cloisite 20 was provided by BYK Additives Company and antioxidant (BASF Irganox B225ED) was added in small quantities, 2.0 wt%, to prevent the polyethylene from oxidizing and thermally cross-linking at

Table 1 Composition of constituents of polyethylene nanocomposites (wt%)

Samples	Matrix	Dose/kGy	Irganox	Cloisite	AgNPs
PE	LDPE/LLDPE	20	2	–	–
PENC1	LDPE/LLDPE	20	2	1	0.5
PENC2	LDPE/LLDPE	20	2	1	1.0

elevated temperatures. Three different formulations containing the LDPE/LLDPE were prepared and are represented in Table 1.

The PVP solution was prepared by adding 2 g PVP in 150 mL of absolute alcohol at 60 °C for 20 min under stirring until complete dissolution. The PVP solution was added 0.5–1.0 wt% of AgNPs. Then we used the ultrasound Unique with power 65 W for 30 min to dispersing and deagglomeration the silver nanoparticles.

Preparation of the Nanocomposites

The LDPE/LLDPE (90/10) pellet were mixed with Irganox B 215 ED in a rotary mixer and maintained under this condition for 24 h. Then the mixture was processed with the addition of clay (MMT 1% by weight) and silver nanoparticles (AgNPs 0.5 and 1% by weight) in a twin-screw extruder Haake co-rotating, model Rheomex PTW 16/25, Fig. 1, with the following processing conditions: the temperature profile (feed to die) was 145–170 °C, with a speed of 100 rpm. After processed, the nanocomposites were granulated in a granulator Primotécnica W-702-3. The PE/MMT-AgNPs (PENC) films were produced in planar sheet extruder and the material was placed directly into the hopper of the extruder with a temperature profile (feed to die) of 150–175 °C, screw speed of 50 rpm and torque of 33–45 Nm. The films were produced with a thickness of ~0.05 mm.

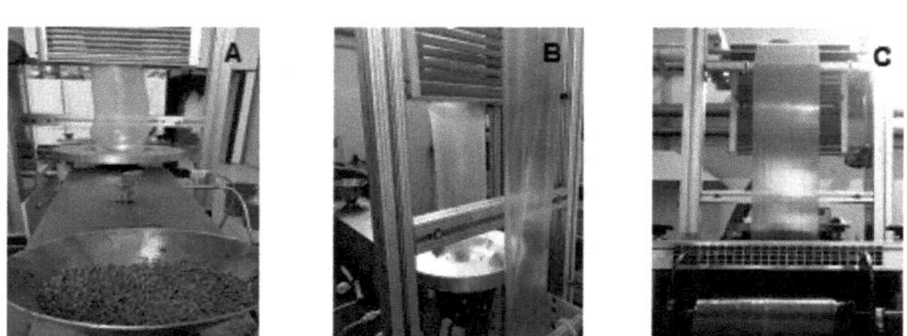

Fig. 1 Haake film blowing unit used to prepare the blown films (**a**). The blown film take-off transports the extruded film in an upward direction (**b**). The film is guided up to the nip rolls where the tube is flattened to create a 'lay-flat' tube of film (**c**)

Characterization

X-ray Diffraction Spectroscopy (XRD)

X-rays diffraction (XRD) measurements were carried out in the reflection mode on a Rigaku diffractometer Mini Flex II (Tokyo, Japan) operated at 30 kV voltage and a current of 15 mA with CuKα radiation (λ = 1.541841 Å).

Scanning Electron Microscopy and Energy Dispersive Spectroscopy

Specimens were examined with a Hitachi TM 3000, coupled with a Bruker Quantax 70 for the collection of EDS data. SEM coupled with backscattered electron detector (BSE) and energy dispersive X-ray spectroscopy (EDS). Sample sections for the EDS analysis were taken at 15 keV, and the acquisition period was 120 s.

Mechanical Test

The tensile tests were carried out on Zwick/Roell Z2.5 materials testing system equipped with TestXpert data acquisition system at 2.5 kN load. Thin films were cut according to the ASTM D882 standard in rectangular strip of 70 × 10 mm. Film thickness were ranged between 10–50 μm and was measured by the optical microscope. Samples at about ten replicates were tested for each film in separation grips of 50 mm. The cross-head speed was set at 0.01 mm s^{-1}. The tensile strength, modulus and toughness were determined by the TestXpert software.

Differential Scanning Calorimetry

Thermal properties of specimens were analyzed using a differential scanning calorimeter DSC 822, Mettler Toledo. The thermal behavior of films was obtained by: (1) heating from −50 to 280 °C at a heating rate of 10 °C min^{-1} under nitrogen atmosphere; (2) holding for 5 min at 280 °C, and (3) cooling to −50 °C and reheating to 280 °C at 10 °C min^{-1}. The crystallinity was calculated according to the Eq. (1):

$$X_C = Px\frac{\Delta H_f \times 100}{\Delta H_0} \tag{1}$$

where: ΔH_f is melting enthalpy of the sample, ΔH_0 is melting enthalpy of the 100% crystalline PE which is assumed to be 280 kJ kg $^{-1}$ [14, 15], and P was the fraction content of PP in the sample.

Results and Discussion

X-ray Diffraction Spectroscopy (XRD)

The X-ray powder diffraction profiles PEs samples are shown in Fig. 2.

In respect to the clay, layer expansion associated with the polyethylene intercalation results in the appearance of a new basal reflection corresponding to larger gallery space [16]. Considering the diffraction peaks related to the (001) reflexions, PENC2 film showed a basal spacing $d_{(001)}$ = 1.4 nm for clay according to the Bragg equation ($2d\sin\theta = n\lambda$, λ = 1.541,841 Å), and crystallite size of 20 nm.

The XRD profile presented in Fig. 2, film PENC2, exhibit characteristic peaks at 2θ of 38.2, 44.4, 64.5, and 77.4 corresponding to scattering from the (111), (200), (220), and (311) crystallographic planes of AgNPs. These diffraction peaks firmly represent a face centered cubic structure of crystalline AgNPs [17–19].

Disappearance of the nanoclay interlayer diffraction peak (d_{001}) indicates possible exfoliation of the nanoclay platelets and broadening of the peak is considered to be the result of partial exfoliation [20]. The results indicate the formation of microstructures predominantly exfoliated in film PENC1.

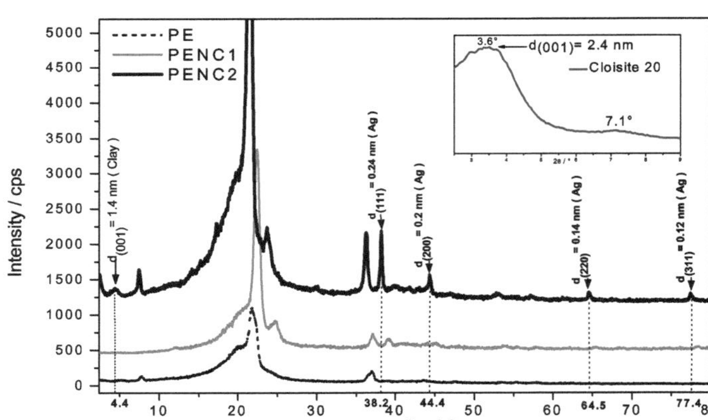

Fig. 2 Typical XRD patterns from polyethylene: LDPE/LLDPE (PE), LDPE/LLDPE/MMT/AgNPs 0.5 wt% (PENC1), and LDPE/LLDPE/MMT/AgNPs 1 wt% (PENC2)

Scanning Electron Microscopy and Energy Dispersive Spectroscopy

The images of Figs. 3 and 4 are related to the PE film with AgNPs and silicon.

The degree of dispersion of nanofillers plays an important role in influencing the properties of the resulting nanocomposites and energy dispersive X-ray spectroscopy (EDX), combined with scanning electron microscopy. EDS on the virgin surface of both nanocomposites of PE matrix, comprising carbon, the presence of silver and silicon (Figs. 3 and 4). The evidence of the elemental analysis suggests that the nanofillers were well dispersed into the bulk of PE matrix without large agglomerates on the whole.

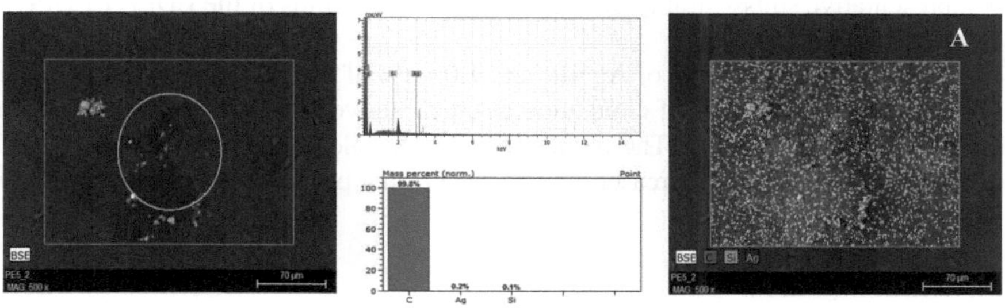

Fig. 3 Elemental maps, magnification 500× (scale = 70 μm) from SEM-EDX of (**a**) The surface of the PENC1 nanocomposite film showing the distribution of silicon (*green*) and silver (*blue*) atoms (color figure online)

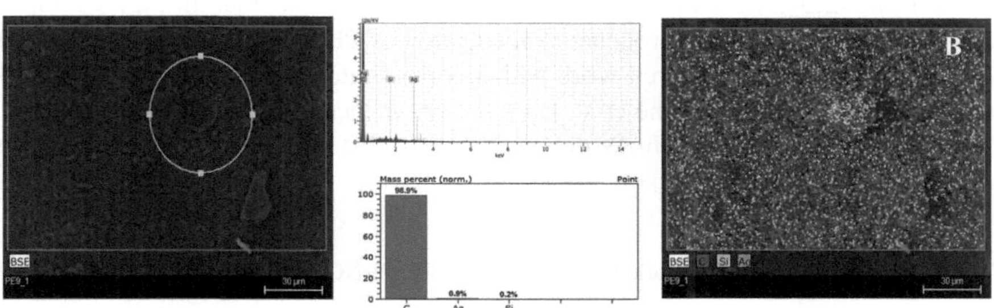

Fig. 4 Elemental maps, magnification 1000× (scale = 30 μm) from SEM-EDX of (**b**) The surface of the PENC2 nanocomposite film showing the distribution of silicon (*blue*) and silver (*green*) atoms (color figure online)

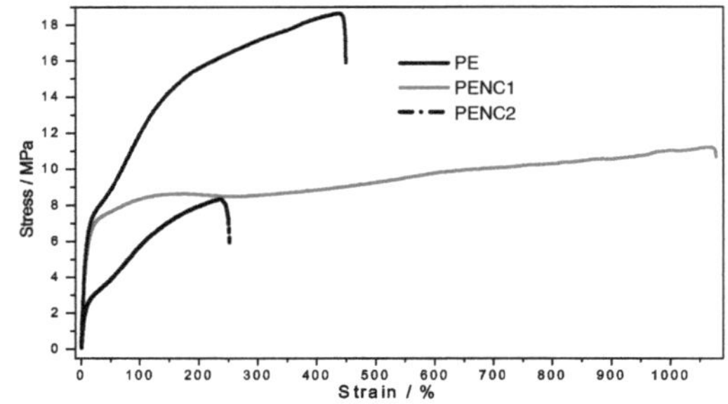

Fig. 5 Stress-strain-curves of the polyethylene samples

Mechanical Test

A representative stress–strain-curve of nanocomposite films of the LDPE/LLDPE is illustrated in Fig. 5.

The addition of 0.5 wt% of AgNPs and 1.0 wt% of clay in polyethylene sample, PENC1, showed increase of elongation at break above 1000% compared to others samples (PE and PENC2). The reduction in the tensile strength in sample PENC2 can be explained by the increased volume of PE that process a stress concentration with AgNPs content.

Differential Scanning Calorimetry

The DSC results are shown in Table 2.

The incorporation of the MMT-AgNPs increased the degree of crystallinity (%) of PE from 32.0–34.1% in sample PENC2. This fact suggest that the AgNPs impart a high efficiency to the heterogeneous nucleation of PE, even at a very low AgNPs content (0.5 wt%). A research work which corroborates with the idea of this work was published by Benchacine et al. [21], the silver incorporation in MMT plays an important role in the crystallinity degree X_C of PE that increases when Ag–MMT is added.

Table 2 DSC parameters for various polyethylene—PE, PENC1 and PENC2

Samples	Melting peak temperature, T_{m1}/°C (±0.1%)	Crystallization peak temperature, T_C/°C (±0.1%)	Melting peak temperature, T_{m2}/°C (±0.1%)	Degree of crystallinity, X_C/% (±0.5%)
PE	113.1	97.9	112.1	32.0
PENC1	115.2	98.0	112.1	33.8
PENC2	111.8	98.0	111.9	34.1

Conclusions

X-ray diffraction spectroscopy of sample PENC2 showed intercalated morphology. The results with MMT indicated the formation of microstructures predominantly exfoliated in film PENC1. The film PENC1 showed increase of elongation at break above 1000%, which suggests the application in packaging. In SEM/EDS microscopy the morphology of the silver/clay particles with homogeneous distribution and some clusters points it were observed.

Acknowledgements The authors acknowledge financial support for this work from CAPES-Process: CSF-PVE's—88887.115684/2016-00, Centre of Science and Technology of Materials—CCTM/IPEN, for microscopy analysis (SEM), the technicians Mr. Eleosmar Gasparin and Nelson R. Bueno, for technical support and multipurpose gamma irradiation facility at the CTR/IPEN.

References

1. Liang JZ, Tan LC, Wang KJ, Li FJ, Zhang SD (2015) Melt elongation flow behaviour of LDPE/LLDPE blends. Polym Test 41:133–139
2. Micic P, Bhattacharya SN (2000) Rheology of LLDPE, LDPE and LLDPE/LDPE blends and its relevance to the film blowing process. Polym Int 49:1580–1589
3. Micic P, Bhattacharya SN, Field G (1997) Rheological behaviour of LLDPE/LDPE blends under elongational deformation. Int Polym Proc 12(2):110–115
4. White JL, Yamane H (1987) A collaborative study of the stability of extrusion, melt, spinning and tubular film extrusion of some high-, low-and linear-low density polyethylene samples. Pure Appl Chem 59:193–216
5. Maryam J, Rahman RA, Ibrahim NA, Abdullah LC, Tan CP (2012) Melt production and antimicrobial efficiency of low-density polyethylene (LDPE)-silver nanocomposite film. Food Bioprocess Technol 5:719–728
6. Hasim SA, Romero MCC, Morris MA, Cummins E, Kerry JP (2015) Effects of a combination of antimicrobial silver low density polyethylene nanocomposite films and modified atmosphere packaging on the shelf life of chicken breast fillets. Food Packag Shelf Life 4:26–35
7. Abbasi AR, Kalantary H, Yousefi M, Ramazani A, Morsali A (2012) Synthesis and characterization of Ag nanoparticles@Polyethylene fibers under ultrasound irradiation. Ultrason Sonochem 19:853–857
8. Savas LA, Hancer M (2015) Montmorillonite reinforced polymer nanocomposite antibacterial film. Appl Clay Sci 108:40–44
9. Girase B, Depan D, Shah JS, Xu W, Misra RDK (2011) Silver–clay nanohybrid structure for effective and diffusion-controlled antimicrobial activity. Mater Sci Eng C 31:1759–1766
10. Giraldo LF, Camilo P, Kyu T (2016) Incorporation of silver in montmorillonite-type phyllosilicates as potential antibacterial material. Curr Opin Chem Eng 11:7–13
11. Oliani WL, Parra DF, Lima LFCP, Lincopan N, Lugao AB (2015) Development of a nanocomposite of polypropylene with biocide action from silver nanoparticles. J Appl Polym Sci 132:42218
12. Fages E, Fenollar O, Sanoguera DG, Balart R (2011) Study of antimicrobial properties of polypropylene filled with surfactant coated silver nanoparticles. Polym Eng Sci 51:804–811

13. Sahoo RK, Mohanty S, Nayak SK (2016) Effect of silver nanoparticles on the morphology, crystallization, and melting behavior of polypropylene: a study on non-isothermal crystallization kinetics. Polym Sci A 58:443–453
14. Mark JE (2007) Crystallization kinetics of polymer. Physical properties of polymers handbook, Springer, New York, EUA, p 636
15. Brandrup J, Immergut EH, Grulke EA (eds) (1999) Polymer handbook. 4th edn. vol 13. Wiley-Interscience, New York, NY, p 1, 445–450
16. Ray SS, Okamoto M (2003) Polymer/layered silicate nanocomposites: a review from preparation to processing. Prog Polym Sci 28:1539–1641
17. Kakkar R, Sherly ED, Madgula K, Devi DK, Sreedhar B (2012) Synergetic effect of sodium citrate and starch in the synthesis of silver nanoparticles. J Appl Polym Sci 126:E154–E161
18. Kim D, Jeong S, Moon J (2006) Synthesis of silver nanoparticles using the polyol process and the influence of precursor injection. Nanotechnology 17:4019–4024
19. Chudasama B, Vala AK, Andhariya N, Mehta RV, Upadhyay VR (2010) Highly bacterial resistant silver nanoparticles: synthesis and antibacterial activities. J Nanoparticles Res 12:1677–1685
20. Ataeefard M, Moradian S (2011) Surface properties of polypropylene/organoclay nanocomposites. Appl Surf Sci 257:2320–2326
21. Benhacine F, Hadj-Hamou AS, Habi A (2016) Development of long-term antimicrobial poly (ε-caprolactone)/silver exchanged montmorillonite nanocomposite films with silver ion release property for active packaging use. Polym Bull 73(5):1207–1227

Production of Concrete Interlocking Blocks with Partial Replacement of Sand in Bulk by Waste Glass Machined

Niander A. Cerqueira, Victor B. Souza, Igor W.D.C. Pereira, Rondinelli C. Ribeiro, Afonso G. Azevedo, Mairyanne S. Souza and Victor T. Bartolazzi

Abstract The common glass machining process used in the market of Brazil's civil construction generates tons of waste through the folder of that waste disposal out by the water used during machining. After drying the product is intended for landfill representing a great environmental liabilities since these deposited glass waste in the environment have a decomposition time greater than four thousand years due to its high chemical and mechanical resistance. Aiming to give a destination that addresses this waste disposal problem in the environment, much has been studied in an attempt to incorporate them into the mix cement to partially replace the sand. In this work it was decided to carry out specific studies using common traits comparing the stroke with sand with a dash with partial replacement of 30% of the sand for the glass machining residue was analyzed resistance compressive resulting in a potential good product.

Keywords Glass · Concrete · Ecological awareness

Introduction

Glasses are now one of the materials most commonly used in construction, more precisely said, the final stage of construction. During years of study, it was possible to develop technologies acquiring numerous advantages over other materials. We

N.A. Cerqueira (✉) · V.B. Souza · I.W.D.C. Pereira · R.C. Ribeiro
A.G. Azevedo · M.S. Souza · V.T. Bartolazzi
Sociedade Universitária Redentor—REDENTOR, BR 356 Km 25, Itaperuna, Rio de Janeiro, Brazil
e-mail: prof.niander@gmail.com

N.A. Cerqueira · A.G. Azevedo · M.S. Souza
Universidade Estadual do Norte Fluminense Darcy Ribeiro—UENF, Rio de Janeiro, Brazil

V.B. Souza
Universidade Federal Fluminense—UFF, Niterói, Brazil

S. Ikhmayies et al. (eds.), *Characterization of Minerals, Metals, and Materials 2017*, The Minerals, Metals & Materials Series,
DOI 10.1007/978-3-319-51382-9_79

can highlight some advantages such as its weight has been reduced at the same time it became tougher, being considered of little mechanical strength, it can now be used in different applications, thanks to the techniques developed as thermal and chemical tempering. With the new technologies applied are achieved high strength and other important properties for the manufacture of different products from glass such as doors, windows, partitions, box for bathrooms, glass bulletproof, car windshield, cups, beverage containers and others, and the doors, windows, partitions and box for bathrooms products more applied in construction [1].

Being a glass machining waste treatment product it is not possible to recycle it and is not viable transport to a location for incineration, thus your waste is disposed of in landfills. Landfills are already reduced capacity due to high amounts of other materials deposited on it and know that they are already at full load and over the years the trend is becoming increasingly overloaded.

The mechanical properties of the glass include high compressive strength values, traction and wear, possessing high mechanical and chemical resistance. We know that over 70% of the material is constituted by silica (SiO_2) which has excellent pazolânicas characteristics (natural or artificial material into contact with calcium hydroxide and water form a binding compound), excellent potential constituent material for the base cement [2].

We know that one of the most used materials in construction containing at its base cement is concrete, which consists of a coarse aggregate mixture and kids, binder and water, and may incorporate additives to achieve specific properties and characteristics. Thinking about it and helping to minimize the environmental impact caused by the disposal of waste glass, and reduce the use of raw material, in this case sand and cost reduction for manufacturing interlocked floor for paving and paving streets.

Aiming the idealization of the use of interlocked floor for paving and paving streets, in order to enjoy the residue of glass machining discarded by Muriaé-MG company, to perform test that indicates a great potential strength of the introduction of waste in concrete. In glass machining process carried out by the company in question, it is made with water aid for glass cooling and cutting parts (drills and grinding wheel), this water is collected by pipes and placed in a water treatment plant. This waste is composed of the factors related to machining, namely: waste glass, waste of cutting drills, glass powder, dust and waste from the grinding wheel.

In the center is made the chemical treatment of water so she can return to nature under normal conditions, this treatment is done by COLLANT-1315 products (composition: chlorides, sulfates, salts and water—Non-flammable and non-toxic), GLASSFLOC (composition: polyacrylamide—not flammable and non-toxic) and GLASSLIQ (composition: cationic metal salts—not flammable and non-toxic), leading to settling of the waste, after it is made to withdraw the waste decanted, being this the product to be embedded in the concrete.

For introduction of the waste to the concrete to produce the interlocked flooring was necessary to dry the decanted residue in an oven, grinding the quantity of waste collected in the company, so that it submit small dimensions approaching the sand

Fig. 1 Block interlocked concrete (type I), with 16 sides. *Source* Alves and Campbell [1]

grains to maintain grain size of the nearest trace of normal concrete without losing its characteristics.

After sizing the particle size of glass waste is made the dosage from the replacement of 30% of the sand trace basis by tailing ground glass product being introduced into the concrete preparation step to give a mixture, filed in the interlocked and compacted floor mold to fill the voids and willing to cure.

Made the samples with glass waste treated for comparison with normal concrete according to standard. After fabrication and waiting seven days to cure the material, they were conducted in the laboratory of the Redentor the compressive strength test in accordance with the resolution of the NBR 9781 [3] and then comparing the results found in testing, checking the feasibility of introducing the waste in concrete composition floor.

This article aims to analyze the mechanical properties of concrete interlocked floor type I (Fig. 1) with the introduction of glass waste in its composition by compression test following the rules according to NBR 9781 [3], and made comparisons of the results obtained from the floor with waste glass and standard concrete floor.

Methodology

Raw Material

To carry out the experimental process the following materials were used natural sand, cement CP IV RS-32 and glass waste. From there, it performed the manufacturing process of the specimens to be tested and analyzed. The waste glass used to carry out this work was collected in water treatment from the glass machining Muriaense company.

Due to the large production company, are produced daily tons of waste that after the time of decantation and drying, it is deposited in landfill. To reduce this disposal will be the study to possibly enter in ISO 9781: 2013 (Annex A), and made comparisons of the results obtained from the floor with waste glass and standard concrete floor base of concrete to interlocked floor. After the collection was necessary in the drying oven (Fig. 2).

Then it was performed manually with the use of grinding sockets obtaining different grain sizes (Fig. 3).

Fig. 2 Reject in a drying process in an oven

Fig. 3 Waste ground

Table 1 Particle size of the residue

Sieve opening (Ø)	% passing by weight (g)
9.5	0
4.75	29.8
2.36	126.1
1.18	170.2
600	182.4
425	455.1
300	29.1
150	6.1
753	1.1
<75	0.1
Total mass (g)	1000

For the studies was made to analyze the particle size of the product. Table 1 below lists the particle size of the waste results.

Proof Bodies

The products showed interlocked blocks segmented formats, with length/width ratio equal to two, sixteen sides (Fig. 4), whose dimensions are available in Table 2. These features allow floors using these blocks are mounted in any position, besides offering better interlocking parts because intertwine with each other on all four sides.

For the production of concrete specimens with the addition of waste glass they were determined from the standard concrete mix paving, making the inclusion of 30% of the fine aggregate replacement in case the sand, the waste glass milled to approach the particle size of the sand. The bodies of evidence were made at the Faculty Redeemer without large scale only in experimental and data analysis.

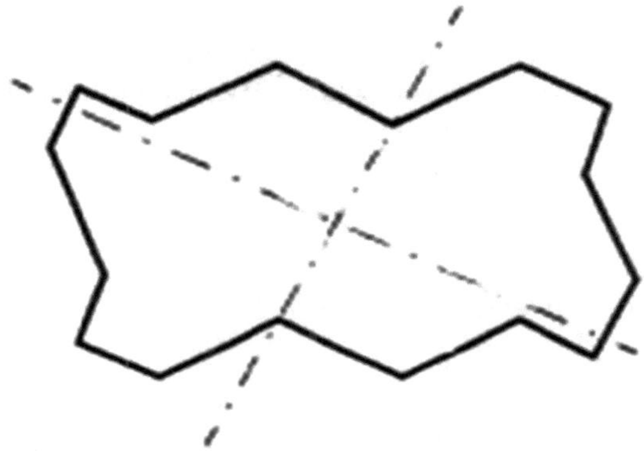

Fig. 4 Top view of the specimen. *Source* NBR 9781: 2013 (Annex A)

Table 2 Dimensions of the analyzed blocks

Width	Length	Height
110 mm	220 mm	100 mm

Fig. 5 Bodies constructed and arranged to support healing

Fig. 6 Test samples after seven days of curing

The curing method applied to the body through natural drying was maintained for seven days without any external intervention. According to Fig. 5.

As this work was only on an experimental basis, we use only four bodies of evidence with the healing time in seven days (Fig. 6).

Compressive Strength Test

All assay development for analysis of compressive strength was following the requirements laid down in ISO 9781 [3]. For the compression test was used one Hydraulic Press of SOLOTEST brand, with a capacity of 100 tons force with

Fig. 7 Hydraulic press SOLOTEST (100 tf, 220 V—60 Hz) Model: 1501220

simple electronic digital indicator and electrohydraulic pump, which allows manual control of the piston forward speed (Figs. 7 and 8). The test was conducted in the soil laboratory of the Faculdade Redentor. The compressive strength test is the result of applying a compressive force to a material, resulting in a reduction of the axial dimensions with the actuation force.

Results and Discussions

The compressive strength test in the samples of floors interlocked with the addition of tailings machined glass, did not show satisfactory values in relation to the specification of the NBR 9781: 2013 (Annex A). According to the standard cement floors with the estimated healing time in seven days, have to withstand the maximum pressure (28 MPa) and according to the data extracted from the three bodies of evidence to reject addition, the lowest value individual presented was (11.036 MPa) and presented the highest floor was (11.496 MPa), performing the calculation samples showed an average of (11.189 MPa). Unfortunately the values

Fig. 8 Sample into position for testing

Table 3 Results of samples

Sample (30% of waste)	Tonnes	Maximum compression (MPa)	Average (Mpa)
1	24	11.04	11.2
2	24	11.04	
3	25	11.50	

found after testing were not satisfactory, since the specification states that for seven day cure it presents the minimum value of 28 MPa in compressive strength testing.

Because this work has been on a trial basis, the production of the floors was not much scale as demanded standard, only four bodies were made of evidence, being a discarded early on to present shortcomings in its structure. Thus, it was not possible to perform the standard deviation calculations (S) and compressive strength estimated by lot ($f_{ck,\ est}$) (Table 3).

Conclusion

Based on the results obtained after the compression strength test, it can be concluded that the use of machined glass tailings coming from the water treatment plant Muriaense company, the concrete floor for paving the interlocked template, replacing 30% in the volume of sand used on the dash basis for this waste, did not show a good result compared to the specification of the NBR 9781 [3].

They were manufactured four bodies of evidence with the addition of waste, such as seven-day cure time, but we performed the test in just three because of a fracture present in its structure before being tested. The other three were referred to as sample 1, sample 2 and sample 3, and the first two tested achieved the maximum compressive strength of 11.04 MPa and the third in 11.50 MPa, and the law has set a 28 MPa minimum for the specimens with seven days of curing.

Several possibilities may have chained the unsatisfactory result in the analysis, and can be marked with added chemicals to treat waste that may have caused a chemical reaction in the composite causing not reach the expected value. Another point to be highlighted is the 30% replacement percentage in volume of sand in the case it would be necessary to make other tests gradually reducing the percentage of the volume until a satisfactory value.

Importantly, this work was carried out on a trial basis, but the rules were followed imposed by the NBR 9781 [3]. Even with values unsatisfactory in this essay, it is possible to further study the possible causes chained this result, it is very important to prioritize the issues related to the environment and to correct destination to the waste that is deposited in landfills and consumption sand as raw material. The local company is interested in providing the waste for collection and this is already of great importance for future studies related to the addition of the waste to the concrete.

References

1. Alves ALJ, Campbell AP (2015) Sustainable employability of waste glass in the manufacture of interlocked concrete floors. XXII congress of mechanical engineering students, Campos dos Goytacazes, RJ
2. Cruz LOF (2003) Floor interlocked concrete: study of the elements, and dimensioning methods. Rio de Janeiro. XVIII, 281 p, 29.7 cm (COPPE/ UFRJ, M.Sc., Civil Engineering, 2003) Thesis. Federal University of Rio de Janeiro, COPPE
3. Brazilian Association of Technical Standards (2013)—ABNT NBR 9781 (ANNEX A) Concrete paving parts: specification and test methods. Rio de Janeiro

Reactive Powder Concrete Production with the Addition of Granite Processing Waste

J.V.B. Lemes, G.E.S. Lima, F.G.D. Machado, L.G. Pedroti, L.A.F. Fioresi, W.E.H. Fernandes, R.C.S.S. Alvarenga and S. Monteiro

Abstract With advances in industrial processes, new products have emerged and become essential in everyday life. These activities generate different types of wastes, which can be harmful to the environment if not properly disposed. Having this problem in sight and considering the importance of concrete in the construction industry, this study analyzes the behavior of Reactive Powder Concrete (RPC) with addition of granite processing waste. Scanning Electron Microscopic (SEM) tests, compressive strength and Young's modulus tests after 1, 3, 7 and 28 days of production and shear and flexural strength after 28 days of production were performed during the analysis. The results indicated that incorporation of this residue into RPC does not affect its performance. RPC was still able to present high compressive and flexural strength.

Keywords RPC · Granite · Granite dust

Introduction

Ornamental rocks are widely used around the world and its processing industry generates a considerable amount of waste. This type of waste, specially granite dust, can harm the environment and, therefore, requires a proper disposal.

J.V.B. Lemes · G.E.S. Lima (✉) · F.G.D. Machado · L.G. Pedroti
L.A.F. Fioresi · W.E.H. Fernandes · R.C.S.S.Alvarenga
DEC—Civil Engineering Department, UFV—Federal University of Vicosa,
Av. PH Rolfs, Vicosa, Minas Gerais 36570-900, Brazil
e-mail: gustavoeslima@yahoo.com.br

J.V.B. Lemes
e-mail: joaovitor.blemes@gmail.com

S. Monteiro
Section of Mechanical and Materials Engineering, IME—Military Institute
of Engineering, Praça General Tiburcio, Urca, Rio de Janeiro 22290-270, Brazil

S. Ikhmayies et al. (eds.), *Characterization of Minerals, Metals, and Materials 2017*, The Minerals, Metals & Materials Series,
DOI 10.1007/978-3-319-51382-9_80

Adding granite processing waste into concrete can reduce the industry's spending in raw materials and the environmental impacts caused by incorrect disposal of this residue.

The Reactive Powder Concrete (RPC) is an ultra-high performance concrete (UHPC), which was created in the 90s through researches done in Canada and France to meet with military demands (BINA [1]). RPC is known for its high compressive strength (over 200 MPa) and low permeability, which is obtained through optimization of particle packing and by containing little water (Kosmatka et al. [2]).

RPC is composed mainly of fine materials, with maximum diameter of particles of 2 mm, such as quartz sand, Portland cement, crushed quartz, active silica, superplasticizer and low water-cement ratio. The absence of coarse aggregate was considered by the inventors to be a key-aspect for the microstructure and the performance of the RPC (Collepardi et al. [3]).

The high compressive strength associated with RPC is fully utilized in prestressed concrete, being used, for example, in beams for bridge structures. Also, RPC's durability allows its use in structural members placed in severe environments, thus reducing the costs of structure maintenance (MENEFY [4]).

This study analyzes the RPC's production and development when the crushed quartz is replaced by granite dust. By assessing the implications of this replacement, a possible allocation for this residue is suggested.

Materials and Methods

Using Vanderlei [5] work as a reference, regular RPC compositions were produced using Portland cement CPV ARI (high-initial resistance), Vicosa's region quartz sand, silica fume and superplasticizer. The water-cement ratio was different in each composition. Besides the usual materials characterization, samples of the quartz sand and active silica were analyzed through X-Ray Diffraction (XRD).

The usual characterization was performed in the Federal University of Vicosa's Laboratory of Construction Materials. The XRD was performed in the X-Ray Laboratory from the Federal University of Vicosa's Physics Department. The used machine was a Bruker D8-DiscoverDavinci X-Ray diffraction system, with Cu radiation and 2θ angle covered between 20–110°. The prepared sample is shown in Fig. 1.

Later, another RPC composition, this time with addition of crushed quartz, was produced and submitted through the same procedures aforementioned. This composition had the smallest water-cement ratio that produced a workable material.

After development and analysis of different RPC compositions, a new one was produced using granite dust instead of crushed quartz. For this composition, the smallest workable water-cement ratio was 0.23. This composition was considered the standard series (S_R). Four compositions (S_1, S_2, S_3, S_4) with different water-cement ratios were produced in order to evaluate if the increase in

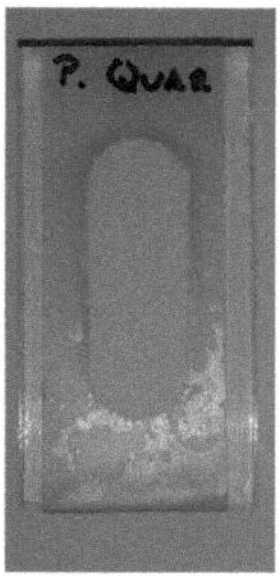

Fig. 1 Sample for X-ray diffraction (XRD) analysis

Table 1 Compositions with addition of granite dust

Series	w/c	Cement (kg)	Sand (kg)	Granite dust (kg)	Silica fume (kg)	Superplasticizer (kg)	Water (kg)
S1	0.25	1.059	1.165	0.248	0.260	0.042	0.267
S2	0.30	1.039	1.144	0.244	0.256	0.042	0.318
S3	0.35	1.020	1.122	0.239	0.251	0.041	0.369
S4	0.40	1.000	1.101	0.235	0.246	0.040	0.400

Table 2 Compositions with crushed quartz

Series	w/c	Cement (kg)	Sand (kg)	Crushed quartz (kg)	Silica fume (kg)	Superplasticizer (kg)	Water (kg)
SR	0.23	1.000	1.101	0.235	0.246	0.040	0.230

workability would compensate the decrease in compressive strength. Each composition produced is shown in Tables 1 and 2.

In each series, 12 specimens were prepared with 50 mm diameter and 100 mm height. They were tested for compressive strength and Young's modulus tests after 1, 3, 7 and 28 days of production. Regarding flexural strength, 3 specimens were prepared with dimensions of 40 × 40 × 160 mm and were tested after 28 days.

During RPC production, all materials (cement, sand, crushed quartz and active silica) were mixed inside a plastic bag until an homogeneous material was obtained. Then, it was placed in a laboratory mortar mixer. The superplasticizer was mixed with cold water (10 °C) because RPC is considerably exothermic and, afterwards, added to the mixer. The mixer worked at low speed for the first minute and, then, at high speed for the following 15 min.

The specimens were prepared according to the Brazilian Standard (ABNT NBR 5738—Concrete—Procedure for molding and curing concrete test specimens [6]). Figure 2 shows the specimens produced for one composition.

After casting, the specimens were taken to a humidified curing chamber, where they stayed for 24 h. As the initial cure finished, they were demolded, properly

Fig. 2 Cylindrical and prismatic specimens

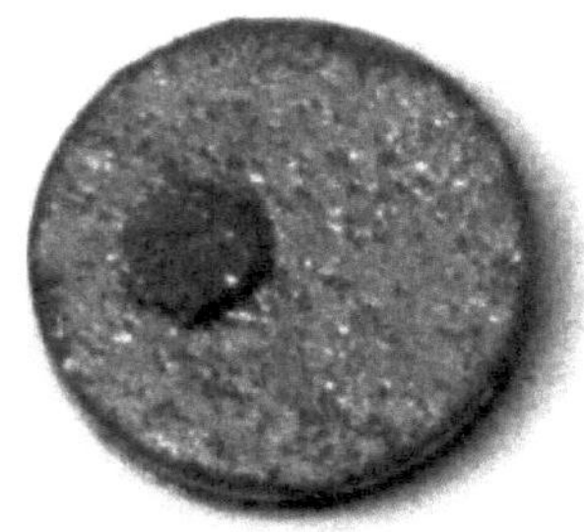

Fig. 3 Scanning electron microscopic (SEM) sample

named and put in a tank with saturated solution of calcium hydroxide, where they remained submerged until the testing day.

Fragments from the standard series (S_R) were removed from the specimens with CNF addition after the tests were performed. Scanning Electron Microscopic (SEM) tests were performed on the aforementioned fragments. The tested sample is shown in Fig. 3.

Results and Discussion

The results of XRD are shown in Figs. 4 and 5.

It is possible to state that crushed quartz and quartz sand are mineralogically similar because the peaks overlap.

It is observed that XRD results of active silica have many peaks in a disorderly way. This happens because active silica is an amorphous material and, therefore, has no defined crystalline structure.

Figure 6 shows the results of Scanning Electron Microscopic (SEM) tests, which is an image of the specimen fragment with 1000× zoom.

It is noticeable in Fig. 6 the homogeneity of RPC, which is one of the main characteristics of this type of material.

Results regarding axial compressive strength tests are shown in Fig. 7.

In the first four compositions, it is noticeable an increase in strength-gain between 7 and 28 days of age, which suggests occurrence of pozzolanic reactions in

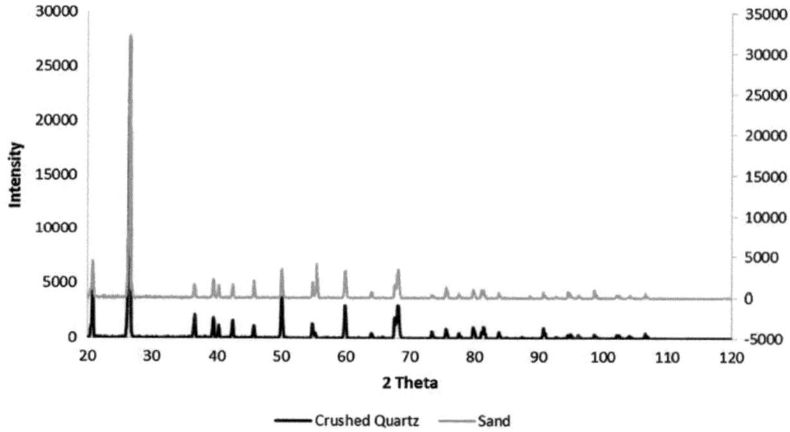

Fig. 4 XRD patterns of crushed quartz and quartz sand

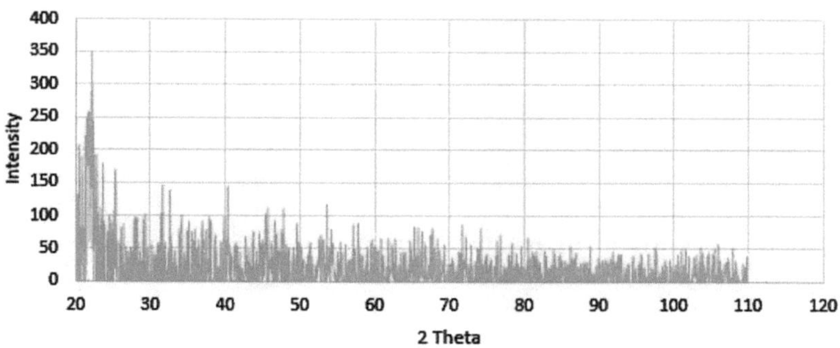

Fig. 5 XRD of active silica

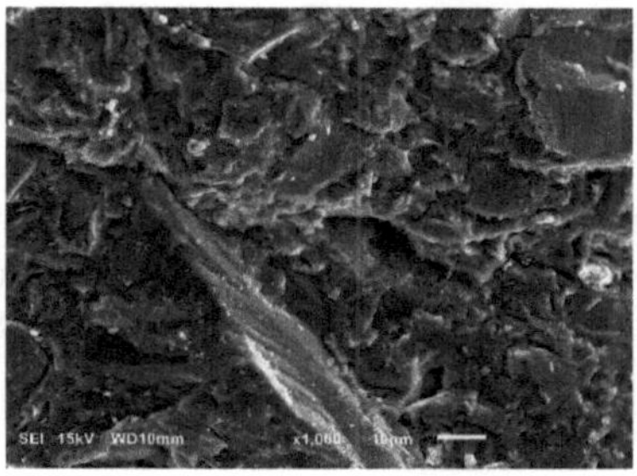

Fig. 6 SEM results of specimen fragment

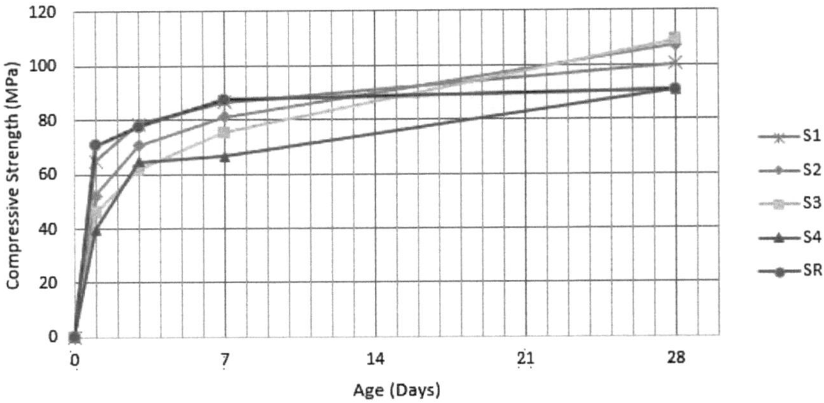

Fig. 7 Compressive strength of cylindrical specimens as a function of age (time)

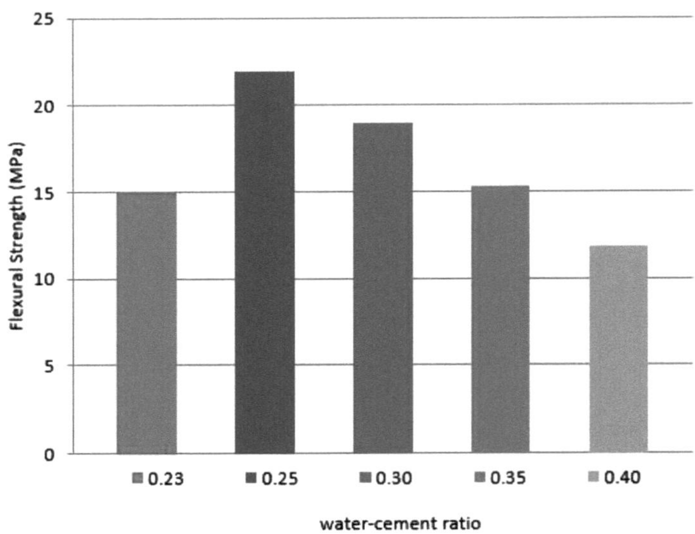

Fig. 8 Flexural strength of prismatic specimens as a function of water-cement ratio

this period. However, the same does not happen with the standard series (S_R), where there is discernible strength-gain only until 7 days of age.

The results of flexural strength tests are shown in Fig. 8.

The results showed strength-gain of series S1 and S2 when compared to the standard series (S_R).

Conclusions

Results of compressive strength tests were above the expected for standard concrete and in the range of high performance concretes (HPC) and not UHPC. This was expected because there was no thermal cure, due to equipment restrictions. Thermal cure increases pozzolanic activity, which results in strength gains for concrete.

As a consequence of a small water-cement ratio, there were difficulties in molding specimens for the standard series (S_R). These problems could have diminished the mechanical properties of this composition's specimens.

The addition of granite dust did not diminish RPC's workability nor compressive strength. The compositions with addition of granite dust presented compressive strength over 100 MPa and flexural strength over 15 MPa in series S1, S2 and S3.

It was noticeable that water-cement ratio directly influences on RPC's mechanical properties. An increase in water-cement ratio caused a decrease in RPC's compressive strength. However, this decrease could be compensated by an increase in RPC's workability.

Furthermore, it was found that is possible to produce RPC with addition of granite processing waste without loss of mechanical properties and workability.

Acknowledgements The authors thank the Brazilian agencies: CNPq, CAPES and FAPEMIG for the support provided to this investigation.

References

1. Bina P (1999) Concretos de Pós Reativos: uma revolução no conceito do concreto. Revista Téchne, n. 38, Jan 1999. Available in http://techne.pini.com.br/engenharia-civil/38/artigo 287168-1.aspx. Accessed on 16 July 2015
2. Kosmatka Steven H, Kerkhoff Beatrix, Panarese Willian C (2003) Design and control of concrete mixtures. Portland Cement Association, Illinois, p 358
3. Collepardi S, Coppola L, Troli R, Collepardi M (1997) Mechanical properties of modified reactive powder concrete. Spec Publ 173:1–22
4. Menefy L (2007) Investigation of reactive powder concrete and it's damping characteristics when utilised in beam elements. Ph.D. thesis, Griffith School of Engineering, Gold Coast Campus, Austrália, 380
5. Vanderlei RD (2004) Análise Experimental do Concreto de Pós Reativos: Dosagem e Propriedades Mecânicas. Ph.D. thesis, Escola de Engenharia de São Carlos da Universidade de São Paulo, São Carlos, 168
6. Brazilian Association of Technical Norms (2016) NBR 5738: Concrete—Procedure for molding and curing concrete test specimens. Rio de Janeiro, 9

Research on the Reason of the Different Type of Chloride Forming in the Process of Blast Furnace Ironmaking

Chuan Hui Li, Jian Liang Zhang, Cui Wang, Bing Ji Yan, Ya Peng Zhang and Hong Wei Guo

Abstract There are a lot of research of the chlorine element's influence on blast furnace, and mainly pay close attention to the harm to gas pipes and equipment. There are three kinds of chloride in the blast furnace gas, including HCl, NH_4Cl, and NaCl (or KCl, ZnCl, etc.). This paper mainly analyzes all kinds of phenomenon of the chlorine element overflowing blast furnaces in Jinan Steel, and puts forward some different viewpoints, which are that the chloride overflowing out of blast furnace in the types of HCl, NH_4Cl, or NaCl (KCl) not only has much to do with raw materials of chlorine load level, at the same time with the blast furnace burden distribution. On this basis, comes up with countermeasures for the different situation, such as adjusting the burden distribution matrix to guarantee the central gas flow smooth and changing the kind of raw materials to reduce the chlorine quantity, which will be helpful to prevent the harmful influence of chlorine element.

Keywords Blast furnace · Chloride · Gas system · Safety production

Introduction

With the popularization of blast furnace dry dedusting equipment and technology, the phenomenon of gas pipe blockage, low pH of the gas condensate water, serious corrosion of the pipes and blade vibration of the top pressure recovery turbine

C.H. Li (✉) · J.L. Zhang · C. Wang · Y.P. Zhang
School of Metallurgical and Ecological Engineering, University of Science and Technology Beijing, Beijing, China
e-mail: jgzmhui@foxmail.com

B.J. Yan · H.W. Guo
School of Iron and Steel, Soochow University, Suzhou, China

S. Ikhmayies et al. (eds.), *Characterization of Minerals, Metals, and Materials 2017*, The Minerals, Metals & Materials Series,
DOI 10.1007/978-3-319-51382-9_81

(TRT) causing accidental shutdown appears in many blast furnaces in China. The phenomenon is more general in the steel plants who use ores by ocean shipping, which influences blast furnaces' safety production.

There were many experts and scholars having researched the phenomenon, and the results showed that the problem was connected with the behavior of chlorine [1–5]. The research mainly forced on the impact on the gas pipes and equipment by chlorine, and developed technology of gas dechlorination, TRT antiscale and so on, in order to alleviate the negative influence of the problems. In addition, some scholars researched the chlorine behavior in blast furnace from side of thermodynamic, got the evolution rule of chlorine, but there was little research about the form of chlorine overflowing from blast furnaces [6–13]. This paper compares different phenomenon of the chlorine element overflowing blast furnaces in Jinan Steel, and puts forward some different viewpoints, which are that the chloride overflowing out of blast furnace in the types of HCl, NH_4Cl, and NaCl (KCl) not only has much to do with raw materials of chlorine load level, at the same time with the blast furnace burden distribution.

Normality of Chlorine Overflowing from Furnace

There is chlorine contained in raw and fuel materials used by blast furnace, and the existing form mainly is chloride, such as NaCl, KCl, $CaCl_2$, and so on. There is also some form of organic chlorine in coal. The temperature of hot metal and blast furnace slag is 1450–1550 °C generally, and it is over 2000 °C in tuyere raceway which is the region in the front of hot-air blast hole at the bottom of blast furnace. Almost all the chlorides are gaseous state under the temperature of over 1450 °C, and the gasification temperature of NaCl, KCl, $ZnCl_2$, NH_4Cl is shown in Table 1.

Shuhui Zhang researched the chlorine exhausting ability of blast furnace slag, and the result showed that the ability was only 3–5% which was low, and most chlorine was exhausted out of blast furnace following with gas in the type of NaCl, KCl (small particle) [14]. The chloride doesn't have adverse effects for the coal gas pipes and equipment, so it is not all the steel plants have the chlorine-harm problem normally.

Table 1 Gasification temperature of different chlorides

	NaCl	KCl	$ZnCl_2$	NH_4Cl
Gasification temperature/°C	1465	1420	733	337 (decomposition)

Anomaly Pattern of Chlorine Overflowing from Furnace

In the Type of HCl

Expression Form of HCl Overflowing from Furnace

Figure 1 shows the change trend of pH of coal gas condensate water obtained from No.1 blast furnace (1750 m^3) from June 1 to August 30 in 2015. The figure shows that the pH of coal gas condensate water has stage differences, which reduces to one to three from the normal value six to seven.

Table 2 shows the different content of Cl^- in the coal gas condensate water under the two different pH state. Compare the values, it can get that the content of Cl^- reduces markedly. If all the Cl^- exists in the form of HCl, the pH will be –lg (1640/1000/35.5) = 1.335, which is a smaller value.

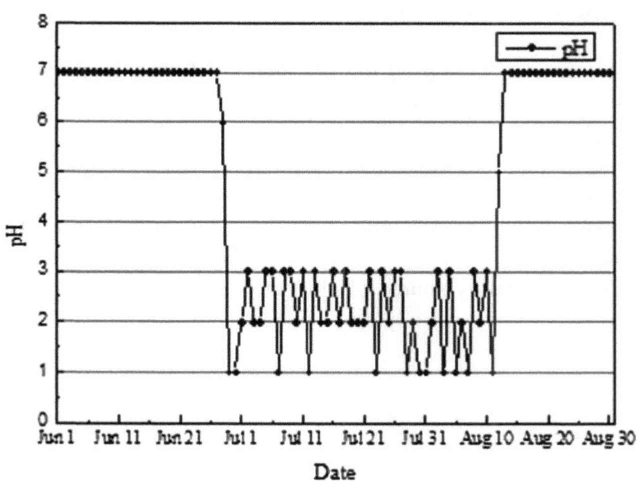

Fig. 1 The change trend of pH of the gas condensate water obtained from the TRT outlet

Table 2 Content of Cl^- for the normal and abnormal pH value

	pH	Cl^-/mg l^{-1}
Normal pH value	7.53	470
Abnormal pH value	2.18	1640
Change in value	−5.35	+1170

Analysis of the Reason of the Abnormal pH

For the pH value in June 29 is the first point which is very abnormal, compare the raw material structure of June 28 and June 29, as shown in Table 3.

Table 3 shows that the change of raw material structure is small, and there is not new kind of ore used. So the decrease of pH in June 29 has nothing to do with raw material. Analysis from another side, it is got that in June 28, in order to improve gas utilization efficiency, the burden distribution matrix was adjusted which was different from the matrix before June 28, and the mode was changed from central coke charging to no-central coke charging. The change is shown in Table 4.

In August 12, for the blast furnace condition is unstable, the burden distribution matrix was adjusted to the mode of no-central coke charging, and then the pH recovered to seven. From the phenomenon, it can give the result that the production of HCl is connected with burden distribution matrix.

Mechanism Analysis of the Production of HCl

In order to notarize the reason of the production of HCl, analyze the content of chlorine in all the materials in and out blast furnace, and calculate chlorine balance as shown in Table 5.

Table 5 shows that chlorine load is 0.523 kg/tHM in the input item, and 0.396 kg/tHM in the output item. The source of chlorine are mainly sinter, lump ore, coal and coke, while the outgoing approach are entering into dust and coal gas. The dechlorination proportion of blast furnace slag is only 4.55%.

If the NaCl is the mainly source of chlorine, the related reactions are shown as below:

Table 3 Raw material structure of June 28 and 29

	Raw material structure (proportion)/%				pH
	High basicity sinter	Newman lump ore	Hainan lump ore	Ukraine pellet	
June 28	75.1	14	2.00	8.90	6
June 29	73.7	14	2.00	10.20	1

Table 4 Burden distribution matrix before June 28 and in June 29

<table>
<tr><td rowspan="3">Before June 28 (central coke charging)</td><td></td><td>Angle/°</td><td>40.5</td><td>38.5</td><td>36.5</td><td>34.5</td><td>31.5</td><td>12</td></tr>
<tr><td>Coke</td><td>Cycle number</td><td>2</td><td>2</td><td>2</td><td>2</td><td>1</td><td>4</td></tr>
<tr><td>Ore</td><td>Cycle number</td><td>4</td><td>4</td><td>3</td><td>2</td><td></td><td></td></tr>
<tr><td rowspan="3">June 29 (no-central coke charging)</td><td></td><td>Angle/°</td><td>41</td><td>39</td><td>36.5</td><td>33.5</td><td>30.5</td><td></td></tr>
<tr><td>Coke</td><td>Cycle number</td><td>3</td><td>3</td><td>2</td><td>2</td><td>1</td><td></td></tr>
<tr><td>Ore</td><td>Cycle number</td><td>3</td><td>3</td><td>2</td><td>1</td><td></td><td></td></tr>
</table>

Table 5 Chlorine balance calculation

Input		Sinter	Pellet	Lump ore	Ore	Coke	Total
	Quantity per hot metal/kg tHM^{-1}	1310	135	235	130	400	2210
	Chlorine content/%	0.012	0.008	0.014	0.100	0.048	–
	Chlorine load/kg tHM^{-1}	0.157	0.011	0.033	0.130	0.192	0.523
	Proportion/%	30.060	2.070	6.290	24.860	36.720	100.00
Output		Gravity dust	Bag dust	Slag	Gas	Hot metal	Total
	Quantity per hot metal/kg tHM^{-1}	23.000	5.000	352.000	1456 m^3/t	1000	–
	Chlorine content/%	1.000	1.500	0.005	5 mg/m^3	0	–
	Chlorine output/kg tHM^{-1}	0.230	0.075	0.018	0.073	0	0.396
	Proportion/%	58.080	18.940	4.550	18.430	0	100.00

$$NaCl + 0.5H_2(g) = Na(g) + HCl(g) \quad T > 1648\,^{\circ}C \text{ (Tuyere Raceway)} \tag{1}$$

$$2NaCl + SiO_2 + H_2O(g) = Na_2SiO_3 + 2HCl(g) \quad T: 740 \sim 1553\,^{\circ}C \text{ (lump and cohesive zone)} \tag{2}$$

$$2NaCl + H_2S(g) = Na_2S + 2HCl(g) \quad T > 1273\,^{\circ}C \text{ (dropping zone)} \tag{3}$$

$$NaCl(s) = NaCl(l) \quad T = 901\,^{\circ}C \tag{4}$$

$$NaCl(l) = NaCl(g) \quad T = 1273\,^{\circ}C \tag{5}$$

NaCl contained in the materials putted into blast furnace from the upper part (ore and coke) can reacts as the Eqs. (1)–(3), and NaCl in the coal can reacts as the Eqs. (1) and (3) [15]. All the reactions will reverse once the chloride shifts to the low-temperature zone in the upper part, and then the NaCl will formed again. In the process of circulating enrichment, some small NaCl particle overflows to dust and coal gas, and then dissolves in the gas condensate water. So, in normal situation, pH of the water solution is about seven.

When the burden distribution matrix is for central coke charging, the center will form a no-ore zone, and the distribution state of the burden for the two mode (central coke charging and no-central coke charging) is shown in Fig. 2. In the figure, the light-grey diagonal-strips shapes represents the burden charge result of ore, the black represents the burden charge result of coke, and the dark-grey between another two kinds of color is the mixing region of ore and coke.

The mode central coke charging makes no ore in center, so that the center temperature is larger than 800 °C, and then the reaction $CO_2 + C = 2CO$ can

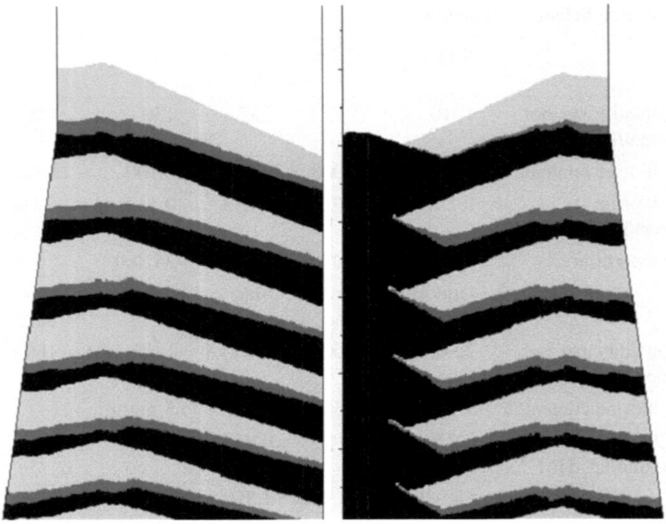

(a) central coke charging (b) no-central coke charging

Fig. 2 Burden surface profile of different charging modes

proceed completely, which leads the gas in center to be CO mainly. So, the reaction condition of $2Na + 2CO_2 = Na_2CO_3 + CO$ is poor. When the sodium vapor flows from the bottom of blast furnace to the up part, a mass of Na reacts with HCl, and then overflows from the center in the type of NaCl.

When the charging mode is no-central coke, the no-ore zone is small, which makes the concentration of CO_2 increase greatly. Then the reaction condition of $2Na + 2CO_2 = Na_2CO_3 + CO$ is good, Na is oxidized again, and the reaction between HCl and Na is weak, which makes much chlorine overflow out of furnace in the type of HCl.

In the Type of NH_4Cl

Expression Form of NH_4Cl Overflowing from Furnace

In July 10, 2015, TRT turbine of Jinan Steel's No. 1 1750 m^3 blast furnace shook seriously, and the four shock monitor points were all about 80 μm which was just the alarm value. From July 10 to 12, the shock distance changed very large from 58 to 120 μm instantaneously, and in July 13, the largest value is 146 μm when the TRT was stopped. The examination result showed that there was much attachment substance enriched on the second stage blade. The attachment substance easily dissolved in water, and easily cleaned. The appearance of the blade before and after cleaning is shown in Fig. 3.

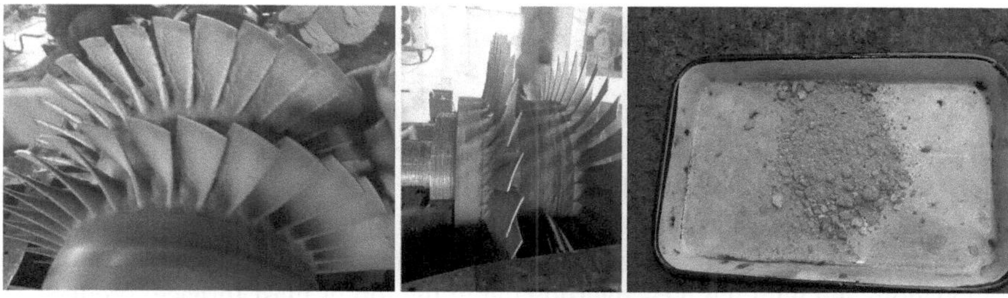

(a) blade before cleaning (b) blade after cleaning (c) attachment substance

Fig. 3 Attachment substance on the blade. **a** blade before cleaning. **b** blade after cleaning. **c** attachment substance

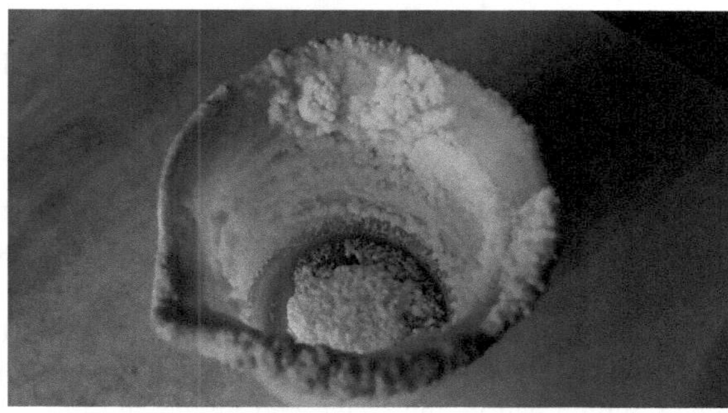

Fig. 4 White crystal precipitating on the cup wall

Analyzed the water solution of the attachment substance that there was about 95% dissolved, and pH of the solution was four. There was white crystal precipitating on the cup wall after the water drying as shown in Fig. 4.

Analyzed the attachment substance by X-ray diffraction, and the result showed that the 99% phase was NH_4Cl. So it can judge that the shock of blade was caused by NH_4Cl.

Mechanism Analysis of the Production of NH_4Cl

NH_4Cl is colourless crystal or white powder in room temperature. When the temperature is lower than 100 °C, the form is solid; when the temperature is higher than 100 °C, the decomposition reaction begins, and decomposes completely at 334 °C. The reaction is shown as follow.

$$NH_4Cl = NH_3 + HCl \tag{6}$$

Most scholars consider that the reason of that there is NH_4Cl in coal gas is that exists following reaction under the specified conditions in blast furnace.

$$14NaCl + 7SO_2(g) + 2NO_2(g) + 10H_2O\ (g) = 7Na_2SO_4 + 14HCl\ (g) + 2NH_3(g)\ T < 374\,°C \quad (7)$$

The reaction shows that the source of NH_4Cl production is the existence of NO_2, while SO_2 and H_2O also need exist in the surroundings. But under the strong reduction atmosphere in blast furnace, the possible of the existence of NO_2 and SO_2 is very low. In the tuyere raceway, NH_3 is produced during the process of coal combustion, and with the NH_3 overflowing to the top of blast furnace, produce the following reaction in the high-temperature zone.

$$NH_3 + C = HCN + H_2 \quad T > 1000\,°C \quad (8)$$

In the low-temperature zone, there are two another reactions as follows.

$$HCN + H_2 = NH_3 + C \quad T < 1000\,°C \quad (9)$$

$$HCN + 3H_2 = NH_3 + CH_4 \quad T < 800\,°C \quad (10)$$

So, the existence of HCN is the main reason for the production of NH_3 [16]. When NH_3 meets HCl, the reaction NH_3 + HCl = NH_4Cl (T < 334 °C) produces overflowing with coal gas. The temperature of the TRT entrance is lower than 70 °C, and under the condition of temperature lower than 100 °C, the front reaction can produce strongly, so a lot of NH_4Cl crystal generates and deposits on TRT blade, which causes the shake and affects the turbine's stability.

Conclusions

(1) Raw and fuel materials used in blast furnace contain some chloride, and in normal situation, the chloride overflows out of furnace in the type of NaCl and KCl, that there is little adverse effect on gas pipes and equipment.
(2) When the central gas flow is not smooth, elements of K, Na, Zn and so on exhausting out of furnace by gas flow will reduce greatly. HCl producing in the bottom of blast furnace has little chance to meet with K, Na, Zn, so that the chloride will overflow out of furnace in the type of HCl, which makes pH of the gas condensate water reduce sharply, and then the gas pipes and equipment is corroded seriously.
(3) The production of NH_4Cl in the gas is caused by the reaction of NH_3 and HCl. In the bottom zone, NH_3 is produced in the process of coal combustion, and transferred from HCN, which has no connection with NO_2 and SO_2.
(4) The way to control chloride in the harmless types not only should control quantity of chlorine in the raw and fuel materials, but also control the charging pattern so that guarantee the central gas flow smooth; then the chloride will not overflow out of furnace in the shape of harm chloride.

References

1. Lei ZC, Zhu WM (2011) Pipe corrosion cause of dry dusting of blast furnace gas and countermeasures. Metall Power 1:22
2. Wang SZ (2006) Influence of chlorine in blast furnace gas on running of gas system. Metall Power 5:22
3. Sun G (2014) Effect analysis of chlorine on blast furnace process. Metallurgical Equipment, Supplement, 50
4. Han XG, Hu BS, Gui YL (2013) Impact of blast furnace smelting process on the chlorine in BF gas and the prevention. J Hebei Polytech Univ Soc Sci Ed 3:36
5. Zhang JL et al (2015) Future prospects and research progress of chlorine in blast furnace system [J]. Iron Steel 7:1
6. Hu BS et al (2011) Behaviors of chlorine in smelting process of blast furnace. Adv Mater Res 396:152
7. Lectard E, Hess E, Lin R (2004) Behaviour of chlorine and alkalis in the blast furnace and effect on sinter properties during reduction. Revue de Metallurgie 1:31
8. Guo HL, Hu BS, Gui YL (2010) Behavior on blast furnace smelting process of chlorine in coal. China Metall 11:12
9. Liu XJ et al (2013) Reaction of chlorine in pulverized coal injection at blast furnace tuyere zone. J Northeast Univ Nat Sci 6:836
10. Lv K et al (2013) Thermodynamic study on the generation of hydrogen chloride in the blast furnace gas. J Hebei Polytech Univ Soc Sci Ed 4:13
11. Zhao Y et al (2004) Technology of chlorine in coal during coal combustion and pyrolysis. J Shandong Univ Sci Technol Nat Sci 2:108
12. Wang C et al (2015) Thermodynamic behavior of chlorine in blast furnace. China Metall 11:9
13. Liu XJ, Lu Q, Zhang SH (2012) Migration principle of chlorine in BF production. Adv Mater Res, 107–110
14. Zhang SH et al (2016) Experiments on dechlorination capacity of blast furnace slag. Iron Steel 5:19
15. Hu BS et al (2011) Effect of chlorine on combustion process of pulverized coal in tuyere zone. Adv Mater Res, 1203–1207
16. Liu X (2009) Analysis and control measures for the over standard cyanide in blast furnace gas washing water. Ind Saf Environ Prot 8:25

Stress and Deformation Analysis of Hot Blast Stove Piping System

Kun Yan and Shusen Cheng

Abstract A three-dimensional mathematical model of hot blast stove pipe shell has been established and used for analyzing the stress distribution and deformation. The results show that stress concentration happens along the intersection line of hot blast main and the hot blast manifold. There is no plastic deformation on the whole pipe region. The shell near the intersection elongates along the vertical direction and shortens along the horizontal direction, and deforms into an oval. The shell of hot blast main rises up, and inclines upward around the end of hot blast main. The hot blast main moves away from the hot blast and the displacement near the middle region is larger, the main deforms into an arch on horizontal surface. The deformation of shell may form a gap between shell and refractory inside and damage the security and stabilization of hot blast stove piping.

Keywords Hot blast stove · Elastic-plastic stress mathematical model · Stress analysis · Deformation

Introduction

A hot blast stove is a huge heat exchange facility generating high temperature air of more than 1200 °C and supply the hot blast to the blast furnace at a constant temperature and flow rate [1, 2]. Therefore, the hot blast stove should be designed to achieve a longer life time than the life of the blast furnace and to operate stably for a long time [3, 4]. However, many researchers investigating the hot blast stove procedure mainly focus on the higher temperature and stable supply of hot blast [5–7], the deformation and damage of hot blast shell is neglected. As the channel to transport the hot blast from hot blast stove to the blast furnace, the hot blast stove piping system is under a high temperature and pressure conditions for a long period

K. Yan (✉) · S. Cheng
School of Metallurgical and Ecological Engineering, University of Science and Technology Beijing, Beijing 100083, China
e-mail: windyk088@163.com

S. Ikhmayies et al. (eds.), *Characterization of Minerals, Metals, and Materials 2017*, The Minerals, Metals & Materials Series,
DOI 10.1007/978-3-319-51382-9_82

and deforms with the hot blast stove operational procedure. In conclusion, stress and deformation analyses of hot blast stove piping system are necessary for security and stabilization of hot blast stove process.

An elastic-plastic stress mathematical model of hot blast stove piping system has been established and used to investigate the stress distribution and deformation along the pipe. Furthermore, the hot blast stove body has been modeled in order to simulate the different constraint conditions suffered at different hot blast manifold, because of the different internal pressure and location of hot blast stove body.

Computational Model

Physical Model

The structure of hot blast stove which is modeled with a three-dimensional model is shown in Fig. 1. The physical model of hot blast stove include three hot blast stove bodies, hot blast manifold, hot blast valve and hot blast main. The height of hot blast stove is 33 m, the length of hot blast manifold and main is 5.5 and 33.5 m, and the diameter of manifold and main is 2.1 and 2.2 m. The hot blast valve is 2.35 m away from the hot blast outlet. The thickness of stove and pipe shell is about 25 mm. The stove body near the end of hot blast main is No.3 stove. The gird density near the intersection line of hot blast stove body, manifold and main has been enhanced as stress concentration happens around these regions. And the average gird is in these region is about 15 mm, less than one fifth of that in stove body shell.

Computational Conditions

An elastic-plastic stress mathematical model of hot blast stove and piping system has been established [8, 9]. The stress strain curve and some other physical

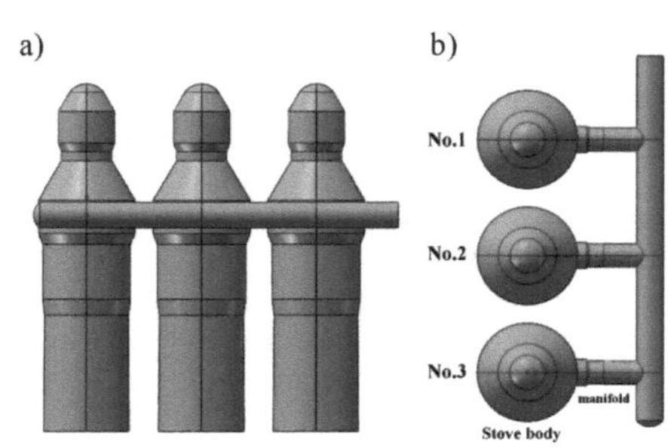

Fig. 1 Physical model of hot blast stove shell. **a** Front view; **b** top view

properties of steel Q235 are based on the experiment results. The computational conditions conclude that:

1. Displacement constraint: Fully-constrained is adopted at the bottom of hot blast stove and the connection position of hot blast main and mix chamber;
2. Pressure conditions: Pressure in single stove or pipe is uniform, and it's 0.38 MPa in blast period and 0.006 MPa in combustion period, expansion pressure of refractory lining is 0.1 MPa;
3. The straining beam is considered as compensation of pipe thickness;
4. No.3 stove is in blast period, the rest stoves are in combustion period.

Result and Discussion

Stress Distribution of Hot Blast Stove Shell

Equivalent stress distribution of hot blast stove pipe is shown in Fig. 2. In this figure, we can see that stress concentration happens along the intersection line of hot blast main and hot blast manifold. And on the rest region of piping system, the equivalent stress is lower than 50 MPa. The highest equivalent stress is about 210 MPa, occurs around the right and left of hot blast outlet. The yield stress of Q235 is about 235 MPa, there is no plastic deformation on the whole pipe region.

Displacement distribution of hot blast stove pipe is shown in Fig. 3. On the hot blast main, we can see that displacement increases along the main from 0 mm to

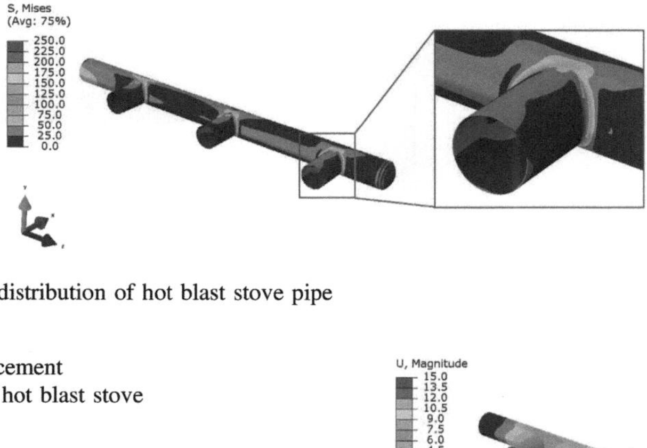

Fig. 2 Stress distribution of hot blast stove pipe

Fig. 3 Displacement distribution of hot blast stove pipe

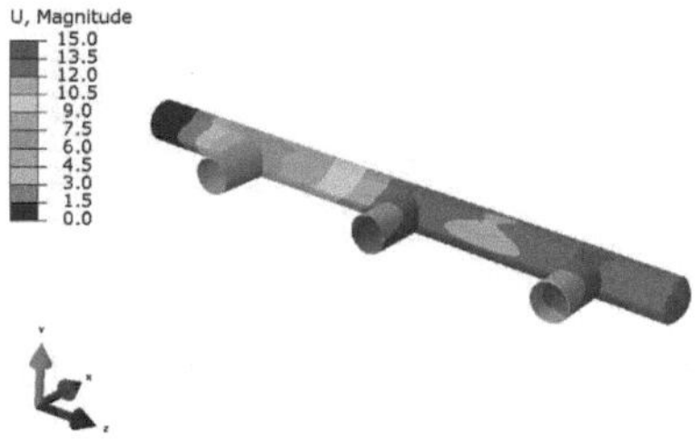

about 13.0 mm. Around the top of intersection line, the displacement is slightly higher than the surrounding region. The highest displacement occurs around pipe intersection line of No.3 stove, about 14.5 mm. On single hot blast manifold, displacement increases from the hot blast stove side to the hot blast main side.

Stress Analyses Along Intersection Line of Hot Blast Main and Manifold

The curve of intersection line of hot blast stove main and manifold (short as pipe intersection line) is shown in Fig. 4. The curve is start at the top of intersection line and spread along the clockwise direction as shown in Fig. 4. In this section, the coordinate of X-axis is the length of the curve. In addition, X direction is paralleled to hot blast manifold length direction, Y direction is altitude direction of stove body, and Z direction is paralleled to hot blast main length direction.

Equivalent stress and normal stress distributions along pipe intersection line are shown in Fig. 5a. The distributions of equivalent stress and normal stress distribution are symmetrical about plane Z = 0. The highest equivalent stress occurs around left and right sides of hot blast outlet, is about 190 MPa, lower than the yield stress of Q235 (235 MPa). The tendencies of normal stresses are similar to the tendency of equivalent, except the X-direction normal stress around the top of pipe intersection line. Except a narrow region around the bottom of pipe intersection

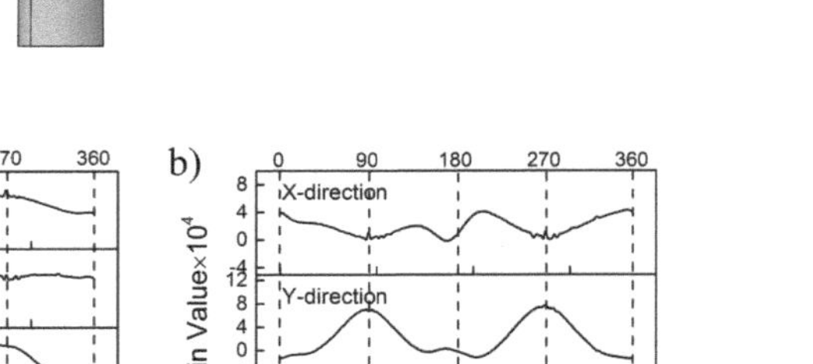

Fig. 4 Schematic of the intersection line of hot blast stove main and manifold

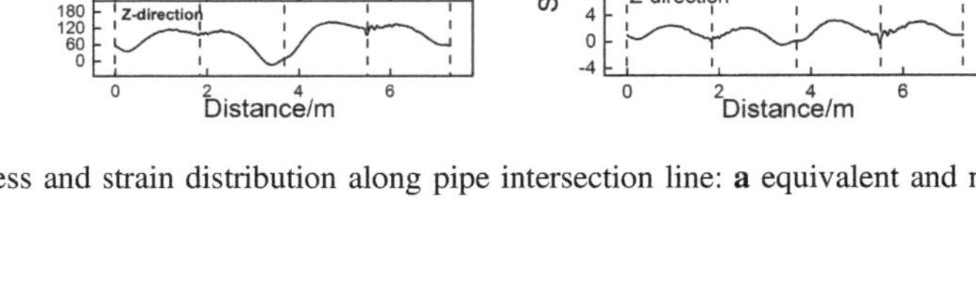

Fig. 5 Stress and strain distribution along pipe intersection line: **a** equivalent and normal stress, **b** strain

line, almost all the normal stress is positive, shows that the shell around these region is under tensile stress.

Normal strain distributions along pipe intersection line are shown in Fig. 5b. The X-direction normal strain is positive and the highest strain is about the 4.3×10^{-4}, around the top of pipe intersection line. Y-direction normal strain is negative around the top and bottom of pipe intersection line and positive around the left and right sides of pipe intersection line. The Z-direction normal strain is positive and fluctuates around 0–3.0×10^{-4}. In conclusion, the shell around pipe intersection line stretches along the X, Z direction, meanwhile, the shell compresses along the Y direction around the top and bottom and stretches along the Y direction around the left and right sides of pipe intersection line.

The displacement distributions along pipe intersection line are shown in Fig. 6a. X direction displacement is positive and increases from the top to the bottom of pipe intersection line, shows that the shell of intersection line moves towards the side of hot blast main. Y direction displacement is positive and decreases from the top to the bottom, shows that the shell of intersection line moves upwards and elongates along vertical direction. Z direction displacement is lower around right side and higher around left side of intersection line, shows the shell of intersection line shortens along horizontal direction.

The deformation of pipe intersection line along the YZ plant is shown in Fig. 6b. To get a better view, the deformation has been magnified 50 times. It can be seen that the shell of pipe intersection line elongates along vertical direction and moves upwards. In addition, the shell shortens along the horizontal direction, the shell around the pipe intersection line deforms into an oval.

Stress Analyses Along Hot Blast Manifold

The curve of hot blast manifold top edge and the start and end point of the curve are shown in Fig. 7.

Equivalent stress and normal stress distributions along hot blast manifold top edge are shown in Fig. 8a. The equivalent stress keeps almost constant along hot

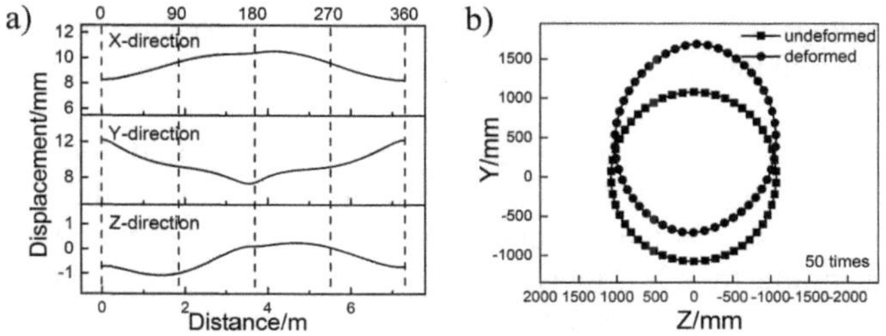

Fig. 6 Displacement and deformation along pipe intersection line: **a** displacement, **b** deformation along the YZ plant

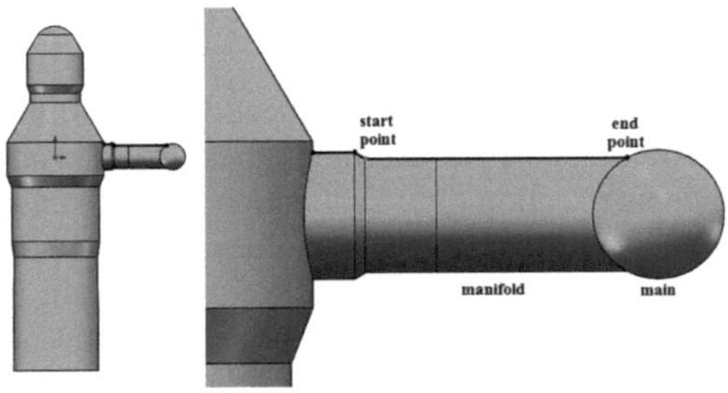

Fig. 7 Schematic of hot blast manifold top edge

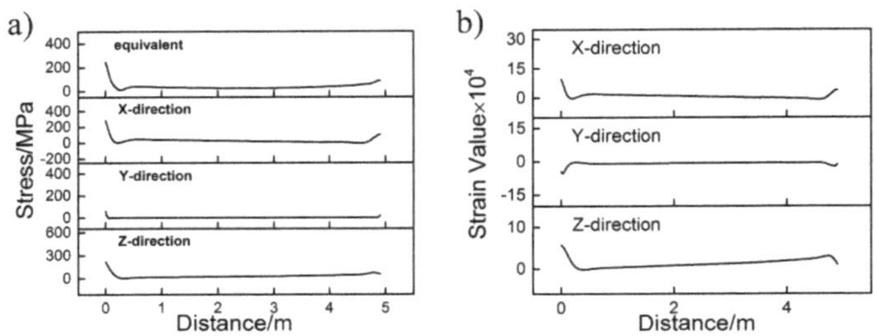

Fig. 8 Stress and strain distribution along hot blast manifold top edge: **a** equivalent and normal stress, **b** strain

blast manifold top edge, about 25–40 MPa, but increases to about 90 MPa around the pipe intersection line. X-direction normal stress decreases along the manifold from about 45 MPa to about −7 MPa, about 0.4 m away from the ending point, then increases to about 102 MPa. Y-direction normal stress keeps almost 0 MPa. Z-direction normal stress increases along the manifold from about 5 MPa to about 75 MPa. The shell along hot blast manifold top edge is under tensile stress along X, Z direction.

Normal strain distributions along hot blast manifold top edge are shown in Fig. 8b. The tendencies of normal strain are coincident with the tendencies of normal stress. The X-, Z-direction normal strain is positive and Y-direction normal strain is negative along hot blast manifold top edge. In conclusion, the shell along hot blast manifold top edge stretches along the X, Z direction and compresses along the Y direction.

The displacement distributions along hot blast manifold top edge and the deformation of manifold XY plant are shown in Fig. 9a. X direction displacement keeps almost constant, about 8 mm, Y direction displacement increases along the manifold from 4.0 to 12.2 mm, and Z direction displacement decreases from −0.2 to −0.7 mm shows that the hot blast manifold shell moves towards the side of hot blast main, and inclines upward around the hot blast main, as shown in Fig. 9b.

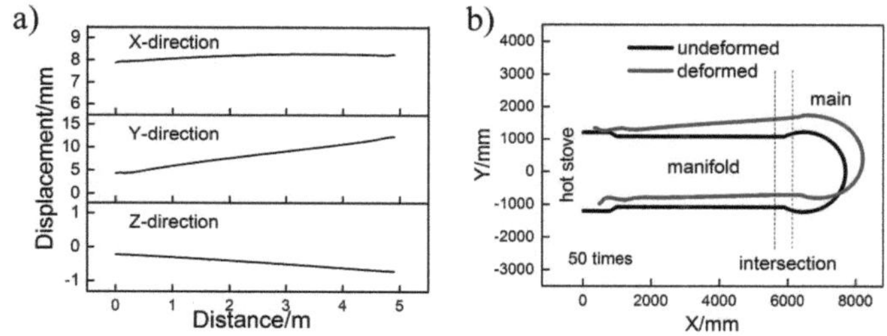

Fig. 9 Displacement and deformation along hot blast manifold: **a** displacement, **b** deformation along the XY plant

Stress Analyses Along Hot Blast Main

The curve of hot blast main top edge and inside edge are shown in Fig. 10.

Equivalent stress and normal stress distributions along hot blast main top edge are shown in Fig. 11a. X-direction normal stress is positive and higher around the pipe intersection. Y-direction normal stress keeps almost 0 MPa. The tendency of Z-direction normal stress is similar to X-direction normal, but the stress is negative around the starting point of main.

Normal strain distributions along hot blast main top edge are shown in Fig. 11b. The tendencies of X-, Z-direction normal strain are coincident with the normal stress. The X-direction normal strain is positive and Y-direction normal strain is

Fig. 10 Schematic of hot blast main top edge: **a** position, **b** top edge, **c** inside edge

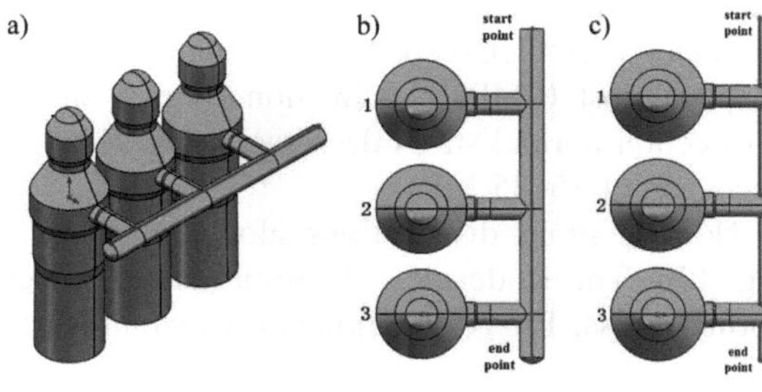

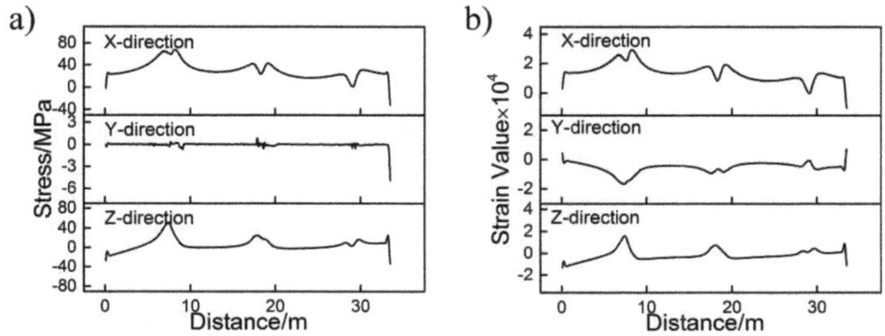

Fig. 11 Stress and strain distribution along hot blast main top edge: **a** equivalent and normal stress, **b** strain

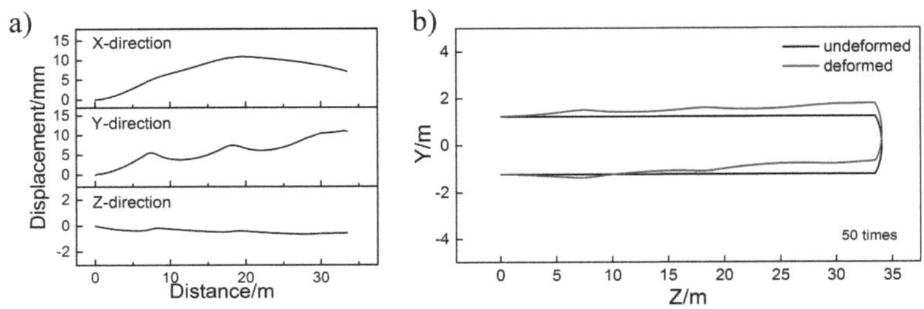

Fig. 12 Displacement and deformation along hot blast main: **a** displacement, **b** deformation along the YZ plant

negative along hot blast manifold top edge, the Z-direction normal strain is negative except the region around the pipe intersection. In conclusion, the shell along hot blast manifold top edge stretches along the X direction and compresses along the Y, Z direction.

The displacement distributions along hot blast main top edge and the deformation of main YZ plant are shown in Fig. 12. X direction displacement increases to 10.9 mm around the pipe intersection line of No.2 stove, and then decreases. Y direction displacement increases along the main from 0 to 11.0 mm. Z direction displacement slight decreases from 0 to −0.5 mm. In conclusion, the hot blast main inclines upward around the end of hot blast main, and the total length of hot blast main keeps almost constant, as shown in Fig. 12b.

Equivalent stress and normal stress distributions along hot blast main inside edge are shown in Fig. 13a. The stress around the pipe intersection is much larger than the other region. Except the pipe intersection region, X-direction normal stress keeps almost 0 MPa, Y-direction normal stress keeps around 15–35 MPa and Z-direction normal stress decreases from 93 to 20 MPa around start of main then keeps about 15–35 MPa.

Normal strain distributions along hot blast main inside edge are shown in Fig. 13b. The tendencies of normal strain are coincident with the tendencies of normal stress. The X-direction normal strain is negative and Y-, Z-direction normal

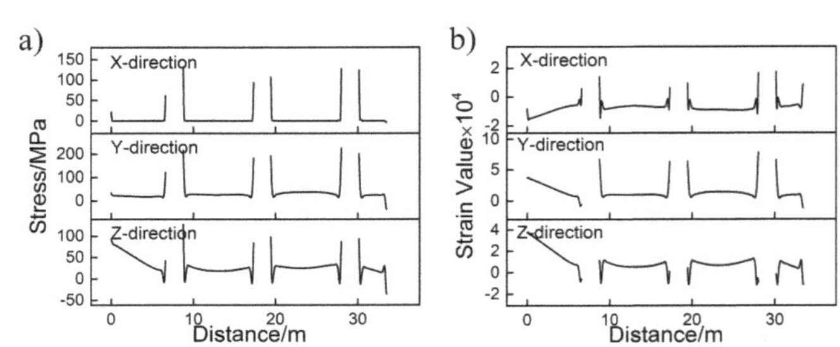

Fig. 13 Stress and strain distribution along hot blast main inside edge: **a** equivalent and normal stress, **b** strain

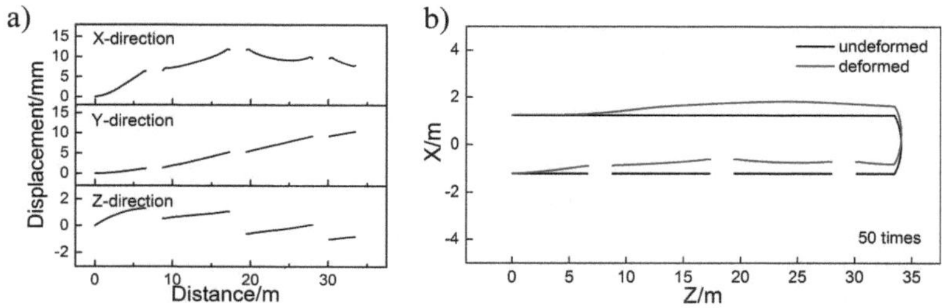

Fig. 14 Displacement and deformation along hot blast main: **a** displacement, **b** deformation along the XZ plant

strain is positive along hot blast main inside edge. The shell along hot blast manifold top edge stretches along main length direction and circle direction of pipe shell, compresses along the thickness direction of pipe shell.

The displacement distributions along hot blast main top edge and the deformation of main YZ plant are shown in Fig. 14. X direction displacement increases to 11.9 mm around the pipe intersection line of No.2 stove, and then decreases. Y direction displacement increases along the manifold from 0 to 10.4 mm. Z direction displacement slight decreases from 0 to −0.8 mm. In conclusion, moves away from the hot blast and the displacement around the pipe intersection line of No.2 stove is larger, the main deforms into an arch on horizontal surface, as shown in Fig. 14b.

Conclusions

Consideration the operation state of hot blast stove and pipe system, an elastic-plastic stress mathematical model of hot blast stove shell has been established in order to investigate the stress distribution and deformation of the shell. As above mentioned work, the followings were obtained.

1. Stress concentration happens along the intersection line of hot blast main and the hot blast manifold. There is no plastic deformation on the whole pipe region. The shell near the intersection elongates along the vertical direction and shortens along the horizontal direction, deforms into an oval.
2. The shell of hot blast main rises up, the displacement along vertical direction near the end of hot blast main is larger, and decreases along the hot blast main direction. The hot blast main moves away from the hot blast and the displacement near the middle region is larger, the main deforms into an arch on horizontal surface.

Future researches will focus on interaction between the shell and refractory inside as the deformation of shell may form gaps between them or force them squeezing each other which is harm for the security and stabilization of hot blast stove piping.

References

1. Kimra Y et al (2010) Three-dimensional mathematical modeling and designing of hot stove. ISIJ Int 7(50):1040–1047
2. Zhong L et al (2004) Computer simulation of heat transfer in regenerative chambers of self-preheating hot blast stoves. ISIJ Int 5(44):795–800
3. Zhang F et al (2012) Dome combustion hot blast stove for huge blast furnace. Int J Iron Steel Res 9(19):1–7
4. Guo H et al (2014) Optimization of the number of burner nozzles in a hot blast stove by the way of simulation. Miner Met Mater Soc 7(66):1241–1252
5. Moon J-W et al (2014) Effect of preheated top gas and air on blast furnace top gas combustion. ISIJ Int 1(54):63–73
6. Muske KR et al (2000) Model-based control of a thermal regenerator. Part 1: dynamic model. Comput Chem Eng 11(24):2519–2531
7. Rieger J et al (2015) Modelling and control of pollutant formation in blast stoves. J Clean Prod 88:254–261
8. Yang G (2013) Introduction to elasticity and plasticity, 2nd edn. Tsinghua University Press, Beijing, pp 41–78
9. Chen M (2007) Elasticity and plasticity. China Science Publishing & Media Ltd, Beijing, pp 226–277

Stress and Deformation Analysis of Top Combustion Hot Blast Stove Shell

Kun Yan and Shusen Cheng

Abstract An elastic-plastic stress mathematical model of hot blast stove shell has been established in order to investigate the stress distribution and deformation of the shell. The results show that stress concentration happens along the intersection line of hot blast stove body and the hot blast manifold. The plastic deformation occurs on the hot blast outlet in blast period. The deformation degree of the hot blast shell near the blast outlet in blast period is much larger than that in combustion period. In blast period, the shell near the blast outlet elongates along the horizontal direction and shortens along the vertical direction, the shell near the outlet deforms into an oval. And in combustion period, its deformation isn't obvious. The repeated deformation of the shell near the outlet between different periods may cause the refractory brick damage and be harmful to the long campaign life of hot blast stove.

Keywords Hot blast stove · Blind flange stress · Elastic-plastic stress mathematical model · Stress analysis · Deformation

Introduction

The maximization, modernization, high-efficiency, longevity of blast furnace is the development trend of the modernized iron making technology [1, 2]. Hot blast stove is the important heat supply facility for blast furnace, the high blast temperature has been attached great importance as an effective action for improving the heat system in the lower zone of blast furnace, reducing the coke ratio and energy consumption of blast furnace [3–5]. However, the increase of blast temperature and pressure will damage the hot blast stove and pipe system as well as enhance the smelting intensity and reduce the consumption. Besides the thermal expansion with the temperature change of refractory, the deformation of hot blast stove shell under

K. Yan (✉) · S. Cheng
School of Metallurgical and Ecological Engineering, University of Science and Technology Beijing, Beijing 100083, China
e-mail: windyk088@163.com

S. Ikhmayies et al. (eds.), *Characterization of Minerals, Metals, and Materials 2017*, The Minerals, Metals & Materials Series,
DOI 10.1007/978-3-319-51382-9_83

high pressure plays an important role to the damage of refractory lining life-span. Therefore, stress and deformation analysis of hot blast stove shell is necessary for security and stabilization of hot blast stove process.

An elastic-plastic stress mathematical model of hot blast stove shell has been established and used to investigate the stress distribution and deformation of the shell [6, 7]. Furthermore, the repeated deformation between different periods (blast period and combustion period) has been investigated as it may be an important reason of refractory lining damage.

Computational Model

Physical Model

The three-dimensional structure of hot blast stove is shown in Fig. 1. The physical model of hot blast stove include the dome, combustion chamber, checker chamber, hot blast manifold, hot blast valve and part of hot blast main. The height of hot blast stove is 33 m, the height of checker chamber is 19.8 m, the diameter of pre-combustion chamber, combustion chamber and checker chamber is 5.2, 9 and 8.2 m. The length of hot blast manifold and main is 5.5 and 33.5 m, the diameter of manifold and main is 2.1 and 2.2 m. The thickness of stove and pipe is about 25 mm. The gird density near the intersection line of hot blast stove body, manifold and main has been enhanced as stress concentration happens along these region.

Computational Conditions

An elastic-plastic stress mathematical model of hot blast stove shell has been established [8, 9]. The main computational conditions conclude that:

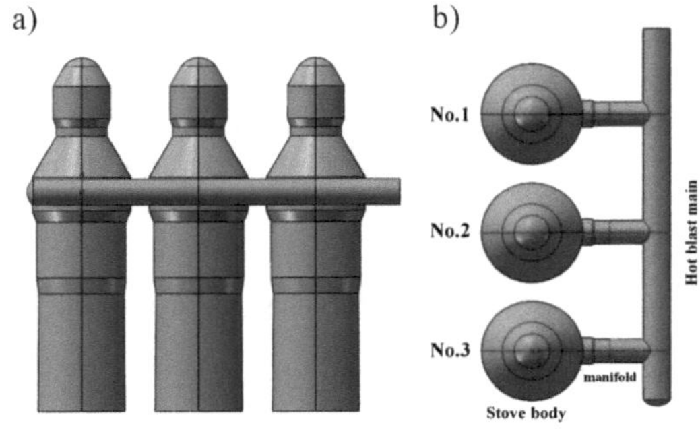

Fig. 1 Physical model of hot blast stove shell. **a** Front view; **b** top view

1. Displacement constraint: Fully-constrained is adopted at the bottom of hot blast stove and the connection of hot blast main;
2. Pressure conditions: Pressure in single stove or pipe is uniform, pressure in stove body is 0.38 MPa in blast period and 0.006 MPa in combustion period, the expansion pressure of refractory lining is 0.1 MPa;
3. The straining beam is considered as a compensation of pipe thickness.
4. The stress strain curve and other physical properties of steel Q235 are based on the experiment results.
5. For case A, No.3 stove is in blast period, the rest stoves are in combustion period. And for case B, No.2 stove changes into blast period, the rest stoves are in combustion period. In this section,

Result and Discussion

Stress Distribution of Hot Blast Stove Shell

Equivalent stress distribution of hot blast stove shell under certain blast pressure is shown in Fig. 2. In this figure, we can see that stress concentration happens along the intersection line of hot blast stove body and the hot blast manifold (short as hot blast outlet). On the hot blast stove shell, high stress region (where equivalent stress is higher than 235 MPa) in blast period is larger than the region in combustion period. The highest equivalent stress is about 460 MPa, occurs around the top of hot blast outlet. The yield stress of Q235 is about 235 MPa, the plastic deformation occurs on a wide-range along the hot blast outlet in blast period. In combustion period, the highest equivalent stress is about 250 MPa, the plastic deformation occurs on a narrow-range along the hot blast outlet.

Displacement distribution of hot blast stove shell under certain blast pressure is shown in Fig. 3. In this figure, we can see that displacement is increase along the height of stove. At the top of stove, the displacement is about 7.9 mm in blast period. In combustion period the displacement is relatively small, at the top of stove, the displacement about 4.7 mm. Along the hot blast outlet, the displacement at left and

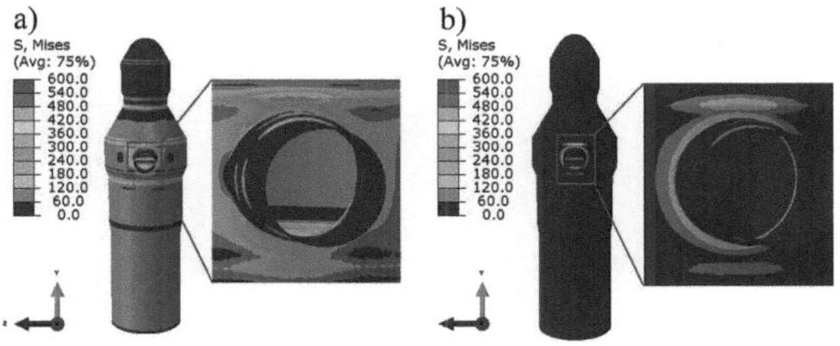

Fig. 2 Stress distribution of hot blast stove shell: **a** in blast period, **b** in combustion period

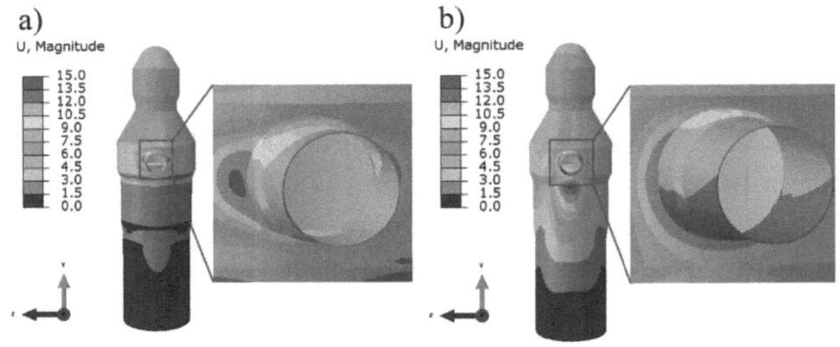

Fig. 3 Displacement distribution of hot blast stove shell: **a** in blast period, **b** in combustion period

right sides is about 12.2 m in blast period, higher than that around top and bottom sides. In blast period, the displacement in the shell of checker chamber is more uniform along the circumferential direction and lower than that in combustion period. And in combustion period, the displacement at front and back sides of stove, especially near the top of checker chamber, is higher than the other region.

Stress Analyses Along Hot Blast Outlet

In Blast Period

The curve of hot blast outlet is shown in Fig. 4. The curve is start at the top of hot blast outlet and spread along the clockwise direction as shown in Fig. 4. In this section, the coordinate of X-axis is the spread length of this curve.

Equivalent stress and normal stress distributions along hot blast outlet are shown in Fig. 5, in this figure, X direction is paralleled to hot blast manifold length direction, Y direction is altitude direction of stove body, and Z direction is paralleled to hot blast main length direction. The tendencies of equivalent stress and normal stress distribution are coincident, and are symmetrical about plane Z = 0. At top and bottom of hot blast outlet, stress is higher than the rest region and the highest equivalent stress is

Fig. 4 Schematic of the intersection line of hot blast stove body and manifold

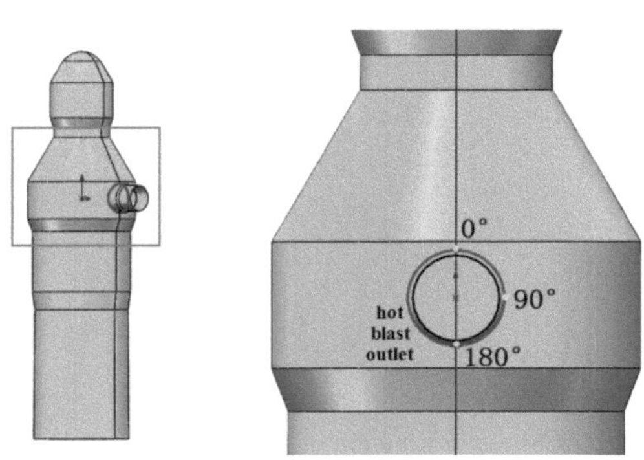

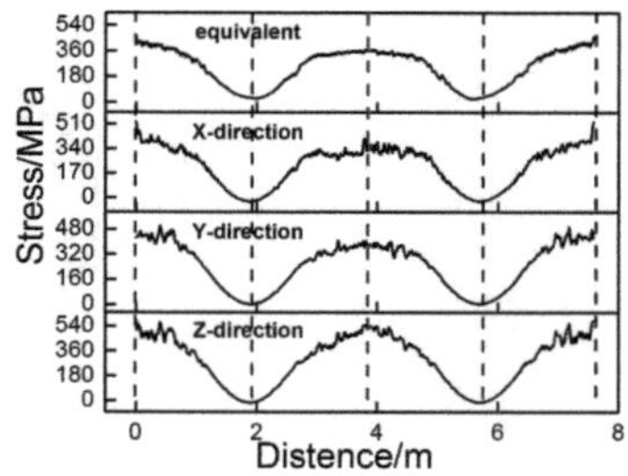

Fig. 5 Equivalent stress and normal stress distribution along hot blast outlet

about 460 MPa, around the top of outlet. The highest X-, Y-, Z-normal stress is about 524, 424, 578 MPa, respectively. All the three direction normal stress around the top and bottom of outlet is positive, shows that the shell around these region is under tensile stress. At left and right sides of hot blast outlet, stress is lower and the lowest equivalent stress is about 25 MPa. The lowest X-, Y-, Z-normal stress are about −32, −1, −17 MPa, respectively. All the three direction normal stress around the left and right sides of outlet is negative, shows that the shell around these region tends to be under slight compressive stress.

Normal strain distributions along hot blast outlet are shown in Fig. 6. From figure a, we can see that all the three direction normal strain concentrates around the top and bottom of outlet. The X-direction normal strain is negative and the Y-, Z-direction normal strain is positive, shows that the shell around the top and bottom of outlet compresses along the X direction and stretches along Y, Z direction. In addition, all the three direction normal strain around the left and right sides of outlet is relatively small, almost equal to 0, while the highest strain is about −0.0043, 0.0026, 0.0048, respectively. From figure b, we can see that the tendencies of Y-, Z-direction elastic and plastic strain along hot blast outlet are coincident, but the tendencies of X-direction elastic and plastic strain are opposite. The X-direction elastic strain is positive and plastic strain is negative. In addition, there is no plastic strain around the left and right sides of outlet, the X-direction total strain around the left and right sides of outlet will be positive, so that the shell around the left and right sides of outlet will stretch along the X direction. In conclusion, the deformation of hot blast outlet is

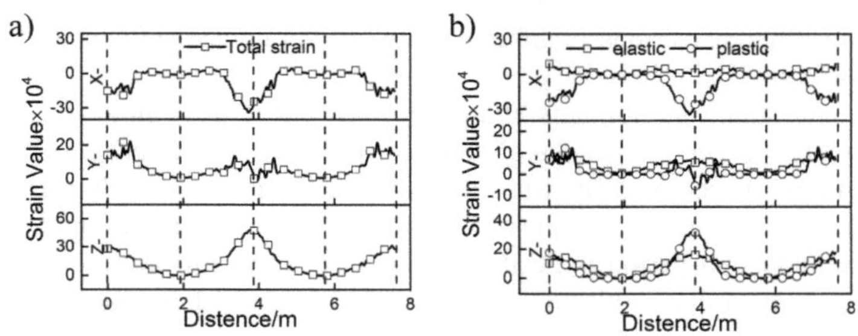

Fig. 6 Normal strain distributions along hot blast outlet: **a** total strain, **b** elastic and plastic strain

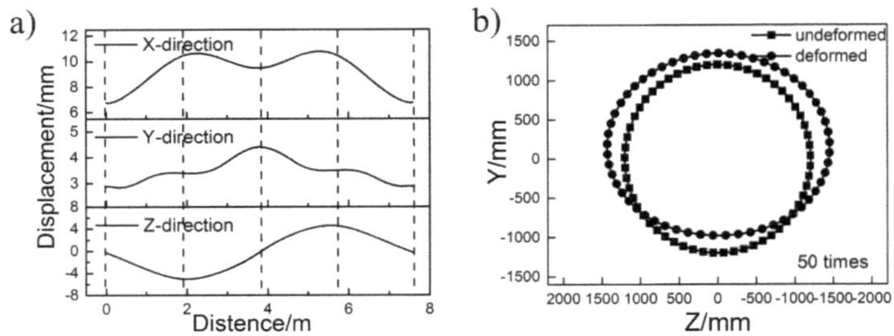

Fig. 7 Displacement and deformation along hot blast outlet: **a** Displacement, **b** deformation along the YZ plant

affected by both elastic and plastic strain, especially the X-direction plastic strain, which dominates the X-direction deformation around the top and bottom of outlet.

Three direction displacements along hot blast outlet are shown in Fig. 7a. X direction displacement is positive, means that the shell of outlet will move towards the hot blast main. X direction displacement around the left and right sides of outlet is higher, about 10.7 mm, and the displacement around top of outlet is lower, about 6.7 mm. Y direction displacement is positive, means that the shell of outlet will move upwards. The shell around the bottom rise higher, about 4.4 mm, and the shell around the top rise lower, about 2.9 mm, means that the shell of outlet will shorten along height direction. Z direction displacement is negative around left side and positive around right side of outlet, the largest displacement is about −5.0, 4.6 mm, respectively, means that the shell of outlet will elongate along the horizontal direction.

The deformation of hot blast outlet along the YZ plant is shown in Fig. 7b. To get a better view, the deformation has been magnified 50 times. It can be seen that the shell of hot blast outlet elongates along the horizontal direction and slightly moves to the left. In addition, the shell rises along height direction and shortens along the vertical direction, the shell near the outlet deforms into an oval.

In Combustion Period

The operating state of hot blast stove circulation with different periods, the inside pressure in different periods is different, and has significant influence to the shell stress distribution.

Stress and strain distributions along hot blast outlet in combustion period are shown in Fig. 8. The tendencies of different direction stress and Y-, Z-direction normal strain are coincident with that in blast period, but the tendency of X-direction normal strain isn't agree with that in blast period. X-direction normal strain in combustion period is positive and all the three direction normal strain is relative lower than that in blast period. In conclusion, the deformation of hot blast outlet in combustion period is mainly affected by elastic strain, and the plastic strain around the hot blast outlet is almost negligible.

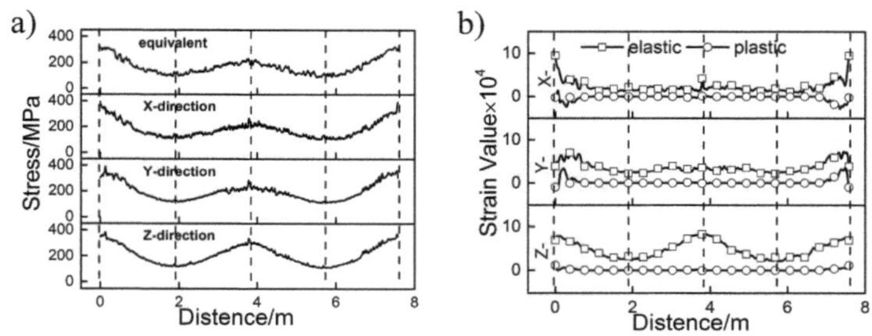

Fig. 8 Stress and strain distribution along hot blast outlet: **a** equivalent and normal stress, **b** elastic and plastic strain

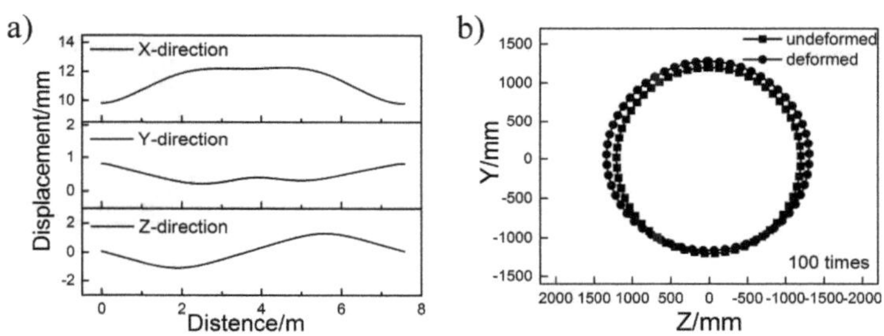

Fig. 9 Displacement and deformation along hot blast outlet: **a** displacement, **b** deformation along the YZ plant

Three direction displacements along hot blast outlet in combustion period are shown in Fig. 9a. All the three direction displacement is relatively smaller than that in blast period. The tendencies of X and Z direction displacement is coincident with that in blast period. However, Y direction displacement around the top of outlet is higher than that around the bottom, means that the shell of outlet will elongate along height direction.

The deformation of hot blast outlet along the YZ plant in combustion period is shown in Fig. 9b. It can be seen that the shell of hot blast outlet elongates along the horizontal direction and the vertical direction, the deformation of shell near the outlet isn't obvious.

Comparison

The normal stress and strain along hot blast outlet in different periods are shown in Fig. 10. All the stress and strain in this figure is that on the top of hot blast outlet. In this figure, it can be seen that all the normal stress and strain in blast period are obvious higher than that in combustion period, part of them are almost twice of that in combustion period. In addition, the X-direction normal strain is negative in blast

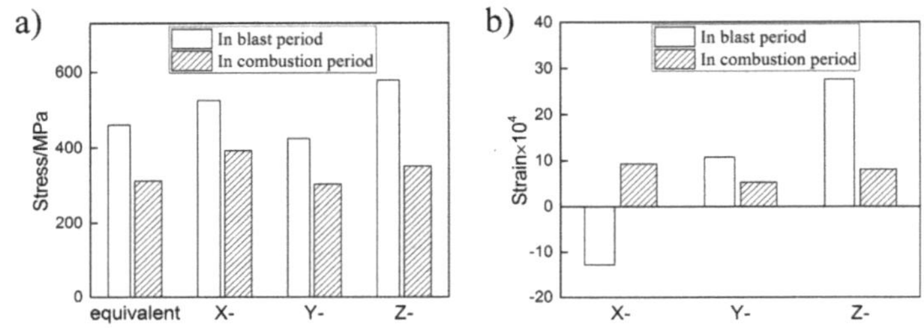

Fig. 10 Stress and strain along hot blast outlet in different periods: **a** stress, **b** strain

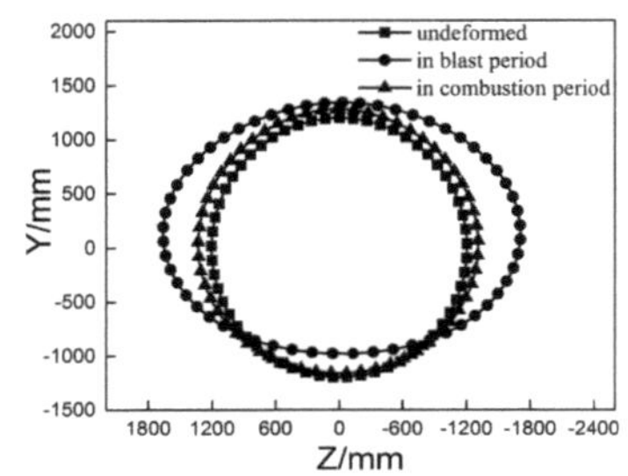

Fig. 11 Deformation of hot blast outlet YZ plant in operation process

period and positive in combustion period, shows that the stress states and deformation along hot blast outlet is obvious difference in different period.

The deformation of hot blast outlet YZ plant in operation process is shown in Fig. 11, and the displacement has been magnified 100 times. It is obvious in this figure that the deformation of shell in combustion period is relatively small, and the center point is almost fixed. In blast period, the shell elongates along the horizontal direction about 7.2 mm and shortens along the vertical direction about 2.0 mm, and the center point rise about 3.5 mm. In operation process, the hot blast stove circulates between blast and combustion period and the shell near the outlet deforms and contacts with the lining refractory repeatedly, this may cause the refractory brick damage and is harmful to the long campaign life of hot blast stove.

Conclusions

Consideration the operation state of hot blast stove, an elastic-plastic stress mathematical model of hot blast stove shell has been established in order to investigate the stress distribution and deformation of the shell, especially the shell of the intersection line of hot blast stove body and the hot blast manifold. As above mentioned work, the following were obtained.

1. Stress concentration happens along the intersection line of hot blast stove body and the hot blast manifold in both blast and combustion period. However, the stress level and deformation degree of the hot blast shell near the blast outlet in blast period is much higher than that in combustion period, and the plastic deformation mainly occurs on the hot blast outlet in blast period.
2. In blast period, the shell near the blast outlet elongates along the horizontal direction and shortens along the vertical direction, the shell near the outlet deforms into an oval. And in combustion period, its deformation isn't obvious. The repeated deformation of the shell near the outlet between different periods may cause the refractory brick damage and be harmful to the long campaign life of hot blast stove.

References

1. Chuan C et al (2014) Hazard control of NO_x in hot stove. Int J Iron Steel Res 3(21):306–311
2. Liu Z et al (2012) Recent progress on long service life design of chinese blast furnace hearth. ISIJ Int 10(52):1713–1723
3. Zhang F et al (2012) Dome combustion hot blast stove for huge blast furnace. Int J Iron Steel Res 9(19):1–7
4. Zhong L et al (2004) Computer simulation of heat transfer in regenerative chambers of self-preheating hot blast stoves. ISIJ Int 5(44):795–800
5. Rieger J et al (2015) Modelling and control of pollutant formation in blast stoves. J Clean Prod 88:254–261
6. Khosoussi S et al (2015) A new approach to the elastic–plastic stress transfer analysis of metal matrix composites. Arch Appl Mech 11(85):1701–1717
7. Çallıoğlu H et al (2015) Elastic–plastic stress analysis of rotating functionally graded discs. Thin-Walled Struct 90:38–44
8. Chen M (2007) Elasticity and plasticity. China Science Publishing & Media Ltd, Beijing, pp 226–277
9. Yang G (2013) Introduction to elasticity and plasticity, 2nd edn. Tsinghua University Press, Beijing, pp 41–78

Study of Calcined Mixtures from Industrial Residues for Production of Agglomerates

L.I.C. Fernandez, L.G. Pedroti, E.B. Ferreira Filho, R.C.S.S. Alvarenga, L.G. Justino and W.E.H. Fernandes

Abstract More and more industries have been concerned with the disposal of the wastes generated in their production processes, once this generates a positive impact, both visual and economic enterprise, no more needing to have resources for transport and final disposal of wastes. In this context, and seen the importance of the Portland cement as a building material, which aims the creation of an ecological cement from the mix of two wastes, the grits, which is a waste generated in pulp production, and the granite beneficiation waste. It proceeded initially with the chemical and physical descriptions of the wastes. Five mixtures, varying the percentages of the wastes and the preparation method of the raw materials, were prepared. After, the blends were sintered in a muffle furnace at 1300 °C for 15 min, forming new compounds that were compared with national cements, according to the Brazilian technical standards. It was concluded that the sintering of the product obtained satisfactory results, presenting good binder characteristics, however, some parameters were outside the limits for common Portland cements such as rapid setting time, large expandability and low compressive strength.

Keywords Grits waste · Granite waste · Ecologic cement

Introduction

The generation of waste from the intense industrial activity, represents nowadays a major environmental threat. Therefore, the proper disposal of waste has become essential to minimize the impacts caused by the disposal of such waste in the

L.I.C. Fernandez (✉) · L.G. Pedroti · E.B. Ferreira Filho · R.C.S.S. Alvarenga · L.G. Justino · W.E.H. Fernandes
DEC—Civil Engineering Department, UFV—Federal University of Viçosa, Av. Peter Henry Rolfs, Viçosa, Minas Gerais 36570-900, Brazil
e-mail: leticia.castrillon@ufv.br

L.G. Pedroti
e-mail: leonardo.pedroti@ufv.br

S. Ikhmayies et al. (eds.), *Characterization of Minerals, Metals, and Materials 2017*, The Minerals, Metals & Materials Series,
DOI 10.1007/978-3-319-51382-9_84

environment [1]. One alternative is the incorporation of the industrial wastes in cementitious products, reducing environmental impacts and costs in the production stage, context in which it was developed this work.

The main constituent components of Portland cement ($CaO = 60–67\%$; $SiO_2 = 17–25\%$; $Al_2O_3 = 3–8\%$; $Fe_2O_3 = 0.5–6\%$; $MgO = 0.8–6\%$; Na_2O and $K_2O = 0.3–1.5\%$; $SO_3 < 3\%$) [2] may be obtained in similar proportions by a mixing the grits waste and cutting granite powder waste.

The grits waste is generated by the production of cellulose by the kraft process, which, according to Gullichen and Fogelholn [3], is the most used process worldwide. This process reuses many of the reagents used in the cooking of the wood, through several unit operations, and one of them is the causticizing, which produces the grits.

The granite cutting residue is obtained from the process in which, after extracted the granitic rocks in shape of boulders, these are transported to sawmills to turn the blocks into sheets or plates using a machine called loom, which use abrasive slurries. After cutting, the waste generated is transported and deposited on the deposition end tanks [4].

In order to study the possibility of inclusion of such industrial wastes in the production of calcined mixtures for the production of binders, this work was developed. Based on the chemical analysis of the waste, it was sought to produce mixtures which had similar characteristics to the commercial Portland cement.

Materials and Methods

Collection, Preparation and Chemical Analysis of Raw Materials

The collection of grits was made in CENIBRA Company, and the waste of ornamental rocks, in the rocks beneficiation step in the Decolores Marble and Granite company from Cachoeiro de Itapemirim.

The preparation of raw materials was divided into the following steps: air drying in the courtyard of the Construction Materials Laboratory of the Federal University of Viçosa and later in the kiln to ensure complete removal of moisture; milling in the Abrasion Los Angeles mill; those which underwent the next step, also passed in the ball mill, to facilitate the sifting process; sifting: the materials which were submitted to the ball mill were subsequently screened through on a 0.074 mm mesh sieve (200 mesh).

The chemical analysis of raw materials was made by Energy Dispersive X-ray Fluorescence Spectrometer (EDX) in the X-ray Laboratory of the Soil Science Department of the Federal University of Viçosa. With the results, it was used the Bogue method to obtain the composition of the clinker phases and percentages of waste mixtures.

Preparation and Sintering Mixtures

Based on the methodology proposed by Costa et al. [5], water was added in the raw mixtures, and hand-molded spheres were prepared with 1–2 cm in diameter. The balls were then kept in a kiln at 100 ± 5 °C for at least 2 h, to then be brought to the furnace (Fig. 1). Because of limitation of the furnace used, the maximum temperature reached was 1300 °C with a 15 min burning level, and the withdrawn temperature was 450 °C. After burning, the spheres were cooled with compressed air to room temperature. Finally, they were broken with a hammer, ground in the ball mill and sieved at the 0.074 mm mesh sieve.

Characterization of the Produced Clinker

The density was obtained following the NBR NM 23:2001 [6]. The Le Chatelier flask was filled up to the level between 0 and 1 cm^3 and left in a thermoregulatory bath for 30 min, so it was made the initial volume reading. It was added a definite amount of cement and made the final reading. The density in g/cm^3 is obtained by dividing the mass of cement by the difference between the initial and final volume.

The determination of the specific surface area was made following the recommendations of the NBR 7224:1984 [7]. Initially, the Blaine equipment cell was filled with cement which provided adequate compaction. So the cell was connected to the manometer tube sealed with fine grease. Then the aspiration was made with the hose, moving the manometric liquid to the top mark of the tube. When the fluid passed the second mark, the chronometer was shot and turned off when it reached the third mark. It was noted the time and the room temperature.

Fig. 1 Balls in the furnace before burning

According to the MB NBR 3433:1991 [8], the water was determined as the normal consistency of cement paste. So 500 g of cement was weighed and an amount of water that has been determined by trial. The mixture was made in a mechanical mixer and molded in the trunk-conical mold and held the penetration of the Tetmajer probe into the paste. The normal consistency is obtained when the probe is located at a distance of (6 ± 1) mm from the base plate.

It was determined the Le Chatelier expansibility according to the MB NBR 3435:1991 [9]. The normal consistency paste was prepared to fill 6 Le Chatelier needles and placed in a moist curing for 24 h. 3 needles were cold cured and the others 3 were hot cured. The heat curing is done by placing the needles in a container with water and raising the temperature up to the boiling point, while the cold curing is made for 6 more days. The results are the difference between the extent of opening of the needle right after molding and after curing.

Under the guidelines of NBR NM 65:2003 [10], it proceeded with the determination of setting time. Again, it was prepared a normal consistency paste and molded the trunk-conical mold. After molding, the readings were done by getting down the Vicat needle over the paste. The initial setting time corresponds to the time interval between the addition of water and the moment that the needle penetrates into the paste to a distance of (4 ± 1) mm from the base plate. For the final setting time, the needle is replaced and the mold is turned upside down. The final setting time is the time elapsed since the addition of water until the time when the Vicat needles penetrates 0.5 mm in the paste.

The compressive strength was determined based on NBR 7215:1996 [11]. The quantities of materials as set out in the standard, were mixed and, immediately after the mortar preparation, the specimens were molded. The standard requires 4 specimens by age, but, because of limited material, only 6 specimens were made, two by age (3, 7 and 28 days). The samples were placed in moist cure until the time of the test. The compressive strength is obtained dividing the breaking strength by the area of the test body.

The chemical analysis of the cements produced were made by Energy Dispersive X-ray Fluorescence Spectrometer (EDX) as mentioned above.

Results and Discussion

Chemical Analysis of Raw Materials

The average results of the raw materials EDX are shown in Table 1.

Table 1 Chemical composition of raw materials

Materials	Compounds (%)				
	CaO	SiO_2	Al_2O_3	Fe_2O_3	SO_3
Grits	81.15	9.90	0.56	0.52	1.07
Granite	2.46	73.94	13.53	2.17	0.04

Preparation and Sintering of Mixtures

According to the proportions of each material in the mixture, a manipulation was done of the raw materials EDX data, to obtain probable chemical composition of the final products (Table 2) to fit in the proportions proposed by Petrucci [12] for the national cements. From these results, the Table 3 was mounted, using the Bogue method for estimating clinker phases.

The mix of 79% grits and 21% granite was the one that fit better in the range of national cements for chemical compounds and clinker phases. It is emphasized that these percentages were obtained by attempts. Moreover, it was specified other two percentages of mix for comparison, and the percentage of 79/21 was used to compare if the preparation of the raw materials had influenced in compressive strength.

It is observed that the highlighted values were those which had fled from the range of clinker phases values for national cements proposed by Petrucci [2].

Characterization of the Produced Clinker

The density results for each produced cement are close to the value of 3.05 g/cm^3 suggested by Coutinho [12] for Portland cements, likewise, the specific surface area was fitted in the range of national cements.

To each produced cement, a relation of water cement (w/c) was determined, corresponding to the amount of water of the normal consistency paste. For the 81/19 mixture it was obtained a 0.56 w/c ratio; for mixing 79/21 the w/c ratio obtained was 0.55 and for the 77/23 a w/c ratio of 0.43. It was noted a high water demand for the mixtures 81/19 and 79/21, when compared to national cement.

Table 2 Chemical compounds obtained from handling of raw materials EDX

Materials	Compounds (%)				
	CaO	SiO_2	Al_2O_3	Fe_2O_3	SO_3
81% grits + 19% granite	66.20	22.07	3.02	0.83	0.86
79% grits + 21% granite	64.62	23.35	3.28	0.87	0.85
77% grits + 23% granite	63.05	24.63	3.54	0.90	0.83

Table 3 Estimating phases of clinker based on raw materials EDX (Bogue method)

Mixtures	Clinker phases (%)			
	C_3S	C_2S	C_3A	C_4AF
81% grits + 19% granite	77.76	4.70	6.60	2.53
79% grits + 21% granite	59.86	21.87	7.23	2.63
77% grits + 23% granite	41.99	39.02	7.87	2.73

Table 4 Synthesis of results

Mixtures (grits/granite)	81/19	79/21	77/23
Specific mass (g/cm^3)	3.05	3.02	3.00
Specific surface area (m^2/kg)	331.01	300.03	278.91
Expansibilities (mm)	24.9	6.2	0.2
Initial setting time (min)	26	27	30
Final setting time (min)	55	66	75

The initial and final setting times obtained from each cements tests showed that the hydration reactions developed quickly, therefore, none of the cements fitted in the national standards.

Soundness tests were executed on the cements by Le Chatelier method, using six Le Chatelier needles, 3 to be cold cured and 3 to be heat cured. However, the heat cured test could not be done, since the materials have been dissolved in the water. The cold cured needles presented a great variability of results, perhaps because of the rapid setting time, which made difficult to mold the needles and, according to standard (MB NBR 3435:1991), the maximum expansibility should not exceed 5 mm and the difference between two individual results should not exceed 1 mm. So only the 77/23 mixture was fitted in the standards. The synthesis of the results is in the Table 4.

The results of the compressive strength tests, for each produced cement, such as the strength growth curve according to the ages are shown in Table 5 and Fig. 2, respectively. As explained above, besides evaluating the behavior of different compositions, for the 79/21 mixture was evaluated the influence of the preparation method of raw materials in the strength, since such activities require a lot of time and effort.

From these results, few observations can be made: firstly, it is noted that increasing grits in the mixing implies the cement increased strength; it was also observed that the preparation method influences the strength, so that the cement produced with ground and sifted raw materials showed better results than the others, which can be explained by the fineness degree, that facilitates molecular interactions between the wastes particles; finally, it was observed a lower water absorption in the cement produced with natural raw materials, allowing a better hydration and molding, which may have interfered in the strength.

Table 5 Results of the compressive strength test

Mixtures	Average strength (MPa)		
	3 days	7 days	28 days
81% grits + 19% granite (grinded)	3.54	4.22	6.47
79% grits + 21% granite (natural)	2.19	3.52	5.87
79% grits + 21% granite (grinded)	2.47	3.24	5.69
79% grits + 21% granite (sifted)	3.82	4.90	6.47
77% grits + 23% granite (grinded)	1.02	1.31	3.45

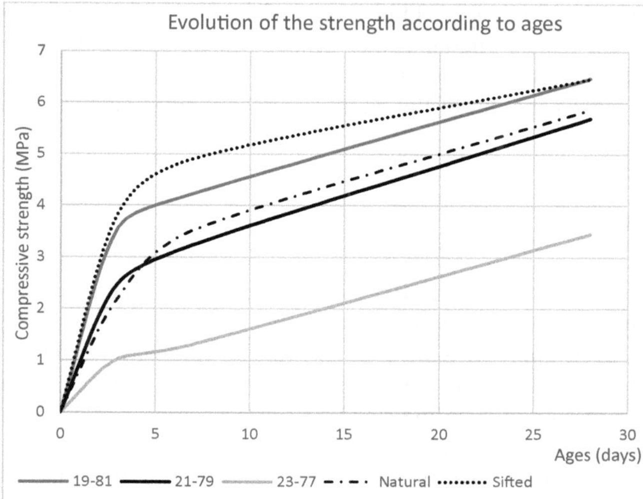

Fig. 2 Evolution of the strength according to ages

The chemical analysis of the cement produced is shown in Table 6. In the Table 7 were again estimated the clinker phases, but now, based on the cement EDX.

Table 6 Chemical composition of cement produced

Materials	Compounds (%)				
	CaO	SiO_2	Al_2O_3	Fe_2O_3	SO_3
81% grits + 19% granite	63.47	23.09	3.69	1.33	1.05
79% grits + 21% granite	60.74	24.46	4.42	1.24	1.15
77% grits + 23% granite	57.85	26.02	5.47	1.19	0.74
79 + 21%—natural	58.62	24.03	4.58	1.49	2.83
79 + 21%—sifted	61.60	25.06	4.23	1.06	0.60

Table 7 Estimates of clinker phases form the cement produced (Bogue Method)

Mixtures	Clinker phases (%)			
	C_3S	C_2S	C_3A	C_4AF
81% grits + 19% granite	53.15	26.19	7.53	4.04
79% grits + 21% granite	26.56	50.17	9.62	3.77
77% grits + 23% granite	−4.89	78.36	13.28	3.62
79 + 21%—natural	14.98	57.67	9.62	4.53
79 + 21%—sifted	28.60	50.35	9.42	3.62

With the results above, it could be observed that the 81/19 mixture was the best fitted one for chemical compounds and clinker phases, what can explain the better compressive strength results for this mix. It is emphasized again that the highlighted values at the table were out of the national cements range according Petrucci [2].

Conclusions

It was cleared that the produced cements have binder characteristics, however, another mixtures should be studied. It is suggested to increase the burning temperature and level, improving the cooling system. This way would be possible to retire the cement from the furnace in higher temperatures and seek the adequacy of the other cement properties in the standards patterns;

The results of density and specific surface area fitted in the national cement ranges, but some characteristic as expansibility and short setting time should not be left aside. The first implies in destructive expansions, and the second harms the workability.

Therefore, the production of binders using the grits and the granite wastes is feasible and promising, however additional studies should be made in order to improve the physical properties of these cements.

Acknowledgements The authors thank the Brazilian agencies: CNPq, CAPES, PIBIC and Funarbe for the support provided to this investigation.

References

1. Lucas D, Benatti CT (2008) Use of industrial waste for the production of cement and clay artifacts utilized in construction. J Agribusiness Environ 1(3):405–418
2. Petrucci EGR (2005) Portland cement concrete. Globo, São Paulo, SP, p 257
3. Gullichen J, Fogelholn CJ (2000) Chemical pulping (Book 6) in papermaking science and technology. Fapet, Helsinki, p 51
4. Silva SAC (1998) Characterization of waste granite sawdust blocks. Study of application potential mortars manufacturing settlement and soil-cement bricks. Sc.M. thesis, Federal University of Espirito Santo
5. Costa EB et al (2013) Clinker portland with reduced environmental impact. Built Environ 13(2):75–86 (Porto Alegre, RS)
6. Brazilian Association of Technical Norms. NM 23 (2001) Portland cement and other powder materials—determination of density, Rio de Janeiro, p 5
7. Brazilian Association of Technical Norms. NBR 7224 (1984) Portland cement and other powder materials—determination of specific surface area, Rio de Janeiro, p 10
8. Brazilian Association of Technical Norms. MB 3433 (1991) Portland cement—determination of water content of the paste with normal consistency. Rio de Janeiro, p 3
9. Brazilian Association of Technical Norms. MB 3435 (1991) Portland cement—determination of soundness by Le Chatelier method. Rio de Janeiro, p 2

10. Brazilian Association of Technical Norms. NM 65 (2003) Portland cement—determination of setting time, Rio de Janeiro, p 4
11. Brazilian Association of Technical Norms. NBR 7215 (1996) Portland cement—determination of the Compressive Strength, Rio de Janeiro, p 8
12. Coutinho JS (2002) Materials science: binders and aggregates. Engineering College, University of Porto, p 161

Study of Synergistic Effect of Light Stabilizer Additive, Conventional and Nanoparticles, Applied to Polyethylene Films Submitted to Ultraviolet Radiation

Patricia Negrini Siqueira Poveda and Leonardo G.A. Silva

Abstract Thermoplastic materials are sensitive to ultraviolet radiation (UV), which can cause structural changes, causing variation in mechanical properties and appearance. UV range between 250 and 400 nm is considered the most aggressive to polymers, and prolonged exposure to it results in breakage and/or crosslinking chains. Light stabilizer additives, blockers or absorbers are added to the polymers for protection in order to extend the life of the final product. The choice of the adequate light stabilizer additive system depends on factors such as resin final application, thickness, geographic location of exposure, presence of fillers in the composition, contact with chemicals, among others. This study evaluates different additive systems: conventional organic, inorganic nanotechnologyc base and the synergy between them when applied to polyethylene considering carbonyl index and UV/VIS results after aging by QUV chamber.

Keywords Polyethylene · Light stabilizer · UV · Nanoparticles · Ageing

Introduction

Aging is a term used in polymer science and engineering when properties of polymer change over a period of time. Changes can be detected on engineering properties as strength and toughness; physical characteristics as density; or

---

P.N.S. Poveda (✉) · L.G.A. Silva
Instituto de Pesquisas Energéticas e Nucleares—IPEN-CNEN/SP, Av. Prof. Lineu Prestes 2242, São Paulo, SP 05508-000, Brazil
e-mail: patricianegrini@usp.br

S. Ikhmayies et al. (eds.), *Characterization of Minerals, Metals, and Materials 2017*, The Minerals, Metals & Materials Series,
DOI 10.1007/978-3-319-51382-9_85

chemical characteristics as aggressive chemical reactivity. These changes can be chemical or a result of interaction with the environment, for example when the oxidation leads to chain scission. Sometimes a series of aging-related phenomena occur simultaneously and/or in an interactive form [1]. The main purpose served by examining data on weathering of materials is to predict any potential changes of physical properties and appearance of those materials. Natural sunlight doesn't offer standard conditions: there are variations in clouds, smog, angle of the sun, rain, industrial environments, etc. Accelerated aging is used to guarantee the test is realized under standard methods [2].

Gulmine et al. studied degradation of polyethylene after artificial accelerated weathering. During degradation polar groups in the polyethylene are generated, as well as chain scission and cross-linking. As effects of these processes increase in crystallinity, density, hardness and surface cracking at late stages of degradation are observed [3]. Tavares et al. evaluated surface mechanical properties of polyethylene subjected to accelerated aging in Weathering Tester (QUV) and Weather-o-Meter (WOM) chambers. Modifications in nano mechanical properties were correlated to the corresponding chemical processes due to aging, which were measured through the carbonyl index profile obtained from ATR–FTIR [4].

Wang et al. [5] investigated mechanical properties of composites and the dispersion of the nanoparticles. The results have been shown that not only the tensile property but also the flexural modulus of the system have been evidently increased by the addition of calcium carbonate (NPCC).

The photodegradation of low density polyethylene (LDPE) containing nano particles of zinc oxide (nZnO) was analyzed by Yang and other [6] by FT-IR to monitor the development of oxidation products of films. Although the absorption of wavelengths below 350 nm by ZnO protect the polymer of photochemistry degradation, it can also induce heterogeneous photocatalytic oxidation of polymer.

Yang and other [7] studied the effect of UV radiation in LDPE films containing different concentrations of titanium dioxide nano particles ($nTiO_2$). When exposed on equipment for accelerated weathering test of type QUV, films resisted 400 h, although TiO_2 presence had promoted the exposition time reduction 50% in mechanical resistance due carbonyl groups formation. Liu and others studied blends of 80 wt% LLDPE and 20 wt% LDPE filled with nanoparticles of SiO_2 and TiO_2 at a ratio up to 5 wt%. The incorporation of nano-TiO_2 contributed to effectively improving the resistance to heating and ultraviolet irradiation of the composites [8].

This study have evaluated the additive systems hindered-amine light stabilizers (HALS), NPCC, nZnO and $nTiO_2$ and the synergy between them when applied to low density polyethylene (LDPE) considering carbonyl index and UV/VIS results after aging by QUV chamber. The samples were molded by cast film and aged in QUV 340 (ASTM G-154 - cycle 01) during until 3000 h, with intervals of 500 h. Tests of FT-IR, carbonyl index and UV/VIS comparing to the non-aged samples were performed to evaluate the performance of additive systems in relation to the degradation behavior of low density polyethylene.

Materials and Methods

Materials

Linear Low Density Polyethylene (LLDPE) butane-1 copolymer, narrow molar mass distribution, without slip additives, supplier Braskem, named ML 2400N, with melt index of 20.0 g/10 min and density of 0.926 g/cm^3. This material was used as a vehicle for the masterbatches production.

Low Density Polyethylene (LDPE), supplier Westlake Chemical Corporation, named EN1817AA, with melt index of 1.7 g/10 min and density of 0.920 g/cm^3, was used as a base resin for the films production.

Cyasorb UV-1164 is 2-[4, 6-Bis(2,4-dimethylphenyl)- 1,3,5-triazin-2-yl]-5-(octyloxy) phenol, produced by Cytec, whose physical form is represented by powder, with molecular weight 509. This material acts as a light stabilizer, offers high inherent UV stability and minimal color contribution.

Cyasorb UV-3529 is 1,6-Hexanediamine, N,N′-bis(2,2,6,6-tetramethyl-4-piperidinyl)-, Polymers with morpholine-2,4,6-trichloro-1,3,5-triazine, produced by Cytec, with physical form in pastilles and molecular weight 1690–1710. It acts as a light stabilizer, offers low volatility and selective reactivity.

Cyasorb UV-5411 is 2-(2′-Hydroxy-5′-octylphenyl)-benzotriazole, produced by Cytec, with physical form in powder and molecular weight 323. This material acts as a light stabilizer, offers minimal color contribution and strong absorption.

Cyasorb UV-531 is 2-Hydroxy-4-n-octoxybenzophenone, produced by Cytec, with physical form in powder and molecular weight 326. It acts as a light stabilizer, offers superb compatibility with polymers and other additives; shows cost efficiency.

Cyanox 2777 is 1:2 blend of 1,3,5-Tris(4-tert-butyl-3-hydroxy-2,6-dimethylbenzyl)–1,3,5-triazine-2,4,6-(1H,3H,5H)-trione and tris (2,4-di-t-butylphenyl) phosphite, produced by Cytec, with physical form in powder and melting point 166–171 °C. It is an antioxidant, offers minimal color and odor contribution; provides low volatility and very good processing stabilization; exhibits low gas yellowing and low water carry-over; shows compatibility with polymeric HALS and UVAs.

Nano calcium carbonate was provided by Nano Materials Technology Pte Ltd., represented in Brazil by Lagos Química, called NPCC-201, with specific density of 2.5 g/cm3, surface area of 40.0 m^2/g and average particle size of 40 nm.

Nano titanium dioxide was supplied by Croda, named Atmer 7354, whose product is a concentrate of additive with 20.0% of active metal oxide content, compatible with polyolefin polymers.

Nano zinc oxide was provided by Croda, named Atmer 7355, whose product is a concentrate of additive with 30.0% of active metal oxide content, compatible with polyolefin polymers.

Preparation of Samples

For the study of the composites, proposed formulations were presented on Table 1, whose compositions were expressed as percentage, in weight.

Formulations were produced in single screw extruder, with screw's diameter of 25 mm and L/D = 30. The heat profile was 120/130/140 °C for feed, compression and dosage zones, respectively, and 140 °C on head. It was used screw rotation of 50 rpm.

To prepare film samples, a casting film extruder Killion/Davis, standard single screw, model KL125, was used with heat profile 120/130/150/155/160 °C, funnel to flat die, producing films with thickness 6 mils or 152 μm.

Analytical Methods

The accelerated aging was conducted in a test chamber UV condensation (QUV), Q-Lab—QUV/se, to simulate the effects of degradation on the surface of specimens, usually generated by exposure to sunlight, rain and dew. The specimens were exposed to radiation generated by UV-B lamp-340, with 8 h of UV cycle to 60 °C/irradiance 0.89 W/m^2 and condensation cycle of 4 h to 50 °C, repeating for 3000 h, with intervals each 500 h to perform analyses of carbonyl index and UV/VIS.

Analyses of Fourier Transform InfraRed (FT-IR) were carried out for verification of the polymer oxidation through the carbonyl index, according to ASTM D 5576, Digilab FTS7000e model, coupled with Universal ATR device ("Sampling Accessory"), in the range of 400–10,000 cm^{-1} with resolution 2 cm^{-1}.

Ultraviolet visible spectrophotometry (UV/VIS) analyses were performed to determine the concentration of compounds that absorb radiation in that range and to verify how much light passes through a sample, using equipment Perkin Elmer, model Lambda 900, in the range of 280–700 nm.

Results

The results of the carbonyl index for the different compositions of samples studied were shown on Table 2.

It was possible to observe the effects of photodegradation on samples #4, #5 and #6 ($nCaCO_3$, $nTiO_2$ and nZnO, respectively) when compared with #13 (control). #13 sample was a conventional and simple formulation with antioxidant and light stabilizer, and then UV radiation was not so aggressive. It was observed nano

Table 1 Formulation of composites (% in weight)

Component/formula	#2	#3	#4	#5	#6	#7	#8	#9	#10	#11	#12	#13
HALS-3529						0.20	0.20	0.20	0.20	0.20	0.20	
UV-1164						0.10						0.30
UV-531	0.30						0.10					
UV-5411		0.30						0.10				
Nano $CaCO_3$—NPCC-530 (100%)			0.30						0.10			
Nano TiO_2—Atmer 7354 (20%)				1.50						0.50		
Nano ZnO—Atmer 7355 (30%)					1.00						0 33	
AO-2777	0.06	0.06	0.06	0.06	0.06	0.06	0.06	0.06	0.06	0.06	0.06	0.06
LDPE	99.64	99.64	99.64	98.64	98.64	99.64	99.64	99.64	99.64	99.24	99.41	99.64

Table 2 Carbonyl Index (CI) of composites

Description	% 0 h	% Δ CI 500 h	% Δ CI 1000 h	% Δ CI 1500 h	% Δ CI 2000 h	% Δ CI 2500 h	% a Δ CI 3000 h
#2: 0.3% UV-531	0.009	0.012	0.021	0.036	0.077	0.509	
#3: 03 94 UV-5411	0.009	0.054	0.460	1.236	1.797		
#4: Nano $CaCO_3$ NPCC 530	0.021	1.555	5.090				
#5: Nano TiO_2—Atmer 7354 (20%)	0.022	1.775	5.879				
#6: Nano ZnO—Atmer 7355 (30%)	0.020	0.406	1.059	1.686	2.408		
#8: 02% HALS—3529 & 0.1% UV-531	0.008	0.009	0.012	0.017	0.020	0.019	
#9: 02% HALS—3529 & 0.1% UV-5411	0.011	−0.004	−0.001	0.003	0.005	0.005	
#10: 02% HALS-3529 & 0.1% Nano $CaCO_3$ NPCC 530	0.012	0.002	0.002	0.005	0.008	0.007	
#11: 02% HALS -3529 & 0.5% Nano TiO_2—Atmer 7354	0.014	0.032	0.039	0.053	0.071	0.099	
#12: 02% HALS—3529 & 0.33% Nano ZnO —Atmer 7355	0.016	0.019	0.022	0.031	0.040	0.047	
#13: Control	0.011	0.287	1.159				

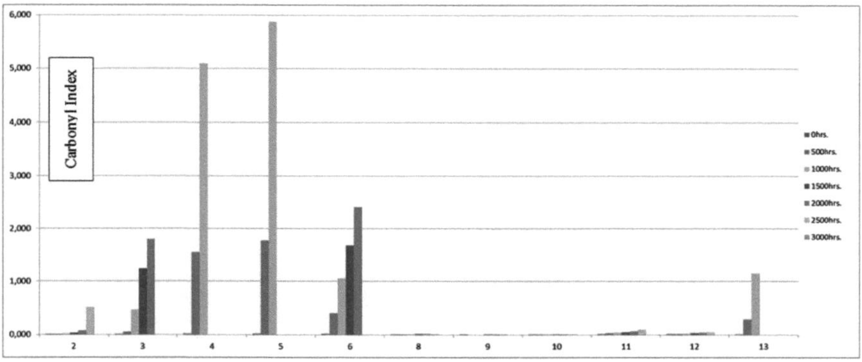

Fig. 1 Carbonyl index of composites, after ageing by QUV until 3000 h

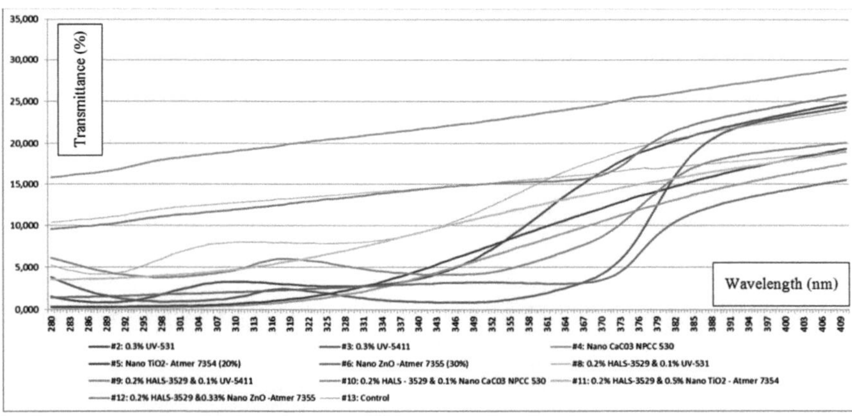

Fig. 2 Ultraviolet visible spectrophotometry (UV/VIS) of composites not aged

composites formulations showed increase in carbonyl index during ageing process. Despite these particles presented light stabilizing characteristic as barrier or blocking, they could generate free radicals, which accelerated the photodegradation process.

Formulations #9 and #10 (HALS 3529 + UV5411 and HALS 3529 + $nCaCO_3$) presented best results in carbonyl index, according to Fig. 1.

The results of ultraviolet visible spectrophotometry (UV/VIS) for composites of samples analyzed were exposed on Figs. 2, 3 and 4.

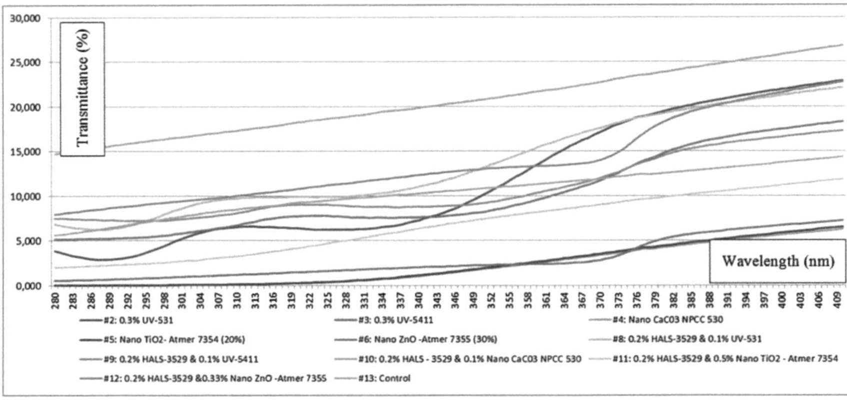

Fig. 3 Ultraviolet visible spectrophotometry (UV/VIS) of composites after 1000 h in QUV

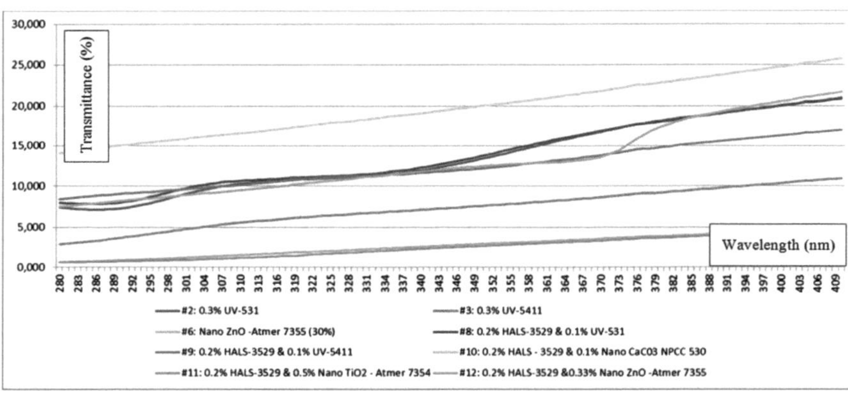

Fig. 4 Ultraviolet visible spectrophotometry (UV/VIS) of composites after 2000 h in QUV

Conclusion

In general, all systems of light stabilizer additives had a positive effect on the LDPE material, with different performances, according to the results showed.

As presented in figures above, it was possible to verify that organic additives (HALS) had its transmittance values increased after the aging period of the samples.

Despite the results of low formation of carbonyl, it was clear that the efficiency of organic additives decreased over time due to the migratory characteristic.

The balance between carbonyl index and UV/VIS showed formulations #10 (HALS 3529 + $nCaCO_3$) the best formulation light stabilizer, among the analyzed nanoparticle systems ($nZnO$, $nTiO_2$ and $nCaCO_3$, individually and/or in blend with HALS).

The advantage of nanoparticles, when compared to organic systems was that they didn't lose their efficiently over time by migration issues. However, it was important to know free radicals formation mechanisms, which interfered negatively in polymer protection as degradation by light exposure.

Acknowledgements The authors wish to thank IPEN, PolyOne and Cytec for all support in this work.

References

1. White JR (2006) Polymer ageing: physics, chemistry or engineering? Time to reflect. Chimie 9:1396–1408
2. Massey LK (2007) The effects of UV light and weather on plastics and elastomers, 2nd edn. William Andrew Publishing, New York
3. Gulmine JV et al (2003) Degradation profile of polyethylene after artificial accelerated weathering. Polym Degrad Stab 79:385–397
4. Tavares AC et al (2003) The effect of accelerated aging on the surface mechanical properties of polyethylene. Polym Degrad Stab 81:367–373
5. Wang W et al (2007) Preparation and characterization of calcium carbonate/low-density-polyethylene nanocomposites. J Appl Polym Sci 106:1932–1938
6. Yang R et al (2010) Degradation products formed during UV exposure of polyethylene-ZnO nano-composites. Polym Degrad Stab 95(9):1533–1541
7. Yang R et al (2011) Spectroscopic studies of photodegradation of polyethylene films containing TiO_2 nanoparticles. J Appl Polym Sci 119(3):1330–1338
8. Liu G et al (2005) Effect of nanoscale SiO_2 and TiO_2 as the fillers on the mechanical properties and aging behavior of linear low-density polyethylene/low-density polyethylene blends. J Polym Environ 13(4):339–348

Study of the Effect of Surface Liquid Flow During Column Flotation of Mining Tailing of the Dos Carlos Dam

Javier Flores Badillo, Juan Hernández Ávila, Isauro Rivera Landero, María Isabel Reyes Valderrama, Eduardo Cerecedo Sáenz, Martín Reyes Pérez, Eleazar Salinas Rodríguez and Mauricio Guerrero Rodríguez

Abstract During column flotation of mining tailings of the Dos Carlos dam, it was observed that the effect of surface liquid flow (J_L) is a variable that determines the rate of rise of packing bubbles and therefore, causes the generation of gas retained within the columnar system. Also overall more stable laminar flows were observed, in a range of superficial flow of liquid in the bottoms stream (J_{LT}) from 0.446 to 0.68 cm/s, showing the stability in the range of superficial gas flow (J_g) from 0.1 to 0.5 cm/s. Observing a uniform bubble size for the three J_{LT} evaluated, with a diameter between 0.05475 and 0.2525 cm, at all the range evaluated of J_g. Furthermore, an uniform performance was obtained, and with the same rate of rise of packing bubbles, showing a greater packing bubbles at J_{LT} = 0.446 cm/s, due to the lesser bubble size observed, in comparison with the values of J_{LT} of 0.68 and 1.017 cm/s.

Keywords Mining tailings · Column flotation · Liquid superficial rate

Introduction

In column flotation, a very useful parameter to understand the system is the superficial speed, either gas (J_g) or liquid (J_L), and which can be compared in columns of different cross section. This flow can be defined as volumetric flow by unit of cross section, being that this flow determines the amount of surface flow of bubbles (S_b), and therefore the cleaning area, it translates as the maximum amount of carries solids or carrying capacity (C_a). This is possible in a bubbly flow regime,

J. Flores Badillo (✉) · J. Hernández Ávila · I. Rivera Landero · M.I. Reyes Valderrama · E. Cerecedo Sáenz · M. Reyes Pérez · E. Salinas Rodríguez · M. Guerrero Rodríguez
Área Académica de Ciencias de la Tierra y Materiales, Universidad Autónoma Del Estado de Hidalgo, Carretera Pachuca-Tulancingo Km. 4.5, C.P. 42184 Pachuca, Hidalgo, Mexico
e-mail: jfb.cmat@gmail.com

S. Ikhmayies et al. (eds.), *Characterization of Minerals, Metals, and Materials 2017*, The Minerals, Metals & Materials Series,
DOI 10.1007/978-3-319-51382-9_86

expressed as the ratio of gas retained within the system columnar (ε_g) and the surface flow of gas (J_g), also being affected by the J_L, [1].

The relationship between the J_g and the ε_g, it can be said that increases according to the concentration of the foaming and an increasing in J_L. This results that the bubble diameter is decreased (D_b) and bubble load is increased (ρ_b), [1].

On the other hand, tailings dam "Dos Carlos" is considered as one of the oldest dam within the mining district of "Pachuca-Real del Monte", which it began operating as such in 1912. Actually, "Dos Carlos" has almost 14.3 millions of tons of tailings [2]. This tailings dam represents an environmental problem due this dam is within the urban area of Pachuca County; this, coupled with the climate that is mostly windy throughout the year, generates constant carries of particulate material thus affecting the surroundings areas, especially the months from February to April, and occasionally November.

In a general manner, the material from the tailings dam has a chemical composition that corresponds to 70.37% SiO_2, 11.35% Al_2O_3, 2.32% Na_2O, 3.67% K_2O, 0.53% CaO, 0.98% MgO, 0.73% MnO, 4.8% Fe_2O_3, 4.32% FeO, 0.53% TiO_2, 0.15% P_2O_5, 55 ppm of Ag, 0.58 ppm of Au, 379 ppm of Ba, 1.1 ppm of Be, 16.8 ppm Cd, 13 ppm of Co, 93 ppm of Cr, 98 ppm of Cu, 9.5 ppm of Mo, 80 ppm of Ni, 600 ppm of Pb, 0.6 ppm of Sb, 11 ppm of Sc, 17 ppm of Sn, 124 ppm of Sr and 1500 ppm of Zn [3–5]. Showing a coarse granulometry of the order of 60% cumulative up to mesh 270 (53 μm), [3, 4].

Previously, it was carried out flotation studies with this kind of tailings (Santa Julia dam), focusing this study on the flotation of silica in a mechanical cell, achieving close to 82.3% recovery of this species. In such study it was used a sample taken from the foam flotation of a concentrate containing 75.16% of silica [6].

Another antecedent was the realization of a flotation in a pneumatic cell using the material of the tailings dam "Santa Julia", where a recovery of 92% of silica after a second wash of the tailings was obtained. It should be noted that during washing of tailings, a loss of 5.9% was noted when HCL was used. Likewise, have conducted studies of flotation of feldspars contained in the tailings, but as these tailings were found similar impurities like in the flotation of silica process, so after crushing material (between −30 mesh and +200 mesh), and subsequent conditioning of the pulp that is comprised of 30% of solids, it proceeds to float the feldspar by using amine acetate, conditioning with sulfuric acid, fuel oil and the addition of pine oil. Then are decanted the flotation tails and are conditioned with sulfuric acid, xanthates, fuel oil and pine oil. Finally the feldspar flotation uses the tails from basic silicates flotation, conditioning with hydrofluoric acid, ammonium salt, fuel oil, and pine oil; fixing an acidic pH [7].

As for the flotation of material from natural deposits of silica sands, there are different methods and reagents depending on the impurities having such sands. For feldspathic sands with quartz, fatty acids are mainly used as collectors in a neutral or slightly basic medium during the primary phase of removal of iron and titanium; then in the secondary phase it is necessary the elimination of mica and feldspar using primary amine collectors in acidic medium, by the addition of sulfuric acid, also that during the removal of feldspar ammonia bifluoride is used as depressant; obtaining a silica concentration of the order of 98.98% with 0.97% of Al_2O_3 and 0.05% of Fe_2O_3, [8].

Separation during secondary phase, is achieved with pH's acids (of the order of 2–2.5), obtained with addition of the H_2SO_4 or in some cases, by using HF in a concentration range of 2000–2500 ppm; the latter is also a depressant agent for the feldspathic portion, also promoting the silica particle surface can be cleaned of metallic oxides, and said modified process gave a silica with a concentration of 98.36%, [9].

Also it is possible the use of cationic reagents based complex sulfates, for silica flotation from sands having quartz and feldspar, by using activators of the silica like Ca^{2+}, Ba^{2+} and Sr^{2+}, thus achieving a selective flotation of silica without using HF [10].

The quality of silica required by chemical and glass industries must have the following composition; 99.8% of SiO_2, with impurities showing values lower than 0.02% of hematite and alumina, as lower values of 0.1% of lime and magnesia [11, 12], and comparing the results found in the above investigations mentioned, the found values are still quite far of industrial requirements, for this reason the present work is focused on silica recovery by flotation column from the tailings dam "Dos Carlos".

Experimental Procedure

Was performed flotation column of tailings, using a concentration of 3% of solids in the system, with 50 ppm of dodecylamine, $C_{12}H_{25}$-NH_2 (analytical grade; Sigma-Aldrich), with the addition of the frother methyl isobutyl carbinol (MIBC) to a concentration of 60 ppm (Alkemin), with values of J_g (0.1, 0.3, 0.5, 0.8 and 1.0 cm/s), J_L of 0.446, 0.68 and 1.017 cm/s, respectively.

From these experiments it was determined the ε_g, D_b and S_b respective, taking samples in each J_g, in the feed streams, tailings and concentrate.

The concentration of samples is determined from flotation processes with the different J_L used, by determining the concentrations of SiO_2, Al_2O_3, Fe_2O_3, Na_2O, K_2O, CaO, MgO, TiO_2, Cu, Zn and S, by EDS technique, using an Scanning Electron Microscope JEOL JSM-6300 equipped with a Noran EDS detector.

The hydrodynamic behavior of the flotation column is simulated using the software COMSOL 5.0, with laminar flow module in 2D, introducing hydrodynamic data flotation of runs at speeds of surface flow of liquid in runs for tailings under the following values, J_{LT} = 0.446, 0.68, 1.017 cm/s. In Table 1 is presented are shown the equation used for the different domains within the flotation column, finding negative values marking a downward vector direction within the flotation column.

Table 1 Equations and values used in hydrodynamics simulation for the column flotation, during its operation at J_{LT} = 0.446, 0.68, 1.017 cm/s, according to their status of domain

Domain status	Equation	Values used
Fluid properties	$\varphi_L\rho_L\frac{\partial U_L}{\partial t}+\varphi_L\rho_L(U_L\cdot\nabla)U_L=\nabla\cdot[-\rho_L+\varphi_L\mu_L(\nabla U_L+(\nabla U_L)^T)]+\varphi_L\rho_L g+F$ $\rho_L\nabla\cdot(U_L)=0,\ \ U_L=U$ $\frac{\partial\varphi_g\rho_g}{\partial t}+\nabla\cdot N\varphi_g\rho_g=m_{gL},\varphi_g\rho_g=rhogeff$ $N_{\varphi_g\rho_g}=\varphi_g\rho_g U_g, U_g=U_L+U_{Slip}$	ρ_L = 997.048 kg/m^3 μ_L = 1 × 10^{-3} Pa s ρ_g = 1.1993 kg/m^3 M_g = 2 × 10^{-2} kg/Mol T = 298.15 K D_b = 0.000065 m
Liquid phase input	$U_L=U_0$ $-n\cdot N_{\varphi_g\rho_g}=0$	U_0, X_1 = 0.00602 m/s; X_2 = 0.00837 m/s; X_3 = 0.01233 m/s, Y = 0 m/s
Gas phase input	$U_L=U_0$ $-n\cdot N_{\varphi_g\rho_g}=\varphi_g\rho_{g0}$	U_0, X = 0.003 m/s, Y = 0 m/s
Without viscous output tension	$P=P_0,[\mu_L(\nabla U_L+(\nabla U_L)^T)]n=0$	P_0 = 0 Pa
Liquid phase output	$U_L=U_0$ $-n\cdot N_{\varphi_g\rho_g}=0$	U_0, X = 0 m/s, Y_1 = −0.00446 m/s; Y_2 = −0.0068 m/s; Y_3 = −0.01017 m/s;
Walls	$U_L=0$ $-n\cdot N_{\varphi_g\rho_g}=0$	
Gravity	$\varphi_L\rho_L\frac{\partial U_L}{\partial t}+\varphi_L\rho_L(U_L\cdot\nabla)U_L=\nabla\cdot[-\rho_L+\varphi_L\mu_L(\nabla U_L+(\nabla U_L)^T)]+\varphi_L\rho_L g+F$	g, X = 0 m/s^2 Y = 9.816 m/s^2

Results and Discussion

It is shown in Fig. 1, the representation of the retained gas in the column flotation, for a J_{LT} of 0.446 cm/s, by the application of a J_g de 0.1, 0.3, 0.5, 0.8 y 1.0 cm/s. In this graph can be seen that the J_g optimal operation for this liquid flow, is in the range from 0.1 to 0.5 cm/s, also observing that in this range there is a light deviation of laminar behavior close to 0.46 cm/s; so that the J_g optimal within this range is of 0.3 cm/s. That value of J_g it can be observed a laminar flow in the 4 zones of flotation column, being the area between the pressure gauges 2 and 3 (central zone), which present a minor increasing of retained gas, so increasing the rate of ascent of the bubbles.

It is observed in Fig. 2 the determination of retained gas in the column flotation, for a J_{LT} of 0.68 cm/s, by applying a J_g of 0.1, 0.3, 0.5, 0.8 and 1.0 cm/s. In this graph is shown the J_g optimal operation to this liquid flow is in the range for 0.1–0.5 cm/s, also observing that in this range there is a light deviation of laminar behavior close to 0.40 cm/s. This laminar flow regime is more evident in the lower and upper peripheral areas at a J_g of 0.3 cm/s, showing a slight decreasing in the amount of retained gas within the central area of column, increasing a bit the rate of rise of bubbles.

It is observed in Fig. 3 the determination of retained gas in the columnar flotation system, for a liquid flow in the tailings stream (J_{LT}) of 1.017 cm/s, by applying of a J_g of 0.1, 0.3, 0.5, 0.8 and 1.0 cm/s. In this graph is demonstrated that the J_g optimal operating to this liquid flow, is present in the range from 0.1 to 0.3 cm/s, observing a deviation from the range of J_g between 0.5 and 0.8 cm/s, subsequently returning to laminar flow and increasing the amount of retained gas in columnar system.

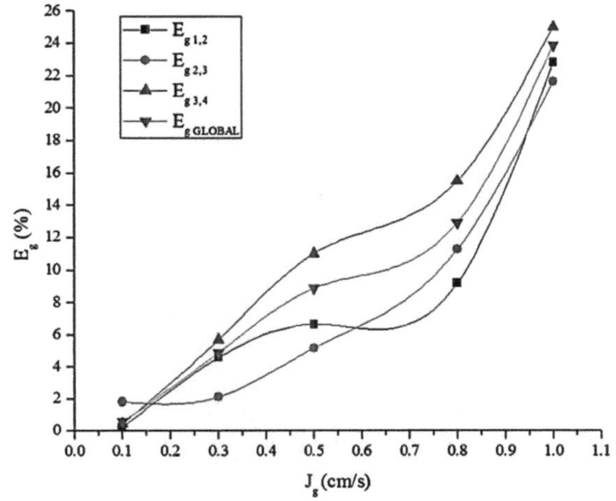

Fig. 1 Determination of ε_g, in column flotation of Dos Carlos tailings, at J_{LT} de 0.446 cm/s, 50 ppm $C_{12}H_{25}NH_2$, as collector and 60 ppm MIBC as frother

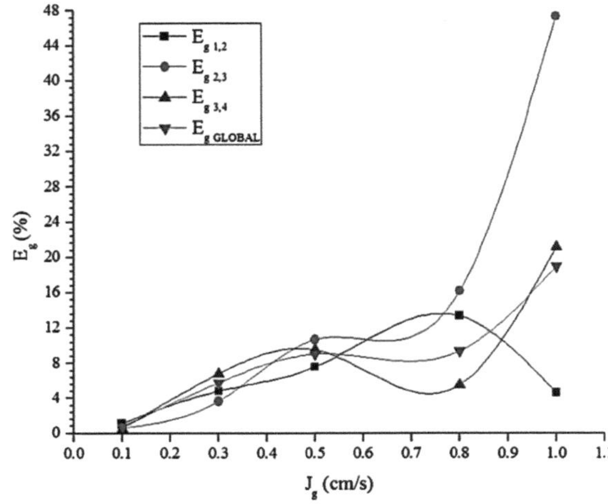

Fig. 2 Determination of ε_g, in column flotation of Dos Carlos tailings, at a J_{LT} of 0.68 cm/s, 50 ppm $C_{12}H_{25}NH_2$, as collector and 60 ppm MIBC as frother

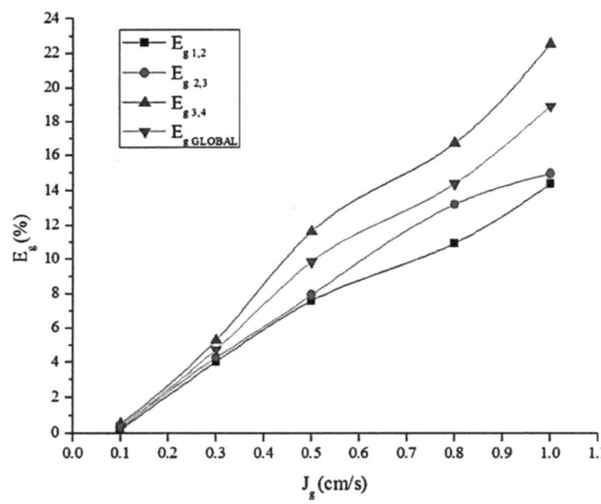

Fig. 3 Determination of ε_g, in column flotation of Dos Carlos tailings, at J_{LT} of 1.017 cm/s, 50 ppm $C_{12}H_{25}NH_2$, as collector and 60 ppm MIBC as frother

This similar behavior of laminar flow regime is presented at the top of the column. With regard to the lower half of the column, from the manometer 1–3, a regime between the lower and the middle area is provided in the range of J_g from 0.1 to 0.5 cm/s, keeping the ascent rate of bubbles constant throughout the flotation column.

It is observed in Fig. 4 the determination of ε_g in the columnar flotation system for the evaluated values of J_{LT} which are of 0.446, 0.68 and 1.017 cm/s, respectively, by applying gas flows J_g of 0.1, 0.3, 0.5, 0.8 and 1.0 cm/s. Laminar flows are observed that are more stables in a range of J_{LT} from 0.446 up to 0.68 cm/s, showing a stability in the range of J_g from 0.1 to 0.5 cm/s. Also is observed in the same range of J_{LT} and with J_g from 0.3 cm/s up to 0.5 cm/s, it has a higher rate of

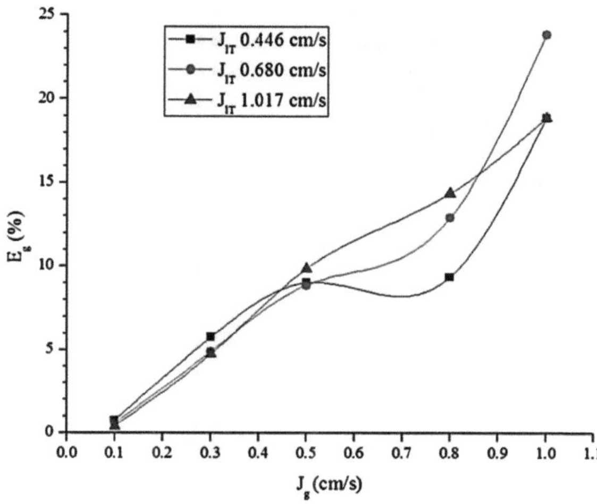

Fig. 4 Determination of $\varepsilon_{g\ Global}$, in column flotation of Dos Carlos tailings, for a J_{LT} of 0.446, 0.68 and 1.017 cm/s, respectively; 50 ppm $C_{12}H_{25}NH_2$, as collector and 60 ppm MIBC as frother

ascent of the bubbles, which would represent a greater recovery of ore than that found at a value of J_{LT} = 1.017 cm/s.

Figure 5 shows the determination of the D_b generated in the disperser for the J_{LT} evaluated and with values of 0.446, 0.68 and 1.017 cm/s, respectively, by applying of values of J_g of 0.1, 0.3, 0.5, 0.8 and 1.0 cm/s. It is noted an uniform bubble size for the 3 values of J_{LT} evaluated with a J_g of 0.5 cm/s, which is between 0.060 and 0.062 cm. At a J_{LT} of 0.446 cm/s and a value of J_g = 0.3 cm/s, a fine bubble size was found, of the order of 0.05475 cm, increasing considerably $\varepsilon_{g\ Global}$ and S_b. At values of J_g minors of 0.3 cm/s, an increased bubble size is observed which goes from 0.12567 cm up to 0.2525 cm, where increases the rate of ascent of bubbles and a decreasing in the amount of global retained in the flotation column.

In Fig. 6 is observed the determination of the S_b generated from the ε_g and the D_b, for the J_{LT} evaluated and with values of 0.446, 0.68 and 1.017 cm/s, respectively, by applying of a J_g of 0.1, 0.3, 0.5, 0.8 and 1.0 cm/s. It is observed in the range for J_g from 0.1 to 0.4 cm/s, is present an uniform behavior and with the same rate of ascent of the bubbles package, showing an increased packing of bubbles at a J_{LT} of 0.446 cm/s, due to the minr bubble size observed, in comparison with the J_{LT} of 0.68 and 1.017 cm/s. From the values of J_g from 0.41 up to 1.0, positive rates of ascent with different slopes are shown, giving as a result an increasing of the ε_g within the columnar system, especially at lower values of J_{LT} such as 0.446 cm/s, where a stagnant of bubble package is presented from values of J_g from 0.5 up to 0.7 cm/s.

Table 2 shows the characterization by EDS of the three streams sampled, showing that the maximum concentrations are observed at J_g in a range from 0.3 to 0.8 cm/s, being possible a higher recovery at J_{LT} from 0.446 and 0.68 cm/s, respectively and at J_g of 0.3 cm/s. The data obtained in these recoveries for the streams of concentrates are of 88.79% SiO_2, 5.83% Al_2O_3 and 5.38% K_2O for J_{LT}

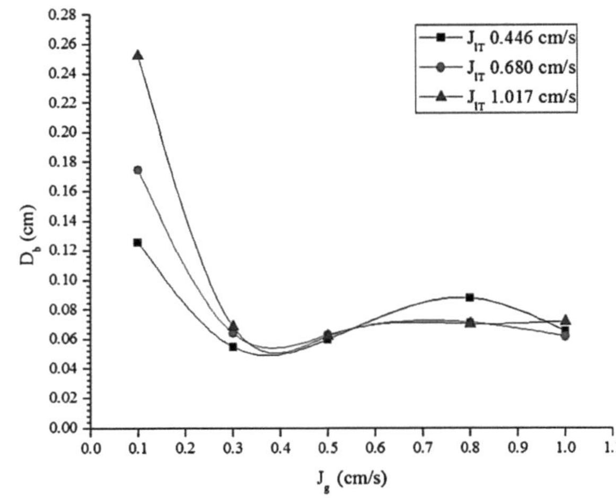

Fig. 5 Determination of D_b, in column flotation of Dos Carlos tailings, for J_{IT} of 0.446, 0.68 and 1.017 cm/s; 50 ppm $C_{12}H_{25}NH_2$, as collector and 60 ppm MIBC as frother

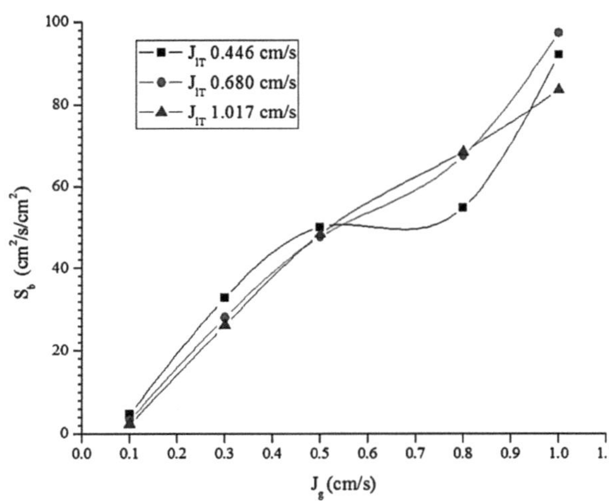

Fig. 6 Determination of S_b, in column flotation of Dos Carlos tailings, for J_{IT} of 0.446, 0.68 and 1.017 cm/s; 50 ppm $C_{12}H_{25}NH_2$, as collector and 60 ppm MIBC as frother

of 0.446 cm/s; as well as 81.66% SiO_2, 8.35% Al_2O_3, 1.58% CaO, 3.62% K_2O, 0.69% MgO and 0.1% S.

In the case of the J_{LT} of 1.017 cm/s, are shown two recoveries higher than 80% SiO_2. The first one in the stream of concentrate which is generated at a J_g of 0.5 cm/s, finding a composition of 82.07% SiO_2, 7.20% Al_2O_3, 3.69% Fe_2O_3, 0.1% TiO_2, 0.26% CaO, 3.58% K_2O and 1.03% MgO. On the other hand, the second recovery was in the stream of concentrate, and it is generated at a Jg of 0.8 cm/s, finding a composition of 82.26% SiO_2, 7.01% Al_2O_3, 3.43% Fe_2O_3, 0.14% TiO_2, 0.22% CaO, 3.08% K_2O, 0.53% MgO and 0.27% S.

In Fig. 7 it is observed the rates obtained with a J_{LT} of 0.44, 0.68 and 1.017 cm/s, in which the rate within the column is lower using a J_{LT} of 0.68 cm/s,

Table 2 Characterization by EDS at weight percentage, of the sampled obtained in the tailings stream, input and concentrate, in a range of J_g from 0.1 to 1.0 cm/s

J_g	Stream	% SiO_2	% Al_2O_3	% Fe_2O_3	% TiO_2	% CaO	% K_2O	% Na_2O	% MgO	% S	% Cu	% Zn
J_{LT} 0.446 cm/s												
0.10	Tailings	82.02	7.01	3.56	0.07	0.30	4.12	0.00	0.30	0.18	0.00	0.00
0.10	Input	78.64	9.48	3.45	0.00	0.41	4.18	0.23	0.82	0.13	0.00	0.00
0.10	Concentrate	68.24	8.48	2.81	0.04	5.93	4.39	1.07	1.78	6.26	0.00	0.00
0.30	Tailings	53.49	7.47	3.19	0.00	1.79	2.20	1.05	1.98	0.51	18.77	7.09
0.30	Input	73.94	8.20	2.30	0.00	2.28	6.70	0.00	0.92	0.46	0.00	0.00
0.30	Concentrate	88.79	5.83	0.00	0.00	0.00	5.38	0.00	0.00	0.00	0.00	0.00
0.50	Tailings	59.53	8.55	2.23	0.00	1.92	2.70	0.07	1.04	0.08	13.87	6.08
0.50	Input	71.44	10.36	6.24	0.04	3.17	4.09	0.37	1.50	0.43	0.00	0.00
0.50	Concentrate	74.62	11.28	0.32	0.01	1.41	6.16	0.00	0.69	1.41	0.00	0.00
0.80	Tailings	77.43	10.18	4.47	0.00	0.38	5.06	0.00	0.84	0.23	0.00	0.00
0.80	Input	73.43	10.38	4.36	0.00	4.76	3.43	0.00	1.27	0.00	0.00	0.00
0.80	Concentrate	55.13	9.25	4.77	0.00	3.86	5.58	9.26	1.07	9.36	0.00	0.00
1.00	Tailings	79.53	8.27	2.08	0.00	3.56	3.99	0.00	0.84	0.08	0.00	0.00
1.00	Input	80.03	8.49	3.21	0.00	2.63	3.77	0.00	0.77	0.00	0.00	0.00
1.00	Concentrate	74.33	9.47	5.39	0.23	4.34	3.87	0.00	0.72	0.28	0.00	0.00
J_{LT} 0.68 cm/s												
0.10	Tailings	59.03	9.24	2.47	0.00	1.52	2.03	0.42	0.73	0.17	13.04	6.03
0.10	Input	67.52	11.63	5.03	0.00	3.95	3.88	1.69	2.11	0.86	0.00	0.00
0.10	Concentrate	71.53	11.66	4.50	0.05	3.29	3.93	0.88	1.62	0.18	0.00	0.00
0.30	Tailings	52.12	8.59	2.98	0.00	1.64	2.11	0.96	0.30	0.51	18.77	7.09
0.30	Input	80.94	9.28	0.00	0.00	2.34	3.78	0.00	0.92	0.46	0.00	0.00
0.30	Concentrate	81.66	8.35	0.00	0.00	1.58	3.62	0.00	0.69	0.10	0.00	0.00

(continued)

Table 2 (continued)

J_g	Stream	% SiO_2	% Al_2O_3	% Fe_2O_3	% TiO_2	% CaO	% K_2O	% Na_2O	% MgO	% S	% Cu	% Zn
0.50	Tailings	59.49	8.47	2.19	0.00	1.79	2.20	0.05	0.98	0.13	14.60	6.08
0.50	Input	70.59	10.83	6.62	0.07	3.28	4.01	0.33	1.20	0.39	0.00	0.00
0.50	Concentrate	74.36	11.79	0.37	0.00	1.44	6.08	0.00	0.68	1.50	0.00	0.00
0.80	Tailings	54.35	13.20	4.96	0.00	4.67	4.30	2.37	1.96	1.95	6.360	2.80
0.80	Input	60.32	10.91	3.28	0.00	2.76	2.93	1.11	1.52	0.50	9.86	3.69
0.80	Concentrate	62.56	10.16	3.76	0.18	2.48	3.14	0.13	1.11	0.18	8.83	4.12
1.00	Tailings	55.11	9.15	4.25	0.00	4.02	5.85	9.05	0.96	9.13	0.00	0.00
1.00	Input	45.30	10.02	4.38	0.01	5.29	7.18	10.89	2.99	10.12	0.00	0.00
1.00	Concentrate	74.78	9.63	0.40	0.57	3.05	2.67	3.88	1.31	0.49	0.00	0.00
J_{LT} 1.017 cm/s												
0.10	Tailings	76.59	10.53	0.00	0.03	0.76	2.76	0.00	0.32	0.00	7.62	0.00
0.10	Input	75.48	7.33	3.48	0.00	2.34	3.94	1.57	0.73	0.22	0.00	0.00
0.10	Concentrate	71.03	8.26	4.62	0.07	3.43	3.89	1.63	0.27	0.76	0.04	0.07
0.30	Tailings	70.92	10.43	4.26	0.27	3.94	3.51	0.00	0.81	0.00	0.00	0.00
0.30	Input	80.20	7.86	2.96	0.00	2.01	3.08	0.00	0.87	0.00	0.00	0.00
0.30	Concentrate	76.83	8.43	4.79	0.28	3.79	3.85	0.00	0.84	0.48	0.00	0.00
0.50	Tailings	73.23	10.57	4.25	0.00	4.83	3.76	0.00	1.26	0.00	0.00	0.00
0.50	Input	73.23	10.57	4.25	0.00	4.83	3.76	0.00	1.08	0.00	0.00	0.00
0.50	Concentrate	82.07	7.20	3.69	0.10	0.26	3.58	0.00	1.03	0.00	0.00	0.00
0.80	Tailings	72.41	10.06	4.77	0.09	3.91	4.27	0.31	1.12	0.17	0.00	0.00
0.80	Input	80.76	7.71	2.97	0.00	2.00	3.89	0.00	0.37	0.00	0.00	0.00
0.80	Concentrate	82.26	7.01	3.43	0.14	0.22	3.08	0.00	0.53	0.27	0.00	0.00
1.00	Tailings	75.94	8.39	2.63	0.00	3.86	3.42	0.00	0.96	0.11	0.00	0.00
1.00	Input	80.08	8.47	3.05	0.00	2.87	3.66	0.00	0.84	0.07	0.00	0.00
1.00	Concentrate	74.37	11.87	5.36	0.24	0.08	4.09	0.00	0.93	0.30	0.00	0.00

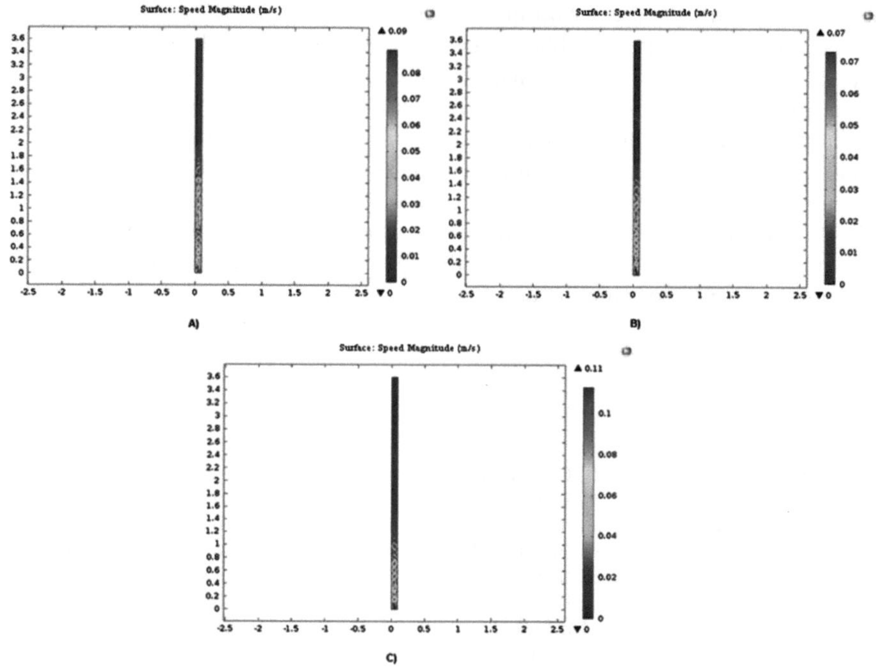

Fig. 7 Hydrodynamics simulation of the flotation column used, for a J_{IT} of **a** 0.446 cm/s, **b** 0.68 cm/s and **c** 1.017 cm/s

finding a quasi-turbulent behavior at the bottom and up to a third of the total length of the flotation column. With a J_{LT} of 0.446 cm/s, were found similar rates and even higher to the liquid inlet rate, due to the generation of flows quasi-turbulent in the lower zone and up to the middle of the flotation column. Meanwhile at J_{LT} of 1.017 cm/s, similar rates are presented and a decreasing in the quasi-turbulent behavior in the lower zone and up to a quart of the total length of the flotation column.

This type of behavior encourages greater recoveries in an environment where there is longer residence time, which is when it has a lower surface speed.

Conclusions

During the flotation column of the material from Dos Carlos dam, it was observed J_L is a variable that determines the rate of ascent of the bubbles package and therefore favors the increasing of retained gas within the columnar system. On the other hand, the effect of J_L in flotation columnar, it was noted that at minors values of J_L (J_{LT} 0.446 to 0.68 cm/s), major rates of ascent of the bubbles package are obtained than that obtained with higher values of J_L (J_{LT} 1.017 cm/s); being possible the highest rate of the bubbles package in major ranges of J_g. It is important to note that at high values of J_L, a major stability of the bubbles flow is present.

According simulation and characterization done by EDS, There is a very particular phenomena, because to a minor surface rate of liquid and especially at J_{LT} of 0.446 cm/s, a turbulence is present with a rate slightly major to the liquid input rate, which generates a zone of major recovery in the lower half of column. That is why this phenomena generates a major recovery instead the others values of J_{LT} tested, finding so a concentration of silica of 88.79%, leaving in the second tails a concentration of 5.83% of alumina and 5.38% of K_2O.

References

1. Finch JA, Dobby GS (1990) Column flotation. Pergamon Press, Gran Bretaña
2. CMRMyP (2005) Resúmen General de presas de Jales. CMRMyP, Pachuca de Soto
3. Hernández Ávila J, Salinas Rodríguez E, Patiño Cardona F, Rivera Landero I, Flores Badillo J, Trápala Pineda NY et al (2012) Tile production using wastes from mining industry of the mining district Pachuca Real del Monte. The Mineral, Metals & Materials Society, USA
4. Hernández Ávila J (2009) Cinética de molienda y recuperación de Ag mediante procesos convencionales y no convencionales de las escombreras de la industria minero-metalúrgica del Estado de Hidalgo. Universidad Autónoma del Estado de Hidalgo, Pachuca de Soto, Hidalgo, México
5. Patiño F, Hernández J, Rivera I, Salinas E, Longoría LC, Gonzalez JM (2007) Characterization and kinetics of the grinding of the Dos Carlos burrows in the state of Hidalgo. International Precious Metals Institute, USA
6. CFM (1961) Experimentación Metalúrgica preliminar corrida sobre muestras de colas de la planta de flotación de Santa Julia de la Comisión de Fomento Minero en Pachuca, Hgo., para la concentración de cuarzo. CFM, Pachuca de Soto, Hidalgo, México
7. CMRMyP (2000) Reporte de flotación de sílice de una muestra proveniente de los Jales de Santa Julia Sur. Compañía Minera Real del Monte y Pachuca (CMRMyP), Pachuca de Soto, Hidalgo, México
8. Klyachin VV, Stepanova MV (1982) Production of high-quality quartz concentrate from quartz-feldspar sands. Glass Ceram 39(10):515–516. Kluwer Academic Publishers, USA
9. Klyachin VV (1969) The selective flotation of quartz-feldspar sands by a new method. Glass Ceram 26(3):174–176. USA
10. El-Salmawy MS, Nakahiro Y, Wakamatsu T (1993) The role of alkaline earth cations in flotation separation of quartz from feldspar. Miner Eng 6(12):1231–1243. United Kingdom
11. Rabone P (1975) Principios de Flotación. México
12. Sutulov A (1963) Flotación de Minerales. Universidad de Concepción, Perú

Study on Bending Test on Concrete Structural Use Crumb Rubber as Substitute in Fine Aggregate

Niander A. Cerqueira, Victor B. Souza, Bruno Padilha, Pâmela Berçot, Afonso G. Azevedo and Victor T. Bartolazzi

Abstract In recent years, concern about environmental issues has become a parameter to consider in the projects in engineering. In order to reduce the impact that the tire improper disposal may result for the environment, we sought a solution to this problem. This article proposes the use of rubber scrap tires as fine aggregate in concrete production. The material produced was analyzed by a bending test, resulting in product quality, in particular for application in situations requiring a more ductile material.

Keywords Evironmental problem · Rubber waste · Bending

Introduction

The tires are residues that are not biodegradable and due to their size are difficult to be stored, so they are configured in a major aggravating the environment. Starting from this issue, CONAMA, the National Environment Council, established criteria for the proper disposal of this material, as well as provided for the prevention of environmental degradation caused by waste tires; through Resolution No. 416 [1].

Worrying about this question, this article aims to search for a possibility to reduce the impacts to the environment through the disposal of waste tires through the use of crumb rubber as a substitute for fine aggregate concrete.

N.A. Cerqueira (✉) · V.B. Souza · B. Padilha
P. Berçot · A.G. Azevedo · V.T. Bartolazzi
Sociedade Universitária Redentor, REDENTOR, BR 356 Km 25,
Itaperuna, Rio de Janeiro, Brazil
e-mail: prof.niander@gmail.com

N.A. Cerqueira · A.G. Azevedo
Universidade Estadual do Norte Fluminense Darcy Ribeiro—UENF,
Rio de Janeiro, Brazil

V.B. Souza
Universidade Federal Fluminense—UFF, Niterói, Brazil

S. Ikhmayies et al. (eds.), *Characterization of Minerals, Metals, and Materials 2017*, The Minerals, Metals & Materials Series,
DOI 10.1007/978-3-319-51382-9_87

For the study was an assay to approximately the particle size of the material through its sieving, enables the construction of the granulometric distribution curve in function of the percentages found in each phase. Subsequently, the construction of the specimen was performed.

For the preparation of the dimensions of the test piece 160 × 40 × 40 mm, it was used as fine aggregate, rubber to replace the sand with concentrations of 5, 10 and 15% of rubber; following the dash 3:2:1; as established by the NBR 12655 [2].

The representative sample was subjected to a bending test after 7 and 14 days elapsed its curing time, in order to measure their strength and a comparison between them.

Literature Review

Particle Size

The particle size is performed in order to classify particles in a sample by measuring the sizes of their grains through the sieve. Its main applicability is the formation of the granulometric curve, which allows the identification of passers material fractions at each stage, allowing the construction of homogeneity for the group in question.

Bending Test

"The mechanical behavior of a material reflects the ratio of the response or deformation at a load or force that is being applied." based on the assertion Callister [3], is a necessary force applied to determine the resistance of a given material. For this study, the definition of the mechanical properties is used in the bending test.

The test is to support the specimen under two supports apart from each other a distance L, and the load of folding or bending applied to the center of the specimen (at L/2 distance of each support). The load should be raised slowly to break the specimen [4].

Such consideration can be checked through the Fig. 1.

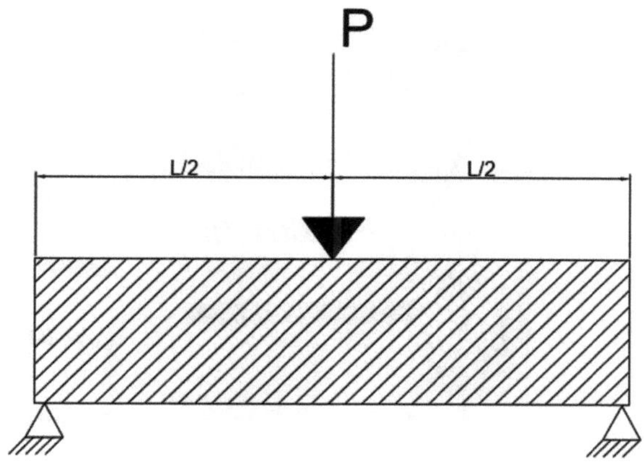

Fig. 1 Representation of the bending test

Materials and Methods

Body of Proof

The construction of the test body was held at 5738 NBR patterns having dimensions of 160 × 40 × 40 mm, a prismatic shape as shown in Figs. 2 and 3.

The specimens were built of following the trace related to concrete, 3:2:1, with concentrations of 5, 10 and 15% crumb rubber as fine aggregate.

Particle Size

The procedure used to determine the granulometric curve of crumb rubber followed determining the following.

(…) The particle size distribution of the aggregate corresponding to the nominal opening in mm mesh sieve normal or intermediate range in which the aggregate has a cumulative percentage retained immediately below or equal to 5% by mass [5].

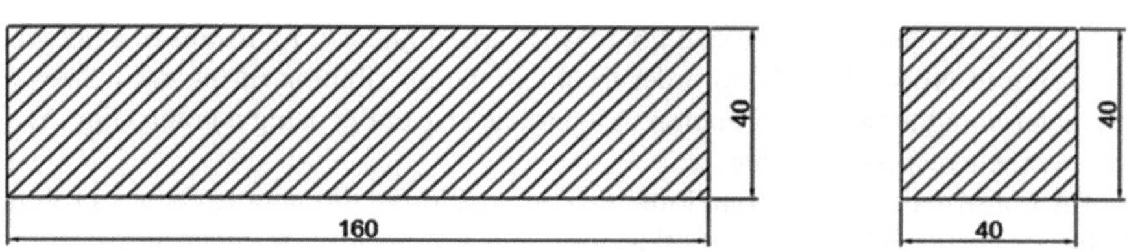

Fig. 2 Representation the specimen dimensions

Fig. 3 Samples (proof of bodies)

Bending Test

The bending test described in this article was conducted at the Laboratory of Mechanical Engineering, where the specimens properly dimensioned according to NBR 5738 [6] were placed in KRATOS machine School Redeemer.

Testing Machine

The bending test machine used in the test specimen for this article is of Kratos equipment shown in Fig. 4, with a maximum of 10,000 kgf cell, and displacement speed of 5.00 mm/min.

Results

Particle Size Distribution Test

Using 1 kg of rubber particle size fibers was carried out test, by means of screening according to NBR 7217 [5], in order to obtain the distribution of the grain sizes of the material; allowing the formation of a particle size curve using a graph (Fig. 5) as well as their respective Table 1.

The construction of the granulometric composition of the sample (Table 1) along with its graphical (Fig. 5) shows that for a 0.6 mm opening resulted in the largest mass fraction 308.1 g, but their pass rate was 17 73%. In contrast the highest passing rate, 99.93%, was recorded in one of the smaller masses, 0.7 g, whose screen had the largest opening 9.5 mm.

Fig. 4 Kratos equipment

Fig. 5 Chart particle size

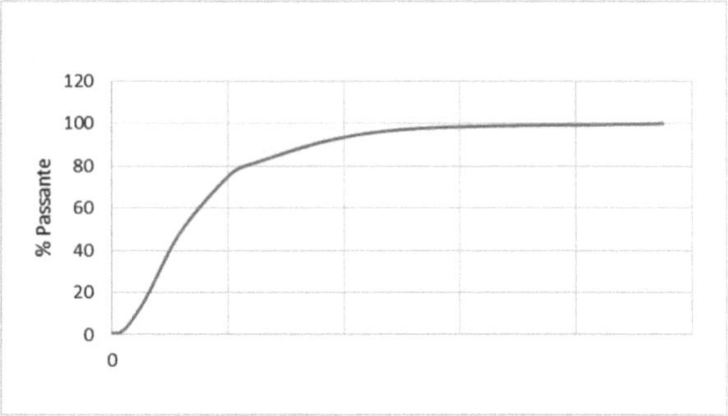

Allowed to be understood that the aggregate remaining material was obtained a passing rate of 0.71% which appeared to one aspect powder function of the homogeneity of size and shape of the grains.

Table 1 Sample composition particle size

Diameter	Weight (W)	% W	Passante (P)	% P
9.5	0.7	0.007	999.3	99.93
4.75	34.1	0.341	965.2	96.52
2.36	159	1.59	806.2	80.62
2	55.7	0.557	750.5	75.05
1.18	265.1	2.651	485.4	48.54
0.6	308.1	3.081	177.3	17.73
0.3	125	1.25	52.3	5.23
0.15	39.8	0.398	12.5	1.25
0.075	5.3	0.053	7.2	0.72
0	0.1	0.001	7.1	0.71

Bending Test

Aiming at the application in construction of concrete with rubber incorporated in its composition, there was a bending test to measure the mechanical properties of the samples.

When the flexural test performed three points, the test body, prismatic shape of dimensions 160 × 40 × 40 mm according to Fig. 2; behaves like a biapoiada beam and a force is applied in a timely manner in the center of the sample according to Fig. 1. According to Beer [7], it is clear that a region of the sample contracts, due to compression, while other region elongates due to traction; generating a stress-strain graph.

It is noteworthy that two (2) specimens with up to 5% of rubber in the fine aggregate 7 days of healing and one (1) other 10% of rubber in the fine aggregate 14 days of healing, they did not hold a minimum effort proof body fixing in Kratos machine and broke before to get some conclusive result. The results are shown in the graphs of Figs. 6, 7, 8, 9 and 10.

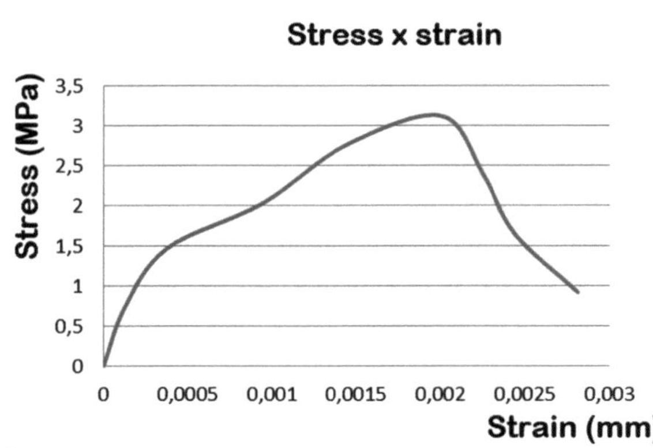

Fig. 6 10% rubber—7 days

Fig. 7 15% rubber—7 days

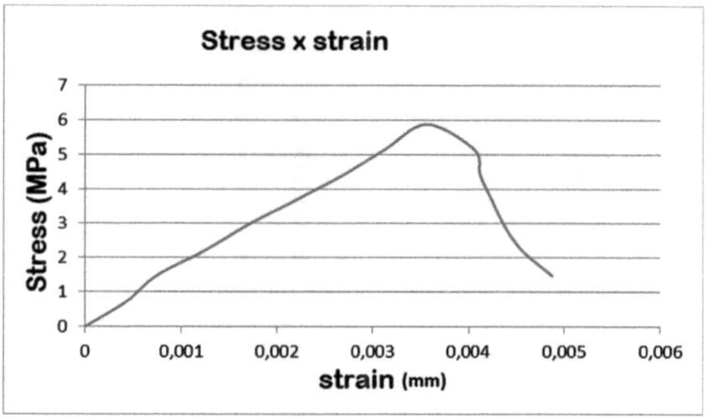

Fig. 8 5% rubber—14 days

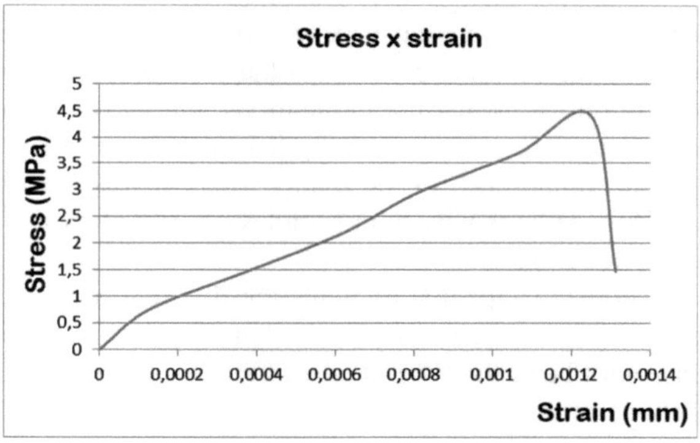

Fig. 9 10% rubber—14 days

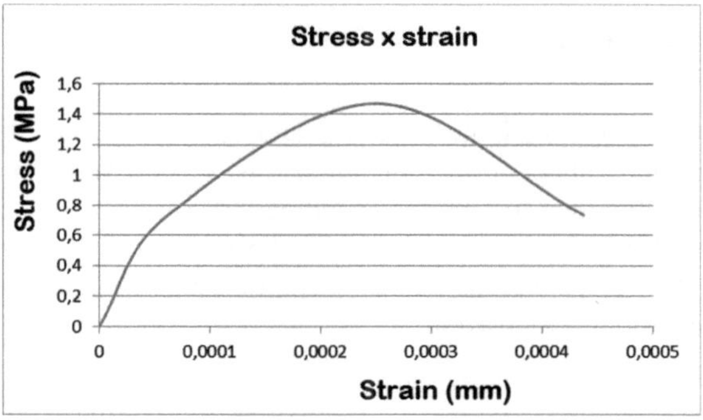

Based on the charts and establishing a relationship between the comparative tests with curing time of 7, and 14 days can realize a significant increase in mechanical properties with time. This fact was explained by Fusco [8] as the increased resistance arising from the concrete maturation so it is with the passage of time and by action of the temperature, since both factors lead to increased cement hydration.

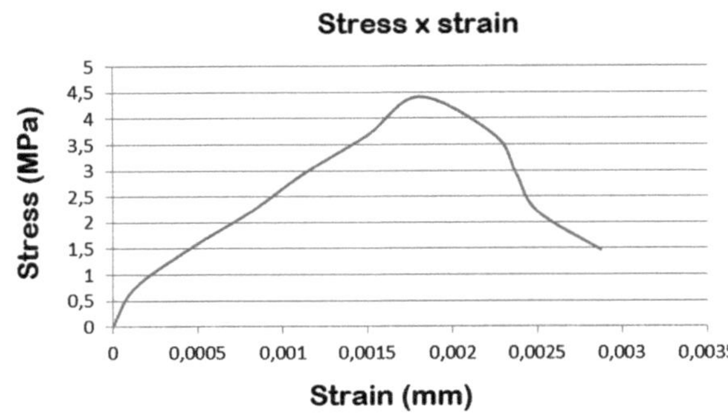

Fig. 10 5% rubber—14 days

Also based on studies Fusco [8], it follows that not all portions of the cast concrete, however, have exactly the same resistance. The effective resistance of each of the concrete portions of the structure will depend on the materials used, the conditions of its mixture in the mixer and transport conditions, launch, densification and cure. The concrete strength is thus a property that may vary in each batch produced and each of the points where it was launched.

This explains why both samples of 10% at 14 days have different results for the same loading.

Another factor that explains the different results found for a specimen with the same age and concentration is that "Due to the variation of the resistance of the internal fibers, the type of cross-section of the specimen influences the graph (...)" [4].

Conclusion

Comparison of the results found showed that at the concentrations studied, the crumb rubber, does not qualify as a good substitute for concrete fine aggregate in situations that require bending stresses.

Nevertheless, in relation to the tests worked, one that performed best was the 15% of rubber as replacement of fine aggregate after 7 days of curing, which acquired 5.884 MPa.

References

1. CONAMA Resolution No. 416 (2009) Preventing environmental degradation caused by waste tires and environmentally sound disposal
2. Brazilian Association of Technical Standards (2006) Portland cement concrete—preparation, control and receiving—procedure. NBR 12655. Rio de Janeiro
3. Callister WD Jr Science and engineering materials—an introduction, 5 ed. Rio de Janeiro, LTC

4. Souza SA (1936) Mechanical testing of metallic materials. Theoretical and practical foundations, 5th ed, São Paulo, Edigard Blücher
5. Brazilian Association of Technical Standards (1987) Added: determination of particle size composition, specification. NBR 7217. Rio de Janeiro
6. Brazilian Association of Technical Standards (2003) Procedure for molding and curing bodies of the test piece. NBR 5738. Rio de Janeiro
7. Beer FP, Johnston ER Strength of materials, 5th edn. McGraw-Hill, São Paulo
8. Fusco PB (2008) Concrete technology structurally applied topics, 1st edn. Editora PINI Ltd, In, São Paulo

Synthesis and Structural Characterization of $BaTiO_3$ Doped with Gd^{3+}

J.P. Hernández-Lara, M. Pérez-Labra, F.R. Barrientos-Hernández, J.A. Romero-Serrano, A. Hernández-Ramírez, A. Arenas-Flores and Pandiyan Thangarasu

Abstract $BaTiO_3$ doped with Gd^{3+}, was synthesized from $BaCO_3$, TiO_2, and Gd_2O_3 as powders using the solid state reaction method ($Ba_{1-x}Gd_xTi_{1-x/4}O_3$). The milling of the precursors was performed in an agate mortar with acetone as control means and compositions x = 0.001, 0.005, 0.05, 0.15, 0.25, 0.35 Gd^{3+} (wt%). Firstly, the powder mixture was decarbonated at 900 °C for 8 h, after it was sintered at 1400 °C for 5 h. Tetragonality of synthesized $BaTiO_3$ doped with Gd^{3+} was analyzed by X-ray diffraction (XRD), scanning electron microscopy (SEM) and Raman spectroscopy (RS). Subsequently, the powders sintered were uniaxially compacted cold until to obtain pellets of about 10 mm in diameter; which again were sintered at 1400 °C for 4 h. XRD patterns and Raman spectra results revealed that the crystal phase of the obtained particles was predominately tetragonal $BaTiO_3$.

Keywords $Batio_3$ · Doping · Tetragonality · Ceramics · Ferroelectrics

Introduction

The perovskite structure has capability to host ions of different size, so a large number of different dopants can be accommodated in the $BaTiO_3$ (BT) lattice which makes BT semiconducting. Doping of BT ceramics is very important to modify

J.P. Hernández-Lara (✉) · M. Pérez-Labra ·
F.R. Barrientos-Hernández · A. Arenas-Flores
Academic Area of Earth Sciences and Materials, Autonomous University of Hidalgo State, Pachuca, Mexico
e-mail: juanp_hernandezlara@hotmail.com

J.A. Romero-Serrano · A. Hernández-Ramírez
Metallurgy and Materials Department, ESIQIE-IPN, Mexico city, Mexico

P. Thangarasu
Faculty of Chemistry, National Autonomous University of Mexico (UNAM), Mexico city, Mexico

S. Ikhmayies et al. (eds.), *Characterization of Minerals, Metals, and Materials 2017*, The Minerals, Metals & Materials Series,
DOI 10.1007/978-3-319-51382-9_88

dielectric, piezoelectric, electrostrictive, and electrooptic properties with corresponding electronic applications [1, 2]. For many years, A- and B-site dopants with different elements [3–6] have been used to modify to modify these properties [7].

Dopants play an important role in the balance of chemical defects in perovskite titanates. The incorporation of isovalent impurities does not have any effect on the population of defects; however, anisovalentes impurities (different valence from that of those who replaces) require the formation of defects compensating counterion for maintaining electrical neutrality.

Substitutional cations of lower charge than the ion to be replaced are commonly called receiving impurities (acceptors). Because they represent a deficiency of positive charge and, therefore carry a charge of negative defect. They require the presence of compensation defects of positive charge such as oxygen vacancies or holes. Substitutional cations of greater charge are called donor impurities and can be compensated by electrons or titanium vacancies or sites covered by M in titanates of the type $MTiO_3$ [8, 9].

When the $BaTiO_3$ is mixed with a dopant donor and is sintered, only a small amount of donor diffuses into $BaTiO_3$ grains. Therefore, the sintered ceramic has a very low concentration of donor in the cores of the grain, but there is considerable solubility of dopant in the regions surrounding the cores where it results in the grain growth [10–13].

The purpose of adding dopant elements electron donors is to create semiconductor regions within the grains of $BaTiO_3$, because an element that plays the role of cation with a different valence to the matrix cations and in substitutional solid solution It could generate electric charge carriers as a way to offset this disequilibrium [14].

As a lanthanide element, Gd exhibits unusual metallurgical properties, apart from its significant use in magnetic resonance imaging (MRI). The inclusion of Gd into the pseudo-cubic host lattice of BT is rarely found in the literature. The main goal of this work was the investigation of structural evolution and electrical properties of $BaTiO_3$ ceramics doped with Gd^{3+} varied within 0.001—0.35 Gd^{3+} (wt%) using the solid-state reaction method.

Experimental Methodology

$BaTiO_3$ samples doped with Gd^{3+} were prepared according to the mechanism $Ba_{1-x}Gd_xTi_{1-x/4}O_3$ using the solid state reaction synthesis method. $BaCO_3$ (Sigma-Aldrich CAS 513-77-9, 99.9%), TiO_2 (Sigma-Aldrich, CAS No. 13463-67-7 99.9%) and Gd_2O_3 (Sigma-Aldrich, CAS No. 278513-25G 99.9%) were dried, weighing and ground for 10 min in an agate mortar with acetone as a control médium and compositions of x = 0.001, 0.005, 0.05, 0.15, 0.25, 0.35 Gd^{3+} (wt%). After that, the powder mixes were sintered in a platinum crucible at 1400 °C for 5 h using a muffle furnace (Thermolyne model 46200). The purity of the products was monitored by X-ray diffraction (diffractometer Equinox 2000 Cu Kα).

For each composition pellets were manufactured for each composition. The powder mixtures were regrounded again and later compacted using uniaxial pressing at 250 MPa in an 10-mm stainless steel die to produce green pellets 3 mm thick. The pellets were sintered at 1400 °C for 4 h in air atmosphere with heating and cooling rates of 5 °C/min.

Once a pure phase was obtained pellets were manufactured for each composition. The morphology studies were performed in a JEOL 6300 SEM. Raman studies were carried out for each sample using a spectrophotometer (Perkin Elmer Spectrum Gx) over the range of 0–1200 wavelength.

Results and Discussion

X-ray Diffraction

Due to the difference of ionic radii between gadolinium ($r(Gd^{3+})$ = 1.02 Å, and titanium r (Ti^{4+}) = 0.68 Å) the lattice will expand to some extent. This may be one of the reasons why the XRD reflections shift to lower angles (Fig. 1) at $2\theta = 65.7$ and $x \geq 0.15$ sintering at 1400 °C.

It can be seen from Fig. 1 that the diffractograms show a double peak at $2\theta \approx 45°$ indicating the presence of tetragonal ferroelectric phase (JCPDS 050626) for the diffractograms when the Gd^{3+} content was $0.001 \leq x \leq 0.25$. This phase disappears for the sample with Gd^{3+} content $x = 0.35$. Determination of the cubic—tetragonal transition temperature was complicated by the presence of a wide metastable persistent region. The phase(s) present after heat treatment was(were) found to be dependent not only on the quenching temperature, but also on the initial phase composition of the sample and thermal history. $Gd_2Ti_2O_7$ phase was identified for $0.001 \leq x \leq 0.25$. Parida et al. [15] studied a series of $Gd_2Ti_2O_7/GdCrO_3$ composites prepared by solid state combustion method using $Gd(NO_3)_3$, TiO_2, Cr_2O_3 as metal source and urea as a fuel. In that work, they found the phase of $Gd_2Ti_2O_7$, in which the main peaks match the phase found in this work.

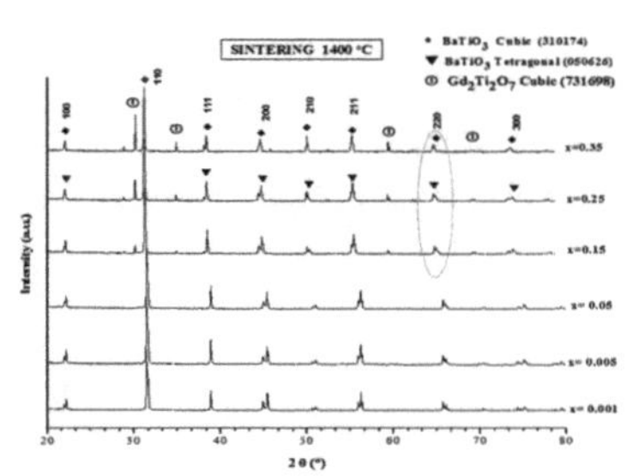

Fig. 1 XRD diffractograms for ($Ba_{1-x}Gd_xTi_{1-x/4}O_3$) powders sintered at 1400° for different values of x

Raman Spectroscopy

Raman spectroscopy RS is a spectroscopic technique used to observe vibrational, rotational, and other low-frequency modes in a system [16] was used to study the phase transition of Gd^{3+} doped $BaTiO_3$ ceramics. Figure 2 shows the Raman spectra for $BaTiO_3$ doped x = 0.001, 0.005, 0.05, 0.15, 0.25, 0.35 Gd^{3+} (wt%) prepared by the solid state reaction. The frequency-covered range is from 100 to 1200 cm^{-1}. The graphs show the characteristic Raman peaks of the tetragonal ferroelectric phase of $BaTiO_3$ [16–18] located at 205, 265, 304, 513 and 717 cm^{-1}. When the Gd^{3+} concentration in $BaTiO_3$ is enough to cross through the tetragonal-cubic phase transition point, all the active Raman modes in the tetragonal phase (P4 mm) are inactive in the perfect cubic phase (Pm3 m) according to forbidding of Raman selected rules. However, the broad A1 (TO) band at 513 cm^{-1} persist into the cubic phase above TC, which is attributed to intrinsic disorder in the cubic phase, and they become broad and weak with increasing Gd^{3+} content [19, 20].

Morphology and Microstructure

Figure 3 shows the X-ray mapping images of O, Ti, Ba, and Gd elements for the sample with x = 0.35 Gd^{3+} (wt%). The X-ray mapping image shows a contrast due to element concentration gradient. For example, in the X-ray image for barium, the bright areas correspond to barium-rich zones and the dark areas correspond to barium-poor zones. The X-ray mapping shows a uniform distribution of Ti, Ba and Gd in a ceramic grain.

The SEM images (Fig. 4) show the Gd^{3+} doped $BaTiO_3$ sintering at 1400 °C for the samples with x = 0.15 and x = 0.35. The images of the samples indicate that Gd^{3+} does not influence drastically the microstructure. All the samples reveal comparable grain sizes, around 1.8–8.4 μm, (Fig. 4a, b). Also can be seen a relatively homogeneous microstructure with higher amounts of inter-granular porosity, this may be due to the behavior of Gd^{3+} in the $BaTiO_3$ ceramics. In Fig. 5 it can be

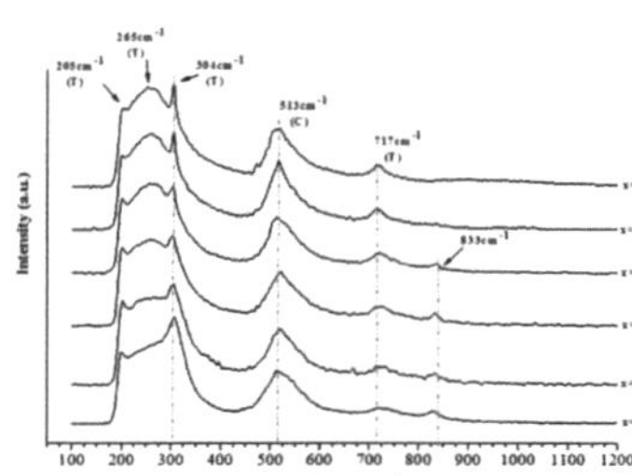

Fig. 2 Raman spectra of sintered powders at 1400 °C, $0.001 \leq x \leq 0.35$

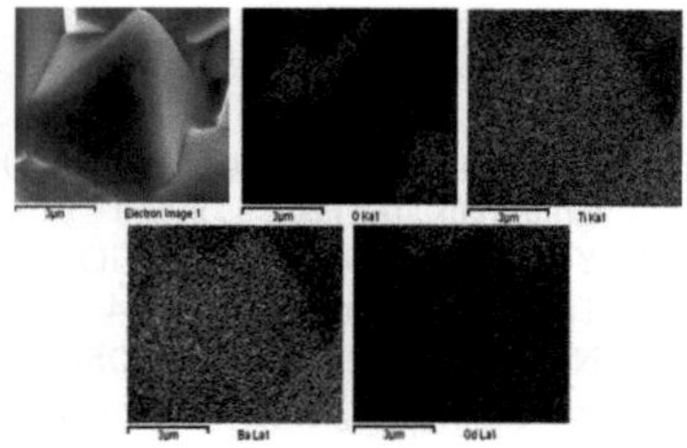

Fig. 3 X-ray mapping images of O, Ti, Ba, and Gd of $BaTiO_3$, doped with different Gd^{3+}. x = 0.35 Gd^{3+} (wt%)

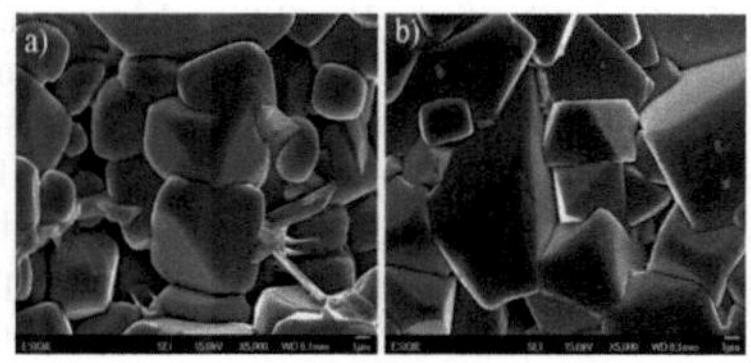

Fig. 4 SEM-EDS micrographs of $BaTiO_3$, doped with different Gd^{3+}. **a** x = 0.15, **b** x = 0.35

Element	Weight %
O K	17.76
Ti K	23.14
Ba L	55.13
Gd L	3.97

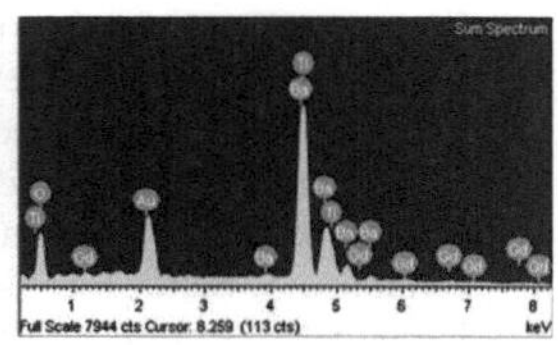

Fig. 5 SEM-EDS analysis of $BaTiO_3$, doped with different Gd^{3+}. x = 0.35

observed the presence of Gd in the EDS spectrum. This confirm again the incorporation of this element into $BaTiO_3$.

Yongping et al. [21] reported that the thermodynamic conditions, such as the partial pressure of oxygen in the sintering atmosphere and temperature, play an important part in the distribution of rare Earth elements with amphoteric behavior in A- site or B- site of $BaTiO_3$ [22, 23].

Conclusions

The $BaTiO_3$ doped with 0.001, 0.005, 0.05, 0.15, 0.25, 0.35 Gd^{3+} (wt%) were investigued by XRD. Raman spectroscopy (RS) and SEM-EDS. It was confirmed by XRD a double peak at $2\theta \approx 45$ indicating the presence of tetragonal ferroelectric phase for the diffractograms when the Gd^{3+} content was $0.001 \leq x \leq 0.25$. This phase disappears for the sample with Gd^{3+} content x = 0.35. The Raman results shown the characteristic Raman peaks of the tetragonal ferroelectric phase of $BaTiO_3$ located at 205, 265, 304, 513 and 717 cm^{-1}. X-ray mapping images confirmed the homogeneous distribution of O, Ti, Ba, and Gd in ceramics.

Acknowledgements The author is grateful to CONACyT-México for financial support.

References

1. Moulson AJ, Herbert JM (1990) Electroceramics: materials, properties and applications. Chapman and Hall, London
2. Yao K, Zhu W (2002) BaTiO3 glass-ceramic thin films for integrated high dielectric media. Thin Solid Films 408:11–14
3. Nemoto H, Oda I (1980) Direct examinations of PTC action of single grain boundaries in semiconducting BaTiO3 ceramics. J Am Ceram Soc 63(7–8):398–401
4. Tangjuank S, Tunkasiri T (2007) Characterization and properties of Sb-doped BaTiO3 powders. Appl Phys Lett 90(7, Article ID 072908)
5. Preis W, Sitte W (2006) Electronic conductivity and chemical diffusion in n-conducting barium titanate ceramics at high temperatures. Solid State Ionics 177(35–36):3093–3098
6. Jana A, Kundu TK (2007) Microstructure and dielectric characteristics of Ni ion doped $BaTiO_3$ nanoparticles. Mater Lett 61(7):1544–1548
7. Vijatovic′ MM, Stojanovic′ BD, Bobic′ JD, Ramoska T, Bowen P () Properties of lanthanum doped $BaTiO_3$ produced from nanopowders. Ceram Int 36:1817–1824
8. Ueoka H (1974) The doping effects of transition elements on the PTC anomaly of semiconductive ferroelectric ceramics. Ferroelectrics 7:351–356
9. Ikushima H, Hayakawa S (1967) Natl Tech Rept 13:209–216
10. Kingery WD (1960) Introduction to ceramics. Wiley, New York
11. Smyth DM (2000) The defect chemistry of metal oxides. Oxford University Press, New York
12. Levinson LM (1988) Electronic ceramics. Marcel Dekker Inc, NY
13. Yanagida H (1996) The chemistry of ceramics. Wiley, Tokyo
14. Shiransaki SI (1985) Structure-property-relationships in perovskite electronic ceramics, fine ceramics. S. Saito. Elsevier, New York, pp 150–161
15. Parida KM, Nashim A, Mahanta SK (2011) Visible-light driven $Gd_2Ti_2O_7/GdCrO_3$ composite for hydrogen evolution. Dalton Trans 40:12839
16. Gardiner DJ (1989) In practical raman spectroscopy. Springer, Berlin. ISBN 978-0-387-50254-0
17. Matsuoka T, Fujimura M, Matsuo Y, Hayakawa S (1972) PTCR Behavior of $BaTiO_3$ with Nb_2O_5 and MnO_2 Additives. J Am Ceram Soc 55:108–116
18. Ikushima H, Hayakawa S (1967) Natl. Tech. Rept 13:209–216
19. Tsuzuku K, Couzi M (2012) In situ investigation of chemical reactions between $BaCO_3$ and anatase or rutile TiO_2. J Mater Sci 47:4481–4487
20. Wada S, Suzuki T, Osada M, Kakihana M, Noma T (1998) Change of macroscopic and microscopic symmetry of barium titanate single crystal around curie temperature. J J Appl Phys 37:5385–5393
21. Yongping Pu, Wenhu Yang, Soutian Chen (2007) Influence of rare earths on electric properties and microstructure of barium titanate ceramics. J Rare Earths 25:154–157
22. Chen Jing Zhi Ang, Zhi Yu, Vilarinho Paula M, Baptista JoãoL (1999) Incorporation of yttrium in barium titanate ceramics. J Am Ceram Soc 83(5):1345–1348
23. Tsur Y, Dunbar D, Randall CA (2001) Crystal and defect chemistry of rare earth cations in $BaTiO_3$. J Electroceram 7:25–34

Texture Analysis and Anisotropic Properties of a Rolled CuZn36 Brass Alloy

A. Vazdirvanidis, G. Pantazopoulos, A. Toulfatzis and A. Rikos

Abstract In this study, the mechanical properties of an annealed CuZn36 rolled brass alloy with 0.2 mm thickness with respect to anisotropy expressed in terms of plastic strain ratio measurements (r-values, normal and planar anisotropy) are related with the development of specific rolling and recrystallization texture components obtained by EBSD. Anisotropic behaviour is primarily controlled by the crystallographic texture and constitutes an indicator of multi-axial mechanical behaviour, affecting directly the drawability and sheet metal formability. Microstructure examination employing optical and Scanning Electron Microscopy of tensile specimens are performed in order to investigate the emergent damage mechanisms in relation to the texture and structure evolution of the final product.

Keywords Brass alloy · Sheet metal formability · Anisotropic properties · Plastic strain ratio · Texture analysis

Introduction

The major applications of brass flat products (CuZn36 alloy) concern sheet metal forming and therefore formability of rolled metal sheets/strips is an important quality parameter for the manufacturing of specially formed parts in electrical and electronic industry as well as in the chemical and manufacturing sector, such as

A. Vazdirvanidis (✉) · G. Pantazopoulos · A. Toulfatzis · A. Rikos
ELKEME—Hellenic Research Centre for Metals S.A, 56th Km
Athens—Lamia National Road, 32011 Oinofyta Viotias, Greece
e-mail: avazdirvanidis@elkeme.vionet.gr

G. Pantazopoulos
e-mail: gpantaz@elkeme.vionet.gr

A. Toulfatzis
e-mail: atoulfatzis@elkeme.vionet.gr

A. Rikos
e-mail: arikos@elkeme.vionet.gr

S. Ikhmayies et al. (eds.), *Characterization of Minerals, Metals, and Materials 2017*, The Minerals, Metals & Materials Series,
DOI 10.1007/978-3-319-51382-9_89

switches, containers, tubes, and other accessories and assemblies. Furthermore, joining is also another fabrication process for the manufacturing of components or final systems that requires critical knowledge concerning the microstructure and the relevant control of brass alloy sheet mechanical properties in order to select and optimize the applied welding/joining parameters.

Texture and microstructure development in face-centered-cubic (fcc) alloys is strongly dependent on stacking fault energy (SFE). There are several research works studying the influence of SFE on rolling texture, microstructure, defect-structure and their impact on final grain size on Cu and fcc alloys [1–5]. High deformation substructure in Cu-30Zn brass consists of thin deformation twins and shear bands while in the high SFE Cu the substructure mainly includes micro-bands and shear bands [1]. A higher SFE results in a smaller separation between the partial dislocations. In Cu-Zn alloys twin and dislocation densities are increased by increasing Zn content and therefore, by decreasing SFE which is consistent to slow rates of recovery [2]. Severe plastic deformation (SPD) and resultant grain size is analyzed and discussed in terms of SFE. Dislocation slip occurs under high SFE and twinning occurs easily under low SFE. SPD in low SFE leads to increased dislocation and twin densities and fine grain size [4]. During deformation of fcc metals cross slip is assumed to occur by pinching of partial dislocations in their slip plane and their subsequent extension on the cross slip plane. Therefore, the requirement of pinching of dislocations controls the dependence of dislocation substructures. Hence a smaller impartial spacing (high SFE) has a greater tendency for cross slip, forming a wavy dislocation substructure. Under large strains at room temperature high SFE forms a cell like dislocation substructure with a dislocation-free core. Lower SFE inhibits cross slip and favours the re-arrangement of dislocation in planar arrays or slip bands [6].

In the present work, a 0.2 mm CuZn36 brass alloy sheet was studied as far as the anisotropic properties and microstructure was concerned. Plastic strain ratios measurements during tensile testing together with microstructure, texture and tensile fracture analysis using metallographic (Light Optical and Scanning Electron Microscopy) and Electron Backscatter Diffraction (EBSD) techniques, were employed as the major analytical techniques to investigate the formability behaviour and its dependence on texture and microstructure of CuZn36 brass alloy sheet.

Experimental Procedure

Microstructure evaluation was conducted in mounted sections parallel to the rolling direction. Metallographic examination was performed employing a Nikon Epiphot 300 inverted optical microscope while SEM evaluation was conducted on Au-sputtered longitudinal sections using a FEI XL40 SFEG scanning electron microscope using Secondary (SE) and Backscattered Electron (BSE) detectors. The microstructure was assessed after immersion etching using $FeCl_3$ solution.

Tensile tests applied to, parallel, transverse and 45° with respect to the rolling direction (RD) specimens were performed according to EN ISO 6892-1 standard, using an Instron 5567 electromechanical testing machine equipped with a load cell of 30 kN maximum load. For the measurement of *R*-values (plastic strain ratio), a transverse type extensometer was used together with an axial one to extract the three components of strain (ε_w, ε_t, ε_l) for 0°, 45° and 90° with respect to rolling direction according to ASTM E517 standard. Moreover, strength coefficient (K) and strain hardening exponent (n) of Hollomon equation were also experimentally determined according to ASTM E646 standard. Cupping tests, following earing (with drawing ratio ~2) standard experimental procedure were also conducted in order to rank the formability performance of the selected Brass samples. Electron Backscatter Diffraction (EBSD) analysis at 70° tilt position, 20 kV accelerating voltage, 10 mm working distance and 0.5 μm step size, was conducted using a high speed camera attached to the SEM vacuum chamber. The data analysis was realized using Orientation Imaging Microscopy analysis software.

Results and Discussion

Mechanical Properties

The *R*-values, given by the following equations are determined as additional important parameters of anisotropic properties, see also Refs. [7–10]. The following equations are used for the definition and calculation of *R*-values:

Plastic strain ratio:

$$R = \frac{\varepsilon_w}{\varepsilon_t} \tag{1}$$

Normal anisotropy (average):

$$Rave = \frac{R_0 + R_{90} + 2 \cdot R_{45}}{4} \tag{2}$$

Earing tendency (planar anisotropy):

$$\Delta R = \frac{R_0 + R_{90} - 2 \cdot R_{45}}{2} \tag{3}$$

Tensile properties and plastic strain ratio are shown in Table 1 and Fig. 1. The tensile properties ($R_{p0.2}$, R_m and A_{50}) along with K, n values are quite reproducible for each direction, 0°, 45° and 90°, showing enhanced material isotropic properties. The high values of R (~1.66–2.14, mean value, R_{ave} ~ 1.83) are consistent to low tendency for localized necking and, therefore, high formability performance. On the other hand, low ΔR values are in agreement to low earing percentage; weak ears

Table 1 List of tensile and anisotropic properties. Elastic modulus (Young's modulus) was preset at 105 GPa

Elastic modulus (GPa)	Direction	Thickness (mm)	Width (mm)	$Rp_{0.2}$ (MPa)	Rm (MPa)	A_{50} (%)	r	n	K (MPa)	ΔR	R_{ave}
105	0°	0.199	12.56	136	314	28	2.143	0.358	627	0.130	1.836
	45°	0.199	12.56	136	320	32	1.771	0.342	609		
	90°	0.199	12.56	137	309	25	1.658	0.354	634		

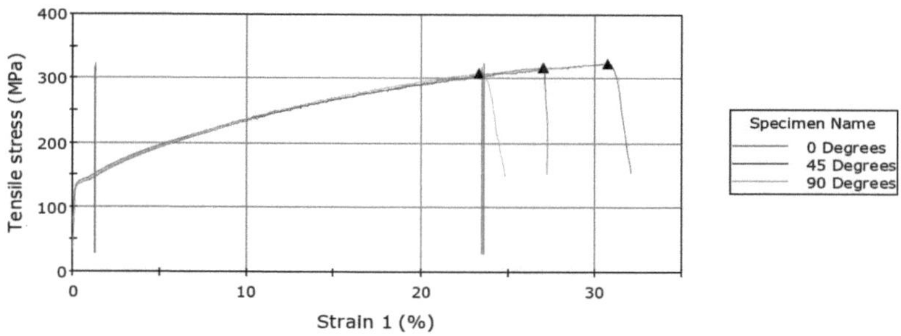

Fig. 1 Stress-strain graphs for characteristic test specimens (0°, 45° and 90°). The realization of tensile test leads to the determination of *R*-values and K, n parameters of Hollomon equation under the following conditions: Plastic zone range: 1–23%, Poisson's ratio = 0.33, Elastic strains were removed

will appear at 0° and 90° directions (positive ΔR). High strain hardening exponent values (0.35) and high $R_m/R_{p0.2}$ ratio (2.25–2.35) indicate high work hardening potential and uniform elongation caused most likely as a result of final recrystallization annealing treatment. Low *R*-values and highly anisotropic properties are obtained in case of hexagonal-close-packed systems with high c/a ratio, such as Zn and ZnTiCu rolled alloys, see also Ref. [11].

Microstructure

The microstructure is uniform across the thickness of the sheet and consists of single a phase equiaxed grains as a result of full recrystallization (recrystallization twins are also discerned) with a mean size of 12 μm. Respective micrograph is given in Fig. 2.

Texture

Texture analysis was accomplished with EBSD technique by scanning of a top view sample (RD-TD) and a metallographic sample in the longitudinal direction (cross-section RD-ND). In both cases a random texture was revealed (see IPF and PF maps in Figs. 3 and 4) which is in accordance with the obtained results of mechanical testing (tensile test results—R and ΔR values). Recrystallization components of texture were also found at very low values (<2%). Removal of recrystallization twins (Figs. 3a, b and 4a, b) with the aid of the software enabled grain size measurement with a high accuracy.

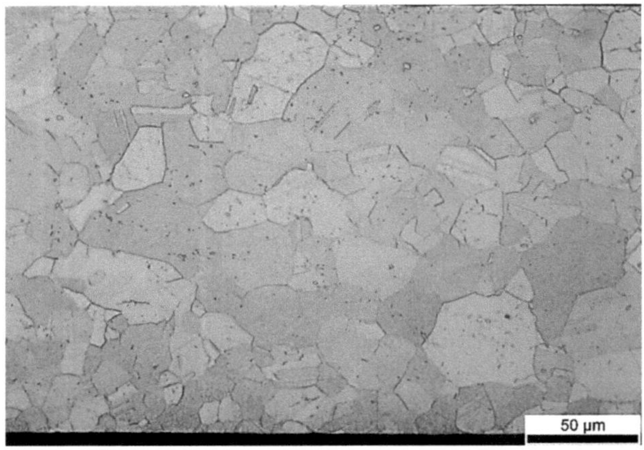

Fig. 2 Microstructure of CuZn36 rolled sheet, longitudinal section

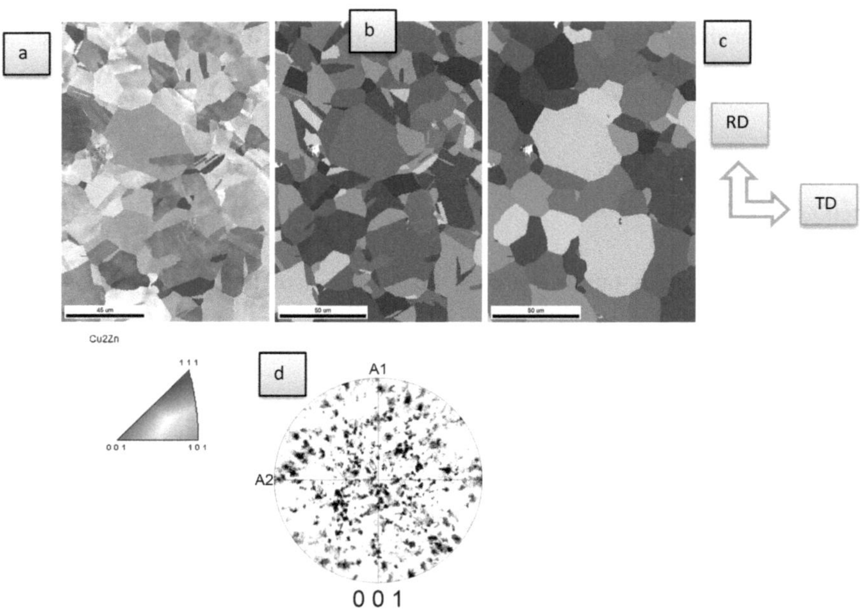

Fig. 3 **a** Inverse pole figure (IPF) map of the brass sheet, top view scan, **b** representation of grain structure and **c** same as (**b**) after removal of recrystallization twins (rotation by 60° around <111> directions for FCC metals). **d** Pole figure (PF) map of the respective scan

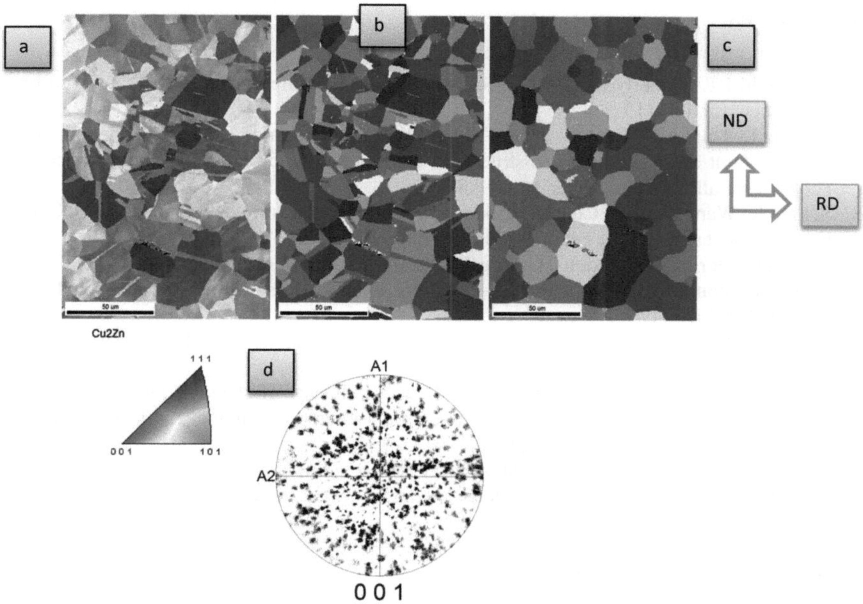

Fig. 4 **a** Inverse pole figure map of the brass sheet, scan of longitudinal section, **b** representation of grain structure and **c** same as (**b**) after removal of recrystallization twins. **d** Pole figure map of the respective scan

Conclusions

The tensile and anisotropic properties together with the texture characteristics and microstructure of a 0.2 mm brass alloy (CuZn36) sheet were investigated, leading to the following conclusions:

1. Uniform tensile properties yielding repeatable values of $R_{p0.2}$, R_m and A_{50} along with K, n parameters at 0°, 45° and 90° were obtained.
2. High *R*-values ($R_{ave} \sim 1.83$) signify improved isotropic and formability properties, accompanied by high thinning resistance and slight earing tendency (0° and 90°).
3. A uniform, equiaxed, fully recrystallized structure with annealing twins was observed across the sheet's thickness. Mean grain size was calculated 12 μm after removal of twins with OIM software.
4. Random texture was exhibited by the sheet by EBSD scans which is in agreement with the results of the mechanical tests signifying high isotropic properties.

References

1. Wakefield PT, Hatherly M (1981) Microstructure and texture of cold-rolled Cu-10Zn brass. Metal Sci. 15:109–115
2. Balogh L, Ungár T, Zhao Y, Zhu YT, Horita Z, Xu C, Langdon TG (2008) Influence of stacking-fault energy on microstructural characteristics of ultrafine-grain copper and copper-zinc alloys. Acta Mater 56:809–820
3. Sekine K, Wang J (1999) Characteristic features of rolling texture development in FCC alloys having low stacking fault energies. Mater Trans, JIM 40:1–6
4. Zhao YH, Horita Z, Langdon TG, Zhu YT (2008) Evolution of defect structures during cold rolling of ultrafine-grained Cu and Cu–Zn alloys: influence of stacking fault energy. Mater Sci Eng, A 474:342–347
5. Engler O (2001) Recrystallisation textures in copper-manganese alloys. Acta Mater 49: 1239–1247
6. Rohatgi A, Vecchio KS, Gray G III (2001) The influence of stacking fault energy on the mechanical behavior of Cu and Cu-Al alloys: deformation twinning, work hardening, and dynamic recovery. Metall Mat Trans A 32:135–145
7. Dieter GE (1988) Mechanical metallurgy, SI Metric Edition. McGraw-Hill, London, UK
8. Hosford WF, Caddell RM (1983) Metal forming: mechanics and metallurgy. Prentice Hall, Englewood Cliffs
9. Marciniak Z, Duncan JL, Hu SJ (2002) Mechanics of sheet metal forming, 2nd edn. Butterworth-Heinemann, Oxford
10. Schey JA (1987) Introduction to manufacturing processes, 2nd edn. McGraw Hill, New York
11. Pantazopoulos G, Toulfatzis A, Vazdirvanidis A, Rikos A (2016) Fundamental aspects of rolled zn alloy sheet formability: structure-property and failure mode relationships. Invited paper at THERMEC 2016, international conference on processing and manufacturing of advanced materials, May 29th–June 3rd, Graz, Austria

The Influence of Titanium Content on the Sinter Ore Phase Structure and the Crack

Dongdong Zhou, Shusen Cheng and Yongqiang Bai

Abstract The strength of sinter could affect the permeability and CO gas utilization of blast furnace, and one of the main impacted factors of sinter strength is crack. With the widely use of vanadium-titanium sinter in China, the blast conditions of some blast furnaces became worsen since the vanadium-titanium sinter has been added. However, there have little public researches for this phenomenon. In this paper, the influence of titanium content on the sinter ore phase structure and crack were studied by DMRX polarizing microscope. In addition, the influence of basicity and oversintering on the sinter ore phase structure were also investigated. The observed results of vanadium-titanium sinter were compared with typical sinter in China, and the influence law of titanium content on the sinter ore crack was obtained.

Keywords Titanium content · Vanadium-titanium sinter · Phase structure · Crack

Introduction

The iron ores with titanium are abundant in China. However, with the prices of imported iron ore increased in recent years, and in order to reduce the economic costs of ironmaking process and improve the level of resources utilization, the vanadium-titanium sinter was used widely in part of the blast furnace in China. For different blast furnace, the titanium content in sinter is distinct, even for the same blast furnace, the change of the titanium content could affect the strength of the sinter. Subsequently, when the vanadium-titanium sinter charged into the blast

D. Zhou (✉) · S. Cheng
School of Metallurgical and Ecological Engineering, University of Science and Technology Beijing, Beijing 100083, People's Republic of China
e-mail: dongdongzhou@live.com

Y. Bai
China Metallurgical Industry Planning and Research Institute, Beijing 100711, People's Republic of China

S. Ikhmayies et al. (eds.), *Characterization of Minerals, Metals, and Materials 2017*, The Minerals, Metals & Materials Series,
DOI 10.1007/978-3-319-51382-9_90

furnace, the permeability and gas utilization rate were reduced simultaneously, worsen the blast conditions and increase the economic costs of ironmaking. On this account, a lot of researches on the sinter ore phase structure and the crack has been made in recent decades.

Xia and Tong [1] studied the broken and crack issues in the sinter of Tanggang and found that the main impacted factor for its strength were the microscopic cracks in the sinter. Qiao and Du [2] analyzed the formation mechanism of the crack in the sinter, it shows that the micro-pore is the main influenced factor of cracks and strength of sinter. Ying et al. [3] used the Vickers indentation test method to analyze the impact of calcium ferrite content on the formation of sinter cracks. Yang and Standish [4] researched the formation mechanism of the micro-pore in the sinter and pellet, the conclusion is that the micro-pore appeared in the round of the limestone and coke in the sinter. Bai et al. [5] used the phase analyzed method to compare the phase structures of vanadium-titanium sinter and ordinary one, and found that the number of pore in the vanadium-titanium sinter is larger than the ordinary one. Sun et al. [6] studied the influence of TiO_2 content on the quality of vanadium-titanium sinter and found that the decrease of perovskite mass percent mainly impacted the quality of sinter. Li et al. [7–11] also investigated the chemical influence on the quality of the vanadium-titanium sinter.

From the above description, it is clear that there has no published literature studied the impact of titanium content on the sinter phase structure and crack. In this paper, the influence of titanium content on the sinter ore phase structure and crack were studied by DMRX polarizing microscope. In addition, the influence of basicity and oversintering on the sinter ore phase structure were also investigated. The observed results of vanadium-titanium sinter has been compared with typical sinter of China, and the influence law of titanium content on the sinter ore crack were obtained.

The Chemical Composition of Sinter

The chemical composition of sintered sample is shown in Table 1. The A1, A2 are the vanadium-titanium sinter with different titanium content, the basicity of A3 is smaller than A1 and A2, while the A4 is the overbake vanadium-titanium sinter, the A5 is the typical sinter of China.

Table 1 The chemical composition of sintered sample

Samples	TFe	FeO	CaO	SiO_2	TiO_2	MgO	Al_2O_3	Basicity
A1	58.79	8.12	11.56	5.94	2.34	2.08	1.69	1.95
A2	52.80	14.44	12.32	5.60	5.88	2.05	2.32	2.20
A3	55.18	9.81	11.23	5.91	3.437	1.87	2.55	1.90
A4	52.80	14.44	12.32	5.60	5.88	2.05	2.32	2.20
B1	55.08	9.47	10.92	5.46	0.52	2.60	2.21	2.00

The Analyses of Sinter Phase Structures

The Influence of Titanium Content

1. A1 sinter ore

   Figure 1 is the phase structures of A1 sinter ore, (a) and (b) are mixed corrosion and melting corrosion structure, separately. The phase structure of A1 sinter ore is uniformly and the main phase structure is mixed corrosion structure, part is melting corrosion structure. The main phase structure is hematite with the skeletal texture, which is non-uniformly distributed in the edge of the sinter ore, as shown in Fig 1d. The particle size of magnetite is also distributed non-uniformly, and it formed the mixed corrosion structure by interweaved with the dicalcium silicate and calcium ferrite (SFCA). The perovskite content is lower and also distributed non-uniformly, which is centralized spread in the hyaline by dendritic or nterlaced structures. Moreover, the binding phase is SFCA and dicalcium silicate. The content of SFCA is lower and presented as

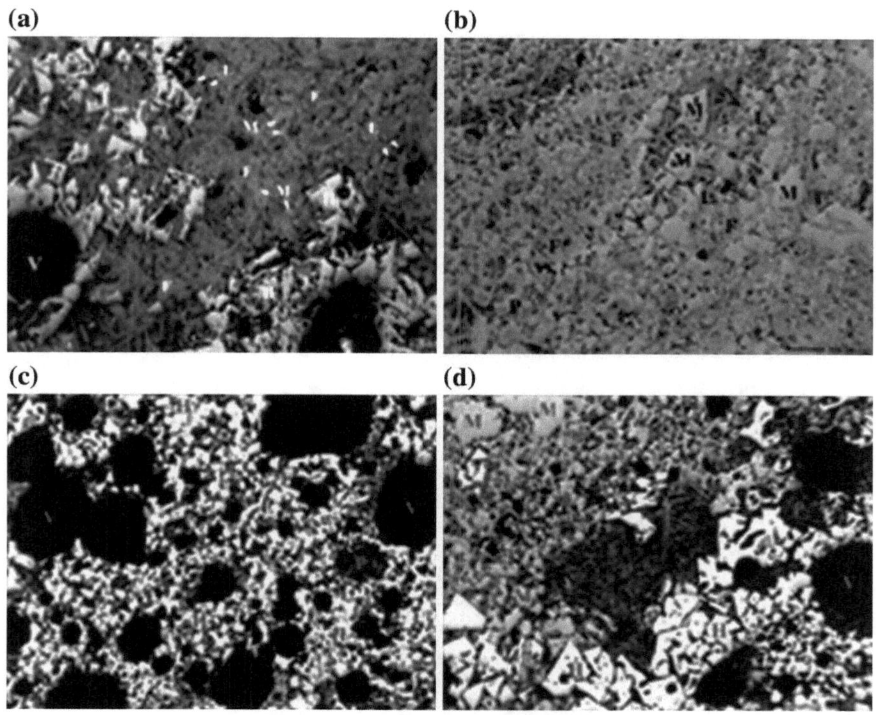

Fig. 1 The phase structure of A1 sinter ore *H* hematite, *M* magnetite, *F* calcium ferrite (SFCA), *L* dicalcium silicate, *P* perovskite, *X* hyaline, *V* void, C-crack

needle-like and clintheriform. Consequently, the content of binding phases is few and the metal phase is large.

Fig 1c shows the pore size distribution of A1 sinter ore, the main pore is the spilehole and the numbers of 51 μm ~ 100 μm pores are larger than that of 10 ~ 50 μm. While the spilehole is concentrate on the positions in which the SFCA and hematite content are high. There have large number pores with different size spread around in the secondary hematite, some pores are distributed around the perovskite. While the particle size of hematite is large and there exists large number micropores and cracks.

2. A2 sinter ore

Figure 2 is the phase structures of A2 sinter ore, (a), (b) and (c) are mixed corrosion, skeletal structure and dendritic perovskite, separately. From those two figures, it seems that the main phase structure in the A2 sinter ore is mixed corrosion, while the skeletal structure appeared in local parts. The crack is internally developed in the sinter. Most of the magnetite is surrounded by SFCA and dicalcium silicate, some local magnetite distributed in the certain position. Therefore, the content of hematite is small and spread non-uniformly, some of them locally concentrated in the edge of the pores and sinter ore, other hematite

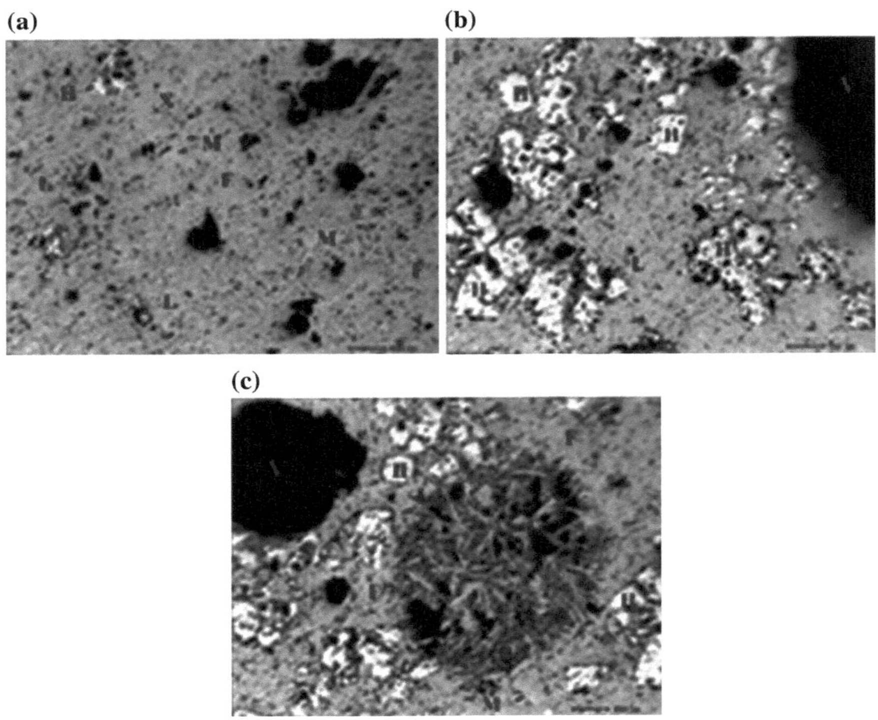

Fig. 2 The phase structure of A2 sinter ore

with large particle size existed as skeletal crystal, a small number of hematite appeared in magnetite. As shown in Fig 1c, the content of perovskite is small and spread non-uniformly, while it also concentrated in the hyaline by dendritic and nterlaced structures. Subsequently, the main constitutions of binding phase are SFCA, dicalcium silicate and hyaline. The SFCA content is large and spread uniformly, while the main structure is needle-shaped and some appeared as foliated columnar. Moreover, the content of dicalcium silicate is also large and appeared in dendritic perovskite as granular and normal particles.

The number of micropores in A2 sinter ore is large, while the large pore is little. Parts of micropores flocked together, some parts of them appeared in the edge of the big pores. The pores are appeared in the corrosion position of magnetite and SFCA, and the dicalcium silicate is surrounded those pores. Eventually, most abreast pores formed to large crack, part of the cracks run through pores, while most of short cracks are formed in the edge of hematite crystalline with skeletal structure.

The Influence of Basicity on the Phase Structure

Compared to A1 and A2, the basicity of A3 sinter ore is small. As shown in Fig. 3, the (a) and (b) are mixed corrosion and dendritic SFCA structure of A3 sinter ore, separately. The phase of A3 sinter ore is spread uniformly and the main structure is mixed corrosion, part of them appeared as porphyritic texture. A small number of SFCA, dicalcium silicate, hyaline and magnetite formed as mixed corrosion structure. The main binding phases are SFCA, dicalcium silicate and hyaline. Moreover, the content of SFCA is large, while the content of dicalcium silicate is small. The large number of pores is distributed in the range of 10–50 μm, some small pores spread centrally, especially appeared in the porphyritic texture of

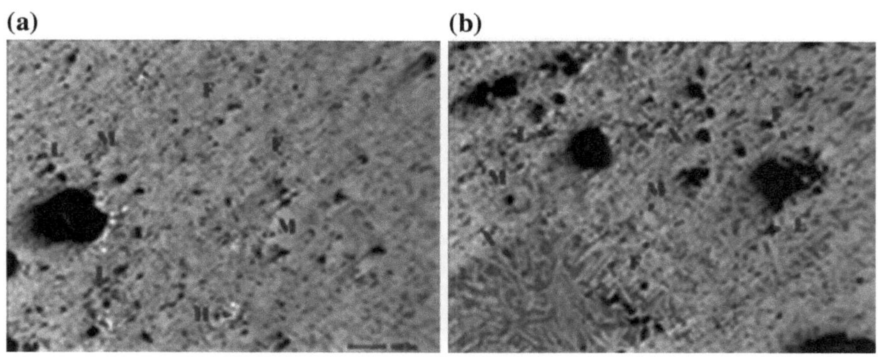

Fig. 3 The phase structure of A3 sinter ore

hematite. Therefore, the large numbers of small cracks are appeared between the hematite particles, some of cracks are run through the whole areas of the hematite.

The Influence of Overbake on the Phase Structure

Compared to A1 and A2, the sintering time of A4 sinter ore is longer. As shown in Fig. 4a–c are mixed corrosion, skeletal structure and perovskite structure of A4 sinter ore, separately. The phase of A4 sinter ore is spread non-uniformly and the main structure is mixed corrosion, part of them appeared as skeletal structure. The content of magnetite and hematite is small and most of them appeared in the pores and edge of the sinter ore. Moreover, the content of perovskite is large and distributed non-uniformly. The main binding phases are SFCA, dicalcium silicate and hyaline, the content of SFCA is small, while the content of dicalcium silicate is large. The main pore in the A4 sinter ore is micropores, most of them spread centrally in the hematite particles, the shape is irregular and size is not spread uniformly. Therefore, the cracks are little and surrounded in the hematite particles.

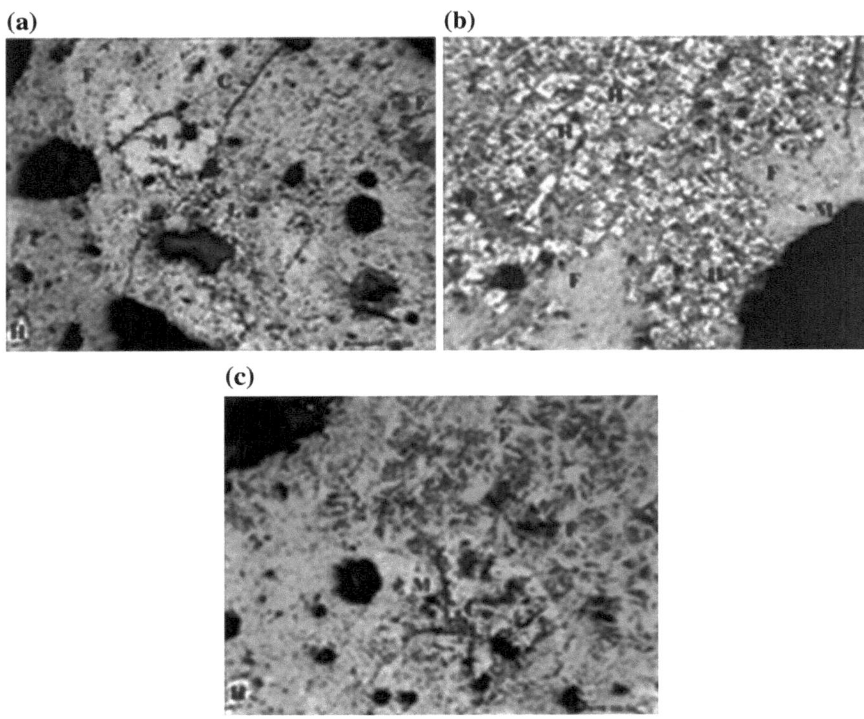

Fig. 4 The phase structure of A4 sinter ore

Compare with the Typical Sinter Ore

Compared to A1–A4, B1 sinter ore is typical sinter ore in China which has none or little titanium content. As shown in Fig. 5a–c are mixed corrosion, skeletal structure and eutectic of magnetite and SFCA of B1 sinter ore, separately. The phase of B1 sinter ore is spread uniformly and the main structure is mixed corrosion, part of them appeared as skeletal structure. The content of perovskite is small and it spread non-uniformly, the main binding phases are SFCA, dicalcium silicate and hyaline. Moreover, the SFCA content is large and spread uniformly, which appeared in needle-shaped. While the content of dicalcium silicate is small and it appeared in granular texture.

Therefore, the numbers of micropores are large in the B1 sinter ore, the long cracks exist in the edge of the sinter ore. Some micropores formed to the crude cracks, surrounding them are the pores with freak shape. In the side of the perovskite are big pores, while there are large numbers of pores and cracks with different shape appeared in the region of regenerated hematite area. In the region of SFCA, hematite and magnetite, there appear many pores and cracks with different

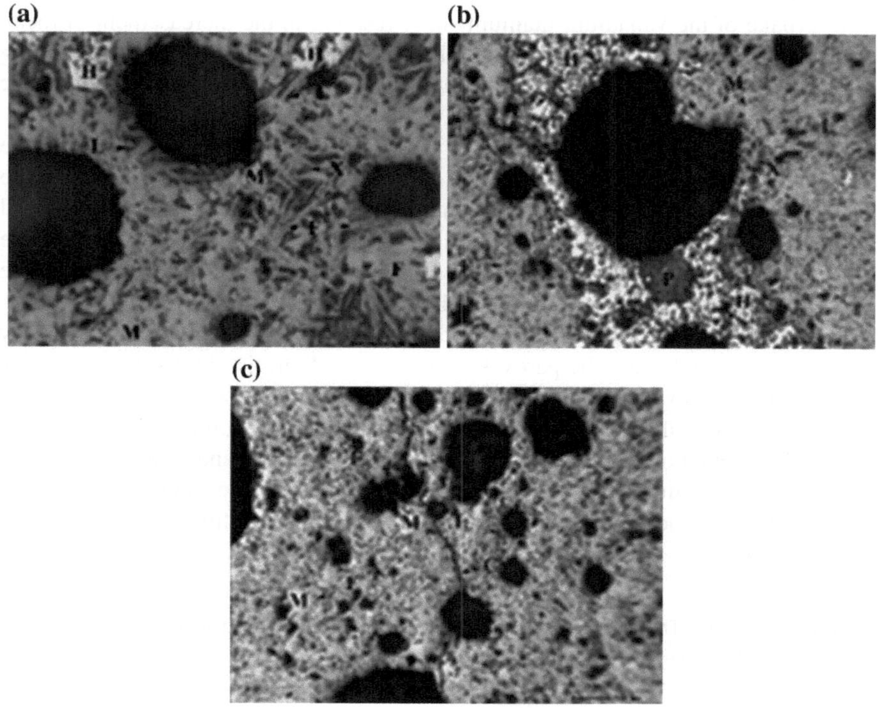

Fig. 5 The phase structure of B1 sinter ore

shapes. Most of small pores are assembled to form big pores with irregular shape. While there are also cracks appeared in the big pores.

Conclusions

1. With the increasement of titanium content in sinter ore, the metal phase content decreased, while the binding phase content increased. The content of SFCA is increased and the hyaline is appeared. Moreover, the nonuniformity of phase structures of sinter ore and particle size of dicalcium silicate are increased with the increasement of titanium content. While the cracks in the vanadium-titanium sinter appeared frequently in the side of the hematite, and the small cracks run through the pores. Moreover, the length of the cracks increased with the increasement of titanium content.
2. With the basicity of the vanadium-titanium sinter decreased, the porphyritic texture in the phase structure and the hyaline in the binding phase are appeared, and the content of SFCA is also decreased. Then most of the cracks with different length and depth are appeared. In order to improve the strength of the vanadium-titanium sinter, the basicity should be increased with the increasement of titanium content.
3. Overbake of the vanadium-titanium sinter resulted in the increasement of metal phase and the content of dicalcium silicate, it also could decrease the binding phase and SFCA content. Consequently, the overbake phenomenon should be avoided during the sintering process.
4. Compared with the typical sinter ore in China, the hematite in the vanadium-titanium sinter is appeared locally and spread non-uniformly, while the binding phase content is also small. Moreover, the cracks, big pores and micropores in the vanadium-titanium sinter are appeared in the center of the sinter and the number is large. However, the cracks in the typical sinter ore are most likely appeared in the edge of the sinter ore.
5. The existed pores in the vanadium-titanium sinter reduced the caking property of crystal particle, and the perovskite in the vanadium-titanium sinter could not play the cohesive action. Eventually, the strength of the vanadium-titanium sinter is worse than the typical sinter ore in China. In order to charge more vanadium-titanium sinter to blast furnace, the basicity and sintering time of vanadium-titanium sinter should be kept in the reasonable range, and then the strength of vanadium-titanium sinter could met the smelting demands of the blast furnace.

Acknowledgements The authors would gratefully acknowledge the financial support from the National Natural Science Foundation of China (No. 61271303).

References

1. Xia ZH, Tong ZX (1997) The study of microscopic crack in the Tanggang sinter. Sintering Pelletizing 1(22):11–13
2. Qiao RQ, Du HG (1999) Forming mechanism of pore in sinter with low fluorine and its effect on the sinter strength. J Iron Steel Res 6(11):1–4
3. Ying ZW et al (2006) Effect of calcium ferrite morphology on anti-fracture ability of sinters. J Iron Steel Res 2(18):55–58
4. Yang YH, Standish N (1991) Fundamental mechanisms of pore formation in iron ore sinter and pellets. ISIJ Int 5:468–477
5. Bai YQ et al (2011) Study V-Ti sinter reduction degradation by mineralogical analysis. Sintering Pelletizing 2(36):1–6
6. Sun YQ et al (2013) Study of the effect of TiO_2 content on the quality of vanadium-titanium in sinter. China Metall 10(23):6–9
7. Bai YQ, Cheng SS (2012) Research on reduction degradation mechanism of low vanadium-titanium sinter. Sintering Pelletizing 1(37):1–5
8. Li ZM et al (2014) Influence of MgO on structure of vanadium-titanium sinter. Sintering Pelletizing 3(39):69–73
9. Li ZM et al (2014) Influence of ordinary ore blending ratio on texture of vanadium-titanium sinter. Iron Steel Vanadium Titauium 4(35):67–70
10. Ding HC, Li FM, Wei XZ (2014) Influence of MgO on structure of vanadium-titanium sinter. J Hebei United Univ 4(36):14–19
11. Peng W, Bi XG, Zhou DJ (2014) Experimental study on suitable MgO content of WISCO sinter. Sintering Pelletizing 3(39):6–11

The Use of Network Simplex Method for Planning the Incorporation of Recycled Paper Mill Sludge in Manufacturing of Ceramic Bodies

Andreiva Lauren Vital do Carmo, Nirlane Cristiane Silva, Anna Paula Sartori, Ana Augusta Passos Rezende, Leonardo Gonçalves Pedroti, Wellington Emílio Hilarino Fernandes and Benício Costa Ribeiro

Abstract The final disposal of sludge from recycled paper mill effluent treatment plants is costly and demands large areas for landfilling. Environmentally sound alternative uses of the sludge, enabling economic valoration of this material appear to be advantageous to industry and to the environment. This paper evaluates the feasibility of the incorporation of paper mill sludge in manufacturing of ceramic bodies. The combination of different proportions of weak clay, strong clay and sludge from recycled paper effluent treatment plant for the production of ceramic bodies was determined by using the Network Simplex method planning of mixtures. Samples were prepared and burnt in oven at temperatures of 750, 850 and 950 °C. Physical, mechanical and environmental tests were carried out, in addition to chemical, microscopic and thermic analysis. The results demonstrated that the flexion resistance increased with the increase in temperature, while the absorption decreased at higher temperatures. The mixture design is efficient combining sludge addition at small fractions with large amount of clay, which demonstrate the viability of sludge incorporation for ceramic bodies manufacture, since the desired characteristics are observed.

Keywords Sludge incorporation · Ceramic bodies · Network simplex

---

A.L.V. do Carmo · N.C. Silva · A.P. Sartori · A.A.P. Rezende ·
L.G. Pedroti · W.E.H. Fernandes (✉) · B.C. Ribeiro
Departamento de Engenharia Civil, Universidade Federal de Viçosa,
Av. P.H. Rolfs s/n, Centro, Viçosa, Minas Gerais, Brazil
e-mail: w.fernandes@ufv.br

S. Ikhmayies et al. (eds.), *Characterization of Minerals, Metals, and Materials 2017*, The Minerals, Metals & Materials Series,
DOI 10.1007/978-3-319-51382-9_91

Introduction

The world industrial growth has been leading to an increase of wastes generated in the mills particularly sludges from effluent treatment plants. Final disposal of sludge is expensive and if not appropriate can contaminate the environment. The search for a sustainable solution depends on the capacity to transform the waste into a useful and economically valiable material to other industrial processes. Several studies related to waste recovery has been conducted aiming their use as a raw material for the production of usable materials [1–5].

Some investigations have been pointed out to solidification/stabilization in solid matrices, such as ceramic clay, cement mortar and concrete in the production of relevant materials in civil construction as a way to reduce disposal costs avoiding negative impacts on the environment [1–5]. Among the materials mentioned, stands out the clay because of its good capacity to absorb waste pollutants, good mechanical strength, in addition to withstand high temperatures and exhibit good stability [2]. Furthermore, the use of sludge as ceramic additives enables the immobilization of heavy metals in the burned matrix, organic matter oxidation and destruction of pathogenic organisms during the firing process [3].

Recycled paper mills generate large amounts of solid waste, including sludge from the effluent treatment plants. An average amount of sludge generated in a recycled paper mill can reach up to 27 kg solids/ton produced [4]. Thus it is crucial to find options for this sludge treatment/reuse. The incorporation of paper mill sludge for the manufacturing of ceramic blocks can bring economical and environmental advantages for both industries. The ceramic industry can increase productivity and save energy and preserve the clay deposits. The paper mills can solve the environmental waste disposal problem reducing treatment and final disposal costs and, also providing a significant gain for the construction sector, promoting quality and productivity of both sectors [5].

The present study aimed to evaluate the incorporation of sludge from a recycled paper mill effluent treatment plant for the manufacturing of ceramic bodies, using a methodology with Network Simplex planning for analysis of the combination of different raw materials mixtures: two types of clay and the sludge.

Materials and Methods

Two types of clays were used as raw material in this study: weak (yellow clay) and strong (gray clay), both collected in the city of Visconde do Rio Branco/MG, Brazil. The sludge used was collected in the effluent treatment plant of a recycled paper mill at Ponte Nova, MG, Brazil.

The clays were dried at 100 °C during 24 h, grounded and sieved prior to the characterization tests. The sludge was arranged on a canvas and dried under ambient conditions during 10 days. After the drying period, it was placed in an oven at 100 °C for 24 h and ground in the Los Angeles Abrasion apparatus for conducting the tests. The sludge was characterized according to ABNT NBR 10004/2004, following the procedures for obtaining leached extract (ABNT NBR 10005/2004), to classify the waste as hazardous or non-hazardous, and obtaining the leached extract (ABNT NBR 10006/2004), aiming to classify as inert or non-inert [6–8].

The clays were characterized according to ABNT NBR 7181/1984, for particle size analysis, ABNT NBR 6508/1984 for determination of specific mass and ABNT NBR 6459/1984, for determining the Atterberg limits [9–11]. It was also carried out X-ray Diffraction Analysis (XRD) using XRD-7000 model of Shimadzu; chemical analysis by Energy Dispersive X-ray (EDX), using the apparatus EDX-700 model of Shimadzu; and thermal analysis (DTA/TGA) using a thermal analyzer of BP Engineering, model RB3000.

The raw materials mixture proportions for the production of the ceramic bodies were determined by the mathematical method of Network Simplex, using the special cubic model, with seven mixing points for different percentages of weak clay (WC), strong clay (SC) and sludge from wastewater treatment (ST). The pseudo-components ranges were 0–100% for WC, 0–70% for SC and 0–70% for ST. Fifteen ceramic bodies were produced for each mixture, including five for each firing threshold temperature (750, 850 and 950 °C), totalizing 105 ceramic bodies.

Prior to production of the ceramic bodies, the raw materials were sieved on 1.2 mm mesh to improve efficiency of the pressing procedure. The pressing was carried out by using a rectangular mold and a load ranging from 8128.4 and 16,256.8 kgf, according to the mixture to be pressed, so that all test bodies had approximate dimensions that kept the specific mass between 1.51 and 1.99 g/cm^3. The average dimensions of the samples were 115.21 ± 0.21 mm $\times$ 11.82 ± 0.49 mm $\times$ 19.60 ± 0.34 mm. An amount of 50 g of mixture was used for making each ceramic body, prepared with 10% moisture content. Subsequently, each ceramic body was identified, weighed using a 0.01 g precision analytical balance and measured with a digital caliper.

The samples were dried in an oven for 24 h at 100 °C and were then weighed and measured to verify a decrease after firing. The burning of the ceramic bodies was then performed in a muffle kiln with different heating stages, initially allowing the temperature to reach approximately 600 °C, remaining then at rest for 1 h and then raised until the firing threshold temperature was reached. Then, the muffle was turned off after 2 h for cooling over a period of 12 h. Again, the samples were weighed and measured. Figure 1 shows the samples produced for each mixture after firing.

After the firing, flexural strength and water absorption tests were performed. From the results, the adjustment of the mathematical model was carried out to evaluate the sludge incorporation in the manufacture of ceramic bodies by generating surfaces responses and statistical graphs using the Statistica software created by Statsoft.

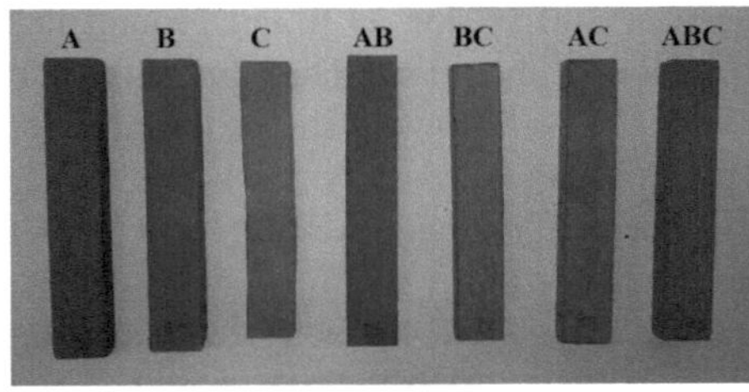

Fig. 1 Samples produced for each mixture after firing

Results and Discussion

Sludge Classification

The sludge originated in the effluent treatment plant of the recycled paper mill was classified as non-hazardous and non-inert.

Particle Size Analysis, Specific Mass and Atterberg Limits

The gray clay presented 39.6% of sand, 19.4% of silt and 41.0% of clay. The yellow clay presented 13.2% of sand, 29.6% of silt and 57.2% of clay. The gray clay had a specific mass of 2.6 g/cm^3 and the yellow clay of 2.7 g/cm^3. The Atterberg limits are showed in Table 1. The gray clay had the largest plasticity index and could be considered a strong clay (SC); and the yellow clay was considered a weak clay (WC).

X-ray Diffraction Analysis

After XRD analysis, a predominance of kaolinite diffraction peaks (C) and the presence of quartz (Q) in minor amounts for both clays were observed, as showed in Fig. 2. The WC also showed a characteristic diffraction peak of gypsum (Gi). The clays that have greater evidence of the presence of kaolinite and minor presence of quartz tend to have better performance in ceramic materials production. Kaolinite is a mineral, which contributes to the development of plasticity on clays, while quartz is a major impurity present in clays.

Table 1 Atterberg limits

Clay	Liquid limit (%)	Plastic limit (%)	Plasticity index (%)
SC	58.9	29.3	29.6
WC	54.9	28.3	26.6

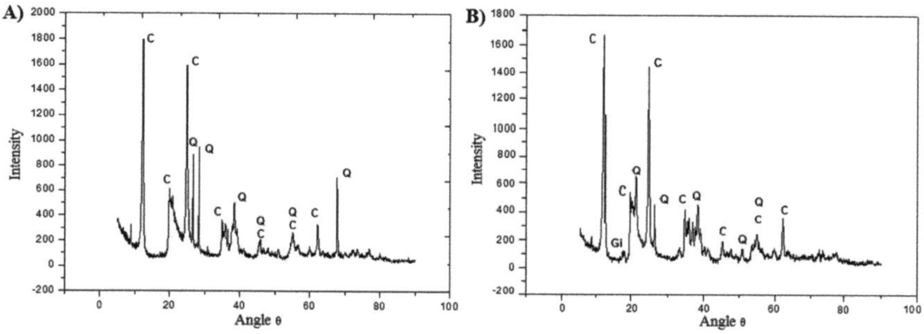

Fig. 2 X-ray diffractograms. Strong clay (**a**) and weak clay (**b**)

Chemical Analysis

Table 2 shows the oxides present in the clays used in this study and their respective percentages. The element SiO_2 is the most present in the samples of analyzed clay, followed by Al_2O_3 and Fe_2O_3.

Thermal Analysis

Figure 3 shows DTA/TGA curves of SC and WC, respectively. The strong clay (Fig. 3a) had a mass loss of 2.6% at the initial temperature, an endothermic peak in the temperature 578.6 °C with 12.2% of mass loss, and an exothermic peak without mass loss at 934.8 °C. The total mass loss of strong clay was 14.8%. The weak clay (Fig. 3b) had a mass loss of 1.53% at the initial temperature, an endothermic peak at 270.95 °C with 2.58% of mass loss, another endothermic peak at 578.60 °C with 11.67% of mass loss, and an exothermic peak without mass loss at 957.60 °C. The total mass loss of weak clay was 15.78%.

Table 2 Chemical analysis of clays

Clay	Element (%)						
	SiO_2	Al_2O_3	Fe_2O_3	SO_3	TiO_2	K_2O	CaO
SC	52.64	37.69	5.20	1.86	1.48	0.56	0.47
WC	50.71	34.36	9.56	1.75	1.74	0.99	0.69

Fig. 3 Thermal analysis. Strong clay (**a**) and weak clay (**b**)

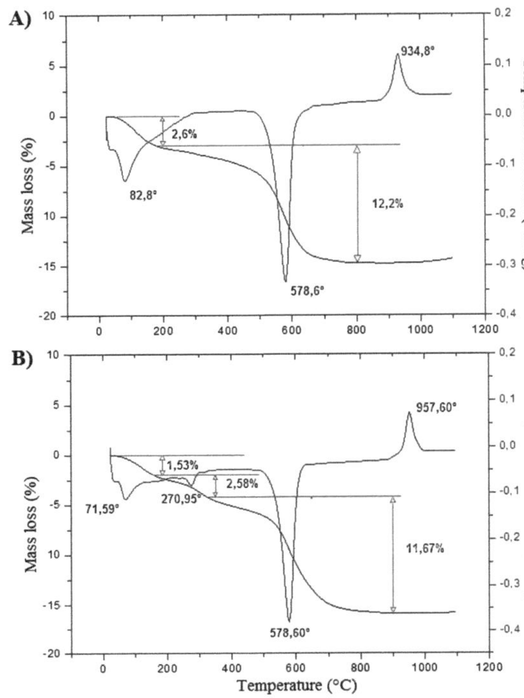

Sludge Incorporation Analysis in Ceramic Bodies

Figure 4 shows the response surface and plotted trace of response estimated for the variable flexural strength in different firing temperature threshold. Graphs A, B and C show that flexural strength increases with increasing temperature. Among the pure mixtures, the WC had higher resistance values. Graphs D, E and F show that, regardless the temperature, the greater the fraction of WC present in the mixture, the greater becomes the flexural strength.

ST showed the lowest flexural strength while increasing its fraction present in the mixture. However, the flexural strength graphs demonstrated that the incorporation of sludge into ceramic bodies, aiming to improve flexural strength performance, it is feasible since carried out using small fractions of sludge and high clay fractions.

Figure 5 shows graphs for water absorption at firing temperature threshold. As opposed to the flexural strength, the water absorption decreased with increasing temperature, and the mixtures with greater fraction of this sludge are those with a

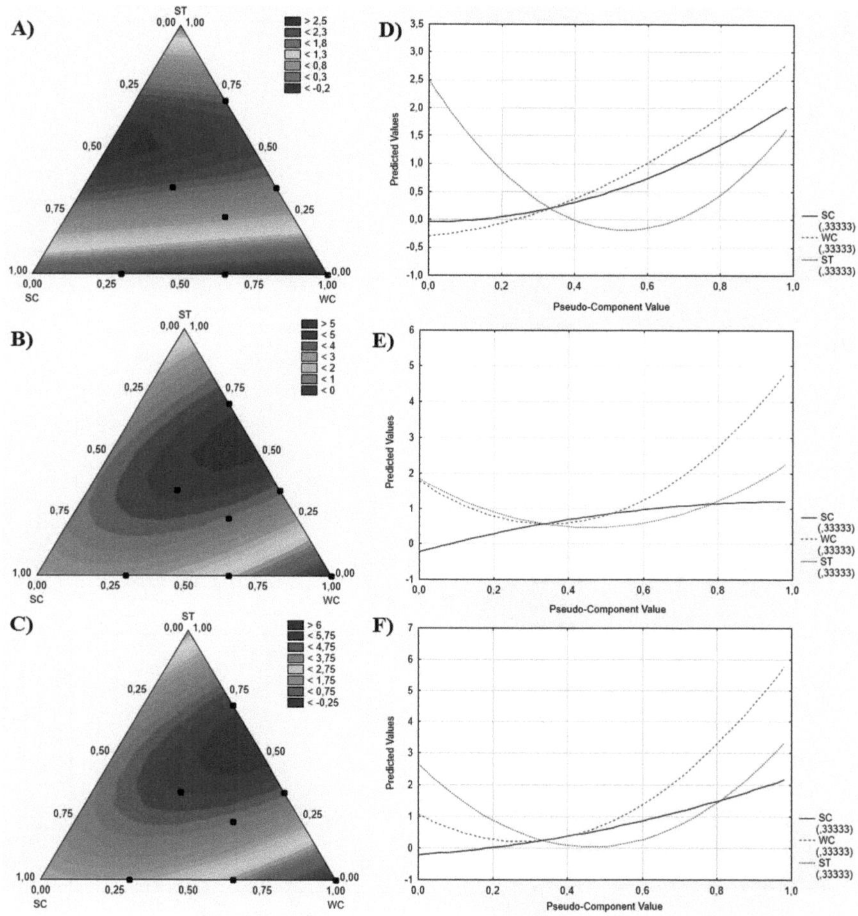

Fig. 4 Flexural strength fitted surface (**a**, **b**, **c**) and flexural strength plot of estimated response (**d**, **e**, **f**) for different firing threshold temperature. **a** T = 750 °C; R^2 = 0.9344; Adj. = 0.9256, **b** T = 850 °C; R^2 = 0.9693; Adj. = 0.9640, **c** T = 950 °C; R^2 = 0.9840; Adj. = 0.9812, **d** T = 750 °C; R^2 = 0.9344; Adj. = 0.9256, **e** T = 850 °C; R^2 = 0.9693; Adj. = 0.9640, and **f** T = 950 °C; R^2 = 0.9840; Adj. = 0.9812. The predicted values for flexural strength corresponds to MPa

higher water absorption percentage. It is emphasized that the ABNT NBR 15270/2005 stipulates values of absorption between 8 and 20% for ceramic sealing and structural blocks [12]. Thus depending on the purpose of the ceramic body, should be prioritized mixtures with higher amounts of clay than sludge in order to get to lower levels of water absorption.

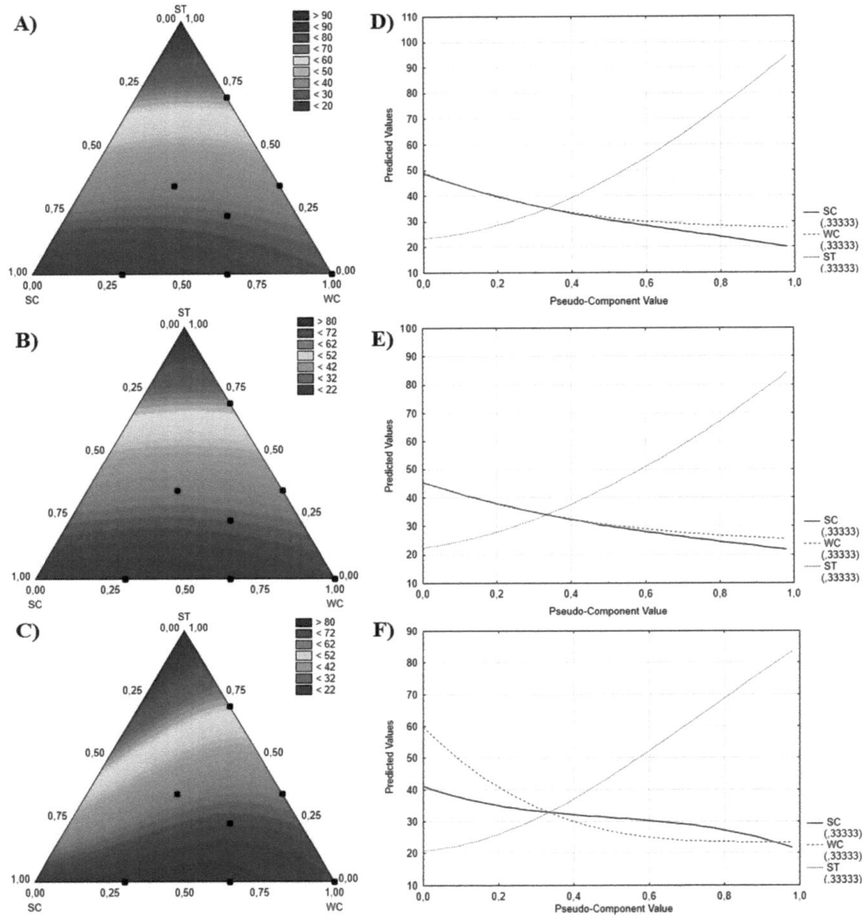

Fig. 5 Water absorption fitted surface (**a**, **b**, **c**) and water absorption trace plot of estimated response (**d**, **e**, **f**) for different firing threshold temperature. **a** T = 750 °C; $R^2 = 0.9992$; Adj. = 0.9991, **b** T = 850 °C; $R^2 = 0.9992$; Adj. = 0.9990, **c** T = 950 °C; $R^2 = 0.9981$; Adj. = 0.9977, **d** T = 750 °C; $R^2 = 0.9992$; Adj. = 0.9991, **e** T = 850 °C; $R^2 = 0.9992$; Adj. = 0.9990, and **f** T = 950 °C; $R^2 = 0.9981$; Adj. = 0.9977. The predicted values for absorption corresponds to %

Conclusions

The combination of different mixtures containing strong clay, weak clay and wastewater treatment sludge employing the Network Simplex planning method enabled the evaluation of waste incorporation feasibility for the manufacture of ceramic bodies. The incorporation of the sludge originated from wastewater treatment of recycled paper mill in ceramic bodies proved to be viable since are observed the required characteristics (flexural strength and water absorption) of the ceramic material to be manufactured.

Acknowledgements The authors acknowledge the dedication of all servers and students DEC UFV, CAPES, Fapemig and Funarbe for the support provided to this investigation.

References

1. Nair KP, Vivek JM, Shibu K (2013) Suitability of sludge as a building material. Int J Sci Eng Res 4(5):23–28
2. Oliveira EMS, Machado SQ, Holanda JNF (2004) Characterization of waterworks waste (sludge) aiming its use in red ceramic. Cerâmica 50:324–330
3. Castro TM et al (2015) Characterization of acoustic ceramic blocks produced with incorporating textile laundry sludge. Engenharia Sanitária e Ambiental 20:47–54
4. Pinheiro RM (2008) Recycling of primary sludge from the effluent treatment plant from the paper industry into ceramic red. M.Sc thesis, Universidade Estadual do Norte Fluminense Darcy Ribeiro, pp 19–24
5. Garcia CM et al (2014) Porosity of expanded clay manufactured with addition of sludge from the brewing industry. Int J Energy Environ Eng 5(4):341–347
6. ABNT (2004) NBR 10004: solid waste—classification. ABNT—Associação Brasileira de Normas Técnicas, Rio de Janeiro, RJ, p 71
7. ABNT (2004) NBR 10005: procedure for obtention leaching extract of solid wastes. ABNT, Rio de Janeiro, RJ, p 16
8. ABNT (2004) NBR 10006: procedure for obtention of solubilized extraction of solid wastes. ABNT, Rio de Janeiro, RJ, p 3
9. ABNT (1984) NBR 7181: soil—granulometric analysis. ABNT, Rio de Janeiro, RJ, p 13
10. ABNT (1984) NBR 6508: grains of soil passing through the sieve of 4.8 mm—determination of density. ABNT, Rio de Janeiro, RJ, p 8
11. ABNT (2016) NBR 6459: soil—liquid limit determination. ABNT, Rio de Janeiro, RJ, p 5
12. ABNT (2005) NBR 15270: ceramic components. ABNT, Rio de Janeiro, RJ, p 15

Use of Alkaline Solid Wastes from Kraft Pulp and Paper Mills, Dregs and Grits in Cement Production

C.M.M.E. Torres, L.G. Pedroti, C.M. Silva, W.E.H. Fernandes, N.G. Viana, R.O.G. Martins, G.E.S. Lima, L.M. Sathler, I.K.R. Andrade and M.A. Caetano

Abstract In Brazil, kraft pulp mills are rapidly expanding and industrial waste generation is increasing. Technical feasibility of using solid wastes from kraft pulp mills as raw material in other industries may be an interesting approach to manage environmental and economic aspects of the industry. This paper proposes the incorporation, in different proportions (0, 2.5, 5 and 7.5%), of alkaline solid wastes from pulp mills, namely dregs and grits, to clinker in the Portland cement industry. The research was carried out in two steps: (i) characterization of dregs, grits and clinker; (ii) laboratory testing (determination of soundness by the Le Chatelier method, determination of setting times, determination of water content of the paste with normal consistency, determination of compressive strength and static modulus of elasticity). The results showed that the addition of dregs and grits to the clinker up to 2.5% were feasible according to the Portland cement Brazilian standards.

Keywords Kraft pulp mills · Dregs · Grits · Waste management · Clinker

Introduction

Bleached kraft pulp mills are rapidly expanding in Brazil. Currently, Brazil is the 4th largest pulp producer of the world, with 17.37 million produced tons. The expectation is that by 2017, Brazilian production exceeds 19 million tons and will occupy the 3rd position in the world ranking [1].

C.M.M.E. Torres (✉) · L.G. Pedroti · C.M. Silva · W.E.H. Fernandes
R.O.G. Martins · G.E.S. Lima · L.M. Sathler · I.K.R. Andrade · M.A. Caetano
DEC—Civil Engineering Departament, UFV—Universidade Federal de Viçosa, Av. PH Rolfs, Viçosa, Minas Gerais 36570-900, Brazil
e-mail: caiomiquelino@yahoo.com.br

N.G. Viana
UniViçosa—União de Ensino Superior de Viçosa, Av. Maria Paula Santana, 3815, Silvestre, Viçosa, Minas Gerais 36570-000, Brazil

S. Ikhmayies et al. (eds.), *Characterization of Minerals, Metals, and Materials 2017*, The Minerals, Metals & Materials Series,
DOI 10.1007/978-3-319-51382-9_92

Aligned to this industrial growth, there is an increase in solid wastes generation. Industrial wastes need appropriate treatment and disposal. Reuse and recycling are highly recommended whenever it is possible.

Kraft pulp mills originate wastes in several parts of the industrial process such as in the water and wastewater treatment plants, wood yard, liquor chemical recovery cycle, biomass boiler, among others. The wastes from pulp mills are classified as non-hazardous and non-inert [2].

Alkaline wastes namely dregs and grits, originated in the causticizing and calcination process of the black liquor chemical recovery cycle, are among the main solid wastes from the kraft pulping process. The dregs are impurities derived mainly from carbon, hydroxides and sulfides having a pH approximately 11. Up to 10 kg adt^{-1} of dregs are generated during the clarification of green liquor. The grits are originated in the slakers and are composed of unreacted lime. The generation of grits is approximately 5 kg adt^{-1}. Both dregs and grits are classified as alkaline wastes and their disposal has been normally to landfills. It is estimated that in 2015 in Brazil were generated 260,500 ton dregs and 868,500 ton grits [3–6].

The use of these wastes as raw material for other industrial purposes is an attractive management option. Studies have been carried out to verify the feasibility for direct application of dregs and grits to soil, ceramic production and paving rural roads [7, 8].

The physico-chemical characteristics of dregs and grits indicate a potential for cement production, since they contain a high amounts of calcium carbonate.

The manufacturing of Portland cement starts with the extraction of limestone rock deposits. The limestones are crushed to obtain appropriate granulometry for mixing with other materials. This mixture is burned in a rotary kiln consisted of an inclined steel cylinder lined with refractory bricks, where temperatures of 1450–1550 °C can be achieved. Chemical reactions undergo in the kiln and the formation of pellets, known as clinker, is formed. Clinker is cooled and milled to a fine powder, usually less than 75 μm [9, 10].

The objective of this research was to study the technical feasibility of the incorporation of alkaline waste called dregs and grits, to clinker in different proportions (0, 2.5, 5 and 7.5%) as raw material in the cement industry. Few studies were found in the literature addressing the use of dregs and grits in the production of clinker [11] and the use of such waste as concrete aggregate [12]. No scientific technical study suggesting the feasibility of using dregs and/or grits as a raw material in the manufacturing of Portland cement was found.

Materials and Methods

Dregs and grits were acquired from a Brazilian bleached kraft pulp mill and stored in big bags. Clinker was obtained from a Brazilian cement company and stored in plastic recipients to avoid contamination. These materials were characterized in the laboratories of the Federal University of Viçosa—UFV and the Federal University of Minas Gerais—UFMG.

The dregs and grits were submitted to an experimental path described by the flowchart of Fig. 1. First, the dregs and grits in nature passed through a drying process in an oven at 100 °C (212 °F) for 24 h and ground in a ball mill. Then the material was broken in the spray, and macerated in hand grail and pistil to complete passage 75 μm sieve (200 mesh). The clinker was ground, macerated to complete passage 200 mesh sieve. All materials were stored in appropriate plastic containers to avoid any risk of contamination.

Trials mass and specific area, fineness, X-ray Diffraction—XRD, Scanning Electron Microscopic—SEM, Energy Dispersive X-ray—EDX and pozolanic activity were performed to determine the physical, chemical and mineralogical characteristics of the materials.

The Energy Dispersive X-ray (EDX) consists of a quantitative analysis of elements present in the materials by fluorescence spectrometry X-rays. EDS has the same purpose as EDX, however with a specific analysis of the material. The used machine was a fluorescence spectrometry X-rays micro EDX 1300Shimadzu.

The Scanning Electron Microscopic (SEM) consists of a microstructural analysis of solid materials with high resolution images. Images were obtained with increases

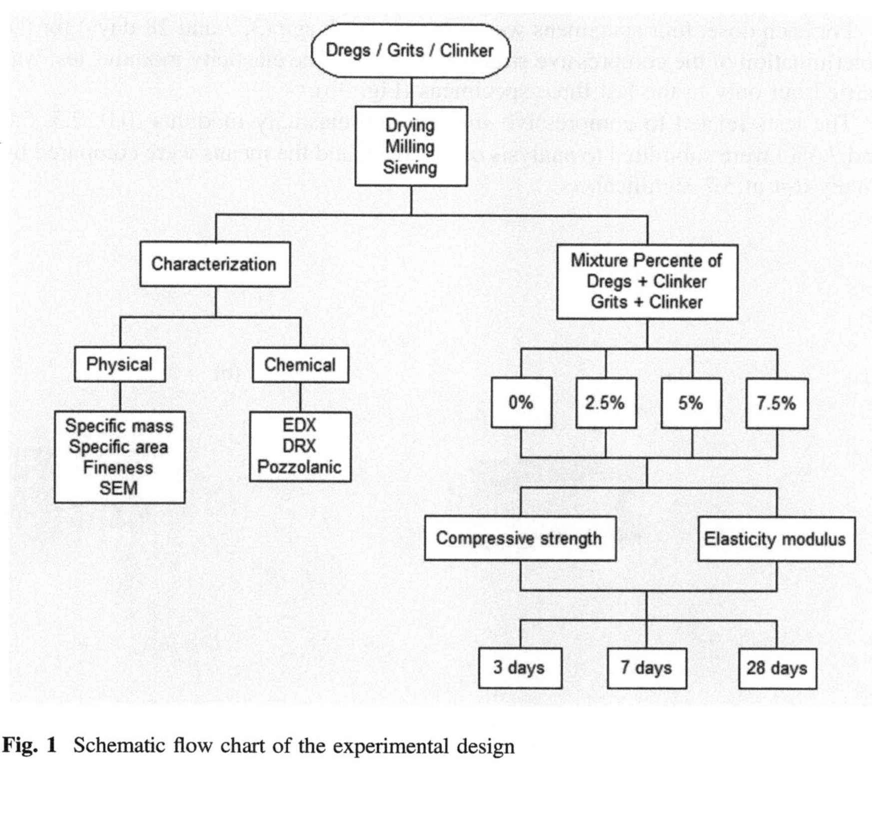

Fig. 1 Schematic flow chart of the experimental design

of 1000–2000 times for the three types of materials (clinker, dregs and grits) and their mixtures. The used machine was a Scanning Electron Microscope Leo 1430VP.

The X-ray Diffraction (XRD) consists in analyzing the crystal structure of the materials tested. The used machine was a Bruker D8-DiscoverDavinci X-ray diffraction system, with Cu radiation and 2θ angle covered between 20 and 110°.

The test pozolanic activity by the method Luxan et al. [13], is to measure the pozzolanic by conductivity variation of a saturated solution of $Ca(OH)_2$ with and without addition of material to be analyzed.

The determination of the specific mass of the material is established by the mass of the volume of a sample according to ABNT NM 23: 2001 [14].

The specific surface area is established to determine the fineness of the materials and indicate the need for grinding thereof was measured following the instructions of ABNT NBR 16372: 2015 [15].

The fineness of the material is determined from the percentage in retained weight of grain larger than 200 mesh ABNT NBR 11579: 2012 [16].

All assays were carried out according to Brazilian standards.

In the present research, 12 specimens of 50 mm diameter and 100 mm in length were produced (Fig. 2), for each incorporation of dregs and grits to the clinker (0.0, 2.5, 5.0 and 7.5%) corresponding to a total of 84 test specimens (Table 1) (ABNT NBR 5738; 2016) [17].

For each dose, four specimens were tested in each age (3, 7 and 28 days) for the determination of the compressive strength (Fig. 3a). The elasticity modulus test was carried out only in the last three specimens (Fig. 3b).

The tests related to compressive strength and elasticity modulus (0.0, 2.5, 5.0 and 7.5%) were submitted to analysis of variance, and the means were compared by Tukey test at 5% significance.

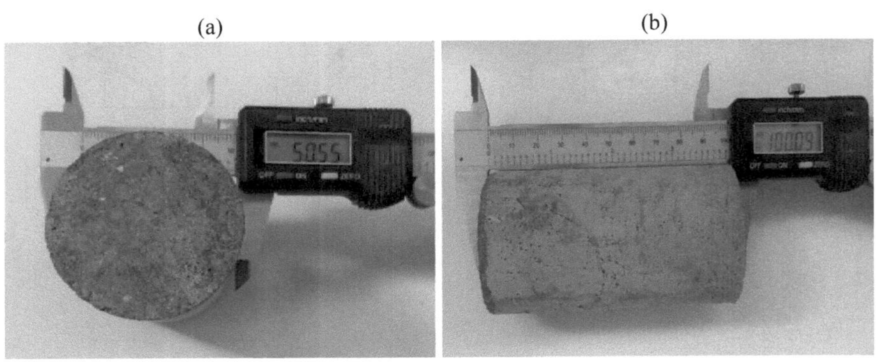

Fig. 2 Specimens **a** 50 mm in diameter and **b** 100 mm in length

Table 1 Dosage materials for experimental clinker

Samples	Series	Trace (%)		Amount of specimens	
		Clinker	Dose	Compressive strength	Elasticity modulus
	CPo—0	100	0	12	9
Dregs	CPd—2.5	97.5	2.5	12	9
	CPd—5.0	95.0	5.0	12	9
	CPd—7.5	92.5	7.5	12	9
Grits	CPg—2.5	97.5	2.5	12	9
	CPg—5.0	95.0	5.0	12	9
	CPg—7.5	92.5	7.5	12	9
Total				84	63

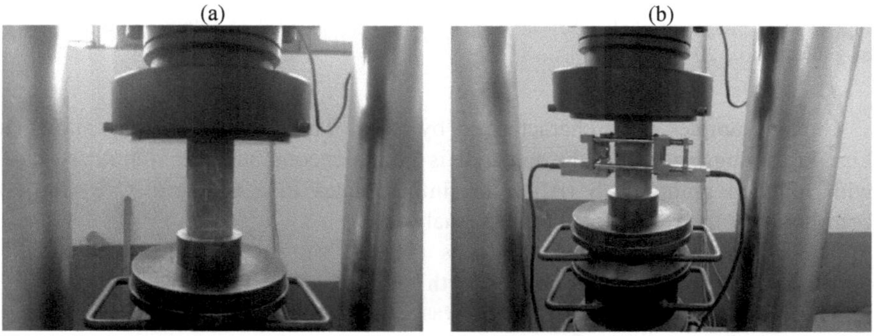

Fig. 3 Test **a** compressive strength and **b** elasticity modulus

Results and Discussion

The chemical composition of clinker, dregs and grits is showed in Table 2. It is observed that these materials have a high calcium oxide content, which influences its incorporation into the clinker mixtures. The dregs presented a higher content of sulfur oxides (SO_3), magnesium (MgO), iron (Fe_2O_3) and sodium (Na_2O) than the grits. Therefore, it can assist the gypsum ($CaSO_4$) in the industrial process.

Table 2 Elemental composition of clinkers, dregs and grits

	CaO	SiO_2	Al_2O_3	Fe_2O_3	SO_3	MgO	K_2O	Na_2O	TiO_2	Cl	Others
Clinker (%)	64.55	20.65	3.02	3.23	1.0	0.33	1.47	1.71	0.27	0.07	3.72
Dregs (%)	68.85	6.67	0.74	3.27	6.44	3.14	0.91	3.64	0.18	0.39	5.78
Grits (%)	83.36	5.21	0.29	1.16	0.97	0.66	0.50	2.77	0.08	0.27	4.73

Fig. 4 SEM images of clinker powder **a** image with magnification of 1000× and **b** image with magnification of 2000×

Clinker samples were characterized by SEM. The obtained images, shown in Fig. 4a, b, suggested that the material is homogeneous and has smooth surfaces with rounded grains, C_3S (Alita) in similar plates to a hexagon, C_2S (Belita), corroborating the results of the EDX analysis [9].

The micrograph of dregs and grits (Fig. 5) showed particles with smaller dimensions compared to the clinker, with rounded grains rich in Ca.

The XRD of the material (Fig. 6) was carried out to support the existence of mineralogical compounds found in the clinker chemical composition, and they present tricalcium silicate ($3CaO{\cdot}SiO_2$)—C_3S; dicalcium silicate ($2CaO{\cdot}SiO_2$)—C_2S; Tricalcium aluminate ($3CaO{\cdot}Al_2O_3$)—C_3A; and iron aluminate tetracalcium phases ($4CaO{\cdot}Al_2O_3{\cdot}Fe_2O_3$)—$C_4AF$ which are the main identified phases in their structures.

Similar results were observed by Souza and Cardoso [7] and Mehta and Monteiro [9] in high concentrations of C_3S (Alita) and C_2S (Belita) between 2θ range 28–40° angles, corroborating the images in the analyzes carried out by SEM.

The XRD of dregs and grits (Fig. 7) was conducted to support the existence of mineralogical compounds found in chemical composition. The calcium carbonate ($CaCO_3$), Pirssote ($Na_2Ca(CO_3)_2{\cdot}2(H_2O)$), calcium hydroxide ($Ca(OH)_2$), and ferrous oxide (FeO) were identified as the major crystal phases. It can also be noted the similarity of dregs and grits XRD patterns.

The pozzolanic activity assay showed a conductivity of 0.31 and 0.87 mS/cm, respectively, for dregs and grits. These results demonstrates that the materials do not have a pozzolanic activity. This demonstrates that neither the dregs nor the grits promotes an interaction improvements in reactivity with cement. However, while cementitious addition of pore fillers, these materials are efficient, with high values of Blaines [9, 10].

Fig. 5 SEM images of dregs powder **a** image with magnification of 1000×, **b** image with magnification of 2000×. With grits powder **c** image with magnification of 1000× and **d** image with magnification of 2000×

Table 3 shows the materials specific mass and specific area. Being identified its fineness levels.

The compressive strength tests (Fig. 8) were conducted with test specimens 3, 7 and 28 days. The mixtures of 2.5, 5 and 7.5% of dregs and grits showed strength below the reference (0% material) based on their estimated results according to their maximum standard deviation. In the 28 days, there are not dregs replacement statistically equal to reference. However, the 2.5% grits substitution was statistically equal to reference.

The results of assays elasticity of static modulus in compressive (Fig. 9) performed with the specimens 3, 7 and 28 days. 2.5, 5 and 7.5% of dregs and grits mixtures were mostly below the reference (0% of materials). In the 28 days, 2.5, 5 and 7.5% of dregs substitution were statistically equal to the reference. However, the 2.5 and 5% of grits replacement were also statistically equal reference.

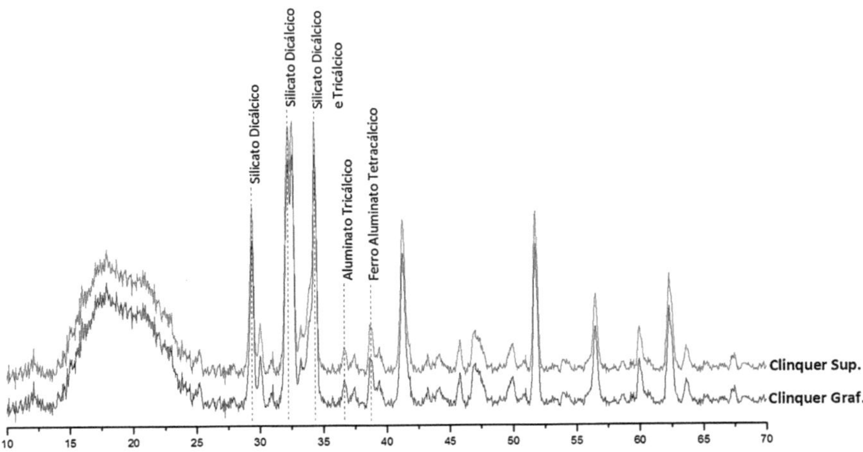

Fig. 6 XRD patterns of clinker

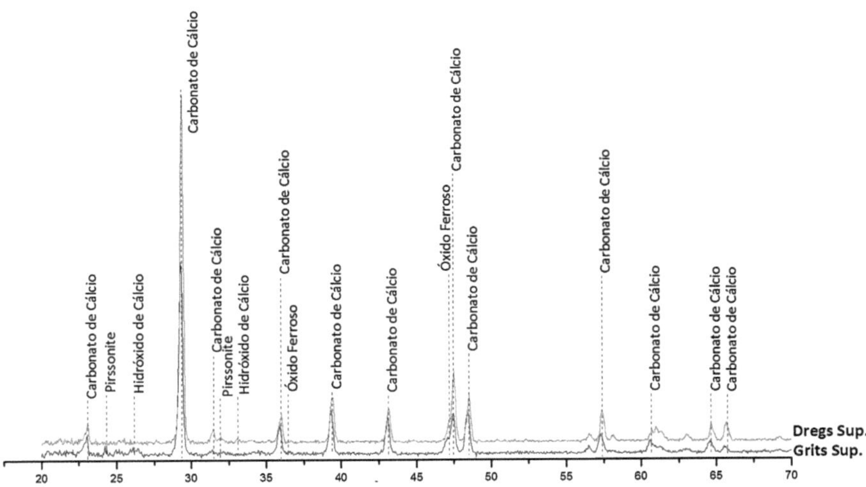

Fig. 7 XRD patterns of dregs and grits

Table 3 Results of tests mass, specific area and fineness

Materials	Specific mass (g/cm^3)	Specific area (m^2/kg)	Fineness index (%)
Clínquer	3.00	246	0.26
Dregs	2.44	1031	0.78
Grits	2.67	972	95.20

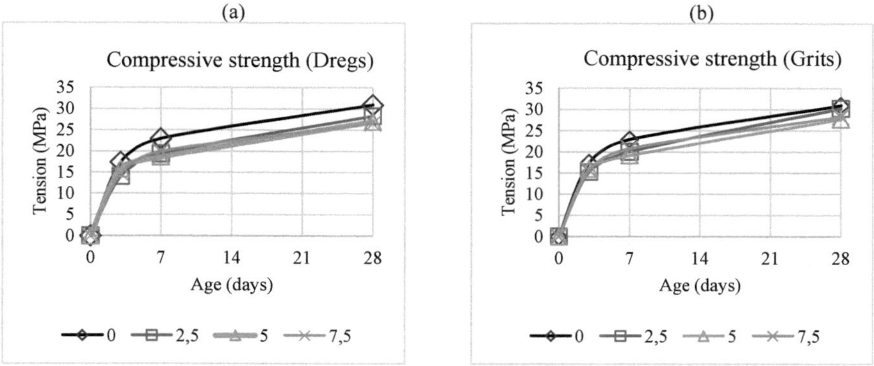

Fig. 8 The results of the compressive strength tests on four of the percentages **a** dregs and **b** grits 3, 7 and 28 days of age

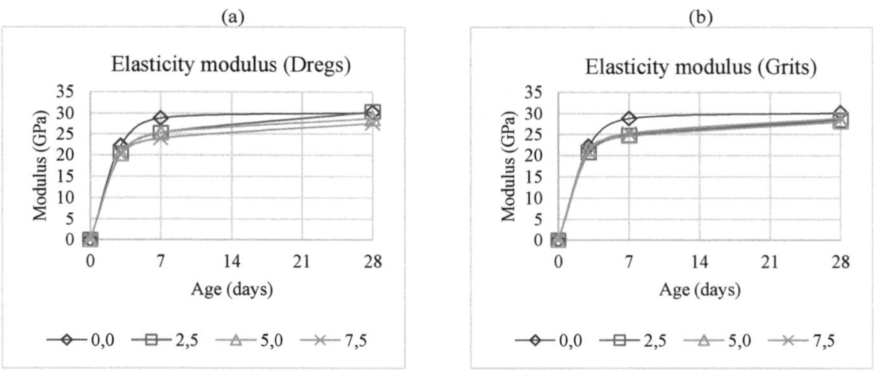

Fig. 9 The results of the elasticity modulus tests on four of the percentages **a** dregs and **b** grits 3, 7 and 28 days of age

Conclusions

The physical-chemical characterization of dregs and grits showed a great potential to use these materials for incorporation into the clinker Portland cement production. The compounds present in dregs and grits were compatible to the production of Portland cement.

The different incorporation of dregs and/or grits (2.5, 5 and 7.5%) to the clinker proved viable for ordinary Portland cement production with addition (CP I-S).

The incorporation of both materials to clinker fulfilled the minimum limits for the compressive strength test and modulus of elasticity established by the Brazilian standard. However, grits replacements have better results compared to the dregs.

Acknowledgements The authors thank the dedication of all servers and students DEF (LCP), DEC (LMC) UFV and the UFMG microscopy center for the images. We also acknowledge the CNPq, CAPES, FAPEMIG and Bioforest for the support provided to this investigation.

References

1. IBÁ—Brazilian tree industry (2016) Statistics data June/2016. Available in: http://iba.org/en/data-statistics/cenarios-iba. Accessed 07 July 2016
2. U.S. EPA (2000) Code of federal regulations, protection of the environment. 40 CFR Parts 260–265—Hazardous Waste Management
3. Biermann CJ (1993) Essentials of pulping and papermaking. Academic Press, San Diego, p 472p
4. Gavrilescu D (2004) Solid waste generation in kraft pulp mills. Environ Eng Manage J 3 (3):399–404
5. Krigstin S, Sain M (2006) Characterization and potential utilization of recycled paper mill sludge. Pulp Paper Canada 107(5):29–32
6. Wolff E (2008) O uso do lodo de estação de tratamento de água e resíduos da indústria de celulose (dregs, grits e lama de cal) na produção de cerâmica vermelha. Ph.D in Saneamento, Meio Ambiente e Recursos Hídricos, Universidade Federal de Minas Gerais, Belo Horizonte—MG
7. Souza TI, Cardoso AV (2009) Utilização de Resíduos Sólidos da Indústria de Celulose Kraft na Fabricação de Cimento: caracterização físico-química. In Congresso brasileiro de engenharia e ciência dos materiais, 18, Porto de Galinhas, 2008. Anais… Porto de Galinhas
8. Machado CC, Pires JMM, Pereira RS, Fontes MPF, Portugal CRM (2006) Chemical and environmental behavior of the solid wastes in forest roads pavements. Sci For 28(70):131–136
9. Metha PK, Monteiro PJM (2006) Concrete—microstructure, properties, and materials, 3rd edn. California, 684p
10. Isaia GC (ed) (2010) Materiais de construção civil e princípios de ciência e engenharia de materiais. Ibracon—Instituto Brasileiro de Concreto, São Paulo, 1773p
11. Castro F, Vilarinho C, Trancoso D, Ferreira P, Nunes F, Miragaia A (2009) Utilisation of pulp and paper industry wastes as raw materials in cement clinker production. Int J Mater Eng Innovation 1(1):74–90
12. Seyyedalipour SF, Kebria DY, Malidarreh NR, Norouznejad G (2014) Study of utilization of pulp and paper industry wastes in production of concrete. J Eng Res Appl 4(1):115–122
13. Luxan MP, Madruga F, Saavedra J (1989) Rapid evaluation of pozzolanic activity of natural products by conductivity measurement. Cem Concr Res 19:63–68
14. Brazilian Association of Technical Norms (2001) NBR NM 23: determination of density, 1st edn. Rio de Janeiro, 5p
15. Brazilian Association of Technical Norms (2015) NBR 16372: determination of fineness by the air permeability method (Blaine method). 2nd edn. Rio de Janeiro, 11p
16. Brazilian Association of Technical Norms (2012) NBR 11579: corrected version: 2013—determination of fineness index by means of the 75 μm sieve (no 200). 2nd edn. Rio de Janeiro. Corrigido, 4p
17. Brazilian Association of Technical Norms (2016) NBR 5738: concrete—procedure for molding and curing concrete test specimens. Rio de Janeiro, 9p

Wood-to-Concrete Joints Using Steel Connectors: Experimental Evaluation

Juliano Correa, Rita de C.S.S. Alvarenga, Beatryz C. Mendes and Márcio Sampaio S. Moreira

Abstract Over the last decades, modern technology to build wood structures have been substantially developed. Building solutions for floor systems of wood-concrete mixed T-beams with wood web and concrete flange stand out among these technologies. This work describes an experimental investigation of the performance of bond connectors made of round steel bars for wood-concrete mixed elements. Symmetric three-member specimens were built. The middle member was made of concrete and the other two were made of wood. The first kind of connector consisted of steel bar designed as straight pins with head, normally driven through the wood. The second kind consisted of arc-shaped connectors directly embedded into the wood or bonded using epoxy adhesive. The specimens were then tested in compression. The compressive strength and the force-displacement curve were obtained, in order to evaluate the failure mode and obtain the stiffness in the ultimate (K_u) and serviceability (K_s) limit states. The stiffness values were calculated in the interval corresponding to the secant straight line between, respectively, 10–40 and 10–70% of the ultimate strength. The obtained results indicate high strength and ductile failure mode for all of the cases. Bond stiffness gains were verified when using the arch-shaped connectors.

Keywords Composite materials connections · Bond strength and stiffness · Steel bar connectors

Introduction

Over the last years, greater incentives to build structures with wood are given, mainly in the developed countries. The increase is motivated by the availability of new application technologies and by the aspect of sustainability of its obtaining and usage.

J. Correa · R. de C.S.S. Alvarenga · B.C. Mendes (✉) · M.S.S. Moreira
DEC—Civil Engineering Department, UFV—Federal University of Vicosa, Av. PH Rolfs, Vicosa, Minas Gerais 36570-900, Brazil
e-mail: beatryz.mendes@ufv.br

S. Ikhmayies et al. (eds.), *Characterization of Minerals, Metals, and Materials 2017*, The Minerals, Metals & Materials Series,
DOI 10.1007/978-3-319-51382-9_93

The social and economic dimensions of the sustainable use of wood can be mainly explained by its high capacity to generate jobs and incomes in the supply chain. In fact, it occurs during the production, processing and manufacturing phases of planted forests, which can happen in rural properties or small and medium-sized businesses.

Wood-concrete mixed solutions for floor systems are one of the newest technologies developed for wood structures. These solutions ensure an easier and faster construction, and reduce waste and costs. Another benefit is the pleasant and welcoming aesthetic appearance of the wood structural members.

When using wood-concrete mixed structures instead of conventional reinforced concrete slabs, the cost of shoring and the structure's self-weight are reduced. They are also less expensive than steel-concrete mixed structures, mainly in areas away from steel mills or areas with greater availability of wood. Compared to simple wood structures, the composite ones present higher load capacity and stiffness, in addition to lower structural vibration. The use of reinforced concrete slabs above wood beams allows better sound insulation, improves the capacity to prevent a fire from spreading, and ensures sealing and mechanical protection. Therefore, it guarantees greater durability, especially for outer slabs and bridges.

Nowadays, one of the most efficient and simplest construction systems consists of simply supported and properly spaced T-beams, whose wood web is mechanically connected to the reinforced concrete slab. According to Correia [1], these composite beams can present gains of 300% in their stiffness and 100% in their modulus of rupture, when compared to conventional wood beams of same dimensions.

The bending behavior of composite members is highly affected by the connection system, which can be considered stiff, semi-stiff and flexible. When mechanical connectors are used, the obtained connection system is semi-stiff or flexible because the connectors are properly spaced, resulting in stress concentrations and deformation in the wood area around the connectors. Furthermore, the connectors can also present large deformation. These effects cause relative displacements between the connected members that cannot be neglected, which modifies the internal stresses distribution, and induces loss of stiffness and strength.

The effective bending stiffness $(EI)_{ef}$ of composite beams is resulted of the association between the different materials. The minimum stiffness $(EI)_0$ is obtained by adding each individual member stiffness. The maximum or apparent stiffness $(EI)_\infty$ or $(EI)_a$ can be obtained by considering a stiff connection. The connection system efficiency is the ratio between the effective stiffness gain and the maximum possible gain of stiffness. This efficiency is equal to one for stiff connection systems and less than one for flexible systems.

When using materials with different stiffness or modulus of elasticity (E_i), the equivalent bending stiffness of the cross section can be simply obtained by considering the additional gain of the each component, as shown by Eq. (1):

$$(EI)_{eq} = \sum_{i=1}^{m} (E_iI_i + \gamma_iE_iA_ia_i^2), \tag{1}$$

I_i, is the moment of inertia of each individual member, with respect to the axis parallel to the bending axis and passing through its centroid. A_i is the area of the member. a_i, is the distance from the centroid of the component part to the neutral axis of the composite section. The coefficient $\gamma_i \leq 1$ defines the contribution of the member to the inertia gain of the composite section.

The Eurocode 5 [2] considers different compositions of composite wood T-sections, and presents the Eq. (1) to obtain the equivalent stiffness. The γ_2 value (for the second considered member) is equal to 1. Then, the Eq. (2) should be taken as

$$\gamma_i = (1 + \frac{\pi^2 E_iA_is_i}{K_iL^2})^{-1}, \quad \text{for } i = 1, 3, \ldots. \tag{2}$$

K_i is the slip modulus and s_i is the fastener spacing used to connect the member i to the member 2. L is the beam span.

Then, the γ_i value is proportional to the K_i value and can endorse the connection efficiency. However, to solve practical problems, the engineer must find a global solution that harmonizes load capacity, ease of construction, and appropriate cost.

For simply supported wood-concrete mixed beams, these expressions can be used when the neutral axis is located in the wood member or while the concrete tensile strength is not reached. Nicolas [3] and Soriano and Mascia [4] present a similar formulation for the analysis of T-beams with wood web and concrete flange, using the cross section homogenization method. When the neutral axis rises and the concrete tensile strength is reached, a new stress distribution occurs. In this situation, iterative methods are required to perform suitable consideration of the physical nonlinear behavior of the material.

Different mechanical connectors can be used to join composite members, such as steel plates, wood notch filled with concrete, pins, nails, bolts or steel bar segments. In recent years, several analyzes of the behavior of different connectors in wood-concrete connections have been published, especially for the international congress, "World Conference on Timber Engineering—WCTE".

Different steel bar connectors have been proposed and evaluated: from the simplest ones, designed as straight segments partially embedded in the wood, to connectors horizontally anchored in the concrete and inserted into the wood at different inclinations, by direct spiking or adhesives. Generally, these connectors allow an easier implementation. Their efficiency and cost depends on the type, size, fixation in the wood and fastener spacing.

When evaluating the behavior of a connection pin normally driven through the interface between the connected parts, a very unfavorable static condition for the pin stiffness is obtained. The use of inclined pin connections allows part of the acting load to be transmitted by normal stresses through the pins, reducing bending stresses and increasing the connection stiffness. The connection's stiffness and strength also depend on the pin adhesion.

The purpose of this study is performing an experimental evaluation of connectors of wood-concrete mixed elements, made of round ribber steel bars, fixed inclined in wood, arranged in arches, and embedded in concrete. For a comparative analysis, connections with straight pins normally fixed into the wood were also evaluated.

Materials and Methods

Symmetric three-member specimens were built and tested. The middle member was made of concrete and the two lateral ones were made of wood. 7-cm-thick beams of medium density sawn wood were used. Their common name is "Angelim Amargoso" and their scientific name is *Votaireopsis araroba*. The material was air dried in order to achieve moisture content between 12 and 15%. After that, the wood pieces were cut in order to obtain a cross section around 6.4 cm × 14 cm.

The timber's parallel compressive strength was obtained by testing six specimens and its modulus of elasticity in bending was obtained by testing six beams with a 480-cm span.

The connectors were made of CA-50 steel bars, with minimum yield strength of 500 MPa and diameter of 12.5 mm. A 12 mm diameter drill bit was used to directly embed the bars into the wood. A 16 mm diameter drill bit was used when fixing with epoxy adhesive. Concrete was produced with crushed stone #8 and plasticizer admixture. Its compressive strength was evaluated by testing six cylindrical specimens with a 10 cm diameter.

Two kinds of connectors were analyzed. The first one consisted of three straight pins normally fixed with epoxy adhesive into holes previously made in wood (LP90). To ensure a better connection between pin and concrete, a pin head was implemented by adding a welded end plate of (38 × 50 × 5) mm with a central hole, as indicated in Fig. 1.

The second kind consisted of arc-shaped connectors. In this case, three different methods were used to fix the bars on the wood. In the first method, the bars were directly embedded into the wood (LPAs). In the second one, they were bonded using epoxy adhesive (LPAC). For the third method, plates were anchored in the compressed side, by including 50-mm diameter and 5-mm thick washers, normally welded to the bar and anchored in the wood (IPL). Two specimens were made and tested for each different connector.

To implement the arches, both of the bars were fixed in wood, and properly inclined in opposite directions. Then, they were folded to get an arc shape and welded together. A reinforcement of 5 mm diameter transverse ties and 10 mm diameter longitudinal bars was also used, as shown in Fig. 2.

Fig. 1 Connections with *steel bars*: view before pouring concrete in the *central part* of the specimens

Fig. 2 Connections with *arc bars* fixed on the wood by direct crimping, epoxy adhesive and using anchor plates and wood carving: view before pouring concrete in the *central part* of the specimens

The arch-shaped connectors were designed to obtain significant gains in individual stiffness of connectors. Furthermore, as the arc shape prevents protruding ends of steel rebar, there is an improvement of safety and of the transportation, erection, reinforcement adding, and concrete pour phases.

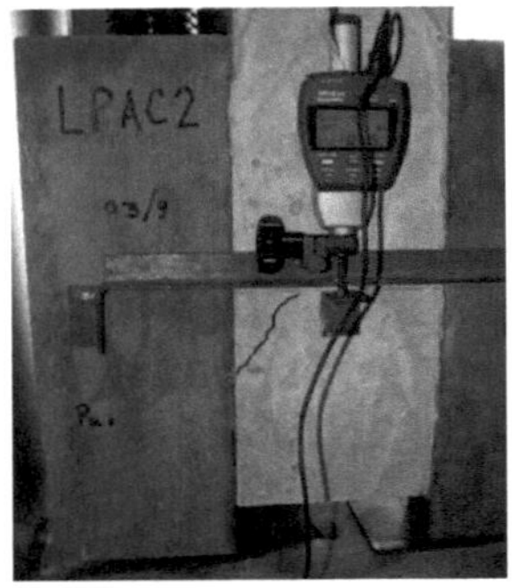

Fig. 3 Test of specimen with *glued arc* connector (LPAC)

Tests were performed using a universal testing machine, applying compressive loads in a constant rate of approximately 15 kN/min and lasting 10–15 min. The relative displacements between the side wood members and the central concrete member were measured, among points on the same straight line, normal to the bond interface and positioned in the center of the connection. Dial gauges within an accuracy of thousandth of millimeter were used, placed on both connection sides, as seen in Fig. 3.

Maximum loads (P_u) and force-displacement curves were obtained. Then, the failure mode and the stiffness in the ultimate (K_u) and serviceability (K_s) limit states were evaluated. The stiffness values were calculated in the interval corresponding to the secant straight line between, respectively, 10–40 and 10–70% of the ultimate strength.

Results and Discussion

The timber's parallel compressive strength obtained with the tests was equal to 48.2 MPa and the modulus of elasticity in bending was 13776 MPa. The concrete compressive strength at 28 days was 35.9 MPa. Maximum loads (P_u), K_s, and K_u stiffness are shown in Table 1.

The force-displacement curves obtained with the tests are represented in Fig. 4. In addition, Fig. 5 shows the force-displacement curve, the polynomial correlation curve and secant straight lines considered to calculate the connection stiffness, when using different kinds of connectors (Fig. 6).

Although the two repetitions for each kind of connector don't allow an appropriate statistical analysis and definitive conclusions, the following trends were observed:

- Strength gain (P_u) in glued arc connections, in relation to the others;
- The lowest ultimate strength value was obtained for the pins in simple arcs (P_u = 115.70 kN)

Table 1 Results of maximum loads (P_u), (K_s), and (K_u) stiffness obtained from the tests

Corpo-de-prova	P_u (kN)	K_s (kN/mm)	K_u (kN/mm)	K_u/K_s (%)
LP90-1	130.9	117.71	71.87	61.1
LP90-2	120.5	125.73	75.02	59.7
Média	125.7	121.72	73.445	60.4
LPAS-1	115.7	117.71	71.87	61.1
LPAS-2	101.9	136.95	123.88	90.5
Média	108.8	127.33	97.87	75.8
LPAE-1	102.7	130.61	118.47	90.7
LPAE-2	150.0	241.29	152.16	63.1
Média	126.4	185.95	135.31	76.9
LPAC-1	167.2	178.91	150.66	84.2
LPAC-2	136.0	192.47	150.45	78.2
Média	151.6	185.69	150.55	81.2

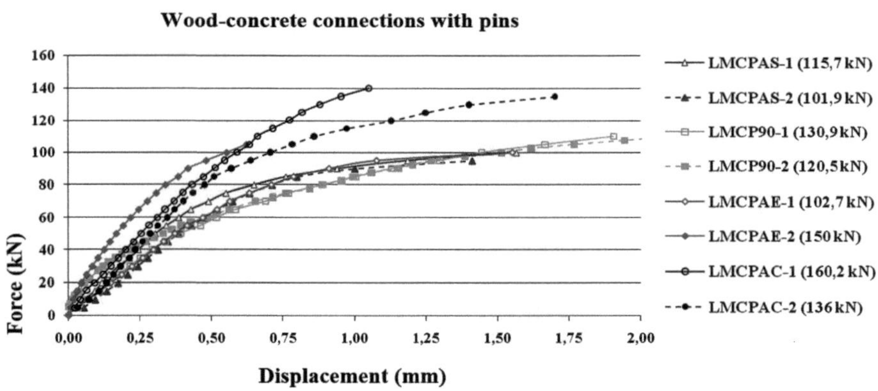

Fig. 4 Force-displacement curve obtained from the tests

- Stiffness gains in glued arc connections (LPAC) and carving connections (LPAE), in relation to the connections mentioned in the previous item;
- Important relative loss of K_u stiffness, compared to the K_s stiffness. This loss was more significant in pin connections (LP90), when happened a reduction of approximately 40%. These were the less stiff connections.
- The connections presented a yield stage before fail. The failure occurred in the concrete by compression and cracks effects.

From the mechanical point of view, results show that the glued arc connections are the best. They had the highest values of strength, stiffness, and K_u/K_s ratio.

The connections presented a yield behavior before failure. Figure 4 shows that the connection made with 90° pins showed higher ductility, with higher displacements under lower compressive loads. The rupture occurred in the concrete by the

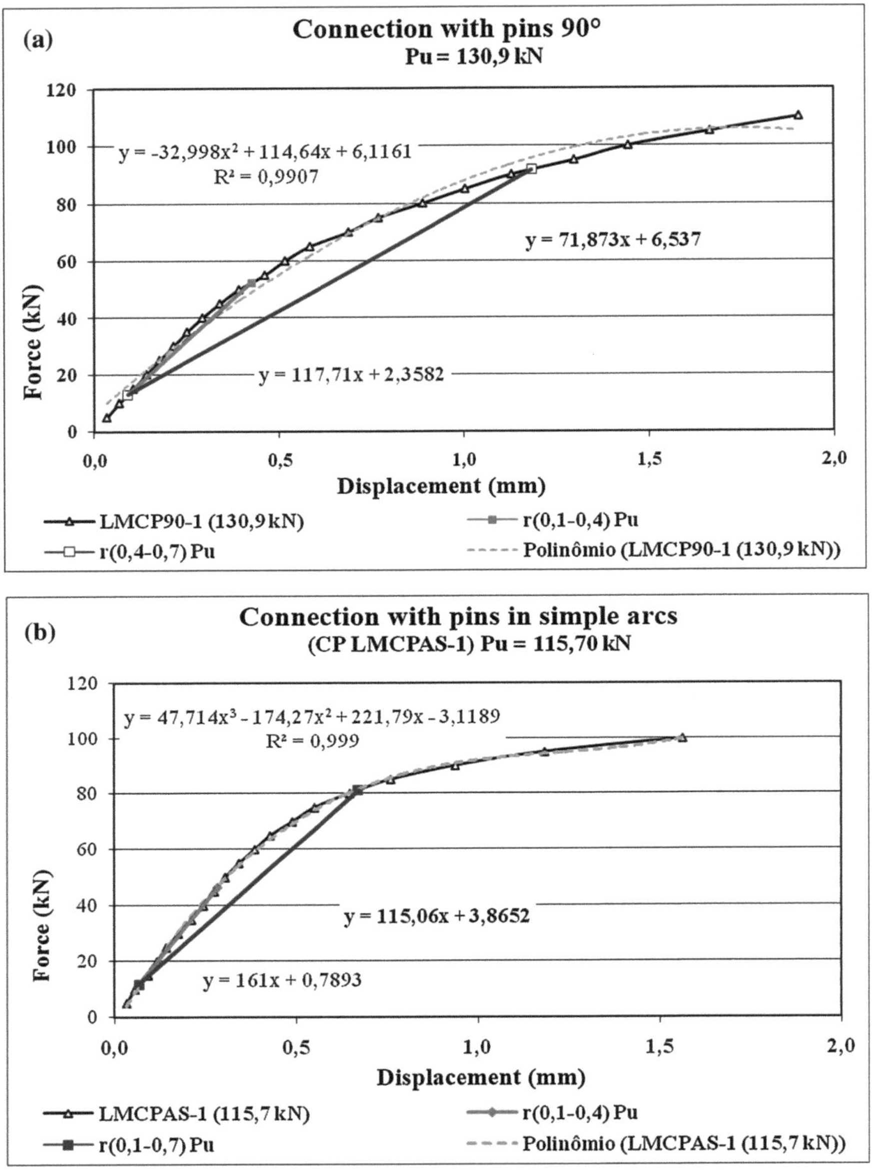

Fig. 5 Force-displacement curves obtained from the connection tests, polynomial correlations, and *straight lines* that define the stiffness: **a** Pins 90°; **b** Pins in simple *arcs*

combined effects of compression and cracking. The steel connectors' compressive strength and the timber's compressive strength parallel to the grain were higher than the concrete compressive strength. In fact, it was supposed that the rupture would occur at the weakest element.

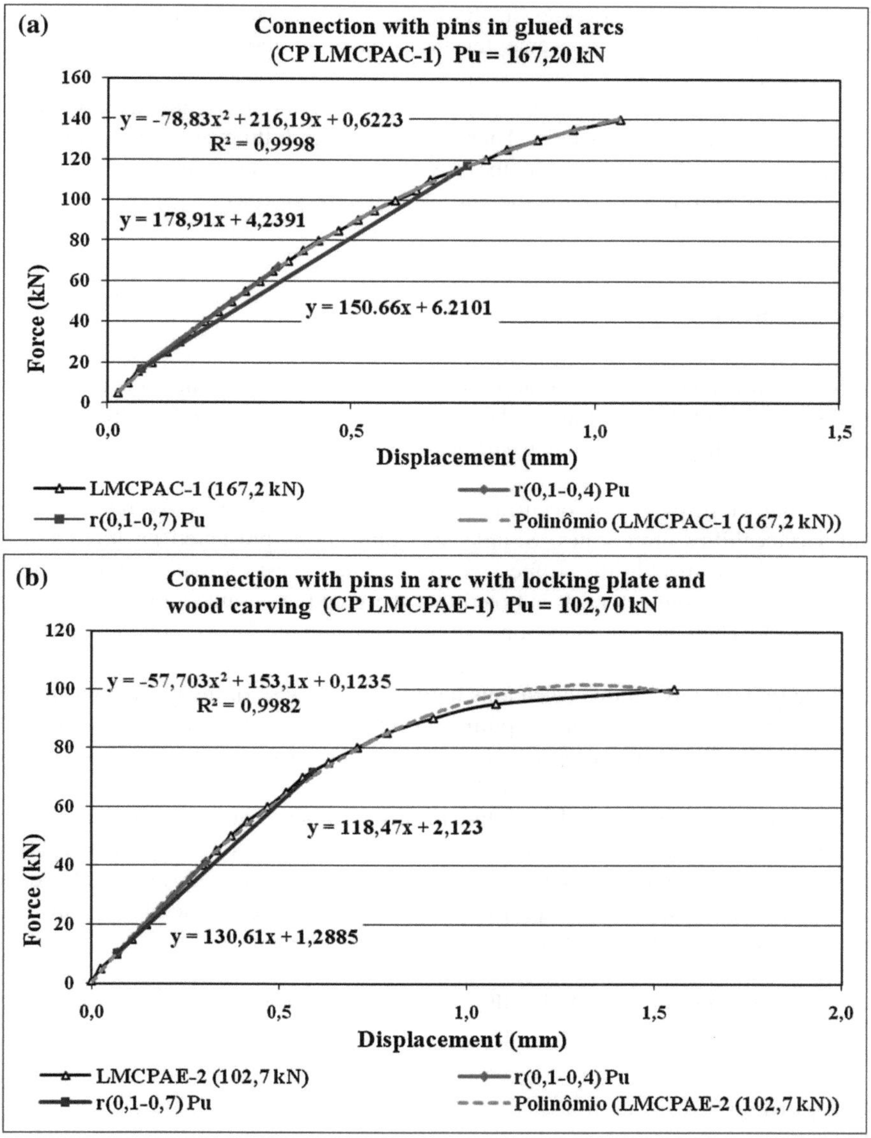

Fig. 6 Force-displacement curves from tests, polynomial correlations and *straight lines* that define the stiffness: **a** Pins in *glued arcs*; **b** Pins in *arc* with locking plate and wood carving

In addition, it is expected that when effectively applied in wood-concrete connections, these connectors can present significant strength gains, compared to the values obtained in isolated connections tests. This is due to the positive influence of the transverse reinforcement. The transmitted stresses are better distributed, decreasing compressive stresses in the concrete region in contact with the connector, and decreasing transverse tensile stresses in the area around the steel connectors.

When using a concrete with higher compressive strength, an increase of the connection strength is expected. Therefore, a greater influence of the connection between wood and steel bars in the stiffness of the composite structure is also expected, as well as a more ductile and clear failure process.

Conclusions

All of the considered connections presented appropriate mechanical behavior, with high strength and stiffness values. The glued arc connections (LPAC) presented the highest strength gain. The higher stiffness values were verified in the connections LPAC and LPAE.

The higher stiffness loss occurred in pin connections (LP90), for both ultimate and serviceability limit states. Simple arc connections presented an intermediate behavior with respect to the stiffness gains, and the lowest ultimate strength among all of the tested connectors. It is noted that the glued arc connection was the most satisfactory. Then, these connections are the most suitable for wood-concrete mixed elements.

Although the limited displacements monitoring because of the need to ensure the dial indicators integrity, the rupture processes showed evidences of ductile behavior. The final rupture occurred in the concrete around the connectors by the combined effects of compression and cracking.

The methodology applied and the results obtained show that the evaluated connectors were easy to build and not expansive, so economically viable. Thus, they are good options to solve floor systems of wood-concrete mixed elements, providing a good mechanical performance of the connected members, and contributing to spread the use of wood as a structural element.

Acknowledgements Thanks to FAPEMIG for the continuous and relevant research support.

References

1. Correia J (2010) Experimental evaluation wood-to-concrete and wood composite members joints using steel-bar and concrete connectors (M.Sc. thesis, Federal University of Viçosa, Viçosa, 2010), 128
2. Comité Européen De Normalisation, Eurocode 5—Design of timber structures (Brussels, 1993)
3. Nicolas EA (2001) Estudo de ligações em estruturas mistas de concreto-madeira (M.Sc. thesis, State University of Campinas, Campinas, 2001), 108
4. Soriano J, Mascia NT (1999) Estruturas mistas em concreto armado e madeira para pontes. Téchne 42:48–50

Author Index

S. Ikhmayies et al. (eds.), *Characterization of Minerals, Metals, and Materials 2017*, The Minerals, Metals & Materials Series,
DOI 10.1007/978-3-319-51382-9

Subject Index

S. Ikhmayies et al. (eds.), *Characterization of Minerals, Metals, and Materials 2017*, The Minerals, Metals & Materials Series,
DOI 10.1007/978-3-319-51382-9

MIX
Papier aus verantwortungsvollen Quellen
Paper from responsible sources
FSC® C105338

If you have any concerns about our products,
you can contact us on
ProductSafety@springernature.com

In case Publisher is established outside the EU,
the EU authorized representative is:
Springer Nature Customer Service Center GmbH
Europaplatz 3, 69115 Heidelberg, Germany

Printed by Libri Plureos GmbH
in Hamburg, Germany